ENERGY SCIENCE, ENGINEERING AND TECHNOLOGY

ADVANCES IN ENERGY RESEARCH

DISTRIBUTED GENERATIONS SYSTEMS INTEGRATING RENEWABLE ENERGY RESOURCES

ENERGY SCIENCE, ENGINEERING AND TECHNOLOGY

Additional books in this series can be found on Nova's website under the Series tab.

Additional e-books in this series can be found on Nova's website under the E-books tab.

ENERGY SCIENCE, ENGINEERING AND TECHNOLOGY

ADVANCES IN ENERGY RESEARCH

DISTRIBUTED GENERATIONS SYSTEMS INTEGRATING RENEWABLE ENERGY RESOURCES

NICU BIZON
EDITOR

Nova Science Publishers, Inc.
New York

For permission to use material from this book please contact us:
Telephone 631-231-7269; Fax 631-231-8175
Web Site: http://www.novapublishers.com

Additional color graphics may be available in the e-book version of this book.

Library of Congress Cataloging-in-Publication Data

Advances in energy research : distributed generations systems integrating renewable energy resources / editor, Nicu Bizon.
p. cm.
Includes index.
ISBN 978-1-61209-991-0 (hardcover)
1. Distributed generation of electric power. 2. Renewable energy sources.
I. Bizon, Nicu.
TK1006.A38 2011
621.31'21--dc23

2011013435

Published by Nova Science Publishers, Inc. † New York

CONTENTS

PREFACE

In this new book, systems and technologies of distributed generation (DG) for electricity generation or Combined Heat and Power (CHP) are discussed. DG generally refers to micro and small-scale (typically 1 kW – 50 MW) electric power generators that produce electricity at a site close to customers or that are tied to an electric distribution system. Distributed generators include, but are not limited to synchronous generators, induction generators, reciprocating engines, micro turbines, combustion gas turbines, fuel cells, solar photovoltaic panels and wind turbines.

Chapter 1 - In this chapter, systems and technologies of distributed generation (DG) for electricity generation or Combined Heat and Power (CHP) were discussed. DG generally refers to micro and small-scale (typically 1 kW – 50 MW) electric power generators that produce electricity at a site close to customers or that are tied to an electric distribution system. Distributed generators include, but are not limited to synchronous generators, induction generators, reciprocating engines, micro turbines (combustion turbines that run on high-energy fossil fuels such as oil, propane, natural gas, gasoline or diesel), combustion gas turbines, fuel cells, solar photovoltaic panels, and wind turbines. There are many reasons for that a customer may choose to install a distributed generator. DG can be used to generate a customer's entire electricity supply; for peak shaving (generating a portion of a customer's electricity onsite to reduce the amount of electricity purchased during peak price periods); for standby or emergency generation (as a backup to Wires Owner's power supply); as a green power source (using renewable technology); or for increased reliability. In some remote locations, DG can be less costly as it eliminates the need for expensive construction of distribution and/or transmission lines.

In second subchapter, with due attention to the existing terms and regulations in developing small-scale generators, these technologies are compared from technical, economic and environmental perspectives. The aim is to present the best scenario to select the existing technologies in applying suitable electricity generator or CHP through a case study. The study was conducted by a computerized code designed and modelled for this purpose. Also considering the presented quantitative and comprehensive analysis and with regard to cooling and heating load and electricity demand (based on base and peak load), the best technical and economic scenario in use can be suggested for implementation.

Chapter 2 - The upraising of the Distributed Generation and Distributed Resources weight within traditional distribution systems is a key topic for discussion and research in the electrical engineering area. This chapter deals with assessing the impact of Distributed Generation upon the existing electrical distribution systems from the reliability point of view.

A main point in the reliability analysis performed in this work is the probabilistic characterization of the random variables involved in the reliability calculations, as time to failure, duration of service restoration and reclosure time. Starting from the actual data collected from a large distribution system, the goodness-of-fit analysis has been performed by means of the Chi-square, Kolmogorov-Smirnov and Geometrical adaptation fit tests, resulting in ranking the most appropriate probability distributions for characterising these random variables.

The reliability assessment has then been performed both for traditional distribution systems and for distribution systems containing Distributed Generation, by means of different types of analysis. In particular, the time-sequential Monte Carlo Simulation and an innovative version of the analytical simulation are used. The latter is based on the combined use of the compound Poisson process and the Fourier analysis for efficiently performing the convolutions among the probability densities of the time-dependent random variables. Numerical tests on various systems, including large real ones, confirm the effectiveness of the approach presented to provide a probabilistic characterisation of the reliability indicators.

Chapter 3 - This chapter reports on the application to field measurements of time series methods developed on the basis of the theory of deterministic chaos. Recent progress in the detection and understanding of nonstationarity is reported. Recurrence plots form an efficient strategy for both visualizing and quantifying nonlinear spatio-temporal dynamics of complex systems. For example, we suggest that the unstable periodic orbits embedded in a chaotic attractor are a useful basis set for the geometry of a recurrence plot of those data. This provides a means to locate unstable periodic orbits in chaotic time series data. Besides that, we present here some results we obtained using the concept of recurrence in the fields of speech processing, signal detection and communications.

Chapter 4 - The chapter deals with a new problem of physical correctness detection in the area of strictly causal system representations. The proposed approach to the problem solution is based on generalization of Tellegen's theorem well known from electrical engineering. Consequently, mathematically as well as physically correct results are obtained. The contribution is mainly concerned with presentation of a new structural approach to analysis and synthesis of linear and non-linear causal systems. It has been proven that complete analysis of instability, conservativity, dissipativity, anti-dissipativity, stability, asymptotic stability and chaoticity reduces to two independent tests: the monotonicity test of abstract state space energy and minimality test of state space dimension (i.e. for zero input that of complete state observability).

Chapter 5 - The use of the variational formulation of power and energy functionals in the analysis of the electric and magnetic circuits makes it possible to appreciate the equilibrium state attained in the circuit at a certain moment. In the present work, a power and energy functional is built for electric and magnetic circuits in static, stationary and quasi-stationary conditions, and its limit is determined. It is shown that the minimum of this functional corresponds to the minimum dissipated (absorbed) power and energy in the circuit. Some applications to electric circuits and magnetic are shown.

Chapter 6 - In a competitive energy marketplace based on the developing Distributed Generation System (DGS), it is necessary to have in real-time complete information for the reliable delivery of energy to all customers, including the industrial end-users. The strategy to plan these demands of electric utilities needs to be based on a high-performance data communication network that can support both existing functionalities and future operational developments. Consequently, the communication network represents the core of the automation system of Electrical Power System (EPS) and its design must be for obtaining reliable and cost-effective network architecture. In this chapter, the standards and challenges of hybrid network architecture are discussed for EPS system automation. So, in next sections some technologies are described, such as mobile communications technologies, power line communications, and wireless communications (wireless sensor networks, WiMAX and

WiFi). The motivation of this chapter is to provide a better understanding of the hybrid network architecture of EPS that can use a secured communication network with some features that are presented in the next chapter.

Chapter 7 - This chapter focuses on different techniques for Securing Communications in network. The intrusion detection system device is one of the most popular equipments for preventing the network security attack (section 7.1). In section 7.2 is proposed a heterogeneous network sensors management service and a method for analyzes the alert conflict in a heterogeneous network sensor environment using different alert fusion technique to understand the security status of the whole network system. The fuzzy cognitive maps to fuse multiple inputs are used. As it is known, cryptography is used to protect data from eavesdropping and to preserve integrity of the data (section 7.3). A problem in the field of information security that has no unitary point of view is that of evaluation of the cryptographic algorithms proposed as public cryptographic standards (section 7.4). Section 7.5 focuses on the link between information security and cryptography that is represented by National Institute of Standards and Technology (NIST) cryptographic standards, Federal Information Processing Standard FIPS 140-2 (Security requirements for cryptographic modules) standard and Common Criteria for Information Technologies Security Evaluation (ISO 15408) standard. In the last period of time a growing attention has been given to chaos-based transmission methods in order to be used for securing communications (section 7.6). Secured chaotic communications are obtained by modulation of the chaotic signal with the informational signal. The evolutions of two chaotic dynamic systems that have the same attractor, but different initial start conditions, will diverge after a certain time, although the evolutions are close to the common attractor. For this is necessary to synchronise these two systems. Some examples of data transmission based on simple discrete-time chaotic systems are presented in section 7.7.

Last section presents a FPGA implementation of the Rijndael-AES Crypto-Algorithm.

Chapter 8 - Load Frequency Control (LFC) is one of the most important issues in electric power system design and operation. The objective of the LFC in an interconnected power system is to maintain the frequency of each area and to keep tie-line power near to the scheduled values by adjusting the MW outputs the LFC generators so as to accommodate fluctuating load demands. LFC provides a mechanism to balance generated power and demand in order to maintain the system frequency at the nominal value and to keep the power interchanges between neighbouring areas at the their scheduled values. In the deregulated environment, it will serve as ancillary service and acquires a principal role to enable power exchanges and provide better condition for electricity trading. LFC goals, i.e. frequency regulation and tracking the load demands, maintaining the tie-line power interchanges to specified values in the presence of modeling uncertainties, system nonlinearities and area load disturbances determines the LFC synthesis as a multi-objective optimization problem and they make a big challenges for controller design. The load frequency controller design with better performance has received considerable attention during past years and many control strategies have been developed for LFC problem.

This chapter presents literature review on the LFC problem and history of new control strategies is outlined. The LFC scheme based on various control methodologies such as linear control, robust and adaptive techniques, artificial computational intelligent based strategies are investigated. Solution of the LFC problem including BES/SMES, wind turbines, FACTs devices and other DG units has also been discussed. Using the obtained results, various comparisons between these methods and the main advantages/disadvantages of them are given. Finally, the recently control methodologies presented for the solution of LFC problem is widely outlined in the deregulated environment.

Chapter 9 - In this chapter are briefly presented the main mitigation techniques of the low frequency fuel cell current ripple: filtering by passive filters on high voltage bus and/or low voltage bus, designing of appropriate structures for inverter system powered by fuel cell, and active control of power converters from inverter system, or the active control of power interface that operate as an active filter. The designing of models used in simulation is also shown, especially for fuel cell model used to evaluate the low frequency fuel cell current ripple. The simulation results are validated by measurements performed in different experiments.

Chapter 10 - The artificial intelligence which is an important branch of computer science or computer engineering includes expert system, fuzzy logic, artificial neural network and genetic algorithm. The application of artificial intelligence in electrical engineering, power electronics and motion control has been rapidly extended recently. The objective of this chapter is to describe how neuro-fuzzy techniques can be applied in the automatic control area, particularly in feedback control. In addition, the fuzzy logic and neural network principles are briefly studied and the applications of neural networks techniques in converter modeling and the control of power electronics converters are described. The results of simulations and experiments for different applications are also given.

Chapter 11 - In this chapter is analyzed a fuel cell inverter system of multistage type. The mono-phase inverter is designed in the first subsection. The boost converter is supplied by a 35 V/2 kW fuel cell stack. The proposed voltage-mode control of the voltage boost stage, which has as fuzzy control variables the voltage on low voltage bus and the fuel cell current, stabilizes the low voltage and mitigates the current ripple, too. Obviously, the main control goal of the full-bridge DC-DC converter is the stabilization of the voltage on high voltage bus, if the stage of boosting voltage is removed. For the mono-phase inverter is proposed a feed-forward control of the switching command circuit that generates a pure sine PWM command as function of the voltage on high voltage bus in order to maintain the RMS level of the AC output voltage. The proposed fuzzy control is compared with classical control techniques. The simulation results confirm the robustness of the proposed control for the fuel mono-phase cell inverter system. The three-phase inverter variant is analyzed, too. Finally, the experimental setup is shown and the obtained results are compared with the simulation results.

Chapter 12 - This chapter present a brief overview of models used in this book. Supplementary information about the relationships of models is here given, which complements the data already presented in the chapters where they were used.

First section shows the used models for Polymer Electrolyte Membrane Fuel Cell (PEMFC) stack and Energy Storage Devices (ESD). Two PEMFC models that were implemented in Matlab Simulink® are presented: a detailed model and its simplified variant when some PEMFC parameters are set to their rated or custom values. The generally accepted models for stacks of batteries or ultracapacitors are briefly shown, too. Some tests that validate the proposed models are presented at the end of this section.

Second section shows generally accepted models for wind and gas turbine. A model of horizontal axis wind turbine with convoluted flow of the wings is presented. Usually, the gas turbine drives a synchronous generator. The model of the synchronous generator is shown, too. Both models are used in simulations presented in chapter named "Case studies concerning DG system implementations".

In the third section are analyzed the multi-stage power topologies that are utilized in the Energy Generation Systems (EGS). There are presented, compared and evaluated different proposed topologies of multi-stage energy conversion for unidirectional or bidirectional power flow from energy source to the load.

Chapter 13 - Most modern electrical and electronic apparatus called as nonlinear loads use AC/DC power supply within their architecture. These devices draw discontinuous current from the AC network. The discontinuous currents include harmonics and cause problems for the power distribution systems. In addition, the harmonic contents of the discontinuous current cause additional losses and dielectric stresses in capacitors and cables resulting in increasing currents in windings of rotating machinery, transformers and electromagnetic interference. With the steadily increasing use of equipments including nonlinear loads, line current harmonics have become a significant problem. So, several international standards such as EN 61000-3-2 now exist limiting the harmonic content due to line currents of equipment connected to AC networks.

In this chapter, parallel to power factor correction (PFC) study, several passive and active topologies applied to single phase power factor correction are analyzed. The major advantages and disadvantages of the topologies are highlighted. In addition, the commonly used control methods for active power factor correction are discussed and compared. The switching losses and soft switching methods of the boost type active power factor correction circuits are also reviewed and discussed.

In the several active power factor correction techniques used to reduce current harmonics and to satisfy the requirements given in international standards, the boost type converters operating in continuous conduction mode have been widely adopted as power factor correction pre-regulators. For this purpose, the boost type power factor correction circuit is simulated and the related waveforms are given to show the importance of active power factor correction methods.

Chapter 14 - The Automatic Generation Control (AGC) mechanism is one of the main control requirements for power generation in an interconnected power system. The most electric power consumers such as AC motors and electrically operated clocks depend on a certain system frequency. Actually, a malfunction of these equipments negatively affects our lives. Hence, AGC is carried out to the power system in order to maintain the constant and the stable system frequency. In this chapter, the main principle of AGC is investigated and its application based on swarm intelligence optimization for a hydrothermal power system is presented. After the explanation of AGC, the gain scheduling Proportional-plus-Integral (PI) control technique and the scheduling algorithm based on swarm intelligence, which is used in this application are introduced. The control technique with tunable weighted cost function is carried out to the power system which includes two control areas having a steam turbine and a hydro turbine connected together through tie lines. Both areas also include Superconducting Magnetic Energy Storage (SMES) units in order to improve the performance of the control. To optimize the gains of PI-controller and the parameters of the proposed cost function, Constriction Factor Approach Particle Swarm Optimization (CFA-PSO) algorithm is used. The usage of this algorithm to optimize the gains of the PI-controllers and the tunable weight coefficients of the cost function was detailed explained with the parts of its program in order to help the researchers which study on optimal AGC for multi area power system. The result of the control was also investigated and discussed. At the end of the simulations, it is seen that the proposed AGC technique improves the dynamic performances of the power system.

Chapter 15 - This chapter aims to provide a non-exhaustive point of view, based on the literature, regarding the premises for building Spatial Decision Support Systems for complex processes monitoring using Web Services-based Infrastructure for Geographic Information, and to identify some challenges and to highlight on the possible solutions in order to overtake the critical point in this areas: syntactic and semantic interoperability from geospatial information used in different fields of geosciences. Starting from the terminological aspects and evolution, has been presented a frame of work covering the European and worldwide initiatives in connection with GI infrastructures building and some application domains associated with the challenges coming from the intention to offer the best solution in order to solve the problems of semantic interoperability and semantic heterogeneity followed by some relevant examples.

Chapter 16 - his chapter is focused on implementation stage of the Distributed Generation (DG) system in some countries: two from the European Union (EU), one candidate to EU and one from outside of EU. If last section of first chapter (named „Stage of Integrating Renewable Sources into DG Systems of the European Union) shows some approach for accessing the available renewable energy of each EU state (that put in evidence the diversity of energy efficiency measures and implementation of renewable energy strategies), then these aspects will be detailed by case studies in this chapter.

First of all, the energy resources of each state (that is subject of a case study) are shown as level of non-renewable energy resources and renewable resources. Then the electricity generation, transportation and distribution is discussed in perspective of integration of the available renewable energy sources.

So, proposed objectives for the DG system, the rehabilitation measures and levels achieved for the implementation stages are presented. In this frame, the energy policy of the each state is discussed, showing the priorities of this energy policy and future energy policy measures. These strategies are based on new technologies and equipments that are applied using new infrastructure (as is the smart grid, for example). Some case studies are sustained by simulation models. Finally, some conclusions are given.

Chapter 17 - Public debate about hydrogen and fuel-cell technology takes place between two extremes: it is perceived as the immediate answer to all questions related to air pollution and noise, climate change and expensive fuels, but also as too dangerous even to think about. There are also obligatory references to hydrogen as the energy carrier of the (far) future, but the option hardly features in the governmental policy debates about the above-mentioned issues. The problem for hydrogen and fuel cells is that they are good for many policy objectives – air quality, climate change, security of energy supply, competitiveness/ innovation – but not the best solution for any of them in the short term. It is the role of science to solve all the issues related to the real hydrogen-based energy applications.

In: Advances in Energy Research
Editor: Nicu Bizon

ISBN 978-1-61209-991-0

Chapter 1

DISTRIBUTED GENERATION SYSTEMS INTEGRATING RENEWABLE ENERGY SOURCES

Naser Mahdavi Tabatabaei[1*], Reza Effatnejad[2], Radu Florin Porumb[3] and Kambiz Rezapour[2]

[1]Seraj Higher Education Institute, Tabriz, Iran
[2]Islamic Azad University, Karaj Branch, Iran
[3]University "Politehnica" of Bucharest, Energetics Department, Romania

ABSTRACT

In this chapter, systems and technologies of distributed generation (DG) for electricity generation or Combined Heat and Power (CHP) were discussed. DG generally refers to micro and small-scale (typically 1 kW – 50 MW) electric power generators that produce electricity at a site close to customers or that are tied to an electric distribution system. Distributed generators include, but are not limited to synchronous generators, induction generators, reciprocating engines, micro turbines (combustion turbines that run on high-energy fossil fuels such as oil, propane, natural gas, gasoline or diesel), combustion gas turbines, fuel cells, solar photovoltaic panels, and wind turbines. There are many reasons for that a customer may choose to install a distributed generator. DG can be used to generate a customer's entire electricity supply; for peak shaving (generating a portion of a customer's electricity onsite to reduce the amount of electricity purchased during peak price periods); for standby or emergency generation (as a backup to Wires Owner's power supply); as a green power source (using renewable technology); or for increased reliability. In some remote locations, DG can be less costly as it eliminates the need for expensive construction of distribution and/or transmission lines.

In second subchapter, with due attention to the existing terms and regulations in developing small-scale generators, these technologies are compared from technical, economic and environmental perspectives. The aim is to present the best scenario to select the existing technologies in applying suitable electricity generator or CHP through a case study. The study was conducted by a computerized code designed and modelled for this purpose. Also considering the presented quantitative and comprehensive analysis

[*] Corresponding Author: n.m.tabatabaei@gmail.com

and with regard to cooling and heating load and electricity demand (based on base and peak load), the best technical and economic scenario in use can be suggested for implementation.

The current stage of integrating renewable sources into DG systems of the European Union (EU) is briefly shown in last subchapter. Energy efficiency measures and implementation target of renewable energy strategies are presented for each EU country, too.

Keywords: Electrical Power System, Distributed Generation, Distributed Energy Resources, Renewable Energy Sources, Combined Heat and Power, small-scale generators.

DISTRIBUTED GENERATION

Distributed Generation

Distributed generation is a latest trend in the generation of electrical power, sometimes combined with generating heat. The Distributed Energy Resources (DER) concept permits, consumers who are generating heat or electricity for their own needs to send surplus of electrical power back into the power grid or share excess heat by a distributed heating grid.

Distributed generation systems can be divided in two parts, as shown in figure 1:

- Combined Heat and Power (CHP)
- Renewable Energy Sources (RES)

CHP is the use of a power generator to generate both heat and electricity. Conventional power plants emit the heat created as a by-product of electricity generation into the environment through cooling towers, as flue gas, or by other means. CHP either captures the excess heat for domestic or industrial heating purposes, very close to the plant, or especially distribute it through steam pipes to heat local housing (district heating).

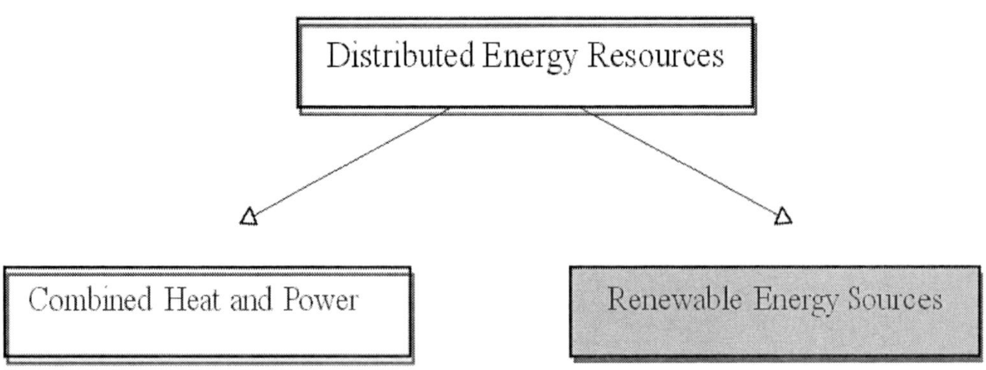

Figure 1. Distributed Energy Resources.

RES capture their energy from existing flows of energy, from on-going natural processes, such as sunshine, wind, flowing water (hydropower), biological processes such as anaerobic digestion and geothermal heat flow. The most common definition is that renewable energy is from an energy resource that is replaced by a natural process at a rate that is equal to or faster than the rate at which that resource is being consumed.

DG is typically connected directly to the distribution network or on the customer side of the meter. Most of these power plants are quite small scale, and because of their size, are often not required to contribute to the system balance. For instance, the grid code of the some excludes generation units of less than 60 MW from the obligation to offer system-balancing capacities to the transmission system operator [1].

Wind Power

Recently, the scale of wind generators has meant that they all qualified as DG. However, offshore wind farms could have total electrical outputs to 1GW in the future and so will require super grid connection. Wind generation present the power system with a number of challenge. Firstly, the output of a wind farm is essentially uncontrollable. This presents an additional challenge to control systems that ensure a balance between supply and demand and constant system frequency. Secondly, most wind generators have different electrical characteristics to conventional synchronous generators and this has to be taken account of when connecting them to the transmission and distribution systems. However, it is possible for wind generation to make a significant contribution to the electricity supply system and it is currently the leading renewable generation technology in many countries [2].

Photovoltaic

Photovoltaic cells convert energy of sun directly into direct current (DC) electricity. A particular advantage of this technology is that it can be very effectively integrated into the structure of buildings. From a distribution system perspective, it is important to note that this technology requires its DC output to be converted to AC at the interface with the grid. This process introduces harmonic distortion to the grid and it is vital the level of this distortion is controlled within limits set by industry standards [3].

Fuel Cell

The fuel cell is essentially a very simple device that was invented in the 19th century. A device allows hydrogen and oxygen to be combined to generate electricity and the reverse of electrolysis process. The development of fuel cells has been constrained primarily by the difficulty of making them cost effective [4].

Biofuels

Biofuel includes:

- Biomass- this can be a wet or dry fuel derived from plant material and animal wastes. It can be combusted to generate electricity and/or heat in plants having a wide range of outputs. Wet wastes can be fermented to produce a gas that can be used in gas engine or gas boilers.
- Biogas- this is essentially the same as the gas derived from wet biomass. The most common example of biogas is landfill gas. However, the same basic process that occurs in a landfill site can be reproduced in an anaerobic digester. This device allows organic materials to decompose in the absence of oxygen in a controlled way, releasing methane.

By dry and gaseous bio-fuels can be used in gas engines (reciprocating) and gas turbines to produce electricity. The recovery heat from this process can be captured to achieve combined heat and power (CHP) [5].

Microgeneration

The term micro generation covers a wide range of electricity generation technologies. There is not yet a well-recognised definition of the term. It is usually used in the context of the generation of electricity at domestic scale. Therefore, it is appropriate to introduce categories of distributed generation. We suggest the following distinction for these categories [6]:

- Micro- DG: ~1 Watt - 5 kW;
- Small - D G: 5 kW - 5 MW;
- Medium-DG: 5 MW - 50 MW;
- Large-DG: 50 MW - 300MW.

The Advantages of DG

In the last decade, the concept of many small-scale energy sources dispersed over the grid gain a considerable interest. Most of all, technological innovations and a changing economic and regulatory environment were that main triggers for this interest. International Energy Agency (IEA) [7] lists five major factors that contribute to this evolution, such as developments in distributed generation technologies, constraints on the construction of new transmission lines, increased customer demand for highly reliable electricity, the electricity market liberalization and concerns about climate change. Especially the last two points seem to offer the most significant benefits, as it is unlikely that distributed generation would be capable of avoiding the development of new transmission lines. At minimum, the grid has to be available as backup supply.

In the liberalized market environment, the distributed generation offers a number of benefits to the market participants. As a rule, customers look for the electricity services best suited for them. Different customers attach different weights to features of electrical energy supply, and distributed generation technologies can help electricity suppliers to supply the type of electricity service they prefer. One of the most interesting features is the flexibility of DG that could allow market participants to respond to changing market conditions, i.e. due to their small sizes and the short construction lead times compared to most types of larger central power plants.

Flexibility in Price Response

Important aspects of the abovementioned flexibility of distributed generation technologies are operation, size and expandability. Flexible reaction to electrical energy price evolutions can be one of the examples, allowing a DG to serve as a hedge against these price fluctuations. Apparently, using distributed generation for continuous use or for peak, shaving is the major driver for the US demand for distributed generation. In Europe, market demand for distributed generation is driven by heat applications, the introduction of renewable and by potential efficiency improvements.

Flexibility in Reliability Needs

Reliability considerations of the second major driver of US demand for distributed generation is quality of supply or reliability considerations. Reliability problems refer to sustained interruptions in electrical energy supply (outages). The liberalization of energy markets makes customers more aware of the value of reliable electricity supply. In many EU countries, the reliability level has been very high, mainly because of high engineering standards. High reliability level implies high investment and maintenance costs for the network and generation infrastructure. Due to the incentives for cost- effectiveness that come from the introduction of competition in generation and from the re-regulation of the network companies, it might be that reliability levels will decrease. However, for some industries, such as chemical, petroleum, refining, paper, metal, telecommunication, a reliable power supply is very important. Such companies may find the reliability of the grid supplied electricity too low and thus be willing to invest in distributed generation units in order to increase their overall reliability of supply.

The IEA recognizes the provision of reliable power as the most important future market niche for distributed generation. Fuel cells and backup systems combined with an UPS (Uninterruptible Power Supply) are identified as the technologies that could provide protection against power interruptions, though it has to be noted that the fuel cell technology is currently not easily commercially available.

Flexibility in Power Quality Needs

Apart from large voltage, drops to near zero (reliability problems), one can also have smaller voltage deviations. The latter deviations are aspects of power quality. Power quality refers to the degree to which power characteristics align with the ideal sinusoidal voltage and current waveform, with current and voltage in balance. Thus, strictly speaking, power quality encompasses reliability.

Insufficient power quality can be caused by failures and switching operations in the network (voltage dips and transients) and by network disturbances from loads (flickers, harmonics and phase imbalance) [8].

Environmental Friendliness

Environmental policies or concerns are probably the major driving force for the demand for distributed generation in Europe. Environmental regulations force players in the electricity market to look for cleaner energy- and cost-efficient solutions. Many of the distributed generation technologies are recognized environmentally friendly. Combined Heat and Power (CHP) technology, allowing for portfolio optimization of companies needing both heat and electrical energy, is one of the examples. Compared to separate fossil-fired generation of heat and electricity, CHP generation may result in a primary energy conservation, varying from 10% to 30%, depending on the size (and efficiency) of the CHP units. The avoided emissions are in a first approximation similar to the amount of energy saving [9]. Furthermore, as renewable energy sources are by nature small-scale and dispersed over the grid. Installing distributed generation allows the exploitation of cheap fuel opportunities. For example, DG units could burn landfill gasses in the proximity of landfills, or other locally available biomass resources. Most government policies that aim to promote the use of renewable will also result in an increased impact of distributed generation [10].

Substitute for Grid Investments

Distributed generation could serve to bypass for transmission and distribution costs (customers investing in DG) or as a substitute for investments in transmission and distribution capacity. Of course, this is possible only to the extent that alternative primary fuels are locally sufficiently available. Otherwise, increased use of distributed generation could result in congestion in other networks, such as the gas transport network.

On-site production could result in cost savings in transmission and distribution of about 30% of electrical energy costs. The smaller the customer size, the larger the share of transmission and distribution costs in the electricity bill (above 40% for households). As such, it is seen as one of the biggest potential drivers for the distributed generation demand [11].

From the viewpoint of the system operators, distributed generation units can substitute for investments in transmission and distribution capacity. In some cases even, it can even be used as an alternative to connecting a customer to the grid in a 'stand alone' application. Furthermore, locating the generation close to the loads could also contribute to reduced grid losses. The IEA estimates the grid losses reduction potential for 6.8% in the OECD countries leading to cost savings of about 10–15%.

The Disadvantages of DG

The question of power quality and distributed generation is not straightforward. On one hand, distributed generation contributes to the improvement of power quality. In the areas where voltage support is difficult, distributed generation offers significant benefits for the voltage profile and power factor corrections.

On the other hand, large-scale introduction of decentralized power generating units may lead to instability of the voltage profile. The bi-directional power flows and the complex reactive power management can be problematic and lead to voltage profile fluctuation. Additionally, short-circuits and overloads are supplied by multiple sources, each independently not detecting the anomaly [12].

Power System Protection

Basic power system protection principles are outlined in standard literature. The primary purpose of power system protection is to ensure safe operation of power systems, thus to care for the safety of people, personnel and equipment [13]. Furthermore, the task is to minimize the impact of unavoidable faults in the system. From an electrical point of view, dangerous situations can occur from overcurrents and overvoltages.

For example, an asynchronous coupling of networks results in high currents. Earth faults can cause high touch voltages and therefore endanger people. The general problem is always voltage and/or current out of limit. Hence, the aim is to avoid overcurrents and overvoltages to guarantee secure operation of power systems.

For the safety of the components it is also necessary to regard device specific concerns, for example oil temperature in transformers, gas pressure in gas insulated components etc. These points are not directly related to electrical values, but, as mentioned, they always come from or lead to un-allowed high voltages or currents.

Another issue is mechanical stress. Whenever power is converted electromechanically, one has to consider not only the electrical but also the mechanical equipment. An example is mechanical resonance of steam turbines due to under frequency. Nowadays, electro-mechanical protection devices are replaced by micro- processor based relays with a number of integrated features. Currents and voltages are suitably transformed and isolated from the line quantities by instrument transformers and converted into digital form. These values are inputs for several algorithms that then reach tripping decisions. Further information about computer relaying can be found in. For the design and coordination of protective relay in a network, some overall rules have become widely accepted:

Selectivity: A protection system should disconnect only the faulted part (or the smallest possible part containing the fault) of the system in order to minimize fault consequences.

Redundancy: A protection system has to care for redundant function of relays in order to improve reliability. Redundant functionalities are planed and referred to as backup protection. Moreover, redundancy is reached by combining different protection principles, for example distance and differential protection for transmission lines.

Grading: For the purpose of clear selectivity and redundancy, relay characteristics are graded. This measure helps to achieve high redundancy whereas selectivity is not disabled.

Security: The security of a relay protection system is the ability to reject all power system events and transients that are not faults so that healthy parts of the power system are not unnecessarily disconnected [14].

Dependability: The dependability of a relay protection system is the ability to detect and disconnect all faults within the protected zone [14].

Different network topologies require different protection schemes. In the following paragraphs, some typical systems should be described shortly. The simplest network structure to protect are radial systems, therefore simple re- lays are used [13]. Normally, time-dependent, graded over current protection is installed regarding redundancy (backup protection). More sophisticated relays are used for the protection of rings and meshed grids. Impedance relays trip due to a low voltage current quotient. Since these relays allow determining the position of the fault on the line, they are also called distance relays. Detailed descriptions are given in. A very common principle for the protection of generators, transformers, bus-bars and lines is differential protection. The trigger criteria is, simply speaking, a certain difference between input and output current. Furthermore, a number of other techniques are used, also device-specific ones.

Protection Issues with DG

The overall problem when integrating DG in existing networks is that distribution systems are planned as passive networks, carrying the power unidirectional from the central generation (HV level) downstream to the loads at MV/LV level. The protection system design in common MV and LV distribution networks is determined by a passive paradigm, i.e. no generation is expected in the network [15].With distributed sources, the networks get active and conventional protection turns out to be unsuitable (Figure 2).

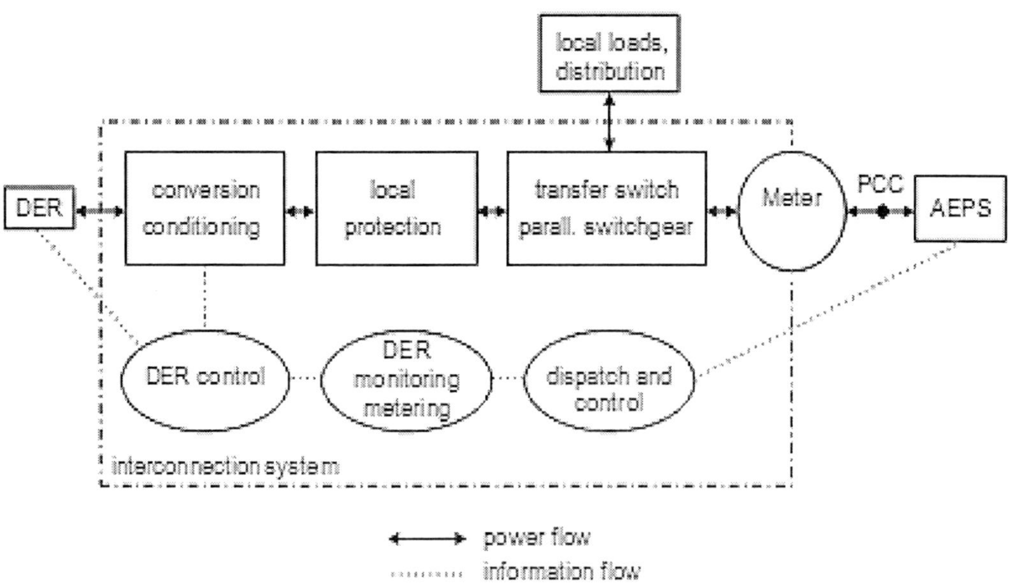

Figure 2. Interconnection system functional scheme [16] (AEPS: Area Electric Power System; DER: Distributed Energy Resource; PCC: Point of Common Coupling).

Connection Issues

In the way the power grid is currently managed, it can be taken as given that the electric power flows from the higher voltage grid to the lower voltage grid. Increased share of distributed generation units may lead to inducing power flows from the low voltage into the medium-voltage grid. A bi-directional power flows asks for different protection schemes at both voltage levels.

Moreover, the added flexibility of DG asks for extra efforts on the grid operation side. As some customers might want to switch to the "island" mode during an outage, they should also meet the requirements for such operation mode. Next to guarantying no power supplied to the grid, they must be able to provide the ancillary services needed. Moreover, once the distribution grid is back into operation, the DG unit must be able to be re-synchronized [12].

Conclusions

Distributed generation is generally perceived to be technology that can help to meet environmental goals, whilst simultaneously being able to provide an economic benefit in many cases and often energy security benefits through primary fuel diversification and improvements in supply reliability. As discussed in this document, a number of issues including market rules, regulation, high gas prices (for CHP) and technical issues continue to impede increased penetration in many States.

Possibly the most important recent changes the legislation that affect DG (and particularly CHP) are the CHP directive and the Emissions Trading System (ETS). The CHP on directive seeks to promote CHP through systematic identification and progressive realisation of the national potential for high efficiency CHP by creating a common definition and removing barriers. It suggests that national targets be set for CHP, and ensures Member States analyse their potential for high efficiency generation, apply consistent definitions and report on progress. The ETS could provide more direct financial assistance for CHP and renewable; where installations achieve emissions reduction over their allowance in the various national allocation plans, thus potentially improving the financial motivation for new installations.

Technical and Economic Analysis and Justification of Application Prevailing Technologies of Distributed Generation For Electricity Generation or Combined Heat and Power

Introduction

A computerized program has been designed in Excel Microsoft program software for technical, economic and environmental modelling of DG prevailing technologies with regard to the existing terms and regulations. In that program, the following technical and cost parameters have been considered: nominal and practical electrical capacity, foreign currency costs of equipment and heat recovery equipments, cost of land for the construction place, consultant cost and connection to gas branching network, customs and others, duration of construction and installation, cost of fixing and maintaining, availability coefficient, technical life timed of the equipments, percent of consuming fuel (gas and gasoil), electrical and thermal efficiency, electrical power density, dropping costs (total equipments and land cost after the end of generative life), high effective efficiency reward (42 percent) and heat using coefficient and economic and thermodynamic parameters such as foreign exchange conversion, devaluation rate, interest rate, energy carriers price (electricity, gas and gasoil), heat value of energy carriers, base rate of purchasing electricity from small scale generators and efficiency of the high electrical effective awards in the communiqués, with regard to the average efficiency of power plants and losses of transforming and distribution network [17; 18; 19; 20].

Generation Cost of Electricity

As it is seen in Figure 3, the cost of electricity generation at present is less under the effects of fuel cost and this cost allocates a low percent (less than 15 percent) in the final cost of electricity generation. Of course, it is worth mentioning that in case of materialization of the real price of energy carriers. For example the price of natural gas will be 680 Rials for per cubic meter (based on 10 thousands Rials equal to 1 dollar, this amount will be equal to 68 Cents), the fuel cost will form 40 percent of the cost of electricity generation. In that case, the economic aspect of the plan and the return of capital will take place in less than three years if these generators to be used Combined Heat and Power (CHP).

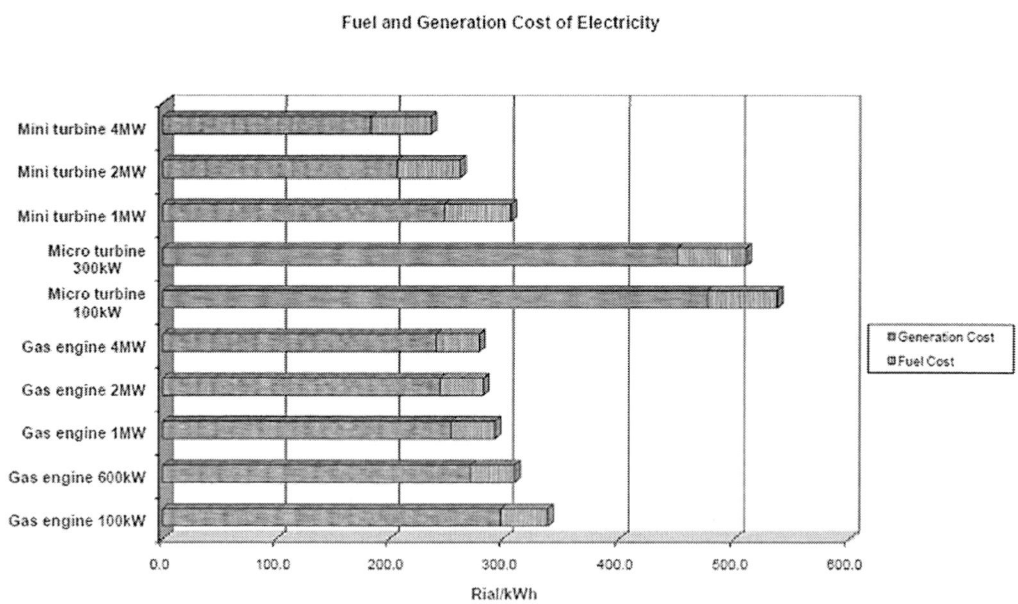

Figure 3. The cost of generation and fuel in the final cost of electricity generation in different DG technologies.

Furthermore, in this bar chart, it is clear that cost of electricity generation and even CHP in micro turbine is uneconomic based on buying electricity generation (368 Rials/kW + one forth of gas price).

Payback Period

The payback period on gas engines with a heat recovery of (100 kW) of which 50 percent of fuel is gasoil and the remaining 50 percent is supposed to be gas , with the prevailing prices (subsidized prices) and generating electricity when they use gas is not economic (Figure 4). Therefore, in order to implement a project, no other alternative fossil fuel but natural gas is suggested. Moreover, this bar chart shows that the electricity generation is not so different from CHP, because the price of natural gas fuel is 150 Rials (15 Cents) per cubic meters. In case of the increase of the fuel cost in a not very far future, with regard to the fuel cost in the cost of electricity generation, only the implementation of CHP will be meaningful and economic. In addition, micro turbines are uneconomic.

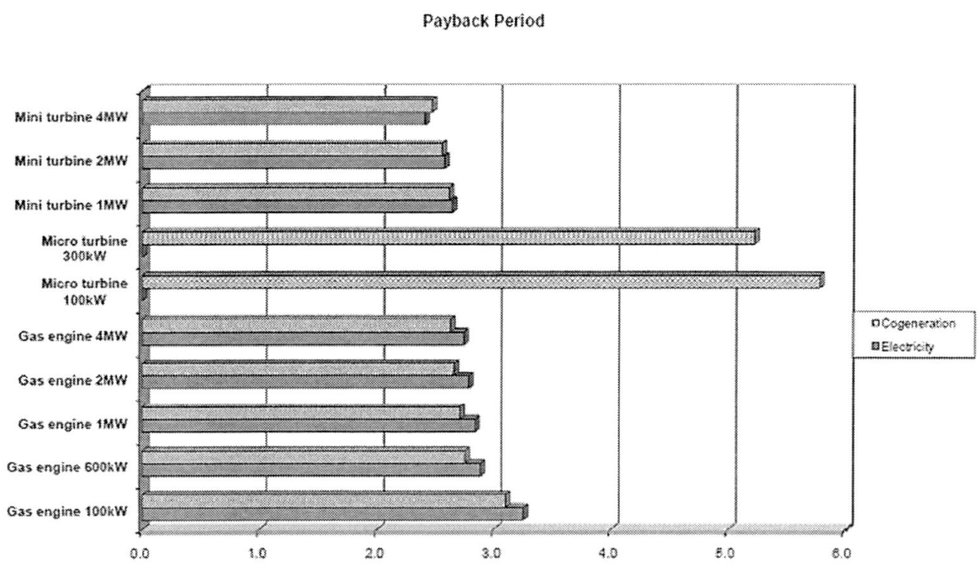

Figure 4. The payback period in using different DG technologies.

Benefit Per Cost

The benefit per cost for different DG technologies of electricity generation and CHP in a small scale is shown in Figure 5. As it is seen in the Figure, either benefits per cost in gas engines at the state of electricity generation or CHP are more than other technologies. In addition, using micro turbine in this study shows an unacceptable benefit to cost during the life of the equipment.

Percent of Annual Profit

One of the other important parameters is the percent of annual profit in using technologies of electricity generation or CHP in small scales. As it is seen in Figure 6, these technologies are compared in considered capacities. As we observe in this figure, still gas engines and mini turbines present a noticeable percent either in electricity generation or co-CHP (More than 35 percent). Furthermore, it is clear in the Figure that the low profit of micro turbine and its inefficiency state from economic point of view is effect the application of this technology.

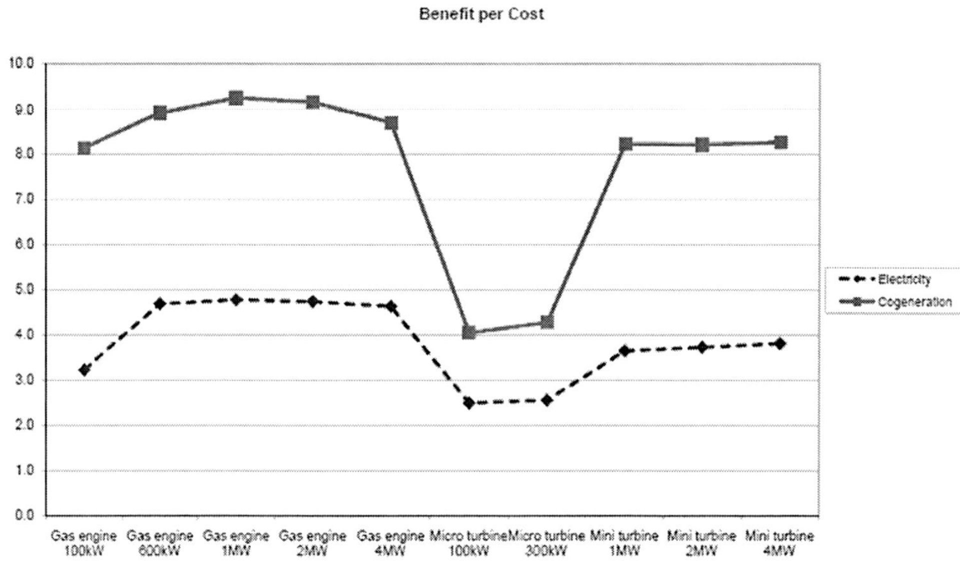

Figure 5. The benefit per cost of different DG technologies.

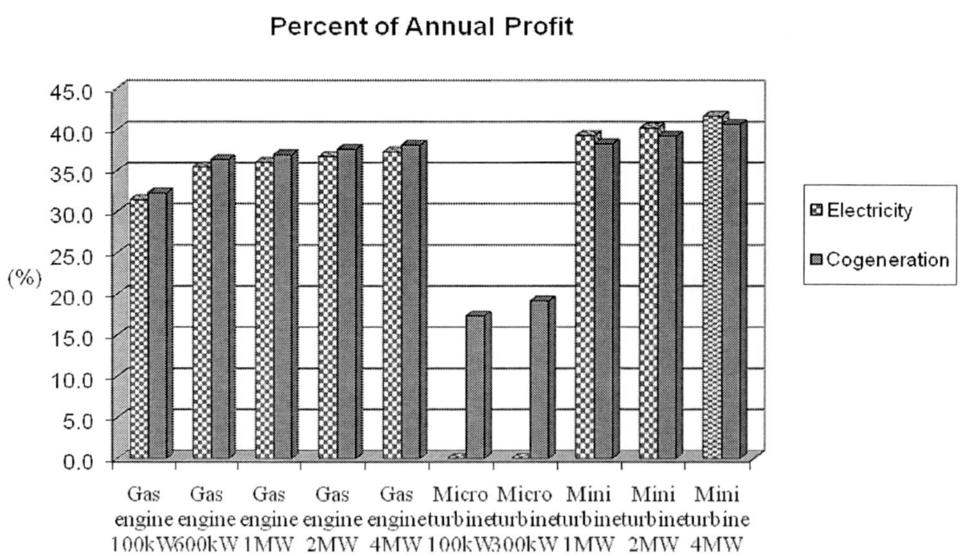

Figure 6. Annual benefit percent of DG different technologies.

Fuel Consumption Rate and Its Annual Saving

As it is shown in bar chart of Figure 7, the fuel consumption rate of gas engines is less. In other words, the rate of its fuel saving rate with regard to the generation capacity is more than other technologies. This important point is one of the strong points for the development of this technology with the possibility of fuel increase price in the near future. This dictates its economic and justified position as compared with micro turbines.

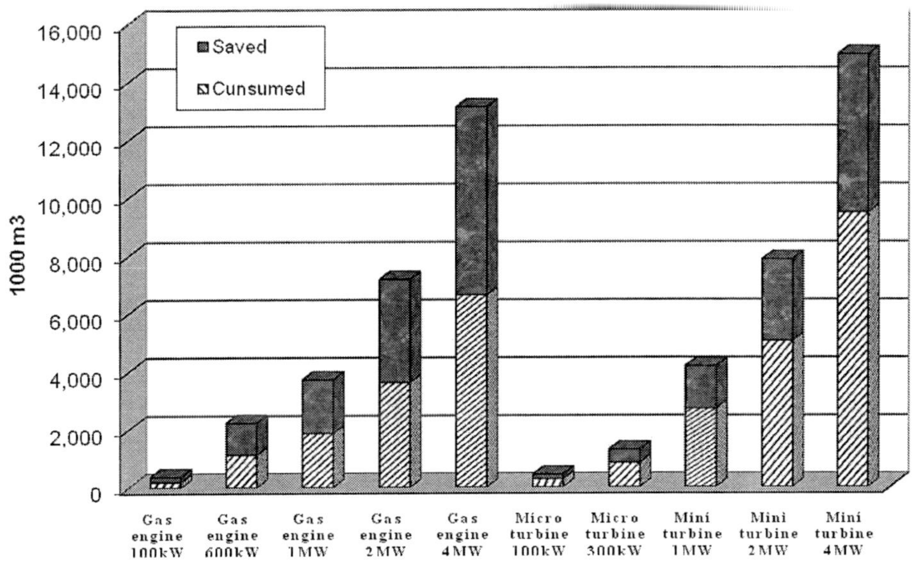

Figure 7. The fuel consumption rate and its annual gas consuming saving in DG different technologies.

Reducing the Dissemination of Environmental Pollutants

The emission of pollutants, in particular NO_x and CO_2, in these technologies are within the accepted limits of global environmental rules, though the emission of these pollutants in technologies of mini turbines and micro turbines are less than gas engines. These comparisons are shown in figures 8 and 9.

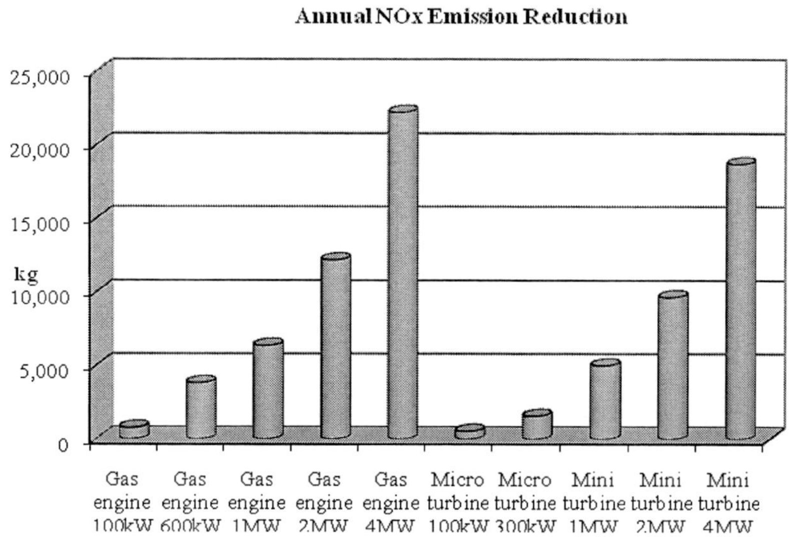

Figure 8. Comparing the emission rate of NO_x in different DG technologies.

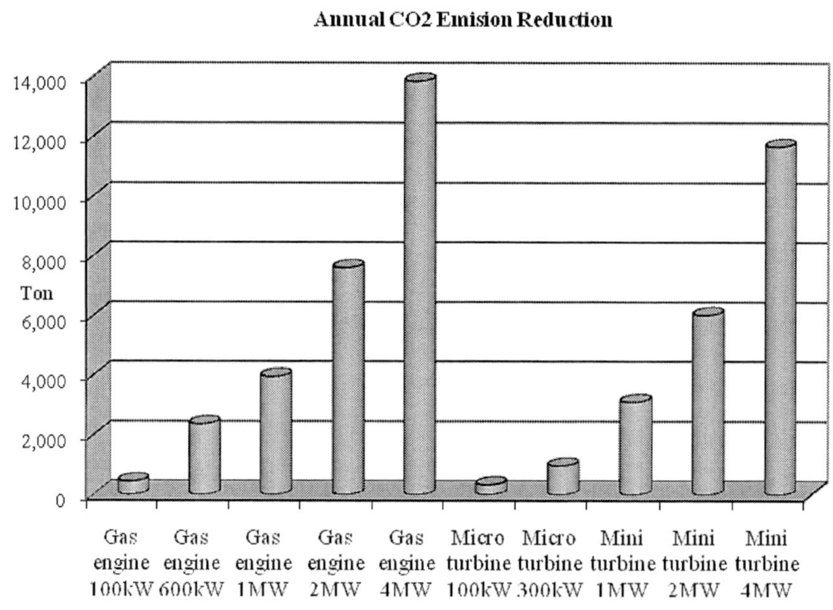

Figure 9. Comparing the emission rate of CO_2 in different DG technologies.

Concluding Remarks on the Selection of Technology

With regard to the analyses, the selection of gas engines and mini turbines are given priority. However, with regard to the generalities of analyzes, in total; we will see that at present also the gas engines will be a better option.

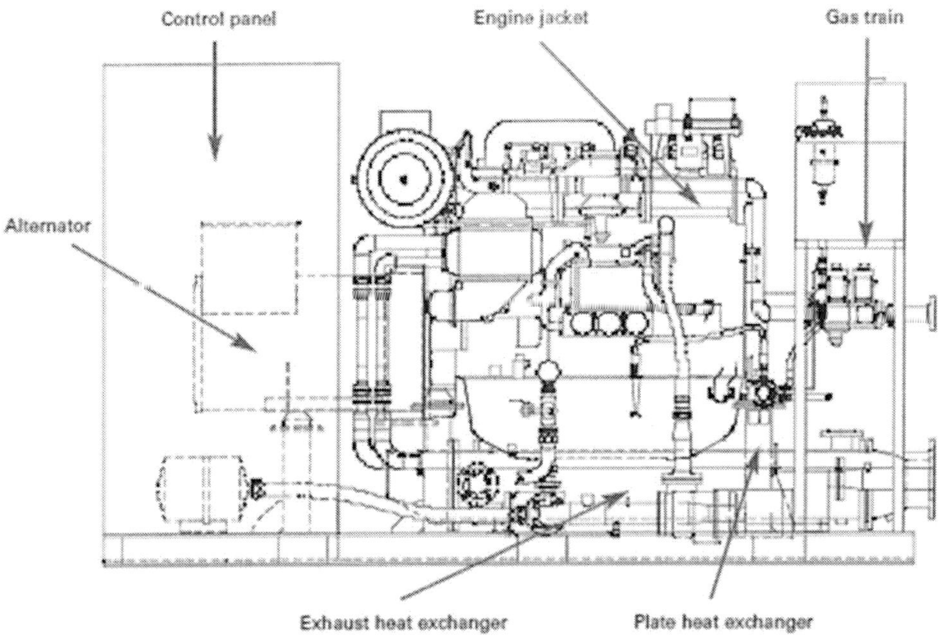

Figure 10. A schematic of gas engine for electricity generation with recovery equipments of heat generated.

In addition to the fact that experience has shown these engines can provide a high level of assurance and accessibility even with a full load , they can be increase DG system abilities by 98 percent by having an additional equipment. On the other hand, the duration of life of these engines is about 20 years if they undergo some trivial fixing and a basic repair. Figure 10 shows a schematic image of this gas engine with a heat recovery.

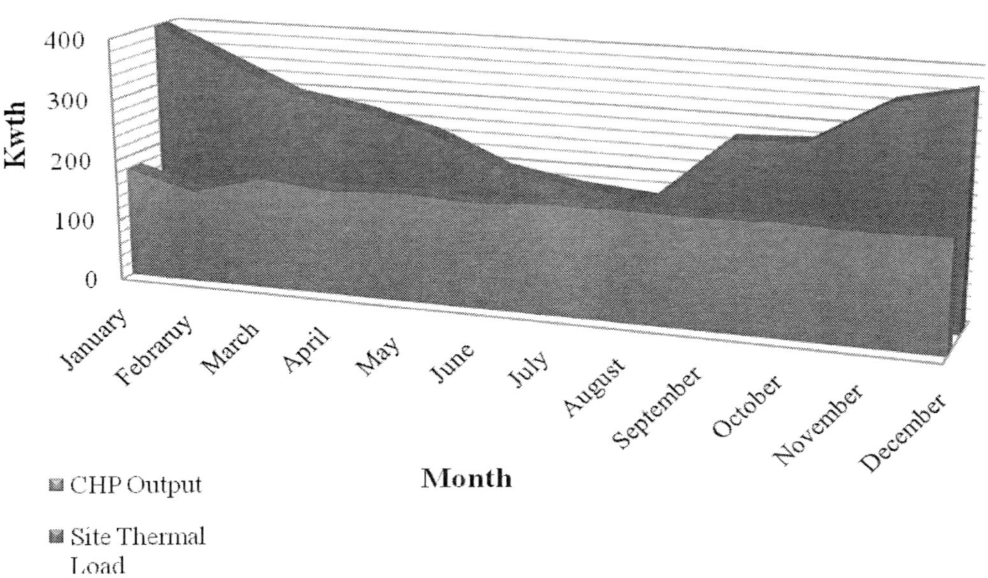

Figure 11. The annual heating (or cooling) demand rate of a sample building with the heat recovery rate of this generator.

Conclusions

As it was said before, using some equipment concurrently or having an additional supportive unit can increase the accessibility and general assurance of DG system by 98 percent. With regard to the fact that at present purchasing electricity from network and selling it (the difference of purchased and sold electricity) and injection of electricity into the nationwide network is the most economic method in returning investment on small scale generators. Therefore, that function of these generators with a full load and injection of its electricity to the network is one of the necessary conditions to enhance assurance of return of capital in less than 2.5 years in this investment.

Moreover, as it was said, with the increase of fuel price, the fuel cost over cost of electricity generation will be a considerable amount. Thus, assurance of investment and return of capital in less than 2.5 year will be materialized fully if a project to be implemented in form of CHP using a heat recovery coefficient higher than 0.8. This needs that the selected generators in full load of generation could meet the heat rate (or cooling) load of the base of building in a year. Figure 11 shows clearly the full use of heat recovery of these kinds of generators.

STAGE OF INTEGRATING RENEWABLE SOURCES INTO DG SYSTEMS OF THE EUROPEAN UNION

States of the European Union have a different approach for addressing renewable energy. The diversity of energy efficiency measures and implementation of renewable energy strategies is presented in sequel.

Table 1. Renewable energy in final energy consumption (2020 target) [21]

	EU Member State	2006	2020 Target	% To cover:	Bar Graph:
1	United Kingdom	1,5 %	15 %	13,5 %	
2	Ireland	2,9 %	16 %	13,1 %	
3	Denmark	17,2 %	30 %	12,8 %	
4	France	10,5 %	23 %	12,5 %	
5	Netherlands	2,7 %	14 %	11,3 %	
6	Spain	8,7 %	20 %	11,3 %	
7	Greece	7,1 %	18 %	10,9 %	
	EU	9,2 %	20 %	10,8 %	
8	Italy	6,3 %	17 %	10,7 %	
9	Latvia	31,4 %	42 %	10,6 %	
10	Belgium	2,6 %	13 %	10,4 %	
11	Cyprus	2,7 %	13 %	10,3 %	
12	Germany	7,8 %	18 %	10,2 %	
13	Luxembourg	1 %	11 %	10 %	
14	Malta	0 %	10 %	10 %	
15	Portugal	21,5 %	31 %	9,5 %	
16	Slovenia	15,5 %	25 %	9,5 %	
17	Finland	28,9 %	38 %	9,1 %	
18	Austria	25,1 %	34 %	8,9 %	

	EU Member State	2006	2020 Target	% To cover:	Bar Graph:
19	Lithuania	14,6 %	23 %	8,4 %	
20	Estonia	16,6 %	25 %	8,4 %	
21	Hungary	5,1 %	13 %	7,9 %	
22	Sweden	41,3 %	49 %	7,7 %	
23	Poland	7,5 %	15 %	7,5 %	
24	Slovakia	6,8 %	14 %	7,2 %	
25	Bulgaria	8,9 %	16 %	7,1 %	
26	Romania	17 %	24 %	7 %	
27	Czech republic	6,5 %	13 %	6,5 %	

Austria

With a share of 70% RES-E of gross electricity consumption in 1997, Austria was the leading EU Member State for many years. Large hydropower is the main source of RES-E in Austria. More recently, a steady rise in the total energy demand has taken place, and a decrease of the share of RES-E has been noted.

Table 2. Renewable energy in the EU- Share of each resource in the renewable electricity generation (in %) [22]

Ranking:	Resource:	2007:	2006:	2005:	Bar Graph:
2	Hydro energy	60,60 %	64,40 %	66,40 %	
3	Wind energy	20,40 %	17,10 %	16,40 %	
1	Biomass	17,00 %	16,70 %	15,60 %	
4	Geothermal energy	1,20 %	1,20 %	1,30 %	
5	Solar energy	0,80 %	0,50 %	0,30 %	

In May 2006, the annual allocated budget for RES-E support was reduced, and annual tariff adjustments have been implemented. Within the new legislation, the annual allocated budget for RES support has been set at EUR 17 million for "new RES-E" up to 2011. This yearly budget is pre-allocated to different types of RES (30% to biomass, 30% to biogas, 30% to wind, 10% to PV and other RES). Within these categories, funds will be given on a "first come – first served" basis.

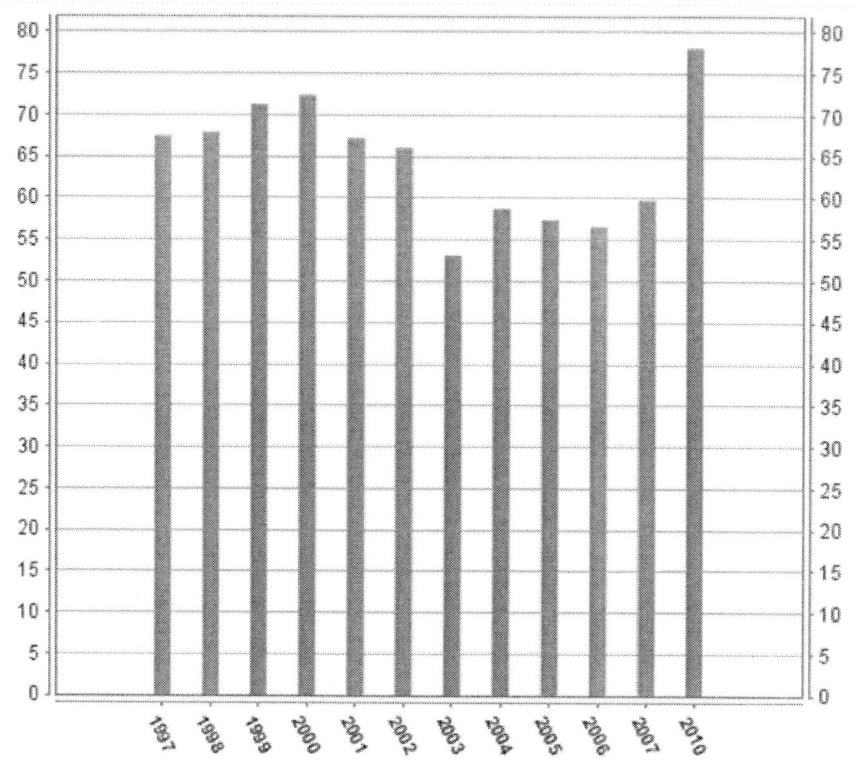

Figure12. Austria - Energy production from renewables between 1997 and 2010 [22].

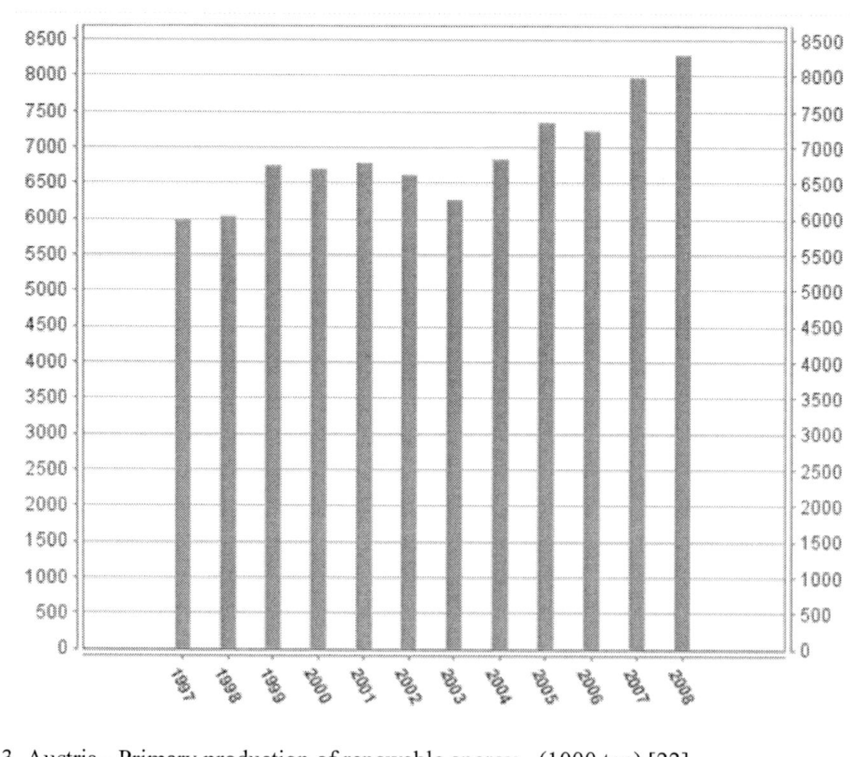

Figure 13. Austria - Primary production of renewable energy - (1000 toe) [22].

Belgium

With a production of 1.1% RES-E of gross electricity consumption in 1997, Belgium was at the bottom of the EU15. Targets differ between the three regions of the country and national energy policies are implemented separately, leading to differing supporting conditions and separate, regional markets for green certificates. Policy measures in Belgium contain incentives to use the most cost-effective technologies. Biomass is traditionally strong in Belgium, but both hydro-power and onshore wind generation have shown strong growth in recent years. RES-HandC is being supported by investment incentives in all three regions. The maximum level of support is as high as 15% in the Walloon region and 20% in both Flanders and Brussels. For biofuels, excise duties have been lowered for diesel that contains at least 3.37% biodiesel and for petrol that contains at least 7% bioethanol. Pure rapeseed oil is completely exempt from excise duties. From March 2007, 4.29% of biodiesel can be incorporated in diesel. Following publication of a federal law on biofuels in 2006, a European call for tenders was launched for selecting a certain number of producers; they will be eligible for fiscal incentives ending in September 2013. An obligation on biofuels is being prepared.

Bulgaria

Bulgaria is approaching its RES-E target for 2010. Large-scale hydro power is currently the main source of RES-E, but its technical and economic potential is already fully exploited. Good opportunities exist for biomass, since 60% of land is agricultural, and about 30% is forest cover.

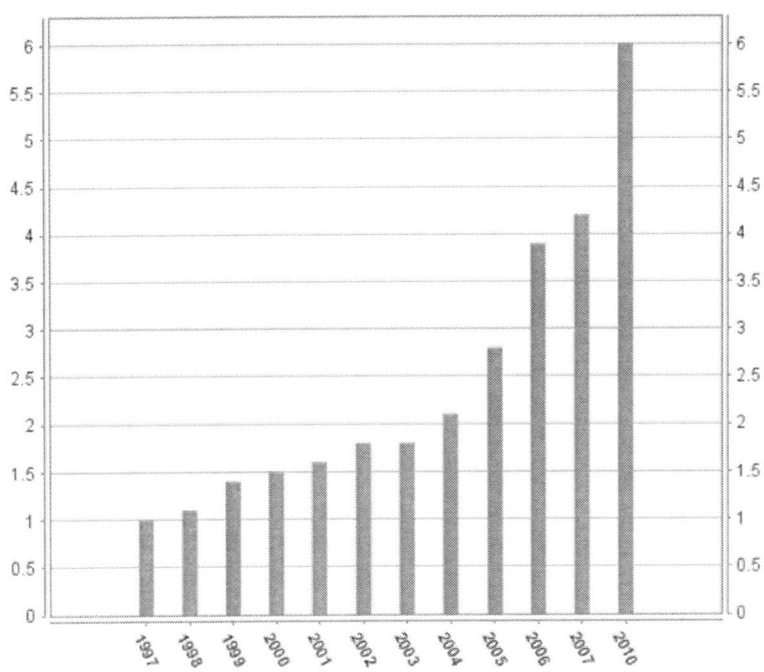

Figure 14. Belgium - Energy production from renewables between 1997 and 2010 [22].

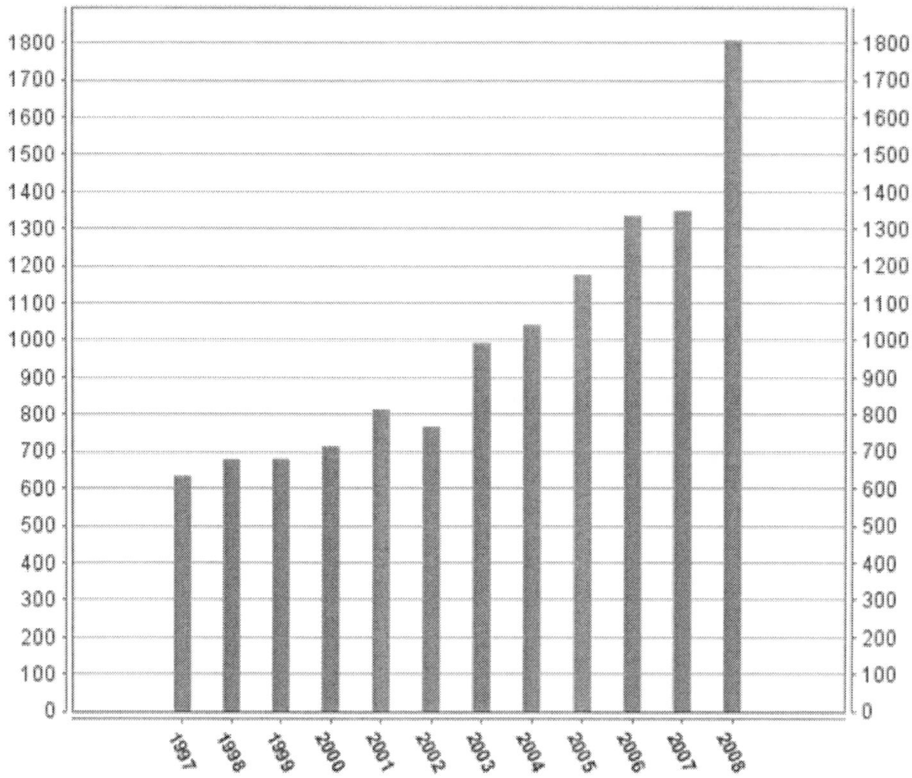

Figure 15. Belgium - Primary production of renewable energy - (1000 toe) [22].

Total wind energy capacity of around 2,200 – 3,400 MW could be installed. Solar potential exists in the East and South of Bulgaria, and 200 MW could be generated from geothermal sources. Energy generated from renewable and alternative energy sources, except for the power generated by hydroelectric power plants with installed capacity of over 10 MW, at preferential prices. A bill is foreseen for 2011 on the market mechanisms for encouraging production of electricity and heating power from renewable energy sources. In order to promote RES, Bulgaria is currently implementing the Bulgarian Energy Efficiency and Renewable Energy Credit Line (BEERECL). RES projects are eligible for a 20% grant. Loans worth more than EUR 12.8 million have already been granted.

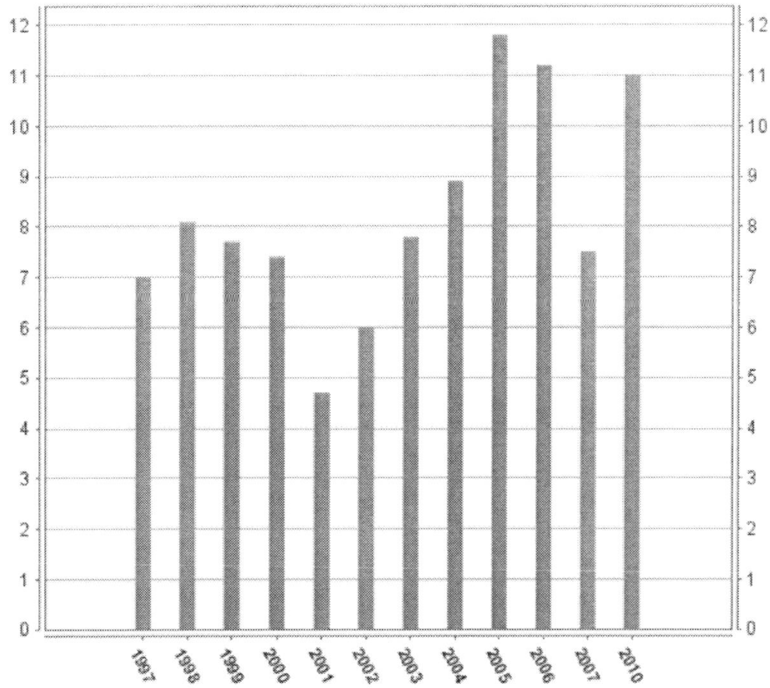

Figure 16. Bulgaria - Energy production from renewables between 1997 and 2010 [22].

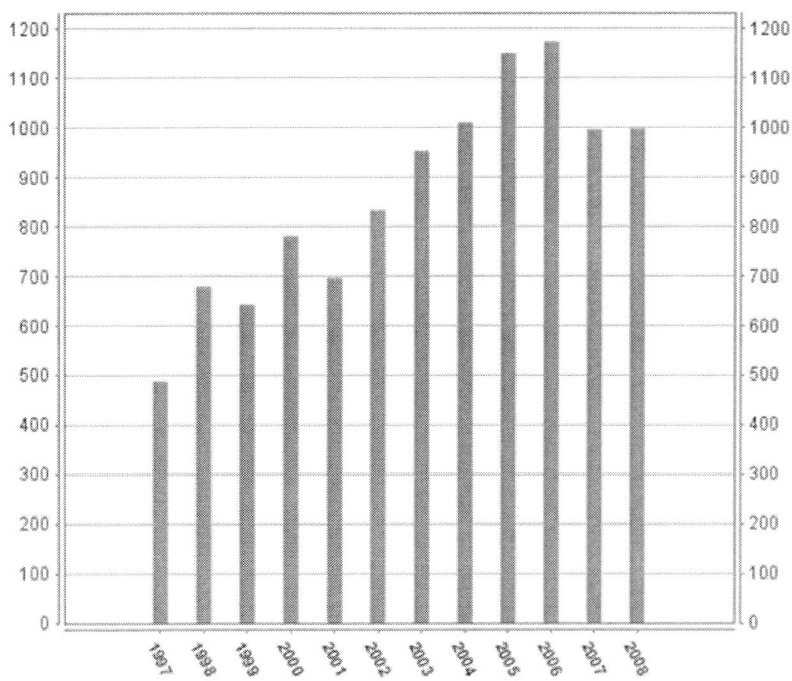

Figure 17. Primary production of renewable energy - (1000 toe) [22].

Czech Republic

The Czech Republic's legislative framework in relation to renewable energy sources has been strengthened by a new RES Act adopted in 2005 and a Government Order regulating the minimum amount of biofuels or other RES fuels that must be available for motor fuel purposes.

The use of biofuels is being encouraged through an air protection Act (2002), which requires that a minimum amount of biofuel (or other fuels produced from RES) is made available to the market for example, as of 1 September 2007, 2% of the total amount of motor-vehicle diesel fuel. Between 2007 and 2012, this should amount to 4.2m tons.

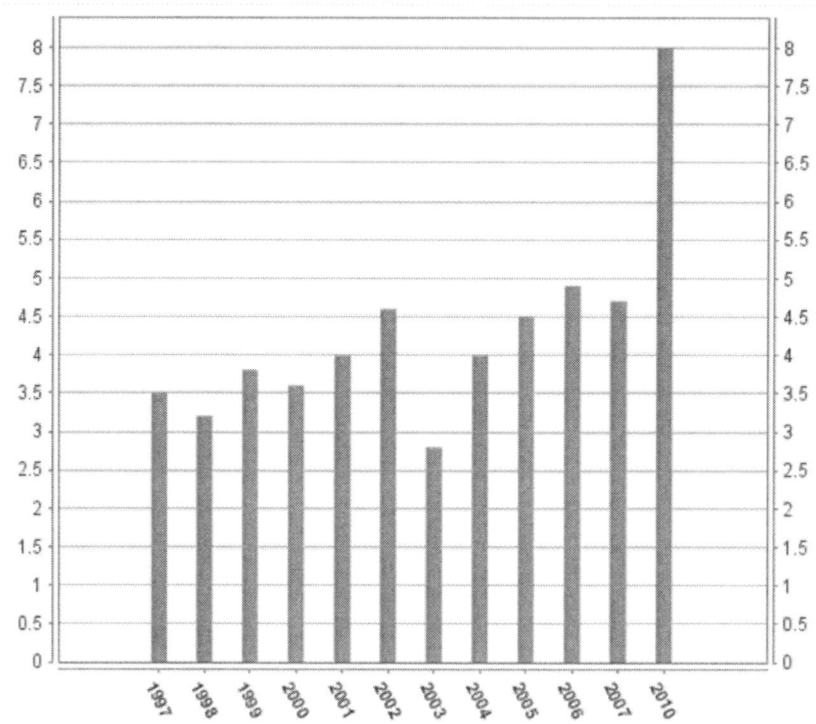

Figure 18. Czech Republic - Energy production from renewables [22].

Denmark

Due to an average growth of 71% per year, Danish offshore wind capacity remains the highest per capita in Europe (400 MW in total in 2006). Denmark is at present close to reaching its RES-E target for 2010. Two new offshore installations, each of 200 MW, are under preparation.

Biofuels have been exempt from the CO2 tax imposed on ordinary petrol and diesel for transport since January 2005; this is currently the main supporting measure for biofuels. In early 2008 a bill will be presented to Parliament, instructing all filling stations to sell at least 5.75% biodiesel and bioethanol as of 2010.

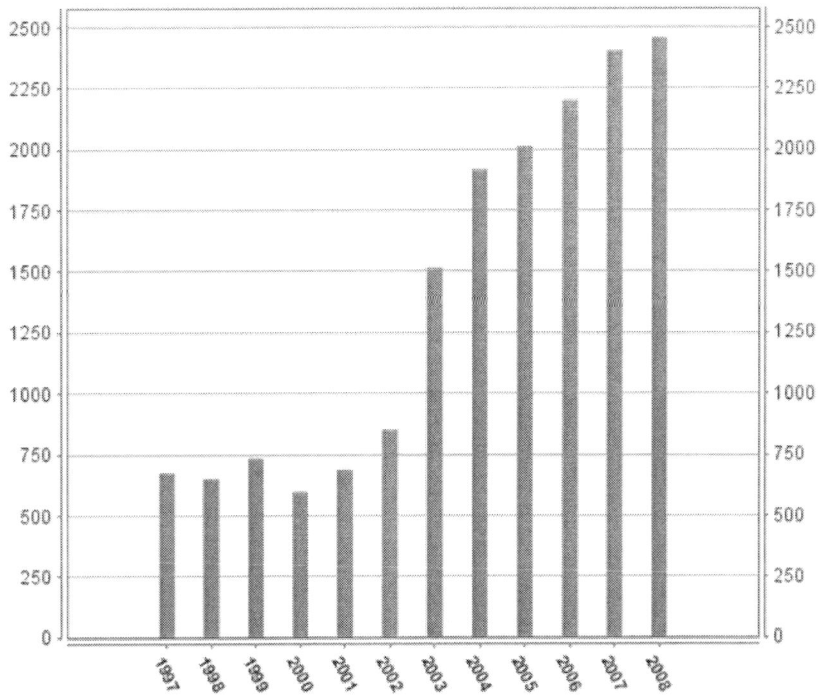

Figure 19. Primary production of renewable energy - (1000 toe) [22].

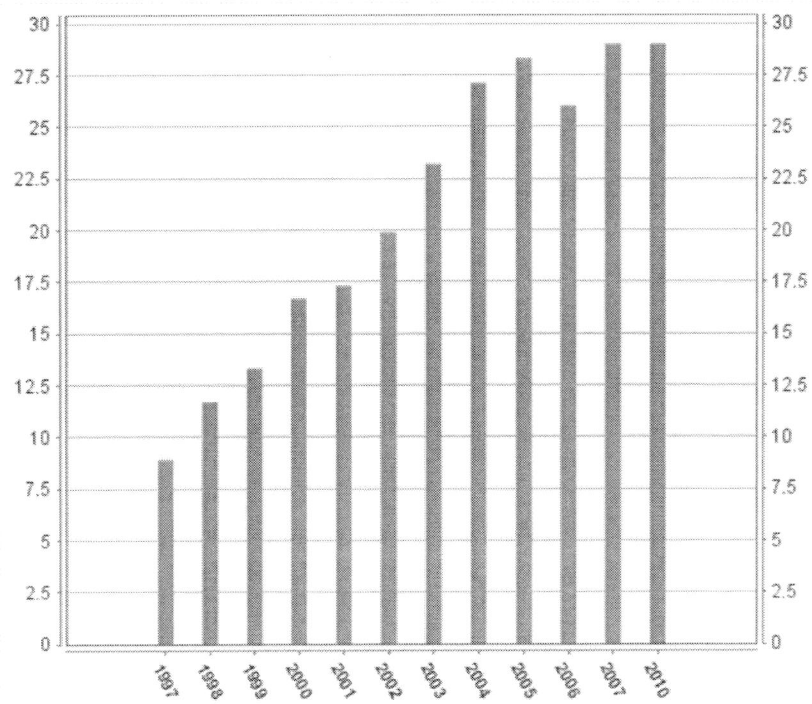

Figure 20. Denmark- Energy production from renewables between 1997 and 2010 [22].

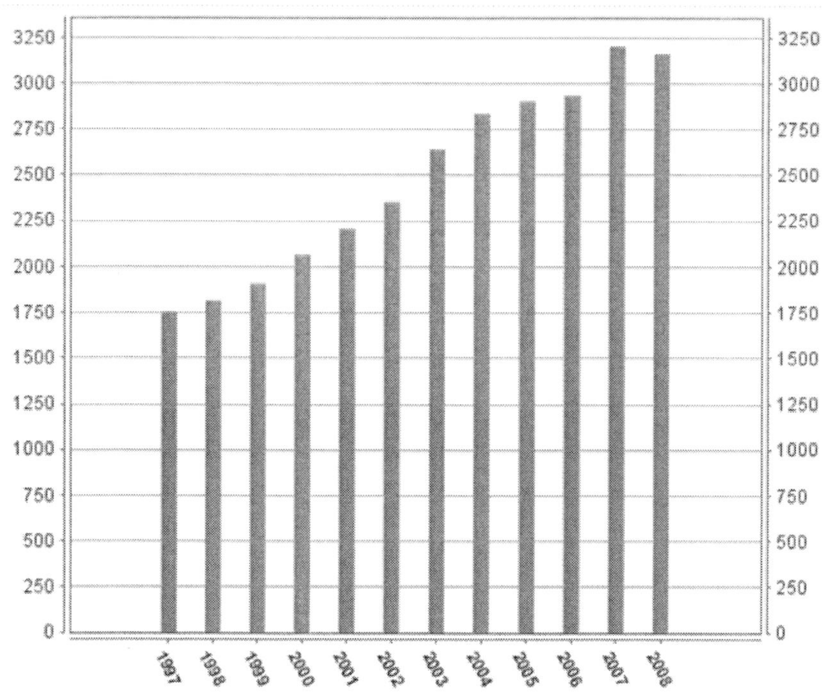

Figure 21. Primary production of renewable energy - (1000 toe) [22].

Germany

Germany is an EU leader in wind utilisation, PV, solar thermal installations and biofuel production. Its onshore wind capacity covers approximately 50% of the total installed capacity in the EU.

A stable and predictable policy framework has created conditions favourable to RES penetration and growth. Feed-in tariffs for RES-E, market incentives for RES-H, and tax exemptions for biofuels have proven to be a successful policy mix leading to a very dynamic market for RES.

In 2006, about 70% of renewable energy was generated from biomass, and 11.8% of electricity was generated from RES. Germany has already exceeded its 2010 biofuel target of 5.75%: in 2006, it achieved 6.3%. From 1 January 2007, firms have been obliged to market biofuels using a quota system: 4.4% for diesel and 1.2% for petrol; this will be increased annually. 2nd generation biofuels, biogas and pure bioethanol (E85) will be granted a decreasing tax incentive until 2015.

Greece

General policies relevant to RES include a measure related to investment support, a 20% reduction of taxable income on expenses for domestic appliances or systems using RES, and a concrete bidding procedure to ensure the rational use of geothermal energy.

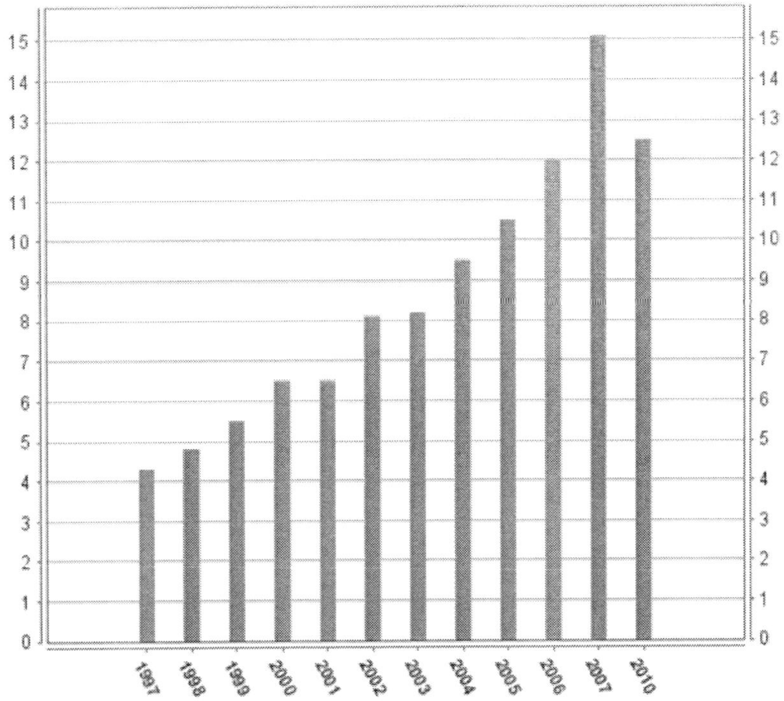

Figure 22. Germany - Energy production from renewables between 1997 and 2010 [22].

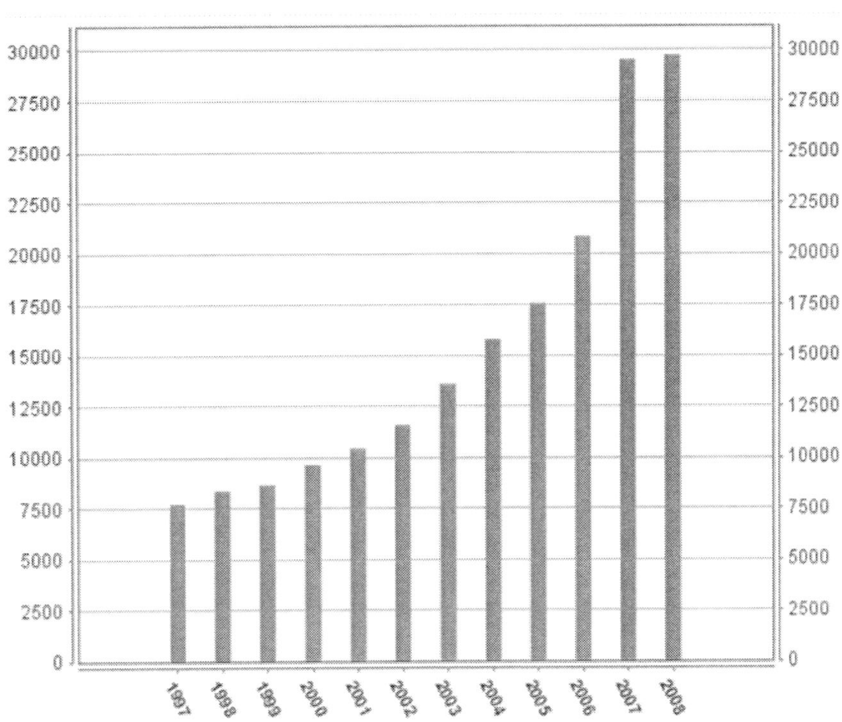

Figure 23. Germany - Primary production of renewable energy - (1000 toe) [22].

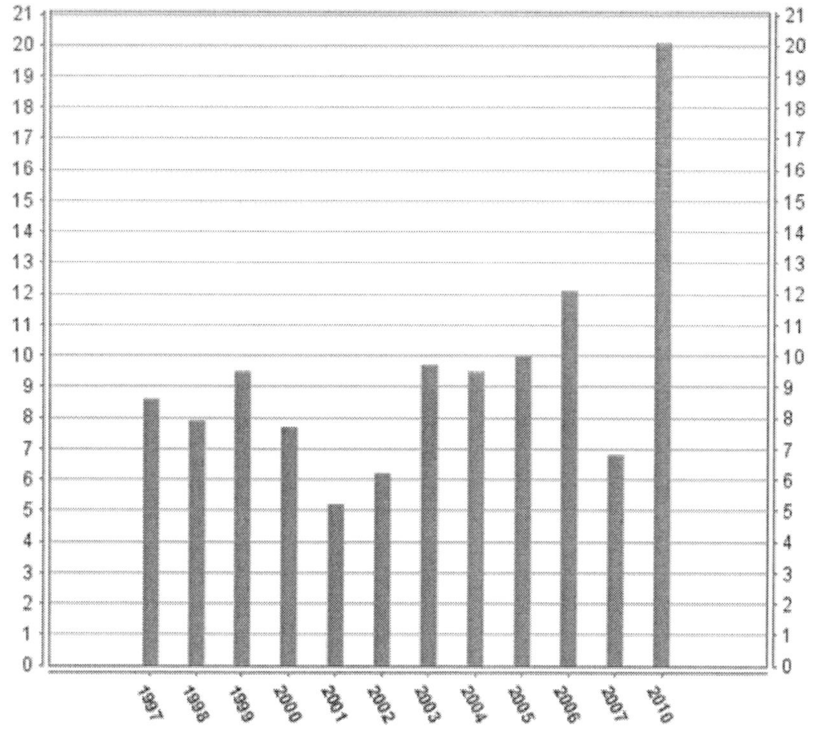

Figure 24. Greece - Energy production from renewables between 1997 and 2010 [22].

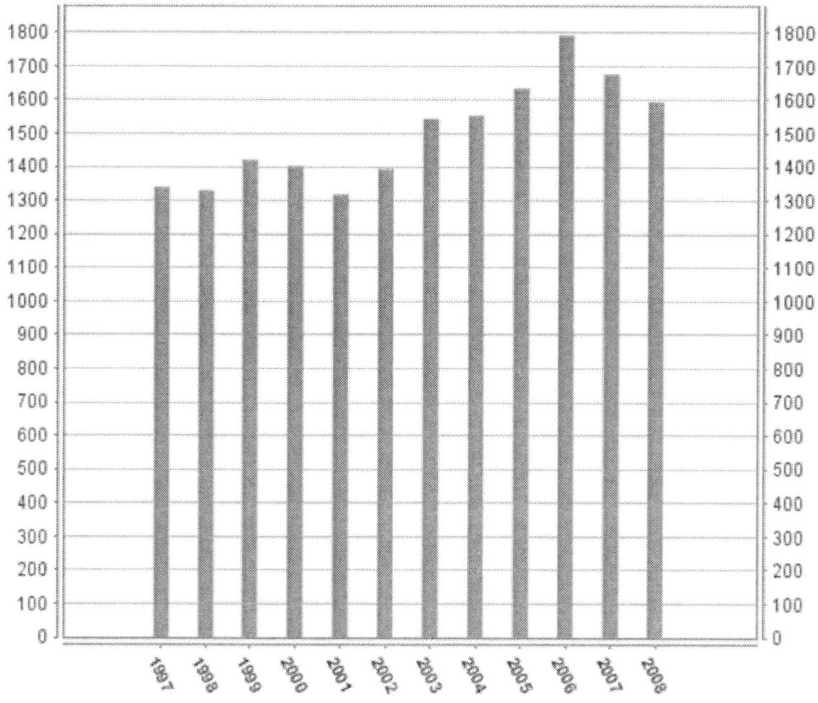

Figure 25. Greece - Primary production of renewable energy - (1000 toe) [22].

Spain

Spain is currently far from its RES-E target. In 2004, hydro power still provided 50% of all green electricity, while onshore wind and biomass had started penetrating the market. PV energy is also promising, with an average growth rate of 54% per year. Proposed changes to the feed-in tariffs and the adoption of a new Technical Buildings Code (2006) show increased support for biomass, biogas, solar thermal electricity, and solar thermal heat. New renewable energy legislation was passed in May 2007 which increased the tariffs for renewables from 50-100% for biomass, and from 16-40% for biogas. RES-H is supported through the new Technical Buildings Code (CTE, 2006). This includes an obligation to cover 30-70% of the domestic hot water demand from solar thermal energy and it applies to all new buildings and renovations. The assumed volume of hot water demand and the geographical location of the building determine the exact percentage that applies. Investments in RES-H are eligible for investment subsidies of 36.4% of the total cost.

France

France has centred its RES approach around feed-in tariffs on the one hand, and a tendering procedure on the other. Hydro power has traditionally been important for electricity generation, and the country ranks high when it comes to biofuel production. France has vast resources of wind, geothermal energy and biomass, and wind power and geothermal electricity have shown growth. In addition, there is potential in the area of solid biomass. In 2006, 3.9 TWh of electricity were produced from biomass, along with 9.2 Mtep for heat production.

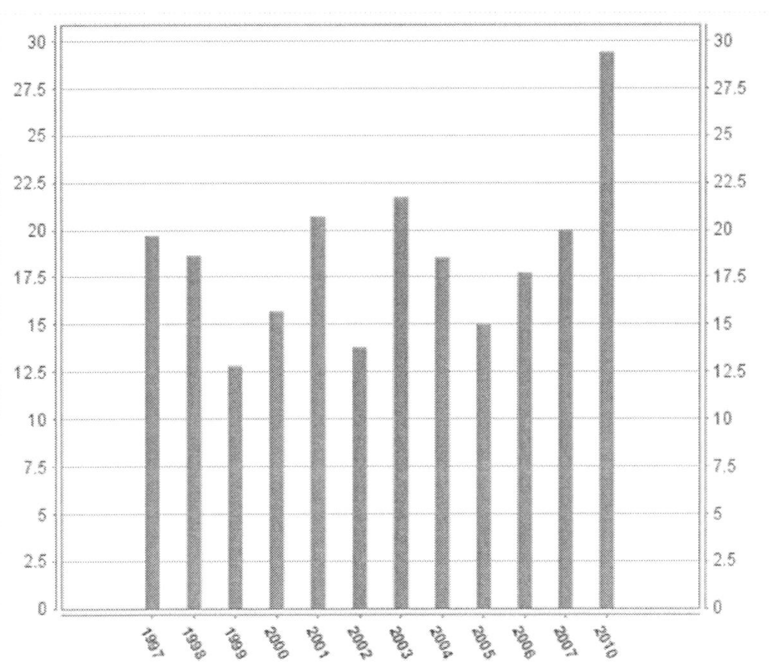

Figure 26. Spain - Energy production from renewables between 1997 and 2010 [22].

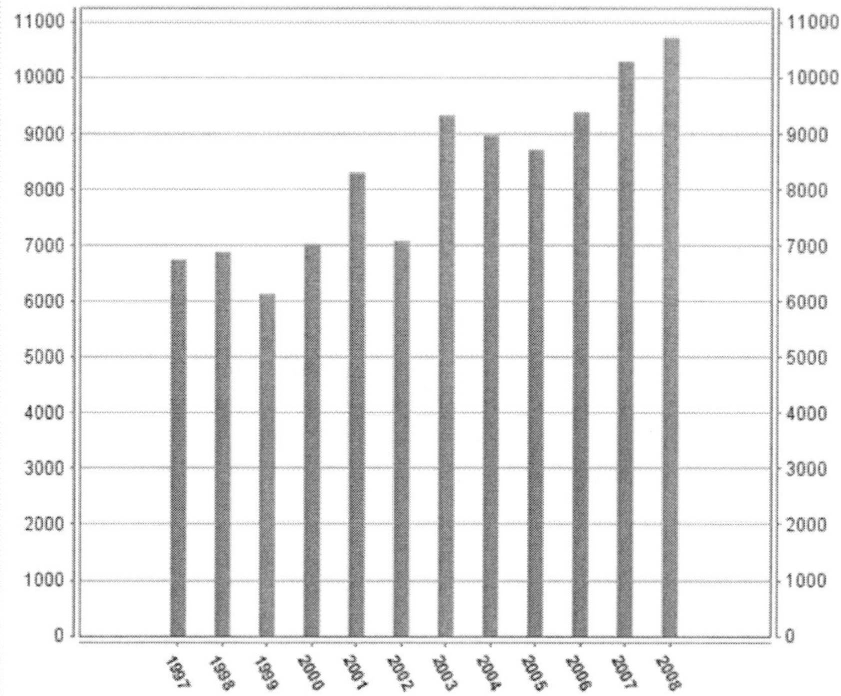

Figure 27. Spain - Primary production of renewable energy - (1000 toe) [22].

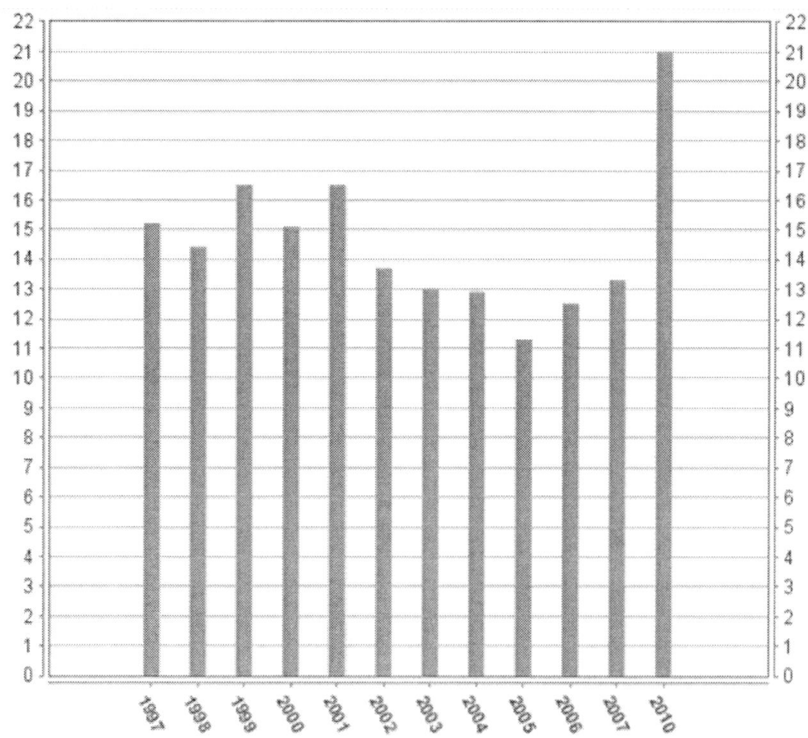

Figure 28. France - Energy production from renewables between 1997 and 2010[22].

Italy

Despite strong growth in sectors such as onshore wind, biogas and biodiesel, Italy is far from the targets set at both the national and European level. At present, the Italian government is working out the details of more ambitious support mechanisms for the development and use of RES. Renewables accounted for 7.6 mtoe (ton of oil equivalent) in 2008.

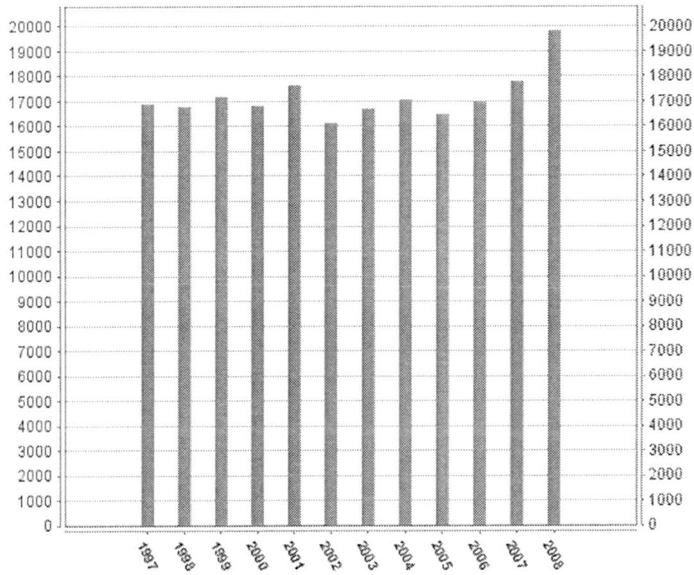

Figure 29. France - Primary production of renewable energy - (1000 toe) [22].

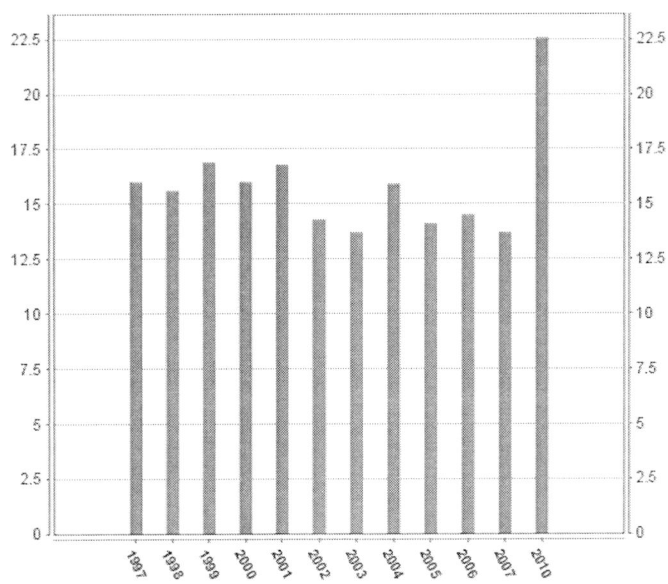

Figure 30. Italy - Energy production from renewables between 1997 and 2010 [22].

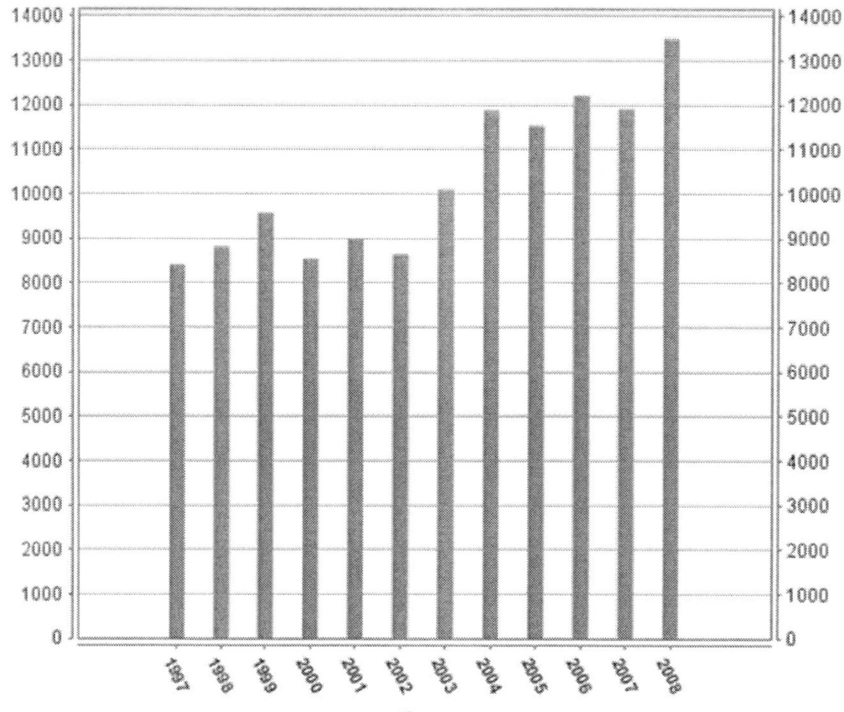

Figure 31. Italy - Primary production of renewable energy - (1000 toe) [22].

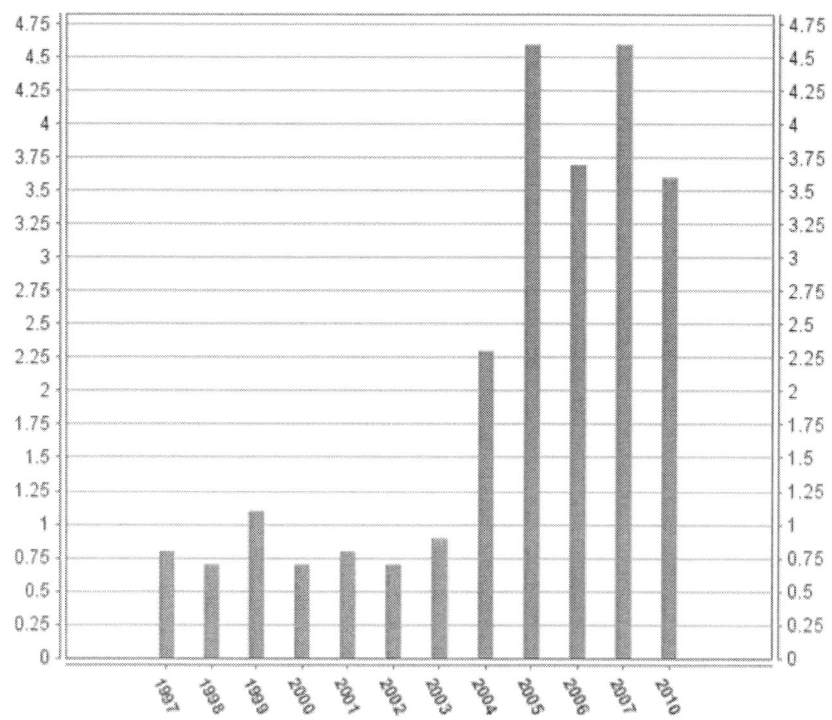

Figure 32. Hungary - Energy production from renewables between 1997 and 2010 [22].

Hungary

Geographical conditions in Hungary are favourable for RES development, especially biomass. Between 1997 and 2004, the average annual growth of biomass was 116%. RES-E 2010 target was achieved in 2006 (5%), with the main contribution being from biomass. However, domestic production was at 4.4%.

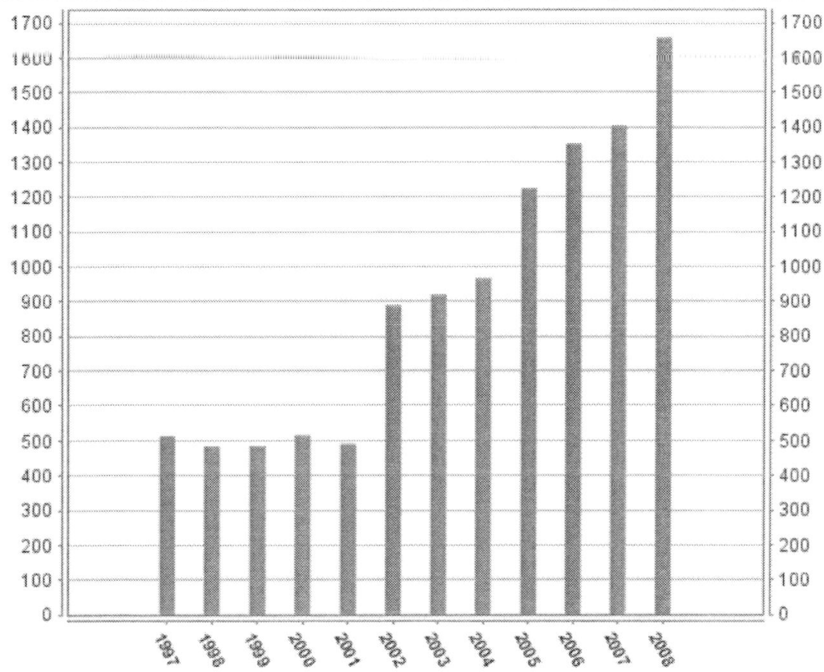

Figure 33. Hungary - Primary production of renewable energy - (1000 toe) [22].

Romania

In terms of RES of gross electricity consumption, Romania is on target. In 2004, the majority of all RES-E was generated through large-scale hydro power. To a large extent, the high potential of small-scale hydro power has remained untouched. Between 1997 and 2004, both the level of production and the growth rate of most RES have been stable.

Provisions for public support are in place, but renewable energy projects have so far not been financed. A quota system with tradable green certificates (TGC) for new RES-E has been in place since 2004. The mandatory quota increased from 0.7% in 2005 to 8.3% in 2010. TGCs are issued to electricity production from wind, solar, biomass or hydro power generated in plants with less than 10 MW capacity.

Legislation on biofuels was transposed into national legislation in December 2005. The list of priorities of the Romanian Energy Efficiency Fund (2002) includes the use of RES for heating. The September 2007 energy strategy includes upgrading and retooling of hydro-power plants with total installed power of 2,328 MW [23].

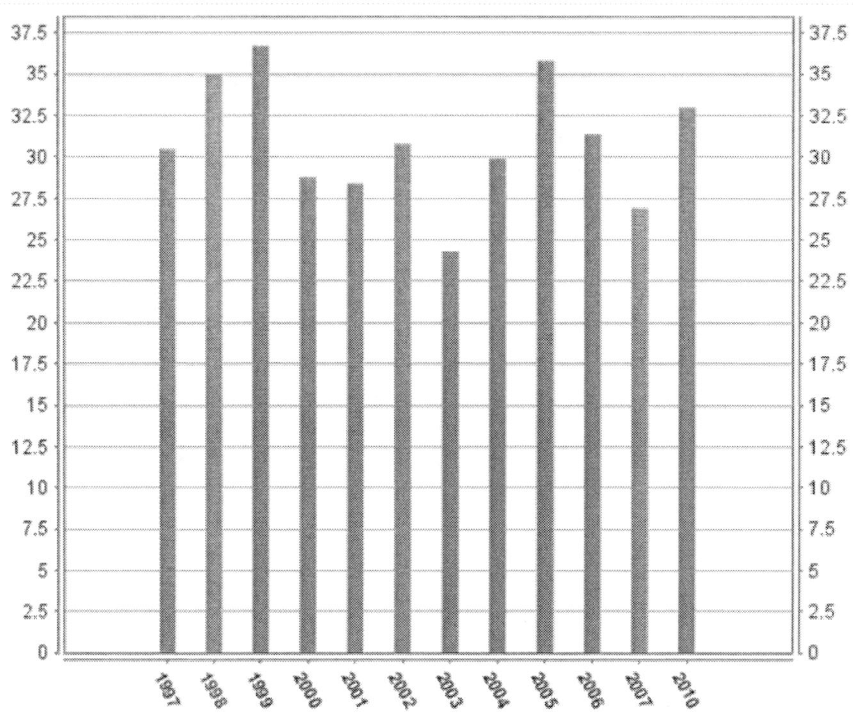

Figure 34. Romania - Energy production from renewables between 1997 and 2010 [22].

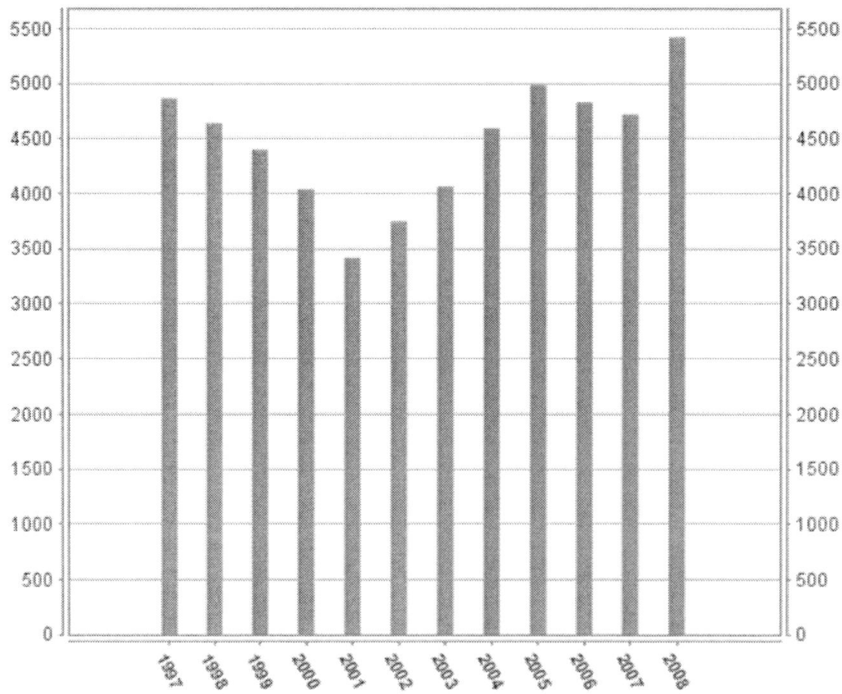

Figure 35. Romania - Primary production of renewable energy - (1000 toe) [22].

Sweden

Tradable Green Certificates were introduced in 2003. The Renewable Energy with green certificates bill that came into force on 1 January 2007 shifts the quota obligation from electricity users to electricity suppliers, and incorporates a new target of 17 TWh by 2016. The environmental premium tariff for wind power is a transitory measure and will be progressively phased out by 2009 for onshore wind. Since 2005, renewable fuels must make up at least 3% of all petrol and diesel consumption for transport operations. A tax relief system is currently in place but will be amended following the implementation of the EU taxation of energy products Directive. Green taxes such as the carbon dioxide tax promote biofuels in an indirect way. In addition, the Swedish government is currently increasing the number of alternative fuel pumps and ensured that 36% of the vehicles it used in 2006 were fuelled, either wholly or in part, by biogas, ethanol or electricity.

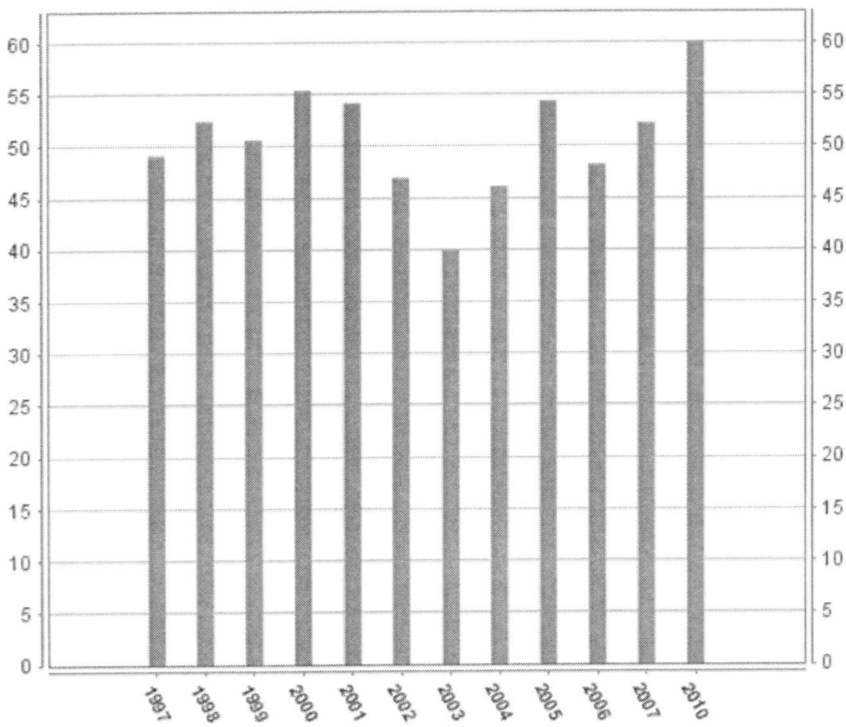

Figure 36. Sweden - Energy production from renewables between 1997 and 2010 [22].

United Kingdom

Obligatory targets with tradable green certificate system (Renewables Obligation on all electricity suppliers in Great Britain to supply a specific proportion of RES-E). The non-compliance 'buy-out' price for 2007-2008 was set at £33.24/MWh (approx €48.20 /MWh), which will be annually adjusted in line with the retail price index. Climate Change Levy: RES-E is exempted from the climate change levy on electricity of £4.3/MWh (approx. €6.3 /MWh).

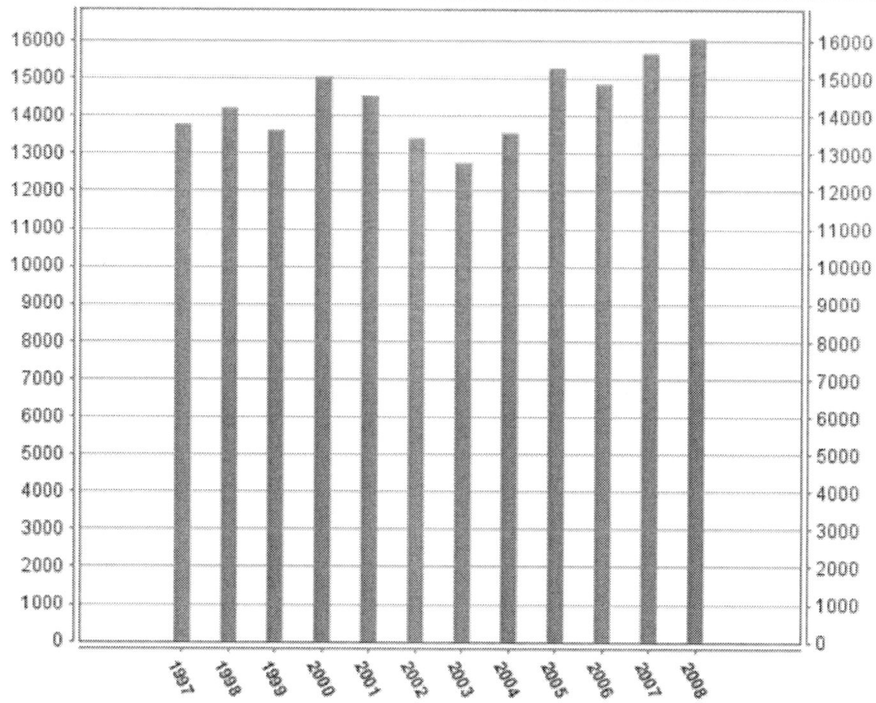

Figure 37. Sweden - Primary production of renewable energy - (1000 toe) [22].

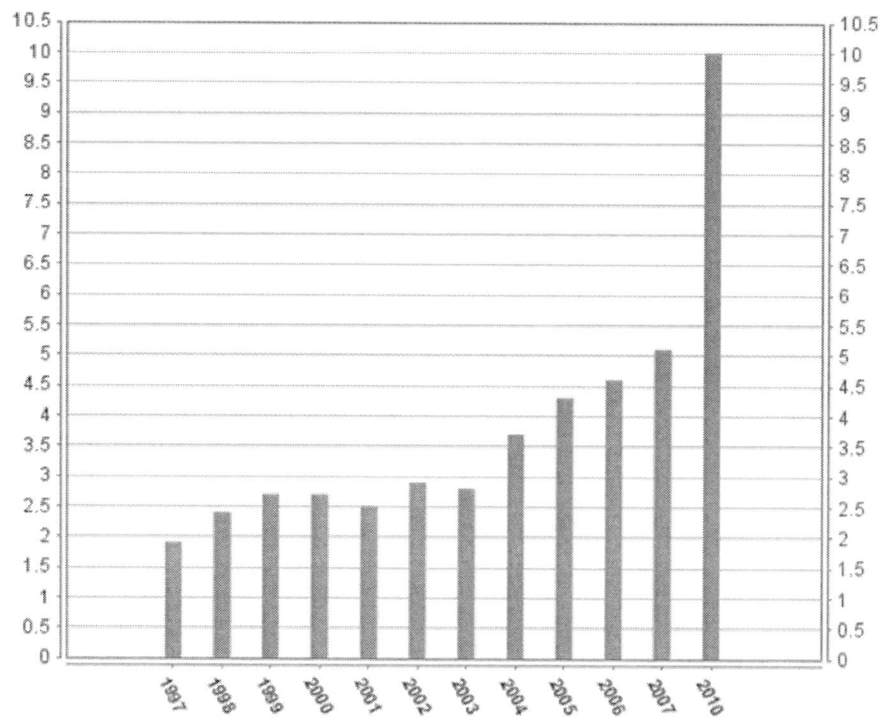

Figure 38. United Kingdom - Energy production from renewables between 1997 and 2010 [22].

In April 2008 the Renewable Transport Fuel Obligation is due to take effect to ensure the UK meets its 2010 target of 5% of transport fuel from biofuels, however, this falls short of the EU target of 5.75% under Directive 2003/30/EC. Certificates can be claimed when biofuels are supplied and fuel duty is paid on them, enabling certificate trading to take place. In 2006, biodiesel and bioethanol received a 20p per litre fuel duty incentive; this will continue until Spring 2010.

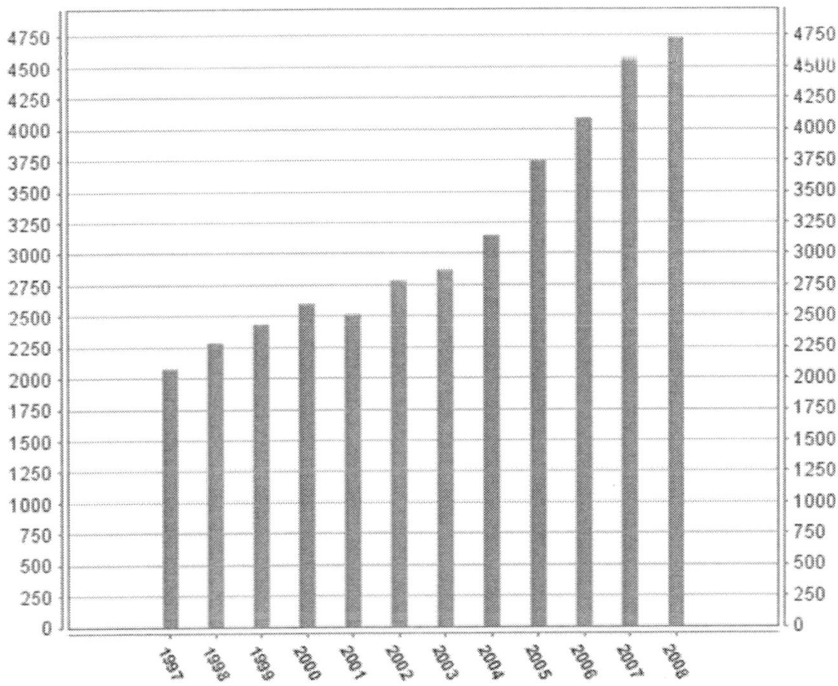

Figure 39. United Kingdom - Primary production of renewable energy - (1000 toe) [22].

RES-H is supported by grant schemes and investment subsidies, biofuels are currently supported by a tax exemption. Following the success of the Renewable (electricity) Obligation, the UK government will introduce a Renewable Transport Fuel Obligation in April 2008. No such scheme is currently planned to promote RES-H.

References

[1] Innovation in Electricity Network. Kunneke University of Technology Delft, 2005.

[2] The Danish Wind Industry Association, www.windpower.org/en/tour.htm.

[3] The Solar Electric Power Association, www.solarelectricpower.org.

[4] Huang X.; Zhang Z.; Jiang J. Fuel cell technology for distributed generation: an overview. *IEEE ISIE* 2006, Montreal, Quebec, Canada.

[5] The European Union Energy, http://ec.europa.eu/energy/res/sectors/bioenergy_en.htm.

[6] Ackermann T.; Anderson G.; Soder L. Distributed generation: a definition. *Electric Power System Research*. 2001, 57, pp 195-204.

[7] International Energy Agency. Distributed generation in liberalised electricity markets. http://www.iea.org.

[8] Eto J.; Koomey J.; Lehman B.; Martin N. Scoping study on trends in the economic value of electricity reliability to the US economy. LBN-47911, 2001, Berkeley, pp 134.

[9] Voorspools K.; D'haeseleer W. The evolution of small cogeneration for residential heating. *International Journal of Energy Research*, 2002, 26, pp 1175-1190.

[10] Voorspools K.; D'haeseleer W. The impact of the implementation of cogeneration in given energetic context. *IEEE Transaction on Energy Conversion*, 2003, 18, pp 135-141.

[11] Dondi P.; Bayoumi D.; Haederli C.; Julian D.; Suter M. Network integration of distributed power generation. *Journal of Power Sources*, 2002, 106, pp 1-9.

[12] Purchala K.; Belmans R.; Exarchakos L.; Hewkes A. D. Distributed generation and the grid integration issues. Imperial Collage London, 2007.

[13] Bergan A. R.; Vittal V. Power systems analysis. Prentice Hall, 2nd Edition, ISBN 0-13-691990-1, 2000.

[14] Ohrstrom M. Fast fault detection for power distribution system. Thesis, Royal Institute of Technology (KTH) Stockholm, 2003.

[15] Ackermann T.; Knyazkin V. Interaction between distributed generation and the distribution network: operation aspects. IEEE/PES Transmission and Distribution Conference and Exhibition Asia Pacific, 2002, Vol. 2, pp 1357-1362.

[16] Friedman N. R. Distributed energy resources interconnection systems: technology review and research needs. Technical Report NREL/SR-560-32459, *National Renewable Energy Laboratory*, 2002.

[17] Wu D. W.; Wang R. Z. Combined cooling, heating and power: a review. *Energy and Combustion Science*, Elsevier, 2006, 32, pp 459-495.

[18] Rezapour K.; Zarbakhsh M. H. Fundamentals of energy conservation and principles of energy management. ISBN 964-6553-17-6, 2nd Edition (with Correction and Additions), 2009.

[19] Guide to decentralized energy technologies. WADE (World Alliance of Decentralized Energy, www.localpower.org), 2002.

[20] Masepohl T. On-site power systems for laboratories. *The National Renewable Energy Laboratory* (NREL), 2003.

[21] Europe's Energy Portal, www.energy.eu.

[22] Eurostat, www.epp.eurostat.ec.europa.eu.

[23] Romanian Energy Regulatory Authority (in Romanian language: Autoritatea Națională de Reglementare în domeniul Energiei), www.anre.ro.

In: Advances in Energy Research
Editor: Nicu Bizon

ISBN 978-1-61209-991-0

Chapter 2

PROBABILISTIC CHARACTERIZATION OF DISTRIBUTION SYSTEMS WITH DISTRIBUTED GENERATION

Radu Florin Porumb [*] ***and Ion Triştiu***
University "Politehnica" of Bucharest, Energetics Department

ABSTRACT

The upraising of the Distributed Generation and Distributed Resources weight within traditional distribution systems is a key topic for discussion and research in the electrical engineering area. This chapter deals with assessing the impact of Distributed Generation upon the existing electrical distribution systems from the reliability point of view.

A main point in the reliability analysis performed in this work is the probabilistic characterization of the random variables involved in the reliability calculations, as time to failure, duration of service restoration and reclosure time. Starting from the actual data collected from a large distribution system, the goodness-of-fit analysis has been performed by means of the Chi-square, Kolmogorov-Smirnov and Geometrical adaptation fit tests, resulting in ranking the most appropriate probability distributions for characterising these random variables.

The reliability assessment has then been performed both for traditional distribution systems and for distribution systems containing Distributed Generation, by means of different types of analysis. In particular, the time-sequential Monte Carlo Simulation and an innovative version of the analytical simulation are used. The latter is based on the combined use of the compound Poisson process and the Fourier analysis for efficiently performing the convolutions among the probability densities of the time-dependent random variables. Numerical tests on various systems, including large real ones, confirm the effectiveness of the approach presented to provide a probabilistic characterisation of the reliability indicators.

[*] Email: raduporumb@yahoo.com, ion_tristiu@yahoo.com

Keywords: Analytical Simulation, Distributed Generation, Electrical Distribution Systems, Goodness-Of-Fit, Reliability Indicators, Sequential Monte Carlo.

NOMENCLATURE

Acronyms

ASAI	Average service availability index
ASIDI	Average system interruption duration index
ASIFI	Average system interruption frequency index
CAIDI	Customer average interruption duration index
CAIFI	Customer average interruption frequency index
CDF	Cumulative distribution function
CEMI	Customers experiencing multiple interruptions
CEMSMI	Customers experiencing multiple sustained interruptions
CTAIDI	Customer total average interruption duration index
DFT	Discrete Fourier Transform
DG	Distributed generation
DER	Distributed energy resources
ENS	Energy not supplied
IDFT	Inverse discrete Fourier transform
LV	Low voltage
MAIFI	Momentary average interruption frequency index
MTBF	Mean time between failures
MTTF	Mean time to failure
MTTR	Mean time to repair
MV	Medium voltage
PDF	Probability density function
PIF	Probability of Islanding Failure
PNS	Power not supplied
RV	Random variable
SAIDI	System average interruption duration index
SAIFI	System average interruption frequency index

Symbols

A	Matrice
C	Load point/system power
D	Total duration of the interruptions (RV)
Dtot	Total duration of service restoration
d	Duration of the interruptions
d	Duration of the interruptions (RV)
E{.}	Expected value

F	Number of faults
f	Frequency of fault occurrence
f	(Subscript) Fault
f	Operator f(.) – Function operator
H	Number of restoration phases
h	Restoration phase
h	(Superscript) Restoration phase
i	Dummy variable (for integer)
j	square root of (-1)
k	Load point (subscript)
L	Node-to-branch incidence matrix
lij	Generic component of L
n	Number of states
M	Number of feeders
m	Raw moment
m	(Superscript) Fault state
N	Number of load points of the system
n	Number of faults / Number of occurrences of fault (RV)
Nf	Deterministic number of occurrences of fault
p	Probability of an event
Pf	Interrupted power
R	Total number of observations
r	Random number
S	Number of system components
T	Duration of the time interval of the analysis
Tm	Equivalent weighted restoration time
t	Dummy variable (for time)
W	Total energy not supplied (RV)
w	Energy not supplied for the whole system (RV)
x	Random Variable
y	Random Variable
z	Random Variable
α	Shape parameter
β	Scale parameter
Φ	Characteristic function
φ	Number of independent restoration phases
Γ	Inverse L
γij	Generic component of Γ
δ	Binary variable
Θ	Set of faults
Θk	Set of faults for which load point k is not supplied
Λmax	Maximum level of significance
λ	Failure rate
μ	Central moment
ν	Degrees of freedom
ξ	Scale factor

ρ	Repair (restoration) rate
σ	Variance (standard deviation)
τ	Restoration time (RV)
Ψ	Second characteristic function
ψ	Observed value for Kolmogorov-Smirnov test
ζ	Observed value for the Chi-square test
ω	Parameter that can have any real value

Distribution System Structure and Operation

Distribution System Structure

Retail sale of electric energy involves the delivery of power in ready-to-use form to the final consumers. Whether marketed by a local utility, load aggregator, or direct power retailer, this electric power must flow through a power delivery system on its way from power production to customer. This delivery system consist of thousands of transmission and distribution lines, substations, transformers and other equipment scattered over a wide geographical area and interconnected so that all function in concert to deliver power as needed to the utility's customers.

As a consequence of the hierarchical structure of voltage levels, a power system can be thought as composed of several distinct levels of equipment, as illustrated in Figure 1 [1]. Each level consists of many units of fundamentally similar equipment, doing roughly the same job, but located in different parts of the system. The present paper deals with the radial type of the electrical distribution system, due to the fact that is the most utilized by nowadays utilities. This concept does not exclude any kind of network topology, as long as it is part of the radial systems (all loads fed by one root node). For performing calculations as close as possible to the reality, the studied networks will include all the elements existing in a real electrical distribution network:

- generators
- branches (lines or transformers)
- circuit breakers;
- disconnects:
- remotely controlled;
- manually controlled
- loads
- control equipment

The design of the electricity distribution system structure is thought to be stratified into layers to simplify its numerical treatment. Therefore, starting from the root node:

the nodes are numbered sequentially in ascending order proceeding from layer to layer (Figure 1), in such a way that any path from the root node to a terminal node encounters node numbers in ascending order;

each branch starts from the sending bus (at the root side) and is identified by the number of its (unique) ending bus.

It was considered a radial network containing n+1 nodes fed at constant voltage at the initial node (root node, node 0), as in Fig 1. Starting from the root (which was not included in the list) and for each i-th node, it was defined path(i) as the ordered list starting from the node 0 and moving to its i-th node [2]. For more convenience, each node belongs to a layer, which represents the position of the node in the network [3]. For the i-th node, layer(i) = dimension(path(i)). As an example, for node #10 in Figure 1, path(10) = {0;1;3;7;8} and layer(10) = 5.

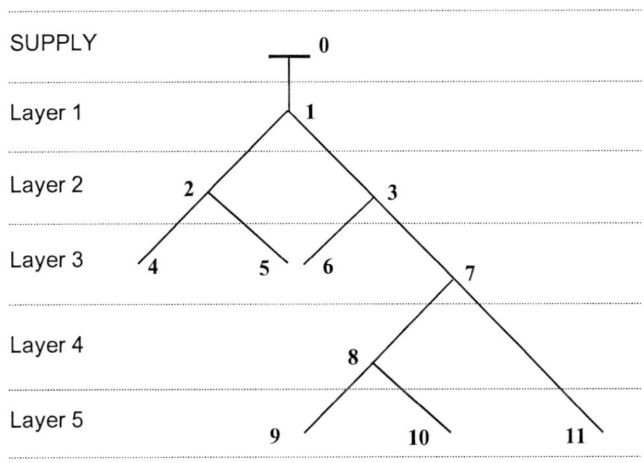

Figure 1. Radial system structure with layer stratification.

The layer representation of the distribution system is used to characterize the distribution system structure by means of the node-to branch incidence matrix.

Faults in the Distribution System and the Restoration Process

A crucial aspect for distribution system reliability is the service restoration process after a fault. In this process, the number of customers affected by an interruption is progressively reduced by performing a set of switching and sectionalising actions aimed at locating the fault, isolating the faulted section and progressively restoring the service to all the customers involved in the interruption. For example, Figure 2 shows for a real distribution system, the evolution in time of the number of customers that experience supply interruption during the execution of manual restoration operations after a fault.

The resulting curve, describing the ongoing process of service restoration, has a multi-step profile. The process of service restoration finishes when the last customer (or group of customers) is energised. The time instant at which all customers are supplied is used to compute the total duration Dtot of service restoration. Furthermore, an equivalent weighted restoration time Tm can be evaluated, corresponding to the weighted average of the duration of supply interruption during the restoration phases, assuming as weights the numbers of interrupted customers at each phase. The pair (Dtot, Tm) provides indicative information on the evolution of the restoration process.

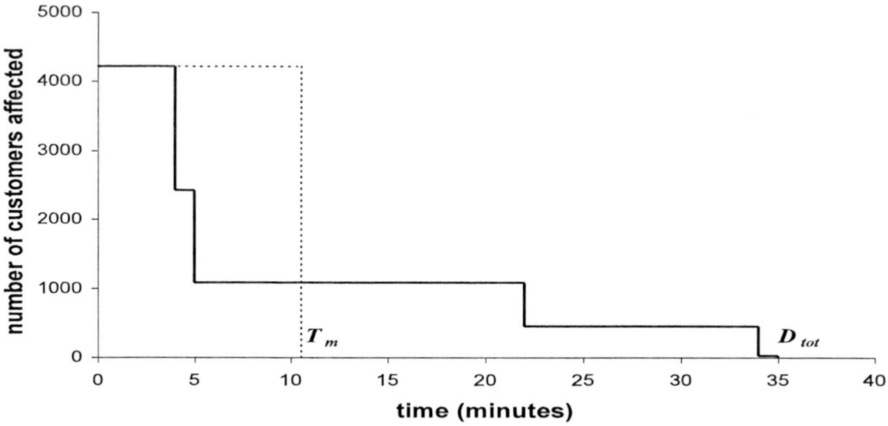

Figure 2. Evolution of the number of customers interrupted during the service restoration process.

Different procedures and techniques have been proposed for assisting the distribution system operators in performing restoration. Switching procedures based on the network structure have been proposed in [4]. Other papers [5,6,7,8] address the factors affecting the restoration times or split the restoration times into their partial time components. The latter technique could be useful in the analysis of rural distribution systems with long distances among the load points.

In this case, the different phases of the restoration process may include ticket creation time, dispatch time, travel time and repair time, according to [9]. The service restoration operations cannot be studied and evaluated uniformly, due to specific properties of each electrical distribution network subdivision. The array of different variables that are inflicting delays for the upper mentioned restoration process phases is too large to be correctly evaluated by analytical methods.

Distributed Generation and Resources

Distributed Generation (DG) and, more generally, Distributed Energy Resources (DER) including DG and storage, have definitely emerged as key options for promoting energy efficiency and use of renewable sources as an alternative to the traditional production of energy. Real-scale applications have been developed more and more rapidly by following the evolution of the technologies, under several constraints imposed by technical and environmental issues. Compared to some forecasts done in the early 80's [10], for which the total installed capacity in the U.S. was expected to be of about 1200 GW in the year 2000 and the impact of DER (wind, thermal solar, photovoltaic, and hydro pumped storage) was expected to be about 2.5% (30 GW), what happened in the year 2000 was a much slower increase of the installed capacity (only slightly higher than 1/3 of the expected growth) reaching about 812 GW [11], while the impact of the above mentioned DER on the installed capacity in the year 2000 was about 2.7% (22 GW). These data do not include the evolution of cogeneration technologies, whose profitability is now emerging for relatively small-scale applications, e.g., using microturbines.

From the technical point of view, the introduction of DER into the distribution systems has significantly changed the characteristics of these systems and has a strong impact on their operation and planning [12,13,14,15]. The introduction of local resources connected to the distribution system primarily affects the distribution system voltage control, short-circuit capability and protection [16,17], that were relatively simple in a radial system supplied by a unique generation point and now is becoming much more complicated. Indeed, positive effects of the DER introduction into distribution systems could result in loss and voltage drop reductions. However, these benefits depend on the level of penetration of the DER, typically intended as the ratio between the amount of peak power provided by the DER and the rated power of the HV/MV supply substation. When this level of penetration increases in such a way that the amount of generation causes significant energy flow reversal in some system branches, the sizing of these branches could become insufficient [18,19].

Although it has emerged more than two decades ago, the DER issue is continuing its way along electrical distribution systems and electricity markets. The present increasing interest in DER raises both from the availability of standardised technologies already in place or under development, and from the flexibility of its application, that merges very well with relatively volatile types of electricity markets that have been set up in different countries. The DER matches the constant need for flexible offer and elastic demand, reliability and price effectiveness, which represent the main trends of interest for the electricity companies and for the consumers. The relatively recent opening of the electricity market even for small consumers (e.g., domestic) has widened the road for DER hardware producers to bring into the light a large array of products that could (at least, in theory) satisfy most types of DER devices demand. Table 1 presents a quick view on the main types of DG, with indications on their power range and market availability.

It was a priori avoided the presentation of costs (of any kind), due to their wide range of values and continuous variation (with an expected decrease in costs as long as the market responsiveness for the new technologies increases). It is an interesting aspect concerns classification of the available technologies for different ranges of rated power, in order to compare their use and possible alternatives. The possibility of finding alternative solutions for the same rated power level depends on the resource availability, which in many cases depends on the location of the site and on other factors such as cost effectiveness, time of return of the investments, environmental impact and possible regulatory incentives. Results of a recent survey put into evidence that nowadays there is a major interest in fuel cells, microturbines, solar and wind DG technologies, while the most interesting DG applications are cogeneration, continuous power and green power. Hybrid systems, allowing for diversified energy mix (e.g., electrical and thermal, such as diffuse cogeneration or solar dishes) can offer advantages in increasing the overall energy efficiency. Moreover, the fuel diversification could be a reasonable option for a country, in order to limit the economical dependence on the resources owned by external producers.

Table 1. Technologies for local generation and storage (up to 100 MW)

Technology		Power [MW]	Fuel Type	Availability
Gas turbines		0.5–100	Natural gas, oil, dual fuel	available
Microturbines		0.025–1	Natural gas, petrol, kerosene	available
Internal Combustion Engines (ICE)		0.005–5	Natural gas, petrol, diesel	available
Stirling engines		0.001–0.02	Natural gas	available
Fuel cells	PA (phosphoric acid)	0.1–0.2	Hydrogen	available
	PEM (proton exchange membrane)	0.003–0.25	Hydrogen	under development
	MC (molten carbonate)	0.25–10	Hydrogen	under development
	SO (solid oxide)	0.001–10	Hydrogen	under development
Biogas		0.1–50	Hydrogen	available
Micro hydro plants		0.025–1	Water	suitable sites
Hydro plants		1–100	Water	suitable sites
Geothermal plants		5–100	Geothermal	suitable sites
Tides/wave motion systems		0.1-1	Tides/waves	suitable sites
Photovoltaic systems		0.001–0.5	Sun	available
Wind systems		0.01–5	Air	suitable sites
Storage		< 5	Various	available

Integration of Distributed Generation into the Electricity Distribution Systems

The development of new technologies for small-scale generation and the trend towards the adoption of distributed resources is modifying the characteristics of distribution systems [20,21,22].

From the reliability point of view, the presence of local generation sources could improve the system performance by allowing temporary islanding, where separate areas are fed by different generators until the repair of the faulted component has been completed. However, feasibility of islanding conditions needs to be accurately verified [23,24,25].

Distribution systems with multiple supply points from HV/MV substations are usually operated with radial configurations, in which each HV/MV substation supplies a radial portion of the system.

Additional local generation units can be located into the system, either connected to the network through a transfer switch to provide backup supply, or connected in parallel to the system during normal operation [26]. Local generators in parallel to the network operate in peak shaving mode if they do not entirely cover the local load, or in net metering mode if they are able to exceed the local load. While a generator in peak shaving mode improves reliability by decreasing the network loading, a generator operating in the net metering mode impacts on the system reliability by inverting some power flows. The local generation should not exceed the contribution of the HV/MV substation, otherwise reverse power flows could be critical for protection coordination and line overloading [22].

In present literature, there have been taken into account, the technical benefits – in a quantitative mode - of DG inclusion into electrical distribution network, as voltage profile improvement index, line-loss reduction index and DG benefit index [27].

One concern is that the presence of DG devices on electric distribution system might contribute significant short-circuit current to faults and destroy the overcurrent protection coordination of reclosers and fuses. This could result in a greater number extended power outages for electric customers. Another concern is that resonant conditions can develop on a section of a distribution feeder that is isolated with DG during single phase-to-ground faults. This may result in an overvoltage condition on the unfaulted phases that can damage the DG devices and feeder line equipment.

Also, there is a problem of a DG device continuing to operate on a feeder that has been isolated from the substation by a recloser or other protection device. This is commonly termed "islanding" in the utility industry. In this situation, the DG device can continue to supply power to the loads on the isolated feeder and become a safety hazard for the line repair personnel.

Reliability Models of Distribution Systems

Continuity of Supply

The liberalisation of the electric power industry has provided end-use customers with the choice of selecting their power delivery options more easily. Power quality, which includes continuity of supply, voltage quality and commercial quality, is a key option for the provision of electricity. A specific regulation concerning power quality issues, aimed at improving the quality of the electricity products and services, is being developed in most countries. The electricity market liberalisation has opened new degrees of freedom to distribution companies and customers for including in the contracts for the electricity supply the provision of services at different levels of quality, but a regulatory action is necessary for maintaining sufficiently high quality levels.

All the disturbances occurring during the system operation that cause an alternation in time or in width of the ideal waveforms are sources of non-quality. For the classification of the disturbances, it was adopted the criterion used in the IEEE standards, referred to the nature of the phenomena (sags, harmonics etc.). In the available literature, has been made a distinction between disturbances linked to the continuity of supply, and those affecting the shape of the waveforms, generally indicated as voltage quality problems [28].

While the voltage quality is related to the sensibility of the customer's devices to disturbances of the wave forms, the continuity of supply deals with the interruption of the electricity service. According to most standards, an interruption is characterised by a voltage magnitude lower than 10% of the rated value. A general classification of the interruptions (IEEE Std. 1159-1995) [29] is based on the duration of the event and leads to the distinction among sustained (over 1 min), temporary (from 3 s to 1 min) and momentary (from 0.5 cycle to 3s) interruptions. The European Norm EN 50160 makes a distinction among long (over 3 min) and short (up to 3 min) interruptions, while a third level of transient interruptions (up to few seconds) is used in some countries. While a classification of the interruptions with three levels of maximum duration is widely used, the limits for the event duration at each level vary from country to country. For example, in Italy the distinction among transients (up to 1 s), short (from 1 s to 3 min) and long (over 3 min) interruptions.

The costs linked with interruptions are simpler to be determined for a commercial or industrial activity than for residential sector. In a competitive electricity market, the value of continuity of supply has another possible interpretation – the customer's willingness to pay to avoid the interruption.

A specific regulation for the continuity of supply is in operation in several countries. In formulating a remuneration mechanism for continuity of supply, the following aspects have been taken into account in most existing regulations:

- different approaches are used: customer-oriented (monitoring the continuity of supply at each customer terminals), or system-oriented (commuting an average performance index to be applied to all the customers in a given area);
- there is a separate recording of major force events (e.g., earthquake, direct lightning, sabotage, strike) and third-party faults (e.g., faults occurred in the transmission system, or in systems managed by other distributors, or caused by customers), which are not included into the list of remunerable interruptions;
- a distinction is made between expected (notified with r days in advance) and unexpected interruptions; this classification is focused on requiring the effective notification to the customers and encompasses other ones where the interruptions are categorised as planned or unplanned;
- there is a classification of the load areas with respect to the number of inhabitants, e.g.:
    - H, with high load concentration or urban
    - M, with medium load concentration or suburban
    - L, with low load concentration or rural
- there is a distinction among the voltage levels at which the customer is connected; for example, customers connected at HV networks and not by the ones caused in the LV network.

In the case of non-compliance with the continuity of supply standards, remuneration should be required by the regulation, either by directly refunding the customers or by creating a fund to be used for quality improvement programmes. The remuneration can be established for single outages, or after computing a global indicator.

One of the main objectives of the application of quality standards is to promote a substantial improvement in the overall continuity of supply. This theoretical concept must be formulated in a quantitative way. One of the many possibilities is to fix a trend of improvement of some indices as a target for a given number of years, thus defining incentives and penalties on the basis of this target. Some national regulations of continuity of supply adopt this approach.

Continuity of Supply and Reliability

The framework for addressing continuity of supply is given by reliability analysis. Reliability indices are used by the utilities as performance limits, and the values of the indices are updated with time in order to force the distribution companies to reach a better performance level.

Distribution systems reliability is one of the most important topics for the electric power industry, due to its high impact on the cost of electricity and its high correlation with customer satisfaction. Reliability assessment of distribution systems is performed in order to evaluate the effectiveness of continuity of supply, resulting in the computation of reliability indices, which use is essential for setting up performance standards for the continuity of supply regulation. A typical distinction for reliability indices is made between local indices, referred to a single load point (e.g., frequency and duration of the interruptions, power and energy not supplied), and global indices, representing the overall reliability of the system [16,30,31].

The probabilistic reliability analysis is particularly useful for assessing the effectiveness of structural changes (e.g., construction or modification of new lines) or operational actions (such as system reconfiguration) which affect the continuity of supply of the distribution systems.

The first step for applying probabilistic reliability analysis is to define the random variables (RVs) involved in the study. However, some indices are defined under the assumption of exponential distribution of the time to failure with constant failure rate λ, so that the corresponding indices can be computed by using only the mean values of the RVs. A more complete view is to consider a set of random variables (e.g., restoration time τ, number of faults occurred in a specified time period n) represented by their PDFs, thus computing the reliability indices by means of analytical or Monte Carlo simulations [32,33,34].

Among the methods for addressing reliability, different conceptual approaches can be used [35]. A first possibility is to adopt a Markov approach, where times to failure and repair times of the system components are assumed to be exponentially distributed, leading to constant failure and repair rates. Anyway, the assumption of exponential distribution in some cases is not realistic. A semi-Markov approach [36] has been proposed in order to work with non-exponential distributions of the repair times. The Markov model is introduced for explaining some general reliability concept. However, a Markov-based model has strong limitations to represent complex restoration schemes (e.g., multi-phase restoration strategies) such as the ones addressed in this work. Then, it is necessary to resort to other techniques, such as Monte Carlo methods and analytical simulation, which are able to represent a broader range of situations and to handle various forms of probability distributions, even multimodal ones.

Reliability of Repairable Systems

Let's consider a system in which two states are identified as 0 (operation state) and 1 (fault state). The probabilities of remaining in or leaving a particular state in a finite time are assumed to be constant. These are defined as the failure rate λ (for the transition between the operation and the faulted state) and the repair rate μ (for the transition between the faulted and the operation state).

At the time instant t:

p0(t) represents the probability that the component is not faulted
p1(t) represents the probability that the component is faulted

At the time instant t + dt:

a) the probability that the system will be in the operation state depends on two independent events: (1) the system was in the normal state at the time instant t and no failure has occurred, and (2) the system was in the faulted state at the time instant t and has been repaired. Hence,

$$p_0(t+dt) = p_0(t)(1-\lambda dt) + p_1(t)\mu dt \tag{1}$$

b) the probability that the system will be in the faulted state depends on two independent events: (1) the system was in the faulted state at the time instant t and no repair has been done, and (2) the system was in the normal state at the time instant t and a fault has occurred. Hence,

$$p_1(t+dt) = p_0(t)\lambda dt + p_1(t)(1-\mu dt) \tag{2}$$

Dividing by dt, it is possible to obtain two differential equations:

$$\frac{dp_0(t)}{dt} = \frac{p_0(t+dt)-p_0(t)}{dt} = -\lambda\, p_0(t) + \mu p_1(t) \tag{3}$$

$$\frac{dp_1(t)}{dt} = \frac{p_1(t+dt)-p_1(t)}{dt} = -\mu p_1(t) + \lambda p_0(t) \tag{4}$$

The solution of these differential equations leads to an exponential behaviour (Figure 3 and Figure 4) with limit state probabilities

$$p_0(\infty) = \frac{\mu}{\lambda+\mu}, \tag{5}$$

$$p_1(\infty) = \frac{\lambda}{\lambda+\mu} \tag{6}$$

A key property is that the limit state probabilities do not depend on the initial state. Therefore, for periods of time relatively long, the system is "without memory".

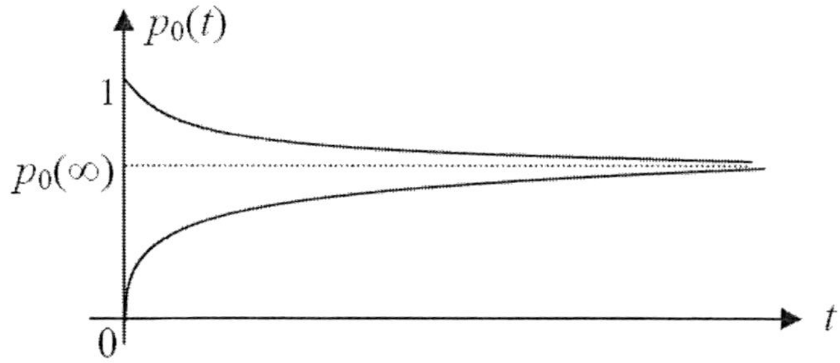

Figure 3. Illustration of p0(t) approaching its limit state.

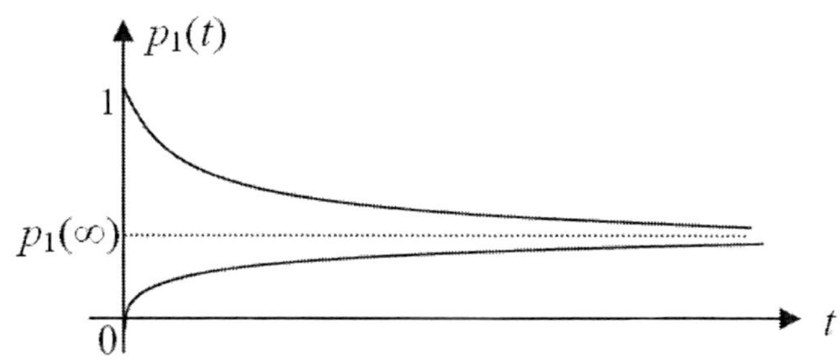

Figure 4. Illustration of p1(t) approaching its limit state.

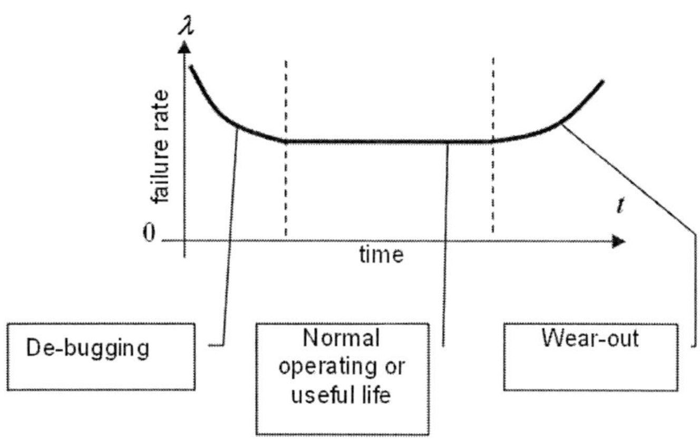

Figure 5. Failure rate curve for a random component.

The assumption of constant failure rate is valid when the system is in its normal life period. This can be seen by looking at the curve of the failure rate (Figure 5), where, with respect to the duration of operation of the component:

The most frequent faults are in initial and final parts of active life

constant failure rate during the intermediary phase, that corresponds to the following assumptions:

the probability that the component function within the [0,t] interval has exponential form ($e^{-\lambda t}$);

the probability that the component is faulted within the [0,t] interval is $1-e^{-\lambda t}$.

For a repairable component, it is possible to define:

Mean Time Between Failures MTBF = T: awaiting time between the faults;
Mean Time To Failure MTTF: awaiting time until the successive fault;
Mean Time To Repair MTTR = τ : awaiting time for repair;
frequency of fault occurrence: f = 1/T.

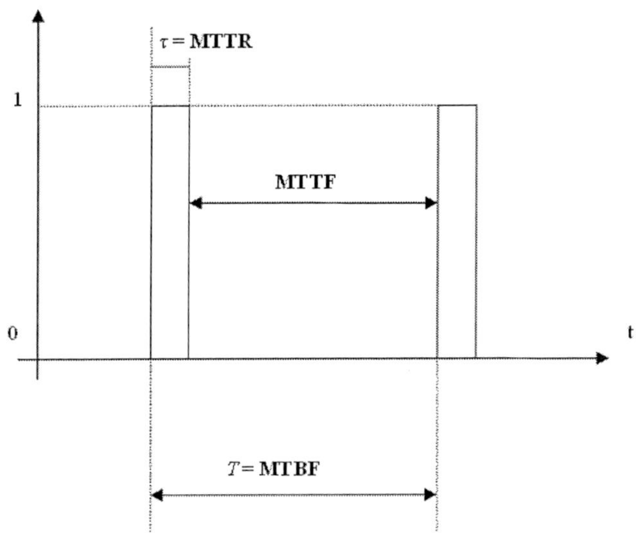

Figure 6. States of a repairable component.

Reliability Model of Distribution System Restoration

General Hypotheses

Let's consider the set Θ representing the whole spectrum of faults that can occur in the system. These faults are distinguished by their type and location. The multi-phase service restoration is assumed to have a random restoration time for each phase, independent of the fault and of the restoration phase for the same fault. The RV λ is used in order to represent failure rate and τ to represent the restoration time.

The reliability analysis is performed by considering number of occurrences of the faults and restoration times to be RVs subject to the randomness of the faults $f \in \Theta$ during the time interval [0,T], under the following hypotheses:

faults are mutually independent;
different occurrences of the same fault are mutually independent;

the number of faults in a given time interval is represented as a RV with Poisson distribution

$$\Pr\{nf = Nf\} = \frac{(\lambda_f T)^{N_f}}{N_f!} e^{-\lambda_f T} \tag{7}$$

where:
λf failure rate;
T length of the period of analysis;
Nf deterministic number of occurrences of fault f.

the restoration time of any restoration phase after a fault is a RVs independent of the fault, of the occurrence of the same fault and of the phase of service restoration, and it is relatively short with respect to the time interval [0,T];
for each occurrence of the fault, the interrupted power Pf is deterministic.

In general, for a better understanding of the purpose of the system analysis, several parameters are considered, some of which are constant and deterministic (e.g., load power), or constant because of their association to an exponential distribution (e.g., failure rate) or probabilistic, as RVs, as shown in Table 2.

Table 2. Nature of the distribution system variables

	Deterministic	Constant	Probabilistic
Load power (C)	X		
Failure rate (λ)		X	
Restoration time (τ)			X
Number of occurrences of fault (n)			X

Representation of the Restoration Times

A key point in the reliability analysis performed in this work is the probabilistic characterization of the restoration times. Generally speaking, there is a multitude of distribution functions which can be used in calculation of τ. They are divided into one-parameter distribution functions and two-parameter distribution functions.

The exponential PDF is sometimes used for its simplicity, but other distributions have been proven to better fit the real behavior of the restoration times.

A very important issue for distribution systems reliability is represented by the service restoration process after a fault occurrence. It was shown that the number of customers affected by a fault and therefore subject to service restoration process, has a decreasing multi-step profile (Figure 2). The evolution of the number of customers interrupted during the service restoration process is characterized by two indices (Dtot, Tm) which can provide valuable information about the evolution of the ongoing process of service restoration.

The impact of the service restoration process on the reliability models is particularly critical. Methods like network modelling and Markov modelling have strong limitations in representing complex restoration strategies [21]. In addition, several literature results [21,30,37,38,39] showed that representing the restoration times by using the exponential distribution is not appropriate. Testing alternative solutions is then worthwhile. If a set of data concerning the occurrence and evolution of faults in a real system is available, an effective assessment of the probability distributions for the fault-related quantities is possible by using suitable goodness-of-fit statistical tests [40].

The Normal, chi-square and Kolmogorov-Smirnov statistics are continuous distributions (in contrast to the Binomial and Poisson discrete distributions). Because continuous statistics are not limited to discrete values, there is almost no probability that a particular precise value will occur. We ask, therefore, about the probability of getting a particular value or less, or the value or more.

The probability of a particular value or less is just the area under the probability density curve to the left of the value; this is the left tail, which is also the cumulative distribution function or CDF. The CDF is just the accumulated sum or integral of the density function; it gives the area to the left of the parameter value.

The probability distributions [41] tested in function of the time t include the (one-parameter) Exponential and Rayleigh distribution, the two-parameter Normal, Weibull, Gamma, Lognormal distributions and the three-parameter Beta distribution.

Goodness-of-Fit Tests

Goodness-of-fit test are statistical tests in which the validity of one hypothesis is tested without specification of an alternative hypothesis. The general procedure consists in defining a statistic, which is some function of the data measuring the distance between the hypothesis and the data (in fact, the badness-of-fit), and then calculating the probability of obtaining data which have a value larger than the value observed, assuming the hypothesis is true. This probability is called the size of the test or confidence level. Small probabilities (say, less than 1 %) indicate a poor fit. Extremely high probabilities (close to 100%) correspond to a fit which is too good to happen very often, and may indicate a mistake in the way the test was applied, such as treating data as independent when they are correlated.

A maximum-likelihood fit to data provides estimates for parameters (and an error matrix). Inserting the estimates into the likelihood function yields a distribution that should model the data reasonably well. How well the data are modelled by that distribution is known as goodness-of-fit. In general, extra work is needed to obtain a quantitative measure of the goodness-of-fit, and there is no universally accepted mathematical definition which is valid in all cases.

Chi-Square Test

The idea behind the chi-square goodness-of-fit test is to see if the sample comes from the population with the claimed distribution. Another way of looking at that is to ask if the frequency distribution fits a specific pattern.

The chi-square test ($\chi 2$)is used to test if a sample of data came from a population with a specific distribution. An attractive feature of the chi-square goodness-of-fit test is that it can be applied to any univariate distribution, for which it is possible to calculate the cumulative distribution function. The chi-square goodness-of-fit test is applied to binned data (i.e., data put into classes). This is actually not a restriction since for non-binned data it is possible to simply calculate a histogram or frequency table before generating the chi-square test. However, the values of the chi-square statistic test are dependent on how the data is binned. Another disadvantage of the chi-square test is that it requires a sufficient sample size in order for the chi-square approximation to be valid.

For a parametric chi-square test in which the same data sample is used for the statistical test, there is a critical range of values (ζmin, ζmax) to which the observed value ζ is compared. The test is successful if $\zeta \leq \zeta$min, unsuccessful if $\zeta > \zeta$max, while for ζmin $< \zeta \leq \zeta$max it has an undefined result. The results of the chi-square test depend on the number of (uniform or non uniform) classes and on the level of significance fixed a priori. The critical range depends on the level of significance used. The maximum level of significance Λmax% corresponding to $\zeta = \zeta$min can be computed to establish the limits of acceptance of the chi-square test results.

$$\chi^2 = \sum_{i=1}^{R} \frac{(x_i - E\{\mathbf{x}\})^2}{E\{\mathbf{x}\}} \tag{8}$$

In the current practice, the following adjustment could be made to the chi-square statistical test, in order to make the value of the statistic smaller. It is hoped that this will give a better estimate of the significance level. The correction is named Yates correction:

$$\chi^2 = \sum_{i=1}^{R} \frac{((x_i - E\{\mathbf{x}\}) - 0.5)^2}{E\{\mathbf{x}\}} \tag{9}$$

In general, the correction is performed only when the number of degrees of freedom is v = 1. For large samples, it produces the same result as a $\chi 2$ without correction. For small samples, where the expected value is between 5 and 10, it is better to make a comparison between the values corrected and those without correction [42].

Kolmogorov-Smirnov Test

The Kolmogorov-Smirnov test (K-S test) tries to determine if two data sets differ significantly (decide if a sample comes from a population with a specific distribution). The K-S test has the advantage of making no assumption about the distribution of data (technically speaking it is non-parametric and distribution free). It enables the graphic view of the data, feature which is very helpful in order to ease the understanding of the data distribution.

An attractive feature of this test is that the distribution of the K-S test statistic itself does not depend on the cumulative distribution function being tested. Another advantage is that it is an exact test (while the chi-square goodness-of-fit test depends on an adequate sample size for the approximations to be valid).

For the K-S test, there is a critical value ψcrit to which the observed value ψ is compared. The test is successful if $\psi \leq \psi$crit. These critical values are found in specific tables. The table valid for a generic probability distribution is reported at p.797 of [42] and in Table 1 of [43]. The critical region contains all cases in which $\psi > \psi$crit.

Despite these advantages, the K-S test has several important limitations:

It only applies to continuous distributions.
It tends to be more sensitive near the centre of the distribution than at the tails.

Perhaps the most serious limitation is that the distribution must be fully specified. If the distribution is fully specified, the result of the test is independent of the distribution. However, if location, scale, and shape parameters are estimated from the data, the critical region of the K-S test is no longer valid. The critical region typically must be determined by simulation or by specific tables. Specific tables have been found for the Exponential distribution (p.798 of [42] and Table 1 of [44]) and for the Normal distribution (p.799 of [42] and Table 1 of [45]).

For other cases in which the distribution parameters are extracted from the data, two limits of the critical region have been set. The lower limit ψmin has been set at the minimum limit for the Normal distribution, while the upper limit ψmin has been provisionally set at the value of the table for a generic distribution. In this sense, the K-S test is not rigorous, but provides only some indicative information. There is a critical range of values (ψmin, ψmax) to which the observed value ψ is compared. The test is successful if $\psi \leq \psi min$, unsuccessful if $\psi > \psi max$, while $\psi min < \psi \leq \psi max$ leads to an undefined result.

Geometrical Adaptation Tests

The results of two graphical tests are shown for assessing the fitness of the ECDF to a reference CDF. By indicating with F the CDF values, two fitness indicators are defined:

$$a_E = ln\left(\frac{1}{1-F}\right) \tag{10}$$

$$a_W = ln\left(ln\left(\frac{1}{1-F}\right)\right) \tag{11}$$

By using the indicator aE, an exponential CDF is represented by a straight line in the (aE, t) plane, while with the indicator aW has the relevant property that a Weibull CDF would be represented by a straight line in the (aW, ln(t)) plane.

Analytical Simulation of the Restoration Process

Introduction to the Analytical Simulation

Once defined the RVs representing the restoration times, it is possible to perform the reliability evaluation by using the analytical simulation technique. The following sections illustrate the basic steps to be performed in the cases of single restoration phase (e.g., for temporary faults) and of multiple restoration phases (e.g., for permanent faults).

Faults with Single Restoration Phase

Let's define the duration dfk of the interruptions at load point k due to fault f ∈ Θ, and the restoration time τfi for the i-th occurrence of the fault. The RV τf represents the restoration time of any occurrence i = 1,..., Nf of fault f. Are computed the expected value and variance of the total duration dk of the interruptions at load point k and of the total energy not supplied w for the whole system. There are provided the expressions of the PDF and CDF of dk for exponentially-distributed restoration times.

The total duration of the interruptions at load point k is expressed by

$$\mathbf{d}_{k} = \sum_{f \in \Phi_k} \mathbf{d}_{fk} \tag{12}$$

where the RV dfk is given by the random sum [46]

$$\mathbf{d}_{fk} = \sum_{i=1}^{\mathbf{n}_f} \boldsymbol{\tau}_{fi} \tag{13}$$

The PDF and CDF of the RV dfk are written in the form

$$f(d_{fk}) = \sum_{N_f=0}^{\infty} f(d_{fk} \mid N_f) Pr\{\mathbf{n}_f = N_f\} \tag{14}$$

$$F(d_{fk}) = \sum_{N_f=0}^{\infty} F(d_{fk} \mid N_f) Pr\{\mathbf{n}_f = N_f\} \tag{15}$$

The PDF $f(d_{fk} \mid N_f)$ is obtained from the convolution of the N_f PDFs of $\boldsymbol{\tau}_{fi}$ since the random sum (13) becomes a deterministic sum of mutually independent terms. Then,

$$E\{\mathbf{d}_{fk} \mid N_f\} = \sum_{i=1}^{N_f} E\{\boldsymbol{\tau}_f\} = N_f E\{\boldsymbol{\tau}_f\} \tag{16}$$

$$\sigma^2_{(\mathbf{d}_{fk} \mid N_f)} = \sum_{f=1}^{N_f} \sigma^2_{\boldsymbol{\tau}_f} = N_f \sigma^2_{\boldsymbol{\tau}_f} \tag{17}$$

resulting in the expected value and variance of the duration dfk respectively written as

$$E\{\mathbf{d}_{fk}\} = \sum_{N_f=0}^{\infty} E\{\mathbf{d}_{fk} \mid N_f\} Pr\{\mathbf{n}_f = N_f\} = \lambda_f T E\{\boldsymbol{\tau}_f\} \tag{18}$$

$$\sigma^2_{\mathbf{d}_{fk}} = \lambda_f T\left(E^2\{\boldsymbol{\tau}_f\} + \sigma_{\boldsymbol{\tau}_f}{}^2\right) \quad E\{\mathbf{d}_k\} = \sum_{f \in \Phi_k} E\{\mathbf{d}_{fk}\} \tag{19}$$

from which the expected value and variance of the total duration of the interruptions at load point k become

$$E\{\mathbf{d}_k\} = \sum_{f \in \Phi_k} E\{\mathbf{d}_{fk}\}; \; \sigma^2_{\mathbf{d}_k} = \sum_{f \in \Phi_k} \sigma^2_{\mathbf{d}_{fk}} \tag{20}$$

The total energy not supplied in the time interval [0,T] is

$$\mathbf{w} = \sum_{f \in \Theta} \mathbf{w}_f = \sum_{f \in \Theta} P_f \, \mathbf{d}_f \tag{21}$$

where the RV wf represents the energy not supplied for all the occurrences of the fault f $\in \Theta$ and the RV df is the duration of interruption of every load point for all the occurrences of the fault, given by the random sum

$$\mathbf{d}_f = \sum_{i=1}^{\mathbf{n}_f} \boldsymbol{\tau}_{fi} \tag{22}$$

For any probability distribution of the restoration times with standard deviation $\sigma^2_{\boldsymbol{\tau}_f}$, we adopt the same scheme of solution to obtain the expected value and variance

$$E\{\mathbf{d}_f\} = \lambda_f T E\{\boldsymbol{\tau}_f\}; \; \sigma^2_{\mathbf{d}_f} = \lambda_f T\left(E^2\{\boldsymbol{\tau}_f\} + \sigma_{\boldsymbol{\tau}_f}{}^2\right) \tag{23}$$

and for the energy not supplied after the fault f $\in \Theta$

$$E\{\mathbf{w}_f\} = \lambda_f T P_f E\{\boldsymbol{\tau}_f\}; \; \sigma^2_{\mathbf{w}_f} = \lambda_f T P_f^2\left(E^2\{\boldsymbol{\tau}_f\} + \sigma^2_{\boldsymbol{\tau}_f}\right) \tag{24}$$

with the corresponding values for the total energy not supplied

$$E\{\mathbf{w}\} = \sum_{f \in \Theta} E\{\mathbf{w}_f\}; \; \sigma^2_{\mathbf{w}} = \sum_{f \in \Theta} \sigma^2_{\mathbf{w}_f} \tag{25}$$

The total duration of the interruptions weighted with respect to the total load power Ctot is

$$\mathbf{d} = \frac{\mathbf{w}}{C_{tot}} \tag{26}$$

with expected value and variance respectively given by

$$E\{\mathbf{d}\} = \frac{E\{\mathbf{w}\}}{C_{tot}},\ \sigma_{\mathbf{d}}^2 = \frac{\sigma_{\mathbf{w}}^2}{C_{tot}^2} \tag{27}$$

In (23)-(25) and (27) the expected values have the same formulations resulting from the classical approach.

Faults with Multiple Restoration Phases

The values of the reliability indices [46,47,48,49] depend on the expected values E{dk} of the interruption duration (dk) at load point k = 1, …, N. Here we recall the computational aspects to obtain E{dk} with a classical approach.

The occurrence of a fault leads to a sequence of mutually exclusive fault states. Each fault state corresponds to a restoration phase, with remote-controlled or manual operations performed for restoring the service. By neglecting the simultaneous faults, it is possible to compute the duration of the service interruption during the restoration phases for any load point. It was assumed that the service restoration after a fault f ∈ Θ to include a number φf of independent restoration phases.

The expected value E{dk} of the duration of the service interruption during the restoration phases at load point k is computed by considering all the restoration phases in which load point k is not supplied. The binary variable $\delta_{kf}^{(m)} = 1$ was introduced, if load point k is not supplied in the phase m of service restoration after the fault f ∈ Θ, otherwise $\delta_{kf}^{(m)} = 0$. Assuming a negligible probability of simultaneous faults, in the restoration phase m = 1,…, φf after fault f ∈ Θ with failure rate λf, corresponding to the restoration time $\boldsymbol{\tau}_f^{(m)}$,, the expected value of the duration of the interruptions is

$$E\{\mathbf{d}_k\} = \sum_{f \in \Theta} \sum_{m=1}^{\varphi_f} \lambda_f \, E\{\boldsymbol{\tau}_f^{(m)}\} \, \delta_{kf}^{(m)} \tag{28}$$

In the presence of permanent faults with multi-phase restoration, it is convenient to evaluate, for each fault f ∈ Θ and for all the load points k = 1,…, N, the binary variable $\delta_{kf}^{(m)}$ introduced by the fault, for m = 1,…, φf. For each fault at a supply node, with a single restoration phase (φf = 1), the variable $\delta_{kf}^{(1)}$ is equal to unity for the load points fed by the faulted supply node, zero otherwise. For each temporary fault, the trip of the circuit breaker isolates all the load points fed by the faulted branch, there is again a single restoration phase (φf = 1) and the variable $\delta_{kf}^{(1)}$ is equal to unity for the load points fed by the faulted branch, zero otherwise. For each permanent fault, with three restoration phases (m = 1, 2, 3), the value of $\delta_{kf}^{(m)}$ depends on the restoration phase:

in the first phase, $\delta_{kf}^{(1)} = 1$ if load point k is not supplied after the trip of the circuit breaker, otherwise $\delta_{kf}^{(1)} = 0$;

in the second phase, $\delta_{kf}^{(2)} = 1$ if load point k is not supplied after the remote-controlled operations, otherwise $\delta_{kf}^{(2)}$ 0;

in the third phase, $\delta_{kf}^{(3)} = 1$ if load point k is not supplied after the manual operations, otherwise $\delta_{kf}^{(3)}$ 0.

After any permanent fault, the evaluation of $\delta_{kf}^{(m)}$ for m = 2, 3 requires the detailed simulation of the operations performed to isolate the fault and to restore the service (Table 3).

Table 3. Evaluation of possible fault types and corresponding load point indices [50]

Type of Fault	f		$\delta_{kf}^{(m)}$		Explanation
Supply point			$\delta_{kf}^{(1}$	1	Fed by the faulted branch
				0	Otherwise
Temporary			$\delta_{kf}^{(1}$	1	Fed by the faulted branch
				0	Otherwise
Permanent			$\delta_{kf}^{(1}$	1	Load point k is not supplied after the trip of the circuit breaker
				0	Otherwise
			$\delta_{kf}^{(1}$	1	Load point k is not supplied after the remote-controlled operations
				0	Otherwise
			$\delta_{kf}^{(1}$	1	Load point k is not supplied after the manual operations
				0	Otherwise

The previous results are extended to faults with φf restoration phases by using the approach presented in [49]. It is assumed that the RVs $\boldsymbol{\tau}_{fi}^{(m)}$ restoration time of the phase m of the occurrence i in [0,T] of the fault f) are mutually independent. It is further assumed that the power $P_f^{(m)}$ is interrupted during the m-th restoration phase of the fault $f \in \Theta$ to be deterministic. The number of occurrences of the fault $f \in \Theta$ in [0,T] is represented by the Poisson-distributed RV $\mathbf{n}_f$ with failure rate λ_f .

The total duration of the interruptions at load point k during the m-th restoration phase of fault $f \in \Theta$ is

$$\mathbf{d}_{fk}^{(m)} = \sum_{i=1}^{n_f} \boldsymbol{\tau}_{fi}^{(m)} \tag{29}$$

The energy not supplied $\mathbf{w}_f$ after all the occurrences of the fault $f \in \Theta$ in [0,T] is

$$\mathbf{w}_f = \sum_{m=1}^{\varphi_f} P_f^{(m)} \sum_{i=1}^{n_f} \boldsymbol{\tau}_{fi}^{(m)} \tag{30}$$

By adopting the same solution scheme used in (II.14 – II.17), the expected value and variance are

$$E\{\mathbf{w}_f\} = \lambda_f T \sum_{m=1}^{\varphi_f} P_f^{(m)} E\{\boldsymbol{\tau}_f^{(m)}\} \tag{31}$$

$$\sigma^2_{\mathbf{w}_f^{(m)}} = \lambda_f T \left[\left(\sum_{m=1}^{\varphi_f} P_f^{(m)} E\{\boldsymbol{\tau}_f^{(m)}\} \right)^2 + \sum_{m=1}^{\varphi_f} P_f^{(m)2} \sigma^2_{\boldsymbol{\tau}_f^{(m)}} \right] \tag{32}$$

The total energy not supplied w is the additive RV

$$\mathbf{w} = \sum_{f \in \Theta} \mathbf{w}_f \tag{33}$$

with expected value

$$E\{\mathbf{w}\} = \sum_{f \in \Theta} \lambda_f T \sum_{m=1}^{\varphi_f} P_f^{(m)} E\{\boldsymbol{\tau}_f^{(m)}\} \tag{34}$$

and, being the faults mutually independent, with variance

$$\sigma^2_{\mathbf{w}} = \sum_{f \in \Theta} \lambda_f T \left[\left(\sum_{m=1}^{\varphi_f} P_f^{(m)} E\{\boldsymbol{\tau}_f^{(m)}\} \right)^2 + \sum_{m=1}^{\varphi_f} P_f^{(m)2} \sigma^2_{\tau_f^{(m)}} \right] \tag{35}$$

Reliability Model of Systems with DG

This section illustrates the basic reliability aspects for distribution systems with local generation and discusses the conditions for successful island operation. It is also presented the overall framework of reliability analysis with local generators, along with the analytical simulation technique used to perform the reliability analysis.

The local generation should not exceed the contribution of the HV/MV substation, otherwise reverse power flows could be critical for protection coordination and line overloading.

The reliability analysis performed in this section assumes that some portions of the system under analysis can be aggregated into a few equivalent networks, each of which containing the components subject to the same interruption in case of fault. These networks are represented by their equivalent reliability parameters [24]. The effect of a DG unit on reliability is investigated by considering the feeder to which the DG unit is connected (Figure 7) and representing two equivalent networks with their failure rates, namely, λU for the upstream network and λD for the downstream network, and with the corresponding load powers CU and CD. The two equivalent networks can be connected to other supply sources during the system restoration after a fault by closing the switches SU and SD. Then, the service restoration after a fault in the downstream network requires opening the circuit breaker BD and performing manual operations. For a fault in the upstream network, the service restoration requires performing manual operations. The presence of the local DG unit can assist the restoration process, by opening the circuit breaker BS and connecting the DG unit to the downstream network.

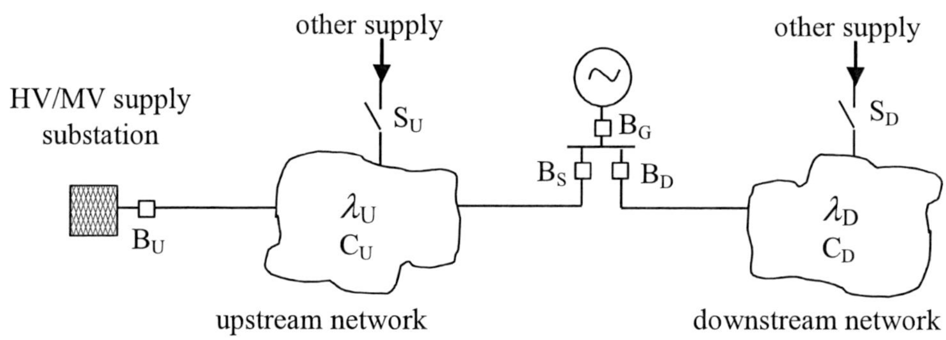

Figure 7. Scheme of the test system for reliability study of a distribution system with DG.

If we consider the probability of islanding failure PIF, the probability pI = 1-PIF of forming the island to supply the power CD depends on the local generator availability, on the probability that CG > CD and on the probability successful transition from parallel to island mode of operation. The reliability model considered takes into account the probability of successful island formation pI, the probability pS of selective operation of the circuit breaker BD with respect to the supply circuit breaker BU for faults occurring in the downstream network, and the probabilities pU (or pD) of successful switching SU for the upstream network (SD for the downstream network) to connect the corresponding network to the other supply when faults in the upstream (downstream) network occurs. All these probabilities are used to compute the probability of successful restoration pR and the interrupted power Cint, by considering different restoration modes, corresponding to independent events, as in Table 4.

Reliability Indices

For a distribution system with K load points, the reliability analysis takes into account two types of indices (local and global) [47]:

Local indices: are defined for each load point k = 1,…,K, and are calculated considering the frequency and duration of the states in which the point k is not supplied. Assuming the power Pk to be delivered to the point k during normal operation, the following local indices [48] are defined:

fk – frequency of the interruptions
dk – duration of the interruptions [min/year]
PkNS = fkPk – power not supplied [MW/year]
EkNS = dkPk – energy not supplied [MWh/year]
dkNS = dk / fk – average duration of the interruptions [min]

Global indices: are defined for the whole electrical network, representing the overall system reliability [29,48]. They are also divided in Sustained Interruption Indices (SII) and Momentary Interruption Indices (MII).

Table 4. Probabilities of successful restoration (restoration mode with initial letter U for faults in the upstream network, D for faults in the downstream network)

Restoration mode	Successful Island Formation	Successfu 1 SU Switching	Successful SD Switching	Successful BD/BU Selective Speration	Interrupted Power Cint (p.u.)	Probability of Successful Restoration pR
U1	yes	yes	---	---	CU	pI pU
U2	yes	no	---	---	CU	pI (1-pU)
U3	no	yes	yes	---	CU + CD	(1- pI) pU
U4	no	no	no	---	CU + CD	(1- pI) (1- pU)
D1	---	---	yes	yes	CD	pS pD
D2	---	---	yes	no	CD	(1- pS) pD
D3	---	---	no	yes	CU + CD	pS (1-pD)
D4	---	---	no	no	CU + CD	(1- pS) (1- pD)

Sustained Interruption Indices

SAIFI (System Average Interruption Frequency Index). This index is designed to give information about the average frequency of sustained interruptions per customer over a predefined area. To calculate the index,

$$\mathrm{SAIFI} = \frac{1}{K}\sum_{k=1}^{K} f_k \tag{36}$$

SAIDI (System Average Interruption Duration Index). This index is commonly referred to as customer minutes of interruption or customer hours, and is designed to provide information about the average time the customers are interrupted. To calculate the index,

$$\mathrm{SAIDI} = \frac{1}{K}\sum_{k=1}^{K} d_k \tag{37}$$

CAIDI (Customer Average Interruption Duration Index). This index represents the average time required to restore service to the average customer per sustained interruption. In order to calculate the index,

$$\mathrm{CAIDI} = \frac{SAIDI}{SAIFI} \tag{38}$$

There is also currently used a variety of SII indices, created from the need to accomplish an exhaustive array of tools used in order to represent the overall electrical power system reliability. These indices will be briefly presented, due to their small significance on current chapter.

CTAIDI (Customer Total Average Interruption Duration Index), representing the total average time in the reporting period the customers were without power

CAIFI (Customer Average Interruption Frequency Index). This index gives the average frequency of sustained interruptions for those customers (counted once, regardless of the number of times interrupted) experiencing sustained interruptions

ASAI (Average Service Availability Index), representing the fraction of time (often in percentage) that a customer has power provided during one year or the defined reporting period

ASIFI (Average System Interruption Frequency Index). This index was specifically created to calculate reliability based on load rather than number of customers, being utilized by utilities that do not have elaborate customer tracking systems

ASIDI (Average System Interruption Duration Index), created using the same reason as ASIFI; but providing information on system average duration of interruptions

CEMIn (Customers Experiencing Multiple Interruptions), designed to track the number (n) of sustained interruptions to a specific customer, aiming at identifying the customer trouble that cannot be seen by using averages

Momentary Interruption Indices

MAIFI (Momentary Average Interruption Frequency Index), is very similar with SAIDI, tracking the average frequency of momentary interruptions instead

MAIFIE (Momentary Average Interruption Event Frequency Index), similar with MAIFI

CEMSMIn (Customers Experiencing Multiple Sustained Interruptions and Momentary Interruptions Events), index designed to track the number (n) of both sustained interruptions and momentary interruption events to a set of specific customers

Substantial differences do exist in the computation of the reliability indices in different countries. The regulation usually applies to "long" interruptions (over 3 min, but in Portugal and UK the limit is 1 min) and, at present or in perspective, to the users of all voltage levels. However, in some countries (e.g., Portugal and Spain) data are recorded only at voltage levels above 1 kV. Furthermore, while the unexpected interruptions are generally subject to regulation, some countries (e.g. Norway, Portugal and UK) also consider the expected interruptions.

Probabilistic analysis is also being used for power quality aspects different with respect to service continuity, such as momentary interruptions, voltage sags, etc.

Relationships between Probabilistic and Traditional Indices

The analysis carried out has considered two reliability indices, defined for the time interval [0,T]:

the total duration of the interruptions, represented by the RVs dk at load point k and d for the whole system;

the total energy not supplied, represented by the RVs wk at load point k and w for the whole system.

For k = 1, …, N, the classical reliability indices are expressed in terms of the expected values of the RVs dk and wk. Let's consider the power Ck delivered to load point k during normal operation.

For the local indices, the probabilistic equivalent of the traditional indices is given by the expected value E{dk} of the duration of the interruptions and the expected value E{wk} = Ck E{dk} of the energy not supplied to load point k.

For the global indices, the Total Energy Not Supplied is

$$ENS_{tot} = \sum_{k=1}^{N} E\{w_k\}, \tag{39}$$

the System Average Interruption Duration Index is

$$\mathrm{SAIDI} = \frac{1}{N}\sum_{k=1}^{N} E\{\mathbf{d}_k\}, \tag{40}$$

and the total duration of the interruptions d_{tot}, defined by using the interrupted power (or alternatively the number of customers) as weighting factor is

$$d_{tot} = \frac{\sum_{k=1}^{N} C_k \, E\{\mathbf{d}_k\}}{\sum_{k=1}^{N} C_k} = \frac{ENS_{tot}}{C_{tot}} \tag{41}$$

Reliability Analysis Techniques for Electricity Distribution Systems

Analytical Simulation for Traditional Distribution Systems

The parameters that are normally associated with reliability evaluation are described by probability distributions. This can easily be appreciated by considering that all components of a given type will not all fail at different times in the future. Consequently, time-to-failure values obey a probability distribution which may, or may not, be known and which describes the probability that a given component fails within a certain specified time. This probability value is a function of the time that is specified or considered. In all practical cases the appropriate probability distribution cannot be determined from knowledge of the geometry of the component, device or system but must be deduced from sample testing or from a data collection scheme associated with the operation of the components, devices or systems.

The probability distributions used in the reliability analysis considered in this work could be defined analytically or in the form of a set of discrete values. Clearly, in the latter case each discrete value is considered in the PDF as a Dirac pulse with given amplitude, and the sum of the amplitudes must be equal to unity. Correspondingly, the CDF is defined as a staircase-like curve.

The analytical simulation method described in the present section [35] works with both continuous and discrete PDFs, and then it is general enough for representing any type of probability distribution. The related computations are simplified in the case the PDFs retain some specific properties from their analytical formulation (e.g., the possibility of finding the analytical representation of the characteristic function, that will simplify the computation of the convolutions, as shown in the sequel).

Let's consider the computation of the energy not supplied in a distribution system with multiple independent faults and multi-phase restoration. The total energy not supplied W is given by the sum of the total energy not supplied for any fault occurred, denoted as Wf for f = 1, …, F. Since the faults are independent, it is possible to write

$$\mathbf{W} = \sum_{f=1}^{F} \mathbf{W}_f \tag{42}$$

For a given fault f, the total energy not supplied Wf depends on the number of occurrences of the fault. Let wf be the RV representing the energy not supplied for a single occurrence of the fault. The number of occurrences of the fault f in the time period of analysis is not known in advance, so that it has to be associated to a RV nf. The total energy not supplied Wf is then defined by the random sum

$$\mathbf{W}_f = \sum_{i=1}^{\mathbf{n}_f} \mathbf{w}_f \tag{43}$$

Note that, when the restoration times are represented by RVs, it is not possible to compute the total energy not supplied by summing up the contributions of the load points, since these contributions are not independent.

Since the occurrence of the same fault in the time period of analysis is considered to be a rare event, it is associated to the RV nf defined with a Poisson distribution, with expected value $E\{\mathbf{n}_f\}$ and standard deviation $\sigma_{\mathbf{n}_f} = \sqrt{\lambda_f T}$. In this case, it is possible to use a key result of the probability analysis, which establishes that it is possible to express the random sum (43) in terms of a compound Poisson distribution, whose characteristic function is given by

$$\Phi_{\mathbf{W}_f}(\omega) = e^{\lambda_f T\left(\Phi_{\mathbf{w}_f}(\omega) - 1\right)} \tag{44}$$

At this point, it is important to recall the links among the expressions of the characteristic functions and the probability distributions.

Characteristic Function of a Probability Distribution

Let z be a stochastic variable and p(z) be the probability density function for z; i.e., the probability of obtaining a value of z between a and b is:

$$\int_a^b p(z)dz \tag{45}$$

The expected value of any function of z, for example g(z), is defined as:

$$E\{g\} = \int_{-\infty}^{\infty} g(z)p(z)dz \tag{46}$$

The expected value of the function $e^{j\omega z}$ is called the characteristic function for the probability distribution p(z), where ω is the parameter that can have any real value and j is the square root of (-1). That is to say, the characteristic function of p(z) is:

$$\Phi(\omega) = \mathrm{E}\left\{e^{j\omega z}\right\} = \int_{-\infty}^{\infty} e^{j\omega z} p(z) dz \tag{47}$$

The characteristic function will generally be a complex function, i.e.,

$$\Phi(\omega) = \mathrm{X}(\omega) + j\mathrm{Y}(\omega) \tag{48}$$

Since

$$e^{j\omega z} = \cos(\omega z) + j\sin(\omega z) \tag{49}$$

the components of the characteristic function are given by:

$$\mathrm{X}(\omega) = \mathrm{E}\{\cos(\omega z)\} = \int_{-\infty}^{\infty} \cos(\omega z) p(z) dz \tag{50}$$

and

$$\mathrm{Y}(\omega) = \mathrm{E}\{\sin(\omega z)\} = \int_{-\infty}^{\infty} \sin(\omega z) p(z) dz \tag{51}$$

Thus, given a probability distribution p(z) it is a straightforward computation to calculate the real and imaginary components of its characteristic function.

The crucial property of characteristic functions is that the characteristic function of the sum of two independent random variables is the product of the characteristic functions of those variables.

It is usually more convenient to work with the logarithm of the characteristic function, also called second characteristic function:

$$\Psi_z(\omega) = \ln(\Phi_z(\omega)) \tag{52}$$

The logarithm of the characteristic function will also be a complex function with real and imaginary components. In this case, the products of characteristic functions become sums of second characteristic functions. This property of characteristic functions can be represented as follows. If $\Phi_x(\omega)$ and $\Phi_y(\omega)$ are the characteristic functions of the independent RVs x and y, respectively, then the characteristic function of a variable that involves taking an observation of x and an observation of y and adding them together to obtain z = x + y is given by:

$$\Phi_z(\omega) = \Phi_x(\omega)\Phi_y(\omega) \tag{53}$$

and hence

$$\Psi_z(\omega) = \Psi_x(\omega) + \Psi_y(\omega) \tag{54}$$

If two variables are not independent, the proposition concerning the characteristic function involves the characteristic function of the conditional probability distribution.

The process of aggregating data, such as combining monthly data to obtain quarterly or annual data, is easily presented in terms of characteristic functions. If the smaller unit data are statistically independent, then the proposition concerning the characteristic function of the sum of random variables applies.

There is another operation that is often involved with combining random variables. Suppose x and y have different probability distributions, but they are treated as coming from the same population. In effect the probability distribution of the combination involves the probabilities that an observation came from the x population or the y population. Let these probabilities be presented as px and py and their probability densities be denoted as f(x) and f(y), respectively. The probability density that an observation from the combined population has a value z is

$$f_z(z) = p_x f_x(z) + p_y f_y(z) \tag{55}$$

If the characteristic functions for f(x) and f(y) are Φ_x and Φ_y then the characteristic function for the combined population is given by:

$$\Phi_z(\omega) = p_x \Phi_x(\omega) + p_y \Phi_y(\omega) \tag{56}$$

In general, the characteristic function is not defined for any probability distribution. However, if it is not possible to define the characteristic function analytically, it is possible to provide a numerical evaluation of the characteristic function by sampling the PDF of the RV in the time domain and successively applying the Discrete Fourier Transform (DFT), as shown in [26].

Analytical Simulation Adopting the Characteristic Functions

The calculation of the RVs representing the reliability indicators can be performed in a simple way by resorting to the use of the characteristic functions.

The procedure is used to represent duration-dependent reliability indices, such as duration of the interruptions and energy not supplied. The link between the total energy not supplied and the total duration of the interruptions is simple. The total duration of the interruptions can be expressed by dividing the total energy not supplied by the total load power Ctot, such that

$$D = W / Ctot \tag{57}$$

Note that in the analysis considered here Ctot is a deterministic variable, so that the PDFs obtained by using the total duration of the interruptions are similar to the ones representing the total energy not supplied, with the only difference of a scale factor.

With reference to the calculation of the total energy not supplied, the general framework is composed of five steps:

1) definition of the PDF of the restoration times $\boldsymbol{\tau}_f^{(h)}$ for any fault f = 1, …, F and for any restoration phase h = 1, …, H, and evaluation of its characteristic function $\Phi_{\boldsymbol{\tau}_f^{(h)}}(\omega)$. This evaluation can be analytical or numerical (by means of the DFT applied to the PDF of the RV)

2) calculation of the characteristic functions of the energy not supplied for a single occurrence of fault f. By considering the multi-phase restoration with Hf phases, the characteristic function is given by

$$\Phi_{w_f}(\omega) = \prod_{h=1}^{H_f} \Phi_{\boldsymbol{\tau}_f^{(h)}}\left(\omega P_f^{(h)}\right) \tag{58}$$

and the second characteristic function is

$$\Psi_{w_f}(\omega) = \prod_{h=1}^{H_f} \Psi_{\boldsymbol{\tau}_f^{(h)}}\left(\omega P_f^{(h)}\right) \tag{59}$$

3) calculation of the characteristic functions of the total energy not supplied for all occurrences of the fault f. This requires applying the result of the compound Poisson process. The characteristic function is given by

$$\Phi_{W_f} = e^{\lambda_f T(\Phi_{w_f} - 1)} \tag{60}$$

and the second characteristic function is

$$\Psi_{W_f} = ln(\Phi_{W_f}) = \lambda_f T(\Phi_{w_f} - 1) \tag{61}$$

4) calculation of the characteristic functions of the total energy not supplied for all occurrences of the faults. This is performed by summing up the contributions of the independent faults.

The sums of RVs would require performing convolutions, which in the characteristic function domain are replaced by products of characteristic functions

$$\Phi_{W} = \prod_{f=1}^{F} \Phi_{W_f} \tag{62}$$

or sums of second characteristic functions

$$\Psi_{W} = \sum_{f=1}^{F} \Psi_{W_f} \tag{63}$$

5) the final step requires to build the PDF of the energy not supplied in the original space of the RV. Starting from the discretized or sampled version of the PDF in the domain of the characteristic functions (with the variable ω), it is necessary to perform the Inverse Discrete Fourier Transform (IDFT).

Monte Carlo Evaluation for Traditional Distribution Systems

General Hypotheses of Monte Carlo Methods

In reliability calculations of electrical distribution networks, both analytical and simulation techniques are used. The analytical techniques are mathematically rigorous and are used to find expected value and variance. These are valuable indices for the system, but they are insufficient in order to give a complete view of the resulting PDF. Monte Carlo (MC) methods consist in experimental simulation of mathematical problems for which random numbers are used for discovering the solution. Various approaches can be followed when performing Monte Carlo simulation on electric power systems. In the available literature, there are distinguishable some different types of Monte Carlo simulations, like

duration sampling approach [37,51] ;
time sequential approach [20,52] ;
system state transition sampling approach [30].

In our test network, the point of interest is constituted by reliability indices. It is well known the fact that MCM is a so called "blind method", which means that it has no "a priori" instrument to guide itself through the iterations for reaching an outcome. The approach is to generate a random variable using a suitable probability density function that could emulate the development of the process.

So, for implementing our tests, we considered that the best type of distribution for our random numbers should be Poisson, due to the fact that its domain of definition covers the "rare event probability", and in our case (reliability indices analysis) the events (interruptions) are rare. Another motivation for our choice was that a convolution made with many discrete

Poisson probability distribution functions (PDFs) has the propriety of keeping a Poisson profile.

The Sequential Monte Carlo Method

The behavior pattern of n identical systems in real time will all be different, including the number of failures, times to failure, restoration times etc. This is due to the random nature of the processes involved. The behavior of a particular system could therefore follow any of these behavior patterns. The time sequential simulation process can be used to examine and predict real behavior patterns in simulated time, to obtain the probability distributions of the various reliability parameters and to estimate the expected or average value of these parameters.

In a time sequential simulation, an artificial history that shows the up and down times of the system elements is generated in chronological order using random number generators and the probability distributions of the element failure and restoration parameters [46].

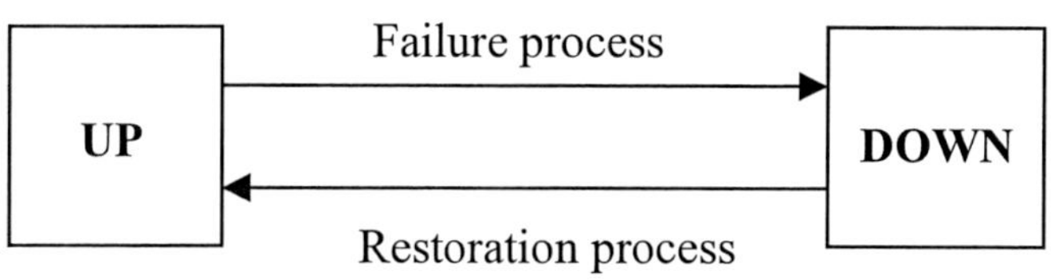

Figure 8. State space diagram of element [20].

A sequence of operating-repair cycles of the system is obtained from the generated component histories using the relationships between the element states and system states. The system reliability indices and their probability distributions are obtained from the artificial history of the system.

The sequential MC method [20] has been used to address the reliability indices computation.

Typically, the number of faults involving a single load point is not very high, so that the load point indices exhibit unusual forms with possible multi-modal shapes. In these cases, the MC method is particularly used. Therefore, some load point indices have been computed. Global indices are less interesting for the MC method application, since the presence of several load points make these indices very close to the Normal form.

The process used to evaluate the distribution system reliability indices using time sequential simulation consists of the following steps [20]:

Step 1 Generate a random number for each element in the system and convert it into time to failure corresponding to the probability distribution of element parameter

Step 2 Determine the element with minimum time to failure

Step 3 Generate a random number and convert this number into the repair time of the element with minimum time to failure according to the probability distribution of the repair time

Step 4 Generate another random number and convert this number into switching time according to te probability distribution of the switching time if this action is possible

Step 5 Utilize the following sequence of procedures and record the outage duration for each failed load point:

determine the type and location of failed element, the failed element number that the failed element is connected to

determine the affected load points connected to the failed feeder and the failure durations of these load points according to the configuration and the protection scheme of the failed feeder

determine the subfeeders which are the downstream feeders connected to the failed feeder and the effects of the failed element on the load points connected to these subfeeders

repeat 2 and 3 for each failed subfeeder connected to the failed feeder are found and evaluated

determine the upfeeder which is the upstream feeder to which the failed feeder is connected ant the effects of the failed element on the load points in the upfeeder

repeat 2 and 5 until the main feeder is reached and evaluated

Step 6 Generate a new random number for the failed element and convert it into a new time to failure, and return to Step 2 if the simulation time is less than one year. If the simulation time is greater than one year, go to Step 9

Step 7 Calculate the number and duration of failures for each load point

Step 8 Calculate the average value of the load point failure rate and failure duration

Step 9 Calculate the system indices

Step 10 Calculate the average values of these system indices

Step 11 Return to Step 2 if the simulation time is less than the specified total simulation years, otherwise output the results

Comparison between Monte Carlo and Analytical Simulation

When dealing with system reliability engineering, the correct questions must be dictated by the reality of the system.

The dilemma of using the Monte Carlo (MC) evaluation or Analytical Simulation refers to the choice of which question to ask. If one chooses to ask about the steady-state availability, analytical methods can and should be used. If one chooses to ask about the time-dependent availability, the MC method must be used.

In the MC method, the difficulty of the integration of any equation (which, as in the solution of integral equations in general, is equivalent to a summation of an infinite series of multiple integrals – the Neumann series) is replaced by a detailed statistical follow-up of the random process leading to that equation.

The mathematical difficulty that prohibits the solution of complex problems limits analytical methods from being applied to a large and vital class of models such as:

time-dependent availability, number of failures, or any other time-dependent properties, except for the case in which all the distribution involved are exponential, i.e. constant failure and repair rates; this in turn means that no aging can be assumed;

no considerations of partial or minimum repair; age reduction and therefore preventive maintenance cannot be analyzed; performance as a function of partial repair and periodic of age-dependent maintenance cannot be studied;

spare parts considerations, resource considerations and any other subjects that are time-dependent.

In short, the mathematical complexity (even for a single component) permits analytical solutions only if we assume that the failure and repair distributions (and any other distribution relevant to the situation such as recycling time) are exponential, or if we assume steady-state eliminates time dependency altogether.

In a system calculation, each component represents a dimension. Thus a system with S components requires a S-dimensional calculation. Each possible state of a component is equivalent to an interval. For a component with n states, thus the number of points to be considered is nS. For example, a system with 30 components each having three possible states, e.g. active, passive and failed (in most realistic systems, components occur in more than just these three states) will have a total of 2,058 * 1014 points. This is the effect that profoundly limits the ability of analytical methods to ever be a solution for advanced calculations of modern complex system predictions.

Analytical Simulation for Distribution Systems with DG

The Analytical Simulation Technique

The reliability analysis performed in this section deals with reliability indices [46] assumed to be random variables (RVs) due to the randomness of the faults occurring in the time interval (0,T).

Reliability analysis is carried out by considering a single feeder and M equal feeders. The fast and efficient technique proposed in [35] is used for performing numerical simulations. This technique adopts a characteristic functions-based approach in which the direct convolution of the PDFs is avoided by resorting to the properties of the compound Poisson process [53] and to the use of the direct and inverse Discrete Fourier Transforms. A key feature of the method is the possibility of handling any type of probability distribution, including multi-modal PDFs, over any time interval of analysis, providing the whole PDF of local and global reliability indices. The results accuracy is checked out by comparing average value and variance of the numerically computed reliability indicators to average value and variance computed with the analytical method illustrated in [35,49].

Restoration modes (Table 4) are scanned by using the index h = 1,..., HU, HU+1,..., HU+HD, where HU = 4 refers to the first four modes and HD = 4 refers to the fifth to last mode. The index h is applied to the interrupted power Pint and to the probability of successful restoration pR. The Gamma RVs representing the switching time τS and the repair time τR are used to form the vector τ = [τS, τR, τS, τR, τS, τR, τS, τR]T. The subscripts U and D represent the upstream and downstream network, respectively. The total energy not supplied is w for a single fault occurrence and W for all fault occurrences.

Results are shown by first taking into account the upstream and the downstream networks separately. The energy not supplied for a single fault occurrence has expected value

$$E\{\mathbf{w}_U\} = \sum_{h=1}^{H_U} p_R^{(h)} P_{i/t}^{(h)} E\{\tau^{(h)}\} \tag{64}$$

$$E\{\mathbf{w}_D\} = \sum_{h=H_U+1}^{H_U+H_D} p_R^{(h)} P_{i/t}^{(h)} E\{\tau^{(h)}\} \tag{65}$$

and second raw moments

$$m_{2wU} = \sum_{h=1}^{H_U} p_R^{(h)} \alpha_{\tau^{(h)}} \left(\alpha_{\tau^{(h)}} + 1\right) \left(P_{int}^{(h)} \frac{E\{\boldsymbol{\tau}^{(h)}\}}{\alpha_{\tau^{(h)}}} \right)^2 \tag{66}$$

$$m_{2wD} = \sum_{h=H_U+1}^{H_U+H_D} p_R^{(h)} \alpha_{\tau^{(h)}} \left(\alpha_{\tau^{(h)}} + 1\right) \left(P_{int}^{(h)} \frac{E\{\boldsymbol{\tau}^{(h)}\}}{\alpha_{\tau^{(h)}}} \right)^2 \tag{67}$$

For all fault occurrences in the time interval (0,T), the total energy not supplied WU has expected value

$$E\{\mathbf{W}_U\} = \lambda_U T E\{\mathbf{w}_U\} \tag{68}$$

and variance

$$\sigma^2_{\mathbf{W}_U} = \lambda_U T\, m_{2\mathbf{w}_U} \tag{69}$$

while WD has expected value

$$E\{\mathbf{W}_D\} = \lambda_D T E\{\mathbf{w}_D\} \tag{70}$$

and variance

$$\sigma^2_{\mathbf{W}_D} = \lambda_D T\, m_{2\mathbf{w}_D} \tag{71}$$

The total energy not supplied W for the whole system with M equal feeders is found by the composition of independent events in the two networks, so that its expected value is

$$\mathrm{E}\{\mathbf{W}\} = \mathrm{M}(\mathrm{E}\{\mathbf{W}_\mathrm{U}\} + \mathrm{E}\{\mathbf{W}_\mathrm{D}\}) \tag{72}$$

and its standard deviation is

$$\sigma_\mathbf{W} = \sqrt{M\left(\sigma^2_{\mathbf{W}_U} + \sigma^2_{\mathbf{W}_D}\right)} \tag{73}$$

The total duration of the interruptions d is easily derived from the above results, with expected value

$$\sigma_\mathbf{W} = \sqrt{M\left(\sigma^2_{\mathbf{W}_U} + \sigma^2_{\mathbf{W}_D}\right)} \tag{74}$$

and standard deviation

$$\sigma_\mathbf{d} = \sigma_\mathbf{W} / \mathrm{C}_\mathrm{tot} \tag{75}$$

An important value characterizing the PDF and the Cumulative Distribution Function (CDF) is the probability p0 that no faults occur during the time interval of analysis (0,T), which does not depend on the effectiveness of the switching operations. The value p0 is the amplitude of the Dirac pulse in the origin of the PDF and the initial step value of the corresponding CDF. For a single feeder where both CU and CD are positive, any fault leads to an interruption, so that

$$\mathrm{p0} = e^{-\lambda_U T} e^{-\lambda_D T} \tag{76}$$

If CD = 0 and CU > 0, the interruptions depend on the occurrence of faults on the upstream network or faults on the downstream network for which the selective operation of BD and BU fails, so that

$$\mathrm{p0} = e^{-\lambda_U T} e^{-(1-p_S)\lambda_D T} \tag{77}$$

If CU = 0 and CD > 0, the interruptions depend on the occurrence of faults on the downstream network or faults on the upstream network for which the island formation fails, so that

$$\mathrm{p0} = e^{-(1-p_I)\lambda_U T} e^{-\lambda_D T} \tag{78}$$

The case of M equal feeders corresponds to $\mathrm{p}_0^{(\mathrm{M})} = \mathrm{p}_0^{(\mathrm{M})}$. The whole PDF of the total energy not supplied is obtained by using the characteristic functions approach, where the Gamma PDFs of switching and repair times are represented by using their characteristic function

$$\Phi_{\boldsymbol{\tau}}(\omega) = (1 + j\omega\beta)^{-\alpha} \tag{79}$$

and second characteristic function

$$\Psi_{\boldsymbol{\tau}}(\omega) = \ln(\Phi_{\boldsymbol{\tau}}(\omega)) = -\alpha \ln(1 + j\omega\beta) \tag{80}$$

For the sake of brevity, the argument ω is omitted in the sequel. For a single fault occurrence, the energy not supplied in a single restoration mode is represented, for h = 1,...HU+HD, by a Gamma PDF with shape factor

$$\alpha_{\mathbf{w}^{(h)}} = \alpha_{\boldsymbol{\tau}^{(h)}} \tag{81}$$

scale factor

$$\xi_{\mathbf{w}^{(h)}} = \mathrm{P}_{int}^{(h)} \mathrm{E}\{\boldsymbol{\tau}^{(h)}\} / \alpha_{\boldsymbol{\tau}^{(h)}} \tag{82}$$

and characteristic functions

$$\Phi_{\mathbf{w}^{(h)}} = (1 + j\omega\xi_{\mathbf{w}^{(h)}})^{-\alpha_{\mathbf{w}^{(h)}}} \tag{83}$$

and

$$\Psi_{\mathbf{w}^{(h)}} = \ln(\Phi_{\mathbf{w}^{(h)}}) \tag{84}$$

A single occurrence of a fault in a feeder of the upstream network (or downstream network) leads to a characteristic function computed as the weighted sum of the characteristic functions of the various restoration modes:

$$\Phi_{\mathbf{w}_U} = \sum_{h=1}^{H_U} p_R^{(h)} \Phi_{\mathbf{w}^{(h)}} \tag{85}$$

$$\Phi_{\mathbf{w}_D} = \sum_{h=H_U+1}^{H_U+H_D} p_R^{(h)} \Phi_{\mathbf{w}^{(h)}} \tag{86}$$

The energy not supplied for all occurrences of the fault in the time interval (0,T) is given by the random sums

$$\mathbf{W}_U = \sum_{i=1}^{\mathbf{n}} \mathbf{w}_U \tag{87}$$

and

$$\mathbf{W}_D = \sum_{i=1}^{n} \mathbf{w}_D \tag{88}$$

where n is the Poisson-distributed RV representing the number of fault occurrences. With n mutually independent RVs with equal PDFs, the above random sums correspond to a compound Poisson process [52], providing the second characteristic functions

$$\Psi_{\mathbf{W}_U} = \lambda_U T\left(\Phi_{\mathbf{w}_U} - 1\right) \tag{89}$$

and

$$\Psi_{\mathbf{W}_D} = \lambda_U T\left(\Phi_{\mathbf{w}_D} - 1\right) \tag{90}$$

The compound Poisson process replaces the direct convolution of the PDFs. The second characteristic functions are additive, so that considering all faults over all feeders leads to the second characteristic function of the total energy not supplied

$$\Psi_{\mathbf{W}} = M\left(\Psi_{\mathbf{W}_U} + \Psi_{\mathbf{W}_D}\right) \tag{91}$$

and to

$$\Phi_{\mathbf{W}} = e^{\Psi_{\mathbf{W}}} \tag{92}$$

The corresponding PDF is then obtained from $\Phi_{\mathbf{W}}$ from the Inverse Discrete Fourier Transform (IDFT). Details of this numerical step are provided in [35]. The PDF of the total duration of the interruptions

$$\mathrm{d} = \mathrm{W} / \mathrm{Ctot} \tag{93}$$

referred to the total load power Ctot connected to the system, can be obtained from the above results. The accuracy of numerically computing the PDF depends on the number of samples and on the maximum duration dmax used for the IDFT, and can be easily verified by comparing expected value and variance of the final PDF evaluating by numerical integration to the analytical results.

Monte Carlo Evaluation for Distribution Systems with DG

In order to achieve the goal of computing the reliability indices for a radial distribution system, a Matlab program has been implemented.

The program is structured to run in steps, facilitating the interventions required for any modifications, also providing a better level of understanding of the algorithm.

Step 1 Data setting of the test-system, for upstream and downstream networks; data are composed of probability of island formation, probability of selectivity of upstream/ downstream disconnects, probability of successful upstream switching, probability of successful downstream switching, restoration time (τ) for Gamma-distributed temporary and permanent faults (min/year), permanent failure rate (λ) for upstream and downstream networks and the number of years for the analysis (in our case, 10 years),

Step 2 The parameters required for Monte Carlo (MC) simulations are initialized. This includes the total extractions parameter and the boundary values which delimit the acceptable values of MC simulation. Also the range of the classes utilized to realize MC histograms is presented, for which two a priori values for each level, and the number of classes inside the range min-max was chosen;

Step 3 For a single load point, a series of routines are executed to calculate:

- power not supplied (PNS), by setting-up the lower and upper limits of PNS values, which would fully contain the calculated MC domain of emerging values of PNS, as well as the class limits within the prior defined domain, and the step (which is a threshold depending on the minimum and maximum values of PNS with respect to the number of classes)
- total duration of the interruptions, by setting-up the lower and upper limits, the limits of classes designated to fully contain the MC calculated domain of emerging values of total outage time and the step (a threshold depending on the minimum and maximum values of total duration of the interruptions with respect to the number of classes)
- number of interruptions, by setting-up the lower and upper limits of number of interruptions, the limits of classes designated to fully contain the MC calculated domain of emerging values of number of interruptions and the step (a threshold depending on the minimum and maximum values of number of interruptions with respect to the number of classes)
- energy not supplied (ENS) by setting-up the lower and upper limits of ENS, the limits of classes designated to fully contain the MC calculated domain of emerging values of ENS and the step (a threshold depending on the minimum and maximum values of ENS with respect to the number of classes)

Step 4 An external cycle for MC calculation is generated, composed of several subroutines:

- creation of the random temporary fault profile, by preparing the vector containing the number of temporary faults of each branch in a single execution, followed by generation of a random number r using the Poisson random distribution and storage of the maximum number of faults to be used afterwards in order to establish the dimension of the matrix A containing the faults instants; then, a set of random numbers is generated from a uniform distribution between

0 and 1 and arranged in increasing order; successively, random restoration times are extracted from a Gamma PDF, to fill the matrix A of the faults instants (normalized with respect to the total duration of the analysis);

- creation of the random permanent fault profile, by preparing the vector containing the number of temporary faults of each branch in a single execution, followed by generation of a random number r using Poisson random distribution and storage of the maximum number of faults to be used, setting the dimension of the matrix A containing the faults instants; then, r random numbers are generated from a uniform distribution between 0 and 1 and arranged in increasing order; random restoration times are extracted from a Gamma PDF to create the matrix A of the faults instants (normalized with respect to the total duration of the analysis)

Another cycle is initialized for computing the reliability indices:

- setting-up of the probabilities to be compared to limit-parameters, by randomly extracting the probabilities of island formation, upward-downward selectivity of the circuit breakers, successful switching operation for upward and downward networks, respectively
- computation of PNS, total interruption time, number of interruptions and ENS for temporary faults at each branch by locating the temporary faults and dividing the whole network into upstream and downstream network. The analysis shall be performed taking into account the location of the temporary fault, either for the upstream or the downstream network, the relation between the probability of island forming, the probability of upstream switching, the probability of downstream (and upstream) switching and probability of selectivity values (limit-parameters) at one side and the values emerged at point (a) at the other side
- computation of the PNS, total interruption time, number of interruptions and ENS for permanent faults at each branch, by locating the permanent faults and dividing the whole network into upstream and downstream network. The analysis shall be performed taking into account the location of the permanent fault, either upstream or downstream, the relation between the probability of island forming, the probability of upstream switching, the probability of downstream (and upstream) switching and probability of selectivity values (limit-parameters) at one side and the values emerged at point (a) at the other side
- computation of MC histograms for PNS, total interruption time, number of interruptions and ENS, by taking into account the PNS summed-up on every branch, the number of total extractions, the number of classes and the class limits of each reliability indicators for each single load point

Step 5 The graphical interface is created, by plotting the results into two different windows (for upstream and downstream networks), for PNS, total interruption time, number of interruptions and ENS.

Numerical Results

Test Systems

Two large real urban distribution systems fed at 6.3 kV and 22 kV have been considered. These system contain many underground cables, some of which with a very long operational life (up to 50 years or more). For characterizing the random variables involved in the restoration process, the 6.3 kV system has been used, with failure data related to three years of operation (2001, 2002 and 2003).

Characterization of the Random Variables

This section deals with assessing the probability density functions of time to failure (time elapsed between two successive permanent faults), reclosure time (duration occurred to perform manual reclosure after a fault), total duration of service restoration Dtot and equivalent weighted restoration time Tm. This assessment is performed on the basis of a set of data representing the faults occurred in a large electrical distribution system.

The evaluations are carried out through various goodness of fit statistical tests, using real data representing the faults occurred in a specific time period of observation on a 6.3 kV urban distribution system.The distributions tested are Gamma, Exponential, Normal (Gauss), Log-normal, Inverse-normal, Beta (max), Weibull and Rayleigh. For all these distributions, an initial table is presented, containing the K-S test error calculated for each type of distribution, thus improving the overall analysis by displaying the most useful distributions, that is the ones with relatively low errors.

Time to Failure

The sample set contains 142 data, with sample mean of 119.9 hours and sample standard deviation of 169.4 hours (141%). The data sample contains also three faults involving pairs of cables fed by the same busbars in which, after the occurrence of a first fault, another fault could have been caused by transient phenomena during restoration and is characterised by almost null time to failure. In addition, there are three very long periods without failures (593, 611 and 1029 hours, respectively).

The goodness-of-fit test of different probability distributions has been carried out by first performing the K-S test, whose results are shown in the Table 5.

The four distributions with lower error, i.e. Weibull, Inverse-Normal, Exponential and Gamma have been studied in more detail by performing the chi-square test, by using classes with uniform width.

Weibull distribution: the calculated shape and scale parameters are $\alpha = 0.721$ and $\beta = 97.34$. The K-S test, performed with the level of significance of 5% showed that the hypothesis is impossible to be verified, due to the observed error (0.0831), which lies in the region between the minimum error for rejection (0.1141) and the maximum value for acceptance (0.0744).

Table 5. Presentation of the K-S test error for every distribution used in the Time to failure assessment

Abbreviation	Type of distribution	K-S test error
w	Weibull	0.0831
i	Inverse-Normal	0.1227
e	Exponential	0.1231
g	Gamma	0.1271
l	Log-Normal	0.1384
b	Beta (max)	0.1983
n	Normal	0.2396
r	Rayleigh	0.3347

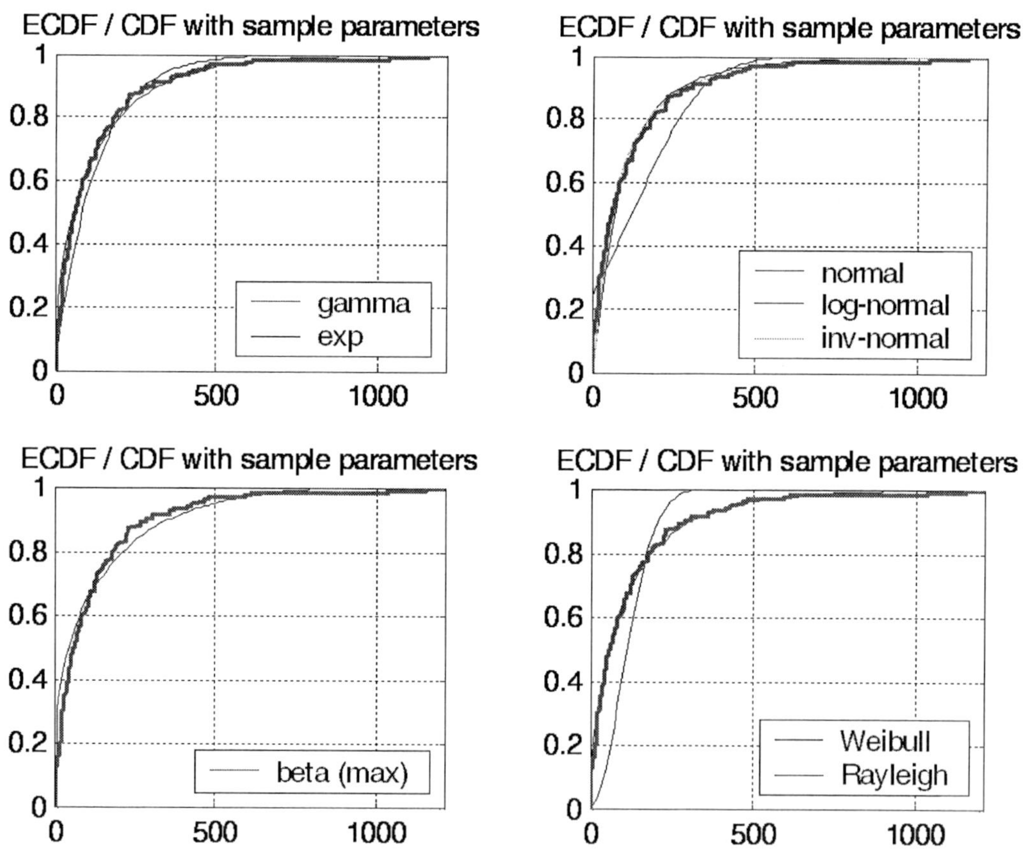

Figure 9. ECDF / CDF comparison for different distributions for time to failure assessment.

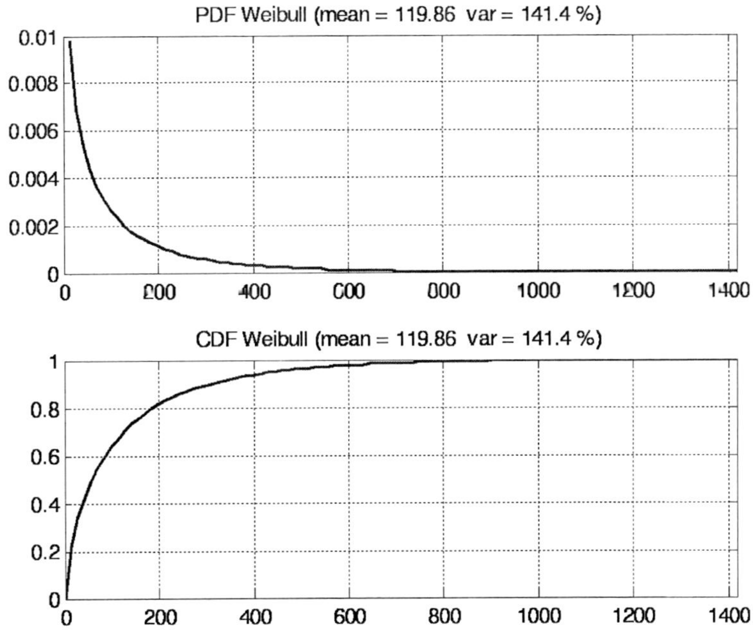

Figure 10. Weibull PDF and CDF with sample parameters.

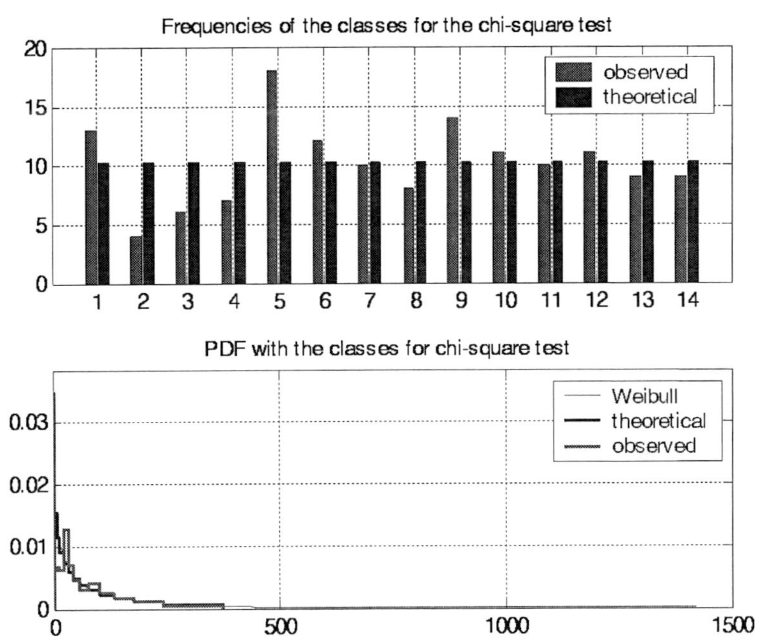

Figure 11. Results of the chi-square test for the time to failure with Weibull distribution and comparison between the ECDF and the Weibull CDF.

A chi-square test has been performed with 14 uniform classes (see Figure 11). The test refers to 11 degrees of freedom (with mean value and variance corresponding to the sample mean and the sample variance, respectively) and results in a maximum error for acceptance of

19.68%. The observed error was 15.94. The hypothesis was accepted with Λ=5% and the maximum level of significance for the acceptance is 14.32%.

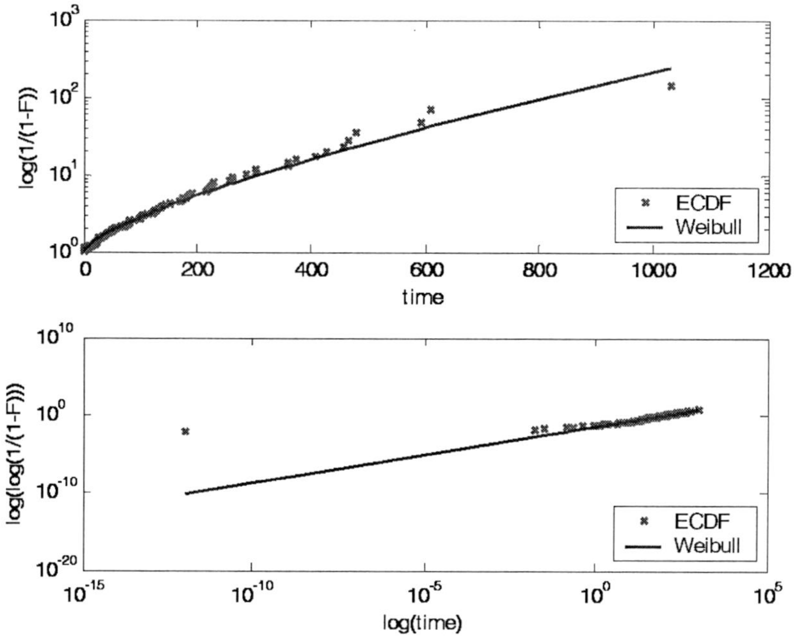

Figure 12. Results of the geometrical adaptation tests for the time to failure with ECDF and Weibull CDF.

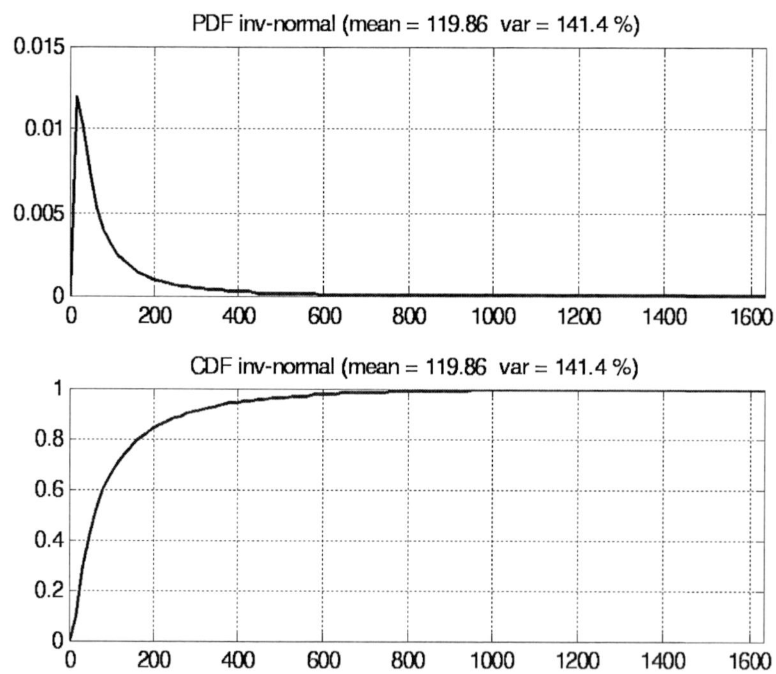

Figure 13. Inverse-Normal PDF and CDF with sample parameters.

Inverse-normal distribution the K-S test, performed with the level of significance of 5% showed that the hypothesis is rejected, due to the observed error (0.1227), which stays outside of the domain situated between the minimum error for rejection (0.1141) and the maximum value for acceptance (0.0744). The chi-square test performed showed that the hypothesis was rejected with Λ%=5%, the observed error (31.32) being bigger than the minimum error for acceptance (22.36), while the maximum error for the acceptance was 19.68 (see details in Figure 14). The maximum significance level for acceptance was Λmax%=0.1%.

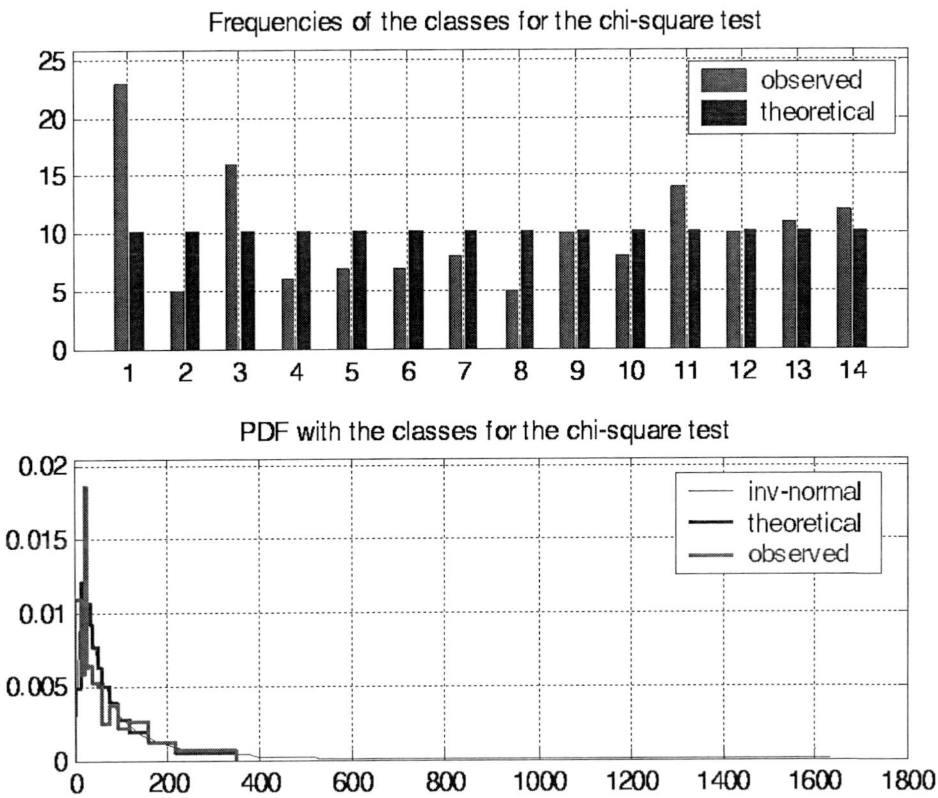

Figure 14. Results of the chi-square test for the time to failure with Inverse-Normal distribution and comparison between the ECDF and the Inverse-Normal CDF.

Exponential distribution the K-S test, performed with the level of significance of 5%, variance σ = 119.86 (100%) and scale parameter β = 1 being calculated, while shape parameter α = 1 was assigned.

The test showed that the hypothesis is rejected, due to the observed error (0.1231), which larger than the critical error (0.089). However, the conceptual importance of the exponential distribution of the time to failure in the reliability study presented required a deeper analysis. For this purpose, the chi-square test with twelve non-uniform classes (Figure 16, with class limits of 0, 15, 28, 40, 50, 65, 75, 110, 150, 200, 300 and > 300 hours).

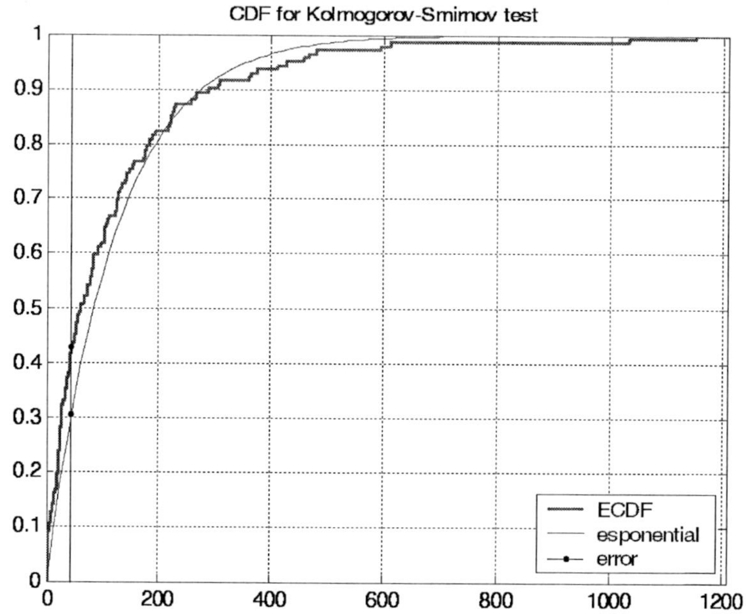

Figure 15. Comparison between ECDF and Exponential CDFs.

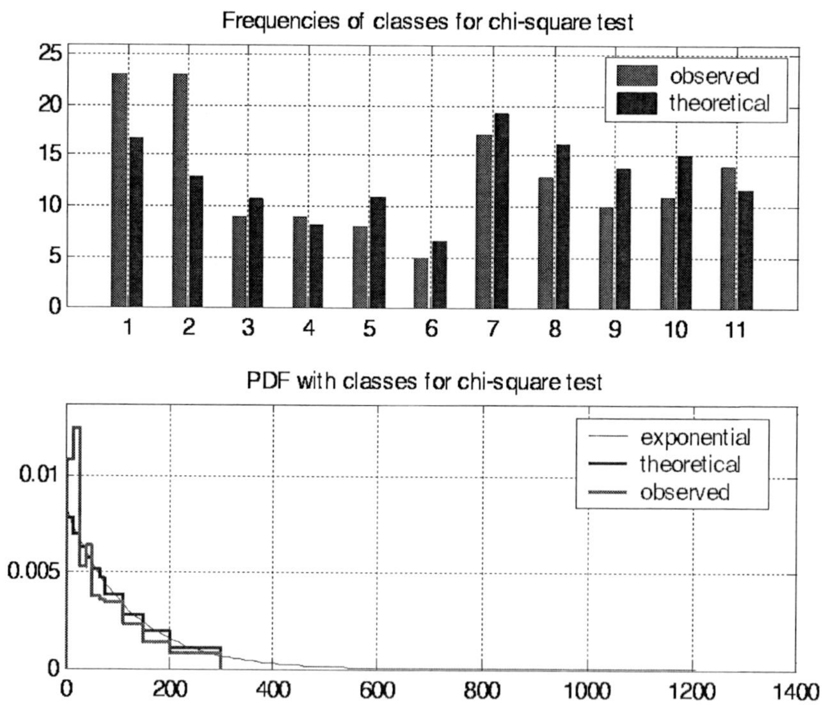

Figure 16. Results of the chi-square test for the time to failure with Exponential distribution and comparison between the ECDF and the Exponential CDF.

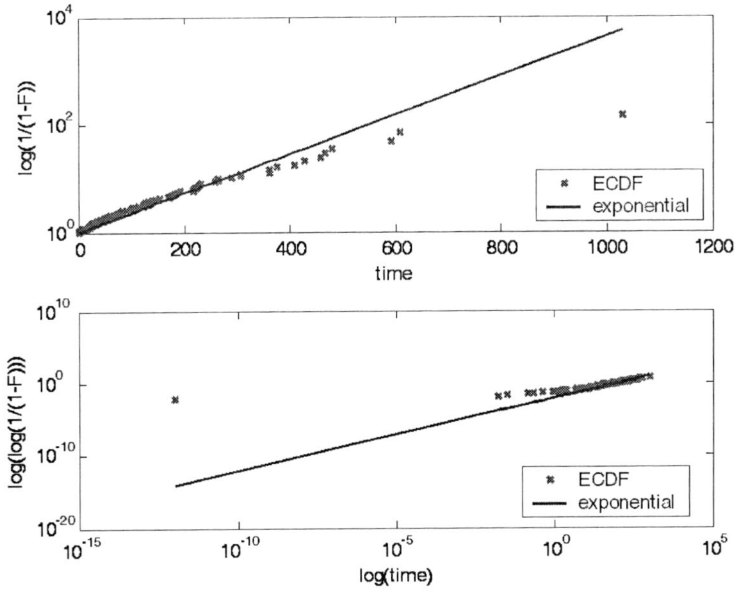

Figure 17. Results of the geometrical adaptation tests for the time to failure with ECDF and Exponential CDF.

It resulted in a maximum significance level Λmax% = 17.78%. From this test, the exponential distribution appears to be satisfactory to characterise the time to failure. Using two Yates error corrections, the observed error (12.68) was smaller than the maximum error for acceptance (16.92) and significantly smaller than the minimum error for rejection (18.31). In order to show more details on this behaviour, see the geometrical adaptation test indicated in Figure 17. The effect of long periods without faults clearly appears.

Gamma distribution: by performing K-S test, the shape parameter can be obtained from the variance of the sample ($\alpha = 0.5$), correspondent to the K-S test error of 0.1271.

The program gives the possibility of obtaining α taking into account the minimum error for K-S test (0.0498) ($\alpha = 0.682$). Beside shape parameter, the mean (119.86) is extracted from the samples, while the variance $\sigma = 169.43$ (141.4%) and scale parameter $\beta = 239.49$ are rigorously calculated. With Λ% = 5%, the hypothesis is rejected, due to the fact that the observed value (0.1271) is bigger than the minimum error for rejection (0.1141) and significantly bigger that the maximum error for acceptance (0.0744).

The chi-square test performed with 11 degrees of freedom (in order to obtain the maximum error for acceptance 19.68), showed that the observed error was 17.92, which means that the hypothesis was accepted with level of significance 5%. The maximum level of significance was Λmax% = 8.36%. Figure 18 graphically illustrates the shape of the Gamma CDF, compared to the ECDF, while Figure 20 shows the results of the geometrical adaptation tests with using the Gamma PDF with $\alpha = 0.5$ and $\beta = 239.49$.

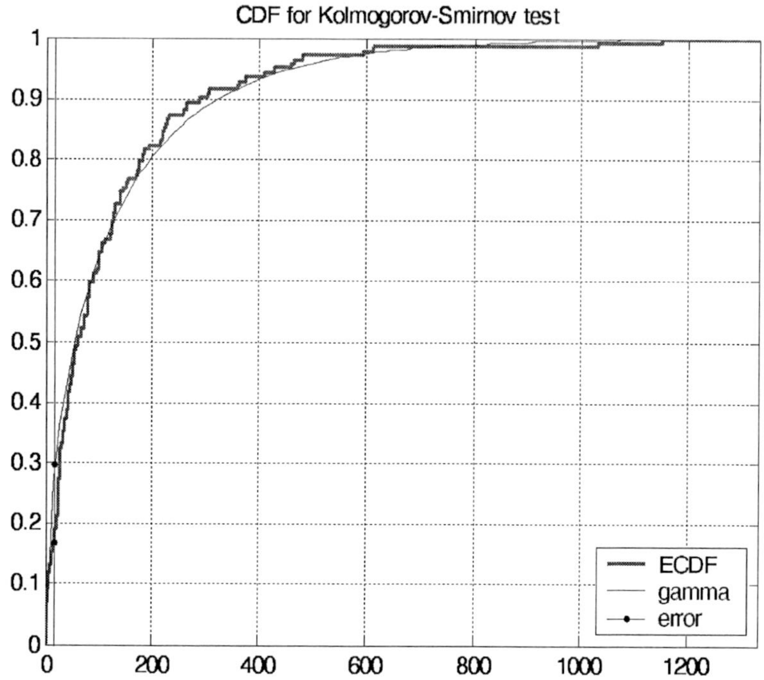

Figure 18. Comparison between ECDF and Gamma CDFs.

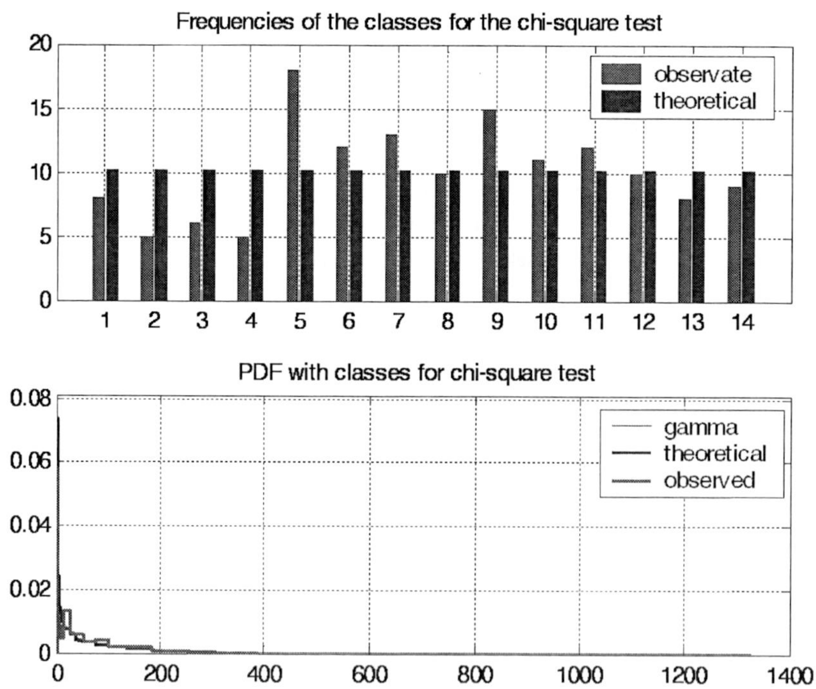

Figure 19. Results of the chi-square test for the time to failure with Gamma distribution and comparison between the ECDF and the Gamma CDF.

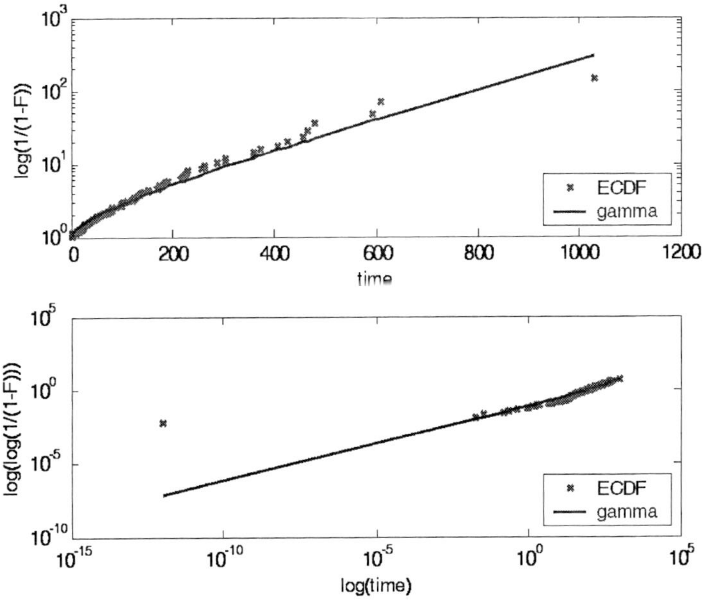

Figure 20. Results of the geometrical adaptation tests for the time to failure with ECDF and Gamma CDF.

Total Duration of Service Restoration

The sample set contains 234 data related to the years 2001, 2002 and 2003, with sample mean of 71.62 minutes and sample standard deviation of 49.62 minutes (69.4%).

The distributions that the program makes possible to be tested are: Gamma, Exponential, Normal (Gauss), Log-normal, Inverse-normal, Beta (max), Weibull and Rayleigh. Table 6 presents the distributions that fits the best like Log-Normal, Inverse-Normal, Gamma, Weibull and Rayleigh.

Log-Normal distribution: the K-S test performed with $\Lambda\% = 5\%$ showed that is impossible to verify if the hypothesis is accepted or rejected, because the observed error (0.0776) was between the maximum error for acceptance (0.0579) and the minimum error for rejection (0.0889).

Table 6. Presentation of the K-S test error for every distribution used in the Total duration of service restoration assessment

Abbreviation	Type of distribution	K-S test error
l	Log-Normal	0.0776
i	Inverse-Normal	0.0906
g	Gamma	0.1060
w	Weibull	0.1125
r	Rayleigh	0.1145
b	Beta (max)	0.1222
n	Normal (Gauss)	0.1745
e	Exponential	0.2321

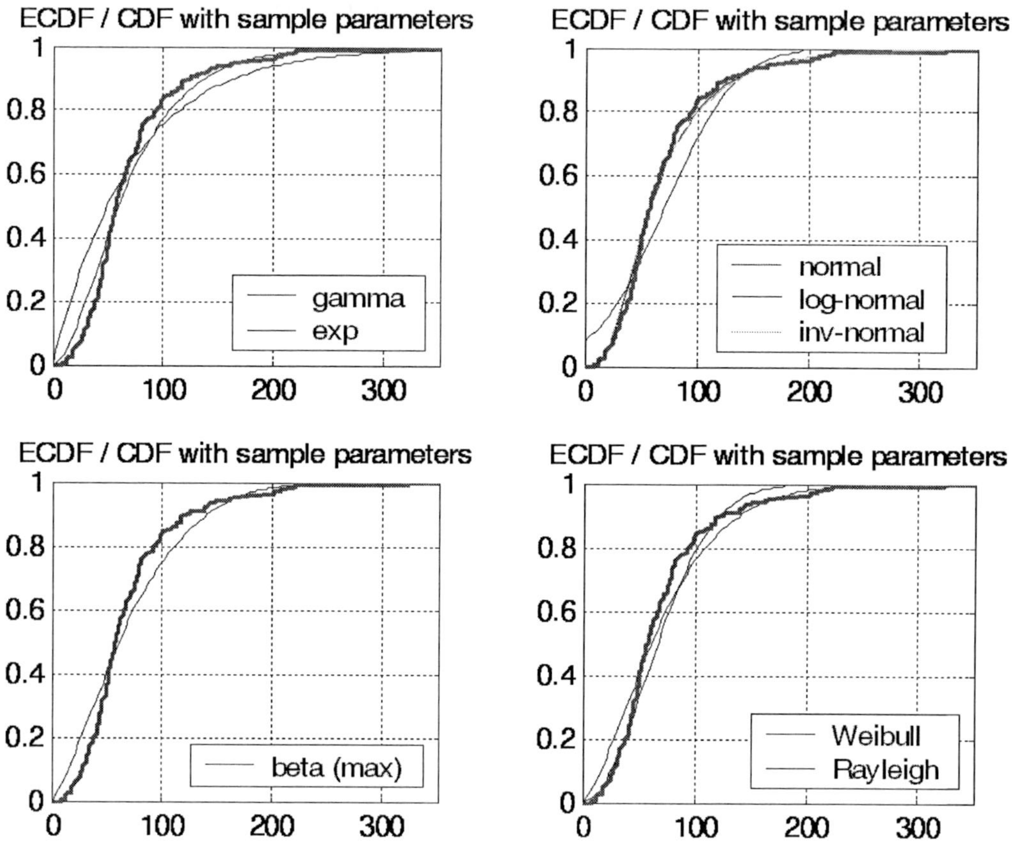

Figure 21. ECDF / CDF comparison for different distributions for total duration of service restoration assessment.

Results of the chi-square test showed that the hypothesis made with Λ% = 5% was impossible to be verified, Λmax% = 4.64% and the observed error stays within the maximum error for acceptance (27.59) and the minimum error for rejection (30.14).

Inverse-Normal distribution: the K-S test prove that the hypothesys was unsatisfactory, the observed error (0.0906) being bigger than the minimum error for rejection (0.0889).

By performing chi-square test with Λ% = 5%, results that is impossible to verify the hypothesis, because Λmax% = 4.64%. The observed error (27.88) stays between the maximum error for acceptance (27.59) and the minimum error for rejection (31.14).

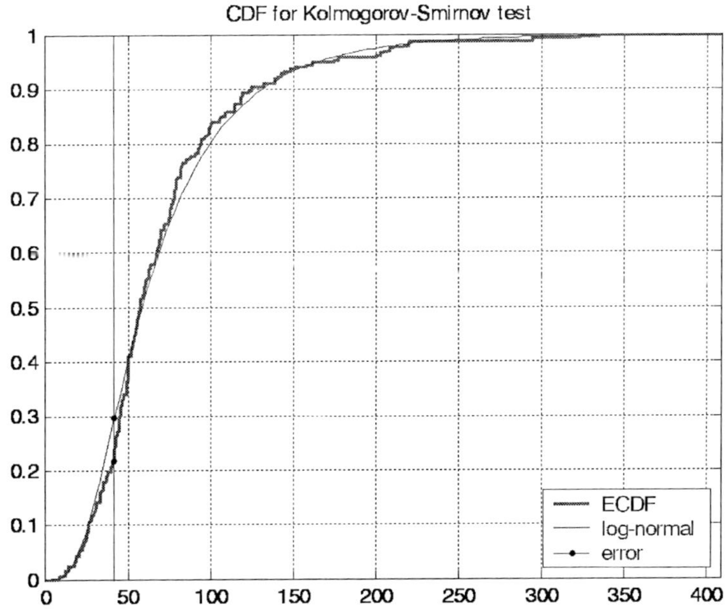

Figure 22. Comparison between ECDF and Log-Normal CDFs.

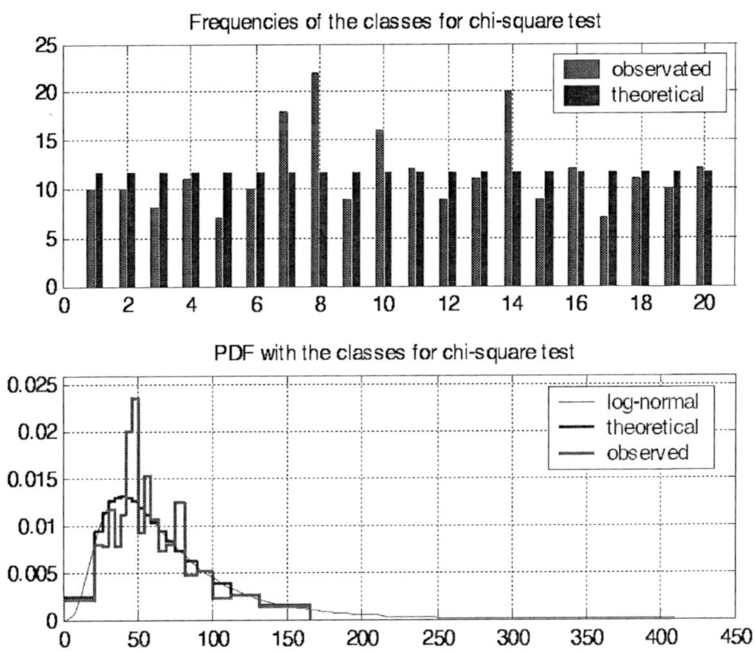

Figure 23. Results of the chi-square test for the total duration of service restoration with Log-Normal distribution and comparison between the ECDF and the Log-Normal CDF.

Gamma distribution: by performing K-S test, the shape parameter can be obtained from the variance of the sample ($\alpha = 2.074$), correspondent to the K-S test error of 0.1060. The program gives the opportunity of obtaining α taking into account the minimum error for K-S test (0.0889)($\alpha = 3.281$).

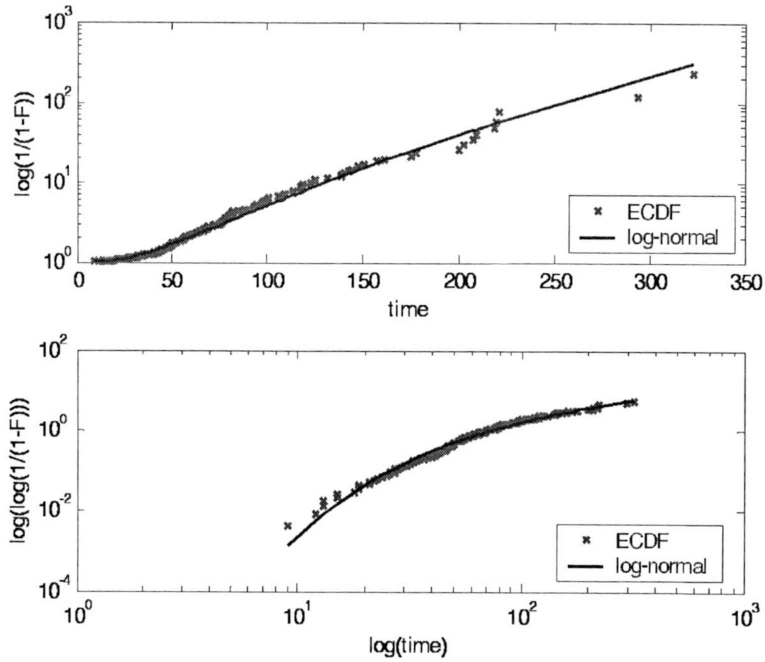

Figure 24. Results of the geometrical adaptation tests for the total duration of service restoration with ECDF and Log-Normal CDF.

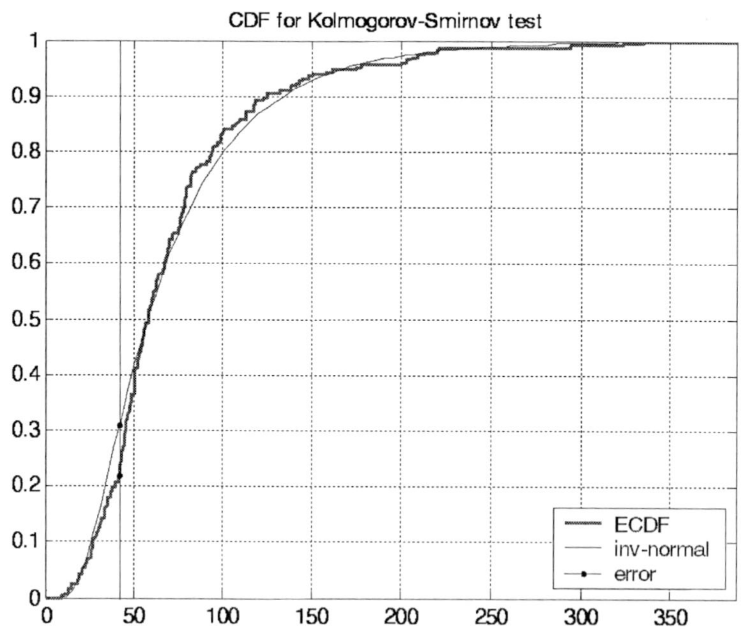

Figure 25. Comparison between ECDF and Inverse-Normal CDFs.

Beside shape parameter, the mean (71.62) is extracted from the samples, while the variance $\sigma = 49.73$ (69.4%) and scale parameter $\beta = 34.535$ are rigorously calculated. With

$\Lambda\% = 5\%$, the hypothesis is rejected, due to the fact that the observed error (0.1060) is bigger than the minimum error for rejection (0.0889) and significantly bigger that the maximum error for acceptance (0.0579).

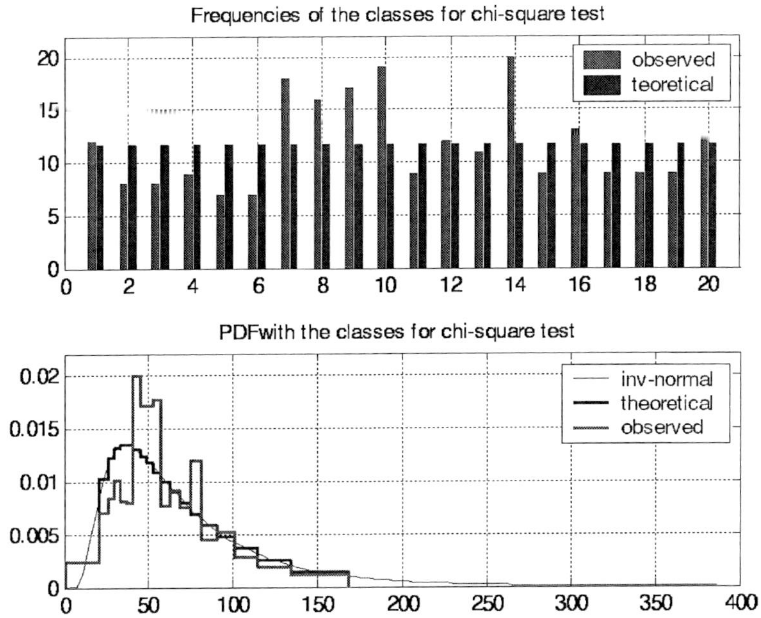

Figure 26. Results of the chi-square test for the total duration of service restoration with Inverse-Normal distribution and comparison between the ECDF and the Inverse-Normal CDF.

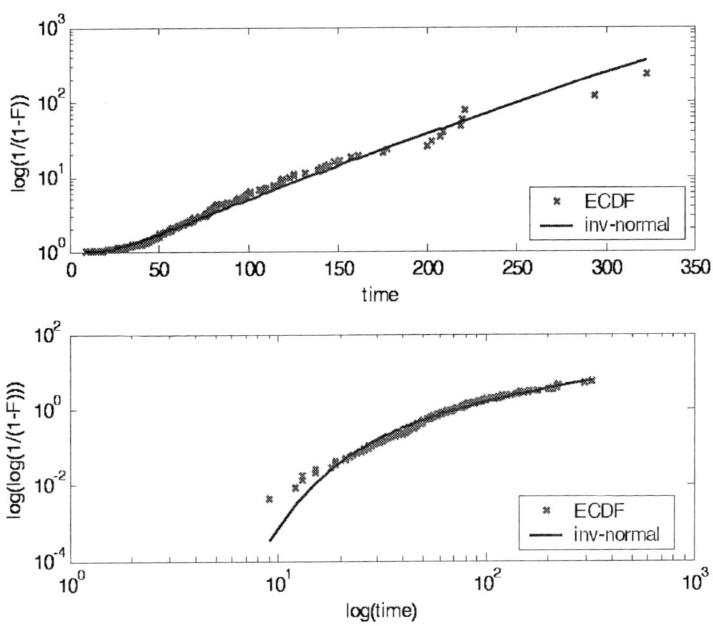

Figure 27. Results of the geometrical adaptation tests for the total duration of service restoration with ECDF and Inverse-Normal CDF.

The performed chi-square test showed that the observed error was 49.93, which means that the hypothesis was accepted with level of significance 5%. Figure 28 graphically illustrates the shape of the Gamma CDF, compared to the ECDF, while Figure 30 shows the results of the geometrical adaptation tests with using the Gamma PDF with $\alpha = 2.074$ and $\beta = 34.535$.

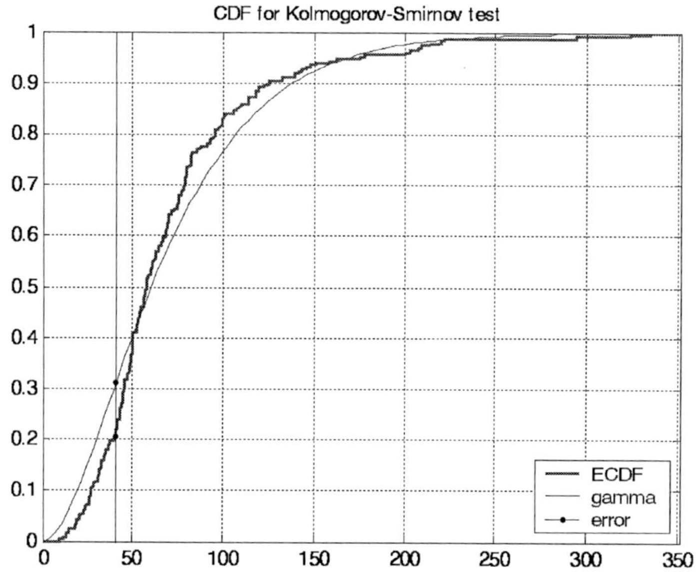

Figure 28. Comparison between ECDF and Gamma CDFs.

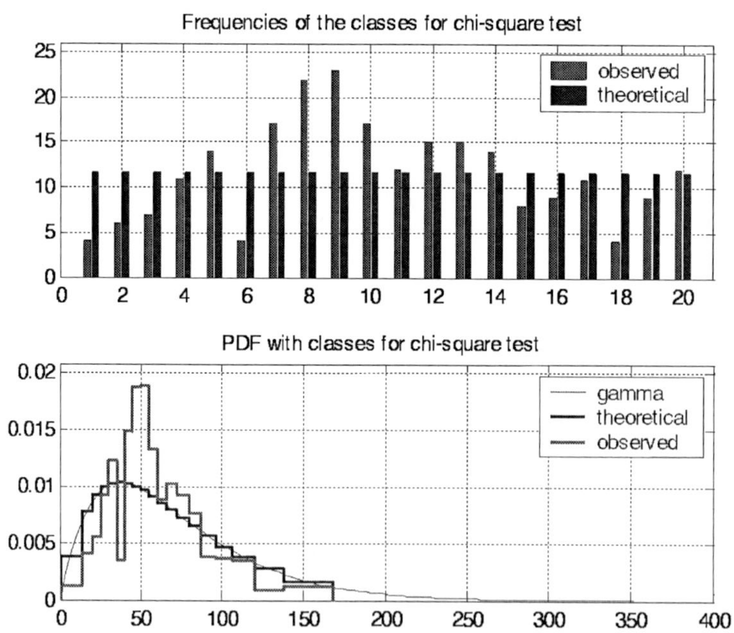

Figure 29. Results of the chi-square test for the total duration of service restoration with Gamma distribution and comparison between the ECDF and the Gamma CDF.

The chi-square test performed with obtained the maximum error for acceptance (5.99) and the minimum error for rejection (9.49), with observed error of 6.39, which means that the hypothesis was accepted with Λ % = 5%, the maximum level of significance being Λmax% = 13.53 %.

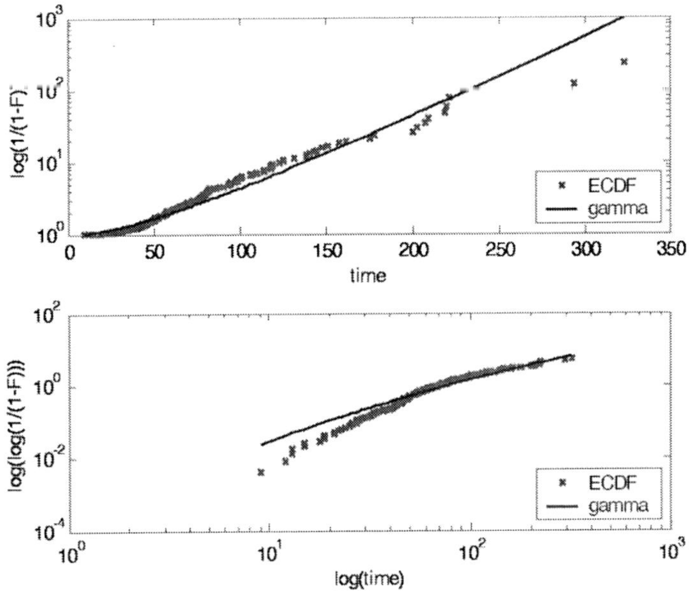

Figure 30. Results of the geometrical adaptation tests for the total duration of service restoration with ECDF and Gamma CDF.

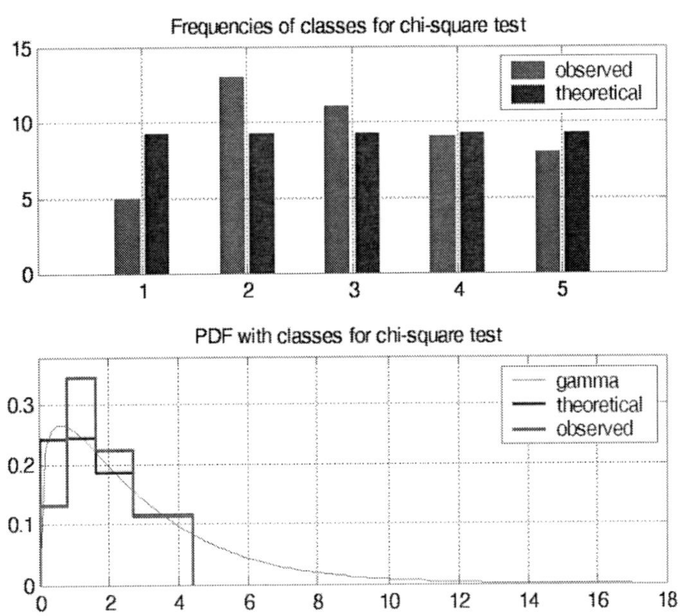

Figure 31. Results of the chi-square test for the Reclosure time with Gamma distribution and comparison between the ECDF and the Gamma CDF.

Figure 31 graphically illustrates the shape of the Gamma CDF, compared to the ECDF, while Figure 32 shows the results of the geometrical adaptation tests using the Gamma PDF. Reliability analysis of distribution system with distributed generation using the Monte Carlo method. The distribution system with distributed generation, whose structure is shown in Figure 7, has been used to carry out the Monte Carlo analysis, with 10000 extractions. The results are shown in Figure 33 for the upstream network and in Figure 35 for the downstream network. Details of the duration of the interruptions are provided in Figure 34 and Figure 36.

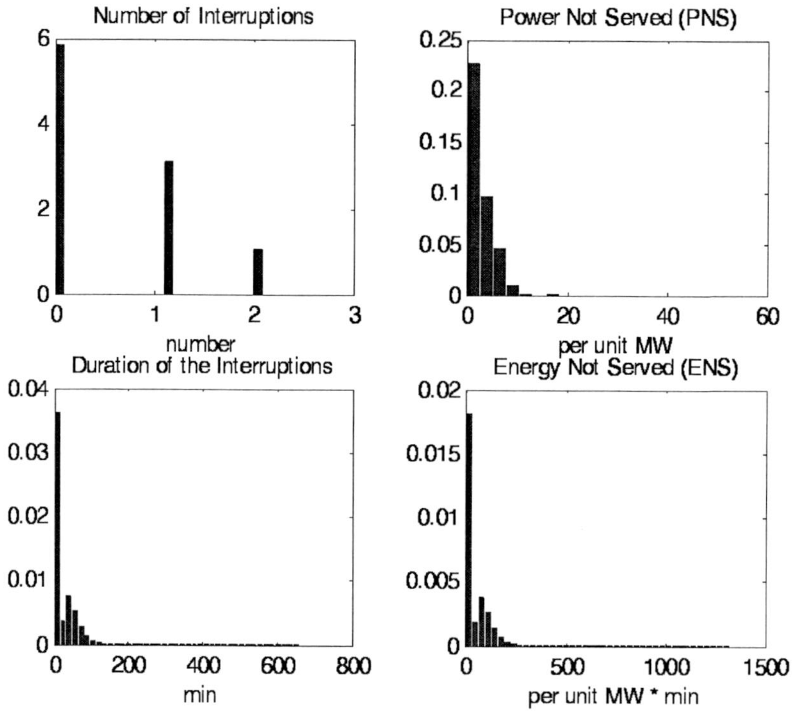

Figure 33. Reliability indicators resulting from the Monte Carlo analysis for the upstream network (Dirac impulse in the origin included).

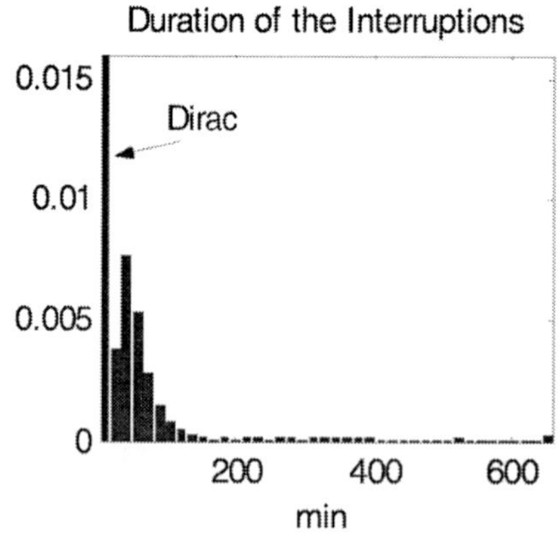

Figure 34. Zoom of the duration of the interruptions for the upstream network.

For the total duration of the interruptions, the Monte Carlo results shown in Figure 33 (for the upstream network) and Figure 35 (for the downstream network)) correspond to the results of the analytical simulation and confirm the validity of the analytical simulation approach.

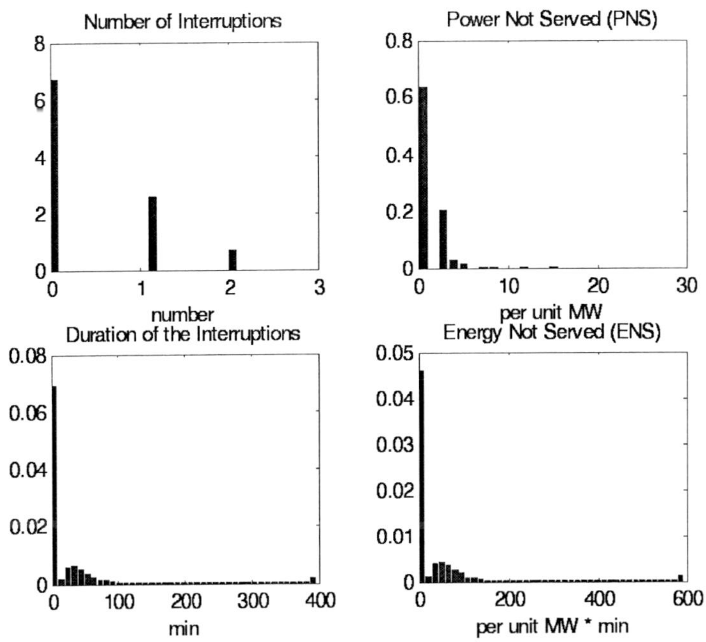

Figure 35. Reliability indicators resulting from the Monte Carlo analysis for the downstream network (Dirac impulse in the origin included).

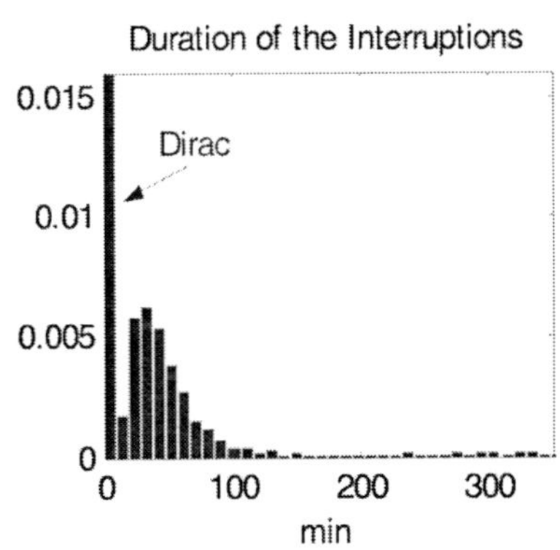

Figure 36. Zoom of the duration of the interruptions for the downstream network.

CONCLUSION

This analysis addressed continuity of supply for electrical distribution systems, investigating the corresponding reliability issues in a probabilistic way. The most challenging aspects concerned the study of the impact of Distributed Generation embedded in distribution systems from the reliability point of view, and the application of efficient computational techniques for handling the probability distributions of the random variables involved in the study in order to obtain the probabilistic formulation of the reliability indicators.

The research efforts went mainly into the direction of evaluating the influence that Distributed Generation has on the traditional electrical distribution systems. The first step for achieving this goal was the probabilistic characterization of the random variables involved in reliability assessment as inputs (number of the interruptions, time to failure, duration of service restoration and reclosure time). In order to perform a proper estimation of upper mentioned random variables, various goodness-of-fit tests (Chi-square, Kolmogorov-Smirnov and Geometrical adaptation) were used, performed on the data gathered from a large real electrical distribution system. This part of the analysis proved to be a particularly useful mean of evaluation of the most suitable probability distributions.

The PDFs of the restoration times allowed for computing the probability distributions of the duration-dependent reliability indices by using different methods (e.g., analytical or Monte Carlo). In particular, the representation of the restoration time by using the Gamma PDFs presented the key advantage of simplifying the numerical treatment of the probability distributions, by enabling the use of analytical expression of the characteristic functions of the PDFs, thus offering the possibility to the distribution system operators to analytically evaluate the PDFs of local and global reliability indices in a fast and general way, even in the case of PDFs with multi-modal shape.Future developments include linking the results of the analysis performed in this chapter to probabilistic evaluations of the economic indicators associated to the interruptions, in order to build a comprehensive probabilistic risk assessment framework for the continuity of supply of electricity distribution systems.

REFERENCES

Lee Willis H., Power Distribution Planning Reference Book, Marcel Dekker, 1997.

Bompard E., Carpaneto E., Chicco G., Napoli R., Convergence of the backward/forward sweep method for the load-flow analysis of radial distribution systems, Electrical Power and Energy Systems 22, 2000, 521-530.

Schwan M., Wellβow W.H. and Koglin H.J., Identification of Probability Distributions of Probabilistic Reliability Indices for Monetary Evaluation and Risk Assessment, Proc. XIV PSCC, Sevilla, Spain, June 24-28, 2002, Session 04, Paper 1.

Dialynas R.N. and Michos D.G., Interactive modeling of supply restoration procedures in distribution system operation, IEEE Transactions on Power Delivery 4, 3 (July 1989) 1847-1854.

Kersting W.H., Phillips W.H. and Clay Doyle R., Distribution feeder reliability studies, IEEE Transactions on Industry Applications 35 (2), March/April 1999, 319-323.

Popović D.S. and Ćirić R.M., A multi-objective algorithm for distribution network restoration, IEEE Transactions on Power Delivery 14 (3), July 1999, 1134-1141.

Nan Miu K., Chiang H.-D., Yuan B. and Darling G., Fast service restoration for large-scale distribution systems with priority customers and constraints, IEEE Transactions on Power Systems 13 (3), August 1998, 789-795.

Lee S.-J., Lim S.-I. and Ahn B.-S., Service restoration of primary distribution systems based on fuzzy evaluation of multi-criteria, IEEE Transactions on Power Systems 13 (3), 1998, 1156-1163.

Douligeris C., Develekos G. and Zografos K., Analysis of the service restoration operations of a large rural utility company, Proc. 39th Rural Electric Power Conference, April 30-May 2, 1995, B3/1-B3/8.

Kirkham H., Nightingale D. and Koerner T., Energy management system design with dispersed storage and generation, IEEE Transactions on Power Apparatus and Systems PAS-100 (7), July 1981, 3432-3440.

U.S. Department of Energy, Energy Information Administration - Annual Energy Review 2002, p.243. Available: www.eia.doe.gov.

Van Zyl S.J. and Gaunt C.T., Control strategies for distributed generators operating on weak distribution networks, Proc. IEEE Bologna Power Tech, Bologna, Italy, June 23-26, 2003, 3, 252-258.

Dondi P., Bayoumi D., Haederli C., Julian D. and Suter M., Network integration of distributed power generation, Journal of Power Sources, 106 (2002) 1-9.

Carpaneto E., Chicco G., Porumb R., Reliability aspects of distribution systems with local generation resources, Proc. CIEM 2003 Conference, Bucharest, România, 22-25 October 2003, Vol. 1, 2.1-2.8

Barsali S., Celli G., Ceraolo M., Giglioli R., Pelacchi P. and Pilo F., Operating and Planning Issues of Distribution Grids Containing Diffuse Generation, Proc. 16th CIRED, Amsterdam, The Netherlands, June 18-21, 2001, paper 4.20.

Barker P.P. and De Mello R.W., Determining the impact of distributed generation on power systems. Part I – Radial distribution systems, Proc. IEEE/PES Summer Meeting (2000) 3, 1645-1656.

Hadjsaid N., Canard J.-F. and Dumas F., Dispersed generation impact on distribution networks, IEEE Computer Applications in Power 12 (1999) 22-28.

Bătrînu F., Chicco G., Porumb R., Postolache P. and Toader C., Current issues on operation and management of distributed resources, Proc. Fifth International World Energy System Conference, Oradea, România, May 17-19, 2004, Vol.3, 598-605.

Lee Willis H., Distributed Power Generation, Marcel Dekker, 2000.

Billinton R., Wang P., Teaching Distribution System Reliability Evaluation Using Monte Carlo Simulation, IEEE Transactions on Power Systems 14, (2), May 1999.

Billinton R. and Li W., Reliability Evaluation of Electric Power Systems Using Monte Carlo Methods, Plenum Press, New York (1994).

Billinton R., Karki R., Maintaining supply reliability of small isolated power systems using renewable energy, IEE Proceedings on Generation, Transmission and Distribution 148, (6), Nov.2001, 530-534

Brown R.E., Modeling the reliability impact of distributed generation, IEEE Power Engineering Society Summer Meeting, 21-25 July 2002, Vol.1, 442-446.

Megdiche M., Besanger Y., Aupied J., Garnier R.and Hadjsaid N., Reliability assessment of distribution systems with distributed generation including fault location and restoration process, 17th CIRED, Barcelona, Spain, May 12-15, 2003, paper 4.74.

Billinton R., Li W., A System State Transition Sampling Method For Composite System Reliability Evaluation, IEEE Transactions on Power Systems 8, (3), August 1993.

Brown R.E., Reliability of distribution systems, Dekker, New York, 2002.

Chiradeja P., Ramakumar R., An approach to quantify the technical benefits of Distributed Generation, IEEE Transactions on Energy Conversion, Vol. 19, No. 4, December 2004, 764-773.

Chicco G., Marin M., Napoli R. and Postolache P., Power quality issues in the regulation of the electricity services, Proc. 4th CEE, Târgovişte, România, 4-5 October 2001, 43-59.

IEEE (The Institute of Electrical and Electronics Engineers, Inc.), IEEE Trial-use guide for electric power distribution reliability indices, IEEE Std 1366-1998, 10 December 1998.

Billinton R., Wojczynski E., Distributional Variation of Distribution System Reliability Indices, IEEE Trans. on Power Apparatus and Systems 104, 11 (1985) 3152-3160.

Endrenyi J., Reliability Modelling in Electric Power Systems, Wiley and Sons, New York, 1978

Schwan M., Wellβow W.H. and Koglin H.J., Identification of Probability Distributions of Probabilistic Reliability Indices for Monetary Evaluation and Risk Assessment, Proc. XIV PSCC, Sevilla, Spain, June 24-28, 2002, Session 04, Paper 1.

Kjølle G. and Sand K., RELRAD – An analytical distribution system reliability assessment, IEEE Transactions on Power Delivery 7, 2 (1992) 809-814.

Carpaneto, E., Chicco, G., Mosso, A., and Ponta, A., Analytical Evaluation of the Probability Distributions of Reliability Indices for Large Distribution Systems, Proc. PMAPS 2002, Naples, Italy, September, 22–26, 2002, pp.31-36.

Carpaneto E., Chicco G., Evaluation of the Probability Density Functions of Distribution System Reliability Indices With a Characteristic Function-Based Approach, IEEE Transactions on Power Systems 19, (2), May 2004, 724-733.

van Casteren J.F.L., Bollen M.H.J. and Schmieg M.E., Reliability assessment in electrical power systems: the Weibull-Markov stochastic model, IEEE Transactions on Industry Applications 36 (3), May-June 2000, 911-915.

Ou Y., Goel L., Using Monte Carlo Simulation For Overall Distribution System Reliability Worth Assessment, IEE Proceedings on Generation, Transmission and Distribution 146, (5), 1999.

Billinton R. and Goel R., An Analytical Approach to Evaluate Probability Distributions Associated with the Reliability Indices of Electric Distribution Systems, IEEE Trans. Power Deliv. 1, 3 (1986) 245-251.

Asgarpoor S. and Mathine M.J., Reliability Evaluation of Distribution Systems with Non-Exponential Down Times, IEEE Trans. Power Syst. 12, 2 (1997) 579-584.

D'Agostino R.B. and Stephens M.A. (ed.), Goodness-of-fit techniques, Dekker, New York (1996)

Papoulis A., Probability, Random variables and Stochastic Processes, McGraw Hill, Singapore (1991).

Trivedi K.S., Probability and Statistics With Reliability, Queuing and Computer Science Applications, New York, Wiley, 2002.

Miller L.H., Table of percentage points of Kolmogorov statistics, J. Am. Stat. Assoc. 51, 113, 1956

Lilliefors H.W., On the Kolmogorov-Smirnov test for exponential with mean unknown, J. Am. Stat. Assoc. 64, 388, 1969.

Lilliefors H.W., On the Kolmogorov-Smirnov test for normality with mean and variance unknown, J. Am. Stat. Assoc. 62, 400, 1967.

Billinton R., Allan R.N., Reliability Evaluation of Engineering Systems, 2nd Edition, Plenum Press, New York, 1992.

Billinton R., Distribution System Reliability Performance and Evaluation, Electrical Power andEnergy Systems 10 (3), July 1998, 190-200.

Carpaneto E., Chicco G., Mosso A., Poggi A. and Ribaldone P., Tools for optimal operation and planning of urban distribution systems, Proc. CIRED 2001, Amsterdam, The Netherlands, June 18-21, 2001, paper 5.22.

Carpaneto, E., Chicco, G., Mosso, A., and Ponta, A., Analytical Evaluation of the Probability Distributions of Reliability Indices for Large Distribution Systems, Proc. PMAPS 2002, Naples, Italy, September, 22–26, 2002, pp.31-36.

Porumb R., Sequential Monte Carlo simulation for computing reliability indices in electrical distribution systems, International Symposium on Electrical Energy (ISEE 2003), Târgovişte, România, November 2003, 6 pages, ISBN: 973-8413-47-8.

Ou Y., Goel L., Subtransmission System Reliability Worth Assessment Using The Monte Carlo Simulation Method, IEEE/Power Engineering Society Winter Meeting, 1999, Vol: 2.

Billinton R., Jonnavithula A., Application Of Sequential Monte Carlo Simulation To Evaluation Of Distributions Of Composite System Indices, IEE Proceedings on Generation, Transmission and Distribution 144, (2), March 1997.

W.Feller, An introduction to probability theory and its applications, 2 Vol., Wiley and Sons, New York, 1971.

In: Advances in Energy Research
Editor: Nicu Bizon

ISBN 978-1-61209-991-0

Chapter 3

NEW APPROACHES IN NONLINEAR DYNAMICS ANALYSIS OF COMPLEX SYSTEMS AND PROCESSES

***Alexandru Serbanescu*[1*] *and Florin-Marian Birleanu*[2]**
[1]Military Technical Academy Bucharest, Romania
[2]University of Pitesti, Romania

ABSTRACT

This chapter reports on the application to field measurements of time series methods developed on the basis of the theory of deterministic chaos. Recent progress in the detection and understanding of nonstationarity is reported. Recurrence plots form an efficient strategy for both visualizing and quantifying nonlinear spatio-temporal dynamics of complex systems. For example, we suggest that the unstable periodic orbits embedded in a chaotic attractor are a useful basis set for the geometry of a recurrence plot of those data. This provides a means to locate unstable periodic orbits in chaotic time series data. Besides that, we present here some results we obtained using the concept of recurrence in the fields of speech processing, signal detection and communications.

Keywords: dynamic systems, phase space, recurrence plot, nonlinear signal analysis.

INTRODUCTION

Comprehension of the physical world depends upon observations, measurements, analysis, and, if possible, prediction of patterns expressed ubiquitously in nature. Linear or nonlinear systems possess regularity and simplicity that are readily amenable to scientific investigations. Yet numerous other [*]systems, especially those in the animate realm, possess complexities that are at best nonlinear and not-predictable. It is not very easy to appreciate if a system is complex.

[*] Corresponding Author: serbalexe1@yahoo.com

There is no specific delimitation between a simple system and a complex one. A system can be complex from different kinds of points of view (structural, functional, dynamic or all three simultaneously).

A communications network, a human cell, a financial market, a city, a bridge, a forest, or a social community – they all form complex systems. Obviously, all interacting with each other and together forming a more complex system, made of nonlinearly interacting units that are hierarchically organized in time and space. Here are some attributes of complex systems:

- A system is complex when it is composed of many parts that interconnect in intricate ways. (Joel Moses, "Complexity and Flexibility"). Metric for intricateness is the amount of information contained in the system.
- A system presents dynamic complexity when cause and effect are subtle, over time [1].
- A system is complex when it is composed of a group of related units (subsystems), for which the degree and nature of the relationships are imperfectly known [2].
- A complex system has a set of different elements connected or related as to perform a unique function not performable by the elements alone [3]. It requires different problem-solving techniques at different levels of abstraction.
- Complexity theory and chaos theory both attempt to reconcile the unpredictability of nonlinear dynamic systems with a sense of underlying order and structure [4].
- Complex and simple systems are disjoint categories that encompass all nature.
- The real world is made up of complex things and the world of simple mechanisms is fictitious and created by science.
- A complex system is non-fragmentable. If it were, it would be a machine. Their reduction to parts destroys important system characteristics, irreversibly.
- A complex system comprises real components that are distinct from its parts. There are functional components defined by the system whose definitions depend on the context of the system. Outside the system they have no meaning. If removed from the system, they lose their original identity.
- Complex systems have models that are analytic or synthetic. But the tools differ. If a synthetic model can replace an analytic model, the system is fragmentable.
- No "largest model". If there were a largest model, all other models could be derived from it and fragmentability would result.
- Causalities in the system are mixed when distributed over the parts. The nature of causality requires closed loops of excluded in the Newtonian Paradigm.
- The important attributes of the system are beyond algorithmic definition or realization.
- Complexity theory targets:
- Understanding the relationship between emergent behavior and intricateness of parts (through the non-fragmentable property).

Every complex system is a changing part of a greater whole, a nesting of larger and larger wholes, leading eventually to the most complex dynamical system of all the systems that ultimately encompasses whatever we mean by order and chaos – the Universe itself.

Complexity theory is a way of thinking about complex systems and predicting their behavior.

Complexity theory studies parts of a system that give rise to the collective behaviors of the (whole) system, how the system interacts with its environment, how interactions give rise to patterns of behavior. Approaches to study complexity allow us to understand the ways of describing complex systems and processes of formation of complex systems through pattern formulation and evolution.

A time series [5] is a collection of observations made sequentially in time. Time series occur in virtually every scientific or experimental domain. Time series account for the fact that data points taken over time may have an internal structure (such as autocorrelation, trend or seasonal variation). Perhaps, the main goals of time series analysis are: identifying the nature of the observed phenomenon, and predicting its future.

The first principle that is common to all these goals is to study similarities between two time series. The utility of similarity search is classification, clustering, association rules and exploratory data analysis. But the fact that the notion of similarity is a "subjective" one makes working with time series very difficult. The definition of similarity depends on the user, the domain and the task at hand. In time series analysis there are different data formats, different sampling rates, noise, missing values etc. We need to be able to handle this "objectively"!

What do we observe? A time series and not the state space in which the phenomenon occurs! Therefore, it seems useful to convert observations (the time series) into state space (or phase space).

Fundamental to science is the presumption that experiments are predictable and repeatable! Thus most scientists were surprised when simple deterministic systems were found that were neither predictable nor repeatable.

Usually, the equations were devised to model a physical process or exhibit a particular behavior. The main lesson is that simple equations can have complicated solutions, including sensitive dependence on initial solutions, strange attractors with fractal structure, irregular basins of attraction and intricate bifurcation sequences.

Having only experimental data, we might want to know whether the dynamics is deterministic and chaotic or nondeterministic and random. In the extreme case (or usually!) we might have only a single sequence of measurements at successive times (a time series).

Most conventional time-series analysis methods implicitly assume the data come from a linear dynamical system, perhaps with many degrees of freedom and some added noise. Stationarity/Non-stationarity of a process is studied by some characteristics of the time series such as: probability distribution, average value, standard/average deviation, higher moments such as: skewness, kurtosis, autocorrelation function, Fourier analysis, Hurst exponent etc.

It is rare that a plot of each value in a time series versus its immediate predecessor reveals one-dimensional structure, even for simple chaotic systems.

However, if the system is deterministic, you would expect each value to depend on some finite number of past values:

$$x_{n+1} = f(x_n, x_{n-1}, x_{n-2}, \dots),$$

since for rapidly sampled time series with small Δt these values capture essentially the same information as a succession of time derivatives.

Furthermore, x may be only one of many dynamical variables that characterize the system, or it may be a combined, transformed or filtered version of them!

It is safe to predict that in the future the ARMA model, including its variations, will continue to play an active role in analyzing time series data due to its simplicity, feasibility and flexibility. But nonlinear features, like for example: non-normality, asymmetric cycles, bimodality, nonlinear relationships between lagged variables, variation of prediction performance over the state space, time irreversibility, sensitivity to initial conditions and others, which have been well-observed in many real time series data, confine ourselves to explore nonlinear models.

The most remarkable and encouraging feature of chaotic data is that it does not matter in principle what variable you use. Most dynamical properties such as the eingenvalues of fixed points and the largest Lyapunov exponent and metric (topological) properties such as attractor dimension are contained within (almost) any variable and its time lags. Furthermore, it is not usually necessary or even desirable to reconstruct the entire state space from the measured variable since the attractor dimension will often be much smaller than the dimension of this space. It is sufficient to construct a new space in which an equivalent attractor can be embedded. This new space should have the properties that every point in it maps to a unique next point by the dynamics and that there is a smooth nonsingular transform between it and the original space. The method is called *state-space reconstruction* and was proposed by Packard *et al.* (1980) and Takens (1981).

State Space Reconstruction Starting from a Time Series

The terminology reconstruction is used for the following reason: from a given state space of dimension D, only one variable, obtained by projection on one axis is taken into account and a new state space is constructed with a different dimension [6]. In special conditions, a theorem and its extensions guarantee the existence of a diffeomorphism between the original and the reconstructed attractors. Both spaces represent the same dynamical system in different coordinate systems.

The *time–delay reconstructed state space* is a space obtained by plotting each value of the time series against some number of previous values. The dimension of this space can be chosen arbitrarily, with a value as large as the number of points in the time series N, in which case the whole data set would be represented by a single point in an N-dimensional space.

The *time-delay embedded space* is a reconstructed state space chosen with a minimum dimension for which the important dynamical and topological properties are preserved. For most purposes this dimension needs only to be the next integer larger than $2D_F$, where D_F is the attractor dimension.

There are several different methods for phase space reconstruction, as:

- derivatives coordinates (consist in using higher order derivatives of the measured time series; but derivatives are susceptible to noise)
- principal value decomposition
- delay coordinates.

Our work will use the method of delay because we can just observe one variable and use it cleverly.

It's easier to do this than observing; for example, three variables (if the correct phase space reconstruction of the dynamical system would require three experimental measurements).

Method of Delays (Time Delay Embedding)

The signal to be analyzed is considered as a time series, *TS* (a set of measurements of a physical variable):

$$TS = \{x_1, x_2, x_3, \ldots, x_N\}\ .$$

From this time series we build the trajectory of the dynamical system in the state space using the method called *time delay embedding*. This method is a very simple one. We compute:

$$\begin{aligned}&\mathbf{R} = \{\mathbf{r}_1, \mathbf{r}_2, \ldots, \mathbf{r}_M\}\\&\mathbf{r}_i = \left(x_i, x_{i+d}, x_{i+2d}, \ldots, x_{i+(m-1)d}\right).\\&x \in TS; \quad i = 1, 2, \ldots, M\end{aligned}$$

In the formula above:

- *R* – is a set of vectors (the r_i) reconstructed with the elements of the time series (the x_i)
- r_i – is a reconstructed state vector, its elements are *m* successive values taken from the time series ST with a delay of *d* positions between them
- *x* – is a value from the time series.

It is easy to observe that if *N* is the number of elements in *TS*, the number of vectors used to represent the signal in the state space depends on *m* and *d* through the following formula:

M=N-(m-1)*d,

where:

- *m* – is the dimension of space and
- *d* – is the time delay.

These two parameters, *m* and *d*, must be chosen appropriately, and in this consists the real difficulty of the method.

- *m* is the dimension of the space in which we reconstruct the trajectory. It can be different from the real dimension of state space represented by the analyzed process. If *m* is not big enough, several portions of the trajectory could overlap.
- *d* is the time delay between the successive components of the reconstructed vectors. It has to be chosen to that the state vector elements are independent. For the time delay embedding method the problem is finding a compromise between redundancy and irrelevance.

The choice of these parameters has to be done in two steps, which will be presented next.

First, we choose *d* using one of the different methods that exist, of which the most famous ones are: *auto-correlation* method, *average moving* method and *average mutual information* method.

Currently, the most used method to choose the *d* parameter is the linear autocorrelation function:

$$C(d)=\frac{\frac{1}{N}\sum_{i=1}^{N-d}(x_i-\tilde{x})(x_{i+d}-\tilde{x})}{\frac{1}{N}\sum_{j=1}^{N}(x_{ji}-\tilde{x})}$$

where $\tilde{x} = \frac{1}{N}\sum_{j=1}^{N}(x_j)$ is the average.

From this function different results can be obtained, such as:

- *d* the value for which *C*(*d*) goes under 1/e for the first time
- *d* is the value for which *C*(*d*) first passes though zero or very near to zero
- *d* is the first point of inflexion of *C*(*d*)
- *d* is the value for *C*(*d*) reaches the first local minimum.

The second interesting method is called *average mutual information* (AMI). This method has a great effectiveness when it comes to finding equilibrium between redundancy and irrelevance. The average mutual information estimates the precision of predicting x_{i+d} from x_i. The average mutual information between the measurements of ‘a’ (system A) and the measurements of ‘b’ (system B) is defined as:

$$I_{AB} = \sum_{a_i, b_k} P_{AB}(a_i, b_k)\log_2\left(\frac{P_{AB}(a_i, b_k)}{P_A(a_i)P_B(b_k)}\right),$$

where:

- $P_{AB}(a_i, b_k)$ – is the probability that measurements of systems A and B are simultaneously a_i and b_k, respectively
- $P_A(a_i)$ – is the probability to measure the value a_i for system A
- $P_B(b_k)$ – is the probability to measure the value b_k for system B.

The AMI formula becomes:

$$AMI(d) = \sum_{i,j=1}^{N} P_{i,j}(d) \log_2((P_{i,j}(d)) / P_i P_j)$$

When using the AMI function, *d* will be chosen as the argument minimizing this function. In certain cases this function does not have local minima, so *d* can be chosen as the value for which AMI(*d*) decreases to 20% of AMI(0).

Secondly, the choice of the dimension of the space, *m*, can be made by different methods, of which the most famous are the method of *decomposition in singulars values* and the method of *false nearest neighbours*. If *m* is not large enough, the reconstructed trajectory will not be able to contain the true dynamics of the original trajectory. But if *m* is too large, the amount of necessary data will have to be greater.

Among the two mentioned methods, the most intensively used is FNN (*false nearest neighbours*). This method is based on the following observations. When *m* is not large enough, we actually perform a projection of points of the real trajectory in a space with less dimensions, which increases (with false neighbors) the number of neighbors for some points. By increasing *m*, the distance between points increases, and the number of false neighbors decreases. When using this method, we must choose for *m* the value for which the proportion of false nearest neighbors becomes zero (or a value that is very close to zero).

A correct reconstruction (in term of similarities with the real trajectory) of the state space trajectory depends on the correct choices for *m* and *d*. Among all the existing methods, AMI and FNN have proved to be the best for most cases.

We can illustrate the method of trajectory reconstruction in state space with the following dynamical system of Lorenz:

$$\begin{cases} x' = 10(y - x) \\ y' = 28x - y - xz \\ z' = xy - \dfrac{2}{3}z \end{cases}$$

We can observe that when it comes to reconstructing the trajectory using the *z* variable, the second wing of the butterfly is absent. Indeed, *z* does not appear in all the three equations of the Lorenz system.

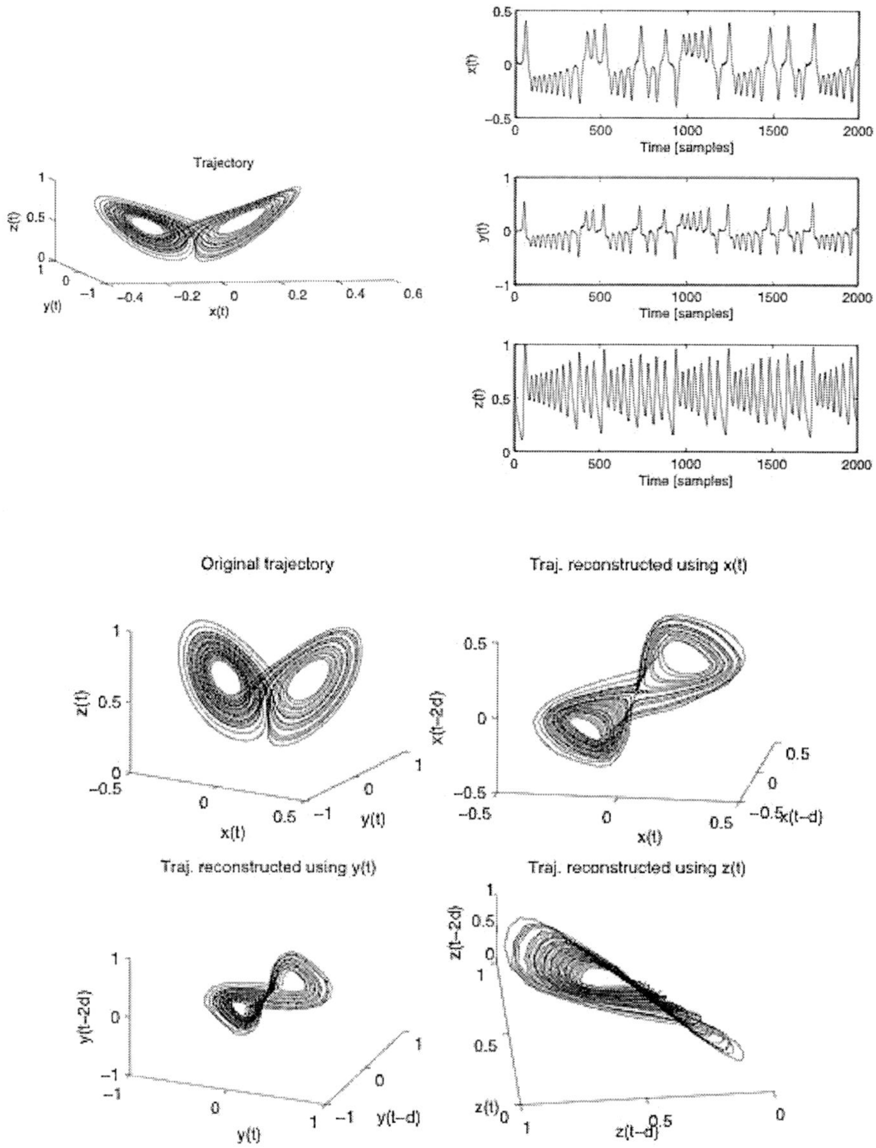

Figure 1. A trajectory for the Lorenz system: the original trajectory, the time variation of the three state variables and the trajectories reconstructed using just one variable at a time.

Recurrence Plot Analysis

Recurrence plots (RPs) were first described by J.P. Eckmann, S.O. Kamphorst and D. Ruelle in 1987 [7]. RP is a relatively new technique for the qualitative assessment of time series. By using it, hidden patterns and structural changes in data can be graphically detected, and similarities in patterns across the time series under study can be seen.

The method developed by Eckmann gives the possibility to represent an m-dimensional state space trajectory in a two-dimensional representation.

Using the reconstructed state space trajectory we can compute the recurrence matrix, whose graphical representation will be called recurrence diagram (or recurrence plot). This computation is done in two steps that are presented below.

Firstly, we must find the *Distance Matrix* (*DM*):

$$DM \in M_{M,M}(R)$$
$$DM = \left\| r_i - r_j \right\|, \quad i,j = 1,2,\ldots,M$$

where $\|\ \|$ is a metric that is used for computing the distance between trajectory points. The most used metric is the *Euclidean norm* ($\|r\|=\sqrt{\sum r_i^2}$). But the *Maximum norm* ($\|r\|=\max\{r_i\}_{i=1,m}$) offers similar results, so it is often used to replace the *Euclidean norm* (because it has the advantage of a faster computation).

The distance matrix is another way (usually not reversible, though) to represent the *m*-dimensional trajectory. It provides a method of representation that is independent of the position and the rotation of the trajectory in its space.

The second step consists of calculating the *Recurrence Matrix* (RM):

$$RM \in M_{M,M}(R)$$
$$RM_{i,j} = H(\varepsilon - DM_{i,j}), \quad i,j = 1,2,\ldots$$

where ε is the threshold used for defining the recurrence neighborhood and *H* is the Heaviside function.

The recurrence matrix has the ability to resume in a compact form the essential characteristics of the analyzed signal. Modification of the amplitude and/or the time scale does not modify the general aspect of the recurrence matrix.

Given a fixed ε, the L_∞*-norm* finds the greatest amount of neighbors; the L_1*-norm*, the least amount of neighbors and the L_2*-norm* (i.e. the *Euclidean norm*) finds an intermediate amount of neighbors. The RP is obtained by plotting the recurrence matrix. Each point (*i,j*) is represented according to the distance between the two trajectory points, $\mathbf{r}_i$ and $\mathbf{r}_j$. In a *Thresholded* RP (TRP), the pixel laying at (*i, j*) is black if the distance falls within a specified threshold ball and white otherwise.

In an *Unthresholded* RP (UTRP), the pixel laying at (*i,j*) is color coded according to the distance between the two trajectory points. In fact, instead of plotting the recurrences with white and black dots, the distances $D_{i,j} = \left\| r_i - r_j \right\|$ can be plotted using some defined color scale.

Both axes of the RP are time axes and show rightwards and downwards (or upwards, by convention). The RP has always a black main diagonal line, the *Line of Identity* (LOI). Furthermore, the RP is symmetric (by definition) with respect to the main diagonal, i.e. $RM_{i,j} = RM_{j,i}$.

A crucial parameter of an RP is the *threshold* ε. Therefore, special attention has to be paid to its choice. If ε is chosen too small, there may be almost no recurrence points and we cannot find anything about the recurrence structure of the underlying system. On the other hand, if ε is chosen too large, almost every point is a neighbor of every other point, which leads to a lot of artifacts. A too large ε includes into the neighborhood also points that are simple consecutive points on the trajectory. This effect is called *tangential motion* and causes thicker and longer diagonal structures in the RP than they actually are. Hence, we have to find a compromise for the value of ε.

Moreover, the influence of noise can entail choosing a larger threshold, because noise would distort any existing structure in the RP. At a higher threshold, this structure may be preserved. Several "rules of thumb" for the choice of the threshold ε have been advocated in the literature, e.g., a few percent of the maximum phase space diameter has been suggested. Furthermore, it should not exceed 10% of the mean distance or of the maximum phase space diameter.

A further possibility is to choose ε according to the recurrence point density of the RP by seeking a scaling region in the recurrence point density. However, this may not be suitable for non-stationary data. For this case it has been proposed to choose ε so that the recurrence point density is approximately 1%.

Structures in a Recurrence Plot

As already mentioned, the initial purpose of RPs was to visualize trajectories in phase space, which is especially advantageous in the case of high dimensional systems. RPs yield important insights into the evolution of these trajectories, because typical patterns in RPs are linked to a specific behavior of the system. Large scale patterns in RPs can be classified in homogenous, periodic, drift, and disrupted ones:

- Homogenous RPs are typical for stationary systems in which the relaxation times are short in comparison with the time spanned by the RP. An example of such an RP is that of a stationary random time series.
- Periodic and quasi-periodic systems have RPs with diagonal oriented patterns – diagonal lines and checkerboard structures. For example, for a periodic system with two harmonic structures and with a frequency ratio of four, the RP contains two diagonal lines and four short lines laying between the continuous diagonal lines.
- A drift is caused by a system with slowly varying parameters, i.e. a non-stationary system. The RP pales away from the main diagonal.
- Abrupt changes in the dynamics as well as extreme events cause white areas or bands in the RP. The RPs allows finding and assessing extreme and rare events easily by using the frequency of their recurrences.

A closer inspection of RPs reveals also small-scale structures which can be typically classified in single dots, diagonal lines as well as vertical and horizontal lines (the combination of vertical and horizontal lines obviously forms rectangular clusters of recurrence points); in addition, even bowed lines may occur:

- Single, isolated recurrence points can occur if states are rare, if they persist only for a very short time or if they fluctuate strongly.

- A diagonal line occurs when a segment of the trajectory runs almost in parallel to another segment for a certain time.
- A vertical (horizontal) line marks a time interval in which a state does not change or changes very slowly. Hence, the state is trapped for some time. This is a typical behavior of laminar states.
- Bowed lines are lines with a non-constant slope. The shape of a bowed line depends on the local time relationship between the corresponding close trajectory segments.

In Table 1 we present the classification of the different patterns with their respective meanings [8].

Table 1. The meaning of patterns in a RP

Pattern	Meaning
(1) Homogeneity	The process is stationary
(2) Fading to the upper left and right corners	Non-stationary data; the process contains a trend or a drift
(3) Disruption (white bands)	Non-stationary data; some state are rare or far from the normal
(4) Periodic / Quasi periodic patterns	Periodicity in the process; the distance between line represents a period
(5) Single isolated point	Strong fluctuation in the process; if only isolated point occur: random process
(6) Diagonals parallel to the LO/	The evolution of state is similar at different time
(7) Diagonals orthogonal to the LO/	The evolution of state is similar at different time but with reverse time; insufficient embedding
(8) Vertical and horizontal lines	Some states do not change or change slowly; laminar states and intermittency
(9) Long bowed line structure	The evolution of state is similar at different time with different velocity

The visual interpretation of the RPs is quite difficult to do and requires experience. There are also other methods to obtain information from RPs – their quantification, which offers more specific techniques for the analysis and the understanding of the considered system.

Recurrence Quantification Analysis as a Set Of Tools for Measuring the Complexity of a System

In order to go beyond the visual impression yielded by RPs, several measures of complexity that quantify the small-scale structures in a RP have been proposed in [9], [10] and are known as recurrence quantification analysis (RQA). These measures are based on the recurrence point density and the diagonal and vertical line structures of the RP. A computation of these measures in small windows (sub-matrices) of the RP moving along the main diagonal yields the time dependent behavior of these variables.

RQA measures are able to identify bifurcation points, especially chaos-order transitions. The vertical structures in the RP are related to intermittency and laminar states. The measures that quantify the vertical structures enable also the detection of chaos-chaos transitions. The simplest RQA measure is the recurrence rate (RR) or per cent recurrences:

$$RR(\varepsilon)=\frac{1}{N^2}\sum_{i,j=1}^{N}\mathbf{R}_{i,j}(\varepsilon),$$

which is a measure of the density of recurrence points in the RP.

The next measures are based on the histogram $P(\varepsilon,l)$ of diagonal lines of length "l":

Processes with uncorrelated or weakly correlated, stochastic or chaotic behavior cause none or very short diagonals, whereas deterministic processes cause longer diagonals and less single, isolated recurrence points. Therefore, the ratio between the recurrence points that form diagonal lines of a length that equals at least $l_{\min}$ and all the recurrence points that form diagonal lines:

$$DET=\frac{\sum_{l=l_{\min}}^{N} l\cdot P(l)}{\sum_{l=1}^{N} l\cdot P(l)},$$

is introduced as a measure for the determinism (or predictability) of the system.

A diagonal line of length l means that a segment of a trajectory is rather close during one time steps to another segment of the trajectory at different time; thus these lines are related to the divergence of trajectory segments. Eckmann et al. have stated that "the length of the diagonal lines is related to the largest positive Lyapunov exponent" if there is one in the considered system.

As mentioned in [11], [8], RPA has become popular in various successful applications in fields of life sciences (as neuroscience and genomics), but also in ecology, physics, earth science and astrophysics, engineering and economy.

Chaos Control – Identifying the Unstable Periodic Orbits for a Chaotic System

In most technical environments chaos is an undesired state of the system which one would like to suppress. Chaos control [12] can help to re-establish at least a regularly oscillating output with judiciously applied minimal perturbations.

Chaotic systems have the outstanding property that the attractor possesses a skeleton of Unstable Periodic Orbits (UPO). In strongly chaotic systems they are even dense.

The set of UPOs in an attractor is a dynamical invariant; their number, distribution, and properties "unfold" the structure of chaotic orbits, and they can be used to calculate other invariants, such as fractal dimension and topological entropy.

Through small perturbations, the unpredictable (undesirable) dynamic is transformed in predictable (desirable) behavior without major changes to the functioning conditions. The method proposed by Ott, Grebogi and Yorke, known today as OGY, allows us to rethink the role of chaos and leads us to interesting applications such as: the control of atria fibrillation and epilepsy.

For a systematic approach of the control problem, the first step is to find the location of the unstable periodic orbits from the experimental time series. When we look at a typical chaotic time series, this seems to be a hopeless task. Nevertheless, a chaotic trajectory reveals information about the periodic skeleton.

Considering a continuous system, it can be described by the following non-linear differential equation:

$$x' = F(x, p)$$

where p is the vector of the control parameters.

A discrete system can be defined by an equation with finite differences:

$$x_{n+1} = f\ x_n; p$$

The long-term behavior of the solutions offered by the systems described in the above equations can be distinguished through regular (fixed point, periodical or quasi-periodical) and/or irregular (chaotic). Usually, by modifying one or more parameters $p_i \in p$, the system dynamics changes from a regular behavior into a chaotic behavior. Throughout this section we will maintain the premise that p is such that, unless altered, the system behaves chaotically.

The whole idea of controlling the chaos is that of changing an unpredictable chaotic behavior into a desired predictable behavior.

For exemplification we will consider the chaotic Lorenz system described by the following equations:

$$\begin{cases} x_1' = -x_1 + x_2 x_3 \\ x_2' = 3(x_2 - x_3) \\ x_3' = -x_1 x_2 + 28 x_2 - x_3 \end{cases}.$$

We can exemplify the process using the following images. If the uncontrolled system behaves as in the picture below, its equivalent trajectory in the state space is given in the Figure 2.

This obvious geometric decomposition of recurrence plot block structure suggests several interesting and potentially useful applications. Not only do RPs of UPOs play the role of geometric basis elements for the structure of an RP of a chaotic attractor, but they are also a useful way to *identify* those UPOs – a task that is algorithmically complex and computationally expensive. To find a UPO, one would simply construct the RP of a trajectory on a chaotic attractor, look for repeated structures, and use that information to index into the trajectory and find the associated state variable values. In the state space, evolution can be represented as a recurrence plot (RP) such as in Figure 4.

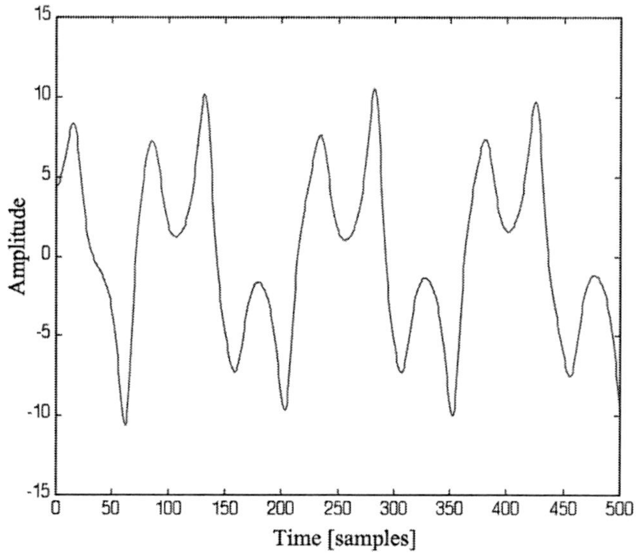

Figure 2. Chaotic Lorenz signal.

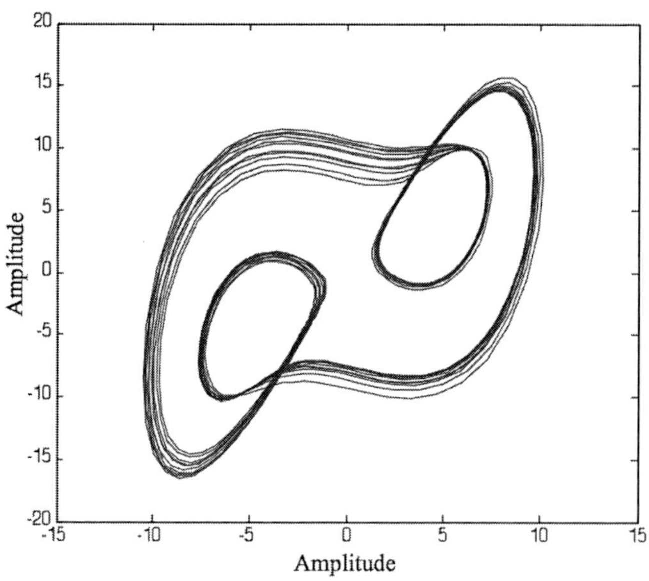

Figure 3. The state space for a Lorenz system.

It can be easily observed that there can be identified a periodic pattern in the RP. This is the observation which rose de idea of using the RP transformation in order to localize UPOs in chaotic time series. But for reaching this desiderate, we have to look more closely to the way the RPs are generated. It has been showed that parallel evolutions in the state space generate in RPs lines of different lengths that are parallel to the main diagonal. The lengths and position of the lines in RP give us information on the times when the system has a similar behavior and at what moments this occurs.

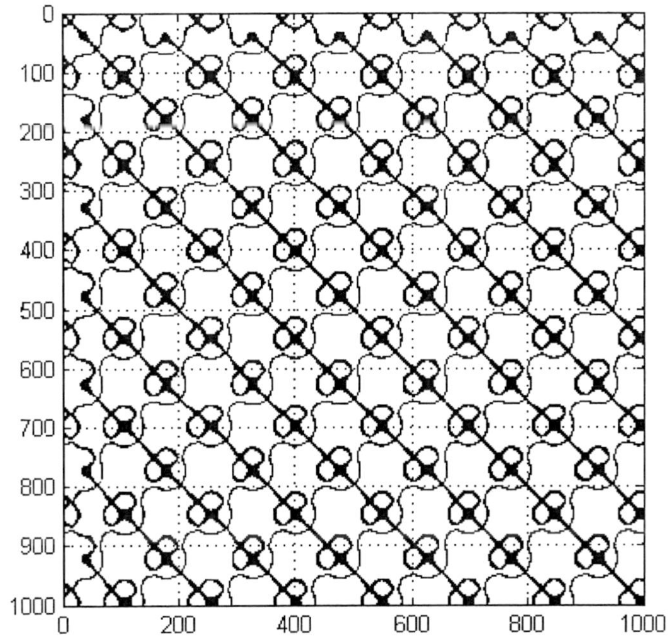

Figure 4. The recurrence plot for a Lorenz system.

Some short lines (or isolated points) in the RP specify that the states the system has at certain moments in time tend never to return. On the other hand, long lines in RP show the system has a "preference" for some specific states. Those are the states in the system evolution that we need to identify/keep in order to extract from the chaotic system behavior its deterministic component, i.e. the UPO. In other words, in order to be able to compute a UPO, we have to retain from the RP just the lines that are parallel with the main diagonal and whose lengths are greater than a specified value (threshold). Filtering the above RP, using a high threshold, we can obtain the following representation:

The basic algorithm for finding UPOs in experimental data using RPs is quite fast. The geometric decomposition also provides a useful way to do a quick, qualitative comparison of two chaotic systems. The RP representation is useful not only for identifying and locating UPOs – a task that is usually computationally demanding – but also for comparing one dynamical system to another: the practical task termed "modeling" by engineers, "scientific discovery" by artificial intelligence practitioners, and "system identification" by control theorists.

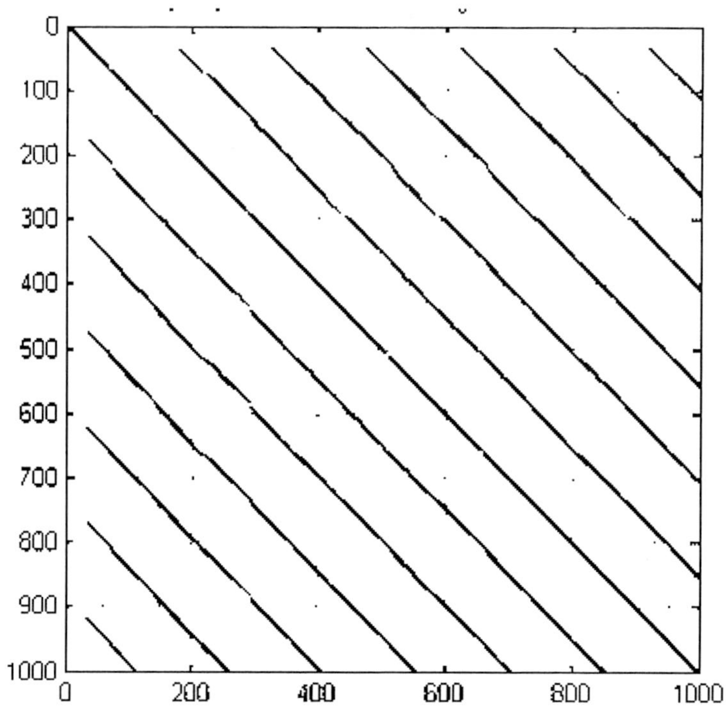

Figure 5. Filtered Lorenz RP using a high threshold.

The above RP corresponds to a quasi-sinusoidal signal obtained from the Lorenz RP.

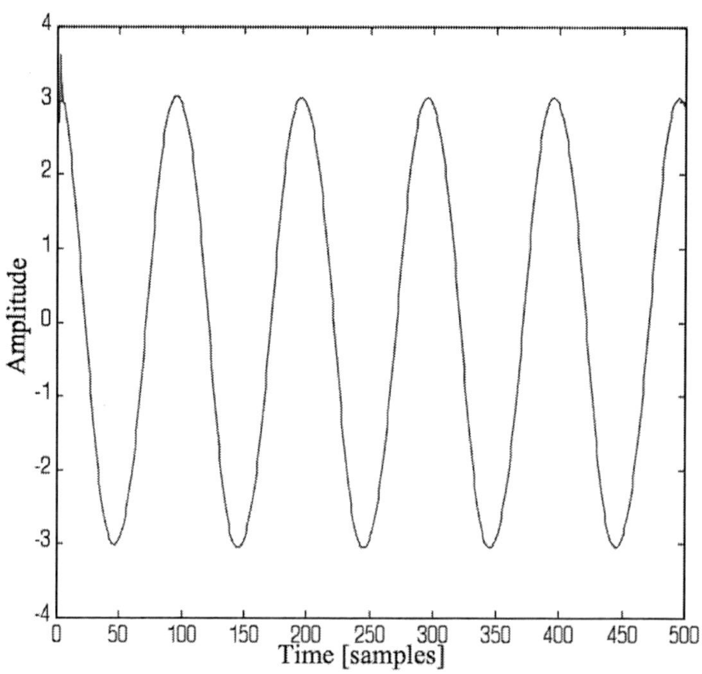

Figure 6. Controlled Lorenz system.

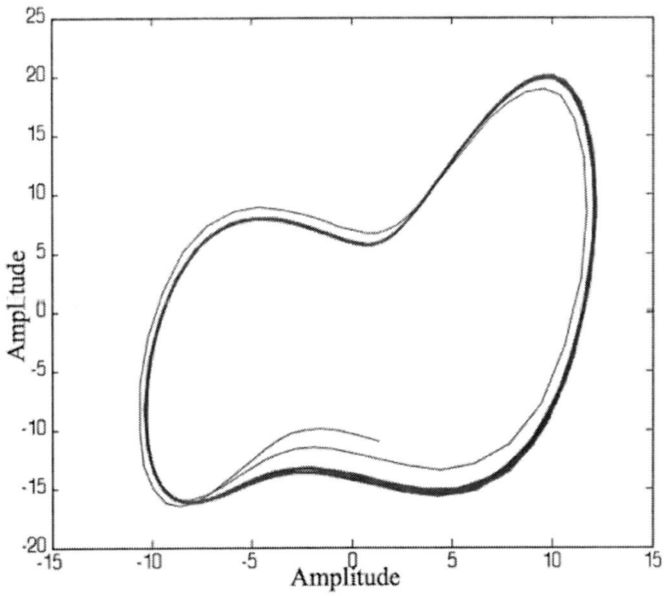

Figure 7. The state space for the controlled Lorenz system.

NONLINEAR SPEECH PROCESSING

Speech processing can be divided into several domains that cannot be completely separated – speech analysis, compression, synthesis and recognition. All those domains where based on the Fant supposition that the speech is deterministic on small periods of time, but it was demonstrated that this type of signal is neither stationary, nor deterministic. All the linear methods used for analyzing the speech (and from this all the imagined operations) were able to provide just approximations of the desired results. Using linear methods, linear filters or transformations put in additional errors in the process of observing the signals. That is why a nonlinear analysis on a nonlinear signal will be able to provide better results than a linear one.

Speech Segmentation

We will show that the RP tool can be easily used for analyzing the speech signal [13] in order to extract the fundamental parameters/characteristics (such as the fundamental frequency and formants). First, one has to split the speech into vowels, consonant and mix parts and to use specific methods for each type of signal. It is also important in the synthesis process to know the type of speech excitation to be used for generating the signal, i.e. periodic, noise-like or mix one. Most actual synthesis methods use to classify speech segments in vowels or consonants, but there are sounds (for example nasal or friction generated sounds) which were formed as a mixture of quasi periodic signals and noise like ones. There are some modern approaches to the synthesis of speech (such as MELP vocoder) that use complex excitation signals determined before the main application emerges, using test like signals. During the analysis stage, by means of a comparing method, the best of those excitations (which are stored in a local database) is chosen and its ID is sent to the receiver. The main problem is that the test signals are different than the analyzed ones (different speakers, different recording conditions etc.) so those methods can provide only approximations of the real excitations. More than that, the involved operations are linear (convolutions) while the speech signal is a dynamical, nonlinear one. A nonlinear approach such as RQA can provide better results.

It has been shown above that the recurrence quantification analysis supplies some easily computable parameters that can be used to determine different signal characteristics. One of these parameters is the recurrence rate defined by the formula:

$$RR = \frac{1}{N^2}\sum_{i,j=1}^{N} R_{i,j} \ .$$

Using the RR value, computed on consecutive segments of speech, we can generate a segmentation signal as shown in Figure 8. It is important to observe that the segmentation signal can be computed in real time (in the same time the speech is produced). More than that, this signal can be used not only for segmentation purposes but also to determine the exact degree of periodicity and noise that have to be mixed in the synthesis process (if the RR signal is normalized). In fact, in order to obtain a very good quality for the synthesized speech, we can use this information as the excitation control signal.

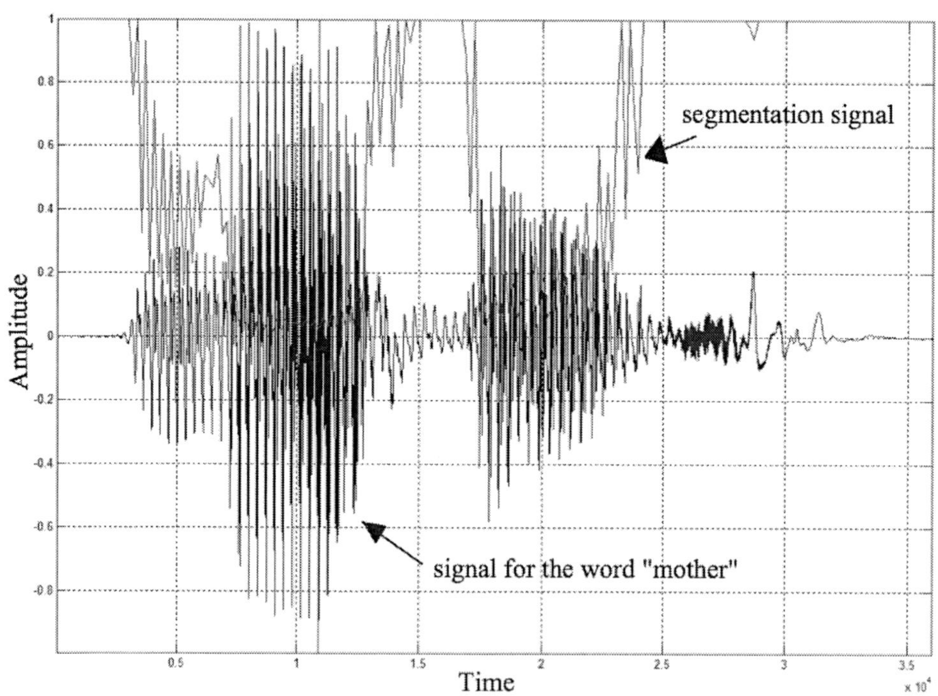

Figure 8. The English word “mother” and the segmentation signal.

Pitch Identification

One of the most important operations in speech processing is the extraction of speech parameters and their variation with time. Having those parameters, it will be very easy to compress the speech, to synthesize or to recognize it. Until now, using linear operations, it was possible just to approximate those parameters. Most of the information which is transmitted using a speech signal is carried out by vowels. Those particular sounds have a quasi-periodic structure. The repetitive part of the vowel is called pitch. The nonlinear speech analysis we propose is able to identify the pitch with a precise controlled accuracy that depends on the analysis window length T_a chosen for the calculation of the RR. To do this, first we calculate the recurrence plot of the speech as shown in Figure 9. It can be observed the periodicity in the RP structure. Each point in the RP coresponds to a return of the system in that state. An area from the RP with a big RR (characterized by a high number of points) is specific to a part of the signal with a rapid variation around a given value. It is specific for the vowels to have high amplitudes and low frequencies, so such signal will produce regions in RP with a small RR. This can be observed in Figure 9 – a vertical/horizontal area without (or with few) points are specific to vowel's pitch.

A very simple method for building a signal for identifying the pitch structures in the vowels is to calculate the RR using an analysis window that fulfils the inequality $\tau_w \leq T_a < \frac{T_0}{2}$ (where τ_w is the reconstruction window and T_0 is the fundamental period of the speech signal).

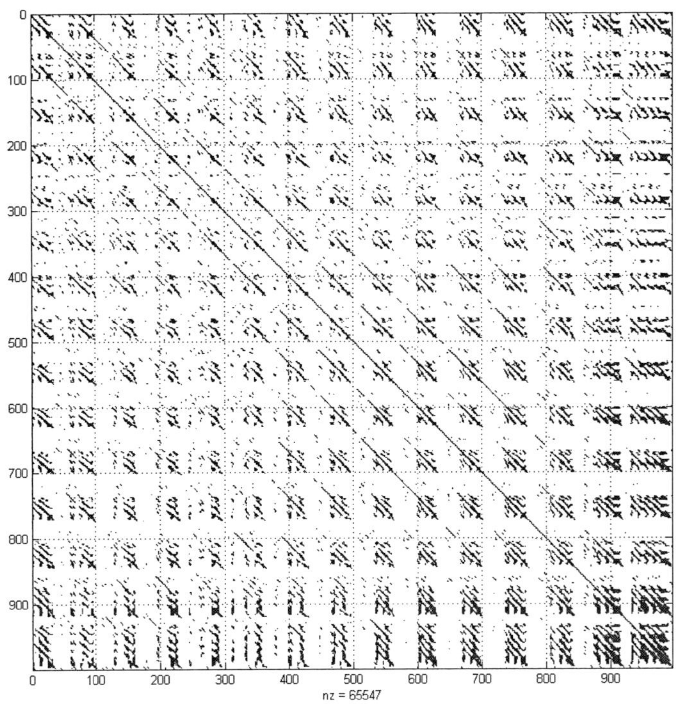

Figure 9. The RP of vowel "a" in the English word "car".

Taking into consideration that the value of T_0 is not known (it has to be determined), we used $T_a = \tau_w$ The resulting signal (which we will call "structure signal") can be observed in the folowing image:

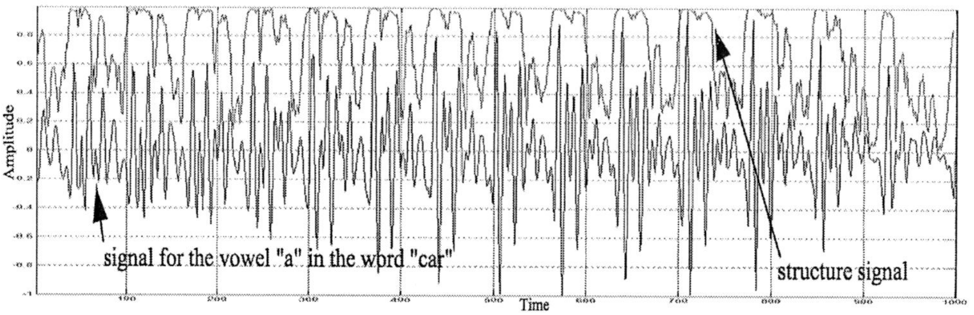

Figure 10. The vowel "a" in the English word "car" and the structure signal.

The obtained structure signal can identify very precisely the pitch evolution, but its shape is sensitive to any variation/modification in the pitch structure. It can be made insensitive to small variation of pitch by filtering the RP in the same way we identified the UPO, but using a small threshold. The obtained RP is represented in the Figure 11 and it has a structure signal corresponding to the one in Figure 12.

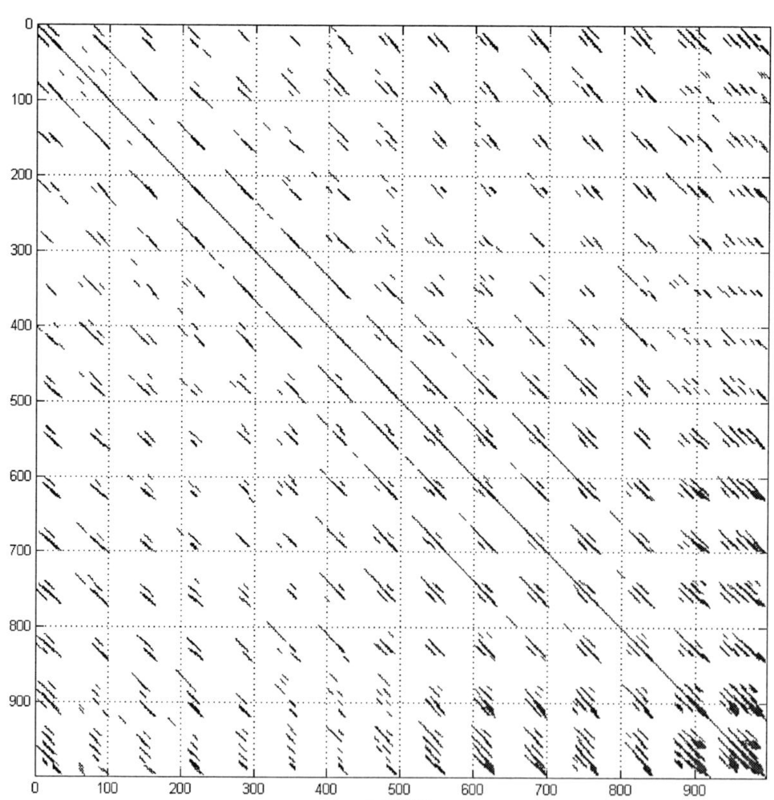

Figure 11. The filtered RP of vowel "a" in the English word "mother".

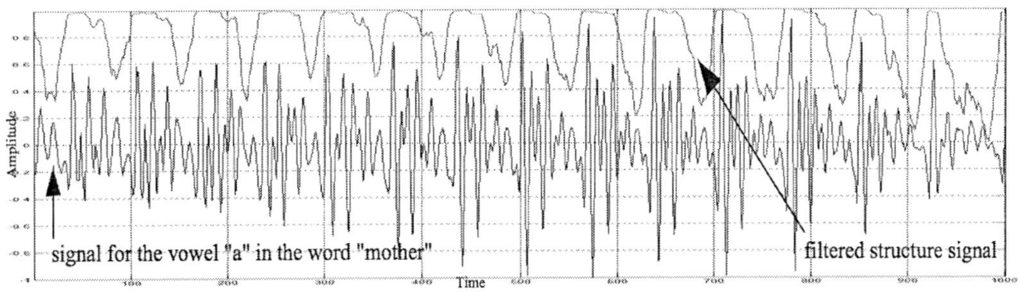

Figure 12. The vowel "a" in the English word "mother" and the filtered structure signal.

The following figure is just a detail of the above one.

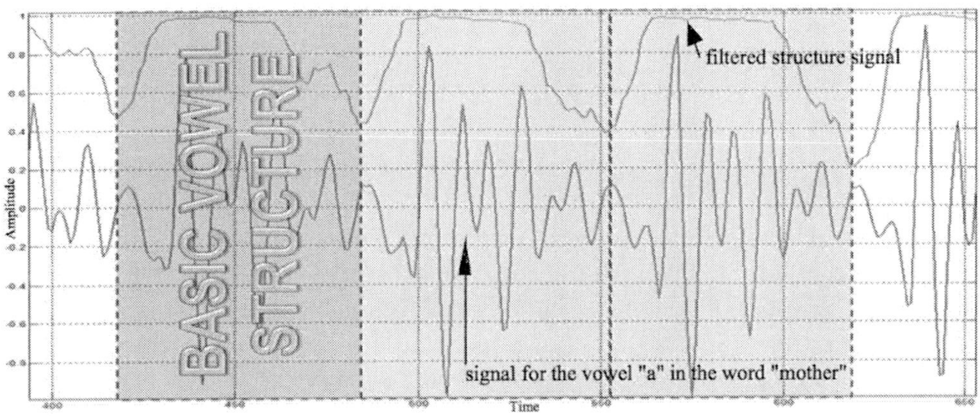

Figure 13. Detail of the vowel "a" in the English word "mother" and the filtered structure signal.

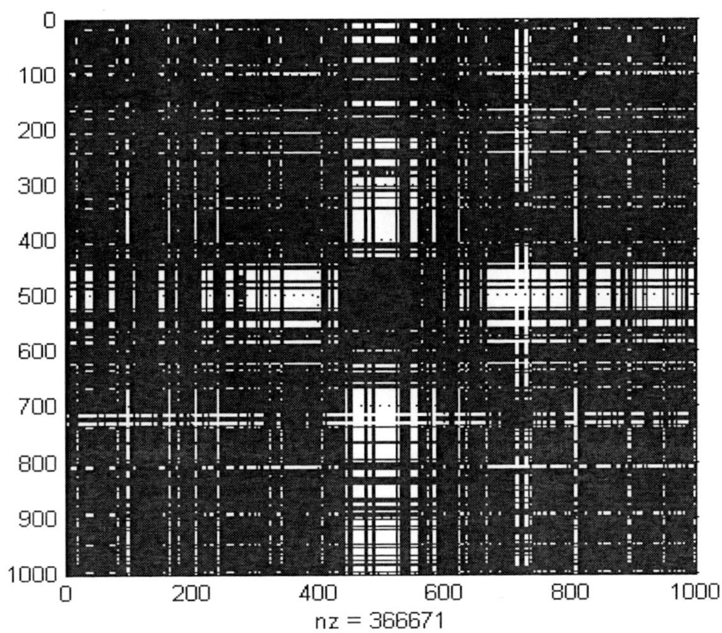

Figure 14. The RP of a consonant.

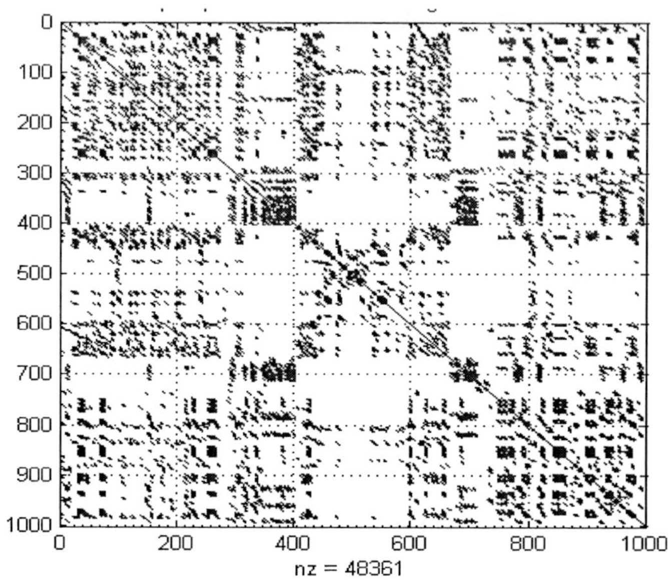

Figure 15. The filtered RP of a consonant.

Having the pitch structure, we can compress the signal, analyze it or use it in a recognition process. For example, one can construct a database containing the pitch structure specific to each vowel. At the emitter level, the same algorithm can be used for extracting the pitch. Next step is to search in the local database for the most similar shape and to send to the receiver just the position in the database of the identified entry and, eventually, for an even greater naturalism, some information about the evolution of this structure (mainly the amplification/attenuation, phase modification at connection points). The receiver, which has the same local database, just extracts the corresponding pitch and synthesizes the signal. The degree of compression achieved this way is very big and the synthesized signal sounds natural.

The consonant case is even simpler – we can build the RP for the consonant and hardly filter it for retaining the dominant characteristics. The process is illustrated in Figure 14 and Figure 15.

DETECTION OF TRANSIENT SIGNALS

Transient signals contain in their parameters (i.e.: the time origin, the duration, the frequency band) information about the process that generated the observed signal. The purpose of transient signal analysis is extracting this information. Especially, precise localization of the time intervals where transients appear is of great interest. Several approaches have been tried, of which the most important are the following: time-frequency representations, time-scale representations and high order statistics. We will present in the following an approach that differs fundamentally from the above mentioned ones. We start by reconstructing the evolution (the trajectory) of the generating process using the signal samples, and then we detect the transients by analyzing its recurrences. The method can be seen as a generalization of the autocorrelation function philosophy, by comparing each portion of the signal to every other portion.

Method

We remind that the main aspect related to transient signal analysis is their detection, meaning that the precise time intervals where they appear are wanted, and not only a raw decision concerning their existence/nonexistence in the analyzed signal. For example, for the signal in Figure 16 we want to obtain the detection curve represented with the dotted line.

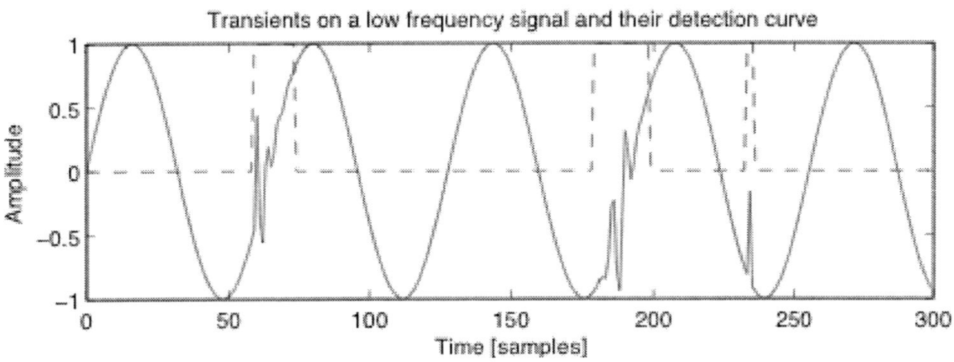

Figure 16. Low frequency signal with transients and the desired detection curve.

Before the actual detection step, a preprocessing of the analyzed signal is often needed. Figure 17 suggestively motivates this statement, by schematically presenting the total signal bandwidth divided between the signal, the transients, and the noise. As the transients band is concentrated in the high part of the spectrum, the low frequency components – that correspond to permanent regimes – can be removed without losing important information (thus simplifying the transients analysis).

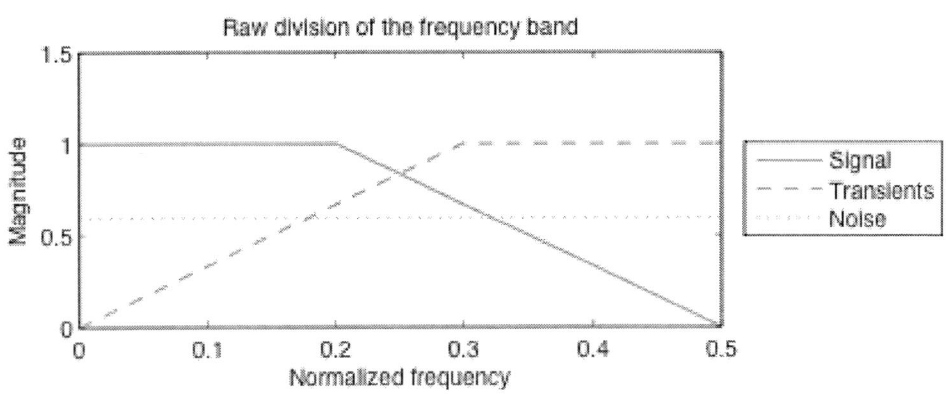

Figure 17. Schematic division of the total available band between the (low frequency) signal, the transients and the noise.

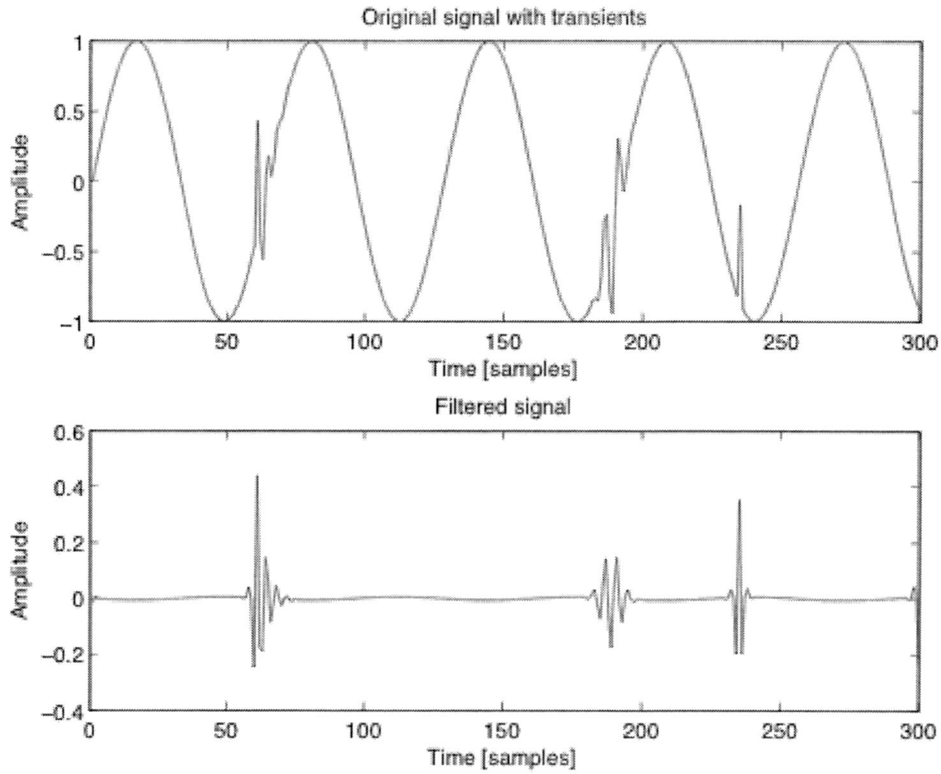

Figure 18. Preprocessing of the signal: (up) the original signal (down) the signal with the low frequency content removed.

We must point out that this separation of transients from the permanent regime signal (low frequency content) is only approximate, if we take into consideration on one hand the fact that ideal frequency filters can not be made, and on the other hand, the fact that the transient signals band is not in fact as the one shown in Figure 17, but spreads over the whole range of frequencies. So, the filtered signal will contain also (much attenuated, it's true) low frequency components, and the shape of the original transients will be slightly modified (as a result of the attenuation of a part of their frequency band). Besides that, the noise that was present in the pass-band filter will stay unmodified – it can not be removed by a filtering operation, due to the fact that this would lead to a loss of transients to be detected. After this preprocessing step, the signal from Figure 16 will turn into a signal that contains only the transients (together with – if it is the case – the high frequency noise), as shown in Figure 18. Having the signal preprocessed like this, a first intuitive idea for revealing the positions and the durations of transients would consist in using a threshold. This detection of the transients that are present in the (preprocessed) signal in Figure 18 can be made visually, due to the fact that their amplitudes are above the level of noise. Thus, by using as threshold a value around the level of noise for the absolute values of the signal samples, an approximation of the ideal detection curve (Figure 16) can be obtained.

The result can be seen in Figure 19.

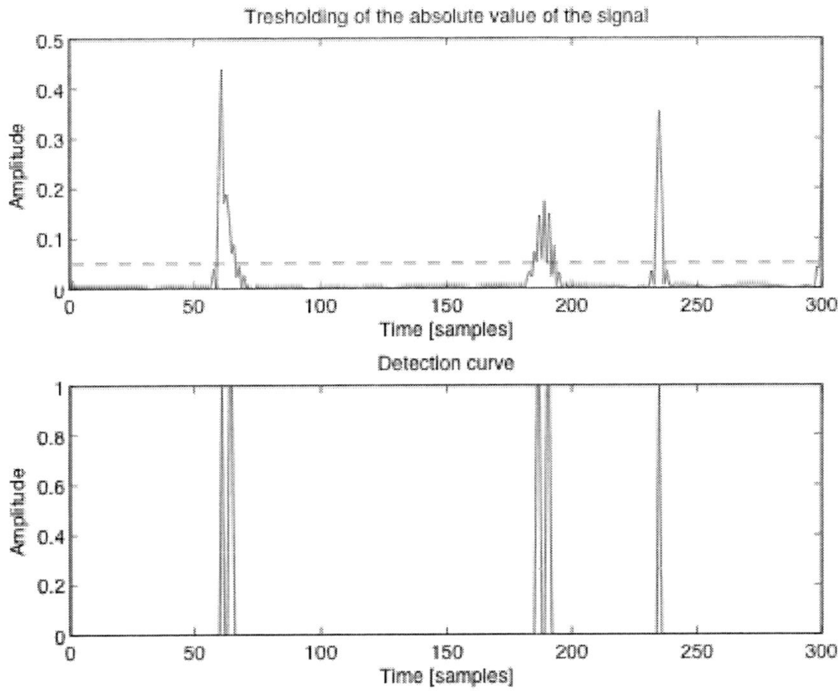

Figure 19. Time-domain detection by using a threshold for the absolute value of the signal.

The previously described solution, besides leading to strongly interrupted detection curves, also raises the problem of noise level estimation, which is necessary for choosing the threshold used in obtaining the detection curve.

This detection method offers, however, results that are surprisingly good for its simplicity. We want to improve it by passing from the time-domain signal to the process state space trajectory. Thus, instead of working with a signal that is composed of one-dimensional samples, we will work with a trajectory that is composed of multi-dimensional vectors (constructed from groups of samples, as shown in the theoretical introduction about the RPA method). In this multi-dimensional space the binarization interval used in the previously described method becomes a ball with its center in the origin of the axes, as shown in Figure 20.

The detection curve obtained this way is less interrupted than in the previous case. This is due to the fact that when the reconstruction is done properly (meaning that the m and d parameters are chosen optimally), the portion of the trajectory that corresponds to an area containing a transient signal becomes a wide curve that leaves the “noise ball” from Figure 20.

We must point out – concerning the detection curve obtained in Figure 5 – the fact that generally there is not a bi-univocal correspondence between trajectory points and signal samples. That is because every point on the trajectory is constructed from samples that are spread over an interval of length:

$$w = (m-1) \cdot d$$

However, if the value of *w* (the analysis window dimension) is sufficiently small compared to the durations of the transients to be detected, a satisfactory time localization of the trajectory points can be done.

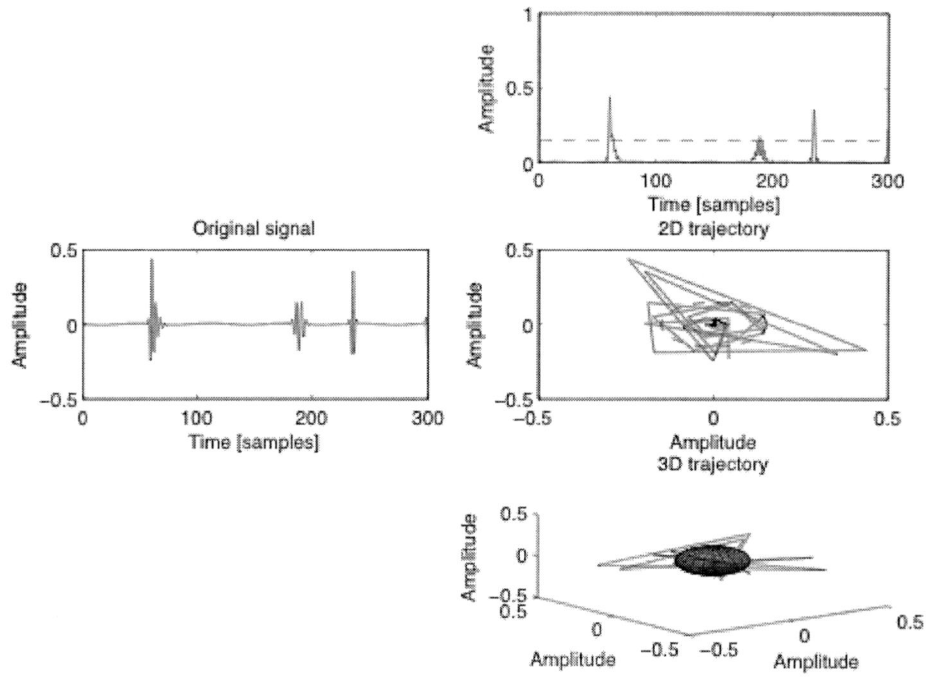

Figure 20. Comparison between time-domain detection and state space detection.

The method we propose goes a little further from the comments made in the previous paragraphs, as it employs the quantification of the recurrences of the state space trajectory. In short, a point of the detection curve is obtained by "summing" all the recurrences between the corresponding trajectory point and all the other points. The steps of the method are the following:

1) *The reconstruction of state space*. We do not yet have an algorithm for automatically choosing the best *m* and *d* parameters to be used for this reconstruction. However, we must impose the restriction that the analysis window dimension, *w*, should be a lot smaller that the durations of the transients to be detected. Ideally, the reconstruction parameters should be chosen such that the obtained trajectory would be completely "unfolded", meaning it should have no self-overlapping and be as wide as possible. Universally accepted methods for doing that are not (yet) known. Experimentally, we noticed that for values of 2 or 3 for *m*, together with a value of 1 or 2 for *d* (in order to ensure a *w* that is small enough), this detection method offers good results, although the trajectory might not be completely "unfolded".

2) *The computation of the recurrence matrix for the reconstructed trajectory*. Two parameters are used at this step: the metric that defines the distances in state space, and the recurrence threshold, □. The role of the metric can be played by the Euclidean norm, as well as by the Maximum norm, which offers similar results with the advantage of a much simplified computation. The choice of the second parameter is not a simple one, as it directly

influences the characterization of the metric relation between two trajectory points as "recurrence". Its choice must also consider the fact that noise "moves" the points from their original positions on the trajectory, thus leading to the loss of some recurrences (as well as to some "false recurrences"). To counteract the effect of noise, the value of □ should be chosen around the level of noise. Hence, we propose the following empirical relation for computing the recurrence threshold:

$$\varepsilon = \frac{\sum_{k=2}^{M} \left\| s_k - s_{k-1} \right\|}{M-1}.$$

This relation is based on the fact that the transients that are present in the signal have a total duration that is very short compared to the observation duration of the signal. The level of noise was estimated by averaging the distances between all the pairs of successive points on the trajectory.

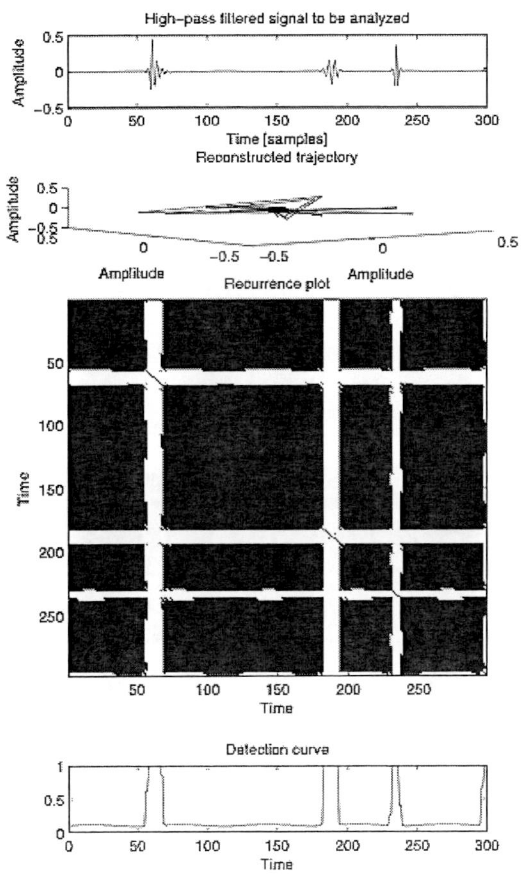

Figure 21. Step by step illustration of the method.

3) *The computation of the detection curve.* The visual inspection of the recurrence plot allows the localization of transients (for relatively low noise levels) in the areas with a small density of recurrences (– because the "noise ball" (having as its radius the recurrence threshold) in Figure 20 is left often during a transient). The global quantification of recurrences for each trajectory point leads to the detection curve, as follows:

$$C_j = \frac{1}{M}\sum_{i=1}^{M}\left(1 - R_{i,j}\right).$$

The detection method described here is illustrated in Figure 21.

It can be noticed that although it allows a visual localization of the existing transients, the obtained detection curve differs from the ideal one. These differences can be reduced by binarizing the former with a properly chosen threshold, possibly after a "smoothing" filtering operation (especially needed when big values for d are used).

Results

The result of applying the method on a synthetic signal can be seen in Figure 21. Figure 22 shows the results of the method for two constructed signals. The method offers similar results for measured signals also, as shown in Figure 23. Figure 24 illustrates the behavior of the method for various levels of noise.

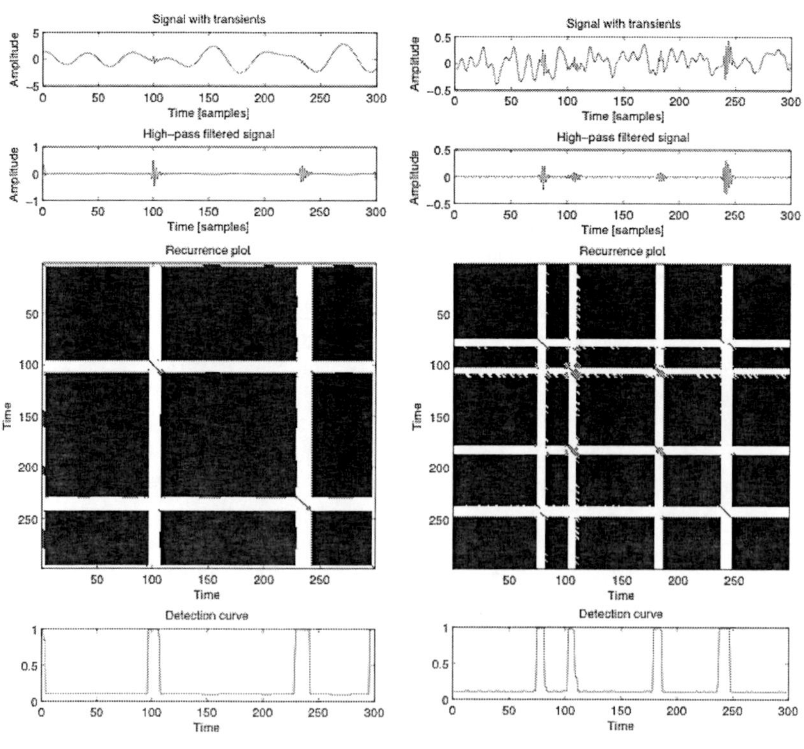

Figure 22. Results of the method for some constructed signals.

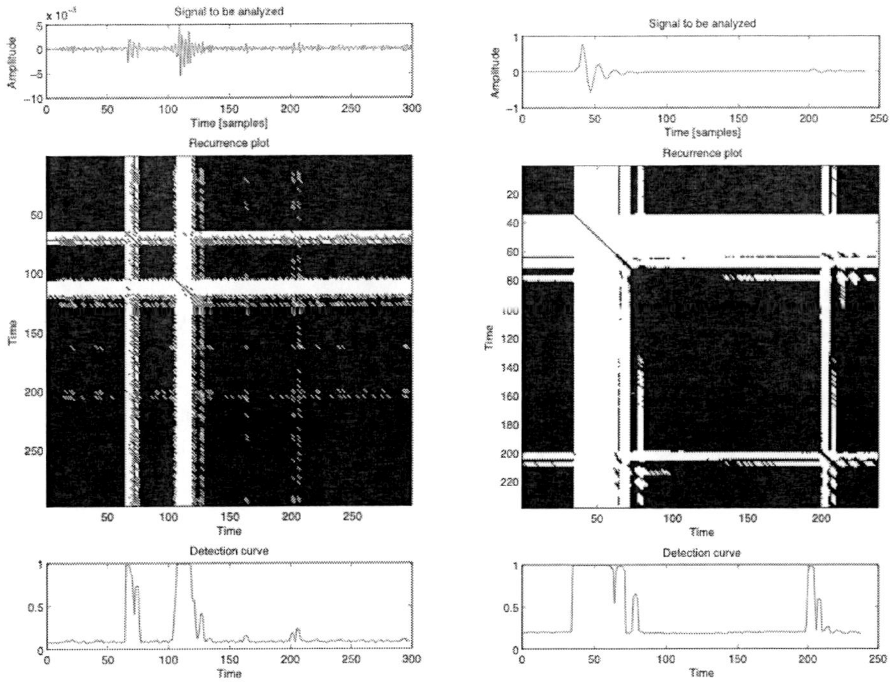

Figure 23. Results of the method for some measured signals.

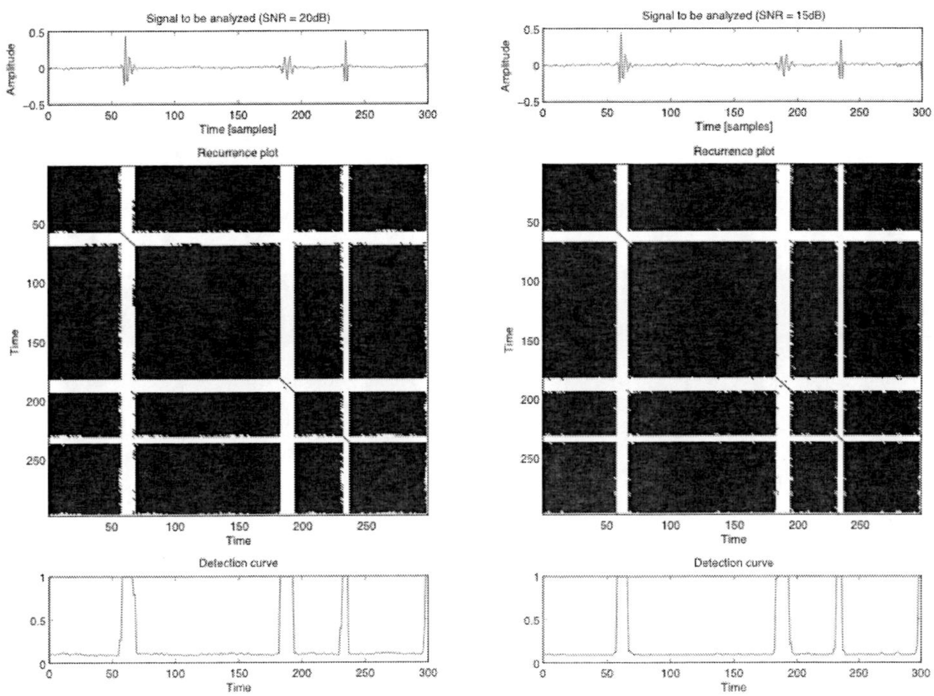

Figure 24. (Continued).

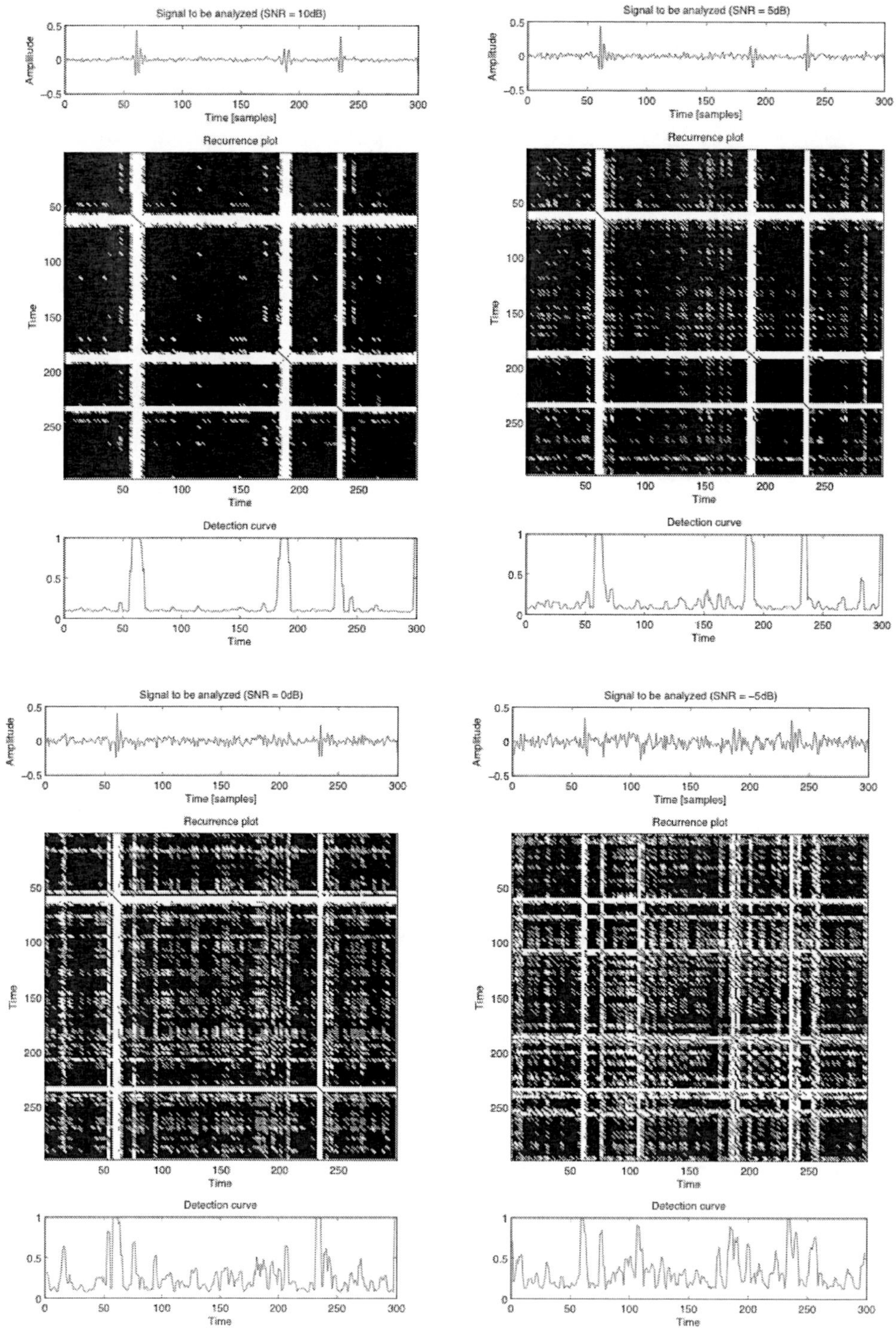

Figure 24. (Continued).

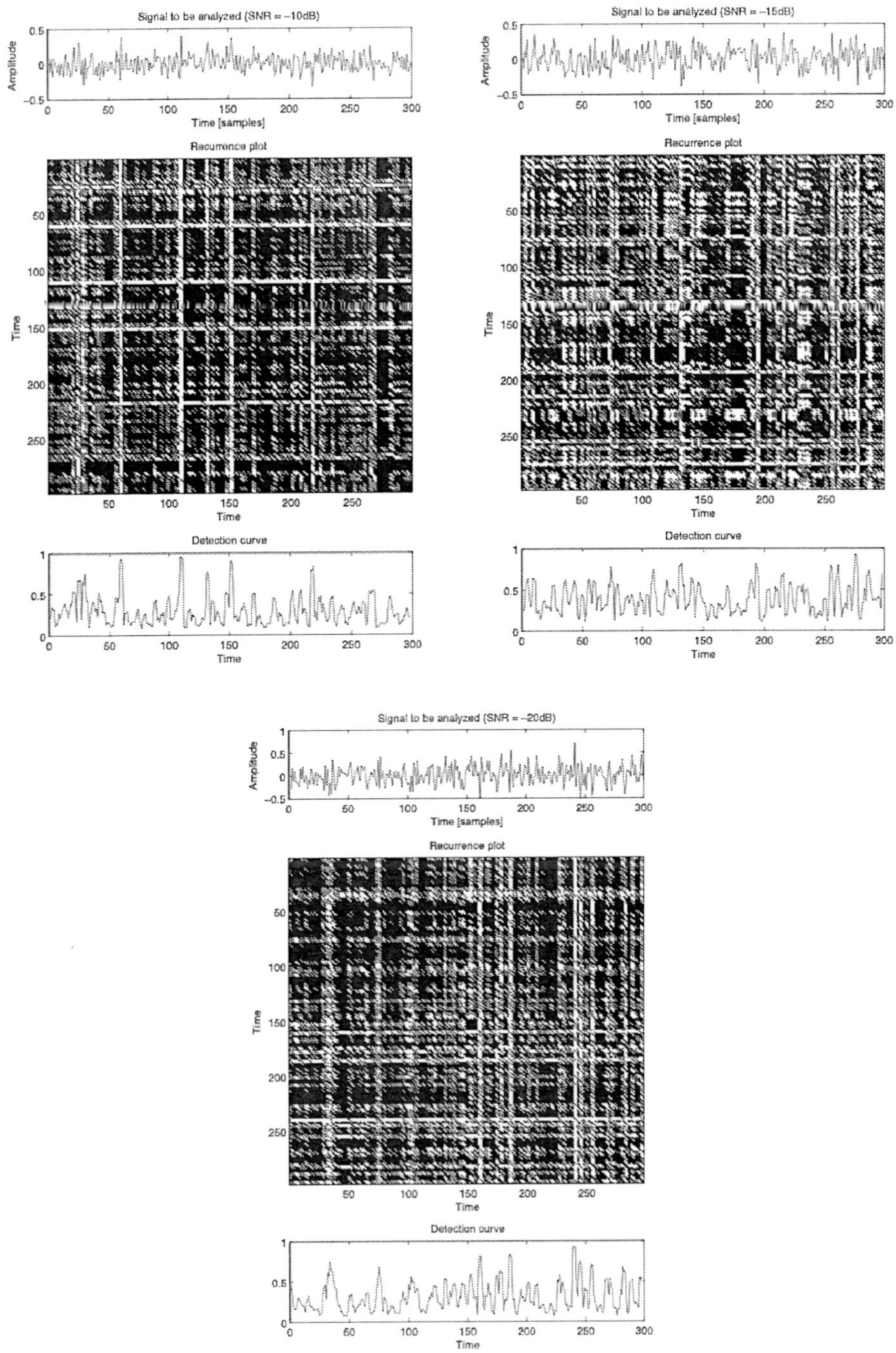

Figure 24. Results of the method for the same signal, but different SNRs.

DISCUSSION

The results presented here support the efficiency of the method. We have shown that satisfactory detection curves can be obtained, regardless of whether the signals are artificially generated or measured from real systems and despite the fact that their shapes are very different. Unlike the time-frequency methods, that have the purpose to characterize the local spectral content of the signals, the method we proposed here gives no information concerning the characterization of the analyzed signals, as it is especially developed for the purpose of localizing the time intervals where transients appear.

As it operates directly with the relationships between the signal samples, the efficiency of the method depends on the level of noise. Hence, the method does not allow the detection of transients that are completely buried in noise – because the detection is based on the fact that during the transients the "noise ball" is left very often. Another disadvantage of the method is that it can not operate directly with the measured signal, but with a preprocessed version of it. This preprocessing step consists in removing the low frequency content of the signal. However, it is not compulsory that this filter be a very good one. The main advantage of the method consists in the lightness of the computations, as no multiplication operations are used (which, on the contrary, are essential for the time-frequency and time-scale methods).

This incipient study on the method is promising, as it shows that the detection of transient signals can be done using some relatively simple computations performed directly with the signal samples, without needing to transform its domain by projecting it on a special basis of elementary signals. Some aspects that we study further regard:

- a thorough study concerning the choice for the state space reconstruction parameters
- finding methods to compute the optimum threshold to be used for the binarization of the detection curve (using, for example, its histogram)
- studying the effect of recursively performing the detection on the obtained detection curve (or, alternatively, on the product between the signal and the obtained detection curve)
- generalization of the method, by working with the distances in state space, instead of the recurrences (in which case the recurrence threshold might be variable or not used at all).

DIGITALLY MODULATED SIGNALS ANALYSIS BY RECURRENCE PLOT

The signals to be analyzed will be modulated by one among six digital modulations methods, i.e. BASK and ASK4; BFSK and FSK4; BPSK and PSK4.

Now the first objective is finding specific patterns of these signals in state space. For example, a sine wave has an elliptic representation in state space. Dots of the trajectory are plotted in Figure 25 and the distances between them are related to the time delay chosen for the acquisition of the measurements.

The size of the ellipse depends on the amplitude: the lengths of the axes increase proportionally with it. The points can be linked together to give an easier global vision of the trajectory.

The BASK modulated signal is presented in Figure 26, in a two-dimensional and in a three-dimensional state space.

The first ellipse, the smaller one, appears first and then it is contained in the bigger one which represents a modification of the signal amplitude. Because the phase of the modulated signal is continuous there is no phase bound in state space when the amplitude changes. Therefore, the transition between the two ellipses is quite simple and can be detected easily. This will not be the case for non digital amplitude modulated signals with discontinuity of the phase as the transition is more complicated and it disrupts the interpretation that could be done.

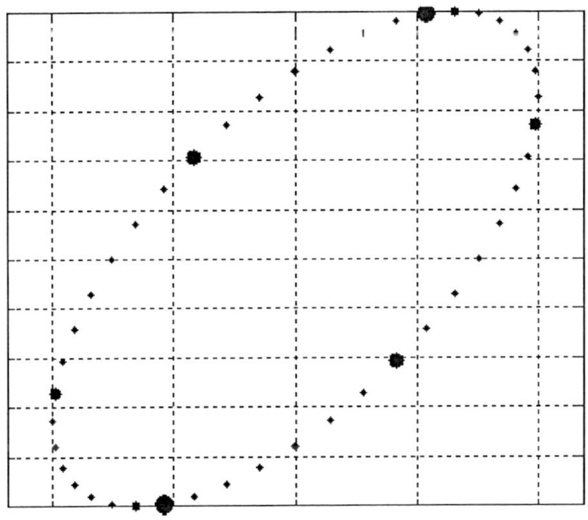

Figure 25. State space trajectory for a sinusoidal signal.

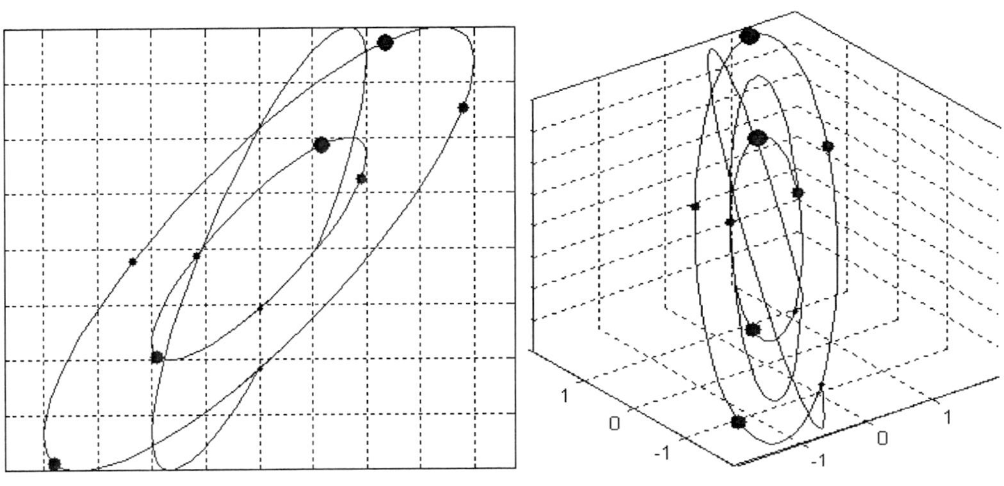

Figure 26. Representation in the state space of a BASK modulation with different dimensions.

In the case of a digital frequency modulated signal, the ellipse changes its orientation in state space. The size of the ellipse is also modified: some closed frequencies may be represented with ellipses with closed orientations and the ellipse will be flattened, i.e. when the ellipse parameters are increasing and decreasing, respectively (or inversely). The state space trajectory of a BFSK modulated signal is presented in Figure 27.

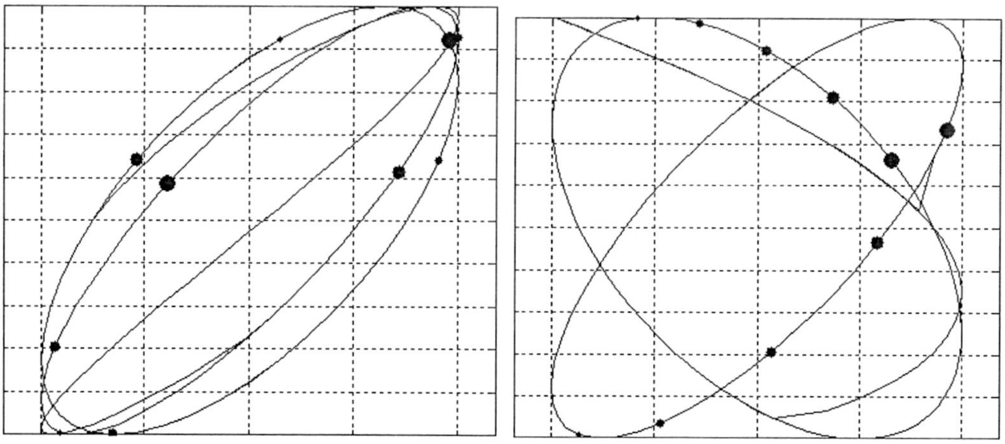

Figure 27. On the left, the difference between the frequencies, the index of modulation, is 11 Hz and on the right, the value is 66 Hz.

In the case of BPSK modulation, the trajectory always comes back on the first ellipse. A transition can be noticed and the direction of rotation is reversed.

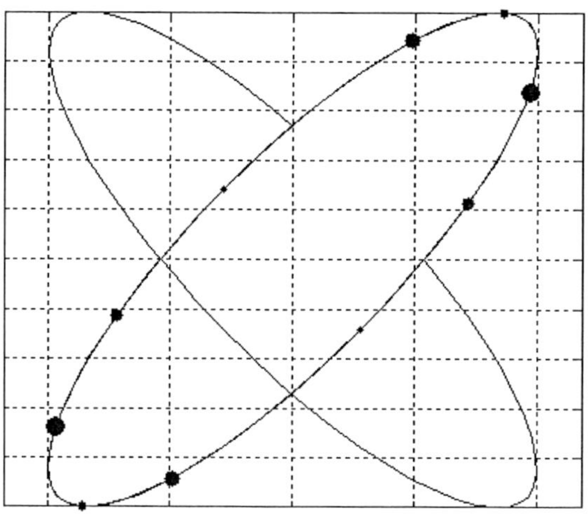

Figure 28. Trajectory in state space of a BPSK modulation.

This can be explained by the fact that a bound of phase of π can be seen as a brutal modification of the amplitude, from 1 to -1 for example. So, just after the changing, the discrete measures give opposite values to those that were expected if nothing had happened. In consequence, the direction of rotation in the ellipse is reversed because all the components of the trajectory vector are the opposite of those of the previous trajectory vector. Hence, specific patterns can be noticed for those modulations.

The number of modifications in the bit stream has also an important influence on the plotting: the transition between two levels of coding can strongly complicate the trajectory leading to an unreadable plot in phase space (Figure 29).

In conclusion, for complex modulated signals, the analysis of state space is not efficient and that is why a recurrence plot analysis will be used.

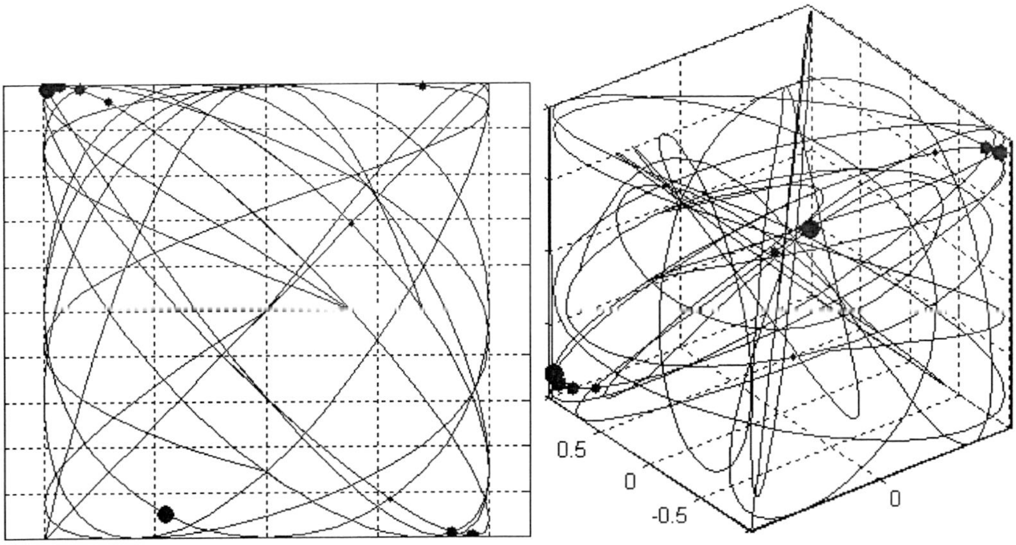

Figure 29. FSK4 modulation in two-dimensional and in three-dimensional space.

For example, the RP of a BFSK modulated signal allows the visualization of the different periods and therefore of the different frequencies contained in the signal. In Figure 30 the two different periods can be easily noticed: the period T_1 appears during the first half of the measure and the period T_2 appears during the rest of the time.

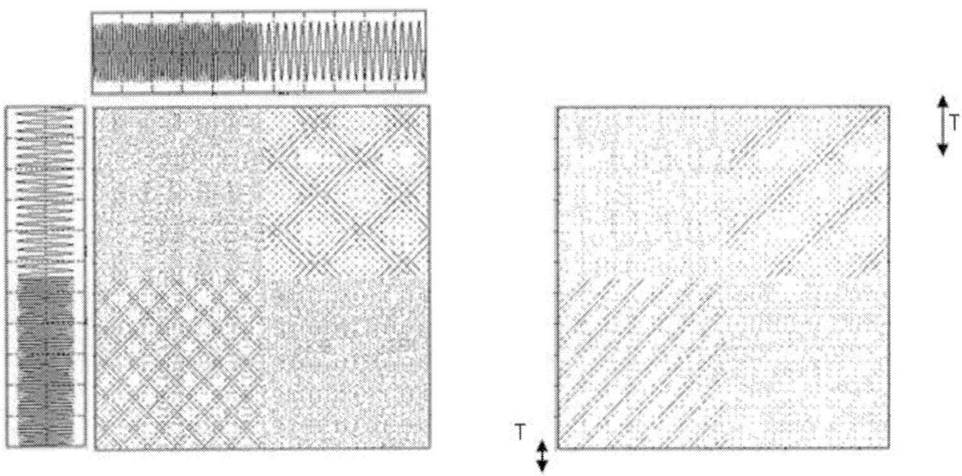

Figure 30. For a BFSK modulation, with m=1 and τ=1 (left side), perpendicular diagonals appear, whereas they are no more present with m=2 and τ=1, (right side).

MATLAB PROGRAM FOR FILTERING THE RP

Besides the two easily noticeable frequencies, the RP contains additional information that we do not need. We removed this unwanted information using an algorithm that we implemented in MATLAB.

- *First step*: The generation of the different digital modulated signals. Some more complicated ones could be tried also, such as Quadrature Amplitude Modulation (QAM).
- *Second step*: The obtained RP is presented in Figure 31. We can identify certain regions corresponding to different informational parts of the signal. Besides those areas that have common characteristics (squares made of lines parallel to the LOI) there are some regions corresponding to different "interferences" – points from one informational region are compared to points from a different informational region. Those regions of interference mst be removed.

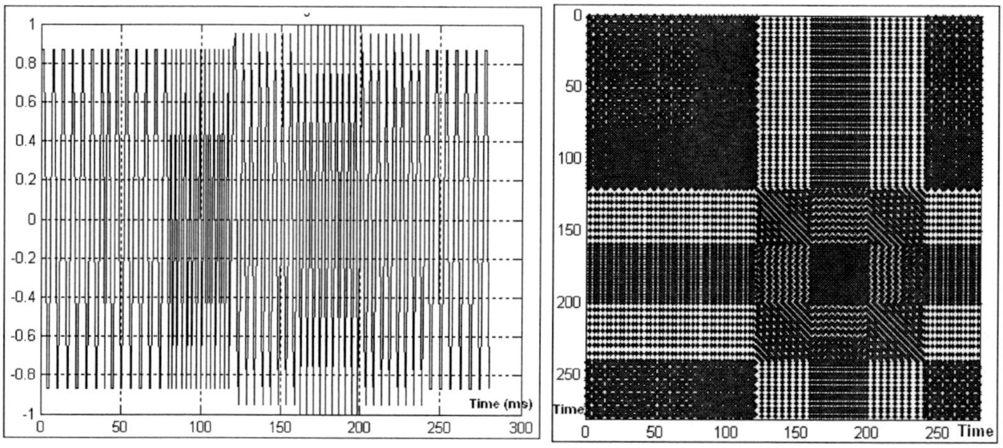

Figure 31. On the left side, the FSK4 modulated signal in reception and on the right side, the RP obtained from this signal without filtering.

The filtering method is explained in the following. Interference might have the effect of random noise signal, which could lead to blackheads on RP. It is possible that two such generated dots have really close positions for a chain of two values. Furthermore, the LOI might not be thin as expected and it might be composed in fact of several lines. This is due to the fact that a given portion of the trajectory is in the neighborhood of some previous one, and does not exactly take the same position. We put an arbitrary threshold on all the diagonal chains of dots whose length is inferior to 3. Hence, the squares on the main diagonal (LOI) remain and most of the interferences are rejected.

A second filtering is performed in order to keep only the square patterns on the main diagonal. It is based on a decimation of the squares and on a modification of the origin point for several iterations. The first origin is (0; 0) and the first time of transition is detected at t_1=80ms. So, all the points over 80 are deleted. The new origin is now the dot of coordinates (80; 80) and the next time transition is found at t_2=120ms. All the points over 120 are deleted and the new origin is now the dot of coordinates (120; 120). The same procedure is used for the entire RP. (Detection of the transition times can be done, for example, by studying the evolution of the recurrence rate).

The duration of the bit is taken as the smallest size of the squares along the main diagonal. For example, the duration of the bit is not t_1=80ms, because the second square is smaller. The period of the signal is the distance between two consecutive diagonal lines in a square pattern. Therefore, this algorithm allows the recognition of the modulation. The

MATLAB program returns the number of intervals, the different parameters and their values. It can be also used for complex signals, e.g. for a modulation in amplitude and frequency or for a QAM.

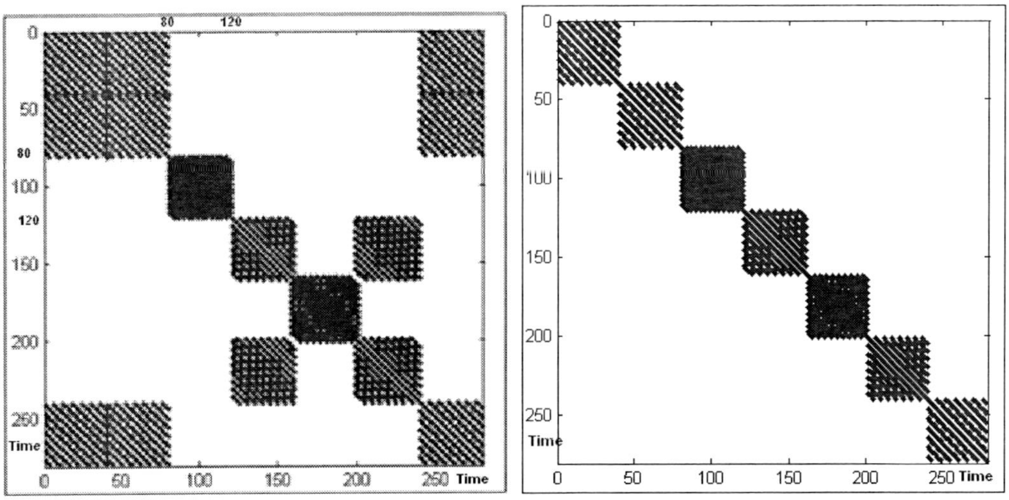

Figure 32. On the left side, the RP after the first filtering and on the right side, the RP after the second filtering.

Final Conclusions

Recurrence analysis turns into a bi-dimensional matrix the entire evolution (a trajectory) of a dynamical system in its phase space. It is very interesting the fact that this 2D image of the dynamics contains much information about the system, regardless of which is the dimension of the phase space in which its state evolves.

In this chapter we tried to show that the new nonlinear analysis and processing techniques based on RPs and RQA can be used not just for the visual inspection of real signals, but also for analyzing and processing them. We show that starting from an RP one can build nonlinear filters that are able to give better results when used on signals produced by real, non-linear systems. These RP-based tools can be used for a variety of purposes, such as: determining the behavior of systems, predicting and controlling system evolution, compressing signals, extracting characteristic parameters from signals, synthesizing signals, comparing signals, etc. The bottom line is that the RP is a good non-linear tool for signals coming from real systems.

References

[1] Senge, P. The Fifth Discipline: The Art and Practice of the Learning Organization; Doubleday Business, 1994.

[2] Sussman, J. The New Transportation Faculty: The Evolution to Engineering Systems; Transportation Quarterly, Eno Transportation Foundation, Washington, DC, 1999.

[3] Maier, M. W.; Rechtin, E. The Art of Systems Architecting, Second Edition; CRC Press, 2000.

[4] Levy, D. L. (2000) Applications and Limitations of Complexity Theory in Organization Theory and Strategy; in Jack Rabin, Gerald J. Miller, and W. Bartley Hildreth (editors), Handbook of Strategic Management, Second Edition; Marcel Dekker, New York, 2000.

[5] Kantz, H.; Schreiber, T. Nonlinear Time Series Analysis, 2nd Edition; Cambridge University Press, 2004.

[6] Mazilu, M.; Serbanescu, A. Reconstruction of the phase space of a dynamic system from experimental measurements; Budva, Montenegro, 2007.

[7] Eckmann, J.-P.; Kamphorst, S.O.; Ruelle, D. Recurrence plots of dynamical systems; *Europhys. Lett.,* 1987, vol. 5, pp. 973-977.

[8] Marwan, N., et al. Recurrence plots for the analysis of complex systems; *Physics Reports*, 2007, vol. 438, pp. 237-329.

[9] Gao, J.; Cai, H. On the structures and quantification of recurrence plots; Physic*s Letters A*, 2000, vol. 270, pp. 75-87.

[10] Marwan, N. Encounters with neighbors: Current developments of concepts based on recurrence plot and their applications; PhD Thesis, Institut fur Physik Fakultät Mathematik und Naturwissenschaften Universität Potsdam, 2003.

[11] Schreiber, T. Interdisciplinary application of nonlinear time series methods; *Physics Reports*, 1999, vol. 308, pp. 1-64.

[12] Ivan, C.; Serbanescu, A. Applications of nonlinear time-series analysis in unstable periodic orbits identification; Proceedings of International Siberian Conference on Control and Communications, Tomsk, Russia, 2009, pp. 93-99.

[13] Serbanescu, A.; Cernaianu L. Applications of Nonlinear Time Series Analysis; Communications 2008, Workshop: Advanced Works on Transient Signal Characterization, Bucharest, Romania.

In: Advances in Energy Research
Editor: Nicu Bizon

ISBN 978-1-61209-991-0

Chapter 4

GENERALIZED TELLEGEN'S PRINCIPLE AND STATE SPACE ENERGY BASED CAUSAL SYSTEMS DESCRIPTION

***Josef Hrusak**[*1]**, Milan Stork**[1] **and Daniel Mayer**[2]*
[1]Department of Applied Electronics and Telecommunications
[2]Department of Theory of Electrical Engineering
University of West Bohemia, Plzen, Czech Republic

ABSTRACT

The chapter deals with a new problem of physical correctness detection in the area of strictly causal system representations. The proposed approach to the problem solution is based on generalization of Tellegen's theorem well known from electrical engineering. Consequently, mathematically as well as physically correct results are obtained. The contribution is mainly concerned with presentation of a new structural approach to analysis and synthesis of linear and non-linear causal systems. It has been proven that complete analysis of instability, conservativity, dissipativity, anti-dissipativity, stability, asymptotic stability and chaoticity reduces to two independent tests: the monotonicity test of abstract state space energy and minimality test of state space dimension (i.e. for zero input that of complete state observability).

Keywords: State space energy, dissipativity, conservativity, stability, convergence, chaoticity, Tellegen's principle.

INTRODUCTION

It is familiar that there are two basic approaches to system modelling. The first one consists in using mathematical formulas and physical principles (a causality principle,

[*] Email: hrusak@kae.zcu.cz, stork@kae.zcu.cz, mayer@kte.zcu.cz

different forms of conservation laws, power balance relations, etc.) in order to describe appropriate system behavior. It has successfully been used in many fields of science and engineering so far. However, there are also situations where physical laws are not known or cannot be expressed in a proper mathematically exact form. In that case the second basic so called cybernetical approach to system modelling can be turned. It is based on identification methods working in terms of experimentally gained data [1], [2]. It is possible to divide the identification methods into two groups: parametric and non-parametric, respectively. If any prior information about a system structure is not available then one of non-parametric procedures has to be chosen for system identification. On the other hand, imagine that a physical structure of an investigated system would be known. In such cases some of available parametric methods can be used and consequently more adequate results from the physical correctness point of view should be obtained [3]. Unfortunately, any reliable explicit knowledge about a physical system structure is more likely an exception than a rule. Therefore, a system structure is mostly chosen *ad hoc* only behalf of heuristic arguments. Subsequently it has to be verified whether obtained quantitative results are not in conflict with obvious qualitative expectations concerning regular system behavior and/or results of additional experiments performed on a real system.

The chapter is organized as follows: The section 1 is devoted to the problem of physical correctness of electrical circuit models. In the section 2, new trends in electronics including a new concept of memristor are discussed. In the section 3, a new concept of the state space energy is introduced and a generalized form of the Tellegen´s theorem called the Liapunov-Tellegen´s principle is investigated. The section 4 demonstrates the way of application of the developed theoretical tools to the problem of the state space energy exchange between two causal systems under a coupling nonlinearity, including a possibility of appearing chaotic dynamics. In the section 5, the abstract form of the state space energy conservation principle is developed and illustrated by nonlinear synchronization of dissipation structures. Section 6 summarizes some knowledge concerning observability, controllability and state minimality of linear system representations, as well as state energy additivity property. In the section 7, the problem of asymptotic stability analysis is used for illustration. The section 8 demonstrates the easy way of nonlinear stability analysis by means of the proposed state space energy based method, while the mechanisms of chaoticity and effects of a memristive feedback are investigated in the final section.

Concept Of Physical Correctness in Electrical Circuits Theory

Certainly, any realizable system has to fulfill some causality and energy conservation requirements. Recall that existence of a state space representation is necessary and sufficient for a system to be causal. On the other hand causality does not imply energy conservation. In electrical engineering validity of Kirchhoff's laws is standardly used as necessary and sufficient condition for energy conservation. Thus for a correctness of a mathematical model of an electrical network we have:

Definition 1: (Physical correctness of an abstract electrical circuit) *Electrical circuit is physically correct if it is not in contradiction with both the voltage and the current Kirchhoff's laws.*

There arise a natural question, how the concept of physical correctness could be generalized beyond the electrical circuits theory.

Tellegen's theorem seems to be a proper starting point, because it has proven to be very general and extraordinary elegant abstract form of energy conservation principle for a class physically correct system representations, in which voltages and currents have been chosen as state variables [4 - 7].

First, let's briefly summarize the essential features of the classical version of Tellegen's theorem and assume that an arbitrarily connected electrical network of b components is given. Let's disregard the specific nature of the individual network components and represent the *network structure* by an oriented graph with n vertices and b branches. Let the set of Kirchhoff's laws constraints is given

$$Ai(t)=0 \qquad Bv(t)=0 \tag{1}$$

where A is a *node incidence matrix*, B is *loop incidence matrix*, and i and v are:

$$i=[i_1,i_2,...,i_n]^T, \quad v=[v_1,v_2,...v_n]^T \tag{2}$$

Let J be the set of all current vectors i and V be the set of all voltage vectors v such that i and v satisfy (1). Then the classical Tellegen's principle can be expressed in the form as follows:

Theorem 1: (Classical Tellegen's theorem)

If $i \in J$ and $v \in V$ then for the *inner product* of both the vectors it holds:

$$\langle i(t), v(t)\rangle = 0 \tag{3}$$

The most important observation is that the *orthogonality condition* (3) is basically *not a physical statement,* but rather one of *geometrical nature.*

In fact, from abstract system theory point of view, it expresses a *neutrality of interactions* between a *gradient field* of an abstract system *state space energy* and the *vector field* of an abstract *state space velocity.* Thus for a *dual product* of a *row gradient vector* with a *column state velocity vector* we could write equivalently:

$$\langle grad_x E(x), f(x)\rangle = 0, \quad grad_x E(x) = \psi(x) \tag{4}$$

where *E(x)* is the *total electromagnetic energy accumulated* in all the electrical circuit components and *f(x)* is the vector field of *state space velocity* defined by the corresponding *state space representation:*

$$\begin{aligned} \frac{dx}{dt} &= f(x), \quad x(t_0) \in X \subset R^n, \quad 1 \le n < \infty \\ y &= h(x), \quad y(t) \in R^n, \qquad\quad 1 \le p < n \end{aligned} \tag{5}$$

The idea of a generalized Liapunov-Tellegen's principle (4) will be further developed in the part 3.

NEW TRENDS IN NANO-ELECTRONICS AND MEMRISTOR FUNDAMENTALS

Roughly speaking, we could characterize many remarkable recent changes in a number of fields including new developments in science and technology as a consequence of dramatic change of the paradigm from various systems in micro-size to nano-size system components, as well as from linearity to nonlinearity. Such revolutionary changes could not remain without any response in engineering sciences. Back in 1971, the theoretically oriented electrical engineer Leon Chua [8] has discovered something missing: the fourth basic electrical circuit element besides the standard trio of linear resistor, capacitor and inductor. He called it the "memristor". The only problem was that as far as Chua or anyone else could see, memristors did not actually exist in reality. At the time they have not only been made, but it seems to be clear that in near future their unique capabilities might revolutionise not only the field of microelectronics, computer architecture, communication systems but even the field of artificial intelligence [9-12].

Remember that in its beginnings the discovery of memristor was a triumph of pure logic. Chua was examining the set of binary relations between four basic quantities that define any electrical circuit. From combinatorics everybody knows that the number of ways by which any set of n objects can be ordered into k-elements groups is determined by (6), and for $n = 4$, $k = 2$ gives

$$\binom{n}{k} = \frac{(n-k)!}{k!} \Rightarrow \binom{n}{k} = \binom{4}{2} = 6 \tag{6}$$

Thus the *four basic state variables*: the charge q, current i, voltage v and flux □ have to be related by *six binary relations.*

Magnetic flux and voltage are related through their definitions by the binary relation:

$$d\varphi = vdt \tag{7}$$

Similarly electrical charge and current are related by the binary relation:

$$dq = idt \tag{8}$$

Three more binary relations correspond to constitutive relations of three traditional circuit elements.

For a resistor we have:

$$dv = Rdi \tag{9}$$

i.e. current passing through a resistor generates voltage.

For a capacitor we have:

$$dq = Cdv$$

(10)

i.e. for a given voltage a capacitor stores a certain amount of charge.

For an inductor we have:

$$d\varphi = Ldi \tag{11}$$

i.e. for a given current an inductor generates a magnetic flux.

Summarizing, that makes *five*. It follows that *one circuit element was missing*. There was a question, what should it do? And consequently: could it be replaced by a combination of resistors, capacitors and inductors? Chua set about exploring both this questions first. It was immediately clear that the missing binary relation has to describe a *new circuit element which relates electric charge with magnetic flux* in some proper way. Because, in fact, moving charges are "currents" and changing fluxes are "voltages" it seems plausible that the *"new device"* would generate a voltage from a current rather *like a resistor*, but in a more complicated *"dynamical way"*. Thus we can see it as a *"resistor that has an ability to remember"* what current had flowed through it before.

Though the memristors were welcome in theory, no real-world device seemed capable of the resistance with the memory effect as described. The situation changed radically in May 2008 with publication of famous papers [10-12], reporting about the accidental discovery of the first physical device with required properties by a research team of Stan Williams at Hewlett-Packard Laboratories in Palo Alto, California.

Recent investigations show that memristors can, similarly as living organisms, *memorise even patterns of events*. It seems to confirm the previous idea of Chua – memristors have something to say about the way *how biological organisms learn.*

The most immediate potential use of memristor seems to be as a powerful replacement for flash memories. Recall that the *fourth passive element* – the *memristor* should in fact be added to the list of three basic electrical circuits elements. Summarizing, the memristor is a *passive two-terminal electronic device,* described by a *nonlinear constitutive relations*

$$v = M(q)i, \quad \text{or} \quad i = W(\varphi)v \tag{12}$$

between the device terminal voltage v and terminal current i. It is just the *inability to duplicate the properties of the memistor* with a combination of the other three *passive* circuits elements, what makes the *existence question* of the *memristor fundamental.*

The resulting *logically complete* set of passive electrical circuit elements is schematically displayed in the Figure 1.

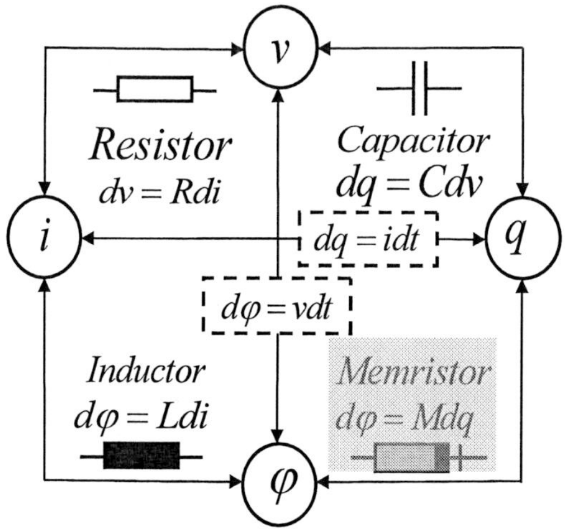

Figure 1. Logically complete set of passive circuit elements

The two nonlinear functions *M(q)* and *W(φ)*, characterizing *two new system properties*, are called the *memristance* and *memductance*, respectively, and are defined by:

$$M(q) = \frac{d\varphi(q)}{dq}, \qquad W(\varphi) = \frac{dq(\varphi)}{d\varphi} \tag{13}$$

They obviously represent the *slope* of scalar functions $\varphi = \varphi(q)$ and $q = q(\varphi)$, respectively, and are often called the *memristor constitutive relations.*

Recall that resistance *R*, is the rate of change of voltage with current; capacitance *C*, is the rate of change of charge with voltage, and similarly *inductance L*, is that of *flux with current.* The point is that any *static nonlinearity* interpreted as a *memristor always appears instantaneously* as a *nonlinear resistor.* However, in fact it *represents* a new *passive element,* which may *relate some state variable to flux without storing a magnetic field.*

Example 1: Let's illustrate the situation by a simple nonlinear electrical circuit shown in the Figure 2.

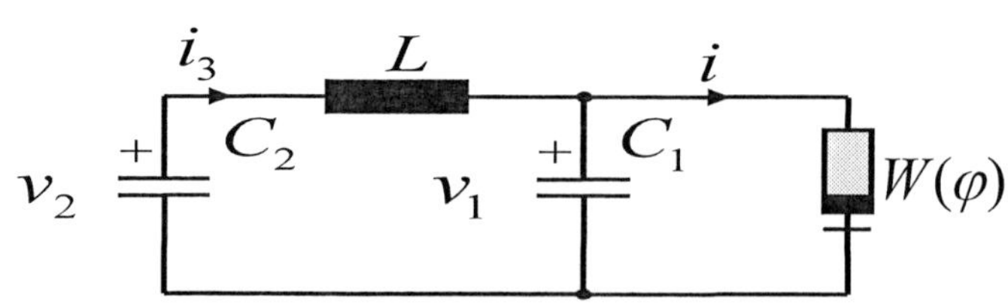

Figure 2. Example of nonlinear oscillator with flux-controlled memductance W(φ).

The circuit according the Figure 2 is described by the following state space representation:

$$\left.\begin{aligned} \frac{d\varphi}{dt} &= v_1 \\ \frac{dv_1}{dt} &= \frac{1}{C_1}\left(i_3 - W(\varphi)v_1\right) \\ \frac{di_3}{dt} &= -\frac{1}{L}\left(v_2 - v_1\right) \\ \frac{dv_2}{dt} &= \frac{1}{C_2} i_3 \end{aligned}\right\} \tag{14}$$

The nonlinearity $W(\varphi)$ can be defined for instance by the function shown in the Figure 3.

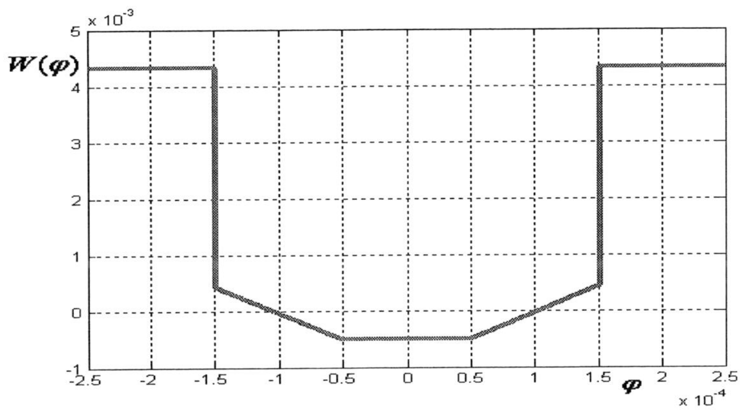

Figure 3. Example of nonlinear function $W(\varphi)$.

The structure of an equivalent 4th order nonlinear state space system representation given by eq. (14) is illustrated in the Figure 4.

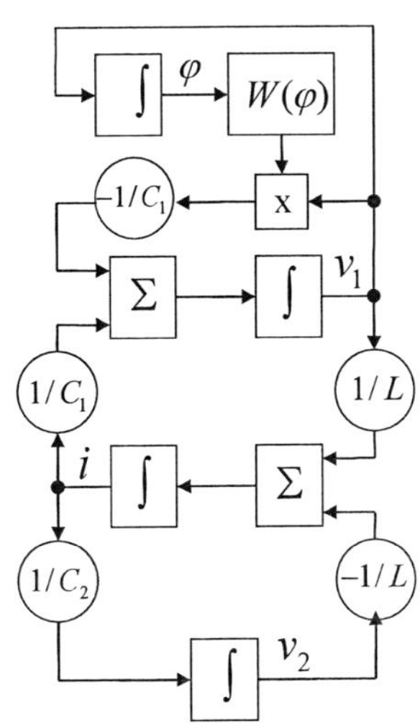

Figure 4. Equivalent structure of a 4th order nonlinear system with memristive feedback $W(\varphi)$.

The simulation results showed that the system described by eqs. (9) with the *topological structure* shown in the Figure 4. may have a chaotic attractor, if parameters are chosen properly. An example of pseudo chaotic behavior with two distinct unstable periodic attractors is also possible. For illustration an example of 3D projection of system trajectories in such a situation is shown in the Figure 5.

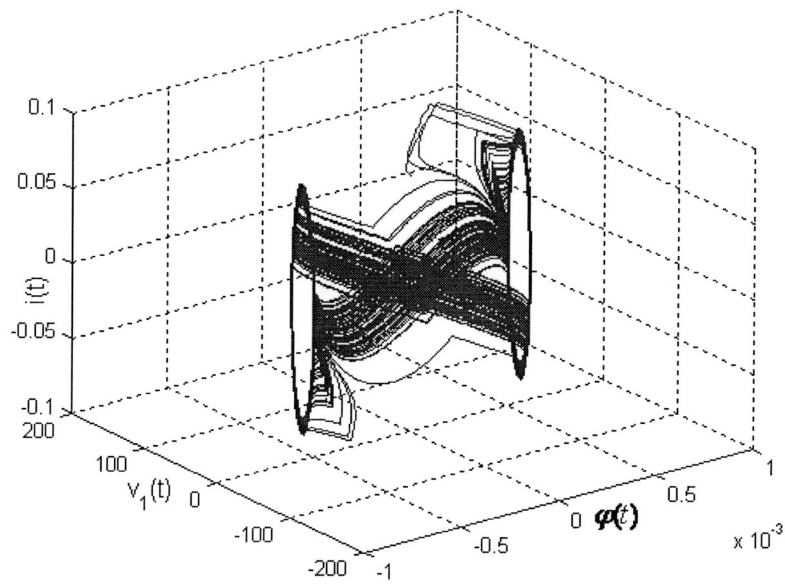

Figure 5. The 3 D projection of pseudo-chaotic system with nonlinear function W(φ).

STATE SPACE ENERGY CONCEPT AND GENERALIZED LIAPUNOV-TELLEGEN`S PRINCIPLE

As an alternative to the well-known physical energy motivated method of Liapunov functions, a new conceptually different approach to stability problems has recently been proposed in [2, 3, 4] and called the signal energy-metric approach. The *crucial idea* is that, in fact, *it is not the energy by itself*, but only a *measure of distance from the system equilibrium to the actual state* $x(t)$, what is needed for stability and in-stability analysis.

Thus a *state space metric* $\rho[x(t), x^*]$, where x^* denotes the *equilibrium state,* has been *defined,* and the basic idea of a new *state energy-metric approach* has then formally been expressed by the following *state space energy definition:*

$$E(x) \;=\; \frac{1}{2}\rho^2\left[x(t), x^*\right] \tag{15}$$

Let's consider a linearly observed *nonlinear* abstract system representation $R\{S\}$ defined by the *state equation in semilinear form*

$$\dot{x}(t) = f[x(t)] + Bu(t),\ x(t_0) = x^0, \qquad (16)$$
$$y(t) = C\,x(t)$$

with an uncontrolled *state velocity vector field* given by:

$$f(x) = A(x)x \qquad (17)$$

1. Let $X \subset R^n$ be a class C_∞ - *manifold*, representing the *state space* of the representation $R\{S\}$ containing the *unique zero input equilibrium state vector*

$$x^* = [x_1^*, x_2^*, \ldots, x_n^*]^T, \qquad x^* \in X \qquad (18)$$

defined by the condition

$$u(t) = 0, t \in [t_0, \infty) : f(x^*) = 0 \Leftrightarrow x^* = 0, \dot{x}(t)\,|_{x=x^*} = 0 \qquad (19)$$

and $x(t)$ be an instantaneous value of the state vector x in a time instant t :

$$x(t) \in X \subset R^n \qquad (20)$$

2. Let ψ be a continuous sufficiently *smooth vector mapping* of the class $C_k, k > 1$

$$\psi : X \to R^n \qquad (21)$$

representing the *state space gradient vector*

$$\psi(x) = grad_x E(x) \qquad (22)$$

of a *state space energy* scalar field, defined *independently of the integration path* by

$$E(x) = \int_{x_1^*=0}^{x_1} \psi_1(\xi_1, 0, \ldots 0)\,d\xi_1 + \int_{x_2^*=0}^{x_2} \psi_2(x_1, \xi_2, 0, \ldots 0)\,d\xi_2 + \ldots$$
$$+ \int_{x_n^*=0}^{x_n} \psi_n(x_1, x_2, \ldots x_{n-1}, \xi_n)\,d\xi_n \qquad (23)$$

For the sake of mathematical correctness we have to require the field ψ to be *nonsingular,* fulfilling the *condition*:

$$\forall x \in X, \quad x \neq x^* : \ \det J_x(\psi) = \det\left[\frac{\partial \psi}{\partial x}\right] \neq 0 \tag{24}$$

and irrotational, i.e. fulfilling the *condition*:

$$\forall x \subset X : \quad curl\left(\psi(x)\right) = 0 \tag{25}$$

or equivalently, satisfying the *Jacobian symmetry condition*

$$\left[J_x(\psi)\right]^T - J_x(\psi) = 0 \tag{26}$$

3. Let f^o be a continuous sufficiently *smooth vector field* of the class $C_k, k > 1$

$$f^o : \ X \to V, \quad V \subset R^n \tag{27}$$

representing the *zero-input state space velocity vector* defined by

$$v(t) = f^o[x(t)] = \left.\frac{dx}{dt}\right|_{u(t)=0}, \qquad t \in [t_o, \infty) \tag{28}$$

4. Let *h* be a continuous sufficiently *smooth vector mapping* of the class C_n

$$h: \ X \to Y, \quad Y \subset R^p \tag{29}$$

representing the *output observation vector of the internal state of a strictly causal system representation*

$$y(t) = h[x(t)], \quad t \in [t_o, \infty) \tag{30}$$

Let *D* be a continuous *scalar mapping* of the class C_2

$$D: \ X \to R \tag{31}$$

defined by the dual product

$$\langle \psi(x), f^{o}(x) \rangle = 0, \tag{32}$$

where $\psi(x)$ is the *gradient field of the state space energy* defined by

$$\psi(x) = grad_x E(x) \tag{33}$$

and $f^{o}(x)$ is the *zero input state velocity vector field* defined by

$$f^{o}(x) = \frac{dx}{dt}\Big|_{u(t)=0} \tag{34}$$

Theorem 2: (Generalized Tellegen's theorem)

If $\psi(x)$ satisfies the relation (24) and $f^{o}(x)$ is defined by (28) then for the *dual product* of both the vectors it holds:

$$\forall x \in X, \forall t, t \in [t_o, \infty) : D(x) = \langle \psi(x), f^{o}(x) \rangle = 0 \tag{35}$$

The proof is obvious by inspection: Because the gradient vector $\psi(x)$ is at any fixed state *x (t) orthogonal* to the state velocity vector $f^{o}(x)$ the state space energy $E(x)$ „ defined by integration of the dissipation function $D(x)$ has to be constant along any trajectory of the uncontrolled system representation $\Re\{S\}$ iff $D(x) = 0$ holds identically.

Remark 1: Theorem 2 should be considered rather as a principle than a proveable statement. It represents an *abstract form of power balance relation* in the case of *zero input power.* Equivalently, we may interpret it as a *state space energy conservation principle in the differential form.*

In fact, the proposed approach is closely related to the well known *Direct method of A.M.Liapunov,* in particular to the *Schulz`s variable gradient method* for solving the nontrivial *problem of Liapunov functions generation* [1-4]. Recall that in spite of obvious conceptual similarity, there are at least *two fundamental diversities* of the proposed approach:

- At first, in contrary to the concept of the state space energy which is implicitly assumed to be unique it is known that unicity of the Liapunov function is not warranted by Liapunov`s axioms.
- At second, in contrary to the Liapunov function concept, concept of the state space energy doesn`t depend on the chosen structure and parametrization of the given system representation.

In order to apply the direct Liapunov`s method a *proper choice* of the *unknown Liapunov function structure and parameters,* (or that of its gradient), has to be made in such a way that a set of axioms required by Liapunov method will be fulfiled. It is not an easy task, because no general constructive method for Liapunov functions generation has been found, yet. Recall that the *class of the Liapunov functions* is characterized axiomatically under the implicit assumption that the information about the *dynamics of the system* is contained in the given vector field $f^o(x)$, which is considered as the *only acceptable fixed model* of the system. Consequently the gradient field $\psi(x)$ is considered as *completely unknown element of a certain class of vector mappings.*

It is obvious from the Generalized Tellegen`s principle (35) that the *operation of the dual product is symmetric.*

It follows that it can be *dualized* in such a sense that instead of the fixed vector field $f^o(x)$ an *equivalence class of vector fields,* generated by the given field will be considered, and instead of the *unknown gradient field representation* one specific, but *universally* acceptable *fixed form* of the gradient vector $\psi(x)$ will be chosen within the *class of vector mappings*, satisfying the *nonsingularity condition* (24) and *set Jacobi symmetry conditions* (26).

The following natural questions arise: Should some forms of the mapping $\psi(x)$ be preferred? If so, what are the most *relevant criteria* for such a decision? The answer to the second question is plausible: besides of *maximal universality the requirement of minimal complexity* should be the dominant features of the proposed paradigm. The *structure symmetry* requirement of the Jacobian $J_x(\psi)$ in the form

$$\left.\begin{array}{l}
\dfrac{\partial \psi_2}{\partial x_1}-\dfrac{\partial \psi_1}{\partial x_2}=0, \\
\dfrac{\partial \psi_3}{\partial x_1}-\dfrac{\partial \psi_1}{\partial x_3}=0, \quad \dfrac{\partial \psi_3}{\partial x_2}-\dfrac{\partial \psi_2}{\partial x_1}=0, \\
------------------------------ \\
\dfrac{\partial \psi_{n-1}}{\partial x_1}-\dfrac{\partial \psi_1}{\partial x_{n-1}}=0, \quad \dfrac{\partial \psi_{n-1}}{\partial x_2}-\dfrac{\partial \psi_2}{\partial x_{n-1}}=0,\ldots, \quad \dfrac{\partial \psi_{n-1}}{\partial x_{n-2}}-\dfrac{\partial \psi_{n-2}}{\partial x_{n-1}}=0, \\
\dfrac{\partial \psi_n}{\partial x_1}-\dfrac{\partial \psi_1}{\partial x_n}=0, \quad \dfrac{\partial \psi_n}{\partial x_2}-\dfrac{\partial \psi_2}{\partial x_n}=0,\ldots, \quad \dfrac{\partial \psi_n}{\partial x_{n-1}}-\dfrac{\partial \psi_{n-1}}{\partial x_n}=0,
\end{array}\right\} \tag{36}$$

reduces the complexity measure given by the number $N=n^2$ of its elements from n^2 to the total number of *independent elements*

$$N = n + \frac{1}{2}n(n-1) \tag{37}$$

where the second term determines the total number of the *off-diagonal elements.*

Using a proper *state space equivalence transformation* and putting

$$\left.\begin{array}{l}\dfrac{\partial\psi_1}{\partial x_2}=\dfrac{\partial\psi_1}{\partial x_3}=.....=\dfrac{\partial\psi_1}{\partial x_{n-1}}=\dfrac{\partial\psi_1}{\partial x_n}=0\\ \dfrac{\partial\psi_2}{\partial x_3}=\dfrac{\partial\psi_2}{\partial x_4}=.....=\dfrac{\partial\psi_2}{\partial x_n}=0\\ -----------------\\ \dfrac{\partial\psi_{n-2}}{\partial x_{n-1}}=\dfrac{\partial\psi_{n-2}}{\partial x_n}=0\\ \dfrac{\partial\psi_{n-1}}{\partial x_n}=0,\end{array}\right\} \tag{38}$$

we can, without any restriction of generality, *eliminate* all the *off-diagonal elements* and further reduce the complexity of the gradient $\psi(x)$ to its *minimal structural complexity*, corresponding to $N=n$ *nonvanishing elements:*

$$\left.\begin{array}{l}\psi_1(x_1,x_2,...,x_n)=\psi_1(x_1)\\ \psi_2(x_1,x_2,...,x_n)=\psi_2(x_2)\\ \psi_3(x_1,x_2,...,x_n)=\psi_3(x_2)\\ \quad\vdots\\ \psi_n(x_1,x_2,...,x_n)=\psi_n(x_1)\end{array}\right\}\Leftrightarrow J_x(\psi)=\begin{bmatrix}\dfrac{\partial\psi_1}{\partial x_1} & 0 & \cdots & 0 & 0\\ 0 & \dfrac{\partial\psi_2}{\partial x_2} & 0 & \cdots & 0\\ \vdots & \vdots & \vdots & \vdots & \vdots\\ \vdots & \vdots & \vdots & \vdots & \vdots\\ 0 & 0 & \cdots & 0 & \dfrac{\partial\psi_n}{\partial x_n}\end{bmatrix} \tag{39}$$

with easy to check *nonsingularity condition*

$$\det J_x(\psi)=\prod_{i=1}^{n}\frac{\partial\psi_i}{\partial x_i}=\prod_{i=1}^{n}\frac{d}{dx_i}\psi_i(x_i)\neq 0 \tag{40}$$

As a consequence, the *nested additive form* (23) of the *total state space energy*

$$E(x)=E_1(x_1)+E_2(x_1,x_2)+E_3(x_1,x_2,x_3)+...+E_n(x_1,x_2,x_3,...,x_n) \tag{41}$$

where the *k-th* term is defined by

$$E_k(x_1,x_2,x_3,...,x_n)=\int_{x_k^*}^{x_k}\psi_k(x_1,x_2,...,x_{n-1},\xi_k)d\xi_k,\quad k\in\{1,2,...,n\} \tag{42}$$

reduces to the *sum of state space energies* for a *complete set* of the *individual first order system components*

$$E(x) = E(x_1, x_2, x_3, ..., x_n) = \sum_{k=1}^{n} E_k(x_k) \tag{43}$$

where *k-th* term defined by

$$E_k(x_k) = \int_{x_k^*}^{x_k} \psi_k(\xi_k) d\xi_k, \quad k \in \{1, 2, ..., n\} \tag{44}$$

determines *the same amount* of the state space energy corresponding to the *k-th interacting subsystem* as *it would be isolated.*

Remark 2: The last observation is of crucial importance from the structure of state space representation point of view, because the class of allowed state space representations is now significantly restricted: any state space representation of this class must namely satisfy the so called *interaction neutrality condition*, which could also be deduced from the generalized Liapunov-Tellegen`s principle.

Remark 3: It is worthwhile to notice that as a by-product *of the Jacobian diagonalization* the n-dimensional mapping $\psi(x)$ reduces to n equivalent independent *one-dimensional nonsingular* mappings $\psi_i(x_i), i \in \{1, 2, ..., n\}$.

Let`s now make the last step in direction of *complexity minimization* requiring an additional property of *a structural nonsingularity*

$$\forall k, k \in \{1, 2, ..., n\} : \frac{d}{dx_i} \psi_i(x_i) = 1 \Rightarrow \forall x \in X \subset R^n : \det J_x(\psi) = 1 \tag{45}$$

as expressed by *gradient renormalization* in the form

$$\left.\begin{aligned} \psi_1(x) &= x_1 \\ \psi_2(x) &= x_2 \\ \psi_3(x) &= x_3 \\ &\vdots \\ \psi_n(x) &= x_n \end{aligned}\right\}$$

(46)

corresponding the most natural choice of gradient mapping $\psi(x)$ as the *identity mapping*, i.e.:

$$\psi(x) = x^T = [x_1, x_2, \ldots, x_n] \tag{47}$$

Thus the *total state space energy* $E(x)$ after integration (3.30) is now expressed as follows:

$$E(x) - E(x^*) = \frac{1}{2}\|x(t)\|^2 = \frac{1}{2}\rho_2^2\left(x, x^*\right)\Big|_{x^*=0} \tag{48}$$

and from the Liapunov-Tellegen`s principle the *conservativity of the state velocity vector field* follows, or equivalently the *state space energy conservation principle* holds:

$$\forall x \in X : \left\langle x^T, f^0(x)\right\rangle = 0 \iff \forall t, t \in [t_0, \infty) : E[x(t)] = E[x(t_0)] \tag{49}$$

Analysis of the State Space Energy Exchange Mechanism

For theoretical analysis of the nonlinear state space energy exchange, as well as for illustrating computer simulation experiments, the structure consisting of two interacting systems as depicted in Figure 6 will be considered.

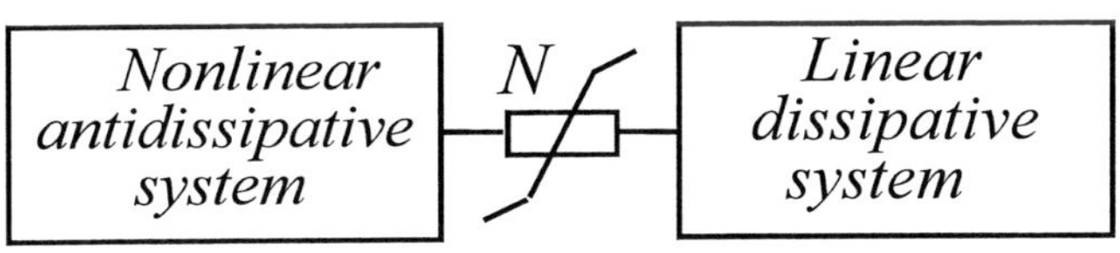

Figure 6. Generalized structure of two nonlinearly coupled systems.

A simple nonlinear electrical circuit consisting of two coupled subsystems is shown in the Figure 7 can be seen as one of its possible physical realizations for motivation. The electrical circuit that is shown in the Figure 7 has two coupled subsystem: first is a subsystem nonlinear of anti-dissipative LC oscillator type and the other is linear and dissipative. The negative impedance converter can be used as a nonlinear negative resistance element, thus giving rise to undamped oscillations in the nonlinear subsystem. Both the LC circuits are coupled by another nonlinear element *N*.

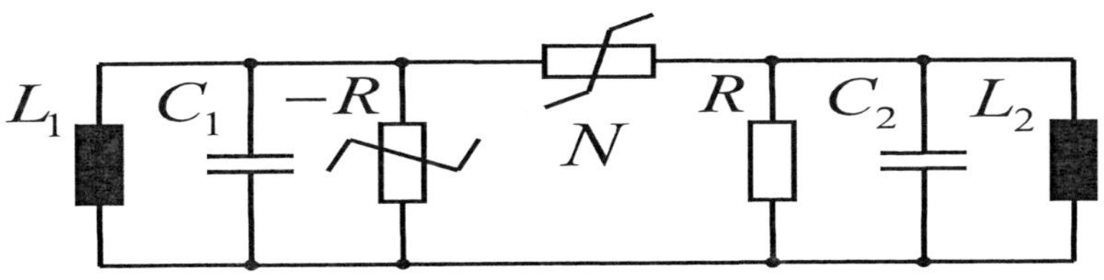

Figure 7. The circuit diagram of nonlinearly coupled oscillators.

Let`s assume that the abstract state equations of the corresponding state space representation are given in the form:

$$\left.\begin{aligned} \dot{x}_1 &= \alpha_1 x_1 - k_1 x_1^3 + \alpha_2 x_2 - N(x_1, x_3, p) \\ \dot{x}_2 &= -\alpha_2 x_1 \\ \dot{x}_3 &= -\alpha_3 x_3 + \alpha_4 x_4 - N(x_1, x_3, p) \\ \dot{x}_4 &= -\alpha_4 x_3 \end{aligned}\right\} \tag{50}$$

where $N(x_1,x_3,p)$ represents a coupling nonlinearity. The *internal structure* of the nonlinearly coupled abstract state space system representations is illustrated by a structure diagram shown in the Figure 8.

Two specific forms of the *coupling nonlinearity N*, as illustrated in the Figure 9, have been chosen for comparison. The first one is the diode-like nonlinearity defined by the expression

$$N = k_{N1}\left[x_1 - x_3 - p + abs(x_1 - x_3 - p)\right] \tag{51}$$

and specified by a threshold parameter p.

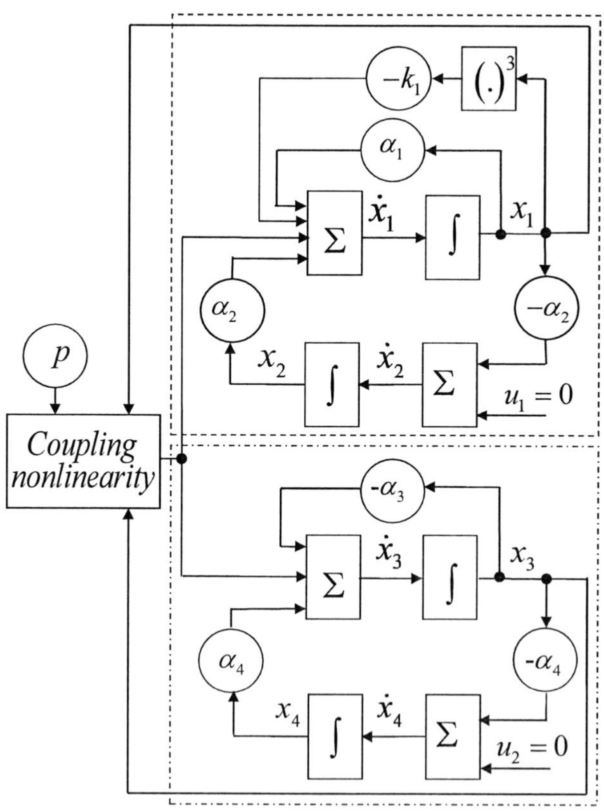

Figure 8. Structure of the state space representation.

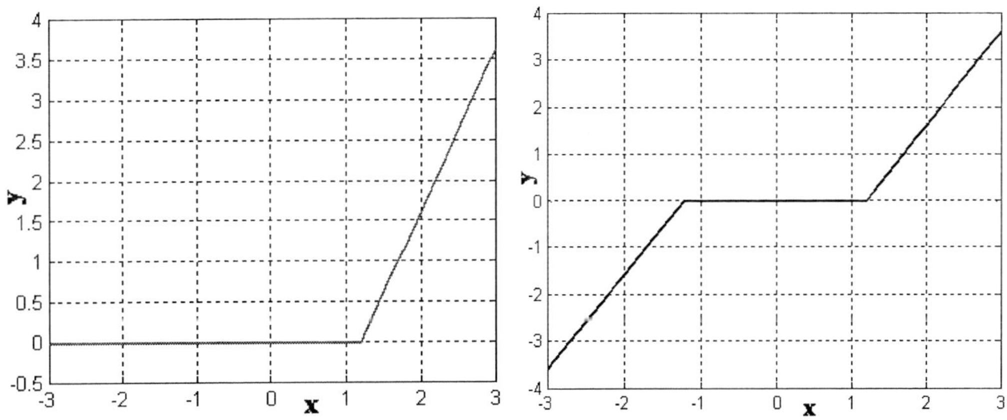

Figure 9. The nonlinear functions: Left – diode-like, right – dead zone-like nonlinearities.

The second one is the dead zone-like nonlinearity defined by the expression

$$\begin{aligned} N &= k_{N2}[(x_1 - x_3 - p) + abs(x_1 - x_3 - p) + (x_1 - x_3 + p) - abs(x_1 - x_3 + p)] = \\ &= k_{N2}[2(x_1 - x_3) + abs(x_1 - x_3 - p) - abs(x_1 - x_3 + p)] \end{aligned} \tag{52}$$

where the lower and upper limits of the dead zone are the same and can be fixed by a free parameter p.

Nonlinear system behavior with diode-like coupling nonlinearity is illustrated by typical results of computer simulation experiments, shown in the Figure 10, where a course of the individual *state vector components* is depicted as an example.

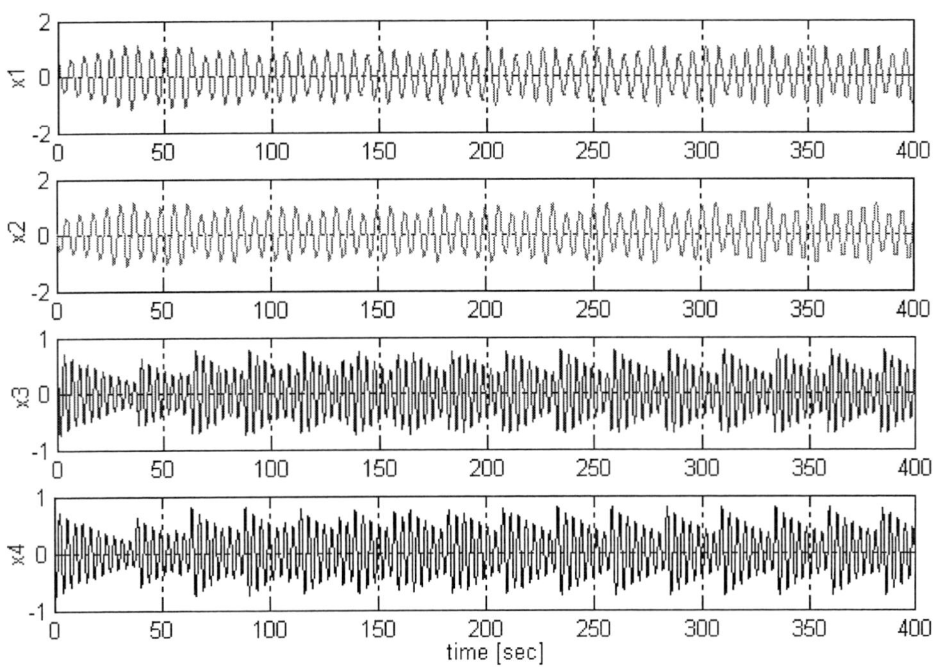

Figure 10. Typical example of the evolution of state variables under diode-like nonlinear coupling.

It is worthwhile to notice that the evolution of the first two state variables in the Figure 10 clearly shows the typical *irregularly repeating switching* intervals of the *anti-dissipative behavior* of the nonlinear subsystem, as demonstrated also in the upper part of the Figure 11. Similarly, the lower part of the Figure 11 illustrates the course of the 3-rd and the 4-th state variables, characterizing the *dissipative behavior* of the linear subsystem under diode-like nonlinear coupling.

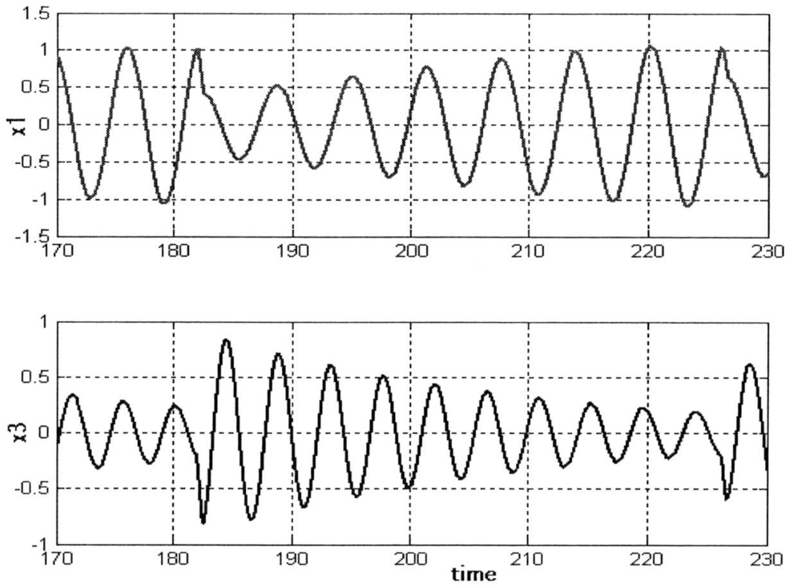

Figure 11. Typical dissipative vs. anti-dissipative behavior of both the isolated subsystems without coupling.

The *state space energy* of the system representation (50) for $x^*=0$ is given by eq. (3.34):

$$E = \frac{1}{2}\sum_{i=1}^{n} x_i^2 \tag{53}$$

where $n=4$ is the order of the system representation (4.1).

Mathematically, the *rate of change of the total state space energy* may be expressed by the *dual product* (35) of the *state space energy gradient field* and the *vector field of the state space velocity*:

$$\left.\frac{dE}{dt}\right|_{\dot{x}=f(x)} = \left.\langle grad_x E, \dot{x}\rangle\right|_{\dot{x}=f(x)} \tag{54}$$

Using eqn. (54) and the state equations (50) we obtain:

$$\left.\frac{dE}{dt}\right|_{\dot{x}=f(x)} = x_1\left[\alpha_1 x_1 - k_1 x_1^3 + \alpha_2 x_2\right] - \alpha_2 x_1 x_2 - \alpha_4 x_3 x_4 + x_3\left[-\alpha_3 x_3 + \alpha_4 x_4\right] - (x_1 + x_3)N(x_1, x_3, p) \tag{55}$$

and so we get

$$\left.\frac{dE}{dt}\right|_{\dot{x}=f(x)} = -P_0(t) = \alpha_1\left(1-\frac{k_1}{\alpha_1}x_1^2\right)x_1^2 - \alpha_3 x_3^2 - (x_1+x_3)N(x_1,x_3,p) \tag{56}$$

where $P_0(t)$ denotes instantaneous value of the *total output dissipation power* of the system including the non-linear coupling.

The *chaotic-like course of energy exchange* obtained by computer simulation for the diode-like coupling nonlinearity is shown in Figure 12 as an example.

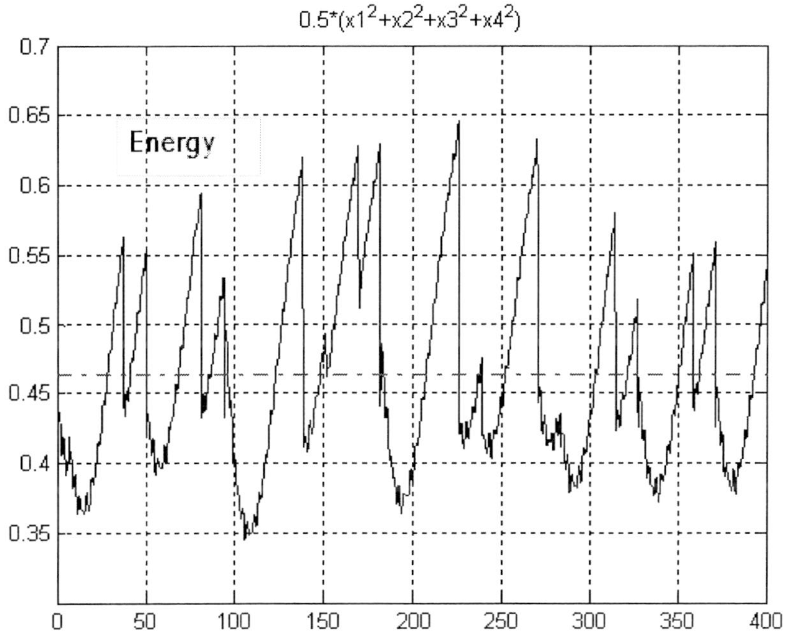

Figure 12. Chaotic-like evolution of the total state space energy

Recall *that the equilibrium states* of the system representation (50) are defined by the set of algebraic equations

$$\left.\begin{aligned} 0 &= \alpha_1 x_1 - k_1 x_1^3 + \alpha_2 x_2 - N(x_1,x_3,p) \\ 0 &= -\alpha_2 x_1 \\ 0 &= -\alpha_3 x_3 + \alpha_4 x_4 - N(x_1,x_3,p) \\ 0 &= -\alpha_4 x_3 \end{aligned}\right\} \tag{57}$$

where the *diode-like nonlinearity* N is described by (51), and the *dead zone-like nonlinearity* by (52).

The evolution of *the partial state space energy*, which illustrates the typical behavior of the *anti-dissipative subsystem* is shown in the upper part of the Figure 13, while in the lower part the corresponding energy changes in the *dissipative subsystem* are shown. In both the cases the *diode-like coupling* nonlinearity was considered.

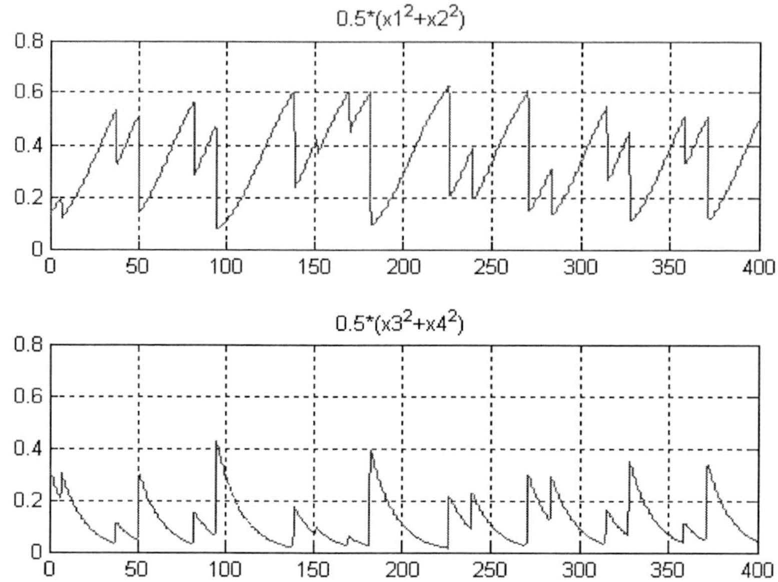

Figure 13. Evolution of the state space energy of the nonlinearly coupled subsystems – diode-like coupling.

Similar results, but for the *dead zone-like coupling nonlinearity,* are displayed in the Figure 14. It is worthwhile to notice that again the evolution of both *the partial state space energies* clearly shows the *irregularly repeating switching intervals, caused by the coupling nonlinearity.*

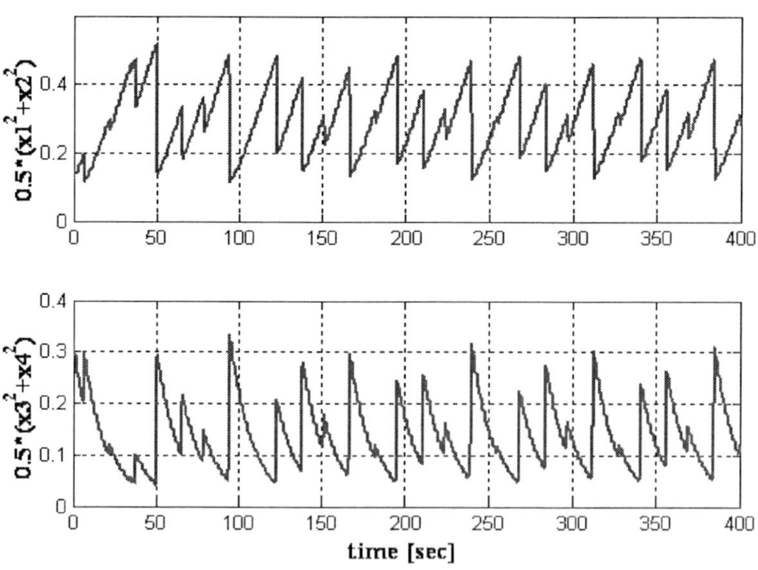

Figure 14. Evolution of the state space energy of the nonlinearly coupled subsystems – dead zone coupling.

It should be emphasized that the *computation* of an instantaneous value of *the state space energy*, as displayed in the Figure 12, may be based either on *direct substitution* into its definition (53), or by an *integration* using the eqn. (56). Conceptually, the second way is obviously more appropriate, because of *state space energy decomposition,* resulting from the corresponding *natural decomposition* of the *output dissipation power* into the following *three components:*

The negative taken value of the first term on the right hand side of (56)

$$P_{0,1}(t) = -\alpha_1\left(1-\frac{k_1}{\alpha_1}x_1^2\right)x_1^2,\ \alpha_1>0,\ k_1>0 \tag{58}$$

expresses the *component* of the instantaneous value of the total *output dissipation power* generated by *the nonlinear anti-dissipative subsystem.*

The negative taken value of the quadratic term on the right hand side of (56)

$$P_{0,2}(t) = \alpha_3 x_3^2 \tag{59}$$

determines the *component* of the instantaneous value of the total *output dissipation power* generated by *the isolated linear dissipation subsystem* and is always positive for $x_3 \in R$, $\alpha_3>0$, and $x_3 \neq x_3^* = 0$.

The last term on the right hand side of (56) depends on the *coupling nonlinearity.* For the *diode-like nonlinearity* (51) the last term in (56) reads:

$$\begin{aligned} P_N(t) &= (x_1+x_3)N(x_1,x_3,p) = k_{N1}(x_1^2-x_3^2)[1+sign(v(t))] \\ &\quad - pk_{N1}(x_1+x_3)[1+sign(v(t))] \end{aligned} \tag{60}$$

and by substitution into (56) we get the value of the *total output dissipation power $P_0(t)$* defined by

$$P_0(t) = P_{1,0}(t) + P_{2,0}(t) + P_N(t) \tag{61}$$

in the final form:

$$\begin{aligned} P_0(t) &= -\alpha_1\left[1-\frac{k_{N1}}{\alpha_1}[1+sign(v(t))]-\frac{k_1}{\alpha_1}x_1^2\right]x_1^2 + \alpha_3\left[1-\frac{k_{N1}}{\alpha_3}[1+sign(v(t))]-\frac{k_1}{\alpha_1}x_1^2\right]x_3^2 \\ &\quad + pk_{N1}[1+sign(v(t))](x_1+x_3) \end{aligned} \tag{62}$$

where

$$v(t) = x_1 - x_3 - p \Rightarrow 1+sign(v(t)) = \begin{cases} 0 \ \ldots x_1 \geq x_3 + p \\ 2 \ \ldots x_1 < x_3 + p \end{cases} \tag{63}$$

Alternatively, for the *dead zone-like nonlinearity* (52) the last term in (56) reads:

$$P_N(t) = 2k_{N2}(x_1^2 - x_3^2)[1 + sign(v_1) - sign(v_2)] \\ - pk_{N2}(x_1 + x_3)[sign(v_1) + sign(v_2)] \tag{64}$$

and by substitution into (60) the corresponding expression for the *total output dissipation power* reads

$$P_0(t) = -\alpha_1\left[1 - 2\frac{k_{N2}}{\alpha_1}[1 + sign(v_1) - sign(v_2)] - \frac{k_1}{\alpha_1}x_1^2\right]x_1^2 \\ +\alpha_3\left[1 - \frac{k_{N2}}{\alpha_3}[1 + sign(v_1) - sign(v_2)]\right]x_3^2 - pk_{N2}[sign(v_1) + sign(v_2)](x_1 + x_3) \tag{65}$$

where

$$v_1(t) = x_1 - x_3 - p\,,\; v_2(t) = x_1 - x_3 + p \tag{66}$$

It is the matter of fact that the state space energy of the isolated dissipative subsystem would monotonically decrease until the corresponding state would reach its equilibrium state.

It is easy to understand that an *irregular interruption* of this process is the *dominating effect of the coupling nonlinearity.* Similarly we could ask *why* the state space energy of the anti-dissipative subsystem does not go to infinity. It is quite plausible that the *cubic feedback nonlinearity* prevents such a situation and *holds the state variables bounded* in a finite region of the state space *X*.

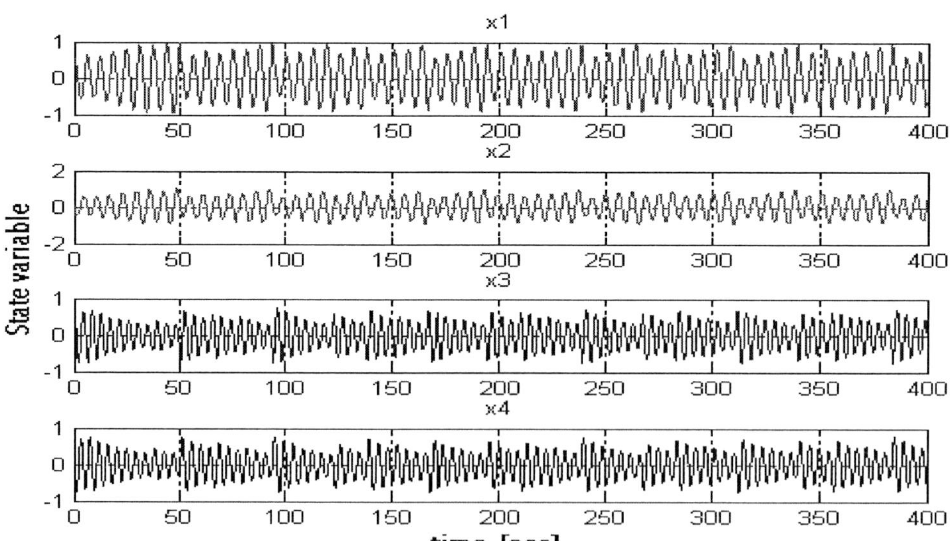

Figure 15. Typical example of the evolution of state variables under dead zone-like nonlinear coupling.

Nonlinear system behavior with dead zone-like coupling nonlinearity is illustrated by results of computer simulation experiments, shown in the Figure 15, where a course of the individual *state vector components* is depicted for comparison with the Figure 10.

The *chaotic-like course of energy exchange* obtained by computer simulation for the *dead zone-like* coupling nonlinearity is shown in Figure 16 for comparison with the Figure 12. It can be concluded that the *nonlinear dynamics* for both types of the coupling nonlinearity is *qualitatively the same.*

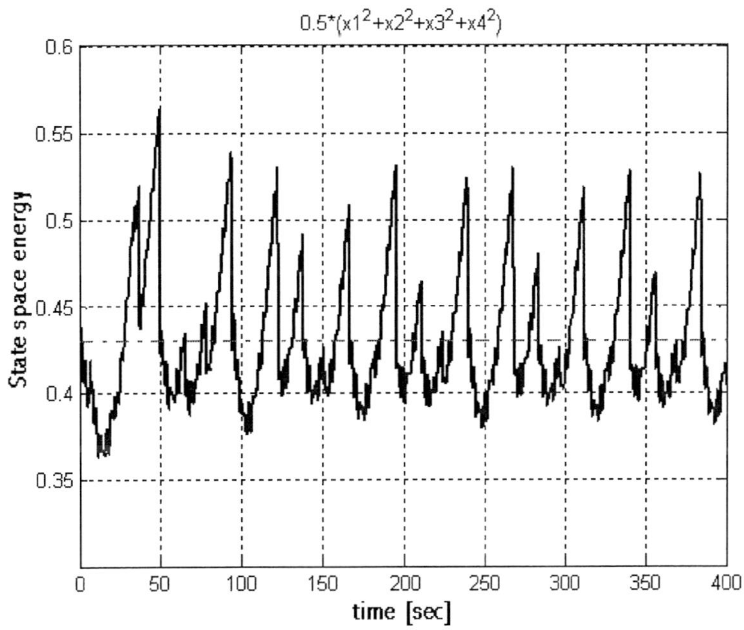

Figure 16. Chaotic-like evolution of the total state space energy for the dead zone nonlinearity.

For the equilibrium states it follows from the general expression (57) for the *diode-like nonlinearity*:

$$\left.\begin{aligned}&\alpha_1\left(1-\frac{k_1}{\alpha_1}x_1^2\right)x_1+\alpha_2x_2=w(t)\\&\alpha_2x_1=0\\&-\alpha_3x_3+\alpha_4x_4=w(t)\\&\alpha_4x_3=0\end{aligned}\right\}\tag{67}$$

where

$$\begin{aligned}&w(t)=k_{N1}v(t)\left[1+sign(v(t))\right]\\&v(t)=x_1-x_3-p\end{aligned}\tag{68}$$

and thus it follows

$$w(t) = \begin{cases} 0 & \ldots v(t) \leq 0 \\ 2k_{N1}v(t) & \ldots v(t) > 0 \end{cases} \tag{69}$$

Summarizing, we get the conditions for equilibrium with the coupling nonlinearity (51) in form:

$$\begin{aligned} &\alpha_2 \neq 0 \Rightarrow x_1^* = 0 \\ &\alpha_4 \neq 0 \Rightarrow x_3^* = 0 \\ &\alpha_4 \neq 0 \Rightarrow x_4^* = \frac{1}{\alpha_4} w(t) \\ &\alpha_2 \neq 0 \Rightarrow x_2^* = \frac{1}{\alpha_2} w(t) \end{aligned} \tag{70}$$

and hence for $v(t) \leq 0$ the *unique zero equilibrium state* $x^*=0$ follows, while for $v(t)>0$ we get

$$x_1^* = 0;\ x_2^* = 2\frac{k_{N1}}{\alpha_2},\ x_3^* = 0;\ x_4^* = 2\frac{k_{N1}}{\alpha_4} \tag{71}$$

It follows from the eqn. (58) that the state space energy accumulated in the *isolated linear* dissipative subsystem decreases monotonically until it reaches the zero value, and simultaneously till the last two state variables reach their zero equilibrium states. Analogously, it follows from the eqn. (57) that *near the equilibrium state* $x_1^* = 0$, the *isolated nonlinear* subsystem would behave as *locally unstable,* or *anti-dissipative,* while for $k_1 > 0$ and $x_1(t)$ *far from the equilibrium* the increased accumulated state space energy will be *bounded by a very strong nonlinear dissipation.*

Notice that the dissipation power $P_{0,1}(t)$ of the isolated nonlinear subsystem vanishes not only in the zero equilibrium point $x_1(t) = x_1^* = 0$, but also in the equilibrium states for which it holds:

$$|x_1(t)| = \sqrt{\frac{\alpha_1}{k_1}} \tag{72}$$

Recall that for analysis of situation with *dead zone-like nonlinearity* the same technique could be used. In order to confirm the *expected chaotic nature* of the system trajectories mathematically, the well known *method of Liapunov exponents,* as a broadly used standard tool, has been used. Hence the four Liapunov exponents have been computed and displayed in the Figure 17.

Because two of computed Liapunov exponents are positive, not only the expected chaotic behavior has been confirmed, but *even hyperchaotic dynamics* had to be considered as proven.

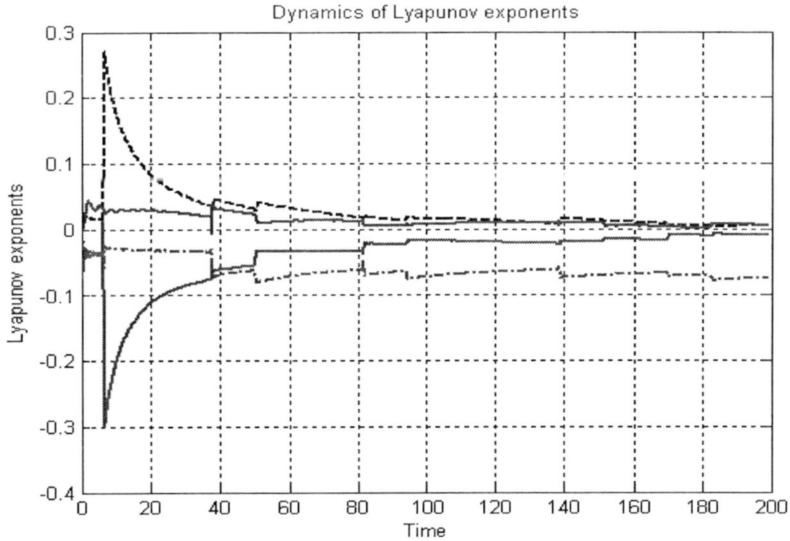

Figure 17. Typical course of the Liapunov exponents.

The chaotic nature of the state space trajectories was observed in variety of computer simulation experiments. As an example of chaotic behavior of the state space trajectories, typical results experimentally obtained by computer simulations are illustrated by 2-D projection in the Figure 18.

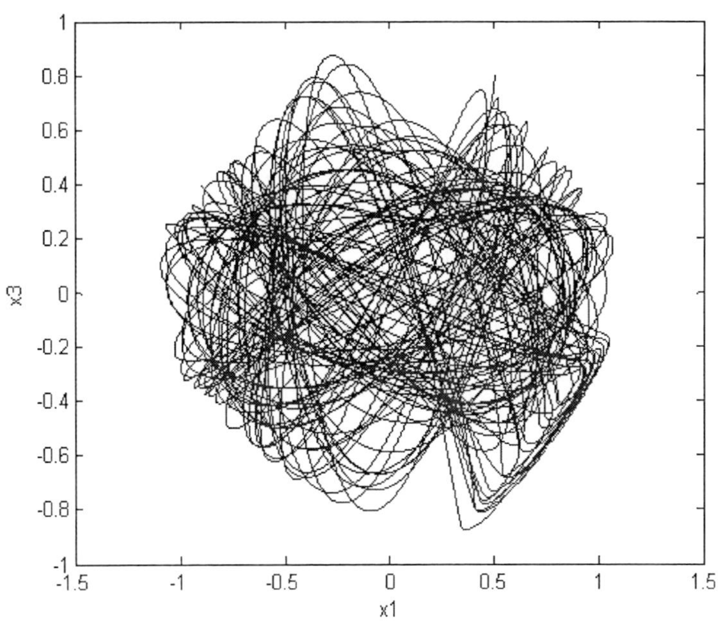

Figure 18. Simulated state space trajectories in the 2-D projection.

In order to be able to verify the theoretical results by a real-world system realization, the state equivalent electronic system implementation has been designed and build for laboratory experiments. Its circuit structure is shown in the Figure 19. The coupling nonlinearity is connected between A-B terminals.

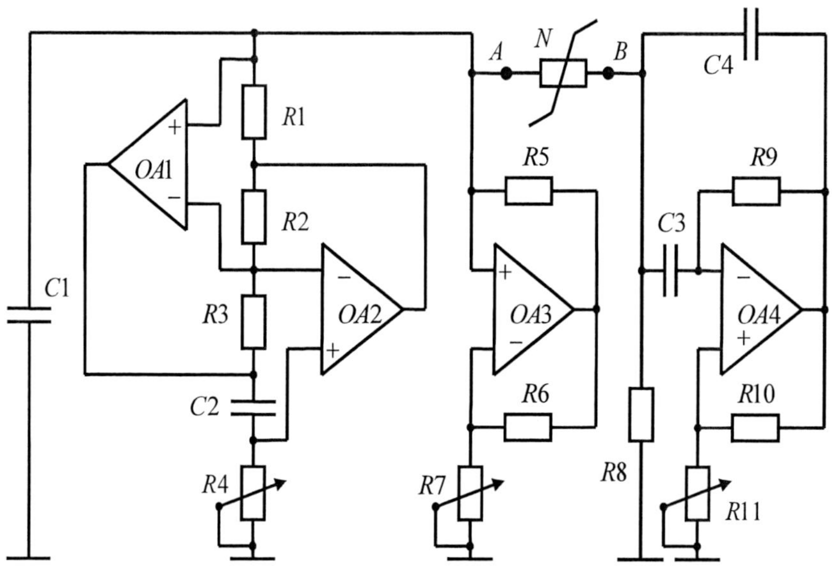

Figure 19. The detailed circuit diagram.

The *anti-dissipative and dissipative nature,* as well as the *effects of the irregular switching* of the circuit state variables, obtained experimentally by real voltage measurement at the points *A and B* on the laboratory circuit implementation is demonstrated in the Figure 20. The 2 channels of waveforms measured on terminals A and B are shown in the Figure 20.

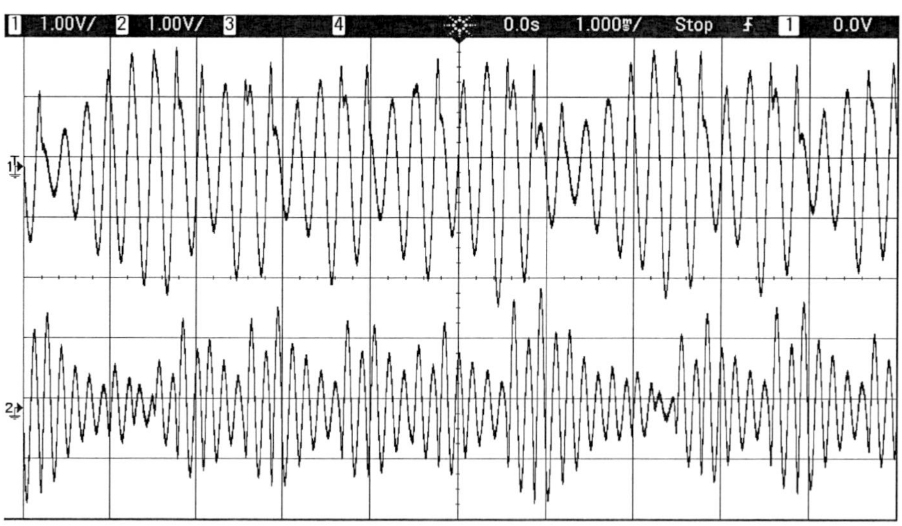

Figure 20. Measured state space trajectories - waveforms measured on terminals A (top) and B (bottom).

The corresponding chaotic nature of the state space trajectories was observed in variety of laboratory experiments. As a typical example of such behavior, observed by measurement on the laboratory circuit shown in the Figure 19 is documented by experimentally obtained results, illustrated by 2-D projection in the Figure 21 for comparison (with the Figure 18 - simulated state trajectories).

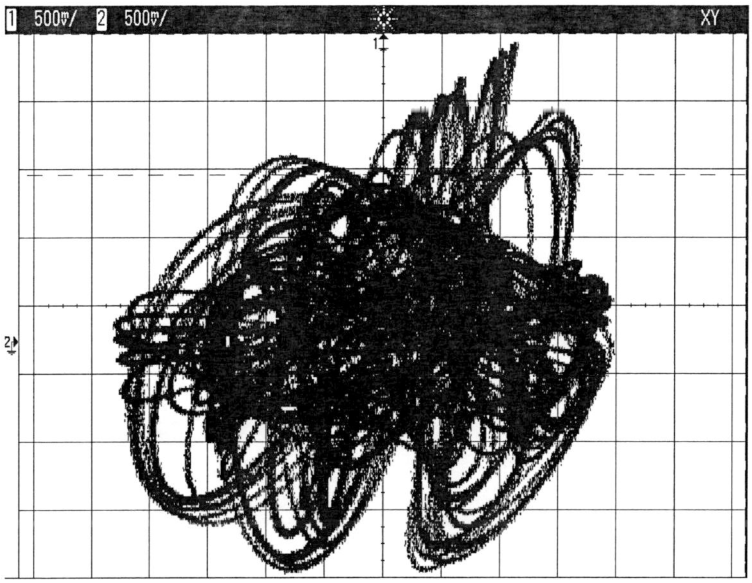

Figure 21. Measured state space trajectories in the 2-D projection - *measured on terminals A, B.*

The electronic circuit consists of 4 JFET operational amplifiers (OA) in one package (TL074). The OA1 and OA2 were used for synthetic inductor. The impedance of the circuit is given by [13]:

$$Z(s) = \frac{sC_2 R_1 R_3 R_4}{R_2} \tag{73}$$

The synthesis inductor and the C_1 is parallel resonance circuit. The OA3 is used as negative impedance converter for negative resistance realization with input resistance (until saturation of OA output):

$$R_{inp} = -\frac{R_5}{R_6} R_7 \tag{74}$$

The rest, circuit around OA4 presents bandpass filter with gain controlled by R_{11}. Used electronic components are specified by the following values of parameters:

C1=C2=C3=C4=33 nF,

R1=R2=R3=1 k, R4=10 kΩ, R5=R6=10 kΩ,

R7=47kΩ, R8=680 Ω, R9=1k5, R10=M12, R11=M33.

Potentiometer R_4 is used for *frequency change of first oscillator,* potentiometer R_7 for the *slope of negative resistance* and potentiometer R11 for *dissipation value of the 2-nd oscillator* [14].

The simulation results and experimental electronic circuit measuring gives good agreement. The theoretical results based on the state space energy approach have been confirmed by electronic circuit construction. This circuit with different modifications can be used in variety of practical applications in which *effects of nonlinear energy exchange* can't be neglected.

As typical application fields especially *chaotic synchronization* in *secured communication systems* and *modern control systems for power electrical networks in non-standard regimes* (such as nonlinear dynamics near blackouts), can be expected in the near future.

CONTROLLED DISSIPATION STRUCTURES AND MECHANISM OF NONLINEAR SYNCHRONIZATION

Recall that according to *Liouville's theorem* of vector analysis, dissipative systems have the important property that any *volume in the state space* strictly *decreases under the action of the system flow.* For a class of nonlinear causal systems with the *state velocity* vector given by a vector field *f*, the property of the (volume-based) *dissipativity* is accordingly defined by:

$$div\, f(x) = \sum_{i=1}^{n} \frac{\partial f_i(x)}{\partial x_i} < 0 \tag{75}$$

Nonlinear systems having at an equilibrium state a dissipative approximate linearization are *locally dissipative* in a vicinity of this equilibrium state, but need not to be globally dissipative.

The *theoretical limiting case* with vanishing divergence:

$$div\, f(x) = 0 \tag{76}$$

implicitly defines a *class of causal systems preserving volume* along *state space trajectories* and thus the corresponding vector fields *f* are naturally referred to as *conservative.*

Let a *class of controlled dissipation structures* is described by a *causally separable nonlinear* system representation $\Re\{S\}$ defined by (16) with the state equation *in semilinear form*

$$\dot{x}(t) = f[x(t)] + Bu(t),\ x(t_0) = x^0, \tag{77}$$

with an *uncontrolled state velocity vector field* given by (3.3) in the form:

$$f(x) = A(x)x \tag{78}$$

where the matrix *A(x), determining the vector field* f, is considered as given. Similarly the matrix B in *the controlled part of the state velocity vector field*

$$v = Bu(t) \tag{79}$$

is assumed to be given.

Example 2: Let's assume that the right hand side of the state equation (77) is given in form (50):

$$\left.\begin{aligned} \dot{x}_1 &= \alpha_1 x_1 - k_1 x_1^3 + \alpha_2 x_2 - N(x_1, x_3, u) \\ \dot{x}_2 &= -\alpha_2 x_1 \\ \dot{x}_3 &= -\alpha_3 x_3 + \alpha_4 x_4 - N(x_1, x_3, u) \\ \dot{x}_4 &= -\alpha_4 x_3 \end{aligned}\right\} \approx \dot{x}(t) = f[x(t), u] \tag{80}$$

with the *output equation* defined by

$$y = x_1 - x_3 \tag{81}$$

where *N*(x_1,x_3,,*u*) represents the dead zone-like *coupling nonlinearity* defined by the expression (52),

$$N(x_1, x_3, u) = k_{N2}\{2y + (y - p).[sign(y - p) - sign(y + p)]\}, \quad p = u \tag{82}$$

and where the *dead zone coupling parameter* p is considered as a *constant control input signal u(t).*

Before starting a *systematic theoretical analysis,* let`s assume that proper values of the *gain parameter* k_{N2}, as well as *the dead zone* parameter $p = u$, characterizing the *nonlinear coupling interaction,* have been *chosen for computer simulation experiments.*

In order to clear the role of coupling parameters, the *total energy evolution* obtained by simulation for *external coupling* with parameters $k_{N2} = 1.3$ and $p = 1.2$, is displayed in the Figure 22.

Remark 4: Notice that for $p = u = 0$ the *dead zone reduces to zero*, and consequently the *coupling nonlinearity N reduces to a linear coupling*, and for $k_{N2} = 0$ the *coupling N* is *completely vanishing.*

Remark 5: Notice that the value of the *interaction parameter* k_{N2} determines the *intensity* of the *external interaction* between the two second order subsystems, while the *value* of the *control signal u* representing the value of the *interaction parameter p*, may be interpreted as a *measure of nonlinearity*.

The components of the *total state space energy* corresponding to both the subsystems are shown in the next Figure 23 for comparison. It is worthwhile to notice the *monotonicity of the state space energy decrease* in the top of the Figure 23, representing *behavior of the strictly dissipative subsystem.*

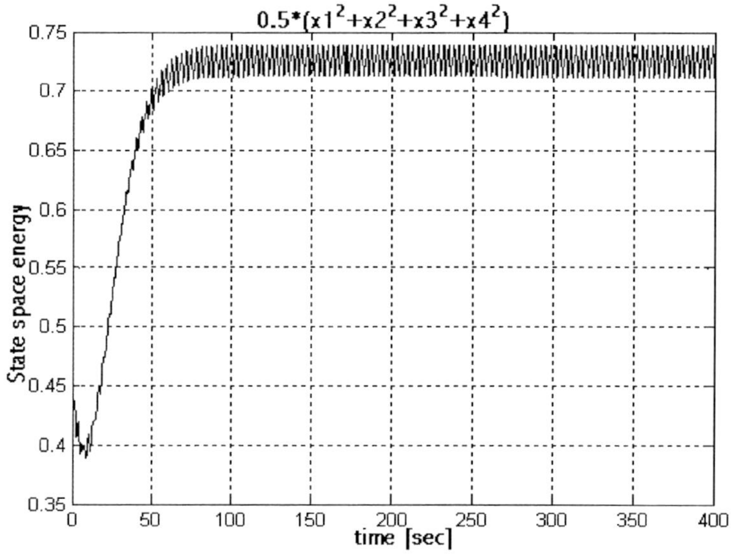

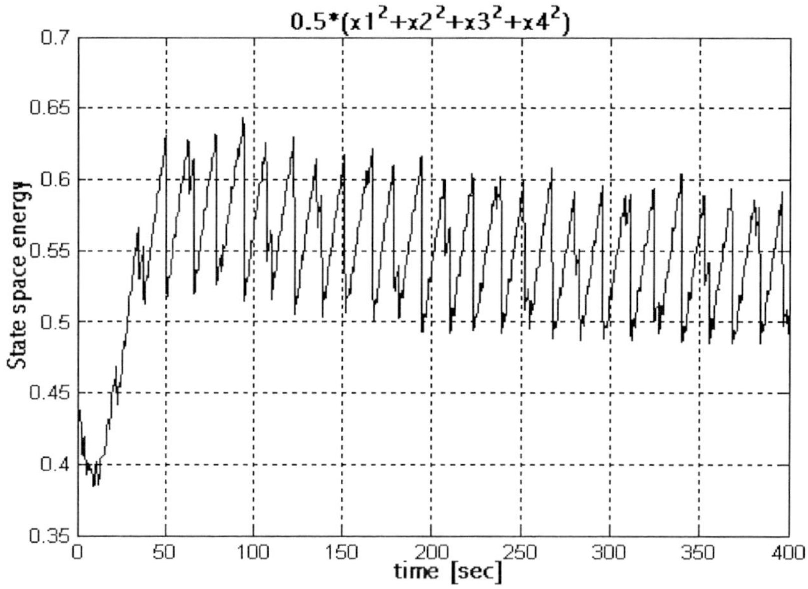

Figure 22. Evolution of the total state space energy: a) top - for vanishing external coupling $k_{N2} = 0$, b) bottom - for nonlinear external coupling $k_{N2} = 1.3$, u = 1.2.

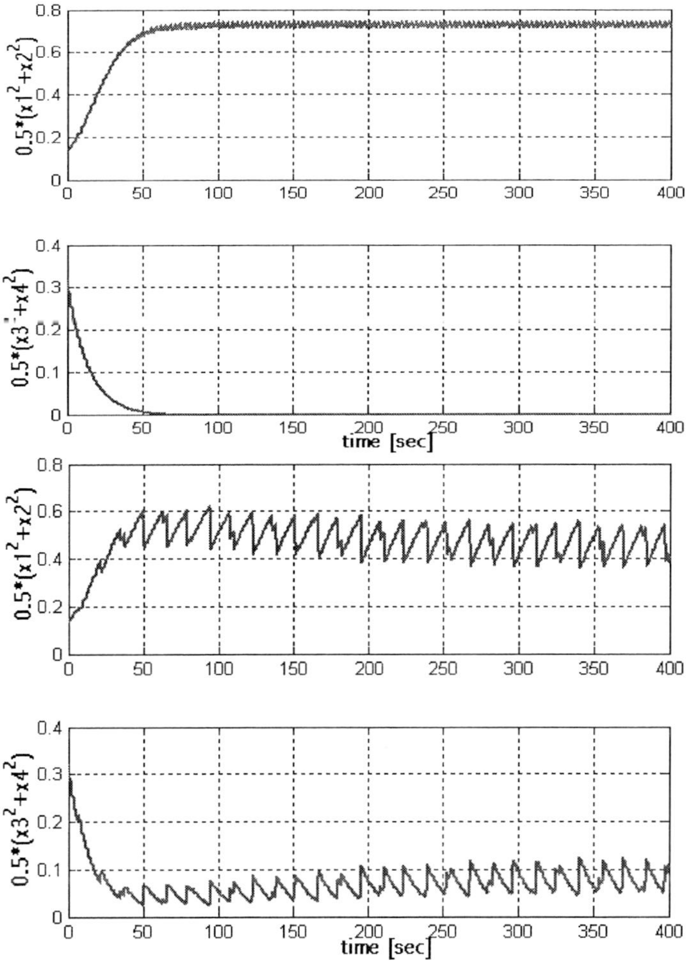

Figure 23. Evolution of the state space energy for individual subsystems: a) top - for vanishing external coupling $k_{N2} = 0$, b) bottom - for nonlinear external coupling $k_{N2} = 1.3$, $u = 1.2$.

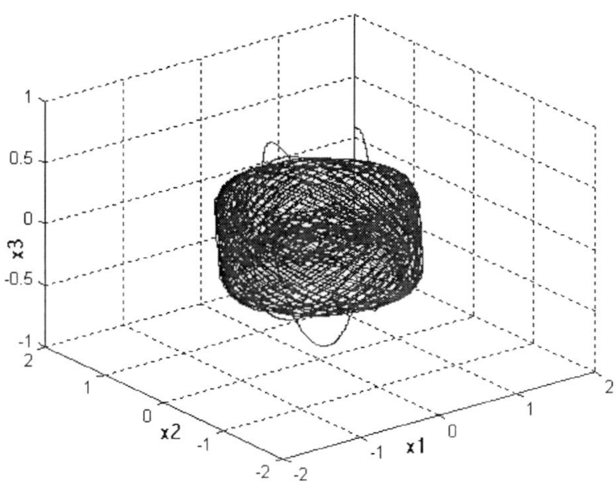

Figure 24. State space trajectories for nonlinear coupling $k_{N2} = 1.3$, $u = 1.2$ in the 3-D projection.

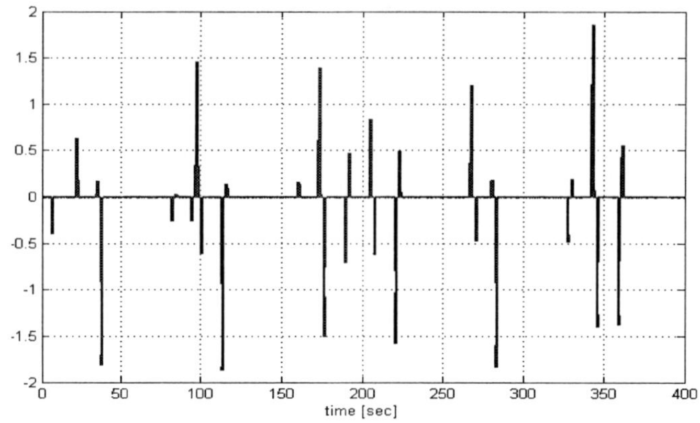

Figure 25. The course of the coupling signal generating the exchange power for external coupling k_{N2} = 5 , u=1.2.

Typical nonlinear behavior of the state space trajectories, obtained experimentally by computer simulations for the same parameter values is illustrated by 3-D projection in the Figure 24. The evolution of the corresponding *coupling signal* is illustrated by the next Figure 25.

The corresponding *exchange power* between these *strongly coupled subsystems* is shown in the right hand part of the Figure 26. The course of the *exchange state space energy,* generated by *radically changed value* of the *coupling parameter* k_{N2}, is shown in the left hand part of the Figure 26.

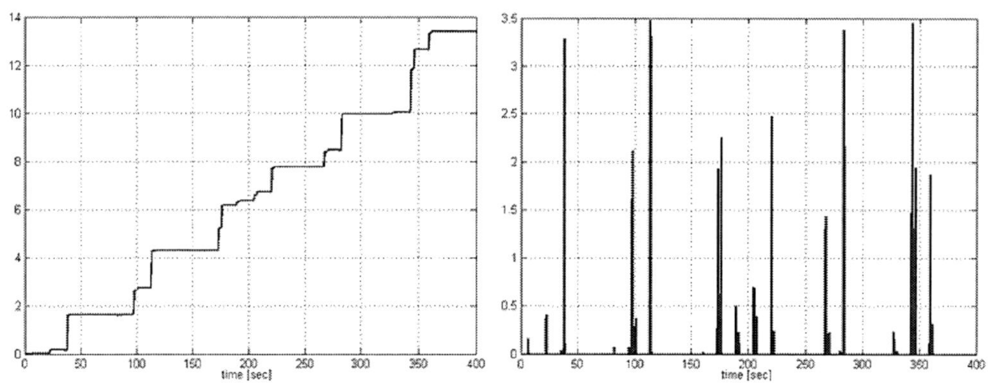

Figure 26. Evolution for k_{N2} = 5 , u =1.2 of: a) the exchange energy, left b) the exchange power, right.

The next three figures illustrate the same quantities for *very weakly coupled subsystems* for choice of the coupling parameter u = 0, i.e *for linear coupling*. The evolution of the corresponding *coupling signal* is shown in the Figure 27. The course of the corresponding *exchange power* between both the weakly coupled subsystems is shown in the right hand side of the Figure 28. and the *exchange state space energy* for the *changed value* of the parameter k_{N2}, is shown in the left part of the Figure 28.

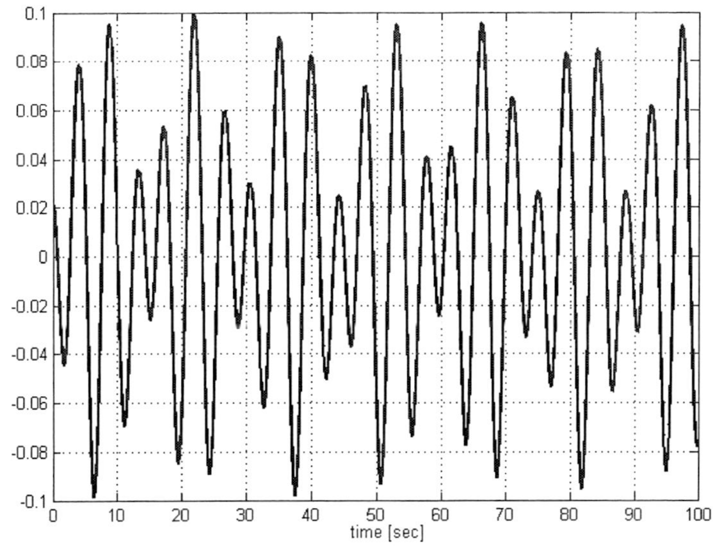

Figure 27. The course of the coupling signal generating the exchange power for parameters $k_{N2} = 0.035$, u = 0.

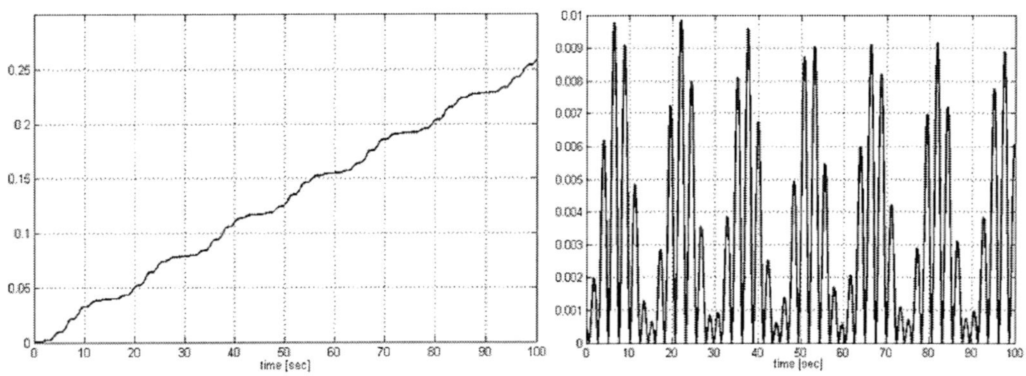

Figure 28. Evolution of the exchange energy and the exchange power for linear coupling $k_{N2} = 0.035$, u = 0.

It is worthwhile to notice the *quantum-like increase* of the exchange state space energy as observed in the Figure 26., and *compare* it with the *smooth monotonous increase* of the *exchange state space energy in the linear case,* as displayed in the Figure 28.

Recall that the value of the *control parameter* u=0 reduces the nonlinearty (82) to *linear coupling interaction*

$$N = 2k_{N2}y = 2k_{N2}(x_1 - x_3) \tag{83}$$

Thus a *linear special case* of the *nonlinear state space representation* (80), (81) follows

$$\left.\begin{aligned}\dot{x}_1 &= \alpha_1 x_1 - k_1 x_1^3 + \alpha_2 x_2 - 2k_{N2} y\\ \dot{x}_2 &= -\alpha_2 x_1\\ \dot{x}_3 &= -\alpha_3 x_3 + \alpha_4 x_4 - 2k_{N2} y\\ \dot{x}_4 &= -\alpha_4 x_3\end{aligned}\right\} \tag{84}$$

$$y = x_1 - x_3 \tag{85}$$

and hence the equivalent form of the state equations (84) follows

$$\left.\begin{aligned}\dot{x}_1 &= \alpha_1\left[1 - \frac{k_1}{\alpha_1} x_1^2 - \frac{2k_{N2}}{\alpha_1}\right] x_1 + \alpha_2 x_2 + 2k_{N2} x_3\\ \dot{x}_2 &= -\alpha_2 x_1\\ \dot{x}_3 &= -2k_{N2} x_1 - \alpha_3\left[1 - \frac{2k_{N2}}{\alpha_3}\right] x_3 + \alpha_4 x_4\\ \dot{x}_4 &= -\alpha_4 x_3\end{aligned}\right\} \tag{86}$$

From the definition (74) the *state space volume based dissipativity condition* in the form:

$$div\, f(x) = \sum_{i=1}^{4} \frac{\partial f_i}{\partial x_i} = \alpha_1 - \alpha_3 - 3k_1 x_1^2 < 0 \tag{87}$$

Using the state equations (86) and computing the corresponding partial derivatives the *divergence of the state space velocity vector field,* given by the eqn. (87) can be obtained by a straightforward way:

$$\left.\begin{aligned}\frac{\partial f_1}{\partial x_1} &= \alpha_1\left[1 - \frac{2k_{N2}}{\alpha_1}\right] - 3k_1 x_1^2\\ \frac{\partial f_1}{\partial x_2} &= 0\\ \frac{\partial f_3}{\partial x_3} &= -\alpha_3\left[1 - \frac{2k_{N2}}{\alpha_3}\right]\\ \frac{\partial f_4}{\partial x_4} &= 0\end{aligned}\right\} \tag{88}$$

Using the *semilinear system description* (77), the *state velocity vector field* given by (76) can be equivalently represented in the form:

$$f(x) = A(x)x \tag{89}$$

with the matrix *A(x)* as follows

$$A = \begin{bmatrix} \alpha_1\left[1 - \frac{k_1}{\alpha_1}x_1^2\right] - 2k_{N2} & \alpha_2 & 2k_{N2} & 0 \\ -\alpha_2 & 0 & 0 & 0 \\ -2k_{N2} & 0 & -\alpha_3 + 2k_{N2} & \alpha_4 \\ 0 & 0 & -\alpha_4 & 0 \end{bmatrix} \tag{90}$$

Remark 6: Notice that the *volume based dissipativity condition* (87) is *invariant* not only with respect to the *internal interaction parameters* α_2, α_4, but even with respect to the *external coupling parameter* k_{N2}.

Remark 7: Notice that the *interaction parameter* k_{N2} determines the *intensity* of an *external linear control interaction* between the two subsystems. The other elements of the skew-symmetric matrix A_A, i.e. the parameters α_2, α_4 define the *internal interactions* of both the subsystems.

The derived *topological structure* corresponding to the matrix (90) with the eqs. (85) and (94), or equivalently, to the eqns. (84), (85), (86), is displayed in the Figure 29.

The *decomposition* of the system *state space representation* (85), (86) from the *state space energy transfer* point of view is *induced* by the standard *decomposition of the matrix A(x)* into its *symmetric* and *skew-symmetric* parts

$$A(x) = A_S(x) + A_A(x) \tag{91}$$

as follows:

$$A_S = \begin{bmatrix} -2k_{N2} + \alpha_1\left[1 - \frac{k_1}{\alpha_1}x_1^2\right] & 0 & 0 & 0 \\ 0 & 0 & 0 & 0 \\ 0 & 0 & -\alpha_3 + 2k_{N2} & 0 \\ 0 & 0 & 0 & 0 \end{bmatrix}$$

$$A_A = \begin{bmatrix} 0 & \alpha_2 & 2k_{N2} & 0 \\ -\alpha_2 & 0 & 0 & 0 \\ -2k_{N2} & 0 & 0 & \alpha_4 \\ 0 & 0 & -\alpha_4 & 0 \end{bmatrix} \quad (92)$$

The elements of the symmetric matrix $A_S(x)$ characterize the dissipation power components $P_{0,1}(t)$, $P_{0,2}(t)$ of the total dissipation power $P_0(t)$ for both the linearly coupled subsystems, respectively:

$$P_0(t) = -\alpha_1\left[1 - \frac{2k_{N2}}{\alpha_1} - \frac{k_1}{\alpha_1}x_1^2\right]x_1^2 + \alpha_3\left[1 - \frac{2k_{N2}}{\alpha_3}\right]x_3^2 > 0 \quad (93)$$

where *for dissipativity the inequality has to be valid almost everywhere.*

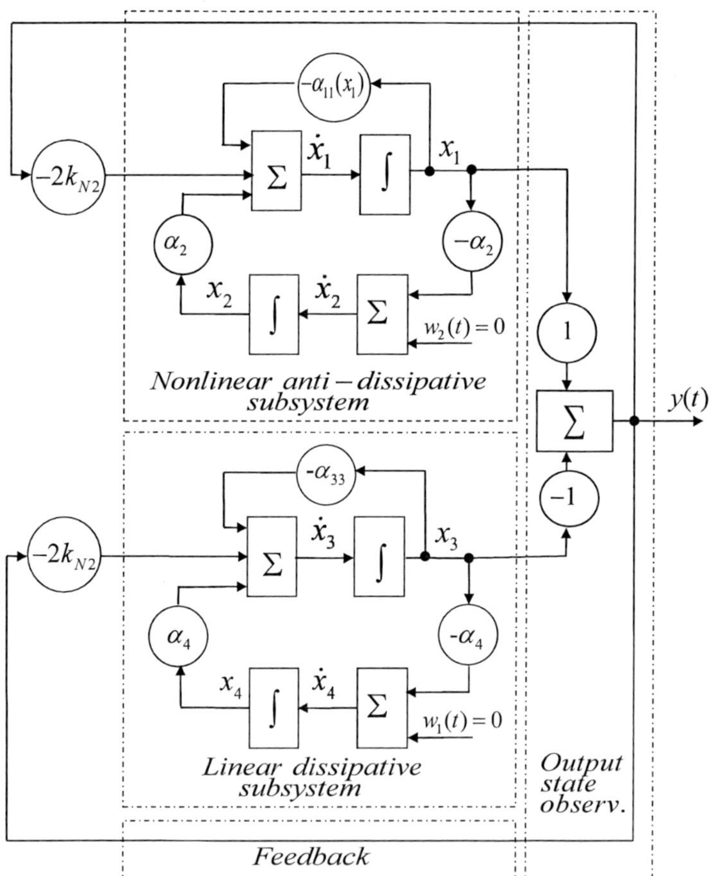

Figure 29. Topological structure of the controlled dissipation structure given by the eq. (85)-(86).

The eq. (93) obviously provides *all the information* needed to derive *necessary and sufficient conditions of the state space energy based dissipativity.* It follows from the eq. (93) that the following two inequalities

$$\alpha_{11}(x_1) = -\alpha_1\left[1 - \frac{2k_{N2}}{\alpha_1} - \frac{k_1}{\alpha_1}x_1^2\right] > 0\ ,\ \alpha_{33} = \alpha_3\left[1 - \frac{2k_{N2}}{\alpha_3}\right] > 0 \tag{94}$$

represent the *state space energy based dissipativity conditions,* which have to be *compared* with the eq.(87)), representing the *standard state space volume based dissipativity criterion.*

At first, let`s assume that the *total dissipation power $P_0(t)$* is required to be *concentrated* in the *first subsystem.* It follows that the *partial dissipation power $P_{0,2}(t)$ generated* by the *linear subsystem*

$$P_{0,2}(t) = \alpha_3\left[1 - \frac{2k_{N2}}{\alpha_3}\right]x_3^2 > 0 \text{ with } \alpha_3 > 0 \tag{95}$$

has to be *completely eliminated by an output feedback.* For the *parameter k_{N2}* of a *linear coupling* we get:

$$\left[1 - \frac{2k_{N2}}{\alpha_3}\right] = 0 \ \Rightarrow\ k_{N2} = \frac{1}{2}\alpha_3\ , \Leftrightarrow\ P_0(t) = P_{0,1}(t) = (-\alpha_1 + \alpha_3)x_1^2 + k_1x_1^4 > 0 \tag{96}$$

and, as a consequence, the state equations (84) *reduce* to the form

$$\begin{aligned}\dot{x}_1 &= (\alpha_1 - \alpha_3)x_1 + \alpha_2x_2 + \alpha_3x_3 - k_1x_1^3\\ \dot{x}_2 &= -\alpha_2x_1\\ \dot{x}_3 &= -\alpha_3x_1 + \alpha_4x_4\\ \dot{x}_4 &= -\alpha_4x_3\end{aligned} \tag{97}$$

Thus *for linear* case $k_1=0$ we conclude from (87) and (96) *that the dissipation power* satisfies:

$$P_0(t) = (-\alpha_1 + \alpha_3)x_1^2 = -[\operatorname{div} f(x)]x_1^2 > 0 \Leftrightarrow \operatorname{div} f(x) < 0 \tag{98}$$

and consequently *both the dissipativity conditions* are *equivalent in the linear case.*

Alternatively, let's assume that the *total dissipation power $P_0(t)$* is required to be *concentrated* in the *second subsystem.*

It follows that the *partial dissipation power $P_{0,1}(t)$*, defined by the expression

$$P_{0,1}(t) = -\alpha_1 \left[1 - \frac{2k_{N2}}{\alpha_1} - \frac{k_1}{\alpha_1} x_1^2 \right] x_1^2 > 0 \text{ with } \alpha_1 > 0 \tag{99}$$

and generated by the *anti-dissipative nonlinear subsystem* has to be *completely eliminated by an output feedback.* For the parameter k_{N2} of the *nonlinear feedback coupling* it follows:

$$\left[1 - \frac{2k_{N2}}{\alpha_1} - \frac{k_1}{\alpha_1} x_1^2 \right] = 0 \Rightarrow k_{N2} = \frac{1}{2}[\alpha_1 - k_1 x_1^2] \text{ , } \Leftrightarrow P_0(t) = P_{0,2}(t) = (-\alpha_1 + \alpha_3)x_3^2 + k_1 x_1^2 x_3^2 > 0 \tag{100}$$

and, as a consequence, the state equations (84) reduce to the form

$$\begin{aligned} \dot{x}_1 &= \alpha_2 x_2 + (\alpha_1 - k_1 x_1^3) x_3 \\ \dot{x}_2 &= -\alpha_2 x_1 \\ \dot{x}_3 &= -(\alpha_1 - k_1 x_1^3) x_1 - [\alpha_3 - \alpha_1 + k_1 x_1^3] x_3 + \alpha_4 x_4 \\ \dot{x}_4 &= -\alpha_4 x_3 \end{aligned} \tag{101}$$

Thus *for linear* case $k_1 = 0$ we conclude from (87) and (100) *that the dissipation power* satisfies:

$$P_0(t) = (-\alpha_1 + \alpha_3)x_3^2 = -[div\, f(x)]x_3^2 > 0 \Leftrightarrow div\, f(x) < 0 \tag{102}$$

and again *both the dissipativity conditions* are proven to be *equivalent in the linear case.*

In the *nonlinear* case $k_1 > 0$ we conclude from (102) that *the total dissipation power satisfies*:

$$P_0(t) = (-\alpha_1 + \alpha_3)x_1^2 + k_1 x_1^4 > 0 \tag{103}$$

whereas the *volume based dissipativity criterion* in the nonlinear case for $k_1 > 0$ requires the condition

$$div\, f(x) = (\alpha_1 - \alpha_3) - 3k_1 x_1^2 < 0 \tag{104}$$

and from (100) it follows that *the total dissipation power satisfies*:

$$P_0(t) = (-\alpha_1 + \alpha_3)x_3^2 + k_1 x_1^2 x_3^2 > 0 \tag{105}$$

whereas the *volume based dissipativity criterion* in the nonlinear case for $k_1 > 0$ requires the condition

$$div\, f(x) = (\alpha_1 - \alpha_3) - k_1 x_1^2 < 0 \tag{106}$$

Similarly results hold in *nonlinear* case for $k_1>0$, too. Hence *both the dissipativity approaches* are *consistent in linear as well as in nonlinear case.* As an example, a *controlled nonlinear dissipation structure,* demonstrating the *essence of a synchronization effect*, was *synthesized* and *studied* using the *state space energy approach* and *verified by computer simulations.*

Remark 8: It is easy to deduce that the total *state space energy exchange* is completely *independent* of the *interactions* defined by the *constant skew-symmetric interaction matrix* $A_A(x)$. Hence it seems natural to call any such coupling as *the neutral interaction.*

Remark 9: In contrary to that, the *nonlinear external interactions* between the both subsystems, characterized by the coupling nonlinearity $N(x_1, x_3, p)$, are *not neutral,* and as such can *cause dramatic changes* in the evolution of the total state space energy, (even the chaotic dynamics, as demonstrated in the Sec. 4.).

Typical *course of the state space trajectories,* demonstrating a typical *mechanism of the nonlinear synchronization*, is shown by 3-D projection in the Figure 30, and further illustrated by the 2-D projections in the Figure 31.

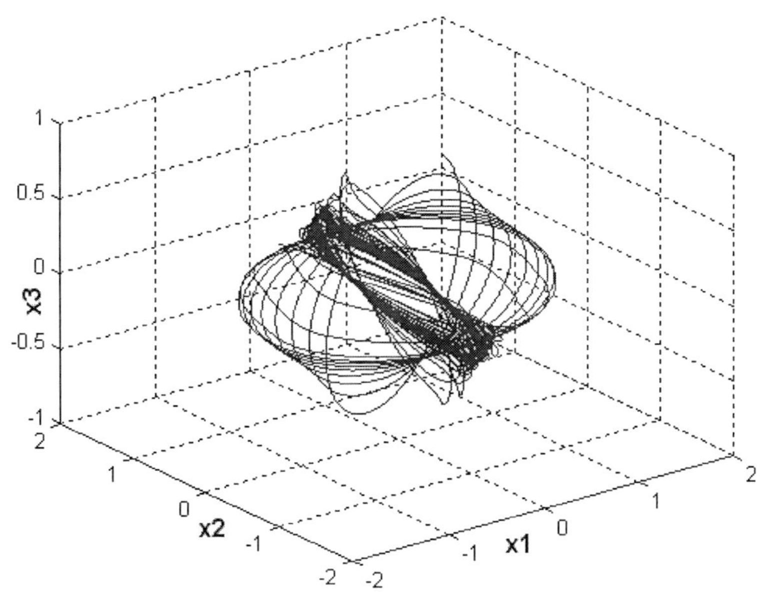

Figure 30. State space trajectories in the 3-D projection for nonlinear coupling $k_{N2} = 1.3$, p =1.2.

For illustration, the *total energy evolution* obtained by simulation for *external coupling* with parameters $k_{N2}= 1.3$ and $p = 1.2$, is displayed in the left part of the Figure 32. The components of the *total state space energy* corresponding to both the subsystems are shown in the right hand side of the Figure 33. Next three figures illustrate the evolution of the coupling variables. It is worthwhile to notice the *asymptotic* course *of the state space energy decrease* in the Figure 34, demonstrating the typical *asymptotic behavior of strictly dissipative systems inclusive of their subsystems.*

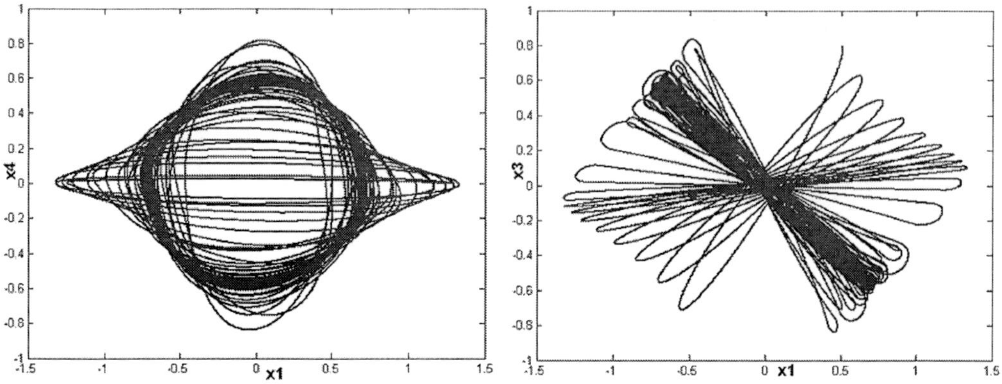

Figure 31. State space trajectories in the 2-D projection for nonlinear coupling k_{N2} = 1.3, p =1.2

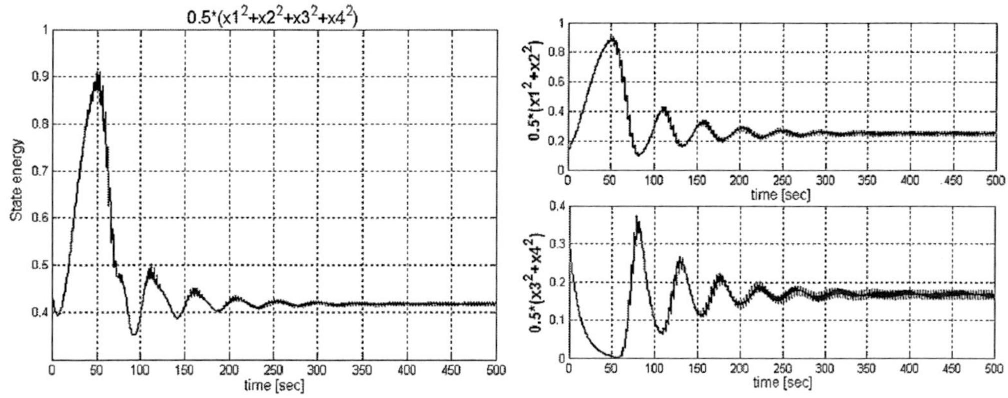

Figure 32. Total state space energy (left) and partial energies of subsystems for parameters k_{N2}=1.3, p = 1.2.

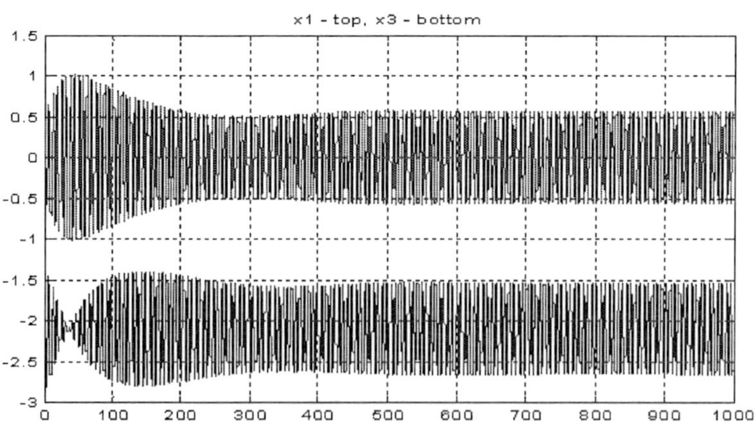

Figure 33. Typical example of the evolution of state variables under nonlinear coupling

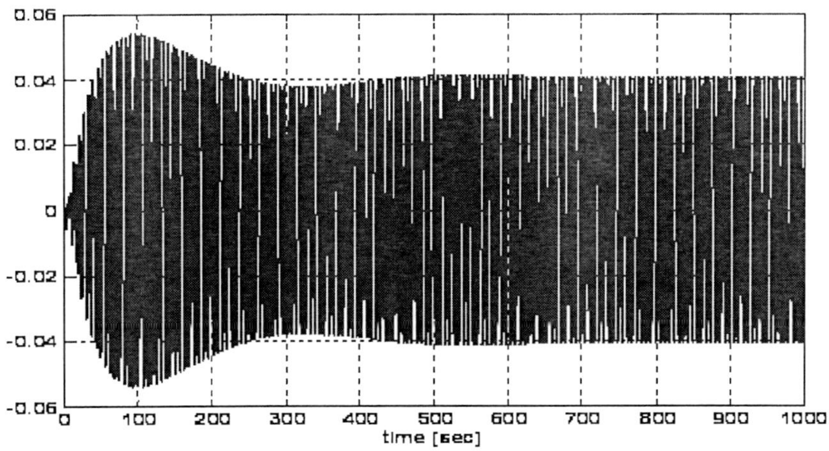

Figure 34. The course of the coupling signal generating the exchange power for parameters $k_{N2} = 0.035$, p=0.

It is easy to see from the Figure 31 and Figure 35 that the asymptotic behavior of synchronized dissipation structure tends to a conservative system. Let's now return to more general dissipativity problems:

$$\Re\{S\} :: \dot{x}(t) = f[x(t)] + Bu(t),\ x(t_0) = x^0, \tag{107}$$

$$y(t) = C\,x(t) \tag{108}$$

and with an uncontrolled state velocity vector field given by :

$$f(x) = A(x)x \tag{109}$$

Let the immediate value of the output signal power be defined:

$$P(t) = \|y(t)\|^2, \tag{110}$$

and the total value of the state space energy, accumulated in a time instance t in the state x(t) be given by:

$$E(t) = \frac{1}{2}\|x(t)\|^2, \tag{111}$$

Putting $u(t) = 0,\ \forall t \ge t_0$ and assuming validity of the state space energy conservation principle in the following differential form

$$\frac{\mathrm{d}\,E(x)}{\mathrm{d}\,t} = -\,P(t) \tag{112}$$

we get by computing the derivative of the state space energy E[x(t)] along the state representation of the given system a signal power balance relation in the following explicit Liapunov equation - like matrix form

$$\frac{dE(x)}{dt} = \frac{1}{2} x^T(t)[A(.) + A^T(.)]x(t) = -\|y(t)\|^2 \tag{113}$$

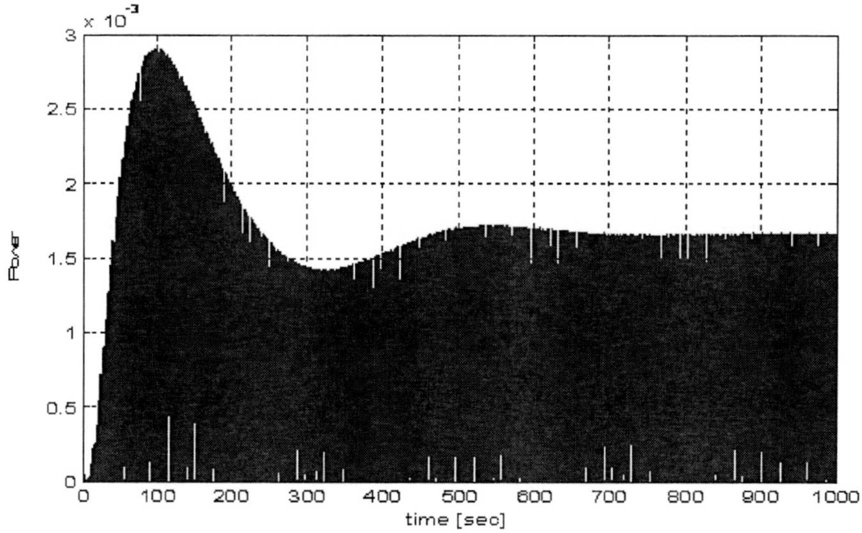

Figure 35. Evolution of the exchange power for subsystems with external coupling $k_{N2} = 5$, p=0.

Remark 10: In contrast to the standard physical concept of system energy, the proposed state space energy concept represented by a class C_2 mapping E: X $\rightarrow$ R^+ is defined independently of any physical variables or parameters having any specific physical meaning. In contrary, it depends exclusively on the abstract system structure and on the corresponding state space topology.

It has been shown in [2, 4] that a special form of a structurally dissipative state equivalent system representation called dissipation normal form exists and its internal algebraic structure can be described by the following matrix:

$$\frac{dE(x)}{dt} = \frac{1}{2} x^T(t)[A(.) + A^T(.)]x(t) = -\|y(t)\|^2 \tag{114}$$

Remark 10: It is important to emphasize that any of the internal or external power-informational interactions may be time-varying or nonlinear with respect to state or input variables.

Remark 11: It follows directly from the expression (113) that the output dissipation power P[x(t)] is unambiguously determined by the symmetric part $A_S(x)$ defined by

$$A_S(x) = \frac{1}{2}[A(x) + A^T(x)] \tag{115}$$

and is completely independent of the skew-symmetric part:

$$A_A(x) = \frac{1}{2}[A(x) - A^T(x)] \tag{116}$$

of the matrix $A(x)$ defined by

$$A(x) = A_S(x) + A_A(x) \tag{117}$$

Remark 12: Modification of the power balance relation and the state space energy conservation principle for the general case of non-vanishing input signals is straightforward.

Remark 13: An abstract form of state space energy conservation principle for a proper chosen equivalent state space representation follows by integration. Hence, in case of zero input:

$$u(t) = 0, t \geq t_0 \tag{118}$$

the total energy accumulated in the system in time t_0 must be equal to an amount of the energy dissipated on the interval $[t_0, \infty)$ by the vector output signal given by the expression:

$$E(t_0) = \int_{t_0}^{\infty} \|y(t)\|^2 \, dt \tag{119}$$

or, in case of scalar output signal, alternatively, by the expression:

$$E(t_0) = \int_{t_0}^{\infty} [y(t)]^2 \, dt \tag{120}$$

The topological structure corresponding to the algebraic structure defined by the matrix (5.40) is displayed in the Figure 36.

It is important to realize that the topological structure of the system representation derived from the algebraic structure (114) is strongly over-parameterized. In many cases it may be reasonable to remove all the redundant diagonal elements of the matrix $A_S(x)$, and concentrate the total dissipation power into a single (scalar) observed output signal y(t).

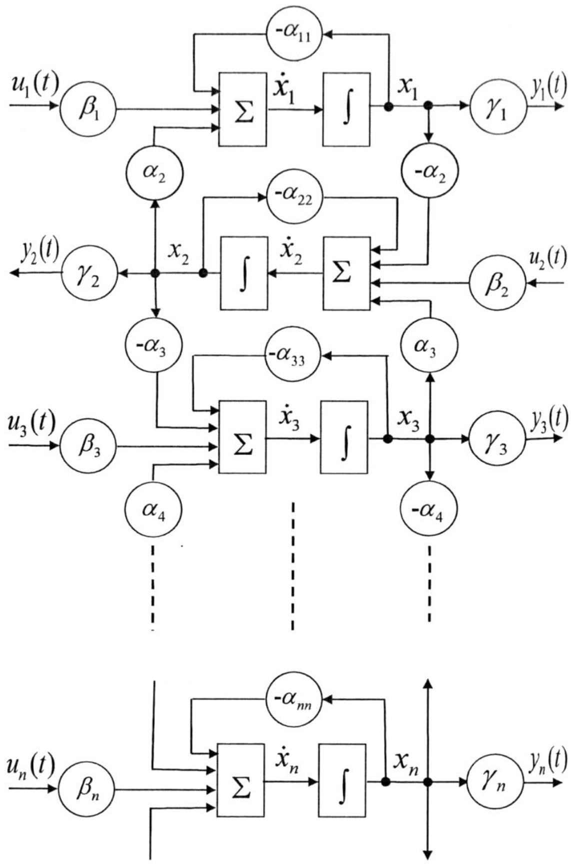

Figure 36. Topological structure of the dissipation normal form with n inputs and with n output channels.

In such a special case of parameter minimal system representation arises, and the corresponding internal algebraic structure of the state space representation reduces to:

$$A = \begin{bmatrix} -\alpha_1 & \alpha_2 & 0 & 0 & 0 & 0 \\ -\alpha_2 & 0 & \alpha_3 & 0 & 0 & 0 \\ 0 & -\alpha_3 & 0 & \ddots & 0 & 0 \\ 0 & 0 & \ddots & \ddots & \alpha_{n-1} & 0 \\ 0 & 0 & 0 & -\alpha_{n-1} & 0 & \alpha_n \\ 0 & 0 & 0 & 0 & -\alpha_n & 0 \end{bmatrix} \tag{121}$$

The correspondingly modified topological structure of the parametrically minimal state space representation is displayed in the Figure 37.

Thus any parametrically minimal system representation defined by a matrix (121) is dissipative if and only if the matrix A has negative trace, i.e. if and only if the single parameter α_1 has a positive value, $\alpha_1>0$.

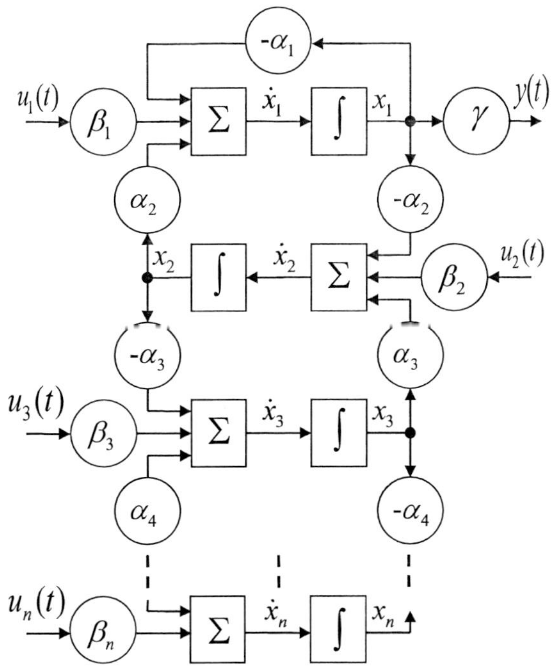

Figure 37. Parametrically minimal structure of the dissipation normal form with single output dissipation channel.

The evolution of state space energy *E[x(t)]* for some typical cases is shown in the next Figure 38.

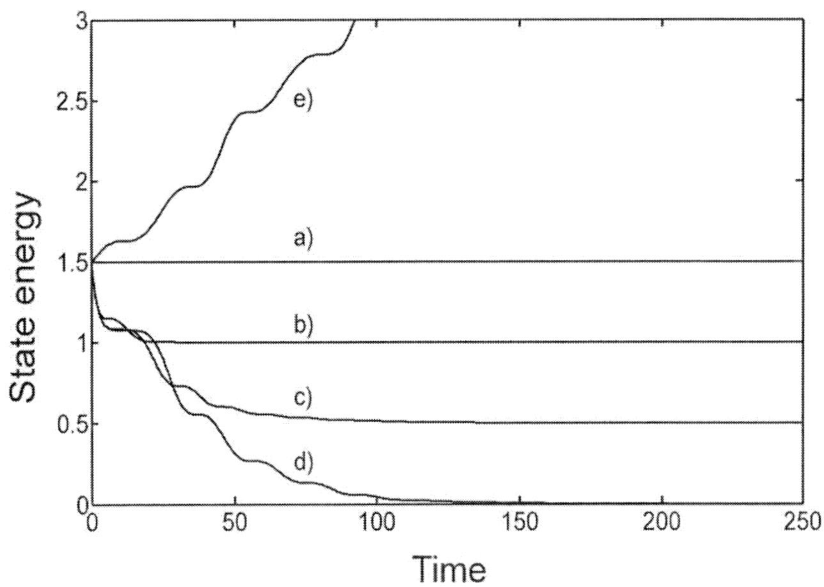

Figure 38. Time evolution of state space energy E[x(t)] for n=6, a) conservative case $\alpha_1 = 0$, b) dissipativity $\alpha_1 > 0$, $\alpha_3 = 0$, c) dissipativity $\alpha_1 > 0$, $\alpha_5 = 0$, d) asymptotic stability $\alpha_1 > 0$, $\alpha_k \neq 0$, k = 2,3,…,n, e) instability $\alpha_1 < 0$.

State Dimension Minimality and State Energy Additivity

Let's consider continuous-time linear time-varying strictly causal state space system representation:

$$\left.\begin{aligned} \Re\{S\}: \frac{dx(t)}{dt} &= A(t)x(t) + B(t)u(t), \quad \dim x(.) = n \\ y(t) &= C(t)x(t), \quad \dim u(t) = r \dim y(.) = p \end{aligned}\right\} \tag{122}$$

where $1 \leq r \leq n$, $1 \leq p \leq n$, and the matrices A(.), B(.), C(.) are supposed to be known. Assume that the given system S has the asymptotic stability property, and that its representation $\Re\{S\}$ has the minimal order n, i.e. it is controlable and observable. In such a case the controlability and observability Grammian matrices $W_c(t)$ and $W_0(t)$ exist, are symmetric and positive definite, and satisfy the following two linear matrix Liapunov-like ordinary differential equations [1]:

$$\left.\begin{aligned} A(t)W_c(t) + W_c(t)A^T(t) + \frac{dW_c(t)}{dt} &= -B(t)B^T(t) \\ A^T(t)W_o(t) + W_o(t)A(t) + \frac{dW_o(t)}{dt} &= -C^T(t)C(t) \end{aligned}\right\} \tag{123}$$

Remark 15: Any such representation induces an equivalence class of minimal, controllable, observable and asymptotic stable state equivalent system representations given by the conditions of state equivalence in the form

$$\bar{A}(t) = \left[T(t)A(t) + \frac{dT(t)}{dt} \right] T^{-1}(t) \tag{124}$$

$$\bar{B}(t) = T(t)B(t) \tag{125}$$

$$\bar{C}(t) = C(t)T^{-1}(t) \tag{126}$$

generated by the group of linear time-varying state equivalence transformations:

$$\bar{x}(t) = T(t)x(t), \qquad x(t) = T^{-1}(t)\bar{x}(t) \tag{127}$$

Remark 16: It is a well known fact that in the linear time-invariant case the Grammian matrices $W_c(t)$ and $W_0(t)$ are closely related to the input and output signal energy functions, respectively [2], [4]. In previous part it has been shown that the state minimality property of any linear strictly causal system representation is equivalent to the property of positive definiteness of both the observability and controllability Grammian matrices if these ones are well defined, i.e. if the representation has the property of asymptotic stability.

From the mathematical correctness point of view there is no reason to require such a property in general case. Hence there is a need to separate the minimality conditions from the asymptotic stability conditions. It is well known that the minimality of linear time-invariant system representations can instead of Grammians be tested independently of stability property using rank conditions of the controllability and observability matrices H_c and H_o defined by the expressions

$$H_c = \left[B,\ AB,\ A^2B,\ \ldots,\ A^{n-1}B\right] \tag{128}$$

$$H_o = \left[C^T,\ A^TC^T, (A^T)^2C^T,\ \ldots,\ (A^T)^{n-1}C^T\right] \tag{129}$$

It is easy to show that for any two state equivalent state representations the following relations hold

$$\overline{H}_c = T\ H_c \tag{130}$$

$$\overline{H}_o = (T^{-1})^T\ H_o \tag{131}$$

and can be used for determination of the state transformation (127)

$$T = \overline{H}_c H_C^{-1} \tag{132}$$

or alternatively

$$T = (\overline{H}_o^{-1})^T\ H_O^T \tag{133}$$

if the structure and parameters of both the equivalent representations are known.

Consider the time-invariant version of the system representations given by Eqn. (122). The finite controllability Grammian matrix $W_c(t)$ at time t is defined as follows:

$$W_C(t) = \int_0^t e^{A\tau}.B.B^T.e^{A^T\tau}\,d\tau \tag{134}$$

and has two important properties [4].

First, the controllability Grammian matrix $W_c(t)$ is symmetric and nonnegative definite

$$W_c^T(t) = W_c \geq 0 \tag{135}$$

and secondly, the columns of $W_c(t)$ span the controllable space, i.e. the image of $W_c(t)$ equals to the image of the controllability matrix H_c :

$$\mathrm{Im}\left[W_c(t)\right] = \mathrm{Im}\left[H_c(A,B)\right] \tag{136}$$

It is easy to show that the state vector defined by the chosen structure of matrices A, B is controllable if and only if $W_c(t)$ is positive definite for some $t > 0$.

If the controllability Grammian matrix is invertible, then the minimum energy input signal u(.) exists and the minimum input signal energy corresponding to the state transfer from the zero state to x(t) is known to be given by

$$\text{Energy}\{u(.)\} = x^T(t)W_c^{-1}x(t), \qquad t > 0 \tag{137}$$

Similarly, using the observability Grammian matrix $W_o(t)$, the output signal energy at time t caused by the initial state $x(t_0)$ is known to be given by the quadratic form:

$$\text{Energy}\{y(.)\} = x^T(t_0)W_o(t)x(t_0), \qquad t_0 < t \tag{138}$$

where the finite observability Grammian $W_o(t)$ at time $t < +\infty$ is defined by the expression [1]

$$W_o(t) = \int_{t_0=0}^{t} e^{A^T\tau}.C^T.C.e^{A\tau}\mathrm{d}\tau \tag{139}$$

and has two important properties.

First, the observability Grammian matrix $W_c(t)$ is symmetric and nonnegative definite

$$W_o^T(t) = W_o(t) \geq 0 \tag{140}$$

Secondly, for positive values of t, the kernel of finite observability Grammian matrix is equal to the kernel of observability matrix H_o

$$\mathrm{Ker}\left[W_o(t)\right] = \mathrm{Ker}\left[H_o(A,C)\right] \tag{141}$$

In the case of standard infinite time observability Grammian, i.e. for t → ∞ the largest observation energy produced by any given value of initial state $x(t_0)$ is given by

$$E_o\left[x(t_0)\right] = x^T(t_0)W_o x(t_0) \tag{142}$$

assuming the couple (A, C) is observable and A is such that asymptotic stability conditions are satisfied. In such case the well known special form of Liapunov's equation

$$A^T W_o + AW_o = -C^T C \tag{143}$$

which in fact expresses a form of the state-output energy transfer balance relation, can be used for determination of the unknown output energy function Eo(t).

In the dual case of standard infinite time controllability Grammian, i.e. for t → ∞ the minimal value of the input energy required to transfer the zero initial state to any state $x(t_1)$ is given by

$$E_I\left[x(t_0)\right] = x^T(t_1).W_c^{-1}.x(t_1) \tag{144}$$

assuming the couple (A,B) is controllable and A is such that asymptotic stability conditions are satisfied. In such a case the well known special form of Liapunov's equation

$$AW_c + A^T W_c = -BB^T \tag{145}$$

which in fact expresses a form of the input-state energy transfer balance relation, can be used as a tool for determination of the unknown input energy function EI(t).

Remark 17: In general, both the energy functions above are generated by symmetric positive semi-definite matrices, which depend heavily on the chosen system representation structure.

From physically motivated energy additivity principle (EAP):

$$E\left[x(.)\right] = \sum_{i=1}^{n} E_i\left[x_i(.)\right], \quad x_i(.) \in R \tag{146}$$

it follows that only such system representation structures can be accepted as physically correct, which are not in contradiction with energy aditivity principle, or equivalently for which the Grammian matrices, generated by state equivalent triplets of matrices {A, B, C}, are diagonal and non-singular, i.e., for which it holds:

$$W = W_o \text{ or } W_c \Leftrightarrow W = \text{diag}\left\{\delta_1, \delta_2, ..., \delta_n\right\}, \ \delta_i \neq 0 \tag{147}$$

ASYMPTOTIC STABILITY ANALYSIS VIA THE STATE SPACE ENERGY CONCEPT

In this part mainly some *elementary consequences* of *dissipativity* and the *state space energy-metric approach* for stability analysis will be presented. We prove that the set of *real parameters* satisfy the following *conditions:*

1. $\alpha_1 > 0$ is necessary and sufficient for dissipativity
2. $\alpha_1 < 0$ is *sufficient for structural instability*
3. $\alpha_1 = 0$ is *necessary and sufficient for conservativity*
4. $\alpha_1 > 0$ is *necessary for asymptotic stability*
5. $\forall\, i,\, i \in \{2,3,\ldots,n\} : \alpha_i \neq 0, \alpha_1 \neq 0$ is equivalent to *parametric and state minimality*
6. $\forall\, i,\, i \in \{2,3,\ldots,n\} : \alpha_i \neq 0, \beta_1 \neq 0$ is equivalent to *structural controlability*
7. $\forall\, i,\, i \in \{2,3,\ldots,n\} : \alpha_i \neq 0, \gamma \neq 0$ is equivalent to *structural observability*
8. $\forall\, i,\, i \in \{2,3,\ldots,n\} : \alpha_i \neq 0, \alpha_1 > 0$ is equivalent to *structural asymptotic stability*

Example 3: In order to demonstrate the way how the proposed concept may be used, let us consider a simple linear time-invariant system for n = 6, and for the input signal u(t)=0, t > t_0, represented by the linear differential equation with constant coefficients

$$y^{(6)} + a_1 y^{(5)} + \ldots + a_4 \ddot{y}(t) + a_5 \dot{y}(t) + a_6 y(t) = 0 \tag{147}$$

with characteristic polynomial

$$A(s) = s^6 + a_1 s^5 + a_2 s^4 + \ldots + a_4 s^2 + a_5 s + a_6 \tag{148}$$

Now, let the parameters $a_1, a_2, \ldots, a_n$ be considered as unknown, and specify the region of asymptotic stability in the parameter space. The state space representation having the required structure (112) is described by:

$$\begin{aligned} \Re(S):\quad & \dot{x}_1(t) = -\alpha_1 x_1(t) + \alpha_2 x_2(t) \\ & \dot{x}_2(t) = -\alpha_2 x_1(t) + \alpha_3 x_3(t) \\ & \dot{x}_3(t) = -\alpha_3 x_2(t) + \alpha_4 x_4(t) \\ & \dot{x}_4(t) = -\alpha_4 x_3(t) + \alpha_5 x_5(t) \\ & \dot{x}_5(t) = -\alpha_5 x_4(t) + \alpha_6 x_6(t) \\ & \dot{x}_6(t) = -\alpha_6 x_5(t) \\ & y(t) = \gamma_1 x_1(t) \end{aligned} \tag{149}$$

Before we use the expression for the state space energy function in the reduced form

$$E(t)=\frac{1}{2}\left\|x(t)\right\|^2, \tag{150}$$

i.e. for the zero input and zero equilibrium state defined by the conditions

$$u(t)=0, t\in[t_0,\infty): f(x^*)=0 \Leftrightarrow x^*=0, \dot{x}(t)|_{x-x^*}=0 \tag{151}$$

Thus, we have to check the necessary and sufficient conditions for existence and uniqueness of the equilibrium state $x^*=0$. From the eq. (149) the following set of linear algebraic equations is obtained:

$$\begin{aligned}
0&=-\alpha_1 x_1(t)+\alpha_2 x_2(t)\\
0&=-\alpha_2 x_1(t)+\alpha_3 x_3(t)\\
0&=-\alpha_3 x_2(t)+\alpha_4 x_4(t)\\
0&=-\alpha_4 x_3(t)+\alpha_5 x_5(t)\\
0&=-\alpha_5 x_4(t)+\alpha_6 x_6(t)\\
0&=-\alpha_6 x_5(t)
\end{aligned} \tag{152}$$

The necessary and sufficient condition for existence and unique equilibrium state $x^*=0$, resulting from the solution of the set of algebraic equations above read:

$$\alpha_2\neq 0, \alpha_4\neq 0, \alpha_6\neq 0 \tag{153}$$

Remark 19: It means that from the existence of a unique equilibrium state point of view, the dissipation parameter α_1, as well as interaction parameters α_3, α_5 can be arbitrary.

Alternatively, using the algebraic structure of the matrix A in the dissipation normal form for n=6

$$A=A_6=\begin{bmatrix}-\alpha_1 & \alpha_2 & 0 & 0 & 0 & 0\\ -\alpha_2 & 0 & \alpha_3 & 0 & 0 & 0\\ 0 & -\alpha_3 & 0 & \alpha_4 & 0 & 0\\ 0 & 0 & -\alpha_4 & 0 & \alpha_5 & 0\\ 0 & 0 & 0 & -\alpha_5 & 0 & \alpha_6\\ 0 & 0 & 0 & 0 & -\alpha_6 & 0\end{bmatrix} \tag{154}$$

we could get the conditions (153) from the equivalent invertability condition of the matrix A:

$$\det A_6 = a_6 = \alpha_2^2 \alpha_4^2 \alpha_6^2 \neq 0, \tag{155}$$

Remark 20: In general case of any finite order n the invertability condition of the matrix A_n can be checked recursively

$$a_n = (-1)^n \det A_n = \alpha_n^2 \det A_{n-2} \neq 0, \tag{156}$$

with the initialization

$$\det A_1 = -\alpha_1, \text{ for } n = 3,5,7,..., \tag{157}$$

for the class of system representations of any odd order, and with the initialization

$$\det A_2 = \alpha_2^2, \text{ for } n = 4,6,8,..., \tag{158}$$

for the class of the conservative system representations.

Remark 21: It is important to emphasize that as a consequence of unique zero equilibrium requirement the order of any physically correct state space representation of a conservative system must be even. On the other hand, for the class of dissipative and/or anti-dissipative systems no such restriction is needed.

Now, let all the parameters $a_1, a_2,..., a_n$ of the characteristic polynomial $A(s)=A_n(s)$ be considered as unknown, and let us specify the region of asymptotic stability in the parameter space. We can compute the instantaneous value of the state space energy using the expression

$$E[x(t)] = \tfrac{1}{2}\rho^2[x(t),0] = \tfrac{1}{2}\|x(t)\|^2 = \tfrac{1}{2}\sum_{i=1}^{n} x_i^2(t) \tag{159}$$

For the derivative of the state space energy function E(x) along the system representation (7.3) we get

$$\left.\frac{dE(t)}{dt}\right|_{\Re\{s\}} = -\alpha_1 x_1^2(t) = -\frac{\alpha_1}{\gamma_1^2} \cdot y^2(t) \tag{160}$$

where γ_1 is a real output scaling parameter $0 < \gamma_1 < \infty$.

Remark 22: It is of crucial importance that the dissipation parameter α_1 is the only element of the matrix A, whose sign separates the system dissipativity from its anti-dissipativity. The critical value of $\alpha_1 = 0$, corresponds to the system conservativity and separates stability of the zero equilibrium state from its instability.

Thus, for real output dissipation power $y^2(t)$ the state space energy conservation principle holds if and only if:

$$P_o(t) = y^2(t) \Leftrightarrow \alpha_1 = \gamma_1^2 \geq 0 \tag{161}$$

or equivalently if and only if the state space energy conservation principle holds

$$\left.\frac{dE(t)}{dt}\right|_{\Re(s)} = -P_0(t), \quad P_0(t) \geq 0 \tag{162}$$

It can be concluded from (7.14) that the total state space energy is constant, i.e. the system is conservative, if and only if the observed output vanishes

$$P_o(t) = y^2(t) = 0 \Leftrightarrow \alpha_1 = \gamma_1^2 = 0 \tag{163}$$

and the system is dissipative if and only if the state space energy monotonously decreases

$$P_o(t) = y^2(t) > 0 \Leftrightarrow \alpha_1 = \gamma_1^2 > \tag{164}$$

Formally, there is no problem to show that the total state space energy monotonously increases, i.e. the system is anti-dissipative, if it holds $\alpha_1 < 0$, or equivalently

$$\left.\frac{dE(t)}{dt}\right|_{\Re(s)} = +P_0(t), \quad P_0(t) \geq 0 \tag{165}$$

but we would have to change the physical meaning of the power $P_o(t) = y^2(t)$ and to interpret it as an input power.

Remark 23: Notice, that if we put $\alpha_5 = 0$, then the state variables x_5 and x_6 become unobservable by the output y ; thus only the first isolated subsystem with the state variables x_i, i = 1,2,3,4, which are observable, will be asymptotic stable, while the second one will oscillate at a constant level of the state space energy. The situation is illustrated by the case c) in the Figure 38.

Similarly, if we put $\alpha_3 = 0$, then the state variables x_i, i = 3,4,5,6 become unobservable by the output y, and only the observable subsystem

$$\begin{aligned}
\dot{x}_1(t) &= -\alpha_1 x_1(t) + \alpha_2 x_2(t) \\
\dot{x}_2(t) &= -\alpha_2 x_2(t) \\
y(t) &= \gamma_1 x_1(t)
\end{aligned} \tag{166}$$

will be asymptotic stable. The situation is illustrated by the case b) in the Figure 38.

In order to be able to interpret the obtained stability results for any given n-th order ordinary differential equation, we need an effective tool to transform any given state space representation, such as (149), but of arbitrary finite order, into the form like (147), and back. One straightforward technique suggests to compute successive derivatives of the output equation up to the order n, combined with successive elimination all of the state variables as follows:

$$y(t) = \gamma_1 x_1(t) \tag{167}$$

$$\dot{y} + \alpha_1 y = \gamma_1 \alpha_2 x_2 \tag{168}$$

$$\ddot{y} + \alpha_1 \dot{y} + \alpha_2^2 y = \gamma_1 \alpha_2 \alpha_3 x_3 \tag{169}$$

$$\dddot{y} + \alpha_1 \ddot{y} + (\alpha_2^2 + \alpha_3^2)\dot{y} + \alpha_1 \alpha_3^2 y = \gamma_1 \alpha_2 \alpha_3 \alpha_4 x_4 \tag{170}$$

$$y^{(4)} + \alpha_1 \dddot{y} + (\alpha_2^2 + \alpha_3^2 + \alpha_4^2)\ddot{y} + \alpha_1(\alpha_3^2 + \alpha_4^2)\dot{y} + \alpha_2^2 \alpha_4^2 y = \gamma_1 \alpha_2 \alpha_3 \alpha_4 \alpha_5 x_5 \tag{171}$$

$$\begin{gathered} y^{(5)} + \alpha_1 y^{(4)} + (\alpha_2^2 + \alpha_3^2 + \alpha_4^2 + \alpha_5^2)\dddot{y} + \alpha_1(\alpha_3^2 + \alpha_4^2 + \alpha_5^2)\ddot{y} \\ +[\alpha_2^2 \alpha_4^2 + \alpha_2^2 \alpha_5^2 + \alpha_3^2 \alpha_5^2]\dot{y} + \alpha_1 \alpha_3^2 \alpha_5^2 y = \gamma_1 \alpha_2 \alpha_3 \alpha_4 \alpha_5 \alpha_6 x_6 \end{gathered} \tag{172}$$

$$\begin{gathered} y^{(6)} + \alpha_1 y^{(5)} + (\alpha_2^2 + \alpha_3^2 + \alpha_4^2 + \alpha_5^2 + \alpha_6^2) y^{(4)} + \alpha_1(\alpha_3^2 + \alpha_4^2 + \alpha_5^2 + \alpha_6^2)\dddot{y} \\ +(\alpha_2^2 \alpha_4^2 + \alpha_2^2 \alpha_5^2 + \alpha_3^2 \alpha_6^2 + \alpha_2^2 \alpha_6^2 + \alpha_3^2 \alpha_6^2 + \alpha_4^2 \alpha_6^2)\ddot{y} \\ +\alpha_1(\alpha_3^2 \alpha_5^2 + \alpha_3^2 \alpha_6^2 + \alpha_4^2 \alpha_6^2)\dot{y} + \alpha_2^2 \alpha_4^2 \alpha_6^2 y = 0 \end{gathered} \tag{173}$$

As a by-product of the state elimination procedure above the following state transformation relations have been obtained:

$$x_1 = \frac{1}{\gamma_1} y, \quad \gamma_1 \neq 0 \tag{174}$$

$$x_2 = \frac{1}{\gamma_1 \alpha_2}(\alpha_1 y + \dot{y}), \quad \gamma_1 \neq 0, \alpha_2 \neq 0 \tag{175}$$

$$x_3 = \frac{1}{\gamma_1 \alpha_2 \alpha_3}(\alpha_2^2 y + \alpha_1 \dot{y} + \ddot{y}), \quad \gamma_1 \neq 0, \alpha_2 \neq 0, \alpha_3 \neq 0 \tag{176}$$

$$x_4 = \frac{1}{\gamma_1 \alpha_2 \alpha_3 \alpha_4}\left[\alpha_1 \alpha_3^2 y + (\alpha_2^2 + \alpha_3^2)\dot{y} + \alpha_1 \ddot{y} + \dddot{y}\right], \quad \gamma_1 \neq 0, \alpha_2 \neq 0, \alpha_3 \neq 0, \alpha_4 \neq 0 \tag{177}$$

$$x_5 = \frac{1}{\gamma_1\alpha_2\alpha_3\alpha_4\alpha_5}\left[\alpha_2^2\alpha_4^2 y + \alpha_1(\alpha_3^2+\alpha_4^2)\dot{y} + (\alpha_2^2+\alpha_3^2+\alpha_4^2)\ddot{y} + \alpha_1\dddot{y} + y^{(4)}\right], \qquad (178)$$
$$\gamma_1 \neq 0, \alpha_2 \neq 0, \ldots, \alpha_5 \neq 0$$

$$x_6 = \frac{1}{\gamma_1\alpha_2\alpha_3\alpha_4\alpha_5\alpha_6}[\alpha_1\alpha_3^2\alpha_5^2 y + (\alpha_2^2\alpha_4^2 + \alpha_2^2\alpha_5^2 + \alpha_3^2\alpha_6^2)\dot{y} + \alpha_1(\alpha_3^2+\alpha_4^2+\alpha_5^2)\ddot{y} \qquad (179)$$
$$+ (\alpha_2^2+\alpha_3^2+\alpha_4^2+\alpha_5^2)\dddot{y} + \alpha_1 y^{(4)} + y^{(5)}], \quad \gamma_1 \neq 0, \alpha_2 \neq 0, \alpha_3 \neq 0, \ldots, \alpha_6 \neq 0$$

which can be expressed in a vector-matrix form

$$x = T\overline{x}, \quad \overline{x} = T^{-1}x \qquad (180)$$

with the corresponding invertible transformation matrix T

$$T = \begin{bmatrix} \frac{1}{\gamma_1} & 0 & 0 & 0 & \cdots\cdots \\ \frac{\alpha_1}{\gamma_1\alpha_2} & \frac{1}{\gamma_1\alpha_2} & 0 & 0 & \cdots\cdots \\ \frac{\alpha_2}{\gamma_1\alpha_3} & \frac{\alpha_1}{\gamma_1\alpha_2\alpha_3} & \frac{1}{\gamma_1\alpha_2\alpha_3} & 0 & \cdots\cdots \\ \frac{\alpha_1\alpha_3}{\gamma_1\alpha_2\alpha_4} & \frac{\alpha_2^2+\alpha_3^2}{\gamma_1\alpha_2\alpha_3} & \frac{\alpha_1}{\gamma_1\alpha_2\alpha_3\alpha_4} & \frac{1}{\gamma_1\alpha_2\alpha_3\alpha_4} & \cdots\cdots \\ \vdots & \vdots & \vdots & \vdots & \vdots \end{bmatrix} \qquad (181)$$

Remark 24: Notice, that the technique above has the advantage that, in principle, it works also for a broad class of nonlinear state representations.

In the linear and time-invariant situations various Laplace transform based techniques may be more appropriate. A simple method based on computation of the characteristic polynomial $A_n(s)$ of the matrix A like that used in (148) will be presented in the sequel.

Let`s consider the matrix A_n as follows:

$$A_n = \begin{bmatrix} -\alpha_1 & \alpha_2 & 0 & 0 & 0 & 0 \\ -\alpha_2 & 0 & \alpha_3 & 0 & 0 & 0 \\ 0 & -\alpha_3 & 0 & \ddots & 0 & 0 \\ 0 & 0 & \ddots & \ddots & \alpha_{n-1} & 0 \\ 0 & 0 & 0 & -\alpha_{n-1} & 0 & \alpha_n \\ 0 & 0 & 0 & 0 & -\alpha_n & 0 \end{bmatrix} \qquad (182)$$

where the conditions

$$\gamma_1 \neq 0, \alpha_2 \neq 0, \alpha_3 \neq 0, ..., \alpha_k \neq 0, \quad k \in \{2,3,4,...n\}, \alpha_1 = \gamma_1^{\ 2}, \tag{183}$$

are the observability conditions of the dissipation normal form (182), and simultaneously represent the set of necessary and sufficient conditions for existence and invertibility of the derived state space transformation.

The n-the degree characteristic polynomial $A_n(s)$of any matrix A_n is defined by the determinant

$$A_n(s) = \det[sI - A_n], \quad s = \sigma + j\omega \tag{184}$$

and can explicitly be expressed in the following standard form:

$$A_n(s) = s^n + a_1 s^{n-1} + a_2 s^{n-2} + a_3 s^{n-3} + ... + a_{n-2} s^2 + a_{n-1} s + a_n \tag{185}$$

It is easy to prove that for the class of matrices A_n of the given structure (182) the characteristic polynomial $A_n(s)$ can be computed recursively, using the following recursive relation

$$A_k(s) = sA_{k-1}(s) + \alpha_k^2 A_{k-2}(s), \quad k \in \{3,4,...,n\} \tag{186}$$

with the initialization for n = 1 and n = 2:

$$A_1(s) = s + \alpha_1, \quad A_2(s) = s^2 + \alpha_1 s + \alpha_2^2 \tag{187}$$

Using it, for instance, for the matrix A_6 in the form (154) we obtain:

$$A_6(s) = s^6 + \underset{a_1}{\alpha_1} s^5 + [\underbrace{\alpha_2^2 + \alpha_3^2 + ... + \alpha_6^2}_{a_2}]s^4 + \underbrace{\alpha_1[\alpha_3^2 + \alpha_4^2 + ... + \alpha_6^2]}_{a_3}s^4 + \underbrace{... + \alpha_2^2\alpha_4^2\alpha_6^2}_{a_n} \tag{188}$$

and by comparing the result (188) with the standard form (185), we obtain a set of algebraic equations from which the unknown elements α_k of the matrix A_n can be computed for given a_i.

For example for the system order n = 6, the set of parameters a_i, $i \in \{1, 2, ..., 6\}$ is obviously given by

$$\begin{aligned} a_1 &= \alpha_1 \\ a_2 &= \alpha_2^2 + \alpha_3^2 + \alpha_4^2 + \alpha_5^2 + \alpha_6^2 \\ a_3 &= \alpha_1(\alpha_3^2 + \alpha_4^2 + \alpha_5^2 + \alpha_6^2) \\ a_4 &= \alpha_2^2(\alpha_4^2 + \alpha_5^2 + \alpha_6^2) + \alpha_3^2(\alpha_5^2 + \alpha_6^2) + \alpha_4^2\alpha_6^2 \\ a_5 &= \alpha_1\alpha_3^2(\alpha_5^2 + \alpha_6^2) + \alpha_1\alpha_4^2\alpha_6^2 \\ a_6 &= \alpha_2^2\alpha_4^2\alpha_6^2 \end{aligned} \tag{189}$$

and can easily be arbitrary reduced or extended for any finite order.

For instance for n = 5 it reduces to

$$\begin{aligned} a_1 &= \alpha_1 \\ a_2 &= \alpha_2^2 + \alpha_3^2 + \alpha_4^2 + \alpha_5^2 \\ a_3 &= \alpha_1(\alpha_3^2 + \alpha_4^2 + \alpha_5^2) \\ a_4 &= \alpha_2^2(\alpha_4^2 + \alpha_5^2) + \alpha_3^2\alpha_5^2 \\ a_5 &= \alpha_1\alpha_3^2\alpha_5^2 \end{aligned} \tag{190}$$

for n = 4 we get:

$$\begin{aligned} a_1 &= \alpha_1 \\ a_2 &= \alpha_2^2 + \alpha_3^2 + \alpha_4^2 \\ a_3 &= \alpha_1(\alpha_3^2 + \alpha_4^2) \\ a_4 &= \alpha_2^2\alpha_4^2 \end{aligned} \tag{191}$$

and for n = 3 it follows:

$$\begin{aligned} a_1 &= \alpha_1 \\ a_2 &= \alpha_2^2 + \alpha_3^2 \\ a_3 &= \alpha_1\alpha_3^2 \end{aligned} \tag{192}$$

For any finite order the analytical solution has been found easily, and can be expressed as follows

$$
\begin{aligned}
\alpha_1 &= a_1 = \Delta_1 , \\
\alpha_2 &= \sqrt{\frac{a_1 a_2 - a_3}{a_1}} = \sqrt{\frac{\Delta_2}{\Delta_1}} \\
\alpha_3 &= \sqrt{\frac{a_1 a_2 a_3 - a_3^2 - a_1^2 a_4}{(a_1 a_2 - a_3) a_1}} = \sqrt{\frac{\Delta_3}{\Delta_2 \Delta_1}} \\
\alpha_k &= \sqrt{\frac{\Delta_k \Delta_{k-3}}{\Delta_{k-2} \Delta_{k-1}}}, \quad k = 4, 5, 6, ..., n
\end{aligned}
\tag{193}
$$

where the set of new parameters Δ_k, k = 1, 2,…,n correspond to the set of diagonal minors of the Hurwitz determinant. Using the state space energy monotonnicity condition (162), together with the state observability conditions (183), it is very easy to prove the resulting asymptotic stability conditions:

$$\alpha_1 > 0, \alpha_2 \neq 0, \alpha_3 \neq 0, ..., \alpha_k \neq 0, \quad k \in \{2,3,4,...n\} \Leftrightarrow \Delta_k > 0,\ k = 1, 2, \dots, n \tag{194}$$

which are obviously equivalent to the set of necessary and sufficient stability conditions representing the well known Hurwitz stability criterion.

Example 4: (Linear asymptotic stability analysis)

For n = 6, the necessary and sufficient observability conditions are given by the condition:

$$\det H_o \neq 0 \tag{195}$$

where H_o is the observability matrix defined by matrices A and C as follows:

$$H_o = \left[C^T,\ A^T C^T, (A^T)^2 C^T,\ ...,\ (A^T)^{n-1} C^T \right] \tag{196}$$

and the calculated observability conditions for the order n = 6 are:

$$\alpha_2 \neq 0, \alpha_3 \neq 0, \alpha_4 \neq 0, \alpha_5 \neq 0, \alpha_6 \neq 0,\ \gamma_1 \neq \neq 0 \tag{197}$$

According to (164) the dissipativity appears if and only if the state space energy monotonously decreases

$$P_o(t) = y^2(t) > 0 \Leftrightarrow \alpha_1 = \gamma_1^2 > 0 \tag{198}$$

Thus the derived set of necessary and sufficient conditions of asymptotic stability reads:

$$\alpha_1 > 0,\ \alpha_k \neq 0,\ k\in\{2,3,...,6\},\ \Delta_k > 0,\ k = 1, 2, \ldots, 6 \tag{199}$$

It has been clearly demonstrated that linear algebraic methods for stability analysis can be seen as a special case of methods based on the proposed state space energy approach. It will be shown in the next section that even nonlinear stability problems can be effectively solved by the same technique. Moreover, the state space energy interpretation makes it possible to gain a better insight into many classical results.

EFFECTS OF MEMRISTIVE FEEDBACK AND MECHANISM OF CHAOTICITY

Example 5: Let a 2nd order linear conservative system with *input v(t):*

$$\ddot{y}(t)+ a_2 y(t) = v(t) \tag{200}$$

Structure of equivalent state representation is assumed:

$$\begin{aligned} \dot{x}_1 &= +\alpha_2 x_2 + u(t) \\ \dot{x}_2 &= -\alpha_2 x_1 \end{aligned} \tag{201}$$

The state space energy *E(x)* can be defined by

$$E(x) = \frac{1}{2}\left(x_1^2 + x_2^2\right) = \frac{1}{2}\rho_2^2\,[x,0] \tag{202}$$

We assume the output is given by the relation $y(t) = x_1$ and a nonlinear control signal $u(t) = -\Phi(y)$ has to be specified in such a way that local dissipativity results, and in addition, the zero equilibrium state will be locally asymptotically stable in a region $D\subset X\subset R^2$.

The state space energy of the representation (7) is observable if and only if $\alpha_2>0$, and for power balance relation we get

$$\left.\frac{\mathrm{d}E(t)}{\mathrm{d}t}\right|_{\Re(s)} = -P(t) = -x_1.\Phi(x_1) \leq 0 \tag{203}$$

It means that the Taylor expansion of the mapping $\Phi(x_1)$ should contain odd terms only, i.e. we can define

$$\Phi(x_1) = \varepsilon\left[\alpha x_1 - \frac{1}{3}\beta x_1^3\right] \tag{204}$$

The structure of the matrix A(x) with feedback is given

$$\mathbf{A}(x_1, x_2) = \begin{bmatrix} -\varepsilon\left[\alpha - \frac{1}{3}\beta x_1^2\right], & \sqrt{a_2} \\ -\sqrt{a_2}, & 0 \end{bmatrix} \tag{205}$$

and the equation (204) with memristive feedback Φ reads

$$\ddot{y}(t) + M(y)\dot{y}(t) + a_2 y(t) = 0 \tag{206}$$

Obviously the term

$$M(y) = \varepsilon\left[\alpha - \beta y^2(t)\right] \tag{207}$$

plays the role of *memristance induced by the static nonlinearity*

$$\Phi(x_1) = \varepsilon\left[\alpha x_1 - \frac{1}{3}\beta x_1^3\right] \tag{208}$$

because for $y(t) = x_1$ the defining relationship holds

$$M(y) = \frac{d\Phi(y)}{dy} = \varepsilon\left[\alpha - \beta y^2(t)\right] \tag{209}$$

Notice that in *linear* case $(\beta=0)$ the *memristance reduces to the resistance* $R=\varepsilon\alpha$, and in *conservative* case ($\varepsilon = 0$) the *memristance reduces to zero,* and consequently the *dissipation power* is $P(t)=0$.

For illustration of the memristance concept a *class of chaotic oscillations* generated by simple example of the *quadratic memristive feedback system representation* with *memristor-like nonlinear dissipation subsystems* will be briefly presented. The similar examples of *nonlinear chaos generating systems* with *memristor-like subsystems* have been described elsewhere [5].

Example 6: (Mechanism of chaoticity in a 4-th order linear system with a memristive feedback).

Let a 4th *order linear system* representation with the *quadratic memristive feedback* is given by:

$$\begin{aligned}\dot{x}_1 &= -\alpha_1 x_1 + \alpha_2 x_2 \\ \dot{x}_2 &= -\alpha_2 x_1 + \alpha_3 x_3 \\ \dot{x}_3 &= -\alpha_3 x_2 + \alpha_4 x_4 \\ \dot{x}_4 &= -\alpha_4 x_3\end{aligned}, \quad \alpha = \begin{bmatrix}\alpha_1 \\ \alpha_2 \\ \alpha_3 \\ \alpha_4\end{bmatrix} = \begin{bmatrix}-1+10{x_2}^2 \\ 1 \\ 1 \\ freq\end{bmatrix}, \quad x(0) = \begin{bmatrix}0 \\ 0 \\ 0.5 \\ 0\end{bmatrix} \tag{210}$$

The chaotic behavior of the *system state space energy* evolution (for $\alpha_4 = 2$) is shown in Figure 39.

The corresponding *3D - projection of state trajectories* for two typical values of the *frequency parameter* α_4, (left chaotic $\alpha_4 = 2$, right quasi-periodic behavior for $\alpha_4 = 2.2$), are shown in the Figure 40.

The existence of chaotic attractor was verified also numerically by calculation of *Liapunov exponents.*

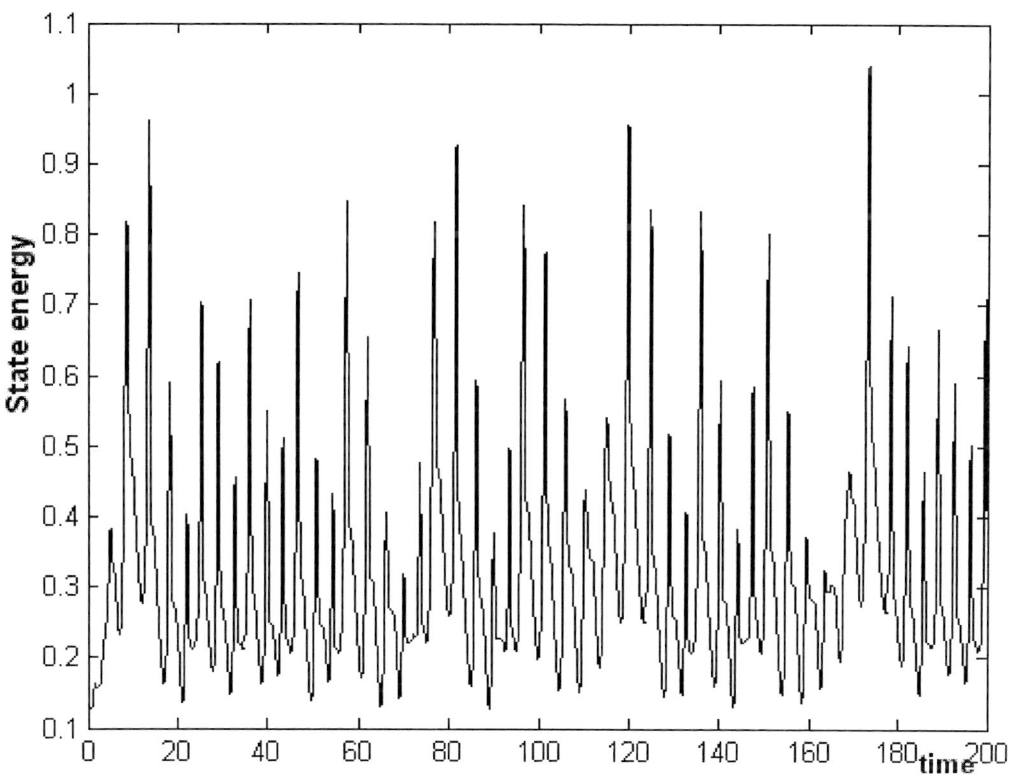

Figure 39. Evolution of the chaotic state space energy.

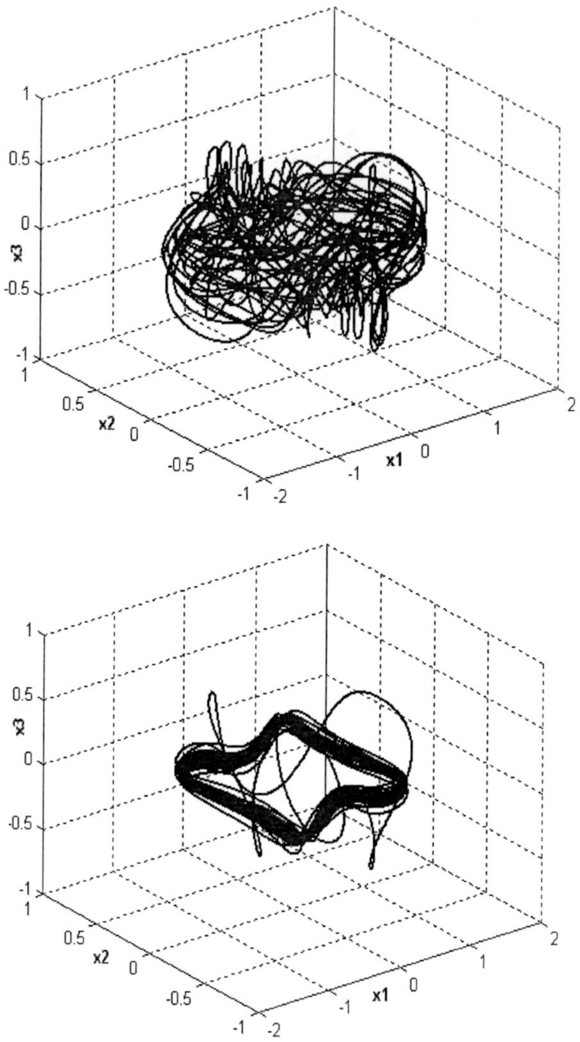

Figure 40. The 3D-projection of the state space trajectories: left α_4=2.0 (chaotic), right α_4=2.2 (quasi-periodic).

Nonlinear Stability Analysis via the State Space Energy Concept

Example 7: (Nonlinear asymptotic stability analysis – Lewis servosystem)

Let us consider well known *non-linear system* called *Lewis servomechanism* given by equation:

$$\ddot{y}(t) + a_1 \left[1 - \beta y(t) \mathrm{sign}[y(t)] \right] \dot{y}(t) + a_2 y(t) = 0 \tag{211}$$

with $\alpha_2 > 0$, and α_1, β as arbitrary real design parameters. Using the state space energy approach for:

$$A(x_1, x_2) = \begin{bmatrix} -a_1\left[1-\beta x_1(t)\text{sign}[x_1(t)]\right], & \sqrt{a_2} \\ -\sqrt{a_2}, & 0 \end{bmatrix} \tag{212}$$

the *unique Liapunov function* $V(x)$ can be determined by *isometric transformations as the state space energy E(x)*. For simplicity we assume $a_2 = 1$ and get:

$$E = \frac{1}{2}\left[\frac{1}{4}a_1^2\beta^2 x_1^4 - a_1^2\beta x_1^3\text{sign}[x_1(t)] - a_1\beta x_1^2 x_2\text{sign}[x_1(t)] + (a_1^2+1)x_1^2 + 2a_1 x_1 x_2 + x_2^2\right] \tag{213}$$

The dissipation power *P(t)* is given by

$$\forall a_2 \geq 0: \ \frac{dE}{dt} = -2a_1 x_1^2\left[1 - \frac{1}{2}\beta x_1 sign[x_1(t)]\right] \tag{214}$$

i.e. the *state* $x^* = 0$ is *asymptotically stable in the region* $D \subset X \subset R^2$, as it is shown in the Figure 41.

For a *special case* ($\beta = 0$), i.e. *linear*, (and for $\alpha_2 = 1$), the *total state space energy reduces to*

$$E\left[x_1, x_2\right] = 0.5(a_1^2+1)x_1^2(t) + a_1 x_1(t)x_2(t) + 0.5x_2^2(t) \tag{215}$$

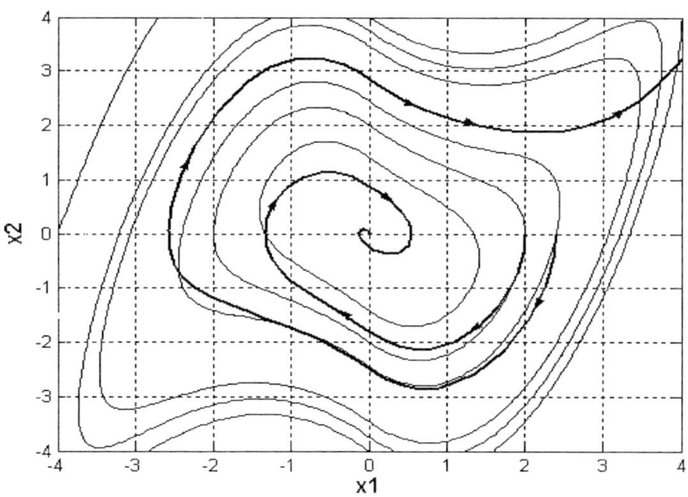

Figure 41. Phase portrait and the typical state space topology generated by the state space energy function (213).

Example 8: (Nonlinear asymptotic stability analysis – Van der Pol equation)

Let us consider a simple *non-linear system given* by the representation:

$$\ddot{y}(t) + \varepsilon\left[\alpha - \beta y^2(t)\right]\dot{y}(t) + a_2 y(t) = 0 \tag{216}$$

Let the matrix C is defined by C=[γ, 0], and the *structure of the matrix A(x)* is

$$A(x_1, x_2) = \begin{bmatrix} -\varepsilon\left[\alpha - \frac{1}{3}\beta x_1^2\right], & \sqrt{a_2} \\ -\sqrt{a_2}, & 0 \end{bmatrix} \tag{217}$$

The system representation is locally observable if and only if $\gamma \neq 0$, $a_2 > 0$, and

$$\left.\frac{dE(t)}{dt}\right|_{\Re(s)} = -P \leq 0, \quad P = \varepsilon\left[\alpha - \frac{1}{3}\beta x_1^2\right] x_1^2 \tag{218}$$

i.e. the *state* $x^* = 0$ is *asymptotically stable in the region* $D \subset X \subset R^2$ almost the same form as Figure 41.

Example 9: (Nonlinear asymptotic stability analysis – generation of the Liapunov function)

Let us consider the same *non-linear system,* but instead of the *structure of A(x),* the state *x*(*t*) is defined by the standard way, i.e.: $x_1 = y$, $x_2 = dy/dt$. Then the *unique Liapunov function V(x)* can be determined by *isometric transformations of the state space energy function E(x),* and for $\alpha = \beta = a_2 = 1$ we get:

$$V(x) = E(x) = \frac{1}{2}\left[\frac{1}{9}\varepsilon^2 x_1^6 - \frac{2}{3}\varepsilon^2 x_1^4 + (1+\varepsilon^2)x_1^2 - \frac{2}{3}\varepsilon x_1^3 x_2 + 2\varepsilon x_1 x_2 + x_2^2\right] \tag{219}$$

and again for the *linear conservative* case (ε=0) it reduces to:

$$V(x) = E(x) = \frac{1}{2}(x_1^2 + x_2^2) = \frac{1}{2}\rho_2^2\left[x, 0\right]. \tag{220}$$

CONCLUSIONS

In this chapter is presented a new structural approach to analysis and synthesis of linear and non-linear causal systems.

Using several examples, it has been proven that complete analysis of instability, conservativity, dissipativity, anti-dissipativity, stability, asymptotic stability and chaoticity can be reduced at only two independent tests: the monotonicity test of abstract state space energy and minimality test of state space dimension

The proposed approach to the problem solution is based on generalization of Tellegen's theorem. The mathematically and physically background is presented, and the correctness of results is shown by simulation, too.

REFERENCES

[1] Khalil, H., K. (1996). Nonlinear Systems, Prentice Hall.

[2] Hrusak, J., Mayer, D., Stork, M. (2004). On System Structure Reconstruction Problem And Tellegen-Like Relation, Proc. of 8th World Multiconf.,SCI, Vol. VIII, Florida, USA, pp. 373-378.

[3] Hrusak, J., Mayer, D., Stork, M. (2006). New Approach to Non-linear Instability and Chaos Based on Generalized Tellegen's Principle, WMSCI 2006. Orlando, Florida, *Int. Institute of Informatics and Systemics*, USA, ISBN 980-6560-67-1, pp. 199-206

[4] Hrusak, J., Mayer, D., Stork, M. (2006). Dissipativity, Minimality and Physical Correctness of State Space Representations. CITSA 2006. Orlando, Florida, USA, *Int. Institute of Informatics and Systemics*, ISBN 980-6560-83-3. pp. 136-141.

[5] Hrusak, J., Mayer, D., Stork, M. (2007). Structural approach to instability and chaos in non-linear causal systems, Systems Theory and Scientific Computation, Athens, WSEAS Press, Greece, ISBN 978-960-8457-96-0. ISSN 1790-5117.pp. 223-228.

[6] Stork, M., Hrusak, J., Mayer, D. (2006). Chaos and Strange Behavior Detection of Continuous and Digital Nonlinear Systems Based on Internal System Energy Autocovariance, WMSCI 2006, Orlando, Florida, *Int. Institute of Informatics and Systemics*, USA, pp. 228-233.

[7] Fradkov, A., L., Pogromsky, A.,Y. (1998). Introduction to Control of Oscillations and Chaos. Singapore, World Scientific Series on Nonlinear Science.

[8] Chua, L. O., (1971). Memristor-The missing circuit element. *IEEE Trans. Circ. Theory*.CT-18, pp. 507-519.

[9] Pershin, Yu., V. and Di Ventra, M., (2008). Spin memristive systems: Spin memory effects in semiconductor spintronics. Phys. Rev. B 78.

[10] Strukov, D., B., Snider, G., S., Stewart, D., R. and Williams, R., S. (2008). The missing memristor found. *Nature* 453, pp. 80-83.

[11] Yang, J., J., Pickett, M., D., Li, X., Ohlberg, D., Stewart, D., R. and Williams, R., S. (2008). Memristive switching mechanism for metal/oxide/metal nanodevices. *Nature Nano-technology* 3, pp. 429 – 433.

[12] Erokhin, V. and Fontana, M. P. (2008). Electrochemically controlled polymeric device: amemristor (and more) found two years ago. arXiv:0807.0333 2008.

[13] Giannakopoulos, K., Deliyannis, T., and Hadjidemetriou, J. (2002). Means for Detecting Chaos and Hyperchaos in Nonlinear Electronic Circuits. Proc. 14th Int. Conf. DSP, Santorini, Greece, pp. 951-954.

[14] Makoto, I., Yang, T. and Chua, L. (1999). Experimental Study of Impulsive Synchronization of Chaotic and Hyperchaotic Circuit. *International Journal of Bifurcation and Chaos*, Vol. 9, No. 7.1393-1424.

In: Advances in Energy Research
Editor: Nicu Bizon
ISBN 978-1-61209-991-0

Chapter 5

MINIMUM DISSIPATED POWER AND ENERGY – TWO GENERAL PRINCIPLES OF THE LINEAR ELECTRIC AND MAGNETIC CIRCUITS IN QUASI-STATIONARY REGIME

Horia Andrei[1]*, Gianfranco Chicco[2] and Fanica Spinei[3]
[1]Faculty of Electrical Engineering, University Valahia Targoviste
[1]8-20 Blv. Unirii, Targoviste, Dambovita, ROMANIA
[2]Dipartimento di Ingegneria Elettrica,
Power and Energy Systems Group
Politecnico di Torino, corso Duca degli Abruzzi 24,
10129, Torino, ITALY
[3]Faculty of Electrical Engineering, University Politehnica Bucharest
313 Splaiul Independentei, Bucharest, ROMANIA

ABSTRACT

The use of the variational formulation of power and energy functionals in the analysis of the electric and magnetic circuits makes it possible to appreciate the equilibrium state attained in the circuit at a certain moment. In the present work, a power and energy functional is built for electric and magnetic circuits in static, stationary and quasi-stationary conditions, and its limit is determined. It is shown that the minimum of this functional corresponds to the minimum dissipated (absorbed) power and energy in the circuit. Some applications to electric circuits and magnetic are shown.

Keywords: linear electric and magnetic circuits, stationary and quasi-stationary regime, absorbed power and energy, variational formulation, Hilbert space, minimum of functionals.

* Email: hr_andrei@yahoo.com

INTRODUCTION

In many domains, such as mechanics, thermodynamics, global climate system, the Earth system, hydrology, and so forth, specific studies have been carried out to express the equilibrium state of the systems in energy terms: [1], [2], [3], [4], [5], and [6].

These studies are mostly based on the concept that for conservative systems it is possible to define suitable functionals expressed in terms of power or energy. Calculating the limits of these functionals represents an important breakthrough in formulating and solving optimization problems. The steady states in the mechanic, thermal, or electric conservative systems generally represent limit states from an energy point of view.

Minimum Energy Principles in Mechanics and Thermodynamics

In the classical mechanics, the analytical principles, underlying analytical mechanics, fall into two categories: differential principles and integral or variational principles. Differential principles examine the mechanical phenomenon at a given time, considering elementary time and space variations around a point.

The main differential principles are the principle of virtual mechanical work, d'Alembert's principle and Gauss's principle or the lowest constraint principle. Integral principles analyze the mechanical phenomenon within a finite space and time interval, reducing the fundamental mechanic problem in determining some values which make certain integrals stationary. In this category there are the Hamilton's principle and the Maupertuis's principle of minimum action. The Maupertuis's principle or the minimum action, formulated in 1745, is expressed as: whenever movement of a material system occurs in nature, the system should work so that the integral of the mass, speed and space product – integral considered on the space and time interval between two successive given positions – is minimal. The quantity is called action, hence the name of the principle. This formulation of Maupertuis's minimum action principle is found in Hamilton's principle, which refers to the state of minimal action in time-related evolution of system, from one quasi-stationary state to another, along the curve $\gamma:[a,b]\rightarrow R^n$, is defined by the minimum of the functional

$$F_M = \int_a^b (\Xi - \Psi)dt \tag{1}$$

also called the „action integral", where F_M is the mechanical energy, Ξ is the kinetic energy (so-called "life power") of the system, and Ψ is the mechanic effect of the power system which works on the system under consideration.

In thermodynamics, the thermodynamic potentials are defined. These potentials are used in calculating the equilibrium of chemical reactions, or for measuring the properties of the substances using chemical reactions. Chemical reactions usually are studied in simple conditions, such as constant pressure and temperature, or constant volume and entropy, and when these conditions are met, the adequate thermodynamic potential is applied. Like in mechanics, the system's potential tends to decrease, and at equilibrium, under those conditions, the potential will reach minimum values. Consequently, thermodynamic potentials can characterize the energy state of a system under given condition.

In particular, if the entropy and the parameters (for instance, volume) of a closed system are kept constant, the internal energy decreases and reaches the minimum value at equilibrium. That comes from the first and second principles of thermodynamics and is called the minimum energy principle. From this principle the following sentences result:

- If the temperature and the parameters of a closed system are kept constant, the pressure and the parameters of a closed system are kept constant, the enthalpy decreases and reaches the minimum value at equilibrium;
- If the temperature, pressure and the parameters of a closed system are kept constant, the Gibbs free energy decreases and reaches the minimum value at equilibrium.

The second principle of thermodynamics can be expressed by another function of state, the entropy. Entropy, which is a measure of molecular disorder of a system, permits to assess whether a state can be reached from another spontaneously. In terms of entropy, the second principle of thermodynamics is stated as follows: In an isolated system the entropy increases in spontaneous processes. Irreversible processes are spontaneous processes and therefore they are accompanied by an increase of the total entropy.

Earth System Processes and the Principles of Minimum and Maximum Entropy Production

The application of non-equilibrium thermodynamics to the planetary energy balance of Earth and its neighboring planets, have been elaborated by [7], [8], [9], and [10]. The principles of minimum and maximum entropy production have been introduced in the context of complex Earth system processes.

Transformations of energy of different forms are governed by the laws of thermodynamics. At the foundation of classical thermodynamics there are the first and second laws. The first law states that the total energy of a system is conserved, while the second law states that the entropy of an isolated system can only increase. The second law implies that the free energy of an isolated system is successively degraded by diabatic processes over time, leading to entropy production. This eventually results in an equilibrium state of maximum entropy. In its statistical interpretation, the direction towards higher entropy can be interpreted as a transition to more probable states.

The formulations from classical thermodynamics can be applied to non-equilibrium systems which are not isolated [11]. By exchanging energy of different entropy (or mass) across the system boundary, these systems maintain states that do not represent thermodynamic equilibrium. In [12] Rinaldo describes how a dissipation minimization procedure can be used to form efficient river networks in model hydrological systems. In these systems, the boundary conditions are fixed. These fixed boundary conditions make a minimization rather than maximization the appropriate procedure to apply, thus obtaining networks comparable with river networks observed in nature. All these models can be put in analogy with the studies of Bejan, [13], in which entropy production minimization can be applied to optimize the design of heat transfer systems. Other studies using relationships similar to the ones adopted for rivers networks refer to exploiting fractal geometry for distribution networks of individual, living organisms and plant communities [14], and [15].

In heat transfer, the existence of a maximum in entropy production associated with heat transport from the tropics to polar regions on Earth can be demonstrated with energy balance models of the climate system [7], [16], [17], [18], [19], [20], and [21].

The maximum entropy production principle has also been confirmed through fluid-dynamics simulations [1], [9], [22], and [23]. A further approach formulated by Shimokawa and Ozawa [22] shows that it is possible to resort to a probabilistic approach, assigning probabilities to multiple steady states and obtaining the maximum entropy production state as the most likely state of the climate system. The maximum entropy principle is also used by Kleidon et al. [9] in order to check consistency of numerical models. Finally, it can be mentioned that controversial arguments and objections to the application of the maximum entropy principle to physical systems have been addressed in [1], [2], and [5].

The differences between the minimum and the maximum entropy principles can be summarized as follows [10]. The minimum entropy principle is applied to linear systems with fixed boundary conditions and few degrees of freedom, in order to deal with situations in which the system perturbations occur away from its steady state. The steady state is one of minimum entropy production relative to any adjacent non-steady state. Conversely, the maximum entropy principle is applied to non-linear systems with many degrees of freedom. In this case, multiple steady states could occur, and it is possible to select the steady state with maximum entropy production relative to the other possible steady states. Constraint such as the conservation of energy, mass and momentum have to be adequately considered in any specific application of the maximum entropy production.

Differential, Integral and Variational Formulation of Classical Electromagnetic Field Theory

Maxwell's Equations of Electromagnetic Field

Specific studies have been dedicated to the electromagnetic field, and to electric circuits, in particular exploiting the consequences of Maxwell's principles of minimum-heat Maxwell [24]. Classical electromagnetic theory is based essentially on Maxwell's equations. In their integral form, Maxwell's equations belong to the category of universal physical laws. Maxwell's equations in integral form are: Faraday's law of electromagnetic induction (2), Ampère's magnetic circuit law amended by Maxwell to include the displacement current (3), Gauss' law for electric field (4), and Gauss' law for magnetic field (5):

$$\oint_{\Gamma} \mathbf{E}\, d\mathbf{r} = -\frac{d}{dt} \int_{S_\Gamma} \mathbf{B}\mathbf{n} d\mathbf{S} \tag{2}$$

$$\oint_{\Gamma} \mathbf{H}\, d\mathbf{r} = \int_{S_\Gamma} \mathbf{J}\mathbf{n} d\mathbf{S} + \frac{d}{dt} \int_{S_\Gamma} \mathbf{D}\mathbf{n} d\mathbf{S} \tag{3}$$

$$\oint_{\Sigma} \mathbf{D}\,\mathbf{n}d\mathbf{S} = \int_{V_\Sigma} \rho_v d\mathbf{V} \tag{4}$$

$$\oint_{\Sigma} \mathbf{B}\mathbf{n}d\mathbf{S} = 0 \tag{5}$$

where the electromagnetic field quantities are:

$\mathbf{E}(\mathbf{r},t)$ electric field strength

$\mathbf{H}(\mathbf{r},t)$ electric field

$\mathbf{D}(\mathbf{r},t)$ electric flux density

$\mathbf{B}(\mathbf{r},t)$ magnetic flux density

$\mathbf{J}(\mathbf{r},t)$ - electric conduction current density.

These quantities are vector functions of position and time. Furthermore, $\rho_v(\mathbf{r},t)$ is the electric charge volume density, and is a scalar function of position and time. We assume that these functions are defined in some domain Ω, and the integral form of the Maxwell equations (2)-(5) gives useful information on global properties of the electromagnetic field around any closed curves Γ, and through any closed surfaces Σ, in Ω.

The differential form of Maxwell's equations is:

$$\nabla \times \mathbf{E} = -\frac{\partial \mathbf{B}}{\partial t} \tag{6}$$

$$\nabla \times \mathbf{H} = \mathbf{J} + \frac{\partial \mathbf{D}}{\partial t} \tag{7}$$

$$\nabla \cdot \mathbf{D} = \rho_v \tag{8}$$

$$\nabla \cdot \mathbf{B} = 0 \tag{9}$$

where the mathematical symbol nabla ("del" operator) ∇ is a vector differential operator.

Electric charge obeys a conservation law. Conservation laws play a significant place in physical theories. To each conservation laws corresponds a certain transformation called symmetry, which makes physical laws invariant under that transformation, Noether's theorem, [25]. For instance, time invariance implies that energy is conserved, translation invariance implies that linear momentum is conserved, and rotation invariance implies that angular momentum is conserved.

The four Maxwell's equations (2)-(5) can be rewritten as two tensor equations: one verified by the pair $(\mathbf{E},\mathbf{B})$, the other one by $(\mathbf{D},\mathbf{H})$. The latter tensor equation is a consequence of the charge conservation law.

The law of charge conservation is implicit in Maxwell's equations (3), (4). In integral form, the law is given by

$$\oint_{\Sigma} \mathbf{J}\,\mathbf{n}d\mathbf{S} = -\frac{d}{dt}\int_{\mathbf{V}_{\Sigma}} \rho_v d\mathbf{V} \tag{10}$$

The differential form of charge conservation law is given by

$$\nabla \cdot \mathbf{J} = -\frac{\partial}{\partial t}\rho_v \tag{11}$$

Using the vector identity $\nabla \cdot (\nabla \times \mathbf{H}) = 0$, from Maxwell' equation (7), one obtains

$$\nabla \cdot (\nabla \times \mathbf{H}) = 0 = \nabla \cdot \mathbf{J} + \nabla \cdot \frac{\partial \mathbf{D}}{\partial t} = \nabla \cdot \mathbf{J} + \frac{\partial \rho_v}{\partial t} \tag{12}$$

To prove the integral form (6), we can choose a closed curve $\mathbf{H}$ through any closed surface Σ in Ω. The circulation of $\mathbf{H}$ a closed curve is called magneto-motive force (abbreviated mmf) and is defined as

$$u_{mm_{\Gamma}} = \oint_{\Gamma} \mathbf{H}\, d\mathbf{r} \tag{13}$$

Using (3), (9), and (10) we conclude that the mmf around that closed path tends to zero when the path reduces to a point. The open surface whose boundary is that path becomes a closed surface. Therefore, the Ampère's law leads to

$$\lim u_{mm_{\Gamma}} = \lim \oint_{\Gamma} \mathbf{H}\, d\mathbf{r} \;\Rightarrow\; \int_{\mathbf{S}_{\Gamma}} \mathbf{J}\mathbf{n}d\mathbf{S} + \frac{d}{dt}\int_{\mathbf{S}_{\Gamma}} \mathbf{D}\mathbf{n}d\mathbf{S} = 0$$

By combining this result with the electric flux law (4), we obtain the charge conservation law (10).

Maxwell's equations are valid for all kind of material media. The price paid for this universal validity is that the set of partial differential equations which results from Maxwell's equations (6)-(9) is incomplete. We need additional relations that are specific to each class of material media, referred to as constitutive relationships. These relations have to be able to describe the interaction of the electromagnetic field with matter, for various different classes of material media. For linear and isotropic dielectrics and magnetic materials, we have the constitutive relationships

$$\mathbf{D} = \varepsilon\mathbf{E}, \;\; \mathbf{B} = \mu\mathbf{H} \tag{14}$$

where $\varepsilon = \varepsilon_0 \varepsilon_r$ is the absolute permittivity of the medium, $\varepsilon_0 = 8.854 \times 10^{-12} C^2 N^{-1} m^2$ absolute permittivity of vacuum, ε_r the relative permittivity of the medium, and $\mu = \mu_0 \mu_r$ is the absolute permeability of the medium, $\mu_0 = 12.566 \times 10^{-7} NA^{-2}$ absolute permeability of vacuum, μ_r the relative permeability of the medium.

The Electromagnetic Field Energy

In the domain Ω in which we have an electromagnetic field, its energy – supposing it does exist – can be transformed into heat, work or other forms of energy. A part of it can leave the domain, through its boundary $\partial\Omega$, still in an electromagnetic form. If the media in Ω are rigid and fixed (at rest) then the work can not be done. We then can admit that we have only two reasons for the decrease of the electromagnetic energy. On the one hand, a part of it can be transformed into other forms through conduction, and, on the other hand, a part can leave the domain through its bounding surface [26].

Energy transformation from the electromagnetic form in other forms, or from other forms into the electromagnetic one, during conduction processes, is governed locally at every point in Ω by the law:

$$p_J = \mathbf{E} \cdot \mathbf{J} \tag{15}$$

where $p_J(\mathbf{r}, t)$ is a scalar function of position and time, named volume density of the instantaneous electromagnetic power, characterizing energy transformation processes. Thus, in the domain Ω we have:

$$\int_{\Omega} \mathbf{E} \cdot \mathbf{J} dv dt + dW_{em} + P_{\partial\Omega} dt = 0 \tag{16}$$

where dW_{em} represents the variation of the electromagnetic field energy, and $P_{\partial\Omega} dt$ represents the energy transferred through the domain boundary. In the time interval dt, equation (16) yields:

$$-\int_{\Omega} \mathbf{E} \cdot \mathbf{J} dv = \frac{dW_{em}}{dt} + P_{\partial\Omega} \tag{17}$$

Starting from relation (17) and Maxwell's equations, the theorem of electromagnetic energy demonstrates that

$$W_{em} = \int_{\Omega} \frac{\mathbf{D}^2}{2\varepsilon} dv + \int_{\Omega} \frac{\mathbf{B}^2}{2\mu} dv = W_e + W_m \tag{18}$$

$$W_e = \int_{\Omega} \frac{\mathbf{D}^2}{2\varepsilon} dv = \int_{\Omega} \frac{\mathbf{DE}}{2} dv = \int_{\Omega} \frac{\varepsilon \mathbf{E}^2}{2} dv \tag{19}$$

$$W_m = \int_{\Omega} \frac{\mathbf{B}^2}{2\mu} dv = \int_{\Omega} \frac{\mathbf{BH}}{2\mu} dv = \int_{\Omega} \frac{\mu \mathbf{H}^2}{2} dv \tag{20}$$

where W_m is the magnetic component of the electromagnetic field energy, and W_e is the electric component of the electromagnetic field energy. On the other hand, the electromagnetic power, transferred to the surroundings is given by

$$P_{\partial\Omega} = \oint_{\Omega} (\mathbf{E} \times \mathbf{H}) \mathbf{n} dS = \oint_{\Omega} \mathbf{Sn} dS \tag{21}$$

where the vector $\mathbf{S} = \mathbf{E} \times \mathbf{H}$ is known as the Poynting vector.

The Variational Formulation of Electromagnetic Field Theory

The electromagnetic field analysis admits a differential formulation and a variational equivalent formulation in the Hilbert space. *The mathematical variational model presumes the establishment of a variational principle capable to supply the equations of the differential mathematical model through the stationary condition of an adequate functional.*

Be the expression:

$$\Im = \iiint_{\Omega} f(x, y, z, V, \frac{\partial V}{\partial x}, \frac{\partial V}{\partial y}, \frac{\partial V}{\partial z}) dx dy dz + \iint_{\Sigma} g(x, y, z) dS \tag{22}$$

called *functional associated of the phenomenon* described through the scalar state quantity $V(x,y,z)$. The quantities $\partial V/\partial x$, $\partial V/\partial y$ and $\partial V/\partial z$ are the partial derivatives of the state quantity. For a phenomenon whose differential model is known, the integrand f of the first integral is a known function on the calculation domain Ω of the quantity V, and the integrand g of the second integral is a known function on the boundary Σ of the domain Ω.

The essence of the variational formulation of a phenomenon consists of the fact that the minimization of the expression (22) *leads to its differential model, i.e. the differential equation is satisfied by the state quantity and its limit conditions.*

For an one-dimensional problem (1D), in which the state size V depends only on the x coordinate, and the second term of the relation (22) doesn't exist, we will deduce properties of the integrand f of the functional.

Be a phenomenon whose unknown state quantity $V(x)$ is defined in the domain $[x_1, x_2]$ and satisfies the limit conditions $V(x_1)=V_1$, $V(x_2)=V_2$.

The functional associated of the phenomenon is expressed through the integral:

$$\Im = \int_{\Omega} f(x, V, \frac{dV}{dx})dx \tag{23}$$

Be $\widetilde{V}$ an approximate solution. In relation to the exact solution $V(x)$, the approximate solution can be written as:

$$\widetilde{V}(x) = V(x) + \delta V(x) \tag{24}$$

where $\delta V(x)$ is the infinitesimal variation of the solution $V(x)$. The stationarity condition of the integral (23), and the minimum value of the functional corresponds to the cancelation of the functional variation $\delta\,\Im$ corresponding to the variation of the solution $\delta V(x)$, can be respectively represented by the relations:

$$\delta\Im = \int_{x_1}^{x_2} \delta f \cdot dx = \int_{x_1}^{x_2} (\frac{\partial f}{\partial x}\delta x + \frac{\partial f}{\partial V}\delta V + \frac{\partial f}{\partial V'}\delta V')dx = \int_{x_1}^{x_2} (\frac{\partial f}{\partial V}\delta V + \frac{\partial f}{\partial V'}\delta V')dx = 0 \tag{25}$$

where $V' = \dfrac{dV}{dx}$ is the derivative of V. The infinitesimal variation δx of the variable x in (25) is null because we search for the variation δf of the integrand f for a given value of the variable x. Integrating by parts the last term of the integrand of expression (25), we obtain:

$$\int_{x_1}^{x_2} \frac{\partial f}{\partial V'}\delta V'dx = \int_{x_1}^{x_2} \frac{\partial f}{\partial V'}\delta(\frac{dV}{dx})dx = \int_{x_1}^{x_2} \frac{\partial f}{\partial V'}\frac{d}{dx}(\delta V)dx = \left[\frac{\partial f}{\partial V'}\delta V\right]_{x_1}^{x_2} - \int_{x_1}^{x_2} \frac{d}{dx}(\frac{\partial f}{\partial V'})\delta V dx \tag{26}$$

The expression (25) then becomes:

$$\delta\Im = \int_{x_1}^{x_2} [\frac{\partial f}{\partial V} - \frac{d}{dx}(\frac{\partial f}{\partial V'})]\delta V dx + \left[\frac{\partial f}{\partial V'}\delta V\right]_{x_1}^{x_2} = 0 \tag{27}$$

Because the variation δV is arbitrary, each term of expression (27) must be null, i.e.

$$\frac{\partial f}{\partial V} - \frac{d}{dx}(\frac{\partial f}{\partial V'}) = 0 \tag{28}$$

and

$$\left[\frac{\partial f}{\partial V'}\delta V\right]_{x_1}^{x_2} = 0 \tag{29}$$

Relation (28) must correspond to the differential equation of the phenomenon, and relation (29) must correspond to the limit conditions of the differential model. If

$$\delta V(x_1) = 0 \text{ and } \delta V(x_2) = 0 \tag{30}$$

then relation (29) is fulfilled. This implies that the values of the state size $V(x)$ at the two limits of the calculation domain are defined, which corresponds to the Dirichlet type limit conditions, called forced limit conditions. If the state size doesn't satisfy Dirichlet forced limit conditions, the variation δV being arbitrary, then the next condition must be fulfilled:

$$\left[\frac{\partial f}{\partial V'}\right]_{x_1} = \left[\frac{\partial f}{\partial V'}\right]_{x_2} = 0 \tag{31}$$

called natural limit condition. This condition must be implicitly fulfilled by the function f.

It is obvious that relation (29) is satisfied also in the case in which at one of the two limits, the differential model implies a natural limit condition, and at the other a forced limit condition.

If the solution in finite element is based on the phenomenon variational formulation, which implies knowing f and g in the functional (25), then the natural limit conditions are automatically incorporated in the formulation and only the forced conditions must be imposed to the solutions.

The functional of the non-stationary electromagnetic field associated to the domain Ω and the volume delimitated by Ω, is:

$$\Im = \int_{\Omega}\{(\int_0^E \mathbf{D}\cdot\mathbf{E} - \int_0^B \mathbf{H}\cdot\mathbf{B}) + (\mathbf{J}\cdot\mathbf{A} - \rho_v V)\}dxdydz \tag{32}$$

where $\mathbf{A}$ is the magnetic vector potential

$$\nabla\times\mathbf{A} = \mathbf{B} \tag{33}$$

and V is the electric potential

$$\mathbf{E} = -\nabla V \tag{34}$$

The minimization of the functional (32) involves the fundamental equations of the electromagnetic field (6)-(9). *On the other hand, the variational formulation implies the particularization of the functional* (32) *according to the specific electromagnetic field problem and with the physical properties of media, and incorporation in the functional expression of the uniqueness conditions of the problem solution.*

The following chapters of this book will refer to the three existence regimes of electromagnetic field which variational formulations are presented here.

The electrostatic field for linear, isotropic and homogenous media, without permanent polarization, is governed by the set of equations

$$\nabla \times \mathbf{E} = 0 \tag{35}$$

$$\nabla \cdot \mathbf{D} = \rho_v \tag{36}$$

$$\mathbf{D} = \varepsilon \mathbf{E} \tag{37}$$

In this case the functional associated of the one-dimensional electrostatic field

$$\Im(V) = \int_{\Omega} \frac{\varepsilon}{2} \left[\left(\frac{\partial V}{\partial x} \right)^2 - \rho_v V \right] dx \tag{38}$$

The stationary electric field, also called the steady electric field of steady current or electro-kinetic field, for linear, isotropic and homogenous conductors, without permanent polarization, is governed by the set of equations

$$\nabla \times \mathbf{E} = 0 \tag{39}$$

$$\nabla \cdot \mathbf{J} = 0 \tag{40}$$

$$\mathbf{J} = \sigma \mathbf{E} \tag{41}$$

where the relation (41) represents the law of electric conduction, here σ is the conductivity of the conductors. In this case the functional associated of the one-dimensional stationary electric field

$$\Im(V) = \int_{\Omega} \frac{\sigma}{2} \left[\left(\frac{\partial V}{\partial x} \right)^2 \right] dx \tag{42}$$

The stationary magnetic field also called the steady magnetic field of steady current, governed by the equations

$$\nabla \times \mathbf{H} = \mathbf{J} \tag{43}$$

$$\nabla \cdot \mathbf{B} = 0 \tag{44}$$

$$\mathbf{B} = \mu \mathbf{H} \tag{45}$$

has the associated functional

$$\Im(\Phi) = \int_{\Omega} \frac{\mu}{2} \left[\left(\frac{\partial \Phi}{\partial x} \right)^2 \right] dx \tag{46}$$

where Φ is the magnetic flux.

The functionals (38), (42) and (46) will be explicitly used in the calculations illustrated in the following chapters.

The Principles of Minimum Absorbed Power and Energy for Electric Circuits

Key contributions to the theory of electric circuits include the definition of "content and co-content" function for the non-linear resistor in the works of [27], [28]. Penfield, in [29], demonstrates the variational principles for "content and co-content" function of the non-linear resistive and reciprocal circuits, in connection with Tellegen's theorem. Smith, in [30], and [31], does an examination of the relationship of electric and magnetic energy storage to the terminal behavior of linear passive networks and puts in evidence the average energy storage by one-port and the minimum energy synthesis. Further contributions to the theory of the minimum and maximum of the powers in the electric circuits can be found in the Romanian technical literature, such as in Ionescu [32], who applied the Hilbert space properties to the analysis of the powers in non-sinusoidal regime, Mocanu, in [33], who introduced the theorem of the minimum power in the resistances for DC circuits, Hantila, in [34], and Fireteanu, in [35], who defined the energy functional for solving the electromagnetic field.

The original contribution of this work refers to the interpretation of the basic circuit laws of electric circuits in static, stationary or quasi-stationary regime in terms of suitable power functionals defined for each type of circuit. The contents illustrated below build on a number of recent contributions of the authors, such as [36], [37], [38], [39], [40], [41], [42], [43], [44], [45], [46], [47], [48], [49].

Using the Hilbert space properties, this work aims at formulating a set of minimum power principles for electric circuits in stationary or quasi-stationary regime, including the case of circuits with distorted voltage and current waveforms. For these circuits, this paper establishes the conditions under which it is possible to formulate the minimum power principles, demonstrates the principles, shows their limits of application and provides a set of illustrative examples.

The common rationale for establishing the minimum power principles is that starting from the definition of the functionals and from topology information it is possible to obtain the conditions under which the circuit variables (e.g., voltages, according to the second Kirchhoff's law) are described. Once identified the relations to obtain the minimum value (when existing) of the functional, it is shown that these relations are verified by the circuit currents that satisfy the first Kirchhoff's law.

Starting from analogies existing between magnetic and electric circuits, the basic concepts of the construction of power and energy functionals for linear electric can be extended to linear electric and magnetic circuits operating in quasi-stationary regime. The results presented in the sequel refer to circuits containing only one type of conservative element, that is, circuits with capacitors or circuits with inductors.

Variational Formulation for Determining the Principle of Minimum Absorbed Power for Dc Circuits

Let us consider a reciprocal DC electric circuit, with K branches and N nodes. All branches are represented with Thévenin equivalents, identifying the voltages and currents as shown in Figure 1 for a generic branch $k = 1, ..., K$, given the conventional direction of the current I_k. The same reference polarity is assumed for the branch voltage ΔV_k and for the voltage U_k at the resistance terminals.

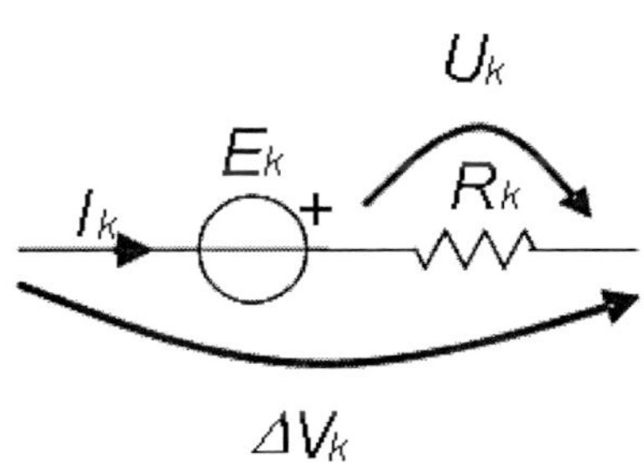

Figure 1. Branch representation. In the conventional indications used for voltages, the arrows point to the negative polarity.

The branch may contain a voltage source E_k with specified polarity. Considering all circuit branches, their voltages and currents form a specific set of vectors belonging to $\mathfrak{R}^{K,1}$, defined as follows (the superscript T denotes transposition):

$$\mathbf{i} = [\mathbf{I}_1, \mathbf{I}_2, ..., \mathbf{I}_\mathbf{K}]^\mathrm{T} \tag{47}$$

$$\mathbf{u} = [\mathbf{U}_1, \mathbf{U}_2, ..., \mathbf{U}_\mathbf{K}]^\mathrm{T} \tag{48}$$

$$\mathbf{e} = [\mathbf{E}_1, \mathbf{E}_2, ..., \mathbf{E}_\mathbf{K}]^\mathrm{T} \tag{49}$$

$$\Delta\mathbf{v} = [\Delta\mathbf{V}_1, \Delta\mathbf{V}_2, ..., \Delta\mathbf{V}_\mathbf{K}]^\mathrm{T} \tag{50}$$

Furthermore, the diagonal branch conductance matrix $G \in \Re^{K,K}$ is defined in terms of the conductance $G_k = 1/R_k$, for $k = 1, ..., K,$ as

$$\mathbf{G} = diag(G_1, G_2, ..., G_K) \tag{51}$$

so that

$$\mathbf{i} = \mathbf{G}\,\mathbf{u} \tag{52}$$

The branch voltages verify the *second Kirchhoff's law*, according to which:

$$\Delta\mathbf{v} = \mathbf{u} - \mathbf{e} \tag{53}$$

This kind of partitioning is one of the key aspects of the formulation of the reasoning proposed for determining the minimum power. It is important to note that the first Kirchhoff's law has not been invoked yet.

Considering the same reference direction of the current and reference polarity of the voltage, shown in Figure 1, according to common notation [24], [33], and [50], let us define, in stationary regime, by using the relation (42), the following power functional $\Im : \Re^{\mathbf{K}} \rightarrow \Re$ in the Hilbert space $\Re^K$ representing one half of the *power absorbed* by all the resistances in stationary regime:

$$\Im \overset{\Delta}{=} \frac{1}{2}\mathbf{u}^{\mathrm{T}}\mathbf{i} \tag{54}$$

By further elaborating (54), taking into account (52) and (53), one obtains:

$$\Im(\mathbf{v}) \overset{\Delta}{=} \frac{1}{2}(\mathbf{e} + \Delta\mathbf{v})^{\mathrm{T}} \cdot \mathbf{G} \cdot (\mathbf{e} + \Delta\mathbf{v}) \tag{55}$$

The components of the vector are the differences between pairs of potentials of the circuit nodes, depending on how the branches are connected in the circuit. Considering constant voltage sources e, the functional (54) can be directly expressed in function of the vector $v \in \Re^{N,1}$ containing the potentials of all the circuit nodes:

$$\mathbf{v} = [V_1, V_2, ..., V_N] \tag{56}$$

To determine the extreme of the functional (54), we apply the function of matrix properties. In stationary regime, the derivative $\dfrac{\partial \Im}{\partial \mathbf{v}}$ with respect to the potentials at the N nodes yields N relations that can be synthetically expressed in the form

$$\frac{\partial \Im}{\partial v} = \frac{1}{2} \mathbf{\Gamma}_{N,K} \, \mathbf{i} = 0 \tag{57}$$

where $\mathbf{\Gamma}_{N,K} \in \Re^{N,K}$ is the branch-to-node incidence matrix which contains the values 0, 1, and -1. Each row of the derivative $\frac{\partial \Im}{\partial \mathbf{v}} \in \Re^{N,1}$ contains the algebraic sum of the currents flowing in the branches converging to the corresponding node, thus verifying for all the N nodes of the circuit the *first Kirchhoff's law*. On these bases, it emerges that $\frac{\partial \Im}{\partial \mathbf{v}} = 0$, namely, the solution obtained also corresponds to an extreme point of the functional. In order to establish the nature of this extreme point, it can be observed that the functional $\Im$ is a quadratic form, and the matrix G is positive defined (assuming that all the resistances are positive).

Thus, the functional is non-negative ($\Im(\mathbf{v}) \geq 0$), whatever the values of the components of vector v might be. Here, the trivial case $\Im(\mathbf{v}) = 0$ is of no practical interest. Thus, the extreme point of the functional is a *minimum*.

According to the definition (54), from the results obtained in (57) we formulate the *Principle of Minimum Power* (*PMP*): *"In a reciprocal and resistive DC circuit, in which the branch voltages obey the second Kirchhoff's law, the condition of minimum power absorbed in the resistances is consistent with the first Kirchhoff's law"*. In other words, *in a reciprocal and resistive DC circuit the currents and voltages are distributed such as the absorbed power in the resistive branches of the circuit is minimum*.

Example 1: Determination of the minimum power functional for a linear DC circuit.

Let us consider the DC circuit shown in Figure 2, with K = 3, and N = 2, where $R_1 = 1\Omega, R_2 = 2\Omega, R_3 = 3\Omega, E_1 = 4\,\text{V}$.

The conductance matrix is $\mathbf{G} = diag(1/R_1, 1/R_2, 1/R_3)$, with $R_k > 0$ for k = 1,.., K.

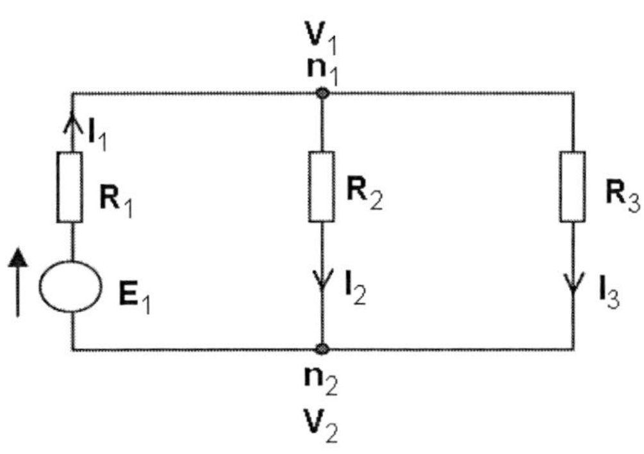

Figure 2. DC circuit with three branches.

The circuit currents and voltages at the resistance terminals (already written in terms of the node voltages) are contained in the vectors

$$\mathbf{i}=\begin{bmatrix} I_1 \\ I_2 \\ I_3 \end{bmatrix};\ \mathbf{u}=\begin{bmatrix} V_2-V_1+E_1 \\ V_1-V_2 \\ V_1-V_2 \end{bmatrix};\ \mathbf{v}=[V_1,V_2] \tag{58}$$

For this DC circuit the branch-to-node incidence matrix $\mathbf{\Gamma}_{N,K}$ is

$$\mathbf{\Gamma}_{N,K}=\begin{bmatrix} -1 & 1 & 1 \\ 1 & -1 & -1 \end{bmatrix} \tag{59}$$

and the power functional, constructed according to (55), is

$$\Im=\frac{1}{2}\{G_1(V_2-V_1+E_1)^2+G_2(V_1-V_2)^2+G_3(V_1-V_2)^2\} \tag{60}$$

By calculating $\frac{\partial \Im}{\partial \mathbf{v}}$, one obtains the following equations, that verify the relation (57) and thus the first Kirchhoff's law at each node (n_1 and n_2):

$$\frac{\partial \Im}{\partial V_1}=-G_1(V_2-V_1+E_1)+G_2(V_1-V_2)+G_3(V_1-V_2)=-I_1+I_2+I_3=\sum_{l_k\in n_1} I_k=0 \tag{61}$$

$$\frac{\partial \Im}{\partial V_2}=G_1(V_2-V_1+E_1)-G_2(V_1-V_2)-G_3(V_1-V_2)=I_1-I_2-I_3=\sum_{l_k\in n_2} I_k=0 \tag{62}$$

For this DC circuit, a PSPICE and MATHCAD algorithm can be used to demonstrate that the functioning point of each resistance of the circuit does not represent the maximum absorbed power point, [51]. Based on the Thèvenin's theorem, the variations of the absorbed powers in each resistances of the circuit P_1, P_2, P_3 depending on the currents I_1, I_2, I_3 are calculated. The dependences between the absorbed power and the current has the expression

$$P_i=(U_{i,0}-R_{e,i}I_i)I_i,\ i=1,2,3 \tag{63}$$

where $U_{i,0}$ is the open voltage and $R_{e,i}$ is the equivalent resistance of the circuit at the terminals of each branch. The power-current *P*(*I*) characteristics are shown in Figures 3, 4 and 5, respectively.

It is remarkable to observe that the absorbed power in each branch has a value lower than the maximum value of the power obtained as one half of the value of the short-circuit current

$\frac{I_{sc}}{2}$

$$P_1 = 3.305\text{W} < P_{1max} = 3.33\text{W}$$
$$P_2 = 2.38\text{W} < P_{2max} = 3\text{W} \qquad (64)$$
$$P_3 = 1.586\text{W} < P_{3max} = 2.67\text{W}$$

On the other side, from Figures 3, 4, and 5, we observe that for each branch the value of the absorbed power P_1, P_2, P_3 (the functioning points) correspond to two values of each current:

$$I_1^{'} = 1.818A, I_1^{''} = 1.515A;$$
$$I_2^{'} = 1.09A, I_2^{''} = 2.909A; \qquad (65)$$
$$I_3^{'} = 0.727A, I_3^{''} = 3.272A.$$

Only one of these values ($I_1^{'}, I_2^{'}, I_3^{'}$) verifies the first Kirchhoff's law. For the branches with resistances, these current values represent the roots of a second degree equation, that is, $I^{','' } = \frac{U_0}{2R_e}(1 \pm \frac{R_e - R}{R_e + R})$.

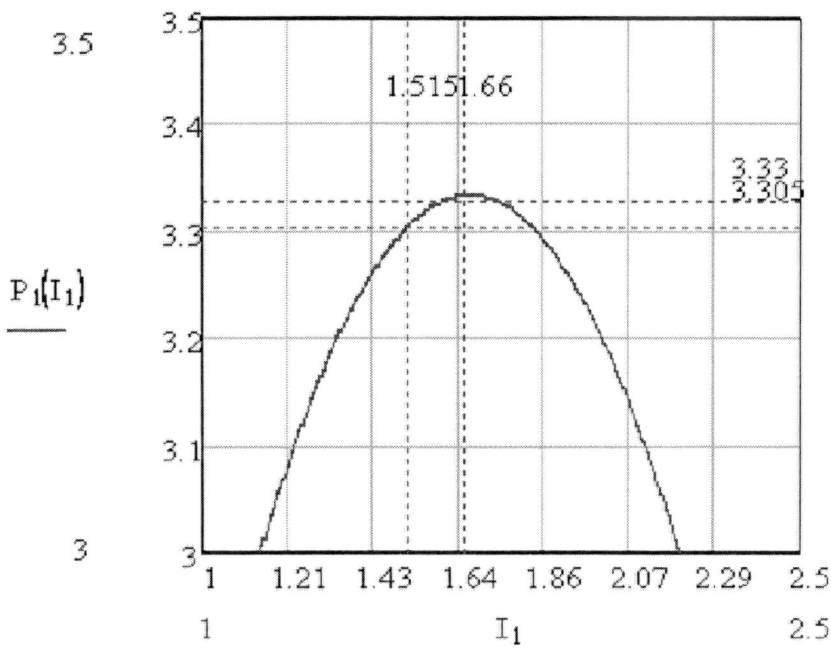

Figure 3. The variation of P1.

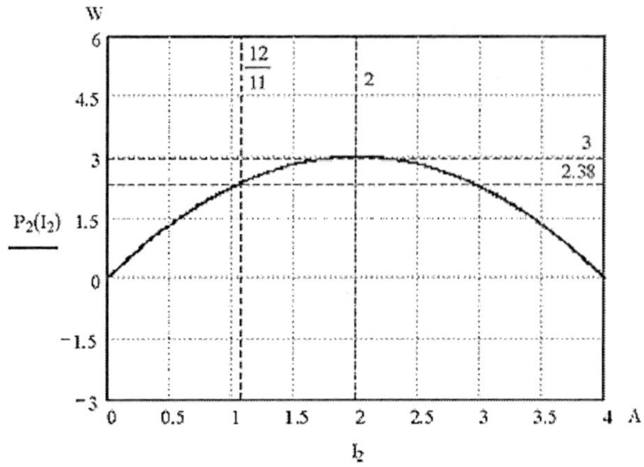

Figure 4. The variation of P2.

The graphic positions of these two values of current compared with the value $\frac{I_{sc}}{2}$ *are depending on the resistance value: if* $R < R_e$ *then* $I' > \frac{I_{sc}}{2}$, *and if* $R > R_e$ *then* $I'' < \frac{I_{sc}}{2}$, as presented in Figure 4 and respectively in Figure 5.

In conclusion, in a DC circuit the currents are distributed on the branches in such a way that they verify, simultaneously, two conditions: the 2nd Kirchhoff law, and the principle of minimum of absorbed power (corresponding to the first Kirchhoff's law).

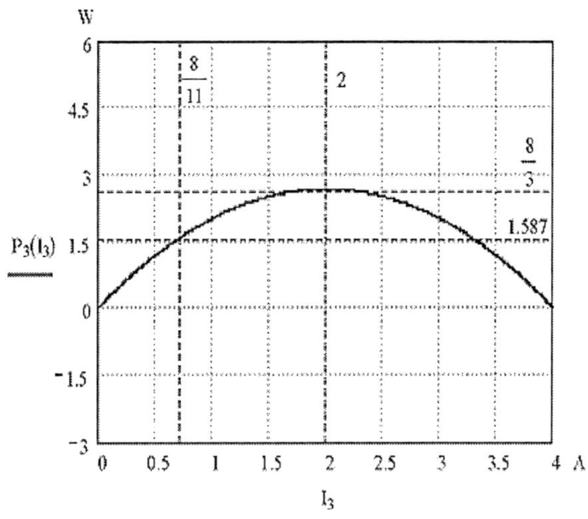

Figure 5. The variation of P3.

Variational Formulation for Determining the Principle of Minimum Absorbed Active and Reactive Power for Linear Ac Circuits

In Andrei and Spinei [45], the minimum energy principle has been demonstrated for electric circuits with capacitors, and for magnetic circuits with coils. More specifically, in a linear circuit with only inductive components (coils) the magnetic potentials and the fluxes linking the coils in the circuit are distributed in such a way to correspond to the minimum magnetic energy of the circuit. Likewise, in a circuit with only capacitive components, electrostatic regime the electric charges and the potentials at the capacitor terminals correspond to the minimum electrostatic energy of the circuit. These results have been demonstrated under the hypothesis that the circuits contain only conservative elements of the same type. In this section we consider the quasi-stationary regime (AC) of a linear electric circuit with K branches and N nodes, and we demonstrate the existence and the limits of applicability of the minimum power principle for such circuit. By using the symbolic method, the voltage at branch $k = 1, \ldots, K$ of the circuit is expressed as:

$$\Delta\overline{V}_k = \overline{V}_{i,k} - \overline{V}_{j,k} = \overline{U}_k - \overline{E}_k = \frac{\overline{I}_k}{\overline{Y}_k} - \overline{E}_k \tag{66}$$

As illustrated in Figure 6, given the reference direction of the complex current $\overline{I}_k$ the same reference polarity is assumed for the complex voltage $\overline{U}_k$ at the admittance terminals and for the complex branch voltage $\Delta\overline{V}_k$. The branch may contain a complex voltage source $\overline{E}_k$ with given polarity. Additional notation is used for the potentials $\overline{V}_{i,k}$ and $\overline{V}_{i,k}$ of the terminals of branch k to be connected to node i and node j, respectively.

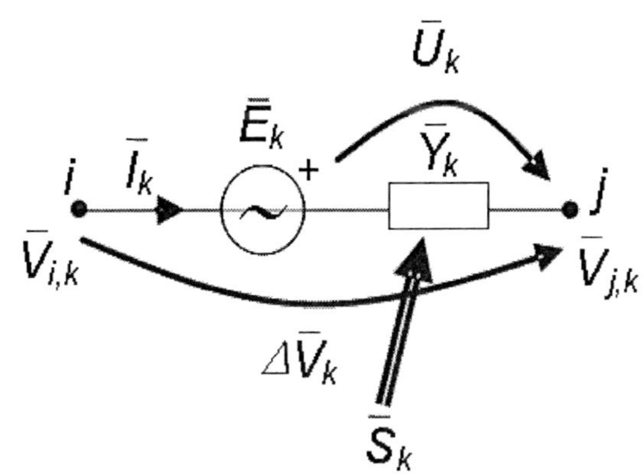

Figure 6. AC circuit branch. In the conventional indications used for voltages, the arrows point to the negative polarity.

Hence,

$$\begin{aligned} &\overline{V}_{i,k} = x_{i,k} + jy_{i,k}; && \overline{V}_{j,k} = x_{j,k} + jy_{j,k}; \\ &\overline{Y}_k = G_k - jB_k; && \overline{E}_k = a_{E,k} + jb_{E,k} \end{aligned} \tag{67}$$

The conjugated complex current in branch k can be expressed as:

$$\bar{I}_k^* = \bar{Y}_k^*(\bar{V}_{i,k} - \bar{V}_{j,k} + \bar{E}_k)^* = [G_k(x_{i,k} - x_{j,k} + a_{E,k}) + B_k(y_{i,k} - y_{j,k} + b_{E,k})] + \\ + j[B_k(x_{i,k} - x_{j,k} + a_{E,k}) - G_k(y_{i,k} - y_{j,k} + b_{E,k})] \quad (68)$$

In the conditions described above, the overall *complex absorbed (entering) power* [24], [33], and [50] related to the admittances of the K branches of the circuit is

$$\sum_{k=1}^{K} \bar{S}_k = \sum_{k=1}^{K} \bar{U}_k \bar{I}_k^* = \sum_{k=1}^{K} (G_k + jB_k)[(x_{i,k} - x_{j,k} + a_{E,k})^2 + (y_{i,k} - y_{j,k} + b_{E,k})^2] \quad (69)$$

The real and imaginary parts of the complex power can be used to define two functionals – according to relation (42) - in Hilbert space: [44], [47], and [52]. The first functional is linked to the real power absorbed by the resistive part of the circuit:

$$\Im_R \equiv \frac{1}{2} Re[\bar{S}] : \Re^{2N} \to \Re; \ \Im_R(x_1, x_2, ..., x_N, y_1, y_2, ..., y_N) = \\ = \frac{1}{2} \sum_{k=1}^{K} G_k [(x_{i,k} - x_{j,k} + a_{E,k})^2 + (y_{i,k} - y_{j,k} + b_{E,k})^2] \quad (70)$$

The second functional is linked to the reactive power absorbed by the reactive part of the circuit:

$$\Im_I \equiv \frac{1}{2} Im[\bar{S}] : \Re^{2N} \to \Re; \Im_I(x_1, x_2, ..., x_N, y_1, y_2, ..., y_N) = \\ = \frac{1}{2} \sum_{k=1}^{K} B_k [(x_{i,k} - x_{j,k} + a_{E,k})^2 + (y_{i,k} - y_{j,k} + b_{E,k})^2] \quad (71)$$

These functionals are a function class C^2 in $\Re^{2N}$ Considering $G_k > 0$ for $k = 1,..., K$, the real function of the complex power is positive defined, that is, $\Im_R(x_1, x_2, ..., x_N, y_1, y_2, ..., y_N) > 0$, for all the pairs (x_i, y_i), $i = 1, ..., N$. Then, *the active power absorbed has a minimum.*

As regards the imaginary functional of the complex power, it is important to analyze the following cases:

i. AC circuits containing *only* resistive-inductive branches, i.e., $B_k > 0$ for $k = 1,...,K$. The imaginary functional $\Im_I = \frac{1}{2} \sum_{k=1}^{K_L} B_k [(x_{i,k} - x_{j,k} + a_{E,k})^2 + (y_{i,k} - y_{j,k} + b_{E,k})^2] > 0$, that is, it is positive defined for all the pairs (x_i, y_i), $i = 1, ..., N$, so that we can state that the *reactive power absorbed by the inductors has a minimum* (see Example 2 below);

ii. AC circuits containing *only* capacitive elements, i.e., $B_k < 0$ for k = 1,…,K. The imaginary functional $\Im_I = \frac{1}{2}\sum_{k=1}^{K_C} B_k [(x_{i,k} - x_{j,k} + a_{E,k})^2 + (y_{i,k} - y_{j,k} + b_{E,k})^2] < 0$, that is, it is negative defined for all the pairs $(x_i, y_i), i = 1,..., N$. In this case, reversing the sign of the reactive power functional it is possible to obtain a positive defined functional and state that the reactive power *generated by the capacitors has a minimum;*

iii. AC circuits with resistive and capacitive components: if a branch contains a capacitor in series to a resistance, it is possible to introduce an auxiliary node in the branch to isolate the capacitor from the resistance. In this case, from the real power functional one obtains that the real power absorbed by the resistances has a minimum, while the reactive power generated by the capacitors has a minimum, as in ii, (see Example 3 below).

iv. in general, when branches of inductive and capacitive nature coexist in the system, the reactive functional F_I is not positive defined, so that it is not possible to conclude that the total reactive power absorbed by the branches has a minimum.

A particular situation occurs when $B_{k,L} = B_{k,C}$ for branch k = 1, …, K, corresponding to the resonance conditions at branch k, so that branch k provides a null contribution to the total (absorbed or generated) reactive power. However, this situation is consistent with the outcomes of the previous sub cases and does not add any further variant.

Consequently, the minimum points of the real and imaginary component of the complex power functionals (when existing) are the solutions of the system which contains 4N equations:

$$\begin{aligned}&\frac{\partial \Im_R}{\partial x_1} = 0, \frac{\partial \Im_R}{\partial y_1} = 0,..., \frac{\partial \Im_R}{\partial x_N} = 0, \frac{\partial \Im_R}{\partial y_N} = 0,\\ &\frac{\partial \Im_I}{\partial x_1} = 0, \frac{\partial \Im_I}{\partial y_1} = 0,..., \frac{\partial \Im_I}{\partial x_N} = 0, \frac{\partial \Im_I}{\partial y_N} = 0\end{aligned} \tag{72}$$

Let us consider the set of branches $\mathbf{\mathit{L}}^{(\mathbf{n})}$ converging to node n, for n = 1,…, N. If we calculate the algebraic sum of the solutions, with the derivatives of $\Im_I$ multiplied by -j, we obtain the expressions:

$$
\begin{aligned}
&\sum_{k \in L^{(1)}} \bar{I}_k^* = 0, \\
&\sum_{k \in L^{(2)}} \bar{I}_k^* = 0, \\
&\ldots\ldots \\
&\sum_{k \in L^{(N)}} \bar{I}_k^* = 0
\end{aligned} \tag{73}
$$

which represent the first Kirchhoff's law expressed in all the N nodes of the circuit.

Consequently, we may state the Principle of Minimum Active and Reactive Power (PMARP): "The minimum of the total active and reactive power absorbed (generated) by the resistive-inductive only (-capacitive only) branches of a linear AC circuit in quasi-stationary state is verified by the circuit currents and voltages that verify the first and second Kirchhoff's laws".

An alternative description of the same principle can be given as follows: "In the linear and reciprocal circuit in quasi-stationary AC regime, the currents and voltages are distributed in such a way that the total active and reactive power absorbed (generated) by resistive-inductive only (-capacitive only) branches of the circuit is minimal".

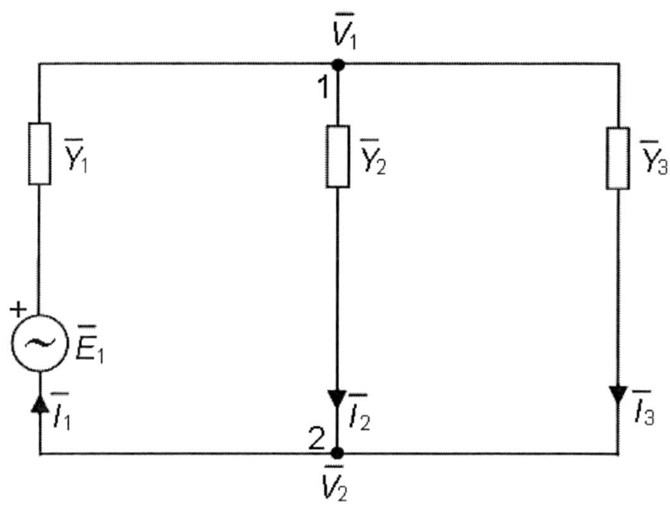

Figure 7. Resistive-inductive AC circuit.

It is however crucial to remark the limits of application of the *PMARP* principle. According to the deduction shown above, based on the sign of the functionals, the *PMARP* principle cannot be applied when inductive and capacitive components are coexisting in the same system. In such conditions, with respect to the power supplied by the generators, the interactions among the inductive and capacitive branches would determine additional reactive power generated by the capacitors going to additional reactive power absorbed by the inductive part of the branches. This conceptual result of general circuit theory is a limiting factor in practical applications with coexisting inductive and capacitive components. More specifically, before applying the minimum power principle one has to carefully check whether neglecting all the inductive (or capacitive) components is acceptable for the purpose of representation of the circuit.

Example 2: determination of the minimum power functional for a resistive-inductive AC circuit.

Let us consider the AC circuit shown in Figure 7, in which the admittances contain only resistive and inductive elements, that is, $G_k > 0$ and $B_k > 0$ for $k = 1,2,3$, satisfying the conditions imposed in case i). The expressions of the complex potentials, complex voltage source, and the expressions of the conjugated complex currents of the circuit are, respectively:

$$\begin{aligned}
&\overline{V_1} = x_1 + jy_1; \overline{V_2} = x_2 + jy_2; \overline{E_1} = a + jb; \\
&\overline{I}_1^* = (G_1 + jB_1)[(x_2 - x_1 + a) - j(y_2 - y_1 + b)]; \\
&\overline{I}_2^* = (G_2 + jB_2)[(x_1 - x_2) - j(y_1 - y_2)]; \\
&\overline{I}_3^* = (G_3 + jB_3)[(x_1 - x_2) - j(y_1 - y_2)].
\end{aligned} \tag{74}$$

The functionals (70) and (71) of the circuit are defined as

$$\Im_R = \frac{1}{2}\left\{G_1\left[(x_2 - x_1 + a)^2 + (y_2 - y_1 + b)^2\right] + G_2\left[(x_1 - x_2)^2 + (y_1 - y_2)^2\right] + G_3\left[(x_1 - x_2)^2 + (y_1 - y_2)^2\right]\right\} \tag{75}$$

$$\Im_I = \frac{1}{2}\left\{B_1\left[(x_2 - x_1 + a)^2 + (y_2 - y_1 + b)^2\right] + B_2\left[(x_1 - x_2)^2 + (y_1 - y_2)^2\right] + B_3\left[(x_1 - x_2)^2 + (y_1 - y_2)^2\right]\right\} \tag{76}$$

Assuming x_1 and y_1 as variables, the minimum of the functionals (75) and (76) is the solution of the system

$$\begin{aligned}
&\frac{\partial \Im_R}{\partial x_1} = -G_1(x_2 - x_1 + a) + G_2(x_1 - x_2) + G_3(x_1 - x_2) = 0 \\
&\frac{\partial \Im_R}{\partial y_1} = -G_1(y_2 - y_1 + b) + G_2(y_1 - y_2) + G_3(y_1 - y_2) = 0 \\
&\frac{\partial \Im_I}{\partial x_1} = -B_1(x_2 - x_1 + a) + B_2(x_1 - x_2) + B_3(x_1 - x_2) = 0 \\
&\frac{\partial \Im_I}{\partial y_1} = -B_1(y_2 - y_1 + b) + B_2(y_1 - y_2) + B_3(y_1 - y_2) = 0
\end{aligned} \tag{77}$$

By summing up the first equation of the system and the third equation multiplied by (−j), and summing up the second equation with the fourth equation multiplied by (−j), we obtain the first Kirchhoff's law expressed at node 1, with $\boldsymbol{L}^{(1)} = \{1,2,3\}$:

$$-\overline{I}_1^* + \overline{I}_2^* + \overline{I}_3^* = \sum_{k \in L^{(1)}} \overline{I}_k^* = 0 \tag{78}$$

We obtain a similar result if the variables of the system are x_2 and y_2, which are the real and imaginary parts of the potential

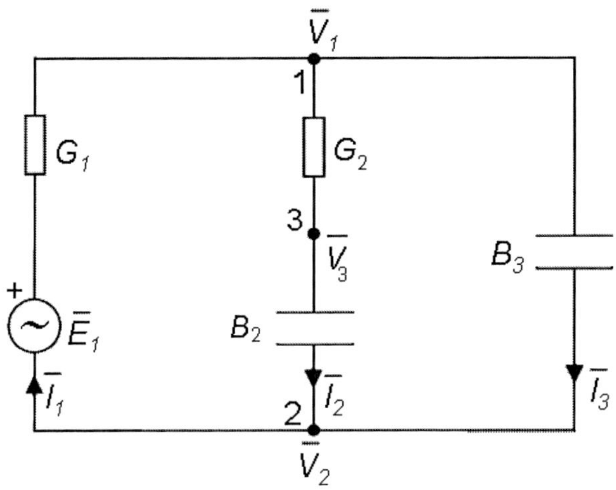

Figure 8. Resistive-capacitive AC circuit.

Example 3: determination of the minimum power functional for a resistive-capacitive AC circuit.

Let us consider the AC circuit shown in Figure 8, in which the branches contain only resistive and capacitive elements. According to the considerations introduced in case ii), let us introduce the auxiliary node 3 with potential $\overline{V}_3$ to isolate the capacitor with susceptance B_2 in the central branch.

The expressions of the complex potentials, complex voltage source, and the expressions of the conjugated complex currents of the circuit are, respectively:

$$\begin{aligned}
&\overline{V}_1 = x_1 + j\,y_1; \overline{V}_2 = x_2 + j\,y_2; \overline{V}_3 = x_3 + j\,y_3;\ \overline{E}_1 = a + jb;\\
&\overline{I}_1^* = G_1[(x_2 - x_1 + a) - j(y_2 - y_1 + b)];\\
&\overline{I}_2^* = G_2[(x_1 - x_3) - j(y_1 - y_3)];\\
&\overline{I}_2^* = jB_2[(x_3 - x_2) - j(y_3 - y_2)];\\
&\overline{I}_3^* = jB_3[(x_1 - x_2) - j(y_1 - y_2)].
\end{aligned} \tag{79}$$

The functionals (70) and (71) of the circuit are defined as

$$\Im_R = \frac{1}{2}\left\{G_1\left[(x_2 - x_1 + a)^2 + (y_2 - y_1 + b)^2\right] + G_2\left[(x_1 - x_3)^2 + (y_1 - y_3)^2\right]\right\} \tag{80}$$

$$\Im_I = \frac{1}{2}B_2\left[(x_3 - x_2)^2 + (y_3 - y_2)^2\right] + B_3\left[(x_1 - x_2)^2 + (y_1 - y_2)^2\right]\Big\} \tag{81}$$

Assuming x_1 and y_1 as variables, the minimum of the functional (80), and of the functional (81) with reversed sign, is the solution of the system

$$\begin{aligned}
&\frac{\partial \Im_R}{\partial x_1} = -G_1(x_2 - x_1 + a) + G_2(x_1 - x_3) = 0\\
&\frac{\partial \Im_R}{\partial y_1} = -G_1(y_2 - y_1 + b) + G_2(y_1 - y_3) = 0\\
&\frac{\partial(-\Im_I)}{\partial x_1} = -B_3(x_1 - x_2) = 0\\
&\frac{\partial(-\Im_I)}{\partial y_1} = -B_3(y_1 - y_2) = 0
\end{aligned} \tag{82}$$

By summing up the first equation of the system and the third equation multiplied by j, and summing up the second equation with the fourth equation multiplied by j), we obtain the first Kirchhoff's law expressed at node 1, with $\boldsymbol{L}^{(1)} = \{1, 2, 3\}$:

$$-\bar{I}_1^* + \bar{I}_2^* + \bar{I}_3^* = \sum_{k \in \boldsymbol{L}^{(1)}} \bar{I}_k^* = 0 \tag{83}$$

We obtain a similar result if the variables of the system are x_2 and y_2, which are the real and imaginary parts of the potential $\bar{V}_2$.

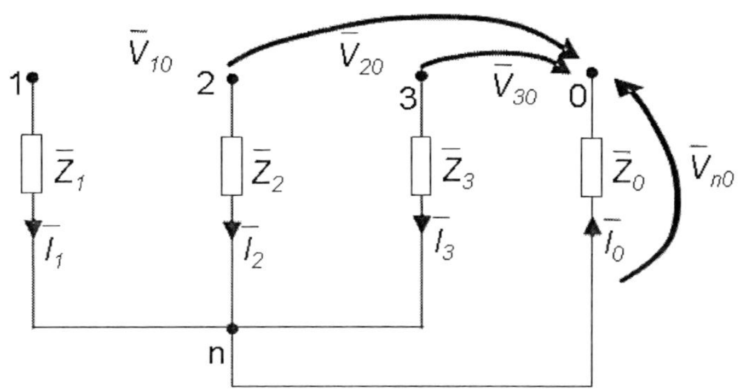

Figure 9. Three-phase circuit with star connection.

Example 4: determination of the minimum power functional for a three-phase circuit with neutral.

Let us consider the three-phase circuit in star connection, with neutral point n, shown in Figure 9, where the branch impedances are modeled as series-connected resistances and inductances, and can be transformed into equivalent admittances $\overline{Y}_k = 1/\overline{Z}_k = G_k - \mathrm{j}B_k$, for k = 0, 1, 2, 3. For the sake of representation, the values of $B_k > 0$ for $k = 1, 2{,}3$ satisfy the conditions imposed in case i). We start from known three-phase symmetrical voltages of the system

$$\overline{U}_{10} = U;\ \overline{U}_{20} = -a - \mathrm{j}b;\ \overline{U}_{30} = -a + \mathrm{j}b \tag{84}$$

and we consider the neutral point displacement voltage

$$\overline{U}_{\mathrm{n}0} = x + \mathrm{j}y \tag{85}$$

Then, the functional (82) for the active power can be expressed as

$$\Im_R = \frac{1}{2}\{\ \{G_1[(U-x)^2 + y^2] + G_2[(-a-x)^2 + (-b-y)^2] + G_3[(-a-x)^2 + (b-y)^2] + G_0(x^2+y^2)\ \} \tag{86}$$

and the functional (83) as

$$\Im_I = \frac{1}{2}\{\ \{B_1[(U-x)^2 + y^2] + B_2[(-a-x)^2 + (-b-y)^2] + B_3[(-a-x)^2 + (b-y)^2] + B_0(x^2+y^2)\ \} \tag{87}$$

The minimum of the functionals (86) and (87) is given by the solutions of the system containing the equations

$$\frac{\partial \Im_R}{\partial x} = 0;\ \frac{\partial \Im_R}{\partial y} = 0;\ \frac{\partial \Im_I}{\partial x} = 0;\ \frac{\partial \Im_I}{\partial y} = 0 \tag{88}$$

so that

$$\begin{aligned}
&\frac{\partial \Im_R}{\partial x} = -G_1(U-x) - G_2(-a-x) - G_3(-a-x) + G_0 x = 0\\
&\frac{\partial \Im_R}{\partial y} = G_1 y - G_2(-b-y) - G_3(b-y) + G_0 y = 0\\
&\frac{\partial \Im_I}{\partial x} = -B_1(U-x) - B_2(-a-x) - B_3(-a-x) + B_0 x = 0\\
&\frac{\partial \Im_I}{\partial y} = B_1 y - B_2(-b-y) - B_3(b-y) + B_0 y = 0
\end{aligned} \tag{89}$$

By summing up the first equation of the system and the third equation multiplied by ($-\mathrm{j}$), and summing up the second equation with the fourth equation multiplied by ($-\mathrm{j}$), we obtain

$$x = \frac{(G_1 - \mathrm{j}B_1)U - (G_2 - \mathrm{j}B_2)a - (G_3 - \mathrm{j}B_3)a}{(G_0 - \mathrm{j}B_0) + \sum_{k=1}^{3}(G_k - \mathrm{j}B_k)} \tag{90}$$

and

$$y = \frac{-(G_2 - \mathrm{j}B_2)b + (G_3 - \mathrm{j}B_3)b}{(G_0 - \mathrm{j}B_0) + \sum_{k=1}^{3}(G_k - \mathrm{j}B_k)} \tag{91}$$

The outcomes of equations (90) and (91) reproduce the well-known relation

$$\overline{U}_{\mathrm{n}0} = x + \mathrm{j}y = \frac{\sum_{k=1}^{3}(G_k - \mathrm{j}B_k)\overline{U}_{k0}}{(G_0 - \mathrm{j}B_0) + \sum_{k=1}^{3}(G_k - \mathrm{j}B_k)} \tag{92}$$

that incorporates the first Kirchhoff's law expressed at node n.

Variational Formulation for Determining the Principle of Minimum Active and Reactive Power for Linear Circuits in Non-Sinusoidal Regime

The concepts illustrated above can be extended to the case of AC circuits operating in non-sinusoidal regime. Let us consider a linear circuit with K branches and N nodes, operating under non-sinusoidal conditions with radian frequency at the fundamental harmonic.

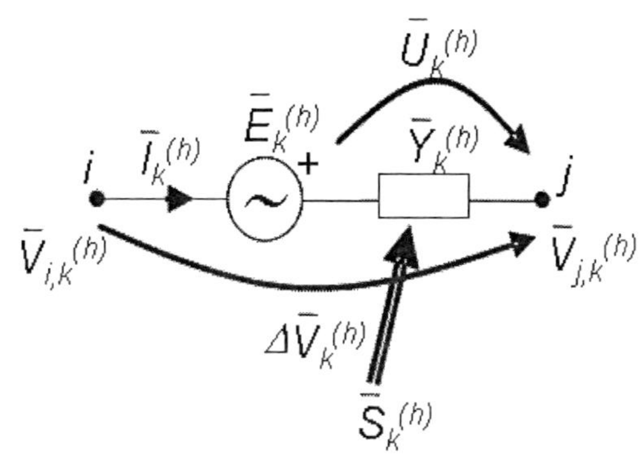

Figure 10. AC circuit under non-sinusoidal conditions.

For each branch of the circuit, the voltage and the current are expressed by using the Fourier series:

$$u_k(t)=\sum_{h=0}^{H} U_k^{(h)}\sqrt{2}\,sin(h\omega t+\varphi_k^{(h)})\;\; u_k(t)=\sum_{h=0}^{H} U_k^{(h)}\sqrt{2}\,sin(h\omega t+\varphi_k^{(h)}) \quad (93)$$

$$i_k(t)=\sum_{h=0}^{H} I_k^{(h)}\sqrt{2}\,sin(h\omega t+\varphi^{(h)}_k-\gamma_k^{(h)}) \quad (94)$$

where $k=1,...,K$ is the number of branches, H is the finite number of harmonics considered, $U_k^{(h)}$ and $I_k^{(h)}$ represents the RMS values of the hth-harmonic of voltage and current, respectively, while $\varphi_k^{(h)}$ and $\varphi_k^{(h)}-\gamma_k^{(h)}$ represent the phase angle of the hth harmonic of voltage and current, respectively. Considering branch *k* of the circuit (Figure 10), by using the complex representation for the hth harmonic of voltage and current, with the same conventional circuit representation adopted in Section for AC circuits in sinusoidal regime, we get

$$\overline{\Delta V}_k^{(h)}=\overline{V}_{i,k}^{(h)}-\overline{V}_{j,k}^{(h)}=\overline{U}_k^{(h)}-\overline{E}_k^{(h)}=\frac{\overline{I}_k^{(h)}}{\overline{Y}_k^{(h)}}-\overline{E}_k^{(h)} \quad (95)$$

In the Hilbert space we can define, similarly with relations (70) and (71), the functionals corresponding to the hth harmonic of real and imaginary parts of the complex absorbed power by all the *K* branches. We obtain the functionals $\mathfrak{I}_{\mathbf{R}}^{(\mathbf{h})}:\mathfrak{R}^{2\mathbf{N}}\rightarrow\mathfrak{R}$ and $\mathfrak{I}_{\mathbf{I}}^{(\mathbf{h})}:\mathfrak{R}^{2\mathbf{N}}\rightarrow\mathfrak{R}$ such that, respectively:

$$\mathfrak{I}_R^{(h)}(x_1^{(h)},x_2^{(h)},...,x_N^{(h)},y_1^{(h)},y_2^{(h)},...,y_N^{(h)})\equiv\frac{1}{2}Re[\overline{S}^{(h)}]=\frac{1}{2}\sum_{k=1}^{K}G_k^{(h)}[(x_{i,k}^{(h)}-x_{j,k}^{(h)}+a_{E,k}^{(h)})^2+(y_{i,k}^{(h)}-y_{j,k}^{(h)}+b_{E,k}^{(h)})^2] \quad (96)$$

$$\mathfrak{I}_I^{(h)}(x_1^{(h)},x_2^{(h)},...,x_N^{(h)},y_1^{(h)},y_2^{(h)},...,y_N^{(h)})\equiv\frac{1}{2}Im[\overline{S}^{(h)}]=\frac{1}{2}\sum_{k=1}^{K}B_K^{(h)}[(x_{i,k}^{(h)}-x_{j,k}^{(h)}+a_{E,k}^{(h)})^2+(y_{i,k}^{(h)}-y_{j,k}^{(h)}+b_{E,k}^{(h)})^2] \quad (97)$$

These functionals are a function class $\mathcal{C}^2$ in $\mathfrak{R}^{2\mathbf{N}}$. Considering $G_k^{(h)}>0$ for $k = 1,\dots, K$, the real functional of the complex power is always positive defined, that is, $\mathfrak{I}_R^{(h)}(x_1^{(h)},x_2^{(h)},...,x_N^{(h)},y_1^{(h)},y_2^{(h)},...,y_N^{(h)})>0$, for all the pairs $(x_i^{(h)},y_i^{(h)})$ with $i=1,...,N$ and $h=1,...,H$; then, *the active absorbed power has a minimum.*

The imaginary functional of the complex power can be analyzed, for all the pairs $(x_i^{(h)},y_i^{(h)}), i=1,...,N$, and each harmonic $h=1,...,H$ as single-phase AC circuits described in Section 3.2 for AC circuits. Similarly as in case i) with resistive-inductive components, the imaginary functional of the complex power is *positive defined*, thus the *total reactive power absorbed by the inductive components has a minimum.* Furthermore, likewise case ii) with resistive-capacitive components, additional nodes are introduced to take into account the series capacitive components, and capacitive components are exploited by changing the sign of the imaginary functional, in order to obtain a *positive defined* functional and to conclude that the *total reactive power generated by the capacitive components has a minimum.* For the general case in which reactive components of different nature coexist at the given harmonic $h = 1,\dots,H$, the functional $\Im_I^{(h)}$ is not positive defined and as such it cannot be stated that the total reactive power absorbed (or generated) by the branches at harmonic h has a minimum.

Consequently, the minimum points of the functionals (96) and (97), when existing, are the solutions of the system containing $4N$ equations:

$$\begin{aligned}
&\frac{\partial \Im_R^{(h)}}{\partial x_1^{(h)}} = 0, \frac{\partial \Im_R^{(h)}}{\partial y_1^{(h)}} = 0,\dots, \frac{\partial \Im_R^{(h)}}{\partial x_N^{(h)}} = 0, \frac{\partial \Im_R^{(h)}}{\partial y_R^{(h)}} = 0 \\
&\frac{\partial \Im_I^{(h)}}{\partial x_1^{(h)}} = 0, \frac{\partial \Im_I^{(h)}}{\partial y_1^{(h)}} = 0,\dots, \frac{\partial \Im_I^{(h)}}{\partial x_N^{(h)}} = 0, \frac{\partial \Im_I^{(h)}}{\partial y_M^{(h)}} = 0
\end{aligned} \tag{98}$$

By calculating the algebraic sum of the solutions, with the derivatives taken with respect to the imaginary parts multiplied by (-j), in the same fashion as seen in AC examples, we obtain the expressions

$$\sum_{\ell_k^{(1)}} \bar{I}_k^{(h)*} = 0, \ \sum_{\ell_k^{(2)}} \bar{I}_k^{(h)*} = 0,\dots, \sum_{\ell_k^{(N)}} \bar{I}_k^{(h)*} = 0 \tag{99}$$

which are the equations of the first Kirchhoff's law for the hth harmonic, expressed at all the N nodes of the circuit.

Consequently, we may state the Principle of Minimum Active and Reactive absorbed Power in linear circuits under Non-sinusoidal conditions (PMARPN): "In the circuits under non-sinusoidal conditions, for each hth-harmonic, the minimum of the total active and reactive power absorbed (generated) by the resistive-inductive only (-capacitive only) branches of a linear circuit in quasi-stationary AC is verified by the solutions in currents and voltages in the circuit, and these are the currents and voltages that verify the first and second Kirchhoff's laws". Another description of the same principle can be stated as follows: "In the linear and reciprocal circuit in quasi-stationary AC regime the currents and voltages are distributed such as the total active and reactive power absorbed (generated) by the resistive-inductive only (-capacitive only) branches of the circuit in non-sinusoidal conditions should be minimal".

It is essential to remark that the same observations regarding the limits of applicability of the minimum power principle indicated in Section for AC circuits hold at every harmonic order. Thus, the minimum power principle applies when no mixture of inductive and capacitive components appears in the circuit at any frequency.

The existence of the minimum power condition is verified if, starting from all inductive (or capacitive) reactive components, no resonance conditions is reached in the circuit at any frequency in the observed range, since the reactive nature of the circuit would change above the resonance frequency. This could be the case of filters containing both inductive and capacitive components, whose equivalent admittance exhibits capacitive behavior at relatively low frequency up to the resonance conditions, and inductive behavior at frequencies higher than the one corresponding to the resonance conditions.

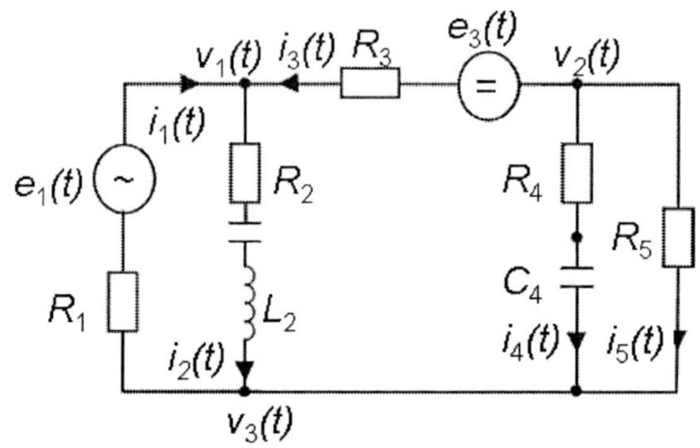

Figure 11. Circuit with five branches under non-sinusoidal conditions and the other one it is a DC voltage source.

Example 5: determination of the minimum power functionals for a circuit with five branches under non-sinusoidal conditions.

Let's consider a linear circuit five branches, shown in Figure 11. One of the two voltage sources contains the fundamental and the third harmonic

$$e_1(t) = 10\sqrt{2}\sin(2\pi 50t) + 2\sqrt{2}\sin(6\pi 50t)[V] \tag{100}$$

$$e_3(t) = E_3^{(0)} = 12[V] \tag{101}$$

For the fundamental, the resonance condition is fulfilled in branch two of the circuit, i.e. $\omega^{(1)} L_2 = \dfrac{1}{\omega^{(1)} C_2}$. In this case, the branch can be redrawn by eliminating the resonant series components (otherwise the principles illustrated cannot be formally applied). If we consider the potentials $\overline{V}_1^{(h)} = x_1^{(h)} + jy_1^{(h)}$, $\overline{V}_2^{(h)} = x_2^{(h)} + jy_2^{(h)}$, $\overline{V}_3^{(h)} = x_3^{(h)} + jy_3^{(h)}$, and the pair's $(x_i^{(h)}, y_i^{(h)}) \in \Re$, $i = 1,2,3$ variables, for the harmonic, h = 0, 1, and 3 of voltage sources we can calculate the minimum of the absorbed power functionals:

a) The absorbed power functional for DC regime of the circuit, when h = 0, can be expressed by

$$\Im_R^{(0)} = \frac{1}{2}\left[G_1 (V_3^{(0)} - V_1^{(0)})^2 + G_3 (V_2^{(0)} - V_1^{(0)} + E_3^{(0)})^2 + G_5 (V_2^{(0)} - V_3^{(0)})^2\right] > 0, \forall V_1^{(0)}, V_2^{(0)}, V_3^{(0)} \in \Re \tag{102}$$

Then, the minimum of the functional is expressed by relations identical with the first Kirchhoff's law at nodes 1, 2, and 3:

$$\frac{\partial \Im_R^{(0)}}{\partial V_1^{(0)}} = -G_1(V_3^{(0)} - V_1^{(0)}) - G_3(V_2^{(0)} - V_1^{(0)} + E_3^{(0)}) = 0 \text{, i.e. } -I_1^{(0)} - I_3^{(0)} = 0 \quad (103)$$

$$\frac{\partial \Im_R^{(0)}}{\partial V_2^{(0)}} = G_3(V_2^{(0)} - V_1^{(0)} + E_3^{(0)}) + G_5(V_2^{(0)} - V_3^{(0)}) = 0 \text{, i.e. } I_3^{(0)} + I_5^{(0)} = 0 \quad (104)$$

$$\frac{\partial \Im_R^{(0)}}{\partial V_3^{(0)}} = G_1(V_3^{(0)} - V_1^{(0)}) - G_5(V_2^{(0)} - V_3^{(0)}) = 0 \text{, i.e. } I_1^{(0)} - I_5^{(0)} = 0 \quad (105)$$

b) The absorbed power functional for AC regime of the circuit, when $h = 1$ is determined taking into account the convention formulated at case iii), because in branch two the resonance condition is satisfied. Thus, at the terminal of the capacitor C_4, we introduce the new node 4, having the potential $\overline{V}_4^{(1)} = x_4^{(1)} + jy_4^{(1)}$, and the variable pair $(x_4^{(1)}, y_4^{(1)}) \in \Re$,. In these conditions, the real part of the complex absorbed power is expressed as

$$\Im_{\mathbf{R}}^{(1)} = \frac{1}{2}\Big\{G_1 [(x_3^{(1)} - x_1^{(1)} + a_{E,1}^{(1)})^2 + (y_3^{(1)} - y_1^{(1)} + b_{E,1}^{(1)})^2] + G_2 [(x_1^{(1)} - x_3^{(1)})^2 + (y_1^{(1)} - y_3^{(1)})^2] + G_3 [(x_2^{(1)} - x_1^{(1)})^2 + (y_2^{(1)} - y_1^{(1)})^2] +$$
$$+ G_4 [(x_2^{(1)} - x_4^{(1)})^2 + (y_2^{(1)} - y_4^{(1)})^2] + G_5 [(x_2^{(1)} - x_3^{(1)})^2 + (y_2^{(1)} - y_3^{(1)})^2] + G_6 [(x_2^{(1)} - x_3^{(1)})^2 + (y_2^{(1)} - y_3^{(1)})^2] \Big\} > 0,$$
$$\forall (x_i^{(1)}, y_i^{(1)}), (a_{E,1}^{(1)}, b_{E,1}^{(1)}) \in \Re, i = 1,2,3,4 \quad (106)$$

and the imaginary part of the complex absorbed power as

$$\Im_I^{(h)} = \frac{1}{2} B_4 [(x_4^{(1)} - x_3^{(1)})^2 + (y_4^{(1)} - y_4^{(1)})^2] > 0, \forall (x_i^{(1)}, y_i^{(1)}) \in \Re, i = 1,2,3,4 \quad (107)$$

where the complex representation of the fundamental harmonic of the voltage source is $\overline{E}_1^{(1)} = a_{E,1}^{(1)} + jb_{E,1}^{(1)}$. The calculation of this functional was used the convention iii), and the minimum of the functional, for fundamental h=1, is obtained when

$$\frac{\partial \Im_R^{(1)}}{\partial x_1^{(1)}} = 0, \frac{\partial \Im_R^{(1)}}{\partial x_2^{(1)}} = 0, \frac{\partial \Im_R^{(1)}}{\partial x_3^{(1)}} = 0, \frac{\partial \Im_R^{(1)}}{\partial x_4^{(1)}} = 0, \frac{\partial \Im_R^{(1)}}{\partial y_1^{(1)}} = 0, \frac{\partial \Im_R^{(1)}}{\partial y_2^{(1)}} = 0, \frac{\partial \Im_R^{(1)}}{\partial y_3^{(1)}} = 0, \frac{\partial \Im F_R^{(1)}}{\partial y_4^{(1)}} = 0 \quad (108)$$

$$\frac{\partial \Im_I^{(1)}}{\partial x_1^{(1)}}=0,\frac{\partial \Im_I^{(1)}}{\partial x_2^{(1)}}=0,\frac{\partial \Im_I^{(1)}}{\partial x_3^{(1)}}=0,\frac{\partial \Im_I^{(1)}}{\partial x_4^{(1)}}=0,\frac{\partial \Im_I^{(1)}}{\partial y_1^{(1)}}=0,\frac{\partial \Im_I^{(1)}}{\partial y_2^{(1)}}=0,\frac{\partial \Im_I^{(1)}}{\partial y_3^{(1)}}=0,\frac{\partial \Im_I^{(1)}}{\partial y_4^{(1)}}=0 \quad (109)$$

relations which are equivalent to the first Kirchhoff's law expressed at node 1, 2, 3, and 4:

$$-\overline{I}_1^{(1)^*}+\overline{I}_2^{(1)^*}-\overline{I}_3^{(1)^*}=0,\ \overline{I}_3^{(1)^*}+\overline{I}_4^{(1)^*}+\overline{I}_5^{(1)^*}=0,\ \overline{I}_1^{(1)^*}-\overline{I}_2^{(1)^*}-\overline{I}_4^{(1)^*}-\overline{I}_5^{(1)^*}=0,\ -\overline{I}_4^{(1)^*}+\overline{I}_4^{(1)^*}=0 \quad (110)$$

c) When h = 3 taking into account the convention formulated at case iv), the branch two is of inductive nature and branch four is of capacitive nature, and coexist in the circuit, then the reactive functional F_I is not positive defined, so that it is not possible to conclude that the total reactive power absorbed by the branches has a minimum.

Example 6: determination of the minimum power functionals for a "spliter".

Let's consider a linear circuit "splitter" which is used in TV cable for power amplifier of signals, shown in Figure 12.

The instantaneous values of non-sinusoidal voltage source have the practical form:

$$e(t)=0.27386+0.27386\sqrt{2}\sin(2\pi 40\cdot 10^6 t)+0.27386\sqrt{2}\sin(2\pi 900\cdot 10^6 t)[\mathrm{V}] \quad (111)$$

If it is consider the potential V_1 variable, for each harmonic of voltage source we can calculate the minimum of the power absorbed functionals.

a) The absorbed power functional for DC regime of the splitter can be expressed by

$$\Im_R^{(0)}=\frac{1}{2}\left[G_1(-V_1^{(0)}+E^{(0)})^2+G_2V_1^{(0)^2}+G_3V_1^{(0)^2}\right]\rangle 0,\forall V_1^{(0)}\in\Re \quad (112)$$

where $G_1=\dfrac{1}{R_1^{'}+R_1^{''}}, G_2=\dfrac{1}{R_2^{'}+R_2^{''}}, G_3=\dfrac{1}{R_3^{'}+R_3^{''}}$ do not depend of the frequency.

Then, the minimum point of functional are obtained when

$$\frac{\partial \Im_R^{(0)}}{\partial V_1^{(0)}}=-G_1(-V_1^{(0)}+E^{(0)})+G_2V_1^{(0)}+G_3V_1^{(0)})=0 \quad (113)$$

relation that is with a identical with the 1st Kirchhoff's law expressed in node 1.

b) If we consider the potential $\underline{V}_1^{(p)}=x+jy^{(p)}$, for the two harmonics, $p=8\cdot 10^5$, and $p=18\cdot 10^6$, then for AC regime of the splitter the functional of the complex absorbed power is

$$\Im_R^{(p)}=\frac{1}{2}\left\{G_1\left[(-x^{(p)}+E^{(p)})^2+y^{(p)^2}\right]+G_2(x^{(p)2}+y^{(p)2})+G_3(x^{(p)2}+y^{(p)2})\right\}\rangle 0,\forall(x^{(p)},y^{(p)})\in\Re \quad (114)$$

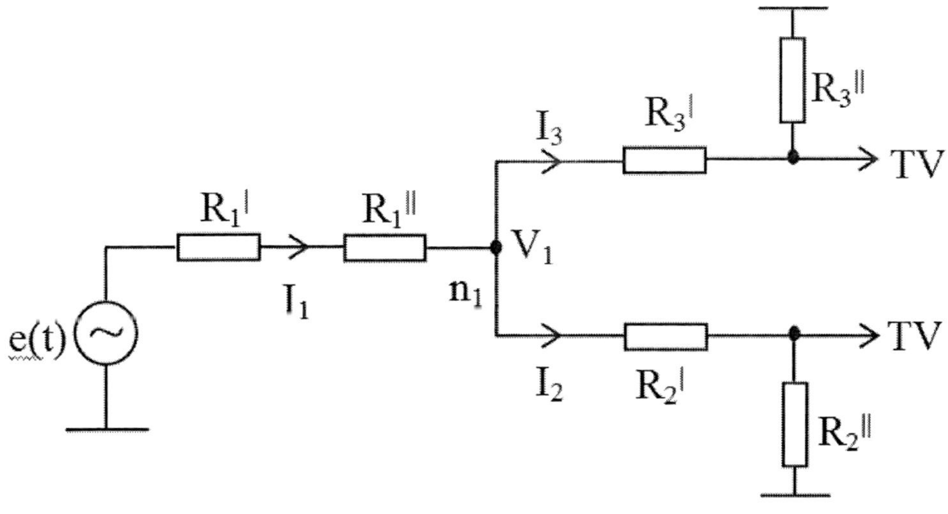

Figure12. The “splitter”.

Then, the minimum of the power functional is the solutions of the system

$$\begin{aligned}
\frac{\partial \Im_R^{(p)}}{\partial x^{(p)}} &= -G_1(-x^{(p)} + E^{(p)}) + G_2 x^{(p)} + G_3 x^{(p)} = 0 \\
\frac{\partial \Im_R^{(p)}}{\partial y^{(p)}} &= G_1 y^{(p)} + G_2 y^{(p)} + G_3 y^{(p)} = 0
\end{aligned} \tag{115}$$

and, if we calculate the algebraic sum of the solutions of system (115), with the second equation multiplied with , results the 1st Kirchhoff’s law expressed in node 1

$$-\underline{I}_1^{(p)^*} + \underline{I}_2^{(p)^*} + \underline{I}_3^{(p)^*} = 0 \tag{116}$$

The imaginary part of power functional is zero, $\Im_I^{(p)} = 0$.

By using a Multisim procedure, shown in Figure 13, a, b, and c, the absorbed active power for different values of the resistances is calculated. The results are shown in Table 1.

The first line contains the nominal values of conductance, and the second line of table contains the adapted values of all the resistances, i.e. each resistance absorbed by the circuit the maximum active power. In this case, the value of the absorbed active power is the big one compared with the other values.

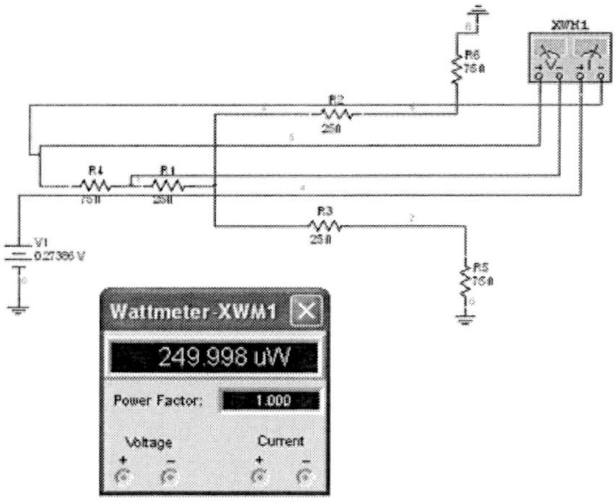

Figure13a. The power absorbed by in DC regime.

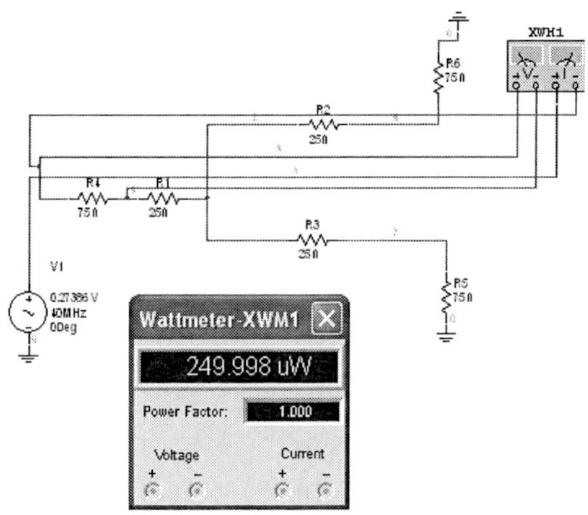

Figure 13b. The active power absorbed by in AC regime.

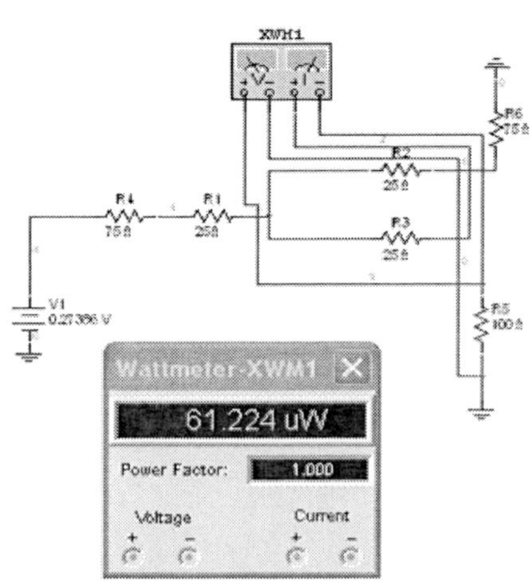

Figure 13c. The power absorbed by in DC regime.

Table 1. Results of simulation

Values of Conductances (S)	Total Absorbed Power in DC and AC (mW)
$G_1 = G_2 = G_3 = 0.01$	499.593
$G_1 = 0.01$; $G_2 = 0.013$; $G_3 = 0.01$	524.818
$G_1 = 0.01$; $G_2 = 0.02$; $G_3 = 0.025$	1.662
$G_1 = 0.01$; $G_2 = 0$; $G_3 = 0.01$	374.832
$G_1 = 0.01$; $G_2 = 0.04$; $G_3 = 0.01$	624.810
$G_1 = 0.02$, $G_2 = 0.01$; $G_3 = 0.01$	749.664

Variational Formulation for Determining the Principle of Minimum Absorbed Energy for Circuits with Capacitors in Electrostatic Regime

The field literature includes valuable works: [53], [54], and [55], which describe results concerning the minimum electrostatic energy in a system of conductors charged with electric charges. We shall now carry out a demonstration of the minimum energetic principle in a circuit made of capacitors [40], [43], [47], [56], and [57].

By applying Kirchhoff's second law for a given branch of the network, shown in Figure 14, connected between two nodes with potentials V_x and V_y, respectively, made of a capacitor C charged with electric charge q from the voltage source E, we get the following expression of the voltage between the two nodes

$$U = V_x - V_y = U_C + U_E = \frac{q}{C} - E \tag{117}$$

The electrostatic energy, according to relation (19), of the system expressed in terms of potentials is

$$W_e = \frac{1}{2} q(V_x - V_y + E) = \frac{1}{2} C(V_x - V_y + E)^2 \tag{118}$$

The electrostatic energy, having values obtained by computing the integral of a function, is an energetic functional $\Im_E$, according to relation (38), defined within the space of the electrical field and having values in the positive real set $\Re_+$

$$\Im_{\mathbf{E}} : S_E \to \Re_+ , \Im_E \equiv W_E = \frac{1}{2} C(V_x - V_y + E)^2 \tag{119}$$

The functional is positive defined, i.e. $\Im_E(V_x, V_y) \geq 0$, whatever the values of the potentials V_x and V_y might be. This means that its extreme points will be minimum points, and therefore the solutions of the system with partial derivatives

$$\frac{\partial \Im_E}{\partial V_x} = 0 \tag{120}$$

$$\frac{\partial \Im_E}{\partial V_y} = 0 \tag{121}$$

will correspond to the minimum electrostatic energy of the network.

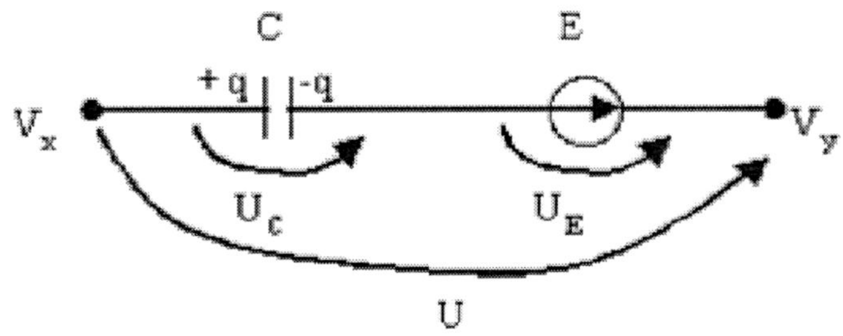

Figure 14. Branch with capacitor.

By computing, for example, the system (120) for all the l_k branches connected the x node, we will obtain a relation that verifies Kirchhoff's first law for circuit with capacitors, expressed in the x node

$$\sum_{l_k \in x} \frac{\partial \Im_{E_k}}{\partial V_{x_k}} = \sum_{l_k \in x} C_k (V_{x_k} - V_{y_k} + E_k) = \sum_{l_k \in x} q_k = 0 \tag{122}$$

A similar equality is to be obtained through computation of the system (121). We can now state the following principle Minimum Energetic Principle in circuits with Capacitors - MEPC: "In the circuits in electrostatic regime the electric charges and the potentials of the capacitors, verify the Kirchhoff's laws, and correspond to the minimum electrostatic energy of the circuit".

Example 7: determination of the minimum energetic functional for a circuit with capacitors.

We consider the circuit with three capacitors, not charged initially, connected like in Figure 15. By applying Kirchhoff's second law we can express the electrostatic energy W_E of the system in terms of potentials V_x and V_y

$$W_E = \frac{1}{2}(q_1 U_1 + q_2 U_2 + q_3 U_3) = \frac{1}{2}\left[C_1(E - V_x + V_y)^2 + C_2(V_x - V_y)^2 + C_3(V_x - V_y)^2\right] \tag{123}$$

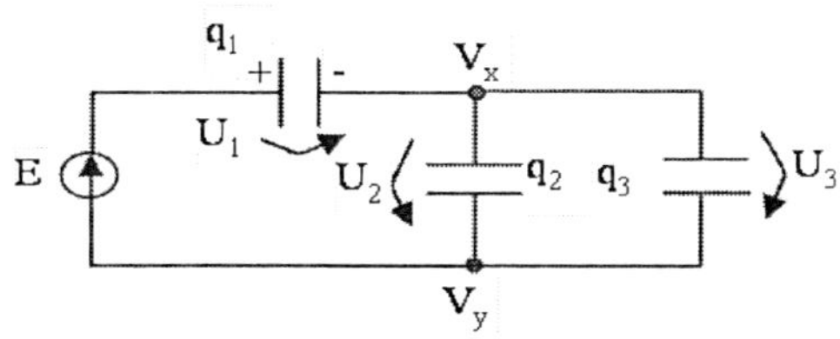

Figure 15. Circuit with capacitors.

The minimization of the energetic functional $\Im_E$ as a dependency on the potential V_x

$$\frac{\partial \Im_E}{\partial V_x} \equiv \frac{\partial W_E}{\partial V_x} = 0 \tag{124}$$

leads to the verifying of Kirchhoff's first law in node x

$$\frac{\partial W_E}{\partial V_x} = -C_1\left(E - V_x + V_y\right) + C_2\left(V_x - V_y\right) + C_3\left(V_x - V_y\right) = -q_1 + q_2 + q_3 = 0 \tag{125}$$

In a similar way, if the optimisation of the energetic functional of the circuit is performed relative to the potentials V_y, Kirchhoff's first law comes to be verified in node y

$$\frac{\partial W_E}{\partial V_y} = C_1\left(E - V_x + V_y\right) - C_2\left(V_x - V_y\right) - C_3\left(V_x - V_y\right) = q_1 - q_2 - q_3 = 0 \tag{126}$$

As a conclusion, the potentials and the electric charges of the circuit are thus distributed as to get a minimum electrostatic energy of the circuit.

THE PRINCIPLE OF MINIMUM ABSORBED ENERGY FOR MAGNETIC CIRCUITS

The analogies between the electric and magnetic circuits as well as the similar definition of the electrostatic and magnetic energy have led to the statement of minimum energetic principle of the linear magnetic circuits in stationary regime [24], [33], [40], [45], [47], [48], [50], and [55].

Variational Formulation for Determining the Minimum Energetic Principle for Linear Magnetic Circuits in Quasi-Stationary Regime

If we consider a coil having N turns, crossed by the conduction current i, situated on a linear, homogenous and isotropic magnetic material, of reluctance R_m, which has the equivalent magnetic circuit shown in Figure 16, then Kirchhoff's second law for magnetic circuits in quasi-stationary regime is

$$U_m + \theta = R_m \Phi_f \tag{127}$$

If $V_{m,x}$ and $V_{m,y}$ are the magnetic potentials of the nodes to which this branch is connected, then the magnetic flux Φ_f can be expressed in terms of these and of the current linkage ($\theta = Ni$)

$$\Phi_f = \frac{V_{m,x} - V_{m,y} + \theta}{R_m} \tag{128}$$

The magnetic energy of the circuit can be expressed as a dependency on the magnetic potentials, according to relation (20)

$$W_m = \frac{1}{2} R_m \Phi_f^2 = \frac{(V_{m,x} - V_{m,y} + \theta)^2}{2R_m} \tag{129}$$

Similarly to the case of the electrostatic energy, the magnetic energy, having values received from the computation of the integral of a function, is an energetic functional $\Im_M$, according to relation (46), defined to the space of the magnetic field S_M, with values in the real positive set $\Re_+$

$$\Im_M : S_M \to \Re_+, \Im_M \equiv W_M = \frac{(V_{m,x} - V_{m,y} + \theta)^2}{2R_m} \tag{130}$$

The energetic functional $\Im_M$ is positively defined, that

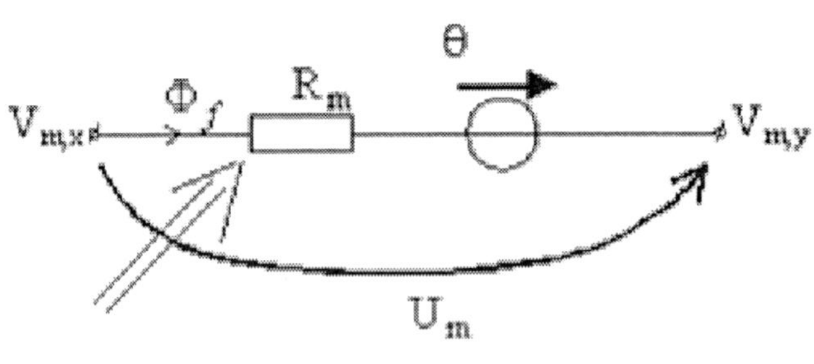

Figure 16. The magnetic circuit branch with a coil.

is $\Im_M(V_{m,x}, V_{m,y}) \geq 0$, whatever the values of the magnetic potentials $V_{m,x}$ and $V_{m,y}$. might be.

Consequently, its extreme points will be minimum points and therefore the solutions of the equation systems with partial derivatives

$$\frac{\partial \Im_M}{\partial V_{m,x}} = 0 \tag{131}$$

$$\frac{\partial \Im_M}{\partial V_{m,y}} = 0 \tag{132}$$

will correspond to the minimum magnetic energy of the circuit. For example by calculating system (131) for all the branches l_k with coils, connected to the node x, we obtain a relation which verifies Kirchhoff's first law for magnetic circuits expressed in node x

$$\sum_{l_k \in x} \frac{\partial \Im_{M_k}}{\partial V_{m,x_k}} = \sum_{l_k \in x} \frac{V_{m,x_k} - V_{m,y_k} + \theta_k}{R_{m_k}} = \sum_{l_k \in x} \Phi_{f_k} = 0 \tag{133}$$

A similar relation is obtained considering the second system (132). We can state the following principle Minimum Energetic Principle for linear Magnetic circuits in quasi-stationary regime - MEPM: "In the linear magnetic circuits in stationary regime, the linking fluxes and the magnetic potentials verify the Kirchhoff's first, respectively second law for magnetic circuits, and correspond to the minimum magnetic energy of the circuit".

Example 8: determination of the minimum energetic functional for a magnetic circuit with coils.

Let us consider a magnetic circuit that contains three coils, connected to the nodes x and y, like in Figure 17.

By applying Kirchhoff's second law for magnetic circuits, we can define the magnetic energy of the circuit W_m in terms of the magnetic potentials $V_{m,x}$ and $V_{m,y}$

$$W_M = \frac{1}{2}\left(R_{m,1}\Phi_{f1}^2 + R_{m,2}\Phi_{f2}^2 + R_{m,3}\Phi_{f3}^2\right) = \frac{1}{2}\left\{\frac{(V_{m,y} - V_{m,x} + \theta_1)^2}{R_{m,1}} + \frac{(V_{m,x} - V_{m,y} + \theta_2)^2}{R_{m,2}} + \frac{(V_{m,x} - V_{m,y} + \theta_3)^2}{R_{m,3}}\right\} \tag{134}$$

where $R_{m,1}$, $R_{m,2}$, $R_{m,3}$ are the magnetic reluctances of the three branches, while θ_1, θ_2, θ_3 are the current linkages of the three coils.

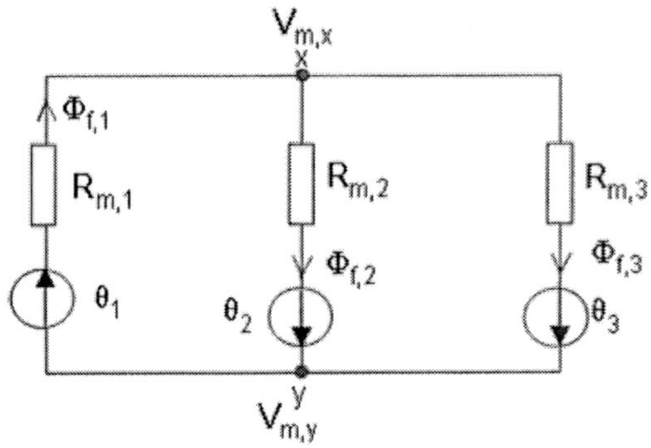

Figure 17. Magnetic circuit with three coils.

The minimization of the energetic functional of the circuit in relation to the magnetic potential $V_{m,x}$

$$\frac{\partial \Im_M}{\partial V_{m,x}} \equiv \frac{\partial W_M}{\partial V_{m,x}} = 0 \tag{135}$$

leads to the verifying of Kirchhoff's first law for magnetic circuit in node x

$$\frac{\partial W_M}{\partial V_{mx}} = -\frac{V_{m,y} - V_{m,x} + \theta_1}{R_{m,1}} + \frac{V_{mx} - V_{m,y} + \theta_2}{R_{m,2}} + \frac{V_{mx} - V_{m,y} + \theta_3}{R_{m,3}} = -\Phi_{f,1} + \Phi_{f,2} + \Phi_{f,3} = 0 \tag{136}$$

In a similar way, if the optimisation of the energetic functional of the circuit is performed in relation to the potential V_y, Kirchhoff's first law for magnetic circuit comes to be verified in node y

$$\frac{\partial W_M}{\partial V_{m,y}} = \frac{V_{m,y} - V_{m,x} + \theta_1}{R_{m,1}} - \frac{V_{mx} - V_{m,y} + \theta_2}{R_{m,2}} - \frac{V_{mx} - V_{m,y} + \theta_3}{R_{m,3}} = \Phi_{f,1} - \Phi_{f,2} - \Phi_{f,3} = 0 \tag{137}$$

In conclusion, the magnetic potentials and the fluxes linking the coils in the circuit get such distribution as to allow the magnetic energy in the circuit to be minimal.

CONCLUSION

The application of the power and energy functionals in the analysis of electric and magnetic circuits in stationary and quasi-stationary regime makes is possible to evaluate the equilibrium state, from an energy point of view, existing in the system at different moments in time.

The formulation of the power functionals in the Hilbert space and the determination of their extremes lead to the conclusion that the equilibrium states in the circuits represent minimum power conditions for these circuits, provided that specific conditions are verified in quasi-stationary circuits. According to these considerations, this paper has formulated a set of minimum power principles, including:

- *PMP: Principle of Minimum Power* for DC circuits;
- *PMARP: Principle of Minimum Active and Reactive Power* for linear AC circuits;
- *PMARPN: Principle of Minimum Active and Reactive absorbed Power* for linear AC circuits in *Non-sinusoidal* conditions;
- *MEPC: Minimum Energetic Principle for* circuits with Capacitors;
- *MEPM: Minimum Energetic Principle for linear Magnetic* circuits in quasi-stationary regime.

For the *PMARP* and *PMARPN* principles, some restrictions apply to the nature of the circuits, since the non-resistive part of the circuit has to be totally inductive or totally capacitive to support the full definition of these principles both on the active and on the reactive power sides. The minimization conditions of the power functionals leads to the first Kirchhoff's law. Thus, the proposed principles and the second Kirchhoff's law together establish a complete characterization of the linear circuits in stationary and quasi-stationary regime. In other terms, the equations of the first Kirchhoff's law can be conceptually interpreted in terms of another set of equations able to represent the behavior of the system from the minimum power point of view, and coincide with the first Kirchhoff's law equations when the system is in stationary or quasi-stationary regime. Besides their conceptual and tutorial worth, the theoretical results shown in this paper can be used to study the transient regimes of linear circuits as transitions from different minimum power states.

Introduction of the principle of minimum absorbed power (PMP) and the minimum energetic principle (MEP) as two natural conditions for the stable functioning of linear electric and magnetic circuits in stationary and quasi-stationary regime has a great importance for understanding of the power transfer in electric and thermal systems.

For this reason, the determining of the functionals in linear circuits is a very important problem, didactically and theoretically, as well as for its practical applications.

REFERENCES

[1] Dewar, R.C. Information theory explanation of the fluctuation theorem, maximum entropy production, and self-organization criticality in non-equilibrium stationary states. *Journal of Physics*, 2003, A36, 631-641.

[2] Ozawa, H.; Ohmura, A.; Lorenz, R.D.; Pujol, T. The second law of thermodynamics and the global climate system - A review of the maximum entropy production principle. *Revue of Geophysics*, 2003, 41, 1018.

[3] Rodriguez-Iturbe, I.; Rinaldo, A. Fractal River Basins: Chance and Self-Organization; New York, USA: Cambridge University Press, 1997.

[4] North, G.H.; Cahalan, R.F.; Coakley, J.A. (1981). Energy balance climate model. *Revue of Geophysics*, 1981, 19, 91-121.

[5] Rodgers C.D. Minimum entropy exchange principle-reply. *Journal of Meteorology Society*, 1976, 102, 455-7.

[6] Clemente-Gallardo, J; Scherpen, J.M.A. Relating Lagrangian and Hamiltonian Formalism of LC Circuits. *IEEE Transactions on Circuits and Systems*, 2003, 50, no.10, 1359-1363.

[7] Lorenz, E.N. Generation of available potential energy and the intensity of the general circulation. Oxford, UK: Pergamon Press, 1960, 86–92.

[8] Kleidon, A. Testing the effect of life on Earth's functioning: How Gaian is the Earth system? *Clim. Change*, 2002, 52, 383–389.

[9] Kleidon, A.; Fraedrich, K.; Kunz, T.; Lunkeit, F. The atmospheric circulation and states of maximum entropy production. *Geophys Res. Lett.*, 2003, 30, 2223.

[10] Kleidon, A; Lorenz, R. D. Non-equilibrium thermodynamics and the production of entropy: life, Earth, and beyond; Heidelberg, GE: Springer Verlag, 2004.

[11] Prigogine, I. Introduction to Non-equilibrium Thermodynamics. New York, USA: Wiley Interscience, 1962.

[12] Rinaldo, A.; Maritan, A.; Colaiori, F.; Flammini, A.; Rigon, R.; Rodriguez-Iturbe, I.; Banavar, J.R. Thermodynamics of fractal networks. *Phys. Rev. Lett.*, 1996, 76, 3364–3367.

[13] Bejan, A. Shape and Structure, from Engineering to Nature. Cambridge, UK: Cambridge University Press, 2000.

[14] Enquist, B.J.; Brown, J.H.; West, G.B. Allometric scaling of plant energetics and population density. *Nature*, 1998, 395, 163–165.

[15] West, G.B.; Brown, J.H.; Enquist, B.J. The fourth dimension of life: fractal geometry and allometric scaling of organisms. *Science* 1999, 284, 167–169.

[16] Paltridge, G.W. Global dynamics and climate – a system of minimum entropy exchange, *Q. J. R. Meteorol. Soc.*, 1975, 101, 475–484.

[17] Paltridge, G.W. The steady-state format of global climate. *Q. J. R. Metorol. Soc.* 1978, 104, 927–945.

[18] Paltridge, G.W. A physical basis for a maximum of thermodynamic dissipation of the climate system. *Q. J. R. Meteorol. Soc.* 2001, 127, 305–313.

[19] Budyko, M.I. Effects of solar radiation variations on climate of Earth. *Tellus* 1996, 211, 611–619.

[20] Sellers, W.D. A global climate model based on the energy balance of the Earth atmosphere system. *J. Appl. Met.* 1969, 8, 392–400.

[21] North, G.R.; Cahalan, R.F.; Coakley, J.A. Energy balance climate models. *Rev. Geophys. Space Phys.* 1981, 19, 91–121.

[22] Grassl, H. The climate at maximum entropy production by meridional atmospheric and oceanic heat fluxes. *Q. J. R. Meteorol. Soc.* 1981, 107, 153–166.

[23] Shimokawa, S.; Ozawa, H. On the thermodynamics of the oceanic general circulation: Entropy increase rate of an open dissipative system and its surroundings. *Tellus*, 2001, 53A, 266–277.

[24] Shimokawa, S.; Ozawa, H. On the thermodynamics of the oceanic general circulation: Irreversible transition to a state with higher rate of entropy production. *Q. J. R. Meteorol. Soc.*, 2002, 128, 2115–2128.

[25] Maxwell, J.C. A Treatise on Electricity and Magnetism, vol. I and II, 3rd ed.New York, USA: Dover Publications, Inc., 1954.

[26] Lanczos, C. The Variational Principles of Mechanics.4th ed. New York, USA: Dover Publication, 1970.

[27] Hammond, P.; Sykulski, J.K. Engineering Electromagnetism – Physical Processes and Computation, London, UK: Oxford University Press, 1994.

[28] Millar, W. Some general Theorems for Nonlinear Systems Possessing Resistance. *Phil. Mag.*, 1951, ser. 7, vol. 42, no. 333, 1150-1160.

[29] Stern, T.E. On the equations of Nonlinear Networks. *IEEE Transactions on Circuits Theory*, 1996, vol. CT-13, no.1, 74-81.

[30] Penfield, P. Jr.; Spence, R.; Duinker, S. Tellegen's Theorem and Electrical Networks. Research monograph no. 58. Massachusetts, USA: M.I.T. Press, 1970.

[31] Smith, W.E. Electric and magnetic energy-storage in passive non-reciprocal networks. *Electronics Letters*, 1967, 3, 389-391.

[32] Smith, W.E. Average energy storage by an one-port and minimum energy synthesis. *IEEE Transactions on Circuit Theory*, 1970, 5, 427-430.

[33] Ionescu, V. Hilbert space applications to distorted waveform analysis (in Romanian: Aplicatii ale spatiilor Hilbert la studiul regimului deformant), *Electrotehnica*, 1958, 6, 280-286.

[34] Mocanu, C.I. Electric circuits theory (in Romanian: Teoria circuitelor electrice), Bucharest, Romania: Editura Didactica si Pedagogica, 1979.

[35] Hantila, I.F.; Vasile, N.; Demeter, E.; Marinescu, S.; Covrig, M. The stationary electromagnetic field in non-linear media (in Romanian: Campul electromagnetic stationar in medii neliniare), Bucharest, Romania: Editura ICPE, 1997.

[36] Fireteanu, V. ; Popa, M. ; Tudorache, T. ; Levacher, L. ; Paya, B.; Neau, Y. Maximum of energetic efficiency in induction through-heating processes, *Proceedings of HES Symposium*, Padua, Italy, 2004, 80-86.

[37] Spinei, F.; Andrei, H. Energetical minimum solution for the resistive network. In: Proceedings of National Symposium of Theoretical Electrotechnics SNET, Politehnica University of Bucharest, Romania, 2004, 443- 450.

[38] Andrei, H.; Spinei, F.; Cepisca, C. Theorems about the minimum of the power functional in linear and resistive circuits, In: Proceedings of Advanced Topics in Electrical Engineering ATEE'04, Politehnica University of Bucharest, Romania, 2004, 22- 18.

[39] Andrei, H.; Chicco, G.; Dragusin, V.; Spinei, F.; Cepisca, C. Equilibrium state of the electric circuits-a minimum dissipated power. *WSEAS Transactions on Systems*, 2005, Issue 2, vol. 4, 2284-2290.

[40] Andrei, H.; Chicco, G.; Dragusin, V.; Spinei, F. Principles of the minimum dissipated power in stationary regime. In: Proceedings of WSEAS, Venice, Italy, 2005, 225-230.

[41] Andrei, H.; Spinei, F. An extension of the minimum energetical principle in stationary regime for electric and magnetic circuits, In: Proceedings of Advanced Topics in Electrical Engineering ATEE'06, Politehnica University of Bucharest, Romania, 2006, 210-213.

[42] Andrei H.; Spinei F.; Cepisca C.; Voicu, N. Contributions regarding the principles of the minimum dissipated power in stationary regime. In: Proceedings of IEEE Conference on Circuits and Systems CAS, Dallas, USA, 2006, 143-47.

[43] Andrei, H.; Spinei, F.; Cepisca, C.; Chicco, G.; Grigorescu, S.D.; Jula, N. On Existance of the Principles of Minimum Dissipated Power for Linear and Nonlinear Electric Circuit, In: Proceedings of WSEAS, Bucharest, Romania, 2006, 186-191.

[44] Andrei, H.; Spinei F. The Minimum Energetical Principle in Electric and Magnetic Circuits In: Proceedings of IEEE European Conference on Circuit Theory and Design ECCTD, Sevilla, Spain, 2007, 906-909.

[45] Andrei, H.; Cepisca, C.; Grigorescu, S.D.; Spinei, F.; Caciula, I.; Andrei, P.C. The Minimum Dissipated Active and Reactive Power: an Equilibrium State of the Circuits under Nonsinusoidal Conditions. In: Proceedings of the 7th WSEAS International Conference on Simulation, Modelling and Optimization SMO'07, Beijing, China, 2007, 284-288.

[46] Andrei, H.; Spinei, F. An extension of the minimum energy principle in stationary regime for electric and magnetic circuits. *Revue Roumaine des Sciences Techniques serie Electrotechnique et Energetique*, 2007, Tome 52, 419-427.

[47] Andrei, H.; Spinei, F.; Caciula, I.; Andrei, P.C. The systematic analysis of the absorbed power in D.C. networks with modifiable parameters using a new mathematic algorithm. In: Proceedings of IEEE AQTR, Cluj Napoca, Romania, 2008, 121-124.

[48] Andrei, H.; Spinei, F.; Jula, N. The Principle of Minimum Dissipated Power and Minimum Energetic Principle-Two general Theorems of the Linear Electric and Magnetic Circuits in Stationary Regime. In: Proceedings of 8th WSEAS International Conference on Systems Theory and Scientific Computation ISTASC'08, Rodos Island, Greece, 2008, 75-87.

[49] Andrei, H.; Chicco, G.; Spinei, F.; Cepisca, C. Minimum energy principle for electric and magnetic circuits in quasi-stationary regime. *Journal of Optoelectronics and Advanced Materials*, 2008, 10 (5), 1203-1207.

[50] Andrei, H.; Spinei, F.; Andrei, P.; Rohde, U.; Silaghi, M.; Silaghi, H. Evaluation of Hilbert Space Techniques and Lagrange's Method for the Analysis of Dissipated Power in DC Circuits. In: Proceedings of IEEE European Conference on Circuit Theory and Design ECCTD'09, Antalya, Turkey, 2009, 862-865.

[51] Chua, L.O. Introduction to Nonlinear Network Theory. Part I. New York, USA: Mc Graw Hill Book Company, 1969.

[52] Spinei, F. Determination of the states of maximum power transferand of power transfer with maximum efficiency in networks with modifiable parameters. Part I: Resistive D.C. Networks, *Revue Roumaine des Sciences Techniques serie Electrotechnique et Energetique*, 1971, 16, 3, 435-446.

[53] Lev-Ari, H.; Stankovic, A. Hilbert Space Techniques for Modeling and Compensation of Reactive Power in Energy Processing System. *IEEE Transactions on Circuits and Systems* 2003, 50 (4), 540-56.

[54] Donati, L. Sulia distribuzione del potenziale neile reti di fili conduttori, Rend. Academia delle Scienze di Bologna, 1910, 14, 136-144.

[55] Tonning, A. Energy density in continuous electromagnetic media", *IRE Transactions Antennas and Propagation*, 1960, vol. AP-8, 428-434.

[56] Penfield, P. Jr.; Haus, H.A. Electrodynamics of Moving Media. Cambridge, Massachusetts, USA: M.I.T. Press, 1967.

[57] Andrei H.; Spinei F.; Cepisca, C.; Chicco, G.; Grigorescu, S.D.; Dascalescu, L. Minimum Dissipated Power for Linear and Nonlinear Circuits. *WSEAS Transactions on Circuits and Systems*, 2006, 5 (11), 1620-1625.

[58] Andrei, H.; Spinei, F.; Cepisca, C.; Chicco, G. On existance of the Principles of Minimum Dissipated Power for Linear and Nonlinear Circuits. In: Proceedings of MMACTEE of WSEAS, Bucharest, Romania, 2006, 186-191.

In: Advances in Energy Research
Editor: Nicu Bizon
*

ISBN 978-1-61209-991-0

Chapter 6

An Overview on Communication Networks for Distributed Generation System

Maaruf Ali*
Oxford Brookes University, UK

Abstract

In a competitive energy marketplace based on the developing Distributed Generation System (DGS), it is necessary to have in real-time complete information for the reliable delivery of energy to all customers, including the industrial end-users. The strategy to plan these demands of electric utilities needs to be based on a high-performance data communication network that can support both existing functionalities and future operational developments. Consequently, the communication network represents the core of the automation system of Electrical Power System (EPS) and its design must be for obtaining reliable and cost-effective network architecture. In this chapter, the standards and challenges of hybrid network architecture are discussed for EPS system automation. So, in next sections some technologies are described, such as mobile communications technologies, power line communications, and wireless communications (wireless sensor networks, WiMAX and WiFi). The motivation of this chapter is to provide a better understanding of the hybrid network architecture of EPS that can use a secured communication network with some features that are presented in the next chapter.

Keywords: Electric power system automation, mobile communications technologies, power line communication, wireless sensor networks, WiMAX, WiFi.

Introduction

With the ever increasing complexity and layout of distributed power generation systems being sited in remote geographical locations, their control, monitoring and cost effective

* Corresponding Author: Dr.Maaruf.Ali@epromagconsultancy.com

management becomes an increasingly expensive and logistically complicated problem. Too often the viable and cheapest option is to use wireless communications. Both traditional and wireless communications for the electric power system will be briefly looked at. Increasingly security is ever becoming an important part of the distributed generation system infrastructure.

The current major wireless standards currently cover:

- IEEE 802.16 d and e (WiMAX – Worldwide interoperability for Microwave Access);
- IEEE 802.11 (WiFi) family of a, b, g, n and s;
- Wireless sensor protocols using part of IEEE 802.15.4 : WirelessHART, ISA 100.11a and Zigbee;
- Mobile communications technologies: GPRS/EDGE/1×RTT, HSPA/EVDO and Long-Term Evolution (LTE)/HSPA+UMTS.

Naturally to cover a wide geographical area quite quickly and cost effectively, a mobile solution is preferred. However, for both data collecting and command and control applications, the three solutions based on IEEE 802.15.4 can all be applied. The current mobile communication network including WiMAX can be used to make the back overhaul link to the power industry from the customer premises. For the generation plant or substation, these standards may be adopted: WirelessHART or ISA100.1s. Any protocol that uses "unlicensed" or shared frequency bands should be avoided to maintain security.

The complexity of the information flow in a smart grid framework is shown in Figure 6.1, below. The quantity of monitored data will vastly exceed that of command and control data, this is so, because of the myriad of devices that have to be monitored.

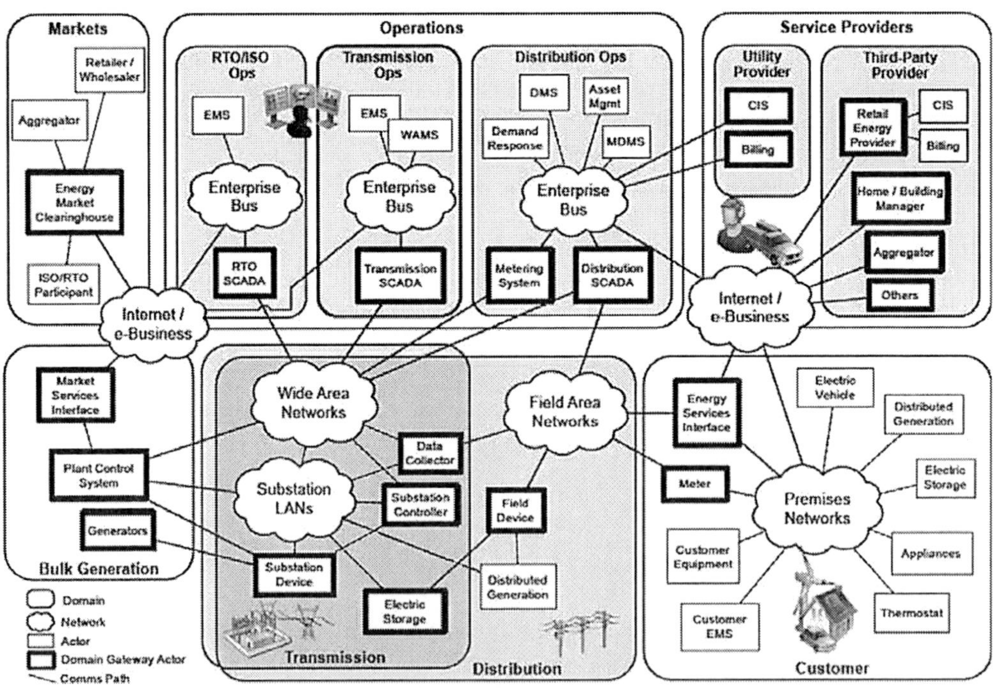

Figure 1. NIST Smart Grid Framework 1.0 (September, 2009) [1].

Clearly the control, command and billing channels need to be highly secure and the information conveyed in a timely manner to be acted upon when demand for action requires it. In some parts of the world, effective monitoring of "stolen energy" is of paramount importance. Figure 1 also identifies five areas of a power generation system: the residence, the distribution system, the substation, the transmission system and the generating station.

The three differing categories of data rates of 'low', 'moderate' and 'high' are summarised in Table 1 below.

Table 1. Summary of Data Rates Terminology [1]

	Data Rate (average)	Latency (average)	Reliability	Security	Distance/ Range	Scalability
Low	<500 Kbps	<250ms	Packet loss is acceptable (It is expected that applications will recover from packet loss at the expense of added delays in communications.)	Clear text communications, integrity checks may or may not be used.	<100m	<100 Nodes / Backhaul Node
Moderate	500 Kbps – 1,500 Kbps	250ms – 1s	acceptable.	may be required, integrity checks are required.	100m – 1000m	100 – 1000 Nodes / Backhaul Node
High	>1,500 Kbps	>1s	Fully reliable communications with error recovery at the data link layer.	Confidentiality and integrity checks are required.	>1000m	>1000 Nodes / Backhaul Node

WIRELESS TECHNOLOGY REQUIREMENT CONSIDERATION

Seven important parameters need to be carefully evaluated in a wireless technology system that is being considered for the communication network in the electric power generation system, these being: latency, bandwidth, resilience, security, scalability, coverage and the life expectancy.

The latency is the round trip delay between the transmitter and the receiver, which includes the media access time, the propagation delay and queuing but not the processing time at the destination.

When considering the bandwidth of the communication link, the effect of channel coding and error coding in reducing the effective goodput of the information transfer capacity must be calculated. The ability of the communication link to recover from noise and natural disasters such as earthquakes is in the domain of the resilience of the link. Security is of ever increasing importance when deciding to adopt the appropriate wireless technology in our uncertain world. Unauthorised access must be prevented and confidentiality of data must be protected. Attacks from 'vampire' nodes can be prevented by using techniques such as mutual authentication. Clearly low powered wireless technologies are only suitable for small scale applications, thus the coverage of the wireless technology has to be appropriate to the application. With a rapidly expanding market, scalability of the technology must be seamless and implementable with ease. For example it must be relatively easy to poll a rapidly increasing number of measuring devices at the correct rate.

The life expectancy of the communication network must match that of the distribution generation system, otherwise it does not make any economic sense to select the communication network in the first place. This is not an easy decision to make with technologies having a short life lifetime often considerably less than the electrical distributed generation system.

Table 2 [1] summarises the major wireless technologies as of January 2010. When updating the communication network, fault tolerance will become more of a critical factor to prevent power outages in times of excessive or peak demands, specially when handling the signalling to implement an efficient load shedding algorithm.

Table 2. Wireless Technologies Summary as of January 2010 [1]

	IEEE 802.16 2004	IEEE 8.2.16e	IEEE 802.11a	IEEE 802.11b	IEEE 802.11g	IEEE 802.11n	Wireless HART	ISA 100.11a	Zigbee	GPRS/EDGE	1×RTT	HDSPA/UMTS	EVDO	LTE/HSPA+
	WiMAX		WiFi				IEEE 802.15.4			2.5G		3G		4G
Latency	L	L	L	L	L	L	L	L	L	M	M	M	M	L
Data Rate	H	H	H	H	H	H	L	L	L	L	L	M	M	H
Resilience	H	H	M	L	L	M	M	M	M	H	H	H	H	H
Security	H*	H	M	M	M	M	H	H	M	H	L	H	L	H
Distance	H	H	L	L	L	M	L	L	L	H	H	H	H	H
Scalability	H	H	M	L	L	M	M	M	L	M	M	H	H	H

Coupled with the properties of the various multi-access wireless technologies, their suitability for a particular application also needs to be determined. Typical application scenarios are shown in Table 3, below, where:

1) Wide Area Networks may be overwhelmed by excessive demand created by an emergency, natural disaster, or large public gathering (e.g., Presidential Inauguration);
2) Technology does not possess necessary geographic coverage area;
3) Technology does not offer sufficient security;
4) Unlicenced Spectrum susceptible to significant intereference;
5) Wide-area technology not suitable for use within a confined area.

Table 3. Wireless Technology Suitability [1]

Examples Scenarios to Wireless Technology Mapping	IEEE 802.16 2004	IEEE 8.2.16e	IEEE 802.11a	IEEE 802.11b	IEEE 802.11g	IEEE 802.11n	Wireless HART	ISA 100.11a	Zigbee	HSPA/EDVO	LTE/HSPA+UMTS
	WiMAX		WiFi				IEEE 802.15.4			3G	4G
Feeder Reconfiguration	Suitable [1]		Not Suitable [2,4]				Not Suitable [2]	Not Suitable [2]	Not Suitable [2,3]	Suitable [1]	Suitable [1]
Within a Customer Premise	Not Suitable [5]		Suitable				Suitable	Suitable	Suitable	Not Suitable [5]	Not Suitable [5]
Customer Premise to Control Centre	Suitable [1]		Not Suitable [2]				Not Suitable [2]	Not Suitable [2]	Not Suitable [2]	Suitable [1]	Suitable [1]
Within a Bulk Generation Plant	Not Suitable [5]		Surveillance and Sensor Aggregation				Sensor Networks	Sensor Networks	Not Suitable [3]	Not Suitable [5]	Not Suitable [5]
Transmission System to Control Centre	Suitable [1]		Not Suitable [2,4]				Not Suitable [2]	Not Suitable [2]	Not Suitable [2,3]	Suitable [1]	Suitable [1]
Bulk Plant to Control Centre	Suitable [1]		Not Suitable [2,4]				Not Suitable [2]	Not Suitable [2]	Not Suitable [2,3]	Suitable [1]	Suitable [1]

RECOMMENDATIONS OF THE COMMUNICATION NETWORK [1]

The communication system, whether it be wireless or otherwise, must have a consistent security policy. Problems in the past have often been with the intra-system communication links which have been carried unencrypted over the air interface. This must be stopped and the channels must be encrypted. Also to stop eavesdroppers from listening in and to combat spoofing, encryption of the links must be maintained and all nodes, particularly gateway nodes must implement authentication. The network must be continuously monitored for security breaches by carrying out periodic vulnerability assessments. In high density urban areas, pairing of devices must be secure and foolproof. It is likely that in the future, due to the economy of scale, 4G systems may eventually win over WiMAX. ZigBee may prove very popular in Home Area Networks (HANs). Wireless Sensor networks should be utilised in substations and generation plants, technologies such as WirelessHART or ISA100.11a are ideal choices. It is very important the wireless devices are protected against denial of service (DOS) types of attacks.

For Home Area Networks (HANs) employing smart system monitoring devices, it is imperative that the network has 'disconnected operational capability'. This is a very likely scenario when the battery in the home device runs out for example or there are other service interruptions between the device and the utility company. The HAN device must have operational capability for a defined period of time in a disconnected state and must be capable of reattaching itself to the utility company without causing any service disruption to the utility network.

IEEE 802.16 D/E WiMAX

Beyond the hype, on ground testing has shown realistic achievable bandwidth at a distance of 8 km of only 1.8 Mbps for reserved traffic. This clearly need to be factored into the link budget calculation when implementing the control links.

Figure 2 illustrates how WiMAX can be utilised to implement a smart grid application.

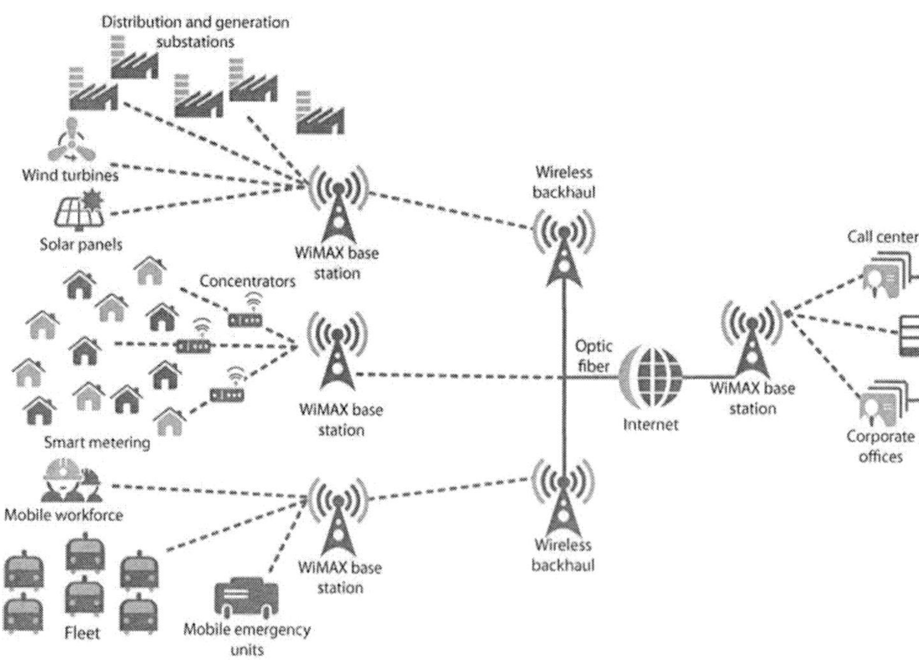

Figure 2. The integration of WiMAX for Smart Grid Applications. [2]

Clearly WiMAX is a very robust system employing OFDM (orthogonal frequency division multiplexing) as its modulation scheme. OFDM is very resilient against multipath conditions, which is the bane of urban environments. OFDM also employs MIMO (multiple input multiple output) antennae technology.

WiMAX has strong security features and these are handled in the MAC (media access layer) privacy layer. However, WiMAX is still vulnerable to attacks which can be divided into those targeting the physical layer and those targeting the MAC layer. Typical physical layer attacks include jamming and scrambling. MAC layer attacks include authentication and management frame spoofing.

IEEE 802.11 A/B/G/N and IEEE 80211S Mesh Networks

These standards refer to wireless local area networks or WLANs. They are the second most popular wireless networks after cellular networks. However, it must be noted that they operate on the unlicensed part of the electromagnetic spectrum frequency of 2.4 GHz and 5 GHz, so interference has to be contended with. IEEE 802.11n supports up to 300 Mbps data rates. Table 4 gives an overview of the main physical layer properties of WLAN.

Regarding security, the Robust Security Network Architecture (RSNA) is used in place of the WEP algorithm. Mutual authentication should also be utilised to increase security.

Table 4. IEEE 802.11 PHY Key Layer Properties [1]

PHY Type	Approval Date	Operating Frequency GHz	Modulation	Typical Throughput Mbps	Net Bit Rate Mbps	Max Indoor Range (m)	Max Outdoor Range (ft)
802.11b	1999	2.4	DSSS	5-7	11	30	91
802.11a	1999	5	OFDM	25-14	54	15	30
802.11g	2003	2.4	OFDM	14	54	30	91
802.11n	2009	2.4/5	OFDM with MIMO	100 (HT20 – 20 MHz mode)	150	91	182

Overview of IEEE 802.11 Mesh Networking

This is the new IEEE 802.11s standard to support mesh networks. Figure 3 shows a typical mesh networking architecture applied to a smart grid. The main difference with this type of network is that not all the nodes have access to the wired network, instead access is gained via anchor access points. Having key access points means that congestion management is a priority to free up bottlenecks. Latency and jitter throughout the mesh must also be taken into account. To enhance radio resource planning and security monitoring, a smart grid should employ a controller-based WLAN architecture.

IEEE 802.15.4 Wireless Sensor Networks

This IEEE standard is mainly concerned with specifying the requirements for power efficient low data rate wireless networks. The ZigBee protocol is the main driver behind this protocol. Other competing standards such as the WirelessHART and ISA100.11a use the physical layer of this standard but have their own respective different layer two MAC definitions. Table 5 gives the main frequency bands for layer one of IEEE 802.15.4. Clearly the lower frequency suffering less attenuation will give longer radio propagation paths than using the 2.4 GHz microwave band.

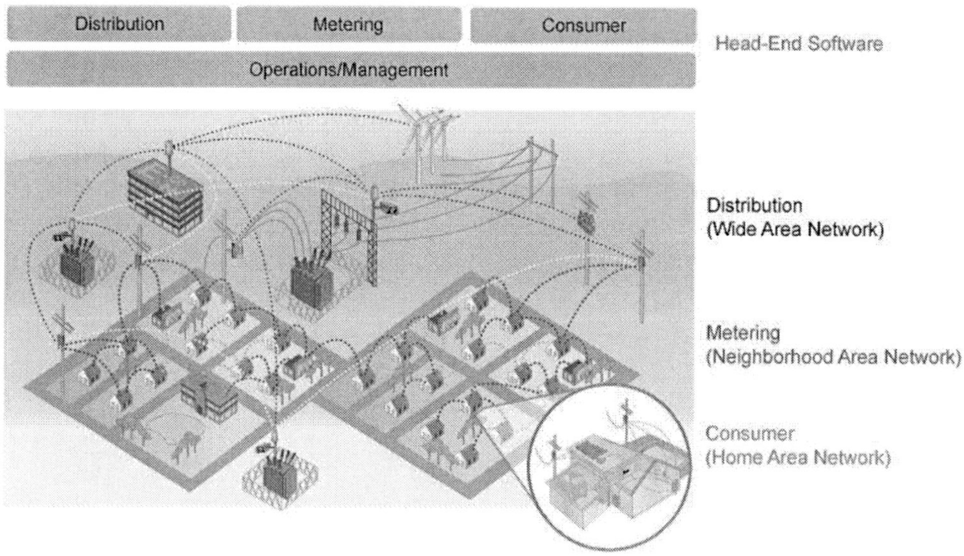

Figure 3. Mesh Networking in Smart Grid Application [3].

Table 5. IEEE 802.15.4 PHY Frequency Band Summary [1]

Frequency Band (MHz)	DSSS Parameters			Data Parameters		
	Chip Rate (kchips/sec)	Modulation	Spreading Factor	Bit Rate (kbps)	Symbol Rate (kbaud)	Symbols
868	300	BPSK	15	20	20	Binary
902-928	600	BPSK	15	40	40	Binary
2400-2483.5	2000	O-QPSK	32	250	62.5	16-ary

The MAC layer for both 802.15.4 and ZigBee are suitable for small networks only.

TSMP (WIRELESSHART)

This uses the same physical layer as IEEE 802.15.4 for the 2.4 GHz version. WirelessHART is popular in the petrochemical industries. It has many useful features such as time synchronised communication, frequency hopping, automatic node joining and network formation, redundant mesh routing and secure message transfer. Figure 4 shows a typical WirelessHART implementation.

The WirelessHART network consists of three main entities: wireless field devices; gateways and the network manager. The wireless field devices may be attached to the grid at key points to monitor energy usage. The gateways interconnect the wireless field devices to the high speed network backbone and the host applications. The management of the network is undertaken by the network manager undertaking such tasks for example as configuration of the network. The topology of the network is of the mesh type which makes it easy to add or remove devices.

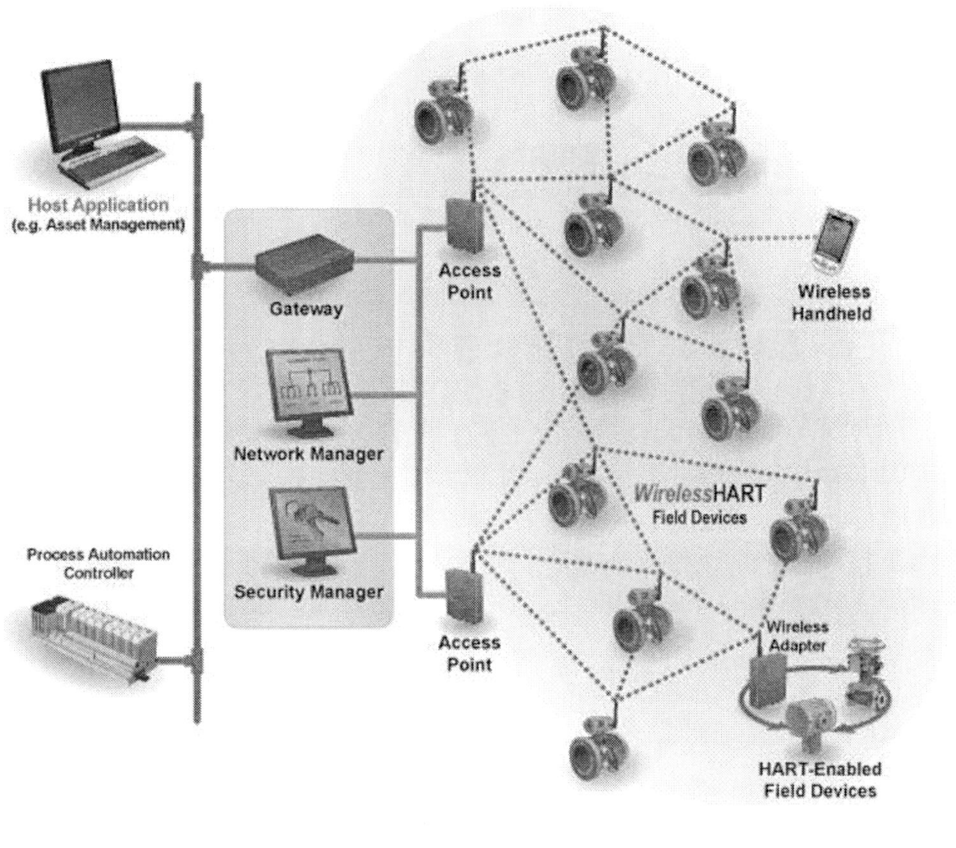

Figure 4. WirelessHART Scenario Application [4].

ISA 100.11A

This is a fairly recently (2009) approved industrial sensor network protocol, which utilises the 802.15.4 physical layer one but has its own MAC layer defined. The over the air data rate is 250 kbps. The network uses mesh topology with slow frequency hopping or frequency diversity. The security of this protocol is robust, employing mutual authentication, optimal encryption and integrity checks. It is a vast improvement over the original 802.15.4 protocol comparable properties with WirelessHART/TSMP.

Table 6. ISA100.11a User Group Importance [5]

User Group Importance	ISA100.11a	ISA100 factory automation study group	Automotive Requirements
Optimised for battery life	High	Medium	Low
Gateway to plant network	High	Medium	Low
Wireless bridge between networks	High	Medium	Low
Discrete I/O	Low	High	High
Fast update rate	Low	High	High
Mesh network	High	Medium	Low

ZIGBEE NETWORK

This uses the same physical and MAC layer of the IEEE 802.15.4 protocol and implements its own network layer protocol. Figure 5 illustrates the utilisation of ZigBee for smart energy monitoring and data collection and dissemination.

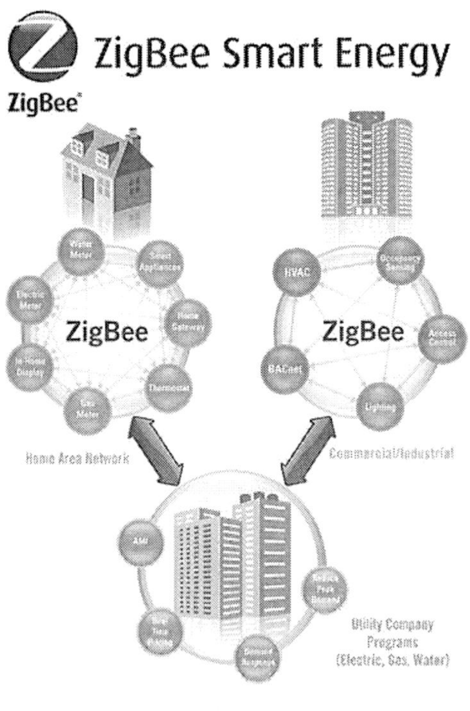

Figure 5. ZigBee Smart Energy [6].

As can be seen from Figure 5, ZigBee can be used in HANs and in the commercial and industrial sector.

GPRS/EDGE/1×RTT 2.5G Cellular Networks

Both GPRS (General Packet Radio Service) and EDGE (Enhanced Data Rates for GSM Evolution) are packet based extension of GSM. 1×RTT (radio transmission technology) is the packet enhancement of the US based cellular IS95 communication network. The '1×' means the use of one 1.25 MHz CDMA channel. Table 7 shows the various key data rates and other features of 2.5G data communication systems.

Table 7. 2.5G Data Communication Systems [1]

Technology	Maximum Data Rate (kbps)	Typical Data Rate (kbps)	Latency (msec)	Encryption – key size
GPRS	60	40-50	700-1000	A5/3 – KASUMI - 128
EDGE	Downlink: 384 Uplink: 100	Downlink: 150 Uplink: 50	500	A5/3 – KASUMI - 128
1×RTT	144	70	500	42 bit PN sequence with ORYX for privacy

The technologies mentioned in Table 7 have mostly been replaced by the 3G cellular communication technology.

HSPA/EVDO 3G Cellular

High Speed Packet Access (HSPA) consists of High-Speed Downlink Packet Access (HSDPA) and High Speed Uplink Packet Access (HSUPA), being based on CDMA technology. Evolution Data Optimised (EVDO) is an evolution of CDMA2000 being an evolution of 1×RTT. A summary of the key technology parameters is given in Table 8.

LTE/HSPA+ UMTS 4G Cellular

The HSPA+ downlink data rate is 42 Mbps for users close to the base station. The uplink stays the same at 5800 kbps. Long Term Evolution or LTE is a 4G technology being based on an all IP network. The downlink data rate for LTE has been increased to 100 Mbps and 50 Mbps for the uplink. LTE supports both FDMA and TDMA with a reduced network latency of 10 ms. The system is also designed to offer reduced operational complexity.

Table 8. Summary of 3G Cellular Data Technologies [1]

Technology	Maximum Data Rate (kbps)	Typical Data Rate (kbps)	Latency (msec)	Encryption – key size
HSPA	Downlink: 7200 Uplink: 5800 (HSUPA) 384 (UMTS)	Downlink: 2000 Uplink: 2000 (HSUPA) 256 (UMTS)	150 ms	KASUMI (f8) & f(9)
EVDO Rev. A.	Downlink: 3100 Uplink: 1800	Downlink: 450-800 Uplink: 300-400	50-150 ms	42 bit PN sequence with ORYX for privacy

Conclusion

The power distribution generation system has already started to adopt wireless based communication system based on 3G and 4G technology. It is likely that to save costs, WiMAX may be replaced by cheaper mass produced 4G network components and eventually adopt the 4G technology in its entirely. SCADA (supervisory control and data acquisition) legacy technology though well entrenched will eventually be replaced by the modern cellular communication systems for power monitoring and control of the power distribution grid.

References

[1] B.A. Akyol, H. Kirkham, S.L. Clements and M.D. Hadley, "A Survey of Wireless Communications for the Electric Power System", January 2010, Report PNNL-19084, US Dept. of Energy. Available from: http://www.pnl.gov/nationalsecurity/technical/secure_cyber_systems/pdf/power_grid_wireless.pdf.

[2] M. Paolini., "Empowering the smart grid with WiMAX", available from: http://www.senzafiliconsulting.com/Blog/tabid/64/articleType/ArticleView/articleId/32/Empowering-the-smart-grid-with-WiMAX.aspx.

[3] http://www.trilliantinc.com/solutions/multi-tier-architecture/.

[4] http://www.hartcomm.org/protocol/wihart/wireless_how_it_works.html.

[5] http://www.isa.org/InTechTemplate.cfm?Section=Article_Index1&template=/ContentManagement/ContentDisplay.cfm&ContentID=75371.

[6] http://www.zigbee.org/Markets/ZigBeeSmartEnergy/Overview.aspx.

In: Advances in Energy Research
Editor: Nicu Bizon

ISBN 978-1-61209-991-0

Chapter 7

TECHNIQUES FOR SECURING COMMUNICATIONS

Charles A. Shoniregun*[*,1], Emil Simion[6], Milan Stork[7], Mircea Andraşiu[5], Paul Burciu[2], Yi-Tung F. Chan[3], Ion Tutanescu[2], Galyna A. Akmayeva, and Andrei-George Oprina[6]

[1]Infonomics Society, UK
[2]University of Pitesti, Romania
[3]Wessex Institute of Technology, United Kingdom
[4]Infonomics Society, Ireland
[5]National Registry Office for Classified Information, Romania
[6]"Simion Stoilow" Institute of Mathematics of the Romanian Academy, Romania
[7]Department of Applied Electronics and Telecommunications
University of West Bohemia, Plzen, Czech Republic

ABSTRACT

This chapter focuses on different techniques for Securing Communications in network. The intrusion detection system device is one of the most popular equipments for preventing the network security attack (section 7.1). In section 7.2 is proposed a heterogeneous network sensors management service and a method for analyzes the alert conflict in a heterogeneous network sensor environment using different alert fusion technique to understand the security status of the whole network system. The fuzzy cognitive maps to fuse multiple inputs are used. As it is known, cryptography is used to protect data from eavesdropping and to preserve integrity of the data (section 7.3). A problem in the field of information security that has no unitary point of view is that of evaluation of the cryptographic algorithms proposed as public cryptographic standards (section 7.4). Section 7.5 focuses on the link between information security and cryptography that is represented by National Institute of Standards and Technology (NIST) cryptographic standards, Federal Information Processing Standard FIPS 140-2 (Security requirements for cryptographic modules) standard and Common Criteria for Information Technologies Security Evaluation (ISO 15408) standard. In the last period of time a growing attention has been given to chaos-based transmission methods in order to be used for securing communications (section 7.6). Secured chaotic communications are

[*] Email: cshoniregun@infonomics-society.eu

obtained by modulation of the chaotic signal with the informational signal. The evolutions of two chaotic dynamic systems that have the same attractor, but different initial start conditions, will diverge after a certain time, although the evolutions are close to the common attractor. For this is necessary to synchronise these two systems. Some examples of data transmission based on simple discrete-time chaotic systems are presented in section 7.7.

Last section presents a FPGA implementation of the Rijndael-AES Crypto-Algorithm.

Keywords: securing communications, intrusion detection, heterogeneous network sensor environment, fusion technique, fuzzy cognitive maps, chaotic communications, information security, cryptography, standards. Intrusion Detection System (IDS), Network Attack, Game Theory, Rational and Irrational Attackers, Network Administrator. Situation Awareness (SA), Alert Conflict, Distributed Intrusion Detection System (DIDS), Fuzzy Cognitive Maps (FCM), Remote Invasion, Virus Invasion. confidentiality, integrity, cryptographic algorithms. cryptographic algorithms, FIPS 140-2, ISO 15408, crypto modules, security evaluation. chaos, dynamic system, attractor, chaotic communications, chaotic encryption, chaotic maps.

A Game Theoretic Approach

Introduction

Generally speaking, researchers are inclined to technical improvement in IDS framework and algorithm that will reduce the misjudgement rate. While, in the DDoS attack and the network worm focused on how to trace back to the attack source or how to build the attack behaviour module on a system.

Therefore, utilizing different point of view to analyze the network interaction between attacker and defender is crucial in defence strategy. The network security problem has been become a very important issue. Nevertheless, the availabilities of many defence mechanisms to prevent the network attack, the intrusion to network security events have never stopped. Therefore, how to efficiently prevent the network security attack is an important research topic area.

The Intrusion Detection System (IDS) device is one of the most popular equipments for preventing the network security attack. Patcha and Park [1] proposed “A Game Theoretic Approach to Modeling Intrusion Detection in Mobile Ad Hoc Networks”. The latter use used game theory to build a model of the relationship between the nodes on Ad Hoc network and analyze each node outcome of the operation. Zou et al., [2] proposed “Code Red Worm Propagation Modeling and Analysis”, simulating and analyzing the Code Red Propagation by using two factors. In order to effectively analyze the interaction between attacker and defender, in the game theory is introduced to the concept of IDS.

IDS Introduction

The IDS model was proposed by Dorothy E. Denning [3]. The IDS is a set of device or technique which is mainly used to detect any inadequate or abnormal activity on the network and immediately generate an alarm and report this alarm to network administrator for further actions. In this model, IDS is consisted of six components Subjects, Objects, Audit records, Profiles, Anomaly records, and Activity rules. The current IDS models are initiated by two international organizations, Common Intrusion Detection Framework (CIDF) [4] and Intrusion Detection Working Group (IDWG) from Internet Engineering Task Force (IETF). It is important to note that this research work utilized the CIDF (see Figure 1).

The CIDF model consisted of four components, Event Generators, Event Analyzers, Response Units, and Event Databases. Firstly, Event Generators gets the event data and transfer the event data to the standard format in order to be convenient for other components to use. Secondly, Event Analyzers receives the event data which was generated from Event Generators and processes comparison and analysis of the intrusion behaviour. Thirdly, Event Databases is used to record the event data and the related information. And fourthly, the Response Units receives the information which was already analyzed and generates the related response action.

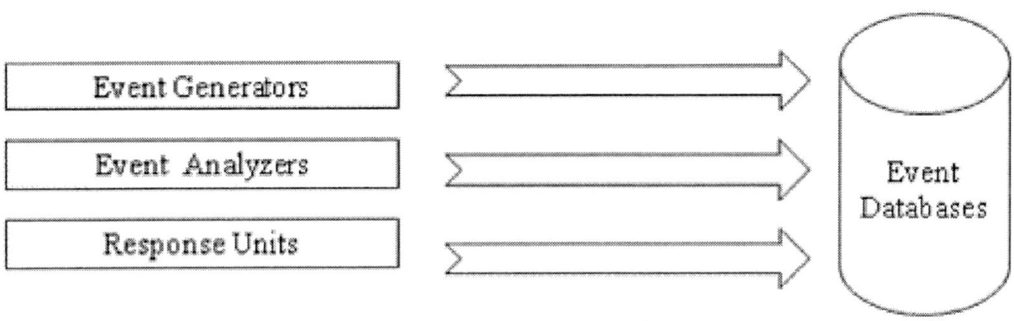

Figure 1. CIDF define IDS Model.

In 2007, Feng-Chi and Shih-Yen [5] states that most of researches emphasize on how to prevent the Internet worm attack and adopt Signature-based technique. In this technique, it has to be controlled by human being to make the system operatability. In 2008, Chan et al., [6] proposed the Cheating defence strategy to prevent the Internet worm attacks. The cheating system used IDS analysis technique to inspect the network traffic and cheat the Internet worm to give up infecting the company network. The cheating system is separated into two parts which are well-known module and unknown module. If the well-known network worm tries to attack the company's network, the cheating system will reject the network connection. On the contrary, if the cheating system detects the potential network worm and it is not listed on the well-known module, the cheating system will add the network pattern into the new blacklist, deny the connection immediately, and send a message to administrator. This research is based on Susceptibility Infestation Recovery (SIR) model [7] and use Kermack-Mckendrick [8, 9] formula to performed the experiment. The result showed that the cheating system can prevent 92.49% worm attack from the Internet.

Deployment of Game Theory Model in IDS Strategy and Attack

The Game theory provides a set of complete analysis [8] and has four basic elements: Player, Strategy, Pay-Off, and Information. In game theory, player will anticipate other players' decision by using other player information and think over the best strategy to take. According to the action sequence, the game can be separated into two categories like the simultaneous-move game (static game) and the sequential-move game (dynamic game). The simultaneous-move game means before players decide to take the action; they do not know other players' decision. Therefore, when the first player made the decision, other players will know the former players' decision and follow this information to process their thinking and the decision. Firstly, Player is the decision maker. The main target of Player is to choose the best strategy and make the maximum benefit. Secondly, Strategy or Action is represented therefore, every player are giving choices (see Table 1).

Table 1. Strategic Form

Player 1 / Player 2	S_a	S_b
S_1	(P_{a1}, Q_{a1})	(P_{b1}, Q_{b1})
S_2	(P_{a2}, Q_{a2})	(P_{b2}, Q_{b2})

In idealization model, we assume that all decision makers (Player) in the game theory are rational and intelligent. The players can pursue their objective and utilize their information to make the judgment and decision on various strategies which possibly cause the cost and the benefit. After then, the player can choose the best strategy and execute it. In order to efficiently analyze the deployment strategy of IDS, the network topology of this research is based on the followings, Network Administrator (Defender) can manage the entire corporation's network architecture, arbitrarily deploy IDS device on the network node, and compare the signature database with the inflow of the sampling network packet in order to eliminate the malicious network packet. While, Attacker tries to attack the network packet with the well-known attack behaviour. The attack behaviour has already been stored in the signature database of IDS device. Using game theory, the Attacker types can be categories into two:

Rational Attacker (RA)

The RA can make the judgment by choosing the best attack strategy. The RA is usually categorized Hacker and can collect the network information.

Irrational Attacker (IA)

The Irrational Attacker randomly chooses one of the attack strategies and executes it. Generally, this kind of Attacker is categorized Virus or Network Worm. The Irrational

Attacker chooses all possible propagation routes to make the attack. In this research work the assumption is that all Attackers are Rational and can make the best judgment by using the public information.

Moreover, Attacker can choose any route to the victim node. If the attack network packet is not detected by IDS device, the attack action will be deemed the successful attack behaviour. Furthermore, Attacker will be based on the size of the network bandwidth to choose the different route. Each route is deemed the link disjointed. When the network bandwidth is bigger; attacker can easily send large quantity of network packet to the victim in order to speedily capture the object node or hide the attack behaviour. The attacks launches on the object node outside network topology and under network domain are connected to the attacked object node and treat the node as the Supernode. This method was utilized to establish the Supernode (see Figure 2).

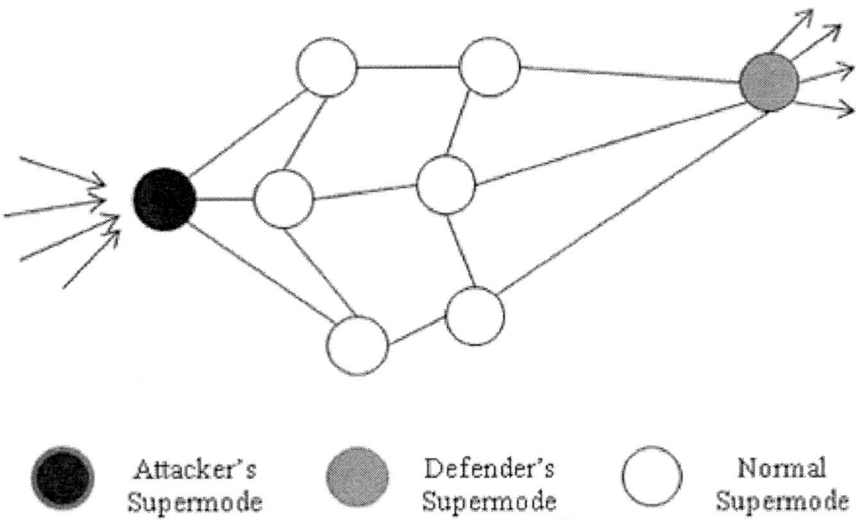

Figure 2. Supernode.

The Network Administrator (Defender) has the limited capacity. It means that Defender has the limited power to deploy quantity of IDS device or collect the sample rate. If rS is the sample rate of IDS, rA will be expressed the attack speed rate which is sent by Attacker. Moreover, l is expressed quantity of IDS device and B is expressed the total sample rate and budget to Network Administrator (Defender). The equation can be showed $\sum^{l}$ rS $\leq$Bl and usually rA $\geq$ rS. Furthermore, we also assume that there are m nodes on the network topology such as K_1, K_2,.., km. If Attacker wants to attack the object node, there are n routes to the destination and the route aggregation is $P_a \rightarrow i$ such as $P_1, P_2, \ldots, P_n$: $P_1, P_2, \ldots, P_n \in P_{a\rightarrow i}$, and f_e(Pi, Kj) represents the link between one of routes P_i from $P_{a\rightarrow i}$, and connection with the node Kj before the link capacity. It is the proportion multiplicator that include the attack packet speed and the sample rate (see next relation):

$$P_d(P_i, K_j) = \frac{r_A}{f_e(P_i, K_j)} * \frac{r_S}{f_e(P_i, K_j)}, K_j \in P_i$$

The Figure 3 shows that if IDS device is deployed on the node Kj where is chosen on the route path Pi, the sample source of IDS will be from the link where the route path is Pi.

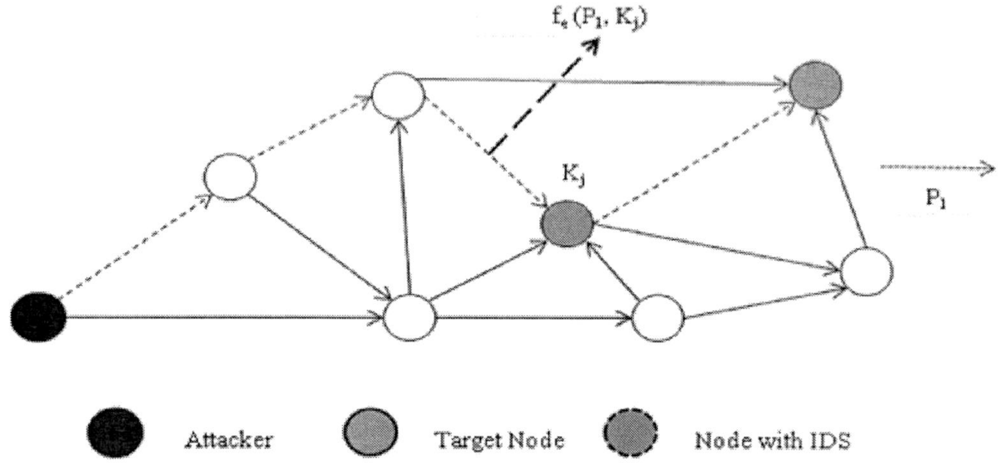

Figure 3. IDS Deployment in Node Kj.

In game theory, making a decision or interaction between players can only be considered from the opponent. In the strategy choice, the Attacker can choose one of the route paths and send the attack packet to the object node from the attack side, that is, $P_1, P_2, \ldots\ldots, P_n \in P_{a\to i}$. Attacker can also utilize other node from the captured route path to control the attack packet direction. On the contrary, Defender can deploy quantity of IDS device and the node selection, that is, the subclass of $\{K_1, K_2, \ldots\ldots\ldots, K_m\}$. Both players have the public information which is included the network topology, the link capacity, the attack packet speed from Attacker, quantity of IDS device from Defender, and the route path selection from Attacker (see Table 2). From these publish information, players can make the judgment by themselves and choose the best strategy.

Table 2. IDS Game Theory Model

Players	Attacker, Defender
Strategies	Attacker: $\mathbf{P_1}, \mathbf{P_2}, \ldots\ldots, \mathbf{P_n} \in \mathbf{P_{a\to i}}$
	Defender: a subclass of $\{\mathbf{K_1}, \mathbf{K_2}, \ldots\ldots\ldots, \mathbf{K_m}\}$
Public Information	Network Topology Link Capacity: $\mathbf{f_e}\ (\mathbf{P_i}, \mathbf{K_j})$, i = 1, 2, ……, m, j = 1, 2, …….n l is quantity of IDS by Defender $\mathbf{r_S}$ is the sample rate of IDS by Defender $\mathbf{r_A}$ is the attack packet of IDS by Attacker Attacker choose the method of the route path

In Equation 2, Pm expressed the detection rate where is located on the node K_m of IDS device. While the h_{ij} expressed the quantity of the nodes before Attacker passes through to the node K_j, which are located on the route path P_i. The C is the cost of successful attacked by

Attacker. Therefore, the payoff function can be alter to join the damage estimation of the by Attacker.

Defender's Payoff:

$$\mathbf{Q_D = (P_i, K_1, K_2, \ldots, K_n) = -P_1 h_{i1} - (1 - P_1)P_2 h_{i2} - \cdots - (1 - P_1)(1 -}$$
P2......1−Pn(hij+C)

$$\mathbf{P_m = \frac{r_A r_S}{f_e^2 \ (P_i, K_{m2})}, m = 1, 2, \ldots\ldots, n; h_{i1} < h_{i2} < \cdots < h_{ij}}$$

Attacker's Payoff:

$$\mathbf{Q_A(P_i, K_1, K_2, \ldots, K_n) = Q_D(P_i, K_1, K_2, \ldots, K_n)}$$

The risk consideration and updating both of players' payoff function, assume that before attacking the object node, Attacker will destroy other nodes. Therefore, considerations were given to quantity of nodes which that passed through by Attacker. When and if the Attacker can successfully attack the object node, how cost effective is the implementation (see Equation 3).

In the following scenario, the Attacker aims at a specific network path and send the attack packet, compare under the same sample rate to deploy one or two IDS devices and compare the difference. The IDS device is deployed on the specific route of the node, that is, l = 1 and the deployment route is on the specific node Pi, and Kj.

Conclusion

The IDS device was exploited using game theory to analyze the relationship between Attacker and Defender and proposed various scenarios. It has been proved that if one powerful IDS device is deployed on one single network path, the detection rate will be high.

Moreover, in a multilink attack, the quantity of the IDS device has to be more than or equal to the number of the attacked path.

Alert Conflict Resolution in Network Security Situation Awareness System

Introduction

The Web Service security and the Network Management are more critical, no matter how secure is an organization transactions, everyone transaction is open to threats. In order to cope with the various threats, organizations deploy various defense mechanisms.

The concept of Situation Awareness (SA) is initially from the aviation security territory. SA deal with complexity, dynamics, high-risk, and situation that requires intervention. The introduction of SA in Network Security Situation Awareness (NSSA) is to response to 'Alert

Conflict', but still the problem remains on solved. Therefore, we proposed Heterogeneous Network Sensors Management Service (HNSMS), a method of analysing the Alert Conflict in a Heterogeneous Network Sensor Environment (HNSE) using different Alert Fusion technique to understand the security status of the whole network system. Thus, we use Fuzzy Cognitive Maps (FCM) and policy to fuse multiple inputs.

Related Work

Endsley (1988) [10] defined SA as the *perception* of elements in the environment within a volume of time and space, the *comprehension* of their meaning, and the *projection* of their status in the near future and further classified SA into three levels which are *Perception*, *Comprehension*, and *Projection* (see Figure 4) [11]. Bass (2000) [12] proposed "Intrusion Detection Systems and Multi-sensor Data Fusion". Yin et al. (2005) [13] proposed "The Design of VisFlowConnect-IP: a Link Analysis System for IP Security Situational Awareness".

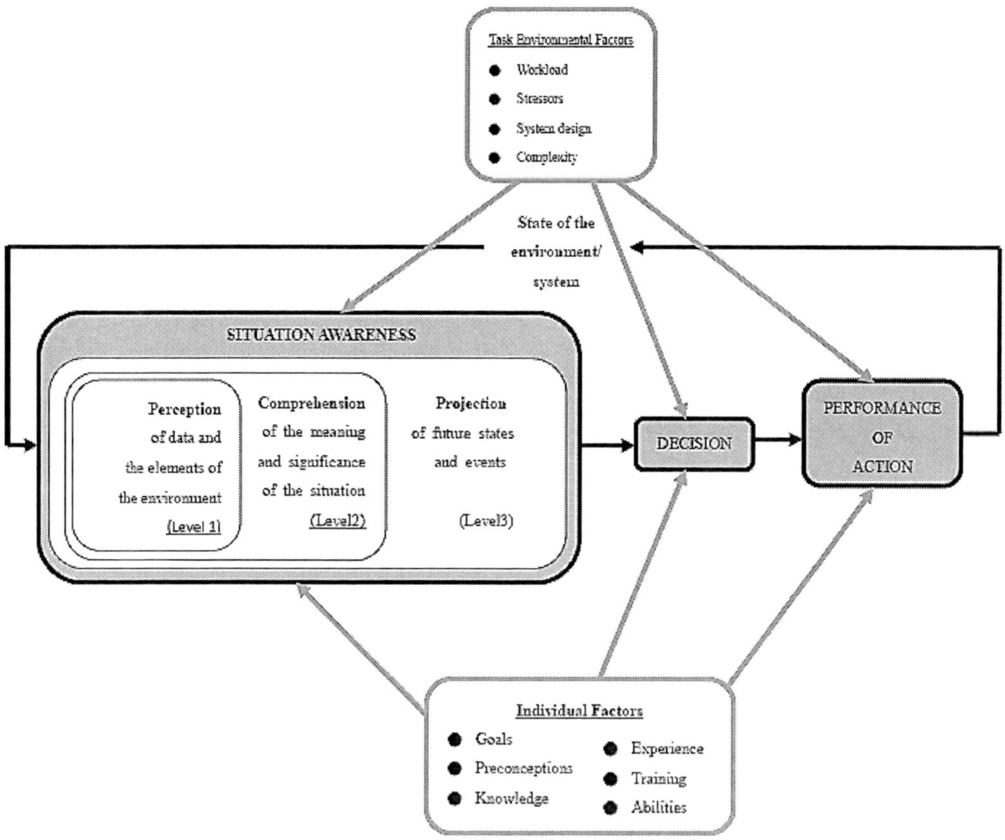

Figure 4. Endsley's the SA Model [10].

Batsell et al. (2005) [14] proposed "Distributed Intrusion Detection and Attack Containment for Organizational Cyber Security". Jibao et al. (2006) [15] proposed "Study of Network Security Situation Awareness Model Based on Sample Additive Weight and Grey

Theory". But none of the existing work has proposed a method of analysing Alert Conflict using HNSE.

In IT Security, the IPO model (Input-Process-Output Model) present the process of the Data Fusion, which is the use of techniques that combine data from multiple sources and gather that information in order to achieve inferences, which will be more efficient that if they were achieved by means of a single source [16] (see Figure 5).

In the Alert Confidence Fusion aspect, the Distributed Intrusion Detection System (DIDS) will be based on different source of the alerts to make a decision whether the system will respond to the attack or not depends on the following [17]:

- The alert confidence information from the remote sensors might be less than local sensors.
- The same network sensors are installed different in place, therefore, the detection ability will be different.
- When different sensors detect the same type of attack, they have different degree of accuracy.

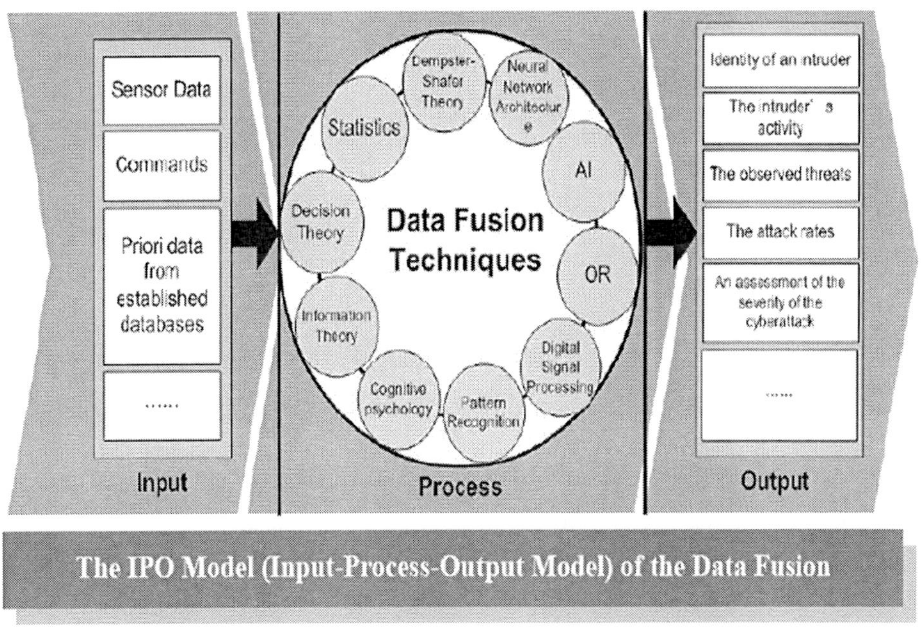

Figure 5. Simple the IPO Model of the Data Fusion.

Yu and Frincke (2005) [17] proposed "Alert Confidence Fusion in Intrusion Detection Systems with Extended Dempster-Shafer Theory", starting that different sensors, different confidence degree, and weight will process the Alert Confidence Fusion. Although the Alert Fusion is efficiency to solve the Alert Conflict by using confidence and weight from different source of the security system, this method does not detect abnormal behavior and respond quickly.

In FCM model, the intrusion detection system used the multi-intrusion sensors at the same time in order to reach the maximum of level of confidence. Siraj et al. (2006, 2004,

2001) used FCM and Fuzzy rule-based to support the causal relationship and inference procedure to detect the potential attack [18, 19, 20]. The conceptual understanding of FCM model is that C_i and C_j has a link value which is e_{ij}. The link value is between -1 and 1. When the link value is 0, this means that C_i and C_j do not have any relationship. When the link value is more than 0, these mean that C_i and C_j have the positive causal relationship. When C_i is increasing, C_j will be multiple increase of e_{ij}. When the modulus of e_{ij} is higher, the impact of the causal relationship is stronger (see Figure 6)

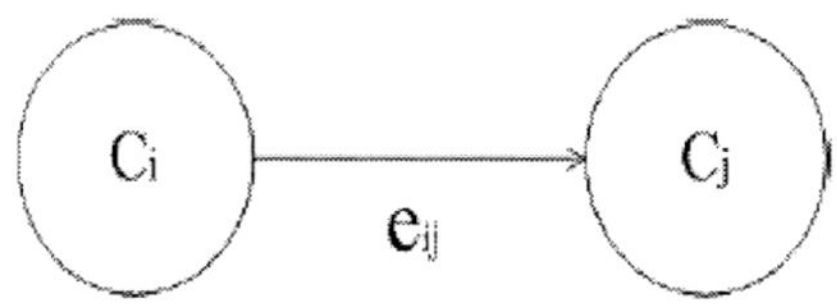

Figure 6. Conceptual Understanding of FCM.

FCM model shows different security events which are Disclosure of Service, Disclosure of Host, System Distress, System Seizure, and System Environment Corruption. It can count the Incident Association Strength (IAS) and represented as a real value in t_{n+1} and the I_k (the influence from e_{ij}) from the evidence of the different security events resource (Ri) (security status of the host). The formula is shown in relation 1.

$$IAS(R_i)(t_{n+1}) = \frac{\sum_{k=1}^{n} I_k t_n \cdot e_{ki}(t_n)}{\sum_{k=1}^{n} e_{ki}(t_n)} \qquad (1)$$

The *IAS* represents the five different security events. The IAS value can be gotten from each security event when the security event is become the quantification of value and the value multiplies the possessed of influence of evidence such as the weight.

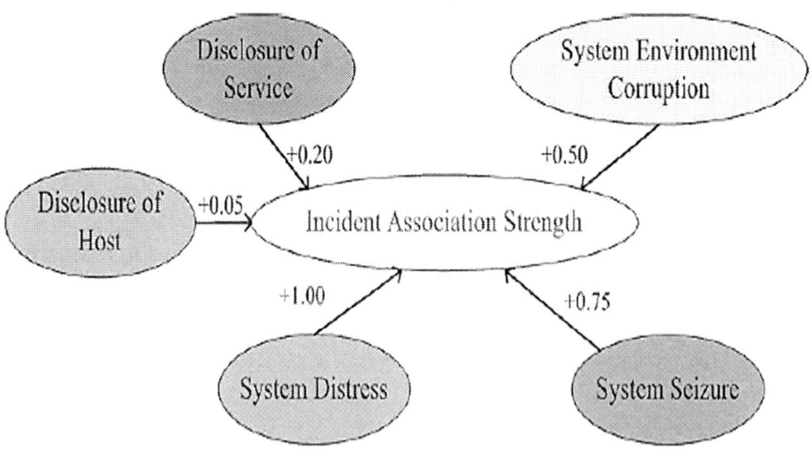

Figure 7. Evidence of combining with different security event.

The Figure 7 shows that the influence of the security event (Disclosure of Service) is 0.2 (positive influence of evidence) and the value is divided by the total amount of all influence of evidence. Therefore, the FCM model can be combined with the evidence of the different security events to estimate whether the whole network environment is safe or not.

Alert Conflict Model

We proposed the HNSMS system that is based on three levels of the SA (*Perception*, *Comprehension*, and *Projection*) and sub divided into nine modules (see Figure 8).

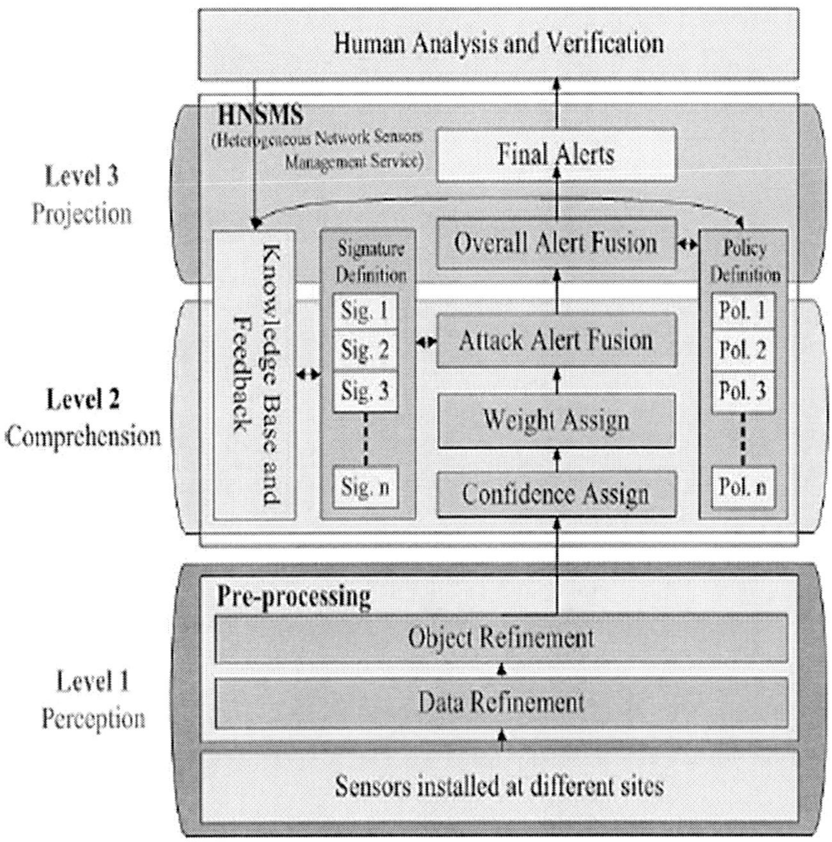

Figure 8.The Framework of the HNSMS System.

The Figure 9 flowchart shows various network alerts from the Multi-sensors that are sent to the HNSMS system.

The alerts have to pass through the Pre-processing and accept the unified format. When the format is accepted by the Pre-processing, the alert will go through the Confidence Assign and Weight Assign to process the Attack Alert Fusion. If it is a critical alert, the alert will be directly reported to the related network administrators.

However, if there is no critical alert, the procedure will continue and until the alerts are confirmed by the Policy Definition. The next stage will process the Overall Alert Fusion.

Finally, when the alerts are confirmed by the network administrators, the secondary data will be stored in the Knowledge Base (Attack Signature Definition and the Security Policy Definition). The 'Attack Signature Definition' is the confidence lever and weight value of the attack for the network sensors while the 'Security Policy Definition' recognizes the attack type in the secondary data.

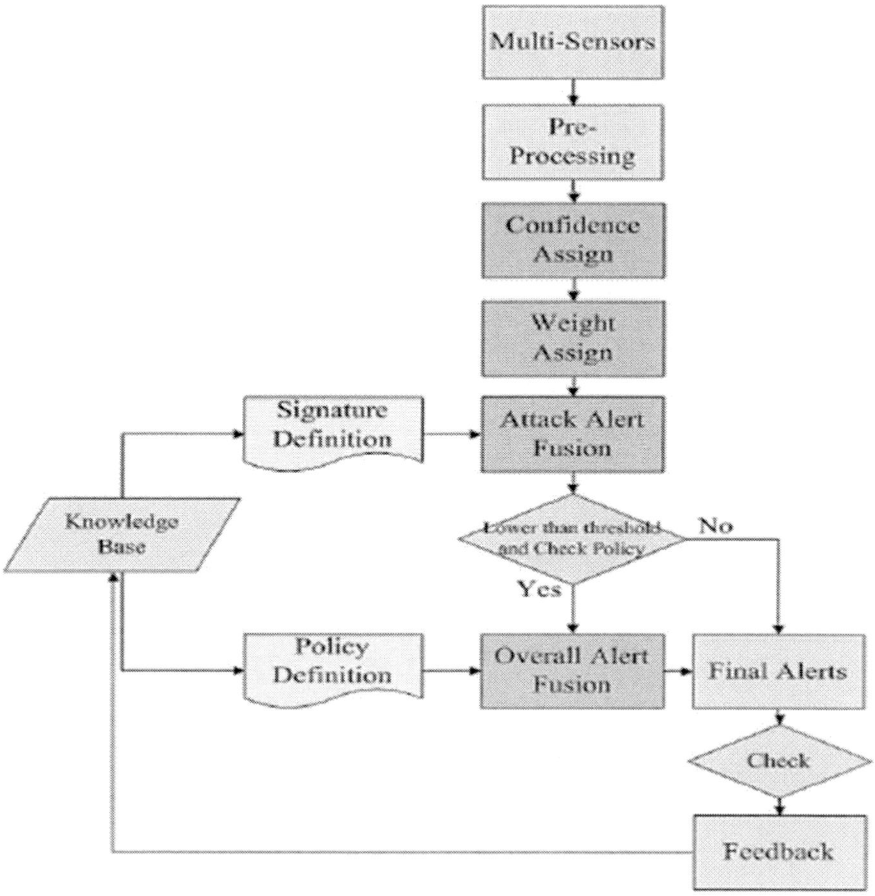

Figure 9. The Flowchart of the HNSMS System.

Weight Assign

The different security equipment can be used to measure and detect the accuracy. Thus, we adopted MaxEnt [14] principle to measure the weight (the condition is known and the Entropy is the biggest object). Therefore, in order to get the proportion of the weight of attack representative from different security equipment, we have to define the scope in advance. We applied the weight assign with the MaxEnt theory (see relation 2), where t_i expressed the time

$$f_i(H,t_i) = \log(H / t_i)$$

Definition of N_k

$$m_{12} = \frac{1}{N_k} \cdot \exp(w_1^k \cdot f_1(H,t_i) + w_1^k \cdot f_2(H,t_i))$$
$$N_k = \sum_H \exp(w_1^k \cdot f_1(H,t_i) + w_1^k \cdot f_2(H,t_i)) \qquad (2)$$

Berger et al. [21] proposed the IIS Algorithm (Improved Iterative Scaling) to estimate the MaxEnt parameter. We used the MaxEnt theory and IIS Algorithm to figure out the best parameter value. In order to be able to discover every attack, we use the FCM model to support the security policy definition. For example, when an invader wants to intrude a specific workstation, the invader has to try to use the account and password from many users to access the workstation illegally. In order to observe this phenomenon, it is impossible to discover the attack from a single feature.

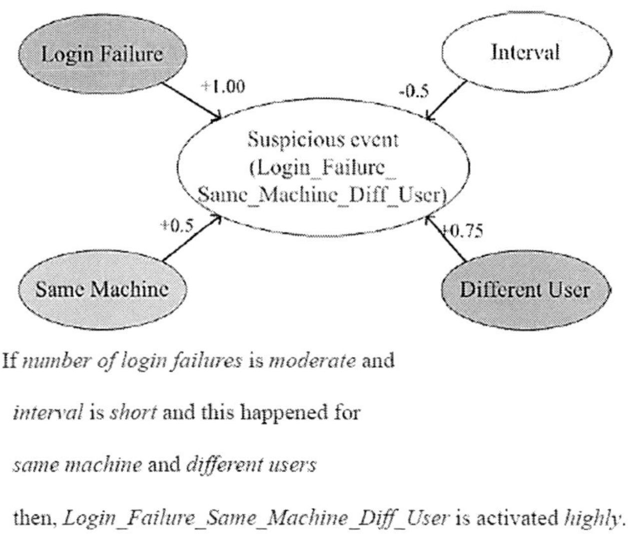

Figure 10. FCM Model Example (Login Failure).

The FCM model is shown in Figure 10. Thus, it has to set up different weighting. For example: "Interval Time" with -0.5. The negative number represents the interval time and the judgement generated by the abnormality and the inverse relationship. In other words, when the interval time is longer, it is impossible to produce the abnormal phenomenon. Furthermore, as to the part of the digital number, it will be indicated how much difference for the influence degree from the positive and negative number. In addition, due to the feature of the multi-attack situation, we have to set up the rule to find out whether the phenomenon is happened or not. In our solution, we used the Fuzzy rule-based to set up the rule (see Figure 10). In other words, when the current situation matches the fuzzy rule, it will give abnormal judgement, and the situation will then be reported to the network administrator for further processing.

Case Analysis

In order to deliberate the efficacy of the HNSMS model, we simulate the situation awareness of the defense in depth (see Figure 8). In the simulation of the security environment, we build four kinds of different sensors. We put the Firewall in the first line of defense. The second line of defense, we put the NIDS (Network Intrusion Detection System). The last line of defense, we put the HIDS (Host Intrusion Detection System) and Anti-Viruses. Through the four kinds of the network security sensors, it will be achieved the efficacy of the defense in depth. As soon as the exogenous factor touches off the alert from sensors, the alert data will be sent to HNSMS server. At this time, the HNSMS system will integrate the alert information, make a judgement for the situation, process the early warning for the information security, deal with the contingency or emergency, and follow the trail. The HNSMS system collects the alert information from different detection sensors and transferred the alert standard format into the consolidation form. We used two case studies to check how the HNSMS operate (see Figure 11 below).

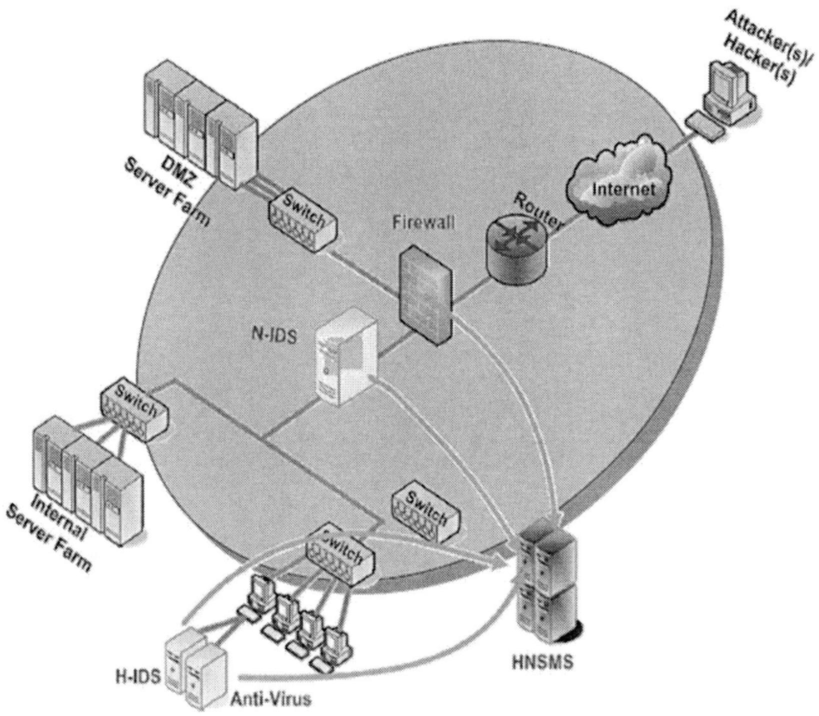

Figure 11. Environment of the Defense in Depth.

Case Study One: Remote Invasion

In this case, when a remote invader tries to access the network and invade a specific workstation, the invader has to try different account and password in the workstation and the invader tries to login to access the resource in this workstation. At this time, because of the login failure, the NIDS and the HIDS will send the alert to the HNSMS server to analyze the alert events (see Figure 12). The alert will be sent to the HNSMS server from two kinds of network security sensors. The reason is that different network security sensor will prevent different security events. From this case, we know that it would be happened certain level of the alert conflict. Therefore, the alert will be sent to the HNSMS server to process the integration. The procedure is shown below.

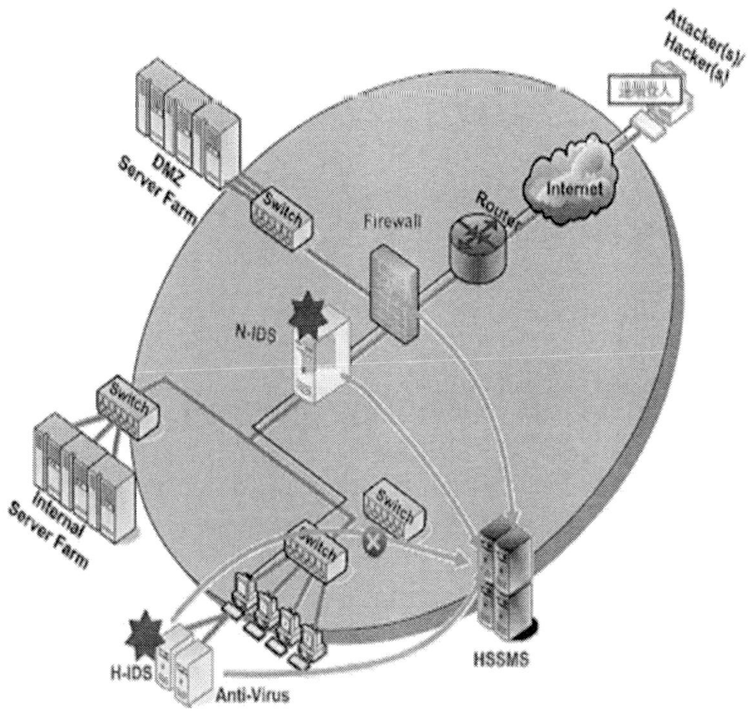

Figure 12. Case Study for the Remote Invasion.

1. In order to process the event analysis, we refer to the information from the knowledge module and this module supports the information. We use the information to give the confidence and the weight on the alert. We also use the related judgement information from the knowledge module to process the Attack Alert Fusion. The Table 3 shows each confidence and weight from the network security sensors in this case study.

Table 3. Related Information on the Knowledge Module (Case Study 1)

id	id	sig_id	Confidence	Weight	Last_modified
1	1	4	0.01	1	2009-04-11 00:00:00
2	2	4	0.2	0.5	2009-04-11 00:00:00
3	3	4	0.4	1	2009-04-11 00:00:00
4	4	4	0.01	1	2009-04-11 00:00:00

2. We fuse the confidence from two kinds of the network security sensors to make the judgement whether the workstation is in secure state or not. We use the concept of

the data fusion from the Dempster-Shafer theory. Moreover, we also use the value of the confidence and the weight from the network security sensors. We use the information which we got and we use it into the formula 2. The result of the Attack Alert Fusion is shown in relation 3.

$$m_{23}(\{A_4\}) = \frac{0.2^{0.5} * 0.4^1}{0.2^{0.5} * 0.4^1 + 0.8^{0.5} * 0.6^1} \cong 0.25 \quad (3)$$

3. Hypothesis: when the confidence value of the Attack Alert Fusion is lower, we can say that the system is in normal state. The situation does not need to report to the network security administrator immediately. The threshold value will be set up by the different situation.
4. In this case, the report from the sensors and fact are not conformed to each other. The reason is that it is common situation when the normal users type wrong account and password. It will be caused the login failure. Therefore, it is impossible to make the judgement through a single timing.
5. Overall Alert Fusion used other auxiliary information to give the judgement whether the system is in secure state or not. The other auxiliary information involved “the number of times for Login Failure”, “Interval Time”, “Users”, and “Machines” (see Figure 10) and we use the security policy which is from the FCM model. We use the estimated criterion in the knowledge module to process the Overall Alert Fusion (see relation 4).

$$\frac{4 * 1 + 4 * (-0.5) + 4 * 0.5 + 4 * 0.75}{1 + (-0.5) + 0.5 + 0.75} = 4 \quad (4)$$

6. From this result, we know that the confidence value is quite high. Therefore, it can be confirmed that the system is in abnormal status and it has to report this situation to the network security administrator. On the other hand, it had better to be provided the automatic control interface, the fast response event interface, and the real time monitored interface for the final alert module. Therefore, because of that, the HNSMS system will be able to produce timely and effective response.
7. Through the friendly users’ interface, the network security administrator can amend and adjust the content information from the knowledge module anytime. For example, when the administrator found that the confidence degree of the Login Failure on the HIDS is not only 40%, the administrator can process the confidence tuning. Therefore, when the HNSMS system processes the alert fusion next time, the operation procedure will be more precise.

From the foregoing case study, when the attacker launched the attack at the first timing, only two security equipments sent out the alert. If the alert information does not process the Attack Alert Fusion via the HNSMS system and the alert does not report the “Login Failure” to the security administrator, the security administrator will have the higher possibility to look at the alert as useless information and the administrator will ignore this attack event.

Although it seems that the alert is the useless information after processing the Attack Alert Fusion in the system, the alert can be executed the further processing such as the Overall Alert Fusion via the security policy and other auxiliary information. It will be a higher possibility to find out the attack behavior and the administrator can make appropriate response to the attack behavior.

Case Study Two: Virus Invasion

When a user opens the E-mail which is infected by the Sasser virus, the infected computer will be sending a lot of network packets into the network. Through sending a lot of malicious E-mail, it can achieve the goal which is disseminating the virus. At this time, because of the malicious procedure invasion, the four kinds of the network security sensors will be operated. In this case, the NIDS and the Anti-viruses will send the alert to the HNSMS server but the Firewall and the HIDS do not send any information. Moreover, when the NIDS and the Anti-virus sent the alert data to the HNSMS server, the server will be analyzing the alert events (see Figure 13).

In this case, the alert will be sent to the HNSMS server from two kinds of the network security sensors. Different network security sensors will prevent different security events. From this case, we know that it would be happened certain level of the alert conflict. Therefore, the alert will be sent to the HNSMS server to process the integration. The procedure is shown below.

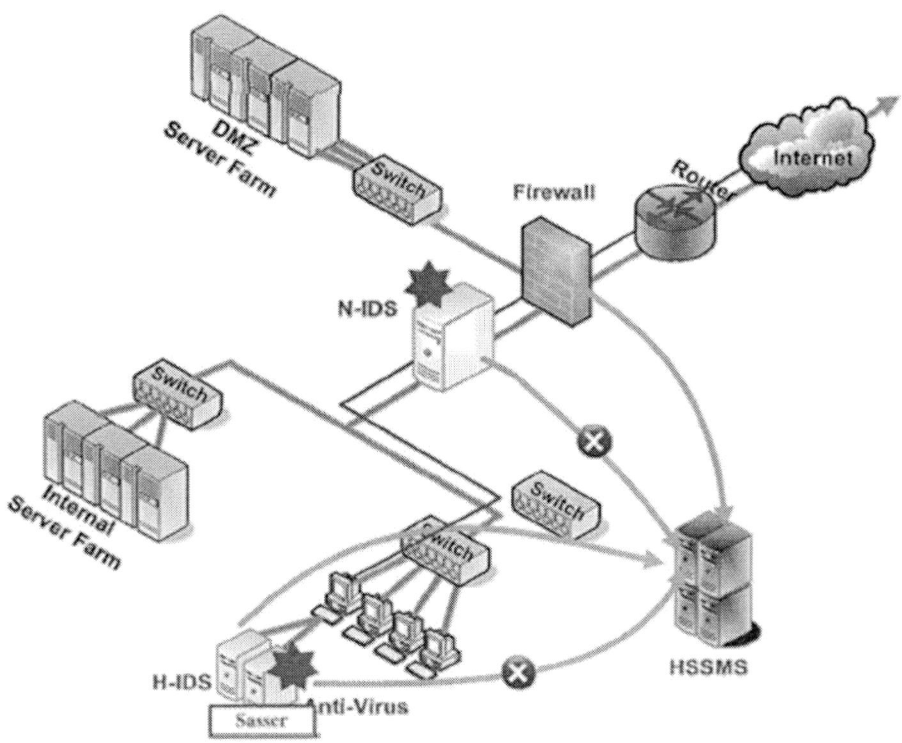

Figure 13. Case Study for the Virus Invasion.

1. In order to process the event analysis, we refer to the information from the knowledge module and this module supports the information. We use the information to give the confidence and the weight on the alert. We also use the related judgement information from the knowledge module to process the Attack Alert Fusion. The Table 4 shows each confidence and weight from the network security sensors in this case study.

Table 4. Related Information on the Knowledge Module (Case Study 2)

id	id	sig_id	Confidence	Weight	Last_modified
1	1	5	0.01	1	2009-04-11 00:00:00
2	2	5	0.2	0.5	2009-04-11 00:00:00
3	3	5	0.2	1	2009-04-11 00:00:00
4	4	5	0.7	1	2009-04-11 00:00:00

2. We fuse the confidence from two kinds of the network security sensors to make the judgement whether the computer system is in secure state or not. We use the concept of the data fusion from the Dempster-Shafer theory. Moreover, we also use the value of the confidence and the weight from the network security sensors. We use the information which we got and we use it into the formula 2. The result of the Attack Alert Fusion is shown in relation 5.

$$m_{24}(\{A_5\}) = \frac{0.2^{0.2} * 0.7^1}{0.2^{0.2} * 0.7^1 + 0.98^{0.2} * 0.3^1} \cong 0.52 \tag{5}$$

3. Hypothesis: when the confidence value of the Attack Alert Fusion is higher, we can say that the system is in abnormal state. The HNSMS has to report this situation to the network security administrator immediately. The threshold value will be set up by the different situation.
4. The system interface had better to be provided the automatic control interface, the fast response event interface, and the real time monitored interface for the final alert module. Thus, because of that, the HNSMS system will be able to produce timely and effective response for the security events.

From the above case study, when the virus is processing proliferation at the first time, only two security systems sent out the alert. The HNSMS system processes the Attack Alert Fusion and reports the abnormal state to the network security administrator.

Conclusion

The HNSMS system processes the Attack Alert Fusion and reports to the administrator. The HNSMS system has to use other auxiliary information to find out the abnormal state. The HNSMS system will execute two modules of the Attack Alert Fusion and the Overall Alert Fusion. Furthermore, as soon as the system receives the information, the system will process the Alert Confidence Fusion. If the alert fusion does not reach the threshold value, the network is in secure state. Thus, the HNSMS system will give up reporting this alert to the administrator. On the other hand, we consider that it would be probably failed to report the alert problem after the alert fusion in a specific timing. Therefore, the HNSMS system has to cooperate with the security policy which is the Policy Definition module. The module is associated with other auxiliary data to recognize the integral security state of the computer system and the network.

A REVIEW OF TECHNIQUES FOR SECURING COMMUNICATIONS USING CRYPTOGRAPHIC ALGORITHMS

Introduction

Cryptography is used to protect data from eavesdropping and also the integrity of the data. There are two types of cryptographic techniques: symmetric and asymmetric. The symmetric techniques involve the usage of the same key in encryption and decryption process or in integrity verification process. The asymmetric techniques are base on the computational difficulty of number theory problems such as factoring the product of two large numbers or discrete logarithms problem.

Communication systems and storage device of data are not reliable in practice because of noise or other forms of introduced interference. The purpose of coding theory [22] is to detect and even correct errors. Usually, coding is defined in terms of source coding and channel coding. Source coding involves changing the message source to a suitable code to be transmitted through the communication channel. The idea of channel coding is to encode the message again after the source coding by introducing some form of redundancy so that errors can be detected or even corrected. Cryptography techniques are used to protect the integrity (including authentication and non-repudiation) and confidentiality of the stored and/or transmitted data. The general scheme of using message source, source encoding and encryption is presented in Figure 14. Source encoding performs, using error detection and correction methods, the reliability of communication through the communication channel.

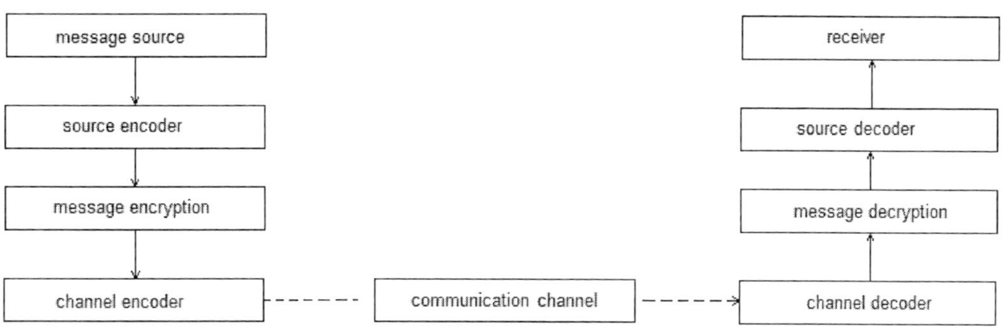

Figure 14. Encoding and encryption data thus a communication channel.

In this chapter is presented a synthesis of cryptographic algorithms used for data protection and integrity through a communication (inclusive storage) channel.

Symetric Algorithms

In the cathegory of symmetric algoritms we find algoritms which ensures condidentiality and/or integrity.

Let's talk about cryptographic primitives which ensures confidentiality: symmetric ciphers. Symmetric ciphers use the same key for encryption and decryption. The information which we desire to protect can be stored information on different types of media and transmitted information on different types of communication chanells. Therefore we need to design two types of ciphers to ensure protection of stored data and protection of transmitted data. Block ciphers are ideal for off-line encryption (storage protection) while stream ciphers are used for encrypting communication. The main difference between this two types of ciphers is that block ciphers have no internal state while stream ciphers have a internal state. Usually the encryption operation in case of usage of stream ciphers is xor-ing the plain text between data and the of stream cipher output. There are methods for transformation a block cipher in a stream cipher and viceversa.

For example, there are several characteristics of block ciphers [23, 24]:

- estimated level of security versus unconditioned security;
- length of effective key K (e.g. in case of AES standard the key size is 128, 192 or 256 bits);
- complexity of output function;
- structure type (e.g. Feistel type in case of DES standard);
- operations used (e.g. AES use subtitutions (S-BOX operation), traspositions (shift row operation), galois field multiplication (mix columns operation) and xor (add round key operation));
- lengh of plaintext block;
- number of rounds (usually depends of the key size and plain text size);
- text inflation: difference between length of plaintext length of cipher text;
- key schedule algorithm;
- balance between encryption and decryption functions;
- complexity of implementation in hardware and software;
- error propagation;
- mode of operations (e.g. Electronic Code Book (ECB), Cipher Block Chaining (CBC), Output Feedback Block (OFB) etc.).

All the above aspect were taken in discussions in AES selection process (1997-2001).

There are several stream ciphers but non of them is standardized at this time. The european project eSTREAM (2004-2008) fail to do this standardization.

Asymetric Algorithms

Asymmetric cryptography allows two or more parties to develop a cryptographic protocol for secure key exchange over an insecure communication channel. There are several asymmetric primitives, based on integer factorizations (IF) of large numbers (e.g. RSA used for encryption and/or signature), discrete log (DL) computing (e.g. El Gamal used for encryption and/or signature, Digital Signature Algorithm (DSA), Diffie-Hellman (DH) protocol that allows two parties that have no prior knowledge of each other to jointly establish a shared secret key etc.) or elliptic curves theory (EC).

In public key cryptography there are two keys for each user: a public key, which is used to encrypt the message and a private key, which is used to decrypt the message. The security of a public key encryption scheme is based on the fact that it is computationally difficult to derive the private key form the public key.

Integer factorization. RSA gets its security from the difficulty of factoring large numbers. The public and private keys are functions of a pair of large (*100* or *200* digits or larger) prime numbers. Recovering the plain text from the public key and the ciphertext is conjectured to be equivalent to factoring the product of two primes.

To generate the two keys, choose two random large numbers, p and q. For maximum security, choose p and q of equal length. Compute the product $n=pq$. Then randomly choose the encryption key, e, such that e and $(p-1)(q-1)$ are relatively prime. Finally, use the extended Euclidian algorithm to compute the decryption key, d, such that

$$ed=1 \bmod (p-1)(q-1).$$

In other words

$$d=e^{-1} \bmod ((p-1)(q-1)).$$

Note that d and n are also relative prime. The numbers e and n are the public key; the number d is the private key. The two primes, p and q, are no longer needed. They should be discarded, but never revealed.

To encrypt a message m, first divide it into numerical blocks smaller than n (with binary data, choose the largest power of *2* less than n). That is, if both p and q are *100* –digit primes, then n will have just fewer than *200* digits and each message block, m_i, should be just under *200* digits long. The encrypted message, c, will be made up of similarly sized message blocks, c_i, of about the same length. The *encryption formula* is simply: $c_i=m_i^e \bmod n$.

To *decrypt* a message, take each encrypted block c_i and compute: $m_i=c_i^e \bmod n$.

Since

$$(c_i)^d = (m_i^e)^d = m_i^{e\,d} = m_i^{k(p-1)(q-1)+1} = m_i m_i^{k(p-1)(q-1)} = m_i \bmod n.$$

There are several attacks on RSA algoritm [25]:

- Chosen ciphertext attack: These are not attacks against the basic algorithm, but against the protocol;
- Common Modulus Attack on RSA: can be applied when RSA implementation gives everyone the same n, but different values for the exponents e and d;
- Low encryption exponent attack against RSA: to avoid this kind of attack the messages must be padded with random values before encrypting them; make sure that m is about the same size as n;

- Low decryption exponent attack against RSA: Another attack will recover d, when d is up to one quarter the size of n and e is less than n. This attack can be avoided if we chose a large value for d;
- Attack on encryption and signing with RSA: It makes sense to sign a message before encryption it, but not everyone follows this practice. With RSA, there is an attack against protocols that encrypt before signing;
- Attack in case of small difference between prime numbers p and q:

In the situation that the prime numbers p and q are close one to each other the following remark allows us to develop an efficient searching algorithm:

$$\left(\frac{p+q}{2}\right)^2 - n = \left(\frac{p-q}{2}\right)^2.$$

For the factorization of n we must test all integers $x > \sqrt{n}$ for which $x^2 - n$ is perfect square; let be this integer denoted by y. We can write

$$\begin{cases} p = x + y \\ q = x - y. \end{cases}$$

Thus the numbers p and q must be of large enough.

Hardware attack: Hardware attacks exploits some hardware parameters such running time, simple power analysis (SPA), differential power analysis (DFA) and fault analysis (FA).

Discrete log theory. ElGamal algorithm is both use for confidentiality and authentication processes. The strength of tha algorithm is based on the difficulty of solving the discrete log problem in a finite field. To generate a public and private key we chose at the beginning a prime number p and two numbers g and x smaller then p. We compute $y=g^x \bmod p$. The public key is $\{y,g,p\}$ and the private key is x. Both numbers g and p can be used in common by a group of users. To encrypt a message M we chose first a random number k relative prime with p-1. We compute after

$$\begin{cases} a = g^k \bmod p \\ b = y^k M \bmod p \end{cases}$$

The pair (a,b) is the cipher text which length is twice then the length of plain text. For decrypting a cipher text (a,b) we compute: $M=ba^{-x} \bmod p$.

Elliptic curves theory. The advantage of using EC [26] versus factorization of large numbers is the key size: the same security margins are achieved using a smaller amount of key bits. EC, used in cryptography, are defined over a general field and the standardized ones are defined over a primes field and over a Galois fields (binary fields). Generally speaking EC, over binary fields are easy to implement in then EC over prime fields. RSA-based protocols have been adapted to EC, replacing the group Z_{pq} with an EC: the ECDH key agreement scheme is based on the DH scheme, the ECDSA is based on the DSA and the ECMQV key agreement scheme is based on the MQV key agreement scheme. As we know there are several cryptographic primitives like symmetric (block and stream ciphers, hash functions etc.) and asymmetric one (RSA, (EC) DSA etc.). EC needs a smaller amount of bits than DL or IF. Some of the application that involve elliptic curve computations are secure communications, e-commerce and digital money.

Hybrid Systems

Hybrid systems are composed by combination of asymmetric and symmetric primitives to provide confidentiality and integrity. Usually confidentiality of information is achieved using a symmetric primitive and the authenticity and non-repudiation are obtained using asymmetric primitives such as RSA, DSA or ECDSA algorithms.

In the case of ensuring confidentiality, the data is usually encrypted by the sender *A* with a symmetric algorithm, the encryption key is encrypted with the public key of the receiver *B*.

Another cryptographic primitive is hash function, which computes a digest of the data. This hash value is signed with the private key of the sender *A*, the receiver *B* will check the signature using the public key or *A*. In Figure 15 is presented the process of signing a document using a PKI infrastructure based on RSA signing function.

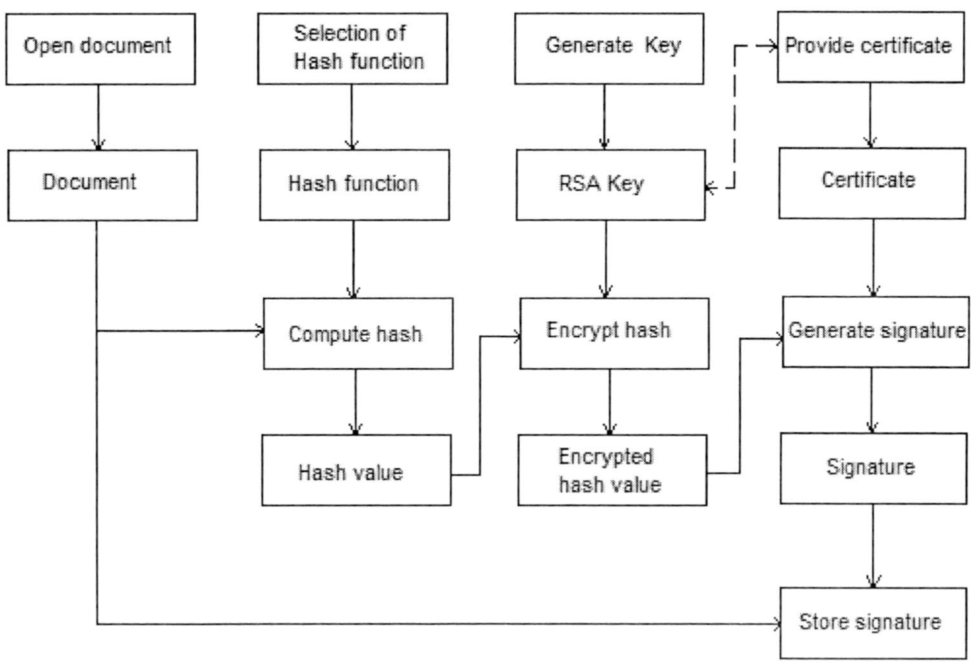

Figure 15. Flowchart of the signing process of an electronic document.

A Proposal for Research and Development of Cryptographic Standards

Introduction

An actual problem in the field of information security, that has no unitary point of view, is that of the methods and techniques used in the evaluation of the cryptographic algorithms proposed as public cryptographic standards. This problem appeared due to the high level of heterogeneity of the cryptographic field and because of the interdisciplinary character of this domain.

The problem is systematically taken into account by the International Association of Cryptographic Research - IACR (http://www. iacr.org/) in more then 60 annual international conferences (http://www.iacr.org/ events/), such as CRYPTO, EUROCRYPT, ASIACRYPT, AFRICA CRYPT with various audience representing governmental institutions, research centers and prestigious universities all over the world [27-42].

Considering those facts this chapter will present some ideas, untouched yet in our country, concerning the evaluation process and cryptographic standards, subject with significance impact in most countries and for Romania as well.

The Character of Interdisciplinary Fundamental Research

The overall objective of this research is the analysis of the main results and of a number of present international standards in the field of information security, explicitly concerning the applied cryptography. As a starting point, the study will employ the mathematical models used for every known standard, developed as a result of national and international projects. By using the comparative analysis of these standards, it is intended to develop a common model that will lead to a national standard in the field. This method was also used by a number of EU and NATO Member States. Such a research can be carried out, from our point of view, by a team consisting of mathematicians, specialists in ITandC, automatics and electronics, with expertise in fundamental fields, that have the capability to develop this standard that is compulsory for the unitary implementation, at a national level, of the main theoretical INFOSEC concepts, possibly in the form of a reference model. The team of mathematicians will study the optimization of the existing models with a view to adapt them to the requirements needed for the implementation in a real system. At this level, a link will be established between the fundamental research and a model suitable for implementation in real systems. In this phase, the team of researchers specialized in optimized implementation will be involved. This team will restrict the model in order to create hardware / software products to comply with the commercial and governmental requirements. Following the testing phase and the analysis by the team specialized in standards and minimum security levels required for each communications level in the national inter-relation data system, there will be a phase of development proposals for models to ensure the confidentiality of information. The models shall be complete from the structural point of view, meaning that it will have a mathematical model, implementation system comprising of analysis schemes, general schemes for adjustment to different systems and analysis of the risk level, as well as the usable API for real applications. This phase will be ensured by the team having vast experience in hardware and software implementation.

The Objectives and the Methodology of Research

Regardless their category, standards are important because they define practices, methods and measures to be followed by the developer, evaluator and customer. For this reason, standards enhance the products reliability and efficiency that results in an improvement of the quality. Standards provide solutions that are accepted by the specific community they address to, and are used by experts in that field. By using standards, organizations are able to reduce costs and to protect their technological investments. For the ITandC field, standards provide interoperability, security and integrity.

Interoperability. Products developed in compliance with a standard are comparable and can be used in conjunction with other products developed in compliances with the same category of standards. By ensuring interoperability between different equipments, standards enable organizations and companies to choose a certain product in order to minimize expenses.

Security. Standards may be used to ensure a certain common level of security. Most managers have no expertise in security issues but, by using a standardized algorithm (e.g. Federal Information Processing Standards) they are certain that this was analyzed by the standardization organization (in this case National Institute of Standards and Technologies, http://www.nist.gov).

Integrity. Standards may be used to ensure integrity of a certain product.

Starting from the way standards are developed, some steps can be made by developing a system to analyze the mathematical models used for the present international standards, the development of an optimized system to be implemented in a source to ensure information confidentiality in commercial and governmental systems. Some comparative studies of technique and methods used in the evaluation of cryptographic public standards can be made, as follows:

- the NESSIE projects (project developed by the European Commission during 2000-2003 in the framework of the IST Information Society Technologies program that proposed the drawing up and evaluation of a number of cryptographic primitives),
- ECRYPT - European Network of Excellence for Cryptology, project under development within the framework of IST-2002-507932 program, both projects being under the coordination of the Katholieke Universiteit Leuven (partners list published at http://www.ecrypt.eu.org/ partners.html,
- CRYPTREC (project regarding the cryptographic evaluation financed by the Japan Government),
- standardization bodies National Institute of Standards and Technologies (NIST – standardization body under the US Government), American National Standards Institute (ANSI – industrial standardization body from the US), GOST (governmental standardization organism within the Community of Independent States, http://www.gost-r.info/).

We also have to take into account the NATO, EU and national standards in the field, whenever possible. The study has to take into consideration the means of approaching the side domains, determined by the implementation of cryptographic algorithms in cryptographic modules and of the modules in cryptographic systems. This approach provides a constructive mediation of the ongoing conflict of the cryptographic policies between the commercial, research and governmental dominions. This research has a multidisciplinary approach, including comparative aspects related to the interrelationship between cryptographic standards and ITandC security standards. The comparative approach will be layered on comparative studies of the standards regarding the cryptographic algorithms, the cryptographic module and ITandC security.

Level of Novelty, Originality and Expected Results

The research is intended to formulate the general framework for the establishment of the requirements (e.g. data format, test vectors, reference implementation) to be met by evaluated and standardized cryptographic algorithms (e.g. by NIST - Cryptographic Toolkit, IEEE, NESSIE, ECRYPT) as follows:

- block ciphering standards;
- stream ciphering standards;
- asymmetric ciphering standards;
- digital signature standards;
- hash function standards;
- message authentication standards;
- entity authentication standards.

The aspects required for accomplish the protection profile of the cryptographic algorithm are distinguished on the Security evaluation, implementation Costs and its Characteristics. In fact, our approach will be structured on the following directions stipulated by NIST regulations:

- development of attached mathematic models, their optimization through adaptation depending on analyses obtained for individual standards existing at global level, evaluation of computational security limits, taking into account the overlapping of the cryptographic algorithms to general attacks types (statistical evaluation, exhaustive research attack), and to particular attacks types (e.g. block cipher type differential cryptanalysis SPN). For a consistent approach we'll take into account the definition of the following concepts: time complexity, space complexity, data complexity, and also space-time compromise. A special attention will be paid to "hard resolvable" problems with applications in cryptographic type standards (factoring matter and discrete algorithm matter, both in Galois field, and elliptic curves);
- software implementation, taking into account the size of the processor word, implementation language, variation of the processing rate in regard to key size, rate

variation in regard to algorithm mode of operation, data processing rate and a comparison of them, in face of particular types of restraints regarding RAM and ROM memory space;

- restrains regarding RAM and ROM, taking into account the memory resources required for cryptographic algorithms implementation. Small resources of implementation space may allow the implementation of the algorithms on specific platforms such as SMART CARDS;
- hardware implementation, taking into account architectural model, implementation methodologies and objectives, implementation within FPGA and ASIC, data processing rate, hardware characteristics, the possibility of distributed computation or collimation of the computations made within specific processing;
- attacks against implementation. Shall be taken into account the following types of attacks assuming cryptanalysis simulation depending on implementation (software or hardware):

FA (fault analysis) – during an arithmetic operation, hardware equipments can produce errors (transient, latent or induced), and through a national exploitation of these errors the private key can be recovered for a series of asymmetric algorithms;

TA (timing attacks) – by calculating execution time we can assume which bytes are zero and which are one;

DPA (differential power analysis) – power consumption depending on the executed instruction, and being monitored we can deduce the source code; in case the instruction sequence depends on key length, than the power consumption can provide information regarding that;

SPA (simple power analysis) – consist in assessing the power spent by the device during cryptographic operation, this type of attack being applied, as a rule, to the devices having exterior voltage source (like SMART CARDS);

EMA (electromagnetic attacks)

- encryption vs. decryption, taking into account the complexity (space - time) of the encryption procedure against the decryption procedure;
- key generation algorithm which, in case of symmetric algorithms, has a special importance from the view of security and of processing rate;
- capabilities and flexibility;
- possibility of implementations collimation at infrastructures level;
- copyrights, taking into account the interoperability character of those.

This research intends to be updatable, in sense that during its development or after its completion, considering international results in the field, if new methods and technologies are developed, these shall be included into a updated version of this project. In fact, this is a standard procedure applied in most international organizations and developed states for motivating both theoretical progress in the field, and the technologic one in implementation and manufacture.

Conclusions

This chapter tried to formulate, a series of requirements for the future development of cryptographic standards in Romania. Having these requirements (relatively) standardized, the evaluation itself becomes more standardized – hence more efficient. Consequently, a metric easily and convincingly can be established for the space of the cryptographic primitives, and a hierarchy based on alternative criteria, other than those concerning the cryptographic strength, can be derived and used subsequently.

INFORMATION SECURITY AND CRYPTOGRAPHY

Introduction

This chapter focuses on the link between information security and cryptography. The link is represented by National Institute of Standards and Technology (NIST) cryptographic standards, Federeal Information Processing Standard FIPS 140-2 (Security requirements for cryptographic modules) standard and Common Criteria for Information Technologies Security Evaluation (ISO 15408) standard [43-48].

To implement the *process of security* of a communication network, system or a device, different procedures must be in our attention. These procedures cover technical and no technical areas in order to minimize *the risk of successful attack* against the system. The risk depends of at least two factors: the threat and the potential vulnerabilities of the system. This risk can't be zero but it can be minimized: the event *will occur in time* but *in finite time* with *small probability*. The threat always will exploit a vulnerability of the system. Thus, the designer of the system must know this weak points and see what potential threat can exploit it and what is the impact on the system. For example, a threat can be an earthquake, which exploits the system availability if there is no proper back-up procedure (back-up of the data, system and personnel). The probability that a threat will exploit a given vulnerability is hard to compute and usually is estimated by the data given by evaluation process. Usually the vulnerabilities are not independent and this computation is based on correlation factors between them. For example, the improper connection of the vehicle power supply to the GPS system will cause a high consumer if the vehicle will be hold into an underground garage and this will affect the life time of the vehicle battery.

Information security (INFOSEC) is focused of protecting information and information systems from unauthorized access, use, disclosure, disruption, modification or destruction. Thus, INFOSEC will cover areas from operational field.

Cryptography deals with design, implemention and evaluating cryptographic algorithms (e.g. NIST AES selection process, SHA-3 competition etc.) in order to be used by products (software and/or hardware) which are intended to protect information or information systems. The cryptographic algorithms are materialised into software, hardware or firmware cryptographic products. For this reason, there are two types of attacks on a cryptographic algorithm: classical attacks and side chanells attacks, where the attacker has the knowledge of the implementation platform.

Before using in information systems, *cryptographic products* also need to be tested and evaluated. One cryptographic evaluation standard is FIPS 140-2. After this evaluation is obtained, from an accredited Laboratory, the system itself needs to be evaluated in order to have a image of the assurance level obtained. Usually these evaluation is made using ISO 15408 (Common Criteria for Information Technology systems) standard.

INFOSEC

INFOSEC domainn is the science that deals with the regulations on electronic information that is processed, stored or transmited and covers the following areas:

Physical security describes measures that prevent attackers (inside or outside) from accessing a facility, resource, or information stored on a physical media and guidance on how to design structures to resist various hostile acts. FIPS 140-2 gives, in section 5, some guidelines for physical protection of cryptographic modules (tamper proof, tamper detection and tamper reponse etc.).

Personal security describes the administrative restrictions on data. Under need-to-know restrictions, even if one a security clearance to access certain information, one would not be given access to such information unless one has a specific need to know.

Procedural security deals with the establishment and enforcement of security procedures. Some of these procedures may be independent of the system involved. Others procedures are invariant to the system. For example, perimeter security controls are usually similar for all type of systems. Procedural security regulates the performance of duties associated with system operation. Procedural security is high depending of the personal security because all users will put in practice the procedures.

Communications security (COMSEC) describes all the measures and controls taken to deny unauthorized persons information derived from telecommunications and ensure the authenticity of such telecommunications. Communications security includes cryptographic security, transmission security (TRANSEC), emission security (TEMPEST), traffic-flow security and physical security of COMSEC equipment.

Computer security (COMPUSEC) is the information security applied to computers. The objective of computer security includes protection of information and property from theft, corruption, or natural disaster.

Information assurance (IA) is the practice of managing information-related risks. More specifically, IA practitioners seek to protect and defend information and information systems by ensuring confidentiality, integrity, authentication, availability, and non-repudiation. IA is the process of ensuring that authorized users have access to authorized information at the authorized time.

A system is protected if all the above areas are properly covered. In the next section we will focus on cryptographic security from the prespective of FIPS 140-2.

INFOSEC Standards

INFOSEC standards can be stratified like in Figure 16: standards for cryptographic algorithms, cryptographic modules and for ITandC security. In this chapter we focus on standards for cryptograhic algorithms, cryptomodules (FIPS 140-2) and ITandC standards (e.g. ISO 15408). The dificult task for an desingner is to ensure the correct implementation of the process specification on the compromise between security, usage and functionality: high security impose restriction on usage.

Cryptographic Standards

Our discussion is based on National Institute of Standards and Technologies (NIST) cryptographic standards. These cryptographic standards can be divided in four classes: symmetric key, public key, secure hash and random number generation.

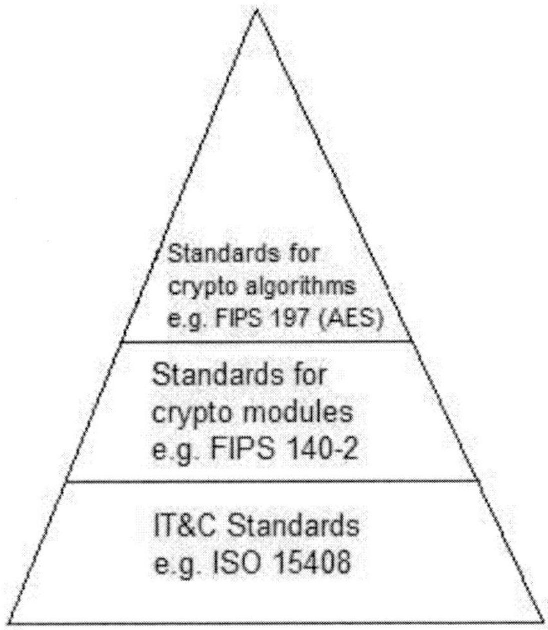

Figure 16. INFOSEC standards stratification.

In symmetric key cryptographic standards we can found for example AES (FIPS 197), DES (FIPS 46-3) for block ciphers standards or HMAC (FIPS 198) for hashing and message authentication code. We remained that simple DES was replaced by AES, 3-DES being in use.

In public key cryptographic standards we can found Digital Signature Standard (FIPS 186-3), Key Establishing Schemes (DHandMQV, FFCandECC SP 800-56A) and Key Management Guideline.

Secure hash is referring to SHA-1, SHA-224, SHA-384, SHA-512 (FIPS 180-2). At this time there exists a draft for SHA-3 that will replace SHA-2.

One standard for random number generation standards is SP 800-90.

The following table gives the theoretical comparable strengths of symmetric and asymmetric cryptographic algorithms:

Table 5. Theoretical comparable strengths of symmetric and asymmetric cryptographic algorithms

Sym Key	80	112	128	192	256
Hash functions (for signatures)	160	224	256	384	512
FFC and IFC	1K	2K	3K	7.5K	15K
ECC	160	224	256	384	512

NIST approved standards are referred by NIST Cryptographic Toolkit.

Some of these standards are allowed to process classified information. For example, AES with 128 bit key can be used to protect SECRET classified information and AES with 192 or 256 bit key can be used to protect TOP SECRET classified information.

FIPS 140-2

Cryptographic controls are provided using cryptographic modules, which may include capabilities such as signature generation and verification, encryption and decryption, key generation, and key establishment.

An undetected error in a cryptographic module design could affect every user in the system for which it is supposed to provide protection. For example, the verification of a chain of public key certificates might not function correctly.

Verifying a chain of public key certificates helps a signature verifier determine if a signature was generated with a particular key. If the function is implemented incorrectly in a cryptographic module, the potential for the dissemination of weak cryptography could be introduced into the system, possibly allowing for signature forgery or the verification of invalid signatures. Therefore, it is important to have cryptographic modules tested before distributing them throughout a system.

The security requirements in FIPS 140-2 cover 11 areas related to the design and implementation of a cryptographic module:

- Cryptographic module specification includes definition of cryptographic boundary, approved algorithms and approved modes of operations;
- Cryptographic module ports and interfaces are referred to the specification of all interfaces and all input data paths. For security level 3 and 4 data ports for unprotected critical security parameters logically or physically separated from others data ports;
- Roles, services and authentication requires, for all security levels, logical separation of required and optional roles and services. For level 2 operators authentication must be role–based or identity–based. To achieve security level 3 and 4 operator authentication must be identity-based;
- Finite state model requires the specification of finite state model, required and optional states, state transition and specification of these transitions;
- Physical security is focusing to tamper evidence, detection and response (e.g. erasing critical security parameters);

- Operational environment is referring to evaluation, for example, of Protection Profile (PP) at (Evaluation Assurance Level) EAL 4;
- Cryptographic Key Management is referring to the key (secret, private and public) manipulation during its life time: generation, pre -activation, activation, usage, storage and deletion;
- EMI/EMC – electromagnetic compliance with Federal standards;
- Self – Tests includes power-up tests and conditional tests;
- Design assurance is referring to configuration management, secure installation, design policy and guidance documents;
- Mitigations of others attacks are referred to specification of mitigation of attacks for which no testable requirements are currently available.

Within most areas, a cryptographic module receives a security level rating of 1 to 4, from lowest to highest, depending on what requirements are met. For other areas that do not provide for different levels of security, a cryptographic module receives a rating that reflects the fulfillment of all of the requirements for that area.

An overall rating is issued for the cryptographic module, that indicates the:

1. Minimum of the independent ratings received in the areas with levels, and
2. Fulfillment of all the requirements in the other areas.

On a vendor's validation certificate, individual ratings are listed as well as the overall rating. It is important for vendors and users of cryptographic modules to realize that the overall rating of a cryptographic module is not necessarily the most important rating. The rating of an individual area may be more important than the overall rating, depending on the environment in which the cryptographic module will be used (this includes understanding what risks the cryptographic module is intended to address). Modules may meet different levels in different security requirement areas; for example, a module may implement identity-based authentication (level 3 or 4) and display tamper evidence (level 2).

At this time the draft for FIPS 140-3 where NIST has updated the standard to reflect changes in technology has a fifth security level. In this draft there is a special section dedicated to software security and specifying requirements to protect against non-invasive attacks. Also the reference to Common Criteria (ISO 15408) and requirements for the use of Common Criteria certified operating systems has been dropped. In this draft NIST improves the requirements for authentication for level 4 at two-factor authentication (at least two of three: something known, something possessed and some physical property). Also a greater importance is given to physical security requirements to defeat non - invasive attacks/side channel attacks (protection to timing attacks (TA), differential power analysis (DFA) etc.).

Cryptographic Module Validation Program (CMVP)

NIST and the Communications Security Esablishment (CSE) of the government of Canada established the CMVP. The goal of the CMVP is to provide Federal agencies with a security metric to use in procuring equipment containing cryptographic modules. The results of the independent testing by accredited laboratories provide this metric. Cryptographic module validation testing is performed using the Derived Test Requirements (DTRs) for FIPS 140-2. The DTRs list of all the vendor and tester requirements for validating a cryptographic module are the basis of testing done by the Cryptographic Module Testing (CMT) accredited laboratories. Figure 17 illustrates the CMV process.

ITandC Assurance Standards (Common Criteria)

Information Technology Security Evaluation Criteria (ITSEC), predecessor of Common Criteria for Information Technology Security Evaluation (abbreviated as Common Criteria or CC), is a structured set of criteria for evaluating computer security within products and systems.

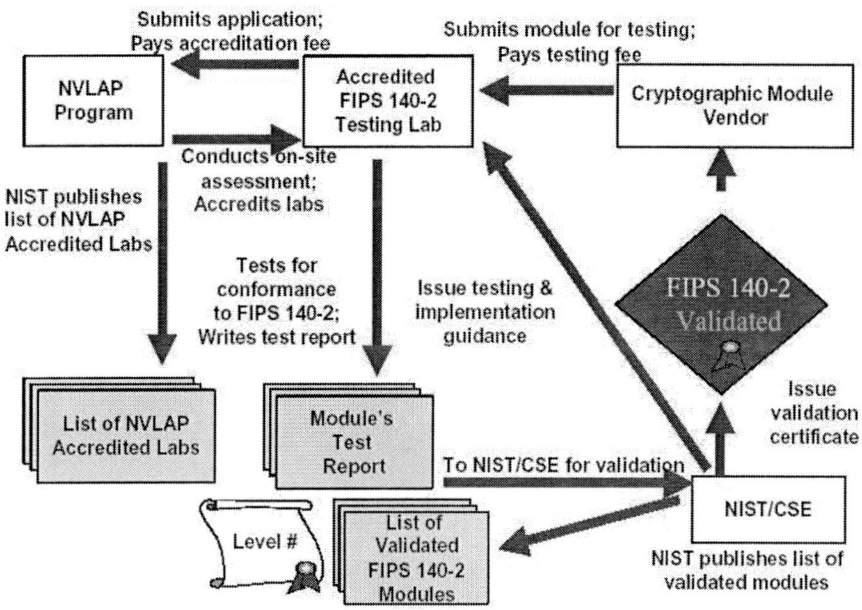

Figure 17. CVMP Process.

The ITSEC was first published in May 1990 in France, Germany, the Netherlands, and the United Kingdom based on existing work in their respective countries. Following extensive international review, Version 1.2 was subsequently published in June 1991 by the Commission of the European Communities for operational use within evaluation and certification schemes. Since the launch of the ITSEC in 1990, a number of other European countries have agreed to recognise the validity of ITSEC evaluations.

Thus Common Criteria is an international standard (ISO/IEC 15408) for computer security certification. It is currently in version 3.1. Common Criteria is a framework in which computer system users can specify their security requirements, vendors can then implement and/or make claims about the security attributes of their products and testing laboratories can evaluate the products to determine if they actually meet the claims. In other words, Common

Criteria provides assurance that the process of specification, implementation and evaluation of a computer security product has been conducted in a rigorous and standard manner. Common Criteria is performed on computer security products and systems and provides similarly-defined evaluation levels, implements the target of evaluation concept and the Security Target document.

Target of Evaluation

Target of Evaluation (TOE) - the product or system that is the subject of the evaluation.

The evaluation serves to validate claims made about the target. To be of practical use, the evaluation must verify the target's security features. This is done through the following:

Protection Profile (PP) - a document, typically created by a user or user community, which identifies security requirements for a class of security devices (for example, smart cards used to provide digital signatures, or network firewalls) relevant to that user for a particular purpose. Product vendors can choose to implement products that comply with one or more PPs, and have their products evaluated against those PPs. In such a case, a PP may serve as a template for the product's ST (Security Target, as defined below), or the authors of the ST will at least ensure that all requirements in relevant PPs also appear in the target's ST document. Customers looking for particular types of products can focus on those certified against the PP that meets their requirements.

Security Target (ST) - the document that identifies the security properties of the target of evaluation. It may refer to one or more PPs. The TOE is evaluated against the SFRs (see below) established in its ST, no more and no less. This allows vendors to tailor the evaluation to accurately match the intended capabilities of their product. This means that a network firewall does not have to meet the same functional requirements as a database management system, and that different firewalls may in fact be evaluated against completely different lists of requirements. The ST is usually published so that potential customers may determine the specific security features that have been certified by the evaluation.

Security Functional Requirements (SFRs) - specify individual security functions which may be provided by a product. The Common Criteria presents a standard catalogue of such functions. For example, an SFR may state how a user acting a particular role might be authenticated. The list of SFRs can vary from one evaluation to the next, even if two targets are the same type of product. Although Common Criteria does not prescribe any SFRs to be included in an ST, it identifies dependencies where the correct operation of one function (such as the ability to limit access according to roles) is dependent on another (such as the ability to identify individual roles).

Evaluation Process

The evaluation process also tries to establish the level of confidence that may be placed in the product's security features through quality assurance processes:

Security Assurance Requirements (SARs) - descriptions of the measures taken during development and evaluation of the product to assure compliance with the claimed security functionality. For example, an evaluation may require that all source code is kept in a change management system, or that full functional testing is performed. The Common Criteria provides a catalogue of these, and the requirements may vary from one evaluation to the next. The requirements for particular targets or types of products are documented in the ST and PP, respectively.

Evaluation Assurance Level (EAL) - the numerical rating describing the depth and rigor of an evaluation. Each EAL corresponds to a package of security assurance requirements (SARs, see above) which covers the complete development of a product, with a given level of strictness. Common Criteria lists seven levels, with EAL 1 being the most basic (and therefore cheapest to implement and evaluate) and EAL 7 being the most stringent (and most expensive). Normally, an ST or PP author will not select assurance requirements individually but choose one of these packages, possibly augmenting requirements in a few areas with requirements from a higher level. Higher EALs do not necessarily imply "better security", they only mean that the claimed security assurance of the TOE has been more extensively validated.

So far, most PPs and most evaluated STs/certified products have been for IT components (e.g., firewalls, operating systems, smart cards). Common Criteria certification is sometimes specified for IT procurement. Other standards containing, e.g, interoperation, system management, user training, supplement CC and other product standards. Examples include the ISO 17799 (or more properly BS 7799-2, which is now ISO/IEC 27002) or the German IT-Grundschutzhandbuch.

Details of cryptographic implementation within the TOE are outside the scope of the CC. Instead, national standards, like FIPS 140-2, give the specifications for cryptographic modules, and various standards specify the cryptographic algorithms in use.

Conclusions

This chapter presented the connections between ISO 15408 (Common Criteria for Information Technologies Security Evaluation), FIPS 140-2 (Security requirements for cryptographic modules) and cryptographic algorithms (NIST Cryptographic ToolKit).

Chaotic Techniques for Securing Communications

Introduction

General speaking, chaos has been avoided due to inherent instability and unpredictability. Systems were devised in such a way as to prevent the chaotic dynamic, or avoid it as much as possible. Some approaches proposed to control chaos by using its own characteristics and proved that chaotic systems can be controlled through the application of small perturbations to one of the parameters. So, the unpredictable dynamic was transformed in predictable behaviour without major changes to the functioning conditions. A development to these was the larger idea of directing general states on chaotic attractors [49-55]. The discovery of feedback mechanisms allowed the synchronization (control) of chaotic systems and started a whole new range of applications in protected communications [56].

In the last period of time a growing attention has been given to chaos-based transmission methods for using in the field of communications. Basically, these methods consist in the information's representation as noise-like waves. The theory that based this research field is the chaotic dynamic systems' theory which deals with the change of a system status in time. The solutions of a chaotic dynamic system can have variations in time for certain values of parameters and for other parameters' values the solutions have a deterministic (regular) behaviour. These variations seem randomly and the behaviour is named chaotic. Therefore, the noise-like waves are the solutions of some dynamic systems which in certain conditions have a chaotic behaviour. There has been significant interest in exploiting chaotic dynamics in communications. A transmission method based on chaotic generator can be a somewhat similar to a FM radio transmission: a message is embedded on a carrier wave, but the wave in this case is rather chaotic than sinusoidal; retrieving the message is simply a question of subtracting the carrier wave from the transmitted signal. The topics explored in the last period of time include chaotic modulation for the transmission of information, chaotic encryption, chaotic spreading codes for multi-user access in spread spectrum systems.

One advantage of using chaos-based communication systems is that it is often easier to generate robust, high-power chaotic signals than conventional ones. It is also harder for an eavesdropper to identify a chaotic signal because it is difficult to distinguish from background noise. Of course, if the chaotic technique is not enough to assure secure communication for higher levels of privacy, the message can be easily encrypted using classical methods of cryptography, or even ones based on chaos. In the last period of time there were tested communications systems using chaos techniques in order to send digital messages at Gbps speeds over one hundred kilometers using commercial optical fibre [57]. An important feature of chaotic systems is that their long-term behaviour is often impossible to predict but their near-term behaviour is quite easy to anticipate, so their immediate evolution can be controlled.

Several chaos-based communication systems were conceived, studied, simulated, implemented and tested in order to evaluate their performances. A series of problems were highlighted: the chaotic carrier systems' mathematical modelling, analysis of adopted models, generation of the chaotic signals, modulation and demodulation methods, synchronisation methods between transmitter and receiver [58]. New chaos-based spreading techniques have been developed including chaotic masking, chaotic modulation, chaos shift keying (CSK), predictive control (or symbolic) modulation a.s.o. The mentioned methods represent techniques that are used to implement communication networks in order to protect the information and also to use efficiently the available spectrum [57].

Chaotic Behaviour

A deterministic system is perfectly predictable given perfect knowledge of the initial conditions and it is in practice always predictable in the short term. Chaos has long-term unpredictable behaviour. This is usually mathematically seen as sensitivity to initial conditions, where the system's dynamics are hard to be predicted from the starting point. Though a chaotic system can have a pattern (an attractor) in the state space, determining where the attractor of the system will be at a distant future moment, given its position in the past, is a problem that becomes exponentially harder as time passes.

The substantiation and development of the dynamic system theory stand out the existence of some systems that have behaviours in time of parameters which can not be predicted, although the dynamics and initial conditions are known. There are very simple dynamic systems that can have very different behaviours at very small changes of the parameters. Thus, it is well known that the solutions of a system of three differential equations can have variations in time for certain values of parameters and for other parameter values the solutions have a deterministic (regular) behaviour. These variations seem randomly and the behaviour is named chaotic [59]. In the time domain, the signal generated by such systems are randomly, their auto-correlation function having a peak in origin and decreasing very fast in the rest. In the frequency domain the signals have a wide spectrum. The chaotic systems generate signals that have much of the random processes properties, but have a determinist structure that makes these systems to be reproducible.

From mathematical point of view a dynamic system consists of a states space (named also phases space or attractor), and a function (named also dynamic) that tells what future state will corresponds to a present state of the system. A determinist dynamic system is completely characterised by its initial state and dynamic. Such a system can have a continuous or a discrete state space and a continuous or discrete dynamic. A continuous time dynamic system is described by a system of differential equations (named also state equation). A discrete time dynamic system is described by a system of equations with finite differences. Let note t the independent variable (time) and $x_1, x_2, \dots, x_n$ dependent variables that characterise the state of the system. The mathematical model of a *continuous* time dynamic system consists of the following first order differential equation's system [60]:

$$\begin{cases} \dot{x}_1 = f_1(x_1, x_2, \dots, x_n; t); \\ \dot{x}_2 = f_2(x_1, x_2, \dots, x_n; t); \\ \dots\dots\dots\dots\dots\dots\dots\dots\dots \\ \dot{x}_3 = f_n(x_1, x_2, \dots, x_n; t). \end{cases} \tag{6}$$

where $\dot{x}_i$ is first order differential with time of x_i. The equations system can be abbreviated so:

$$\dot{x} = f(x,t), \tag{6}$$

where $x \square R^n$ is a system state vector of which components are $x_1, x_2, \dots, x_n$, and $f(x, t)$ is a field vector of which components are $f_1, f_2, \dots, f_n$. A *discrete* time dynamic system is based on a mathematical model that uses the recurrent relations

$$\begin{cases} x_1^{(N)} = f_1\left[x_1^{(N-1)}, x_2^{(N-1)}, \dots, x_n^{(N-1)}\right]; \\ x_2^{(N)} = f_2\left[x_1^{(N-1)}, x_2^{(N-1)}, \dots, x_n^{(N-1)}\right]; \\ \dots\dots\dots\dots\dots\dots\dots\dots\dots\dots\dots\dots \\ x_n^{(N)} = f_n\left[x_1^{(N-1)}, x_2^{(N-1)}, \dots, x_n^{(N-1)}\right] \end{cases} \tag{7}$$

which offer the values of parameters $x_1, x_2, \ldots, x_n$ after N periods of time. The above equations can be put in the abbreviated form:

$$x^{(N)} = f\left[x^{(N-1)}\right], \tag{8}$$

where $x^{(N)}$ is a state vector having components $x_1^{(N)},\ldots,x_n^{(N)}$ and f a field vector having the components $f_1, \ldots, f_n$. The last two equations are applied when $x_1,\ldots, x_n$ do not vary continuously.

Chaotic Communications

In order to produce chaotic signals, there are different generators using: natural noise sources (resistors, diodes), shift registers (digital filters) and nonlinear dynamic systems with chaotic behaviour. There are continuous time generators and discrete time generators. As a continuous time chaotic generator the Chua circuit is well-known. Other such generators are: RLC circuit with nonlinear resistance, nonlinear RC circuit, LC/RC oscillators with chaotic behaviour. As discrete time chaotic generators are generally used digital filters [61].

For realising a protected communication it is necessary to modulate the chaotic signal with the informational signal. The general structure of a chaotic carrier communication system is presented in Figure 18. For a proper operation it is necessary to have absolutely identical chaotic generators in transmitter and in receiver. The most used modulation methods are: chaotic masking, chaotic parametrical modulation, chaotic direct modulation, Chaos Shift Keying (CSK) modulation, Differential Chaos Shift Keying (DCSK) modulation, etc. [62].

The chaotic dynamic systems have an unpredictable behaviour on long term. A chaotic dynamic system has a specific compact state space, named attractor. The evolutions of two identical chaotic dynamic systems (with the same attractor), but starting in the state space from two close points, will diverge after a certain time, although the evolutions "stay" inside the same attractor. For this is necessary to synchronise the two systems. The most used methods of synchronisation are: master - slave synchronisation, error's feedback synchronisation, observer's method, inverse system's method, method using synchronisation pulses, dead-beat synchronisation [63].

The basic idea of digital chaotic communications is that the bits (in case of binary modulation) or symbols (M-ary modulation) are mapped to sample functions of chaotic signals originating from one or more chaotic attractors. In order to avoid periodicity, the symbols are mapped to the actual non-periodic outputs of chaotic circuits and not to parameters of certain known sample functions.

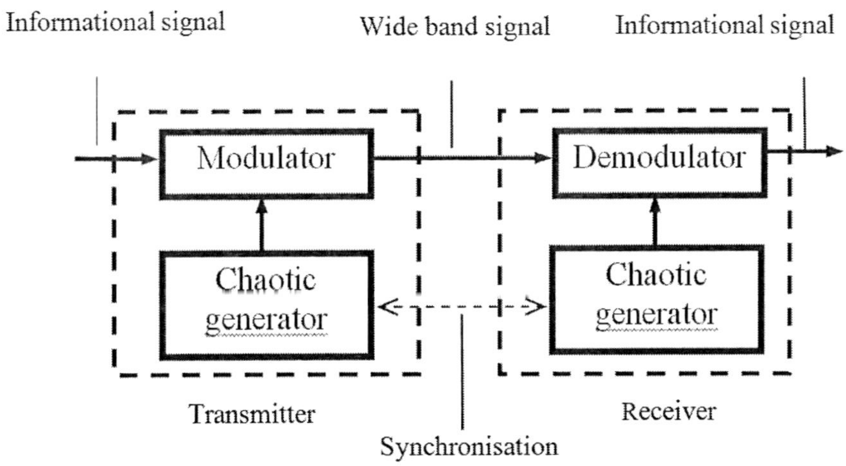

Figure 18. The general structure of a chaotic carrier communications system.

Chaotic Encrypted Communications

There were developed techniques that can be used in order to additionally increase the communication security. The more complex is the transmitted signal, the higher is the security of the communication system. These techniques combine the classic encryption with the synchronisation of chaotic systems, aiming to enhance the security level. The intruder needs to reconstruct the key signal which is different from the transmitted signal. It seems very difficult to create reconstruction methods in order to obtain the key signal. In order to hide the contents of a message using chaotic signals different techniques have been developed. Basically, for increasing the interception resistance of the chaotic carrier communications, three classes of methods are used:

- use of digital filters having a chaotic behaviour together with nonlinear mapping functions,
- use of digital filters having a chaotic behaviour together with binary shift registers, shift cipher and auto-key ciphers,
- use of chaotic systems involving two chaotic signals: first for the chaotic encrypter and decrypter synchronisation and the other for informational signal encryption using a classical encryption algorithm (by example, shift ciphers).

A different approach uses two chaotic signals [63]: one of them is used for the chaotic encrypter and decrypter synchronisation, the other signal is used for informational signal encryption using a encryption scheme, by example with shift ciphers. The general block diagram of the chaos-based cryptosystem is represented in Figure 19.

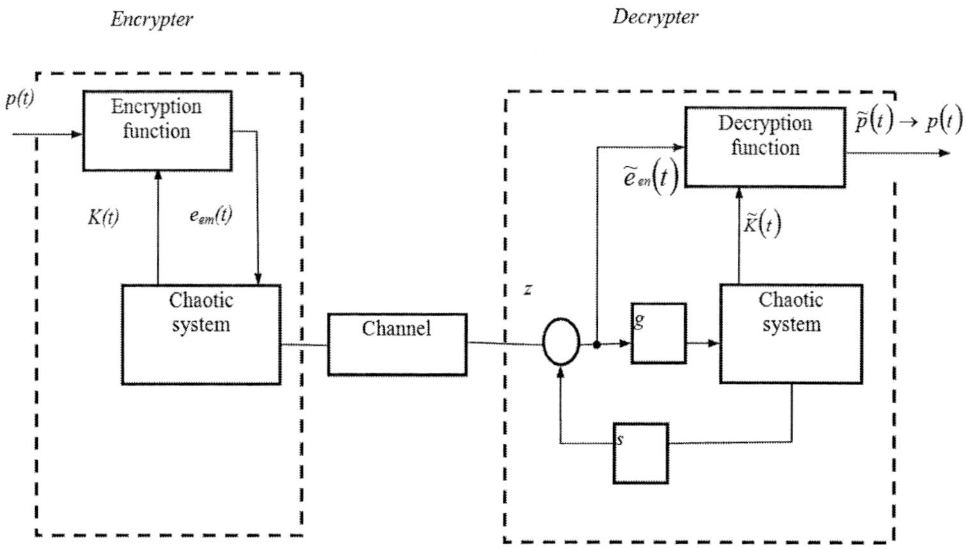

Figure 19. The block diagram of the chaos-based cryptosystem.

The encrypter consists of a chaotic system and an encryption function *e(t)*. The cryptographic key *k(t)* is one of the state variables of the chaotic system. The transmitted signal *s(t)* is another state variable of the chaotic system. It is sent through a public channel to the decrypter and used to synchronise the decrypter. The decrypter consists of a chaotic system and a decryption function *d( )*. It should be noted that both the encrypted signal *y(t)* and the key signal *k(t)* are not sent to the decrypter. It is different from traditional discrete cryptosystems where both the encrypted signal and the key should be transmitted to the decrypter. The communication channel's noise *n(t)* is added to *s(t)*, so the decrypter receives the sum signal *z(t)= s(t) + n(t)*. Only when the decrypter and the encrypter are synchronised, the decrypter can find the encrypted signal and the key signal. Then, the decryption function *d( )* is used to decrypt the encrypted signal.

The crucial points in chaotic cryptography are continuous-value information and the usage of continuous- value systems, which may operate in continuous or discrete time. In the case of discrete time chaotic systems, their states at certain moments are described by chaotic maps. Chaotic maps and cryptographic algorithms have some similar properties: sensitivity to initial conditions and parameters, random like behavior and unstable orbits with long periods, depending upon the precision of the numerical implementation. Encryption rounds of a cryptographic algorithm lead to the desired diffusion and confusion properties of the algorithm. Similarly, iterations of the chaotic map spread the initial region over the entire phase space, while the parameters of the chaotic map may represent the key of the encryption algorithm.

Conclusion and Future Trends

The generation of chaotic signals has several fields of application as communication systems, information security, the simulation of phenomena or processes from external circumstance etc. Because of properties they have, the chaotic communication systems are a viable alternative to the existing communication systems. They provide a good level of security because of chaotic signals used for transmission; the more complex is the transmitted signal, the higher is the security of the communication system. Using the chaotic wide frequency band signals, these systems are noise-resistant.

In order to enhance the interception resistance of chaotic carrier communications, there were proposed other methods which use classical cryptographic blocks in the same structure with the chaotic systems. Many times the cryptosystems have feedback loops to implement specific encryption functions and to adapt encrypter and decrypter using adaptive algorithms.

The research in chaotic techniques for securing communications is to be continued for improving the performances, especially for increasing of transmission speed in real conditions of operation.

Discrete Chaotic Systems Synchronization and Using for Secure Data Transmission

A number of methods have been proposed for synchronizing chaotic systems. The most widely used methods are derived for continuous synchronization schemes. In a continuous synchronization scheme, chaotic systems are coupled to each other continuously such that synchronization errors converge to zero. In this sub-chapter, chaos synchronization in coupled discrete-time dynamical systems is presented. Especially, practical impulsive synchronization scheme for 3 discrete time chaotic systems is shown. Simulation results finally demonstrate the effectiveness of the method. Experimental results show that chaotic and hyperchaotic systems can be synchronized by impulses sampled from one or two state variables. The impulsive synchronization can be applied to almost all chaotic and hyperchaotic systems even in the case when continuous synchronization systems fail to work. The example of data transmission based on simple discrete-time chaotic systems is also presented.

Introduction

In the past time, synchronization of the chaotic systems has attracted considerable attention due to its great potential applications in secure communication, chemical reactions and biological systems. The first idea of synchronization between two chaotic systems with different initial conditions was introduced by Pecora and Carroll [64]. Since then, many different methods have been applied theoretically and experimentally to synchronize the chaotic systems, such as linear and nonlinear feedback control, adaptive control, back stepping control, variable structure control, impulsive control, etc. Among these methods, the impulsive control provides an efficient method to deal with the dynamical systems. Additionally, in synchronization process, the response system receives the information from the drive system only at the discrete time instants, which reduces the amount of synchronization information transmitted from the drive system to the response system and makes this method more efficient in a great number of applications. Most of the researchers just concern the synchronization between two identical chaotic systems with known parameters or identical unknown parameters. However, in reality there is hardly find the case where the structures of the drive and the response systems can be assumed to be identical. Moreover, the parametric uncertainties of the drive and response systems are always different and time-varying. Therefore, it is significant to investigate the synchronization between two little different or different chaotic systems in the presence of time-varying parametric uncertainties.

A chaotic system is extremely sensitive to tiny variations of parameters. But parameters of some systems in practical circumstances cannot be exactly known in advance. The effect of these uncertainties will destroy the synchronization and even break it. Therefore, it is important and necessary to study synchronization in such systems with parametric uncertainties.

Impulsive synchronization had been applied to several chaotic spread spectrum secure communication systems, and had exhibited good performance [65, 66, 67]. Recently, the detailed experiments and performance analysis of impulsive synchronization were carried out for the purpose of applying impulsive synchronization to chaotic communication systems. The following experimental results were reported in [68]:

a) The accuracy of synchronization depends on both the period and the width of the impulse samples.
b) The minimum impulse width for synchronization increases as the impulse period increases.
c) The hyperchaotic systems can be synchronized by transmitting two kinds of samples through a single channel via a time-division scheme.

These experimental results showed that chaotic systems can be impulsively synchronized by using impulse samples derived from some of the state variables of the driven system. This is because some chaotic systems can be decomposed into two parts. One part tends to make the synchronization unstable and the other part tends to make the synchronization stable. If we construct impulsive controllers to stabilize the unstable part we can synchronize chaotic systems by using samples from some state variables of the driven system.

The stability of impulsive synchronization is closely connected to the values of the Lyapunov components of the variational systems. In a unidirectional directional synchronization scheme, during some time periods, the driving signal can cause the synchronization error to increase and during some other time periods it can cause synchronization error to decrease. These two kinds of time periods can be distinguished by monitoring the eigenvalues of the variational synchronization error systems. When the driving signal in a time period is detected to be more likely to increase synchronization errors, we do not use it to drive the driven system. By doing this, we can synchronize two chaotic systems much more efficiently than a by continuous synchronization scheme. This kind of synchronization scheme is called selective synchronization. The selective synchronization scheme can synchronize chaotic systems which cannot be synchronized by a continuous synchronization scheme under similar conditions.

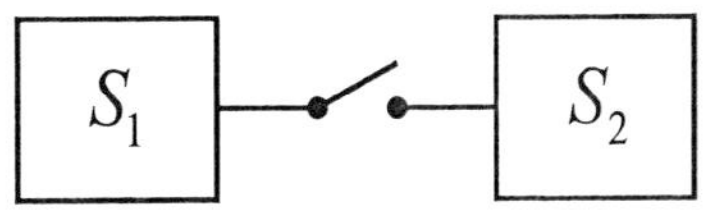

Figure 20. Block diagram of impulse synchronization between 2 systems.

Chaotic System Synchronization Methods

In this section, we only consider the impulsive synchronization between two chaotic systems with unidirectional coupling. In a unidirectional coupled synchronization scheme (Figure 1), we transmit impulses sampled from one state variable of the driving system S_1 (master or transmitter) to the driven system S_2 (slave or receiver). To avoid clutter and without loss of generality, we study the case when impulse samples are equidistant. Consider the following general form of a continuous dynamical system whose state variables can be separated into two parts x_{n+1} and y_{n+1} by (9):

$$S_1: \left.\begin{aligned} x_{n+1} &= f(x_n, y_n) \\ y_{n+1} &= g(x_n, y_n) \end{aligned}\right\} \text{ transmitter} \qquad (9)$$

Let us assume that samples of the state variable x_n are sent to the driven system. During the time interval when the n-th sample is send from drive system to driven system (Fig. 20, switch on), the driven system can be described by:

$$S_2: \left.\begin{aligned} \bar{x}_{n+1} &= x_n \\ \bar{y}_{n+1} &= g(x_n, \bar{y}_n) \end{aligned}\right\} \text{ receiver} \qquad (10)$$

During the time interval when the switch is off, the driven system can be described as:

$$S_2: \left.\begin{aligned} \bar{x}_{n+1} &= f(\bar{x}_n, \bar{y}_n) \\ \bar{y}_{n+1} &= g(\bar{x}_n, \bar{y}_n) \end{aligned}\right\} \text{ receiver} \qquad (11)$$

The synchronization error systems are given by:

$$s_{n+1} = \frac{\partial g(x_n, y_n)}{\partial y_n} s_n \qquad (12)$$

or by equation (13):

$$\begin{bmatrix} r_{n+1} \\ s_{n+1} \end{bmatrix} = \begin{bmatrix} \dfrac{\partial f(x_n, y_n)}{\partial x_n} & \dfrac{\partial f(x_n, y_n)}{\partial y_n} \\ \dfrac{\partial g(x_n, y_n)}{\partial x_n} & \dfrac{\partial g(x_n, y_n)}{\partial y_n} \end{bmatrix} \begin{bmatrix} r_n \\ s_n \end{bmatrix} \tag{13}$$

where $r_n = x_n - \overline{x}_n$ and $r_n = x_n - \overline{x}_n$.

The Lyapunov exponents for a particular driven trajectory are called conditional Lyapunov exponents. Then the discrete version of Pecora and Carroll Theorem is given as follows [69]:

Theorem 1: The systems S1 and S2 will synchronize if the conditional Lyapunov expoments of the difference system (11) are all negative.

The most often way to synchronize the two systems is coupling. There are many ways of coupling. In this sub-chapter, the following coupling is used:

$$\begin{aligned} &x_{n+1} = f(x_n, y_n), \quad y_{n+1} = g(x_n, y_n) \\ &q_n = \overline{x}_n + K(x_n - \overline{x}_n) \\ &\overline{x}_{n+1} = f(q_n, \overline{y}_n), \quad \overline{y}_{n+1} = g(q_n, \overline{y}_n) \end{aligned} \tag{14}$$

or equivalently:

$$\begin{aligned} &x_{n+1} = f(x_n, y_n), \quad y_{n+1} = g(x_n, y_n) \\ &\overline{x}_{n+1} = f(\overline{x}_n + K(x_n - \overline{x}_n), \overline{y}_n) \\ &\overline{y}_{n+1} = g(\overline{x}_n + K(x_n - \overline{x}_n), \overline{y}_n) \end{aligned} \tag{15}$$

where K is a constant. The system (6), (7) is x_n coupled and it is important to note that system S_1 is not affected by S_2.

Theorem 2: The systems x_n, y_n and $\overline{x}_n, \overline{y}_n$ will synchronize if the conditional Lyapunov expoments of the difference system (11) are all negative and K is sufficiently close to 1.

Discrete Chaotic System Description

In this sub-chapter we tested the chaos synchronization of three discrete-time chaotic systems: Lorenz, Fold and Rössler. The discrete-time chaotic systems have some advantages over the continuous systems. They are chaotic in wider range and are can be easily constructed by computer, microcontroller or by programmable logic. The Lorenz discrete-time chaotic system is given by:

$$x_{n+1} = (1+ab)x_n - bx_n y_n$$
$$y_{n+1} = (1-b)y_n + bx_n^2$$
$$a = 1.2, \quad b = 0.8 \tag{16}$$

The example of state space trajectories are shown in Fig. 21 and evolution of y_n in Fig. 22. Maximal Lyapunov exponent is 0.335.

The second system is "Fold" discrete-time chaotic system (sometimes also called Fold bifurcation, Saddle-node bifurcation or Tangent bifurcation). This system is simple for construction and equations of this system are given by:

$$x_{n+1} = y_n + ax_n$$
$$y_{n+1} = x_n^2 + b$$
$$a = -0.1, \quad b = -1.7 \tag{17}$$

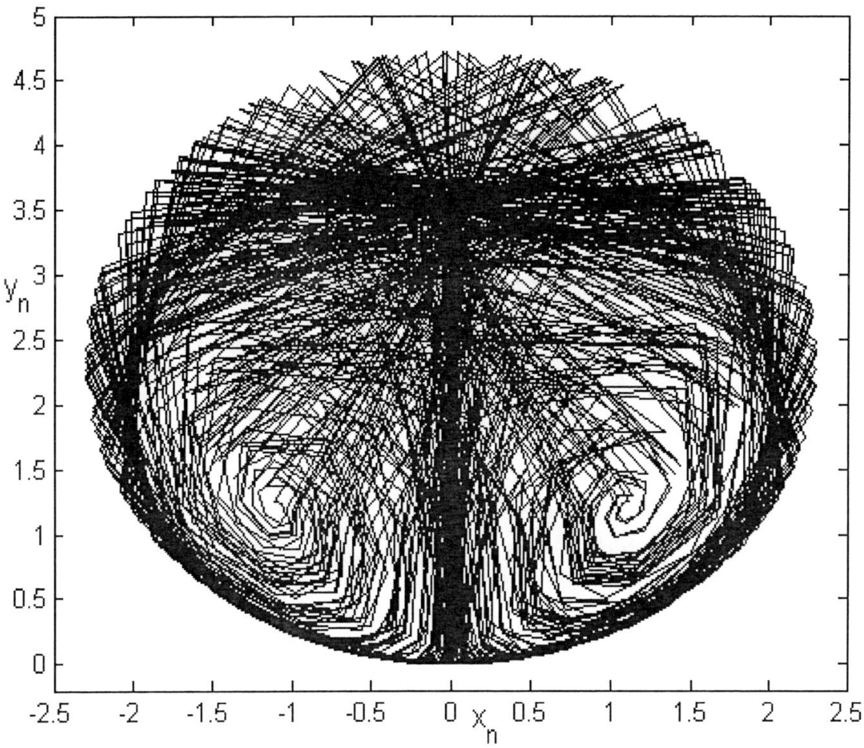

Figure 21. The projection of state space trajectories for Lorenz discrete-time chaotic system for a=1.2; b=0.8; x_0=0.1; y_0=0.1.

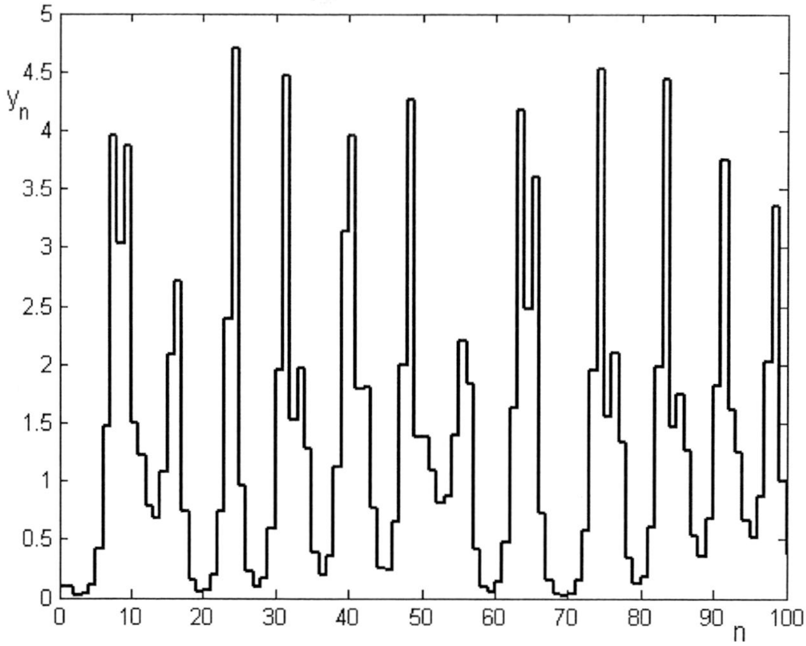

Figure 22. Evolution of y_n for Lorenz discrete-time chaotic system for a=1.2; b=0.8; x_0=0.1; y_0=0.1

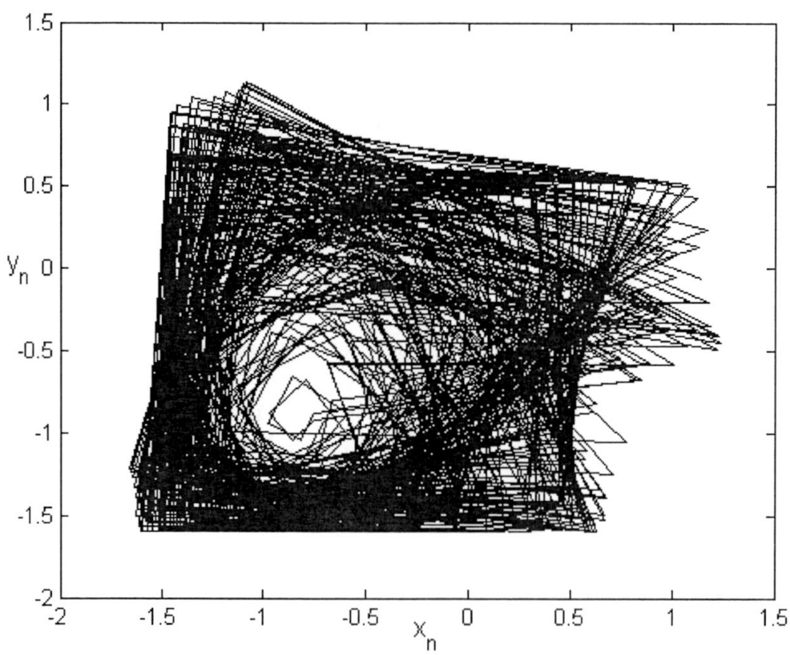

Figure 23. The projection of state space trajectories for fold discrete-time chaotic system for values: a=-0.09; b=-1.6; x_0=0.1; y_0=0.1.

The state space trajectories are shown in Figure 23 and evolution of x_n in Figure 24. Maximal Lyapunov exponent for this system is 0.02.

The Rössler discrete-time hyperchaotic system is described by:

$$\begin{aligned}
x_{n+1} &= ax_n(1-x_n) - b(z_n + c)(1-2y_n)\\
y_{n+1} &= dy_n(1-y_n) + ez_n\\
z_{n+1} &= f\left[(z_n + c)(1-2y_n) - 1\right](1-gx_n)\\
a &= 3.8; \quad b = 0.05; \quad c = 0.35; \quad d = 3.78;\\
e &= 0.2; \quad f = 0.1; \quad g = 1.9;
\end{aligned} \tag{18}$$

The state space trajectories of this system are shown in Figure 25, evolution of x_n, y_n and z_n are shown in Figure 26, 27, and 28. Maximal Lyapunov exponent for this system is 0.41 (This system has 2 positive exponents).

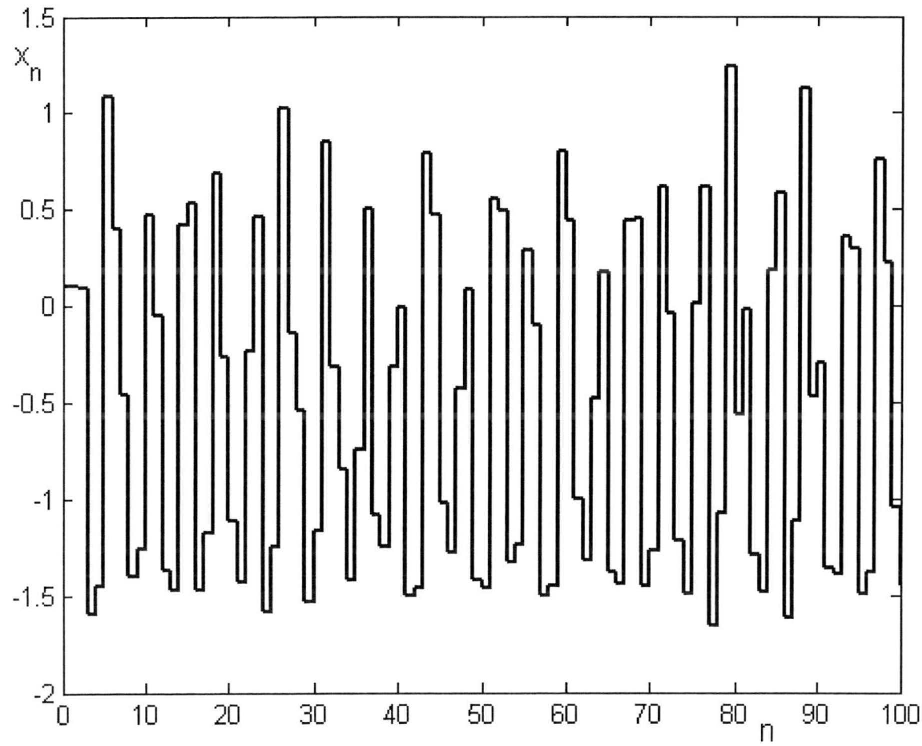

Figure 24. Evolution of x_n for fold discrete-time chaotic system for a=-0.09; b=-1.6; x_0=0.1; y_0=0.1.

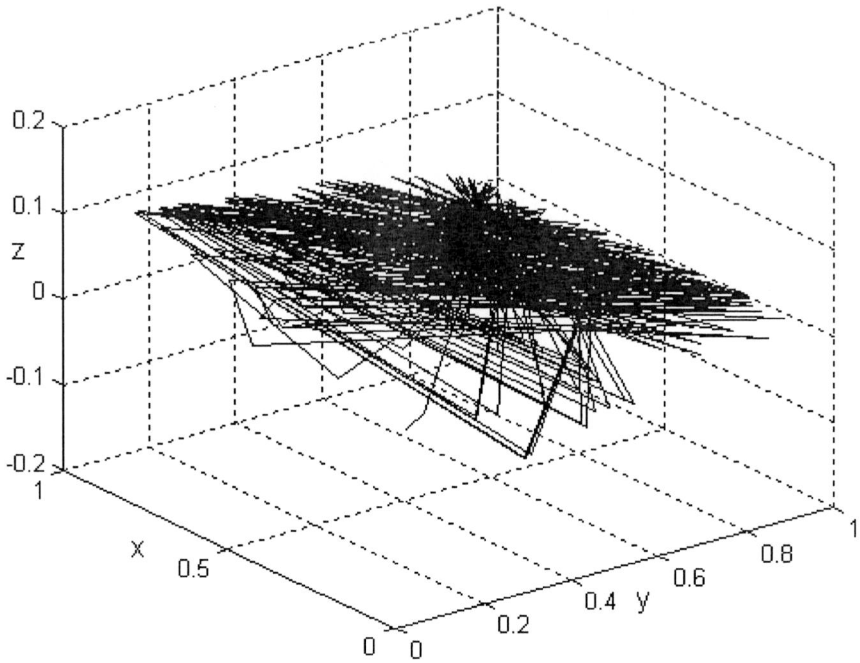

Figure 25. The projection of state space trajectories of Rössler discrete-time hyperchaotic system for values: a=3.8; b=0.05; c=0.35; d=3.78; e=0.2; f=0.1; g=1.9; x_0=0.1; y_0=0.1; z_0=0.

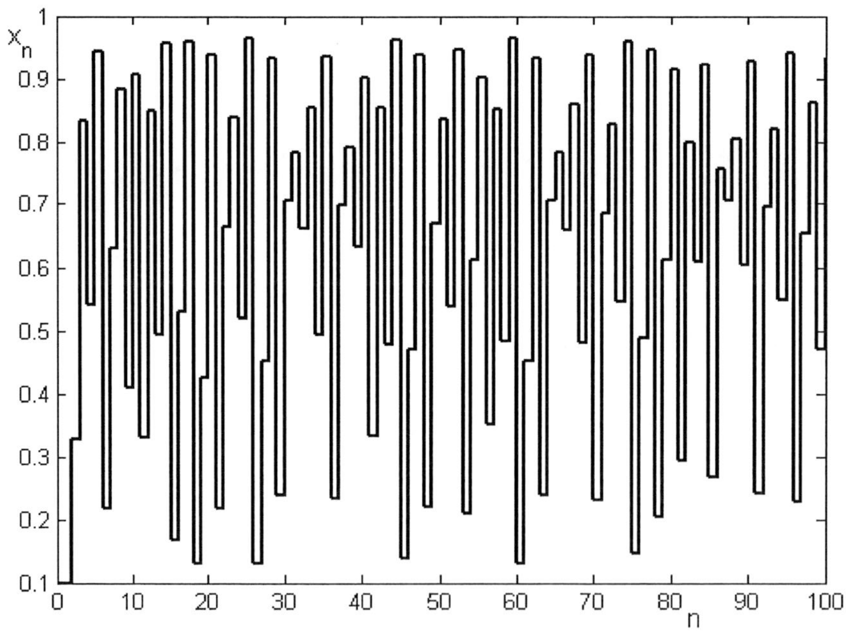

Figure 26. Evolution of x_n for Rössler discrete-time hyperchaotic system. Constants and initial values are the same as in Figure 25.

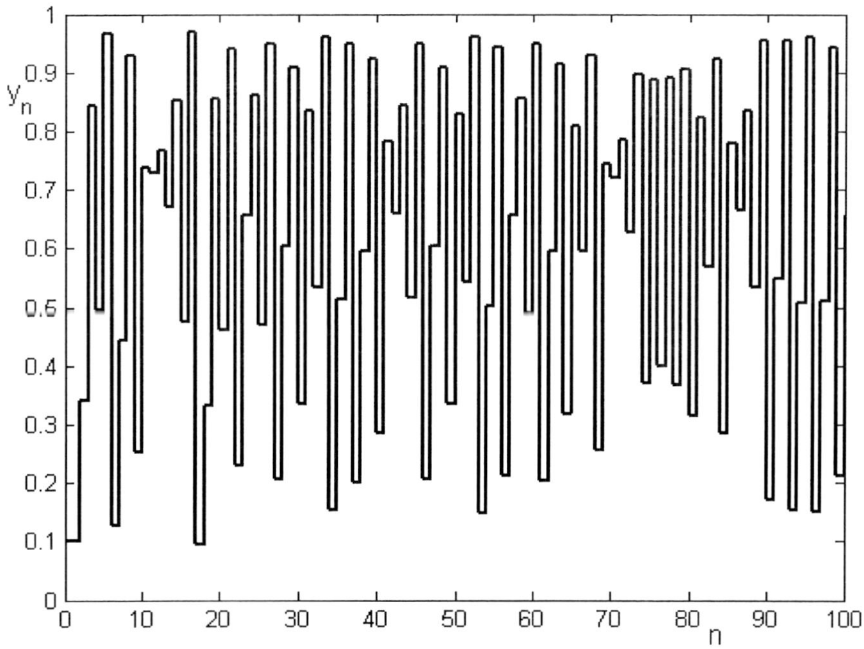

Figure 27. Evolution of y_n for Rössler discrete-time hyperchaotic system. Constants and initial values are the same as in Figure 25.

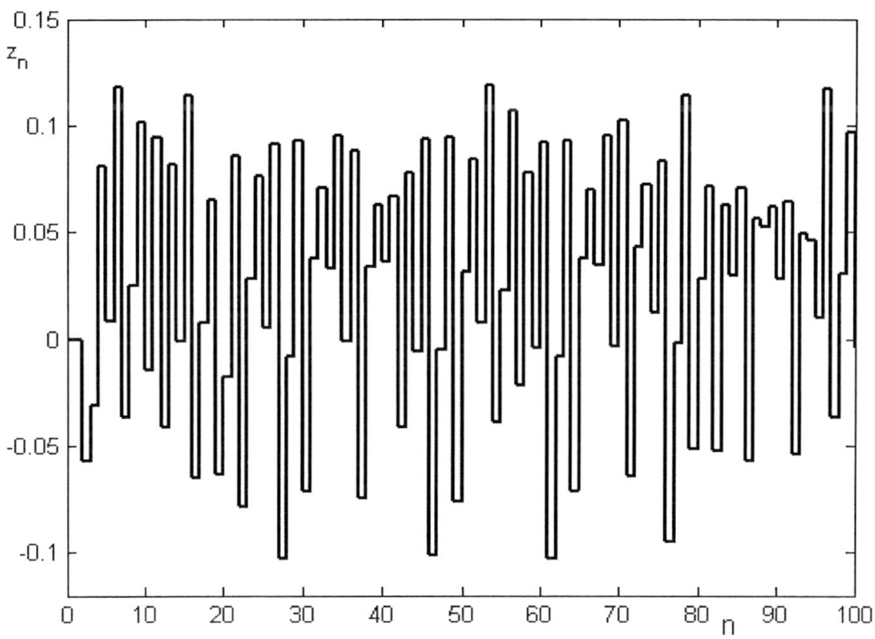

Figure 28. Evolution of z_n for Rössler discrete-time hyperchaotic system. Constants and initial values are the same as in Figure 25.

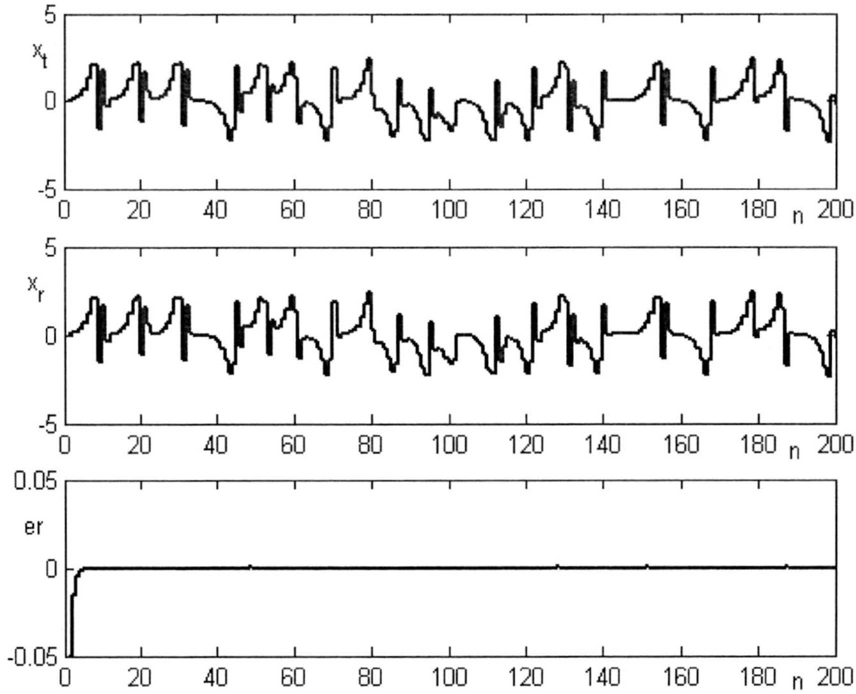

Figure 29. Synchronization of 2 discrete-time Lorenz systems with different parameters and initial values. Transmitter state evolution x_t (top), receiver state evolution x_r (middle) and error = x_t - x_r (bottom).

Chaotic Systems Synchronization

In this section, simulation examples of discrete/time chaotic systems are presented. Only unidirectional coupling cases are shown. The minimal number of controlled variables has to be equal to the number of positive Lyapunov exponents of the system. If the hyperchaotic systems have two positive Lyapunov exponents, then, at least two driving signals are needed to synchronize them [70, 71, 72, 73].

The example of synchronization of 2 Lorenz systems (S_1 and S_2) with different parameters and initial value is shown in Figure 29, where systems parameters are:

S_1: a=1.25; b=0.75; x_0=0.05; y_0=0.05;

S_2: a=1.22; b=0.73; x_0=0.10; y_0=0.01;

Driving signal for Lorenz systems synchronization is given by d_n (generated by driving system):

$$d_n = (1+ab)x_n - bx_n y_n - \varepsilon x_n \tag{19}$$

where ε is positive constant (0<ε<1). In presented example ε=0.3. Slave system equations are:

$$\bar{x}_{n+1} = d_n + \varepsilon \bar{x}_n$$
$$\bar{y}_{n+1} = (1-b)\bar{y}_n + b\bar{x}_n^2 \qquad (20)$$

The example of synchronization of 2 Fold systems (S_1 and S_2) with different parameters and initial value is shown in Figure 30, systems parameters are:

S_1: a=-0.10; b=-1.70; x_0=0.10; y_0=0.10;

S_2: a=-0.11; b=-1.71; x_0=0.05; y_0=0.05;

Driving signal for fold systems synchronization is given by d_n (generated by driving system):

$$d_n = x_n^2 + b - \varepsilon y_n \qquad (21)$$

where ε is positive constant (0<ε<1). In presented example ε=0.2. Slave system equtions are:

$$\bar{x}_{n+1} = \bar{y}_n + a\bar{x}_n$$
$$\bar{y}_{n+1} = d_n + \varepsilon \bar{y}_n \qquad (22)$$

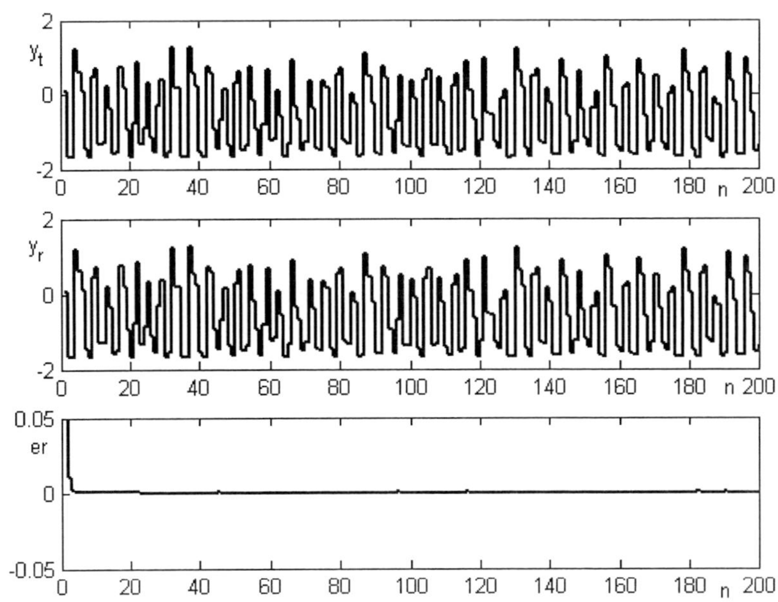

Figure 30. Synchronization of 2 discrete-time fold systems with different parameters and initial values. Transmitter state evolution y_t (top), receiver state evolution y_r (middle) and error = y_t - y_r (bottom).

The example of synchronization of 2 Rössler discrete-time hyperchaotic system (S_1 and S_2) with different parameters and initial value is shown in Figure 31. Systems parameters are:

S_1: a=3.8; b=0.05; c=0.35; d=3.78; e=0.2; f=0.10;

g=1.9; x_0=0.10; y_0=0.10; z_0=0;
S_2: a=3.7; b=0.07; c=0.34; d=3.80; e=0.18; f=0.12;

g=2.0; x_0=0.05; y_0=0.05; z_0=0;

Driving signals for Rössler discrete-time hyperchaotic system synchronization are given by ${}^1d_n$ and ${}^2d_n$ (generated by driving system):

$$\begin{aligned} {}^1d_n &= ax_n(1-x_n)-b(z_n+c)(1-2y_n)-\varepsilon x_n \\ {}^2d_n &= dy_n(1-y_n)+ez_n-\varepsilon y_n \end{aligned} \tag{23}$$

where ε is positive constant (0<ε<1). In presented example ε=0.1. Slave system equations are:

$$\begin{aligned} \overline{x}_{n+1} &= {}^1d_n+\varepsilon\overline{x}_n \\ \overline{y}_{n+1} &= {}^2d_n+\varepsilon\overline{y}_n \\ \overline{z}_{n+1} &= f\left[(\overline{z}_n+c)(1-2\overline{y}_n)-1\right](1-g\overline{x}_n) \end{aligned} \tag{24}$$

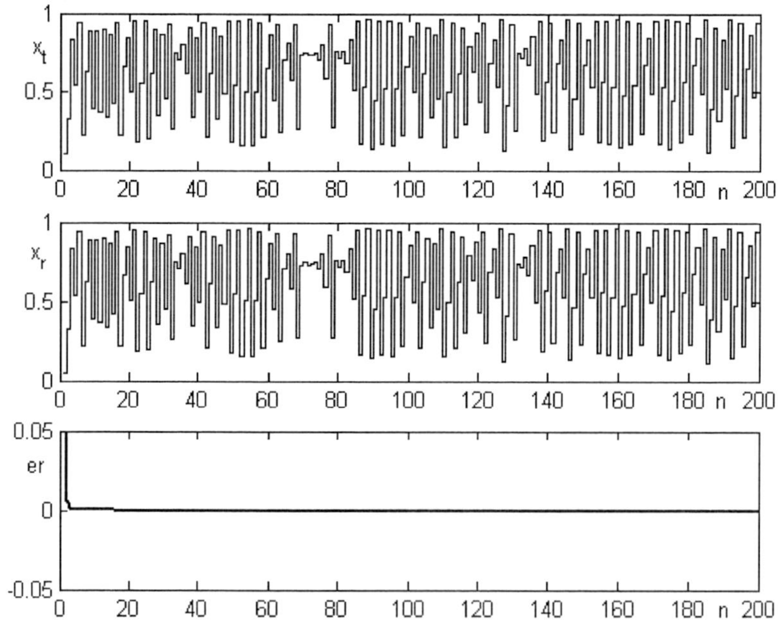

Figure 31. Synchronization of 2 discrete-time Rössler systems with different parameters and initial values. Transmitter state evolution x_t (top), receiver state evolution x_r (middle) and error = x_t - x_r (bottom).

Example of Data Transmission

This part is devoted to application of discrete-time chaotic systems synchronization for digital data transmission. The two Fold chaotic systems (17) are used for this purpose. The principle of transmission is based on transmitter chaotic system parameter b changing according logical signal H and L [74, 75,76,77]:

b=-1.7 for H, b=-1.65 for L (25)

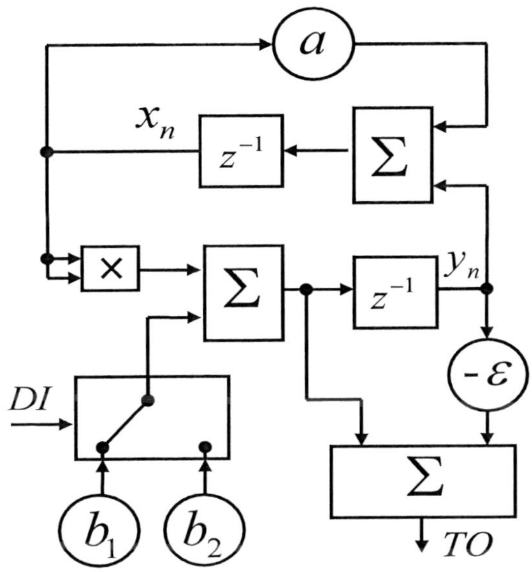

Figure 32. Transmitter block diagram. DI – digital input, TO – transmitter output

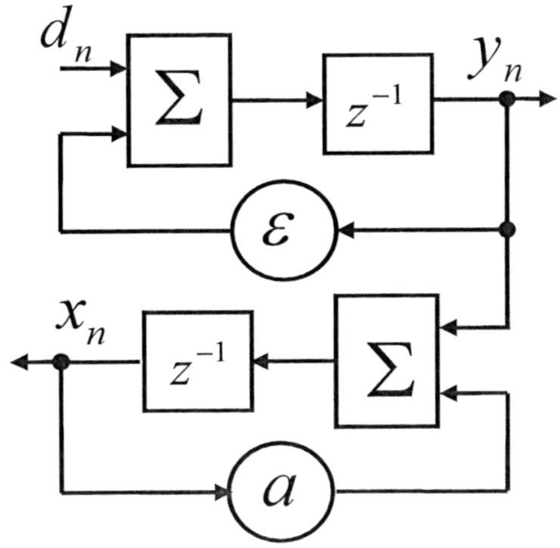

Figure 33. Receiver block diagram.

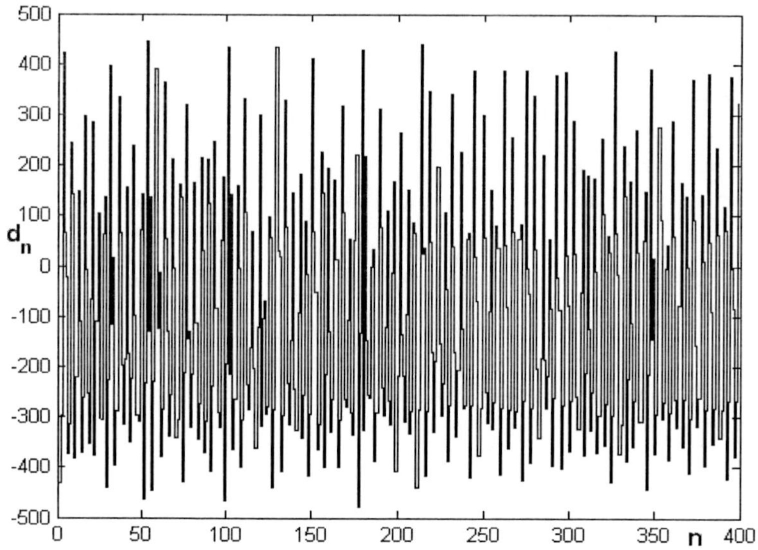

Figure 34. Transmitted samples (10 bit word).

State space variable y_n is multiplied by constant and rounds the value to the nearest integers (converted to n-bit data word) and transmitted to receiver. In receiver, first of all received signal is divided by constant and used for synchronization. On the end, the digital signal is recovered by signal processing. In this example, digital data word is transmitted:

w = [0 1 0 1 0 0 0 0 0 1 0 1 1 1 0 1 0 1 0 1]

Block diagram of the transmitter and receiver are shown in Figure 32 and Figure 33. The transmitted signal (binary coded in 10 bit word) is shown in Figure 34, autocorrelation value of transmitted signal in Figure 35 and raw received signal, low-pass filtered signal and recovered digital signal are in Figure 36.

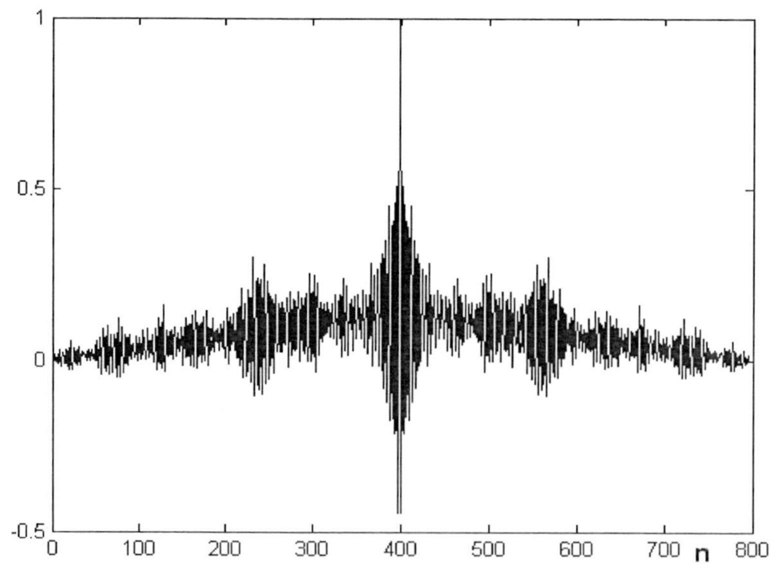

Figure 35. Autocorrelation of transmitted signal.

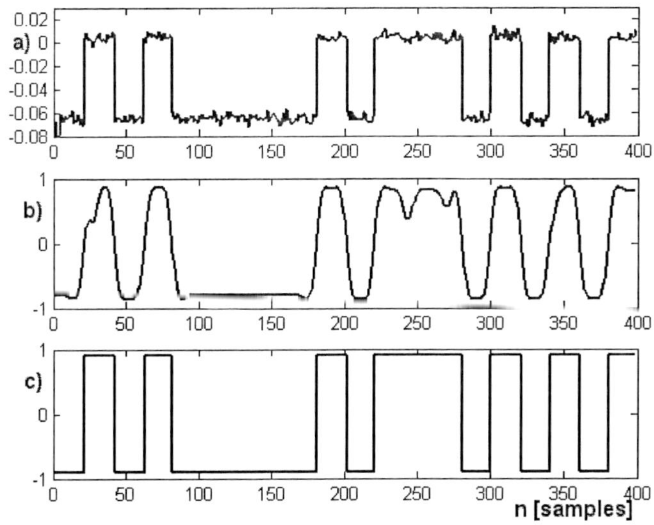

Figure 36. Example of data transmission: a) Raw recovered signal, b) Low-pass filtered signal, c) Recovered digital signal: w = [0 1 0 1 0 0 0 0 0 1 0 1 1 1 0 1 0 1 0 1].

Conclusion

In this sub-chapter example of 3 discrete-time chaotic systems synchronization was shown. Also the data transmission based on chaotic system was simulated. All presented systems can be more easily realized by means of microcontroller or programmable array than continuous chaotic systems. Moreover, the discrete-time systems aren't so sensitive for chaotic behavior area as continuous systems. In future, a more advanced system for data transfer will be developed based on chaotic systems.

Acknowledgments

This work has been supported from Department of Applied Electronics and Telecommunication, University of West Bohemia, Plzen, Czech Republic and project No.: FT-TA 3/058.

New Approach of Rijndael-AES Crypto-Algorithm

This paper presents a new approach of Rijndael-AES crypto-algorithm, based on the addition of MixColumns transformation, which originally belongs to the AES standard algorithm, before/after/beforeandafter the Rijndael-AES encryption/decryption block and to the secret encryption/decryption key, resulting in a new symmetric algorithm, with 3 different versions (M-AES, AES-M, M-AES-M), which the author of this paper generically called the AES PLUS algorithm. This paper also presents some aspects regarding the hardware (FPGA) implementations of the standard and proposed algorithms. The NIST statistical tests, both applied to the AES-128 algorithm and to the new one, proved that the M-AES-M version has a better statistical behavior than the original AES-128 standard algorithm.

Introduction

The specifications for Rijndael-Advanced Encryption Standard (AES) crypto-algorithm [78, 79] stipulate a final encryption/decryption round without its most complex transformation, which is the MixColumns transformation. This is a finite field or Galois Field (2^8) multiplication between the data block (organized in a 4x4 binary matrix, with 8 bit numeric coefficients, which is called State) and a constant matrix operand (e.g. in case of the AES-128 version of the standard algorithm it is organized in a 4x4 binary matrix, with 8 bit constant coefficients). The author of this paper did not find any clear explanation of this fact, except the symmetry of encryption and decryption blocks, so he had the idea of adding the missing transformation before/after/before and after the Rijndael-AES encryption/decryption block, together with the same transformation of the secret encryption/decryption key, resulting in a new symmetric algorithm, with 3 different versions (M-AES, AES-M. M-AES-M), which the author of this paper generically called the AES PLUS algorithm.

This paper also presents the hardware (FPGA) implementations of the standard and the proposed algorithm, all in iterative loop configurations that will provide the corresponding cryptographic modules.

Finally, the resulted cryptographic modules will be statistically tested, using a special designed FPGA test unit, together with the NIST statistical test suite.

The improved algorithm called AES-128 PLUS, which is the modified AES-128 version of AES, will be presented, but the other versions of AES can be similarly modified.

New Architectural Approach

The architecture of the AES block encryption/decryption algorithm contains, as shown in Figure 20, an initial round where a 128 bits plaintext/ciphertext is applied on its input, consisting of 4 columns with 4 bytes each, or else 4 words with 32 bits each. The plaintext/ciphertext is XORed with the initial key, also represented as a data block consisting of 4 words with 32 bits each.

The algorithm specification (NIST FIPS Publication 197, 2001) mentions 3 possible versions for the AES encryption/decryption algorithm, respectively $N_k = 4, 6, 8$ (the number of 4 bytes columns, or 32 bits words, of the key) corresponding to AES-128, AES-192 and AES-256 versions, also uniquely $N_b = 4$ (the number of 4 bytes columns, or 32 bits words, of the plaintext/ciphertext), namely 128 bits. The architectures presented by Figure 37 refer to the AES-128 version. For each of the 3 constructive versions we have a specific round number, as shown in Table 6.

After the initial round, AES-128 contains 9 encryption/decryption rounds. Each of these rounds has a corresponding round key supplied by the key schedule mechanism which starts the key generation process with the secret initial key.

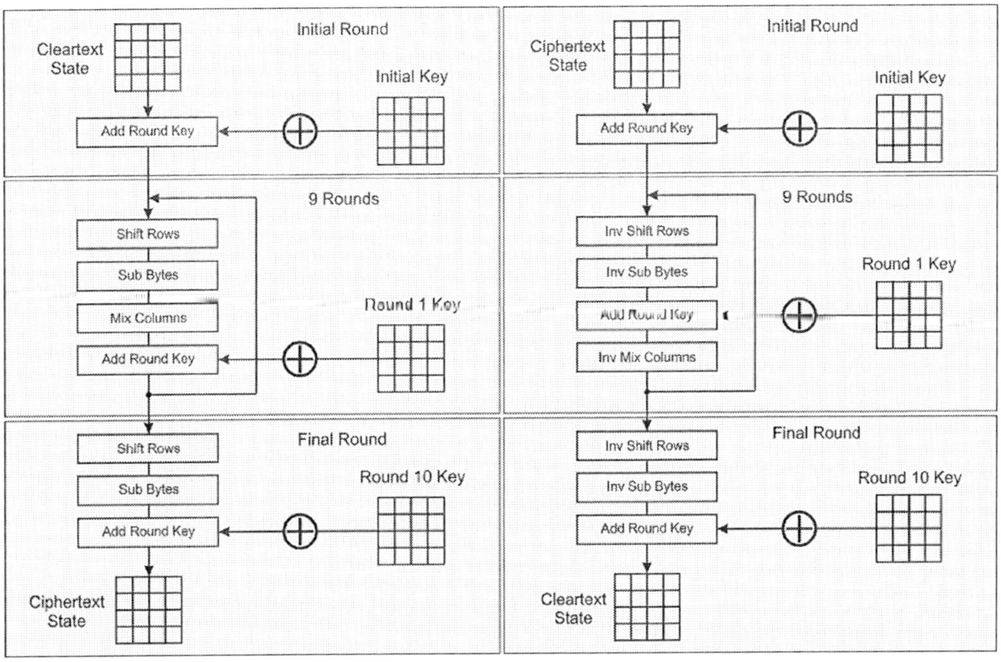

Figure 37. Basic AES-128 encryption/decryption architecture (encryption block - left picture, decryption block - right picture).

For the encryption process, every round consists of 4 main operations: ShiftRows (every row of the State is shifted by a specific number of positions), SubBytes (the substitution of every byte of the State using a special substitution table), MixColumns and the AddRoundKey (the round key is XORed with the resulted State of the previous operation, namely MixColumns). This sequence of operations is 9 times iterated.

Table 6. AES version characteristics

AES Version	Key Length (N_k words)	Block Size (N_b words)	Number of Rounds (N_r)
AES-128	4	4	10
AES-192	6	4	12
AES-256	8	4	14

MixColumns is a Galois Field (2^8) multiplication between the State, each of its columns being polynomials over GF(2^8), and a fixed polynomial given by:

$$a(x) = \{03\}x^3 + \{01\}x^2 + \{01\}x + \{02\}, \text{ or: } \begin{bmatrix} s'_{0,c} \\ s'_{1,c} \\ s'_{2,c} \\ s'_{3,c} \end{bmatrix} = \begin{bmatrix} 02 & 03 & 01 & 01 \\ 01 & 02 & 03 & 01 \\ 01 & 01 & 02 & 03 \\ 03 & 01 & 01 & 02 \end{bmatrix} \begin{bmatrix} s_{0,c} \\ s_{1,c} \\ s_{2,c} \\ s_{3,c} \end{bmatrix}, \text{ for } 0 \le c < N_b.$$

The final round differs from the others by the lack of the MixColumns operation. Thus, this round only consists of ShiftRows, SubBytes and an AddRoundKey operation. Finally, the result of this last round is the ciphertext State, which is a matrix with 4 words, exactly like the plaintext State.

For the decryption process, every round consists of 4 main operations: InvShiftRows (every row of the State is right shifted by a specific number of positions, which is the inverse transposition of those made in the encryption case), InvSubBytes (the substitution of every byte of the State using an inverse substitution table), InvMixColumns (which produces the two variants of implementation) and the AddRoundKey (the round key is XORed with the resulted State of the previous operation). This sequence of operations is also 9 times iterated.

The InvMixColumns function is a Galois Field (2^8) multiplication between the operand (described above), each of its columns being polynomials over GF(2^8), and a fixed polynomial given by:

$$a^{-1}(x) = \{0B\}x^3 + \{0D\}x^2 + \{09\}x + \{0E\}, \quad \text{or:} \quad \begin{bmatrix} s'_{0,c} \\ s'_{1,c} \\ s'_{2,c} \\ s'_{3,c} \end{bmatrix} = \begin{bmatrix} 0E & 0B & 0D & 09 \\ 09 & 0E & 0B & 0D \\ 0D & 09 & 0E & 0B \\ 0B & 0D & 09 & 0E \end{bmatrix} \begin{bmatrix} s_{0,c} \\ s_{1,c} \\ s_{2,c} \\ s_{3,c} \end{bmatrix}, \quad \text{for}$$

$0 \le c < N_b$

The hardware (FPGA) implementations of the decryption use a decomposition of the inverse affine matrix, in two matrix terms [80]:

$$\begin{bmatrix} 0E & 0B & 0D & 09 \\ 09 & 0E & 0B & 0D \\ 0D & 09 & 0E & 0B \\ 0B & 0D & 09 & 0E \end{bmatrix} = \begin{bmatrix} 02 & 03 & 01 & 01 \\ 01 & 02 & 03 & 01 \\ 01 & 01 & 02 & 03 \\ 03 & 01 & 01 & 02 \end{bmatrix} \begin{bmatrix} 05 & 00 & 04 & 00 \\ 00 & 05 & 00 & 04 \\ 04 & 00 & 05 & 00 \\ 00 & 04 & 00 & 05 \end{bmatrix}.$$

The final round differs from the others by the lack of the InvMixColumns operation. Thus, this round only consists of InvShiftRows, InvSubBytes and an AddRoundKey operation. Finally, the result of this last round is the plaintext State, which is a matrix with 4 words, exactly like the input ciphertext State.

The key schedule uses a key expansion routine which provides the encryption algorithm with N_b (N_r + 1) words, namely 44 key words, or 11 round keys, in the case of AES-128 encryption version. The resulting key schedule consists of a linear array of 4 byte words, denoted [w_i], with i in the range $0 \le i < N_b(N_r + 1)$. The expansion routine uses some specific functions, like SubWord(), which takes a 4 byte input word and applies the substitution S-box to each of the 4 bytes to produce an output word, and RotWord(), which takes a word [a_0,a_1,a_2,a_3] as input, performs a cyclic permutation, and returns the word [a_1,a_2,a_3,a_0]. The round constant word array Rcon[i] contains the values given by $[x^{i-1}, \{00\}, \{00\}, \{00\}]$, with x^{i-1} being powers of x, where x is denoted as {02} in the GF(2^8) field. Every following word, w[i], is equal to the XOR of the previous word, w[i-1], and the word N_k positions earlier, w[i-N_k]. For words in positions that are a multiple of N_k, a transformation is applied to w[i-1] prior to the XOR, followed by a XOR with a round constant, Rcon[i]. This transformation consists of a cyclic bytes shifting of a word (RotWord()), followed by application of a table lookup to all four bytes of the word (SubWord()).

The inverse key schedule uses the same key expansion routine as in the encryption case, with the difference from encryption of a reverse order of keys. Another difference from encryption key expansion is the reverse order of the applied round constant (Rcon[i]) values [81].

For the purpose of security improvement, the author of this paper had the idea to add the missing transformation to the input/output and to both the input and output of Rijndael-AES encryption/decryption block, together with the same transformation of the secret encryption/decryption key. A new symmetric algorithm with 3 different versions (M-AES, AES-M, M-AES-M) is obtained, which the author generically called the AES-128 PLUS algorithm [82]. The 3 versions are represented in Figure 38 (the encryption algorithm) and Figure 39 (the decryption algorithm).

It can be easily noticed that MixColumns transformation, which is applied to the input of AES-128 encryption (called M-AES version), will have a corresponding InvMixColumns transformation applied to the output of the AES-128 decryption.

Similarly, the MixColumns transformation, which is applied to the output of the AES-128 encryption (called AES-M version), will have a corresponding InvMixColumns transformation applied to the input of the AES-128 decryption. For the case of M-AES-M version, the MixColumns transformation will be applied both to the input and output of an AES-128 encryption/decryption block.

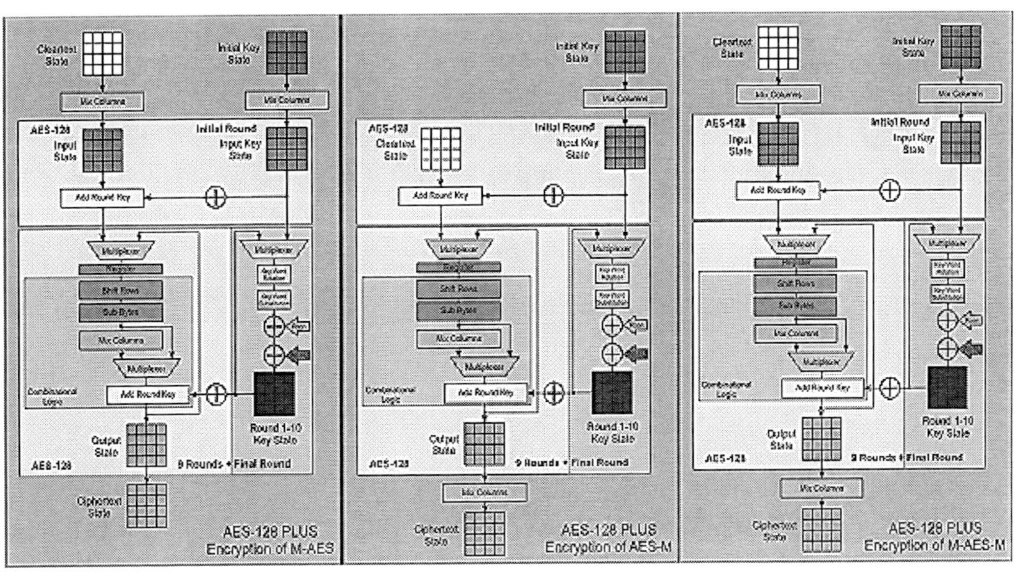

Figure 38. AES-128 PLUS encryption algorithm [82].

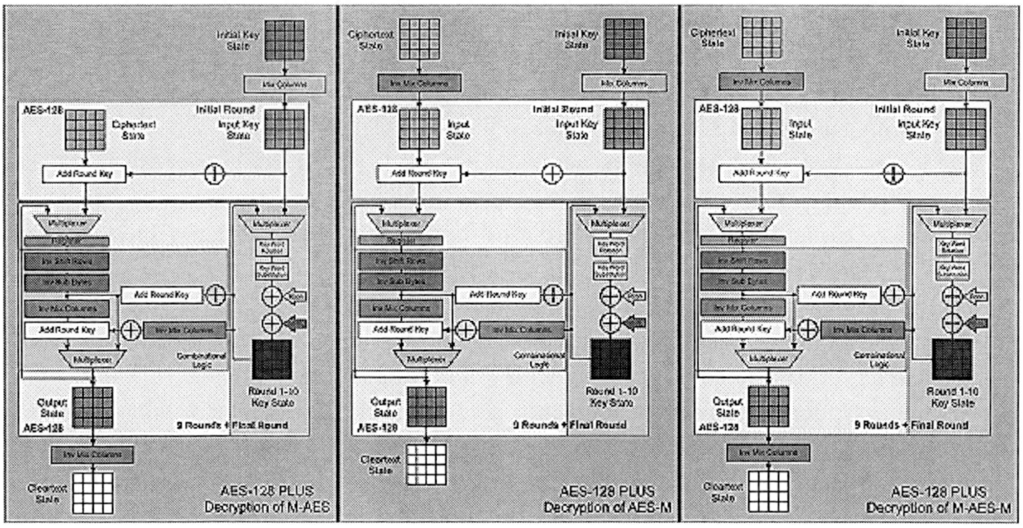

Figure 39. AES-128 PLUS decryption algorithm [82].

Hardware (FPGA) Implementation and Security Aspects

In order to prove the functionality of the new AES-128 PLUS algorithm, the 3 versions, including both the encryption and decryption, together with the AES-128 algorithm, were implemented using a reconfigurable hardware device, namely a Xilinx FPGA platform model XC4VFX12 - FF668-10C, which is a low resource platform, with 5472 slices, 320 I/O blocks, 10944 LUT components with 4 inputs each.

The implementation used the VHDL programming language, which nowadays is a well-established commonly used language for FPGA devices. The design and simulation software was Xilinx ISE 10.1, a complete version of this software being provided by the producer's website for evaluation purpose. These implementations refer to the Electronic Code Book (ECB) mode of operation on AES-128 and AES-128 PLUS encryption/decryption.

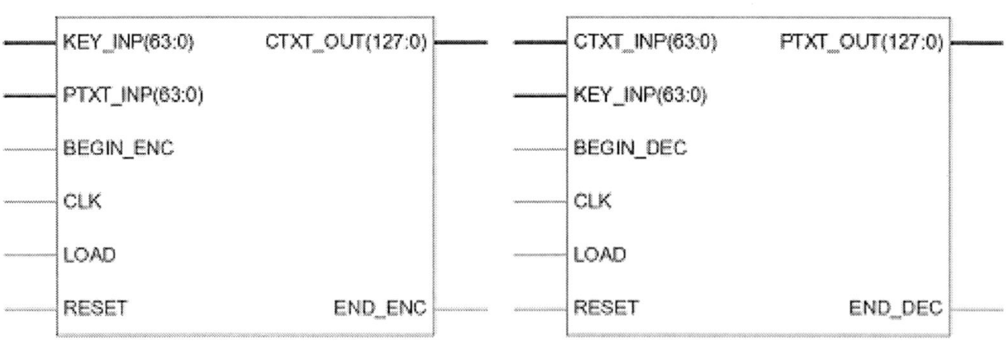

Figure 40. The encryption/decryption blocks of AES-128.

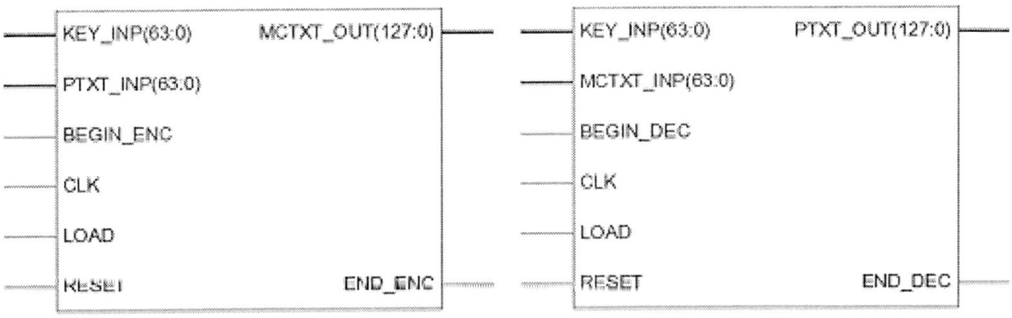

Figure 41. The encryption/decryption blocks of AES-128 PLUS.

In Figure 40 (AES-128) and Figure 41 (AES-128 PLUS), the encryption and decryption blocks are represented, showing the main signals used by the corresponding implementations.

In theory there are two main types of hardware implementation of the block ciphers: iterative and pipelined implementations [80]. Some characteristics of the iterative architectures are: knowing that the block ciphers are themselves of iterative nature (*n* iterations of the same algorithm are made in a single encryption/decryption, and n clock cycles are used in a single encryption/decryption), the iterative structure is a natural choice for implementation, occupying small areas of the FPGA devices, with the disadvantage of low throughput. Iterative architectures will implement a reduced number of rounds, usually one, in an independent way. The iterative architectures would be useful for applications where hardware area is limited and speed is not critical, which is the case of our target FPGA device. A pipelined version of AES implementation was not included because the target of this paper was not speed, but the functionality of this new algorithm.

The evaluation of the security that is provided by this implementation must take into consideration the specific security requirements for cryptographic modules which are stipulated by *FIPS Pub 140-3 (draft)* [83]. In fact, the security of a cryptographic module stands on the security that is provided by the cryptographic algorithm itself. In *Report on the development of the Advanced Encryption Standard (AES)* [79], a complex evaluation of the 5 finalists of the NIST's international contest concludes that using S-boxes as non-linear components, 'Rijndael appears to have an adequate security margin, but has received some criticism suggesting that its mathematical structure may lead to attacks. On the other hand, the simple structure may have facilitated its security analysis during the timeframe of the AES development process.' Even if some critiques were formulated (e.g. 'the key schedule does not have high diffusion'), Rijndael-AES was highly appreciated for the provided level of security. Regarding attacks on implementations, the same paper remarks that 'the operations used by Rijndael-AES are among the easiest to defend against power and timing attacks'.

Another way to evaluate the security that is provided by discussed implementations is the application of NIST statistical test suite (version 1.7 includes 15 tests) on the implemented algorithms (resulted cryptographic modules), as given by *NIST Special Pub 800-22* [84].

The support idea of these tests is based on the supposition of equivalence between the output of each implemented encryption block (AES-128 and AES-128 PLUS) and a pseudorandom number generator. In case of AES-128, this equivalence is proven by the AES development process and its evaluation, as mentioned in *Report on the development of the*

Advanced Encryption Standard (AES) [85]. This paper includes results of complete statistical tests application on the well-known pseudorandom generator BBS (Blum-Blum-Schub), on the AES-128 encryption module and also on the 3 versions of AES-128 PLUS encryption module.

Table 7. The Statistical Test Charts Evaluation

Test no.	BS	AES-128	M-AES	AES-M	M-AES-M
(1)	2	4	1	3	5
(2)	3	4	1	2	5
(3)	3	2	1	4	5
(4)	4	1	2	3	5
(5)	1	5	4	3	2
(6)	4	1	2	3	5
(7)	3	2	1	4	5
(8)	3	4	1	2	5
(9)	3	4	1	2	5
(10)	4	5	2	1	3
(11)	2	3	1	4	5
(12)	3	5	1	4	2
(13)	2	1	5	4	3
(14)	2	4	5	3	1
(15)	4	3	1	2	5
Total no.	43	48	29	44	61

The special built test unit is based on the same Xilinx FPGA device, which applies an iteration loop (850,000 iterations) to every evaluated algorithm (an implementation using VHDL programming language which integrates each algorithm in an iteration loop). The initial cleartext message consists of 128 ‘0’ bits, same as the initial key (the initial key takes this form permanently). The algorithm’s ciphertext output is connected to the module’s input for a new iteration. After all 850,000 iterations, 4 binary files will result, 103 MB each, consisting of a continuous catenation of 128 bit ciphertext output blocks. The 15 statistical tests use 100 series of 1,024,000 bits each, that is the content of each single binary file. Every test provides 100 test points, included in statistical test charts, which are considered by *NIST Special Pub 800-22* [84] enough for the test to be relevant.

After the application of the statistical test suite, firstly an evaluation of statistical test charts was made (how close to the diagonal are the test points), resulting in a classification which is based upon the synopsis of evaluated algorithms for each statistical test (e.g. for Frequency Test the hierarchy is: M-AES-M - 5 points, AES-128 - 4 points, AES-M - 3 points, BBS - 2 points, M-AES - 1 point). The complete results of this evaluation are presented in Table 7, where the test numbers are: (1) Frequency (Monobit) Test, (2) Frequency Test within a Block, (3) Cumulative Sums (Cusums) Test, (4) Runs Test, (5) Tests for the Longest-Run-of-Ones in a Block, (6) Binary Matrix Rank Test, (7) Discrete Fourier Transform (Spectral) Test, (8) Non-overlapping Template Matching Test, (9) Overlapping Template Matching Test, (10) Maurer's "Universal Statistical" Test, (11) Approximate Entropy Test, (12) Random

Excursions Test, (13) Random Excursions Variant Test, (14) Serial Test, (15) Linear Complexity Test.

Table 8. The Number of "FAILURE" Conclusions [82; 86]

Test no.	BBS	AES-128	M-AES	AES-M	M-AES-M
(1)	0	2	2	1	1
(2)	1	0	1	0	1
(3)	0	4	4	1	2
(4)	1	2	0	2	1
(5)	1	0	1	0	0
(6)	0	3	2	2	3
(7)	2	3	3	1	1
(8)	153	148	139	158	181
(9)	4	0	0	1	0
(10)	0	1	2	0	0
(11)	0	2	2	1	1
(12)	1	4	2	3	7
(13)	6	11	9	8	10
(14)	5	1	1	1	3
(15)	1	2	0	1	0

Table 9. NIST Tests Diagrams Of The Evaluated Algorithms [82]

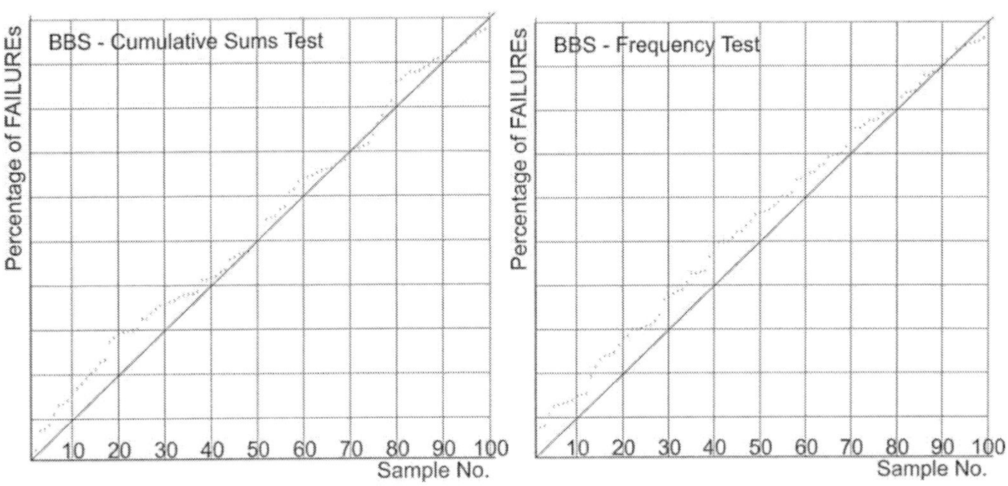

Secondly, an evaluation of experimental results of the statistical tests was made, this time counting the number of "FAILURE" conclusions for each test and algorithm. Table 8 presents this evaluation with the same order of statistical tests as before. Table 9 presents the results for two test charts, respectively for Frequency (Monobit) Test and Cumulative Sums (Cusums) Test.

The abscissa of every NIST test diagram represents the sample number and the ordinate represents the percentage of "FAILURE" results, which will not exceed 1%. The diagonal is expressed by a linear equation of the form $y = 0.99x$, which is very close to the bisectrix. If the points of a test diagram are situated on the bisectrix this means that the algorithm is cryptographically secure, which is not practically possible.

Conclusion and Future Trends

The conclusion of these evaluations is that M-AES-M's behavior is better than AES-128's. The other 2 versions (M-AES and AES-M) of the proposed AES-128 PLUS algorithm are close to the standard AES-128 algorithm. In only one of the tests, namely Non-overlapping Template Matching Test, M-AES-M's behavior is worse than AES-128's, but with an acceptable percentage of "FAILURE" conclusions, that is 1,25%.

The future work aims the pipeline implementation of the AES standard crypto-algorithm and of its new proposed version, AES-128 PLUS, using low resource FPGA platforms and trying to gain higher levels of throughput.

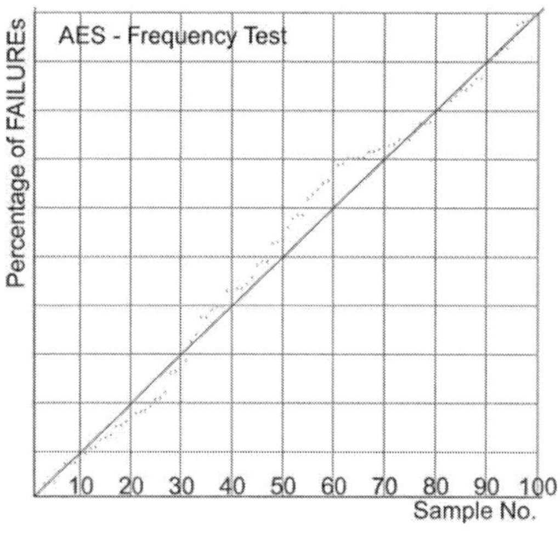

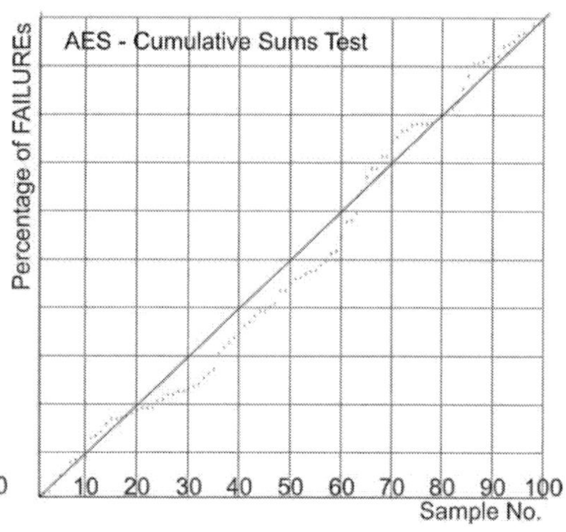

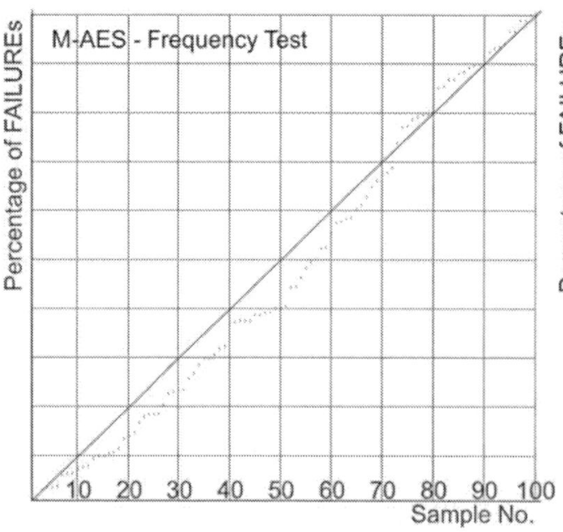

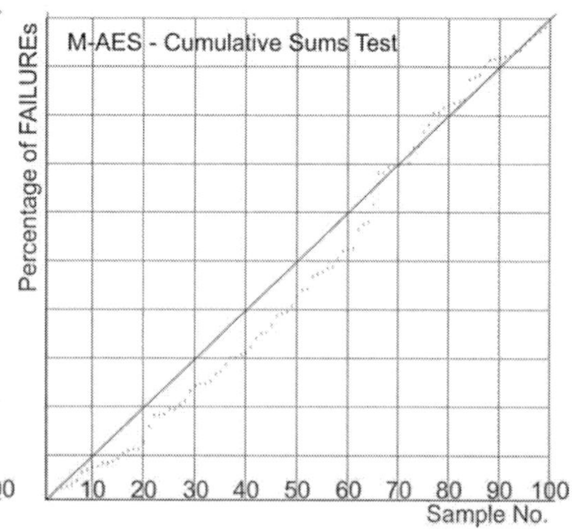

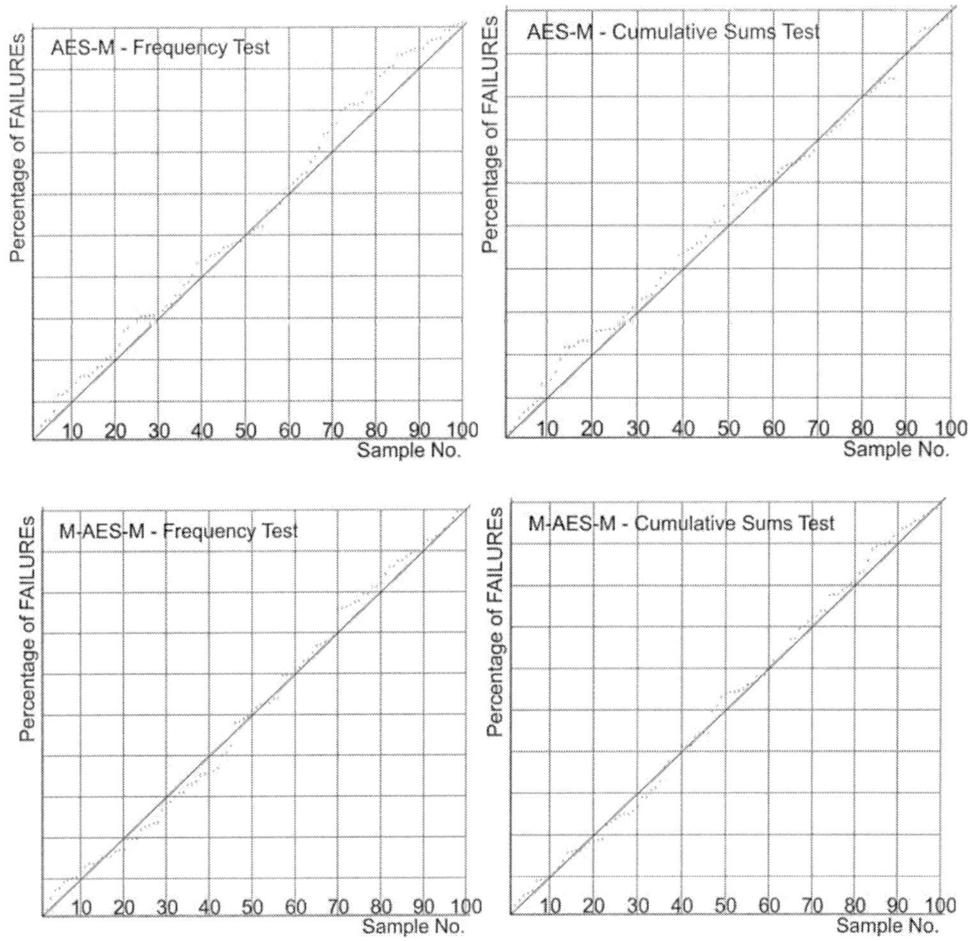

REFERENCES

[1] Patcha A., Park J.-M., "A Game Theoretic Approach to Modeling Intrusion Detection in Mobile Ad Hoc Networks", in the Proceedings of 2004 *IEEE Workshop on Information Assurance and Security*, 2004.

[2] Zou C. C., Gong W., Towsley D., "Code Red Worm Propagation Modeling and Analysis", *ACM on Computer and Communications Security*, 2002.

[3] Denning D. E., "An Intrusion-Detection Model", *IEEE Transactions on Software Engineering*, 1987.

[4] Tung B., "CIDF", http://gost.isi.edu/cidf/, (Access Date: 12th April 2010).

[5] Feng-Chi L., and Shih-Yen H., "Restraint method on the Internet worm self-reproduction and dissemination in the intranet", Hui-Chung C., Information and Communication Institute, 2007 Digital Technology and Innovation Management Seminar, Huafan University, Taiwan, 2007.

[6] Chan Y. F., Shoniregun A. C., Akmayeva A. G., "The network worm vulnerabilities and the Internet host attacks ", in the Proceedings of the 5th International Conference on Information Technology and Applications (ICITA 2008), 2008.

[7] Hui Cheng, "Research on the Internet Worm", (PhD Thesis), Tsinghua University, China, 2003.

[8] Daley D. J., Gani J., "Epidemic Modelling: An Information", Cambridge University Press, Cambridge, UK, 1999.

[9] Dolgoarshinnykh, R.G., "Epidemic modeling: SIRS models". University of Chicago, 2003

[10] Endsley M., "Design and evaluation for situation awareness enhancement", In proceedings of the Human Factors Society 32nd Annual Meeting, Human Factors Society, 1988.

[11] Situational awareness, from Wikipedia, the free encyclopedia.

[12] Bass T., "Intrusion Detection Systems and Multisensor Data Fusion", Communications of the ACM, April 2000.

[13] Yin X., Yurcik W., Slagell A., "The Design of VisFlowConnect-IP: a Link Analysis System for IP Security Situational Awareness", *The third IEEE International Workshop on Information Assurance* (IWIA), 2005.

[14] Batsell G. S., Rao S. N., Shankar M., "Distributed Intrusion Detection and Attack Containment for Organizational Cyber Security", http://www.ioc.ornl.gov/projects/documents/containment.pdf , 2005, Accessed on 12th June 2006.

[15] Jibao L., Huiqiang W., Liang Z., "Study of Network Security Situation Awareness Model Based on Simple Additive Weight and Grey Theory", 2006.

[16] Data Fusion, from Wikipedia, the free encyclopedia, http://en.wikipedia.org/wiki/Data_fusion , Accessed on 26the May 2008.

[17] Yu D., and Frincke D., "Alert Confidence Fusion in Intrusion Detection Systems with Extended Dempster-Shafer Theory", 43rd ACM Southeast Conference, Kennesaw, GA, USA, March 2005.

[18] Siraj A., "A Unified Alert Fusion Model for Intelligent Analysis of Sensor Data in an Intrusion Detection Environment", the Faculty of Mississippi State University, Mississippi, USA, Aug. 2006.

[19] Siraj A., Vaughm B. R., Bridges M. S., "Intrusion Sensor Data Fusion in an Intelligent Intrusion Detection System Architecture", In Proceedings of the 37th Hawaii International Conference on System Sciences, IEEE, 5-8 Jan. 2004.

[20] Siraj A., Bridges M. S., Vaughn B. R., "Fuzzy Cognitive Maps for Decision Support in an Intelligent Intrusion Detection Systems". In IFSA World Congress and 20th NAFIPS International Conference, 25-28 Jul. 2001.

[21] Berger A., Pietra D. S., Pietra D. V., "A Maximum Entropy Approach to Natural Language Processing", *Computational Linguistic*, March 1996.

[22] Ling S. and Xing C. (2004). Coding Theory, Cambridge University Press.

[23] Nechvatal J. et. al. (2000). Report on the Development of the Advanced Encryption Standard (AES), National Institute of Standards and Technology.

[24] Daemen J. and Rijmen V. (2002), The design of RIJNDAEL: AES - the Advanced Encryption Standard (Information Security and Cryptography), Springer-Verlag.

[25] Schneier B. (1996). Applied Cryptography, Second Edition: Protocols, Algorthms, and Source Code in C, John Wiley and Sons, Inc.

[26] Hankerson D., Menezes A., Vanstone S. (2004), Guide to Elliptic Curve Cryptography, Springer-Verlag New York, Inc.

[27] Alexander W. D., Chris J. M. (2006). User's guide to Cryptography and Standards, Artech House.
[28] ISO standards: http://www.iso.ch/.
[29] IEC standards: http://www.iec.ch/.
[30] ITU standards: http://www.itu.int/ITU-T/.
[31] ANSI standards: http://webstore.ansi.org/.
[32] BSI standards: http://www.bsi-global.com/.
[33] NIST standards: http://www.nist.gov/ , http://www.csrc.nist.gov/.
[34] 3GPP standards: http://www.3gpp.org/.
[35] ETSI standards: http://www.etsi.org/.
[36] IEEE standards: http://standards.ieee.org/.
[37] IETF standards: http://www.ietf.org/.
[38] SECG standards: http://www.secg.org/.
[39] PKCS standards : http://www.rsasecurity.com/rsalabs/pkcs/index.html.
[40] NESSIE evaluation project: http://www.cryptonessie.org/.
[41] ECRYPT evaluation project: http://www.ecrypt.eu.org/.
[42] CRYPTREC evaluation project: http://www.ipa.go.jp/security/enc/CRYPTREC/index-e.html.
[43] Barker, E.B., Barker, W.C., and Lee, A.(2005). Guide line for implementing cryptography in the federal systems- Second Edition (SP 800-21), Gaithersburg, USA: *National Institute of Standards and Technology* (NIST).
[44] Common Criteria for Information Technology Security Evaluation, *ISO* 15408.
[45] Federal Information Processing Standards Publication (FIPS) 140-2 (2002). Security requirements for cryptographic requirements, Gaithersburg, USA: National Institute of Standards and Technology.
[46] ISO standards: http://www.iso.ch/.
[47] NIST standards: http://www.nist.gov/, http://www.csrc.nist.gov/.
[48] Security requirements for cryptographic requirements, *FIPS 140-2*.
[49] Stavroulakis P. (2006). *Chaos Applications in Telecommunications*, CRC Press, Taylor and Francis Group.
[50] Larson, L.E., Liu, J.M., Tsimring L.S. (2006). Digital Communications using Chaos and Nonlinear Dynamics, Springer.
[51] Alvarez, G., Hernandez, L., Munoz, J., Montoya, F., Li, S. (2005). Security Analysis of communication system based on the synchronization of different chaotic systems", *Physiscs Letters*, A345 (4), pp. 245-250.
[52] Shujun, L. (2003). Analysis and new Designs of Digital Chaotic Ciphers, PhD thesis, School of Electronic and Information Engineering, Xi'an Jiaotong University.
[53] Kocarev, L., Szczepanski, J., Amigo, J. M, Tomosvski, I. (2006). Discrete chaos - I: Theory. *IEEE Transactions on Circuits and Systems I,* 53(6), pp. 1300–1309.
[54] Lazzouni, A.S., Bowong, S., Kakmeni, F.M., Cherki, M. (2007). An adaptive feedback control for chaos synchronization of nonlinear systems with different order, *Communications in Nonlinear Science and Numerical Simulation*, 12(4), pp. 568-583.
[55] Şerbănescu, A., Tutănescu, I., Sofron, E., Scheianu, D., Irimia, C., (2002). Applications of Chaotic Dynamic Systems, International Conference ECIT 2002, Iaşi, Romania.

[56] Tutănescu, I. (2005). An Alternative Transmission Method in Communications, Automatic and Informatics 2005 International Conference, Sofia, Bulgaria, 3-5 October 2005, published in CD of Conf. Proceedings.

[57] Rîncu, C.I., Iana, G.V., Serban, G., Tutănescu, I. (2007). Hardware implementation of a tent map-based chaotic generator, Proceedings of ELECO'2007 Conference, Bursa, Turkey.

[58] Şerbănescu, A., Tutănescu, I., Sofron, E., Scheianu, D., Irimia C. (2002) Applications of Chaotic Dynamic Systems, International Conference ECIT 2002, Iaşi, Romania.

[59] Frey, D.R. (1998). On Adaptive Chaotic Encoding, IEEE Transactions on Circuits and Systems-I: Fundamental Theory and Applications, Vol. 45, No. 11, November 1998, pp. 1200-1204.

[60] Götz, M., Kelber, K., Schwarz, W. (1998). Discrete-Time Chaotic Encryption Systems, Parts I, II and III, *IEEE Transactions on Circuits and Systems - I: Fundamental Theory and Applications*, Vol. 44, No. 10, October 1997, pp. 963-970/ Vol. 45, No. 2, Febr. 1998/ Vol. 45, No. 9, Sept. 1998, pp. 983-988.

[61] Chua, L.O., Lin, T. (1998). Chaos in digital filters, *IEEE Transactions on Circuits and Systems*, Vol. 35, June 1998, pp. 648-658.

[62] Şerbănescu, A. (2000). Integrated communications systems, Volume I – Wide band communications using dynamic chaotic systems, Military Tehnical Academy, Bucharest.

[63] Tutănescu, I., Irimia, C., Şerbănescu, A. (2000). Design and simulation of a chaotic based cryptosystem, International Conference Communications 2000 Proceedings, Bucharest, 7-9 Dec. 2000, pp. 381-385.

[64] Pecora, L., M., Carol, T., L. (1990). Synchronization in chaotic systems, *Phy. Rev. Lett.*, vol. 64, no. 8, pp. 821-824.

[65] Brucoli, M., Cafanga, D., Camimeo, L. (1999). Synchronization method of hyperchaotic circuits based on a scalar transmitted signal, Proc. European Conf. Circuit Theory and Design, Stresa, pp. 365 368.

[66] Chua, L., O., Kocarev, L., Eckert, K., Itoh, M. (1992). Experimental chaos synchronization in Chua's circuit, *Int. J. Bifurcation and Chaos* 2(3), pp. 705-708.

[67] Chua, L., O., Itoh, M., Kocarev, L., Eckert, K. (1993). Chaos synchronization in Chua's circuit, *J. Circuits Syst. Comput.* 3(1), pp. 93-108.

[68] Itoh, M., Yang, T., Chua, L., O. (1999). Experimental study of impulsive synchronization of chaotic and hyperchaotic circuits, *Int. J. Bifurcation and Chaos* 9(1), pp. 1393-1424.

[69] Pyragas, K. (1993). Predictable chaos in slightly perturbed unpredictable chaotic systems, *Phys. Lett.* A181, pp. 203 210.

[70] Itoh, M. (1999). Spread spectrum communication via chaos, *Int. J. Bifurcation and Chaos* 9(1), pp. 153-213.

[71] Panas, A., Yang, I., Chua, L., O. (1998). Experimental results of impulsive synchronization between two Chua's circuits, *Int. J. Bifurcation and Chaos* 8(3), pp. 639-644.

[72] Stjanovski, T., Kocarev, L., Parlitz, U. (1997). Digital coding via chaotic systems, *IEEE Trans. Circuit Syst.* 44(6), pp. 562-565.

[73] Wu, C., W., Yang, T., Chua, L., O. (1996). On adaptive synchronization and control of nonlinear dynamical systems, *Int. J. Bifurcation and Chaos* 6(3), pp. 455-471.

[74] Chang-Pin, Li., Chen, G. (2003). A note on Hopf bifurcation in Chen's system, *Int. J. Bifur. Chaos*, 13, pp.1609–1615.

[75] Chang-Pin, Li., Chen, G. (2004). Chaos in the Chen's system with a fractional order, *Journal: Chaos, Solitons and Fractals*, 22, pp. 443–450.

[76] Yan, Y., Chang-Pin, Li. (2005). On synchronization of three chaotic systems, *Journal: Chaos, Solitons and Fractals*, 23, pp.1683–1688.

[77] Peng, Z., Li, X. Liao, Y., Li, C. (2006). Generalized projective synchronization for the chaotic Lorenz system and the chaotic Chen system, Journal of Shanghai University (English Edition), Shanghai University Press, ISSN 1007-6417 (Print) 1863-236X (Online), Volume 10, No: 4.

[78] Daemen, J. and Rijmen, V. (1999). AES proposal: Rijndael, AES Algorithm Submission, available via http://www.nist.gov/CryptoToolkit.

[79] National Institute of Standards and Technology (2000). Report on the development of the Advanced Encryption Standard (AES).

[80] Rodriguez-Henriquez, F., Saqib, N. A., Diaz-Perez, A. and Koc, C. K. (2006). Cryptographic algorithms on reconfigurable hardware, Springer Science+Business Media, LLC.

[81] Burciu, P. and Sima, I. M. (2009). AES-128 encryption/decryption implementation on reconfigurable hardware (FPGA), Proceedings of the 3rd International Conference on Electronics, Computers and Artificial Intelligence, Vol. 2, Pitesti, Romania.

[82] Burciu, P. (2009). Design And Optimization Methods For The Hardware Implementation Of Information Encryption Algorithms On Digital Communications, Doctoral Thesis, University of Pitesti, Pitesti, Romania.

[83] National Institute of Standards and Technology (2007). Security requirements for cryptographic modules, FIPS Pub 140-3 (draft).

[84] National Institute of Standards and Technology (2008). A statistical test suite for random and pseudorandom number generators for cryptographic applications, NIST Special Pub 800-22.

[85] National Institute of Standards and Technology (2001). *Announcing the Advanced Encryption Standard* (AES), FIPS Pub 197.

[86] Burciu, P. (In press). A Different Approach Of The Rijndael-AES Algorithm, The Annals of Maritime University of Constanta, Constanta, Romania.

In: Advances in Energy Research
Editor: Nicu Bizon

ISBN 978-1-61209-991-0

Chapter 8

A REVIEW ON LOAD FREQUENCY CONTROL STRATEGIES

Hossein Shayeghi[1*] and Naser Mahdavi Tabatabaei[2]

[1] Technical Engineering Department,
University of Mohaghegh Ardabili, Ardabil, Iran
[2] Electrical Engineering Department,
Seraj Higher Education Institute, Tabriz, Iran

ABSTRACT

Load Frequency Control (LFC) is one of the most important issues in electric power system design and operation. The objective of the LFC in an interconnected power system is to maintain the frequency of each area and to keep tie-line power near to the scheduled values by adjusting the MW outputs the LFC generators so as to accommodate fluctuating load demands. LFC provides a mechanism to balance generated power and demand in order to maintain the system frequency at the nominal value and to keep the power interchanges between neigh*bouring areas at the their scheduled values. In the deregulated environment, it will serve as ancillary service and acquires a principal role to enable power exchanges and provide better condition for electricity trading. LFC goals, i.e. frequency regulation and tracking the load demands, maintaining the tie-line power interchanges to specified values in the presence of modeling uncertainties, system nonlinearities and area load disturbances determines the LFC synthesis as a multi-objective optimization problem and they make a big challenges for controller design. The load frequency controller design with better performance has received considerable attention during past years and many control strategies have been developed for LFC problem.

This chapter presents literature review on the LFC problem and history of new control strategies is outlined. The LFC scheme based on various control methodologies such as linear control, robust and adaptive techniques, artificial computational intelligent based strategies are investigated. Solution of the LFC problem including BES/SMES, wind turbines, FACTs devices and other DG units has also been discussed. Using the obtained results, various comparisons between these methods and the main

* Corresponding Author: hshayeghi@gmail.com

advantages/disadvantages of them are given. Finally, the recently control methodologies presented for the solution of LFC problem is widely outlined in the deregulated environment.

Keywords: LFC, AGC, Restructured Power System, Robust and Adaptive Control, Intelligent/Soft Computing Control Strategy, Power System Control.

NOMENCLATURE

ACE	Area Control Error
AGPM	Augmented Generation Participation Matrix
AFRC	Automatic Frequency Ratio Control
AGC	Automatic Generation Control
AI	Artificial Intelligence
ANFIS	Adaptive Neuro Fuzzy Inference System
ANN	Artificial Neural Network
BES	Battery Energy Storage
CES	Capacitive Energy Storage
DAMC	Data Analysis and Monitoring Center
DAC	Data Acquisition Center
DG	Distribution Generation
DISCO	Distribution Companies
FD	Figure of Demerit
GAs	Genetic Algorithms
GENCOs	Generation Companies
GDB	Governor Dead-Band
GRC	Generation Rate Constraint
HVDC	High Voltage Direct Current
IGBT	Insulated Gate Bipolar Transistor
ISE	Integral Square Error
ITAE	Integral of Time of Absolute Error
LFC	Load Frequency Control
LMI	Linear Matrix Inequality
LQG	Linear Quadratic Gaussian
MSF	Multi Stage Fuzzy
PI	Proportional plus Integral
PID	Proportional, Integral, and Derivative
PSO	Particle Swarm Optimization
P_{Tie}	Net tie-line Power Flow
P_T	Turbine power
P_v	Governor valve position
PV	Photovoltaic
RBF	Radial Biased Function
SA	Simulated Annealing
SMES	Superconducting Magnetic Energy Storage

SVC	Static Var Compensator
SSSC	Static Series Synchronies Compensator
TCPS	Thyristor Controlled Phase Shifter
TRANSCO	Transmission Companies
VCA	Virtual Control Area
VSC	Variable Structure Controller
VSS	Variable Structure System

INTRODUCTION

Power systems are used to convert natural energy into electric power. They transport electricity to factories and houses to satisfy all kinds of power needs. To optimize the performance of electrical equipment, it is important to ensure the quality of the electric power. A good quality of the electric power system requires both the frequency and voltage to remain at standard values during operation. In actual power systems, the load is changing continuously and randomly due to stochastic nature of operation techniques. Hence, system may experience deviations in voltage, frequency and scheduled exchanges to other area, which may yield undesirable effects [1]. Thus, a control system is essential to cancel the effects of the random load changes and to keep the frequency and voltage at the standard values. Although the active power and reactive power have combined effects on the frequency and voltage, the control problem of the frequency and voltage can be decoupled. The frequency is highly dependent on the active power while the voltage is highly dependent on the reactive power. Thus, the control issue in power systems can be decoupled into two independent problems. One is about the active power and frequency control while the other is about the reactive power and voltage control. The active power and frequency control is referred to as load frequency control [2]. LFC or AGC is one of the most important issues in electric power system design and operation for supplying sufficient and reliable electric power with good quality.

Generation in large interconnected power system comprises of thermal, hydro, nuclear, and gas power generation. Nuclear units owing to their high efficiency are usually kept at base load close to their maximum output with no participation in system automatic generation control. Gas power generation is ideal for meeting varying load demand. However, such plants do not play very significant role in AGC of a large power system, since these plants form a very small percentage of total system generation. Gas plants are used to meet peak demands only. Thus, the natural choice for AGC falls on either thermal or hydro units.

A typical large-scale power system is composed of several areas of generating units. In order to enhance the fault tolerance of the entire power system, these generating units are connected via tie-lines. The usage of tie-line power imports a new error into the control problem, i.e., tie-line power exchange error. When a sudden active power load change occurs to an area, the area will obtain energy via tie-lines from other areas. But eventually, the area that is subject to the load change should balance it without external support. Otherwise there would be economic conflicts between the areas. Hence, each area requires a separate load frequency controller to regulate the tie-line power exchange error so that all the areas in an interconnected power system can set their setpoints differently. In summary, the LFC goals for a power system are:

- Ensuring zero steady state error for frequency deviations
- Minimizing unscheduled tie line power flows between neighboring control areas
- Getting good tracking for load demands and disturbances
- Maintaining acceptable overshoot and settling time on the frequency and tie-line power deviations

Based on the above objectives, a control signal made up of tie line flow deviation added to frequency deviation weighted by a bias factor called ACE is used as the control signal in the LFC problem. ACE serves to indicate when total generation must be raised or lowered in a control area.

Nowadays, the electric power industry is in a transition from vertically integrated utility scenario, where a single utility owned and operated the generation, transmission and distribution systems and provided power at regulated rates to the deregulated scenario, where, competitive companies sell unbundled power at lower rates. Furthermore, various kinds of apparatus with large capacity and fast power consumption such as testing plants for nuclear fusion, steel factories, etc increase significantly. When these loads are concentrated in power systems, they may cause a serious problem of frequency oscillations. Thus, it is very important to consider how the control services of system frequency should be implemented. In a deregulated environment, any power system controls, such as the LFC as an ancillary service acquires a principal role to maintain the electric system reliability at an adequate level and is becoming much more significant today in accordance with the complexity of interconnected power system [3; 4]. Thus, stabilization of frequency oscillations in an interconnected power system becomes challenging when implemented in the future competitive environment. A new frequency stabilization service which emphasizes not only efficiency, reliability and economics, but also advanced and improved controls for satisfying the requirements of power system operation, is much in demand.

Valuable research contributions from time to time for LFC problem has been augmented, like LFC regulator designs to cope with parameter variations uncertainties, load characteristics, excitation control and parallel ac/dc transmission links. In this field, the application of neural networks, fuzzy logic and computational intelligent based optimization techniques such as GA and PSO is the most recent advancement to tackle the difficulties associated with the design of LFC controllers for the power systems with nonlinear models and/or insufficient knowledge about the system required for its accurate modeling. Apart from advances in control concepts, there have been many changes during the last decade or more, such as deregulation of power industry and use of superconducting magnetic energy storage, wind turbines, photovoltaic cells as other sources of electrical energy to the system. Due to these, the control philosophies associated with LFC problem have changed to accommodate their dynamics and effects on the overall system dynamic performance. Generally, LFC designs methods can be categorized as: i) Classical methods, ii) Adaptive and variable structure methods, iii) Robust control approaches and IV) AI based methods.

Types of power system models for solution of the LFC problem and the history of the recently various control strategies with their salient features is outlined in this study.

POWER SYSTEM LFC MODELS

The LFC problem has been dealt with extensively for more than four decades. The power systems are usually large-scale systems with complex nonlinear dynamics. However, the major part of the work reported so far has been performed by considering lineanized models of two/multi area power systems [5]. A comprehensive litratures review about the earlier studied in the field of LFC problem has been presented by [6]. The first attempt in the area of LFC problem has been to control the frequency of a power system via the flywheel governor of the synchronous machine. This technique was subsequently found to be insufficient and a supplementary control was included to the governor with the help of a signal directly proportional to the frequency deviation plus it's integral. This scheme constitutes the classical approach for the solution of the LFC problem. ACE serves to indicate when total generation must be raised or lowered in a control area. Since ACE is directly influenced by random load variations, this criterion can be treated statically by saying that the standard deviation of ACE should be small [2]. In order getting zero steady state error for the frequency and tie line power flow deviations, the ACE signal as a control input in LFC scheme is considered as linearly combination of these variables. As a result, an ACE is defined as:

$$ACE_i = \sum_{i=1, j\neq i}^{N} a_{ji} \Delta P_{tie,ij} + B_i \Delta F_i \tag{1}$$

where, ΔF_i, $\Delta P_{tie,ij}$ and B_i are frequency, tie-line power flow deviations and the frequency response characteristic, respectively. Based on the experiences with actual implementation of AGC schemes, modifications to the definition of ACE are suggested from time to time to cope with the changed power system environment.

In [7] has been illustrated that supplementary controller designs based on tie-line bias control strategy is caused that the ACEs are regulated to zero effectively. Figures 1 and 2 show the block diagram of the LFC system in the traditional environment [8; 9].

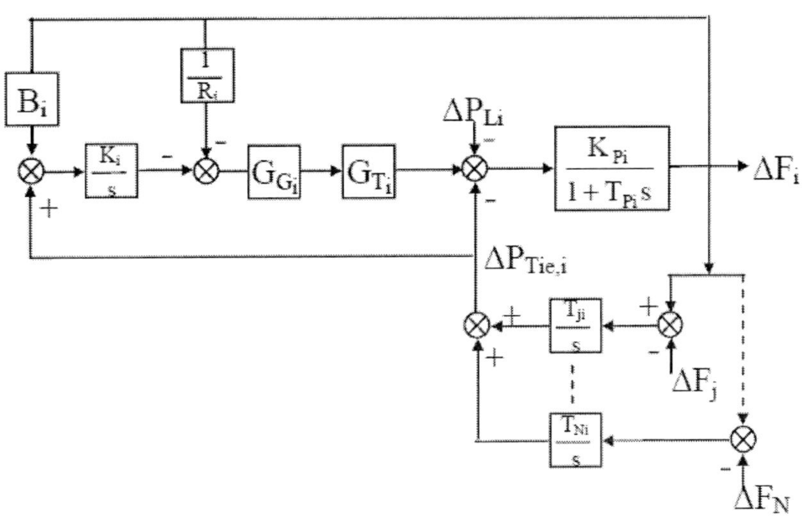

Figure 1. Block diagram of i-th control area in LFC scheme.

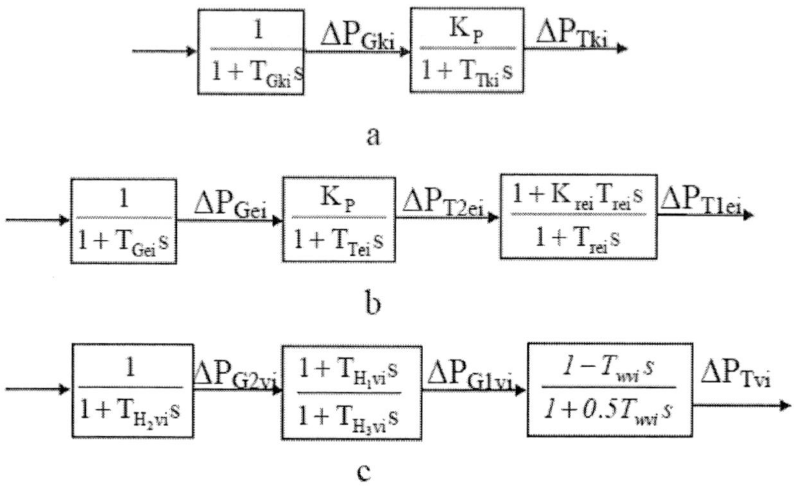

Figure 2. Governor and turbine model ($G_{Gi} \times G_{Ti}$) a) Thermal unit b) Thermal reheat unit c) Hydro unit.

A. LFC Model in Deregulated Environment

In the deregulated power systems, the vertically integrated utility no longer exists. However, the common LFC objectives, i.e. restoring the frequency and the net interchangers to their desired values for each control area, still remain. Since many presently regulated markets are likely to evolve into a hybrid scheme, and some deregulated markets are already of this type (e.g., Norway), the effects of deregulation of the power industry on LFC have been addressed in [9; 10]. Generally, three structures is considered for LFC system with progressing from, no LFC markets to a fully distributed market as follows:

- Free LFC
- Charged LFC
- Bilateral LFC

Free LFC is the present situation in the traditional environment, where a vertical integrated utility issues real power control signal to units it owns to perform LFC task.

In the charged LFC, the TRANCO no owns generation, but has an obligation to perform LFC task. In this scheme, the TRANCO must purchase active power from GENCOs and resell it to the DISCOs, on a short term, real time basis. The control error calculation is identical to the free LFC structure, because the control area bounding is the same as (Figure 3). Charged LFC is subjected to some technical issues such as computation of ACE, allocation of control error to GENCOs and accounting [10].

In the bilateral LFC scheme, DISCOs have the freedom to have a contract with any GENCO in their and other areas and GENCOs may or may not particulate in their own or other areas for AGC task. In this structure, a separate control process and then a control area associated for each load matching contract is needed. This control area is virtual which come into existence at the start of a load matching and terminate at its end (Figure 4) [9]. The boundary of the virtual control area encloses the GENCO and the DISCO associated with the contract.

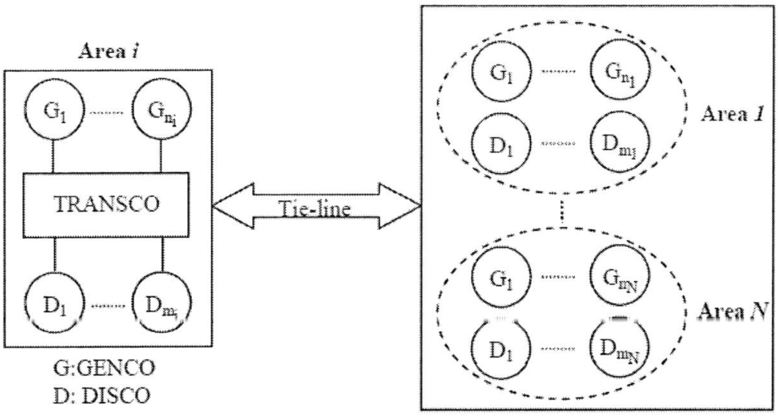

Figure 3. Charged LFC structure.

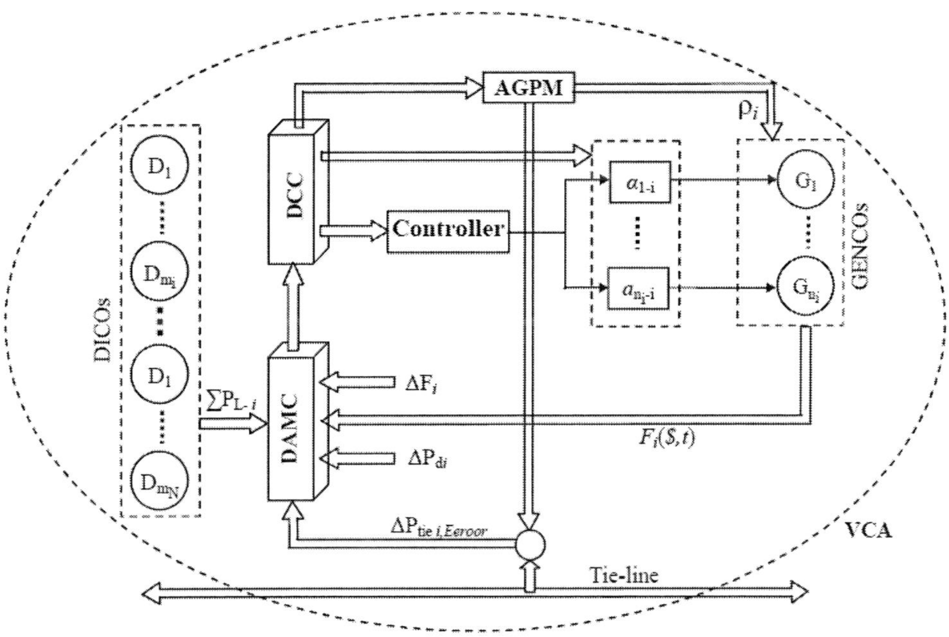

Figure 4. Restructured LFC Scheme.

The deregulated power system consists of GENCOs, TRANSCOs and DISCOs with an open access policy. Thus, various combinations of possible contracted scenarios between DISCOS and GENCOS are possible. All the transactions have to be cleared by the independent system operator or other organizations. Due to these, a study on simulation and optimization in an LFC system after deregulation has been carried out by [11]. The concept of DISCO participation matrix is proposed that helps for implementation of the contracts. Later, a generalized dynamical model for LFC scheme has been developed in the deregulated environment by [12]. The proposed generalized model helps the visualization of the contracts and introducing the added new information signals on traditional LFC scheme dynamic which carry information as to which GENCO has to follow a local demanded by which DISCO.

Here, a brief review of his proposed model for LFC systems is given and a test system is simulated to illustrate its effectiveness.

Global analysis of the power system markets shows that the frequency control is one of the most profitable ancillary services at these systems. This service is related to the short-term balance of energy and frequency of the power systems. The most common methods used to accomplish frequency control are generator governor response (primary frequency regulation) and load frequency control. The goal of LFC is to reestablish primary frequency regulation capacity, return the frequency to its nominal value and minimize unscheduled tie-line power flows between neighboring control areas. From the mechanisms used to manage the provision this service in ancillary markets, the bilateral contracts or competitive offers stand out.

Based on the idea presented in (Donde and Pai 2001), the concept of an '*AGPM*' to express the possible contracts following is presented here [12]. The AGPM shows the participation factor of a GENCO in the load following contract with a DISCO. The rows and columns of AGPM matrix equal the total number of GENCOs and DISCOs in the overall power system, respectively. Consider the number of GENCOs and DISCOs in area i be n_i and m_i in a large scale power system with N control areas. The structure of AGPM is given by:

$$AGPM = \begin{bmatrix} AGPM_{11} & \cdots & AGPM_{1N} \\ \vdots & \ddots & \vdots \\ AGPM_{N1} & \cdots & AGPM_{NN} \end{bmatrix} \tag{2}$$

where,

$$AGPM_{ij} = \begin{bmatrix} gpf_{(s_i+1)(z_j+1)} & \cdots & gpf_{(s_i+1)(z_j+m_j)} \\ \vdots & \ddots & \vdots \\ gpf_{(s_i+n_i)(z_j+1)} & \cdots & gpf_{(s_i+n_i)(z_j+m_j)} \end{bmatrix}$$

for $i, j = 1, \cdots, N$ and

$$s_i = \sum_{k=1}^{i-1} n_i \,,\; z_j = \sum_{k=1}^{j-1} m_j \quad \& \quad s_1 = z_1 = 0$$

In the above, gpf_{ij} refers to '*generation participation factor*' and shows the participation factor of GENCO i in total load following requirement of DISCO j based on the contracted scenario. Sum of all entries in each column of AGPM is unity. The diagonal sub-matrices of AGPM correspond to local demands and off-diagonal sub-matrices correspond to demands of DISCOs in one area on GENCOs in another area. Block diagram of the generalized LFC scheme in a restructured system is shown in Figure 5. Dashed lines show interfaces between areas and the demand signals based on the possible contracts. These new information signals are absent in the traditional LFC scheme.

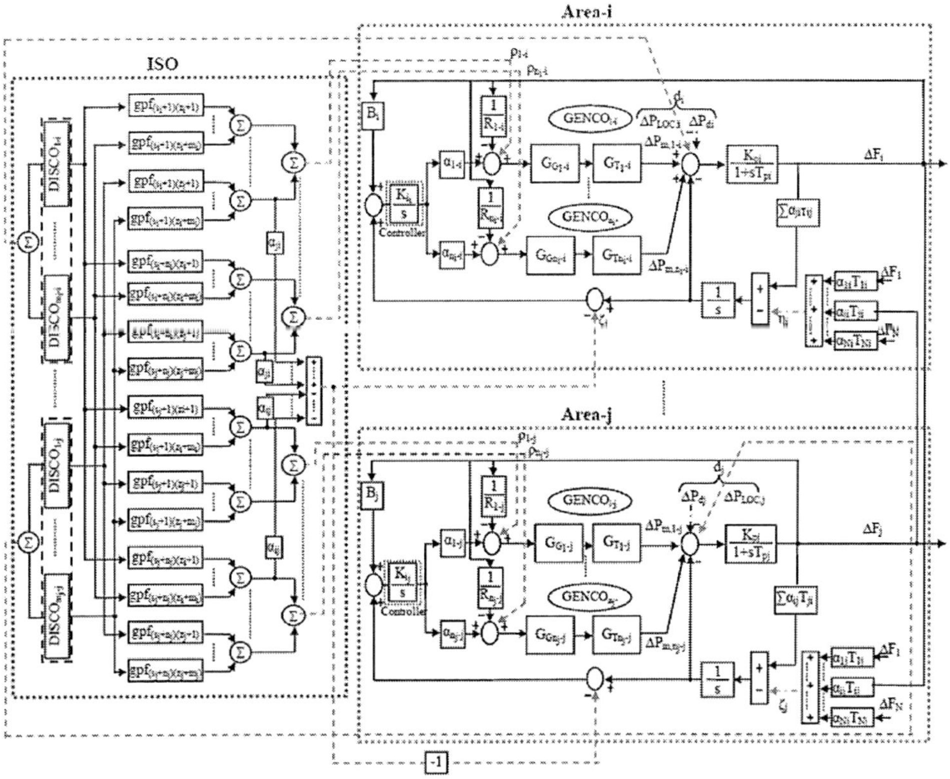

Figure 5. The generalized LFC scheme in the restructured system.

As there are many GENCOs in each area, ACE signal has to be distributed among them due to their ACE participation factor in the LFC task and $\sum_{j=1}^{n_i} \alpha_{ji} = 1$.

We can write:

$$d_i = \Delta P_{Loc,i} + \Delta P_{di}\ , \quad \Delta P_{Loc,i} = \sum_{j=1}^{m_i} (\Delta P_{Lj} + \Delta P_{ULj}) \tag{3}$$

$$\eta_i = \sum_{\substack{j=1 \\ j \neq i}}^{N} T_{ij} \Delta f_j \tag{4}$$

$$\zeta_i = \Delta P_{tie,i,sch} = \sum_{\substack{k=1 \\ k \neq i}}^{N} \Delta P_{tie,ik,sch} \tag{5}$$

$$\Delta P_{tie,ik,sch} = \sum_{j=1}^{n_i} \sum_{t=1}^{m_k} apf_{(s_i+j)(z_k+t)} \Delta P_{L(z_k+t)} - \sum_{t=1}^{n_k} \sum_{j=1}^{m_i} apf_{(s_k+t)(z_i+j)} \Delta P_{L(z_i+j)} \tag{6}$$

$$\Delta P_{tie,i-error} = \Delta P_{tie,i-actual} - \zeta_i \tag{7}$$

$$\rho_i = [\rho_{1i} \cdots \rho_{ki} \cdots \rho_{n_i i}]^T, \quad \rho_{ki} = \sum_{j=1}^{N}[\sum_{t=1}^{m_j} gpf_{(s_i+k)(z_j+t)} \Delta P_{Lt-j}] \tag{8}$$

$$\Delta P_{m,k-i} = \rho_{ki} + apf_{ki} \sum_{j=1}^{m_i} \Delta P_{ULj-i} \tag{9}$$

$\Delta P_{m,ki}$ is the desired total power generation of a GENCO k in area i and must track the demand of the DISCOs in contract with it in the steady state.

The small signal analysis is justified for studying the systems response for small perturbation one. However, the implementation of LFC strategy based on a linearized model on an essentially nonlinear system does not necessarily ensure the stability of the systems [5]. The destabilizing effect of governor dead-band nonlinearity on the conventional AGC system was demonstrated by [13]. It was shown that the GDB nonlinearity tends to produce continuous oscillations in the area frequency and tie-line power transient response. Figure 6 shows the nonlinear model of governor for consideration back lash or GDB.

One of the importance constraints in the LFC problem is GRC, i.e. practical limit on the rate of change in the generation power of each GENCO. The results in [7; 14] indicated that GRC would influence the dynamic responses of the system significantly and lead to larger overshot and longer settling time. In order to take effect of the GRC into account, the linear model of turbine $\Delta PV_{ki}/\Delta PT_{ki}$ in Figure 6 is usually replaced by a nonlinear model of Figure 7 (with ±δ limit). Also, a limiter, bounded by ±δ limit was used within the PI controller for governor system to prevent the excessive control action.

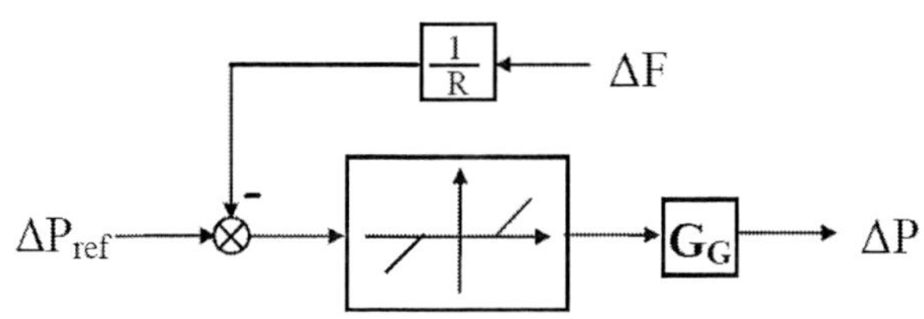

Figure 6. A nonlinear model of governor with back lash.

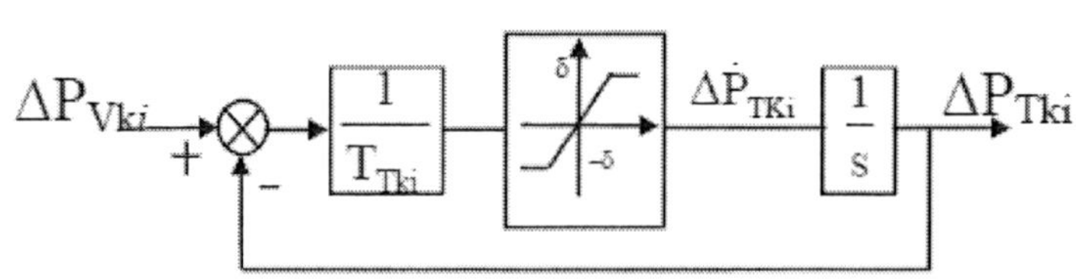

Figure 7. Nonlinear turbine model with GRC.

B. Case Study

A two-area restructured power system as shown in Figure 8 is considered as a test system to demonstrate the effectivness of the proposed LFC model in the pervious subsection. It is assumed that each control area has two GENCOs and DISCOs. The system parameters are given in [15] and it is assumed that α_{ij} is 0.5 for each GENCO.

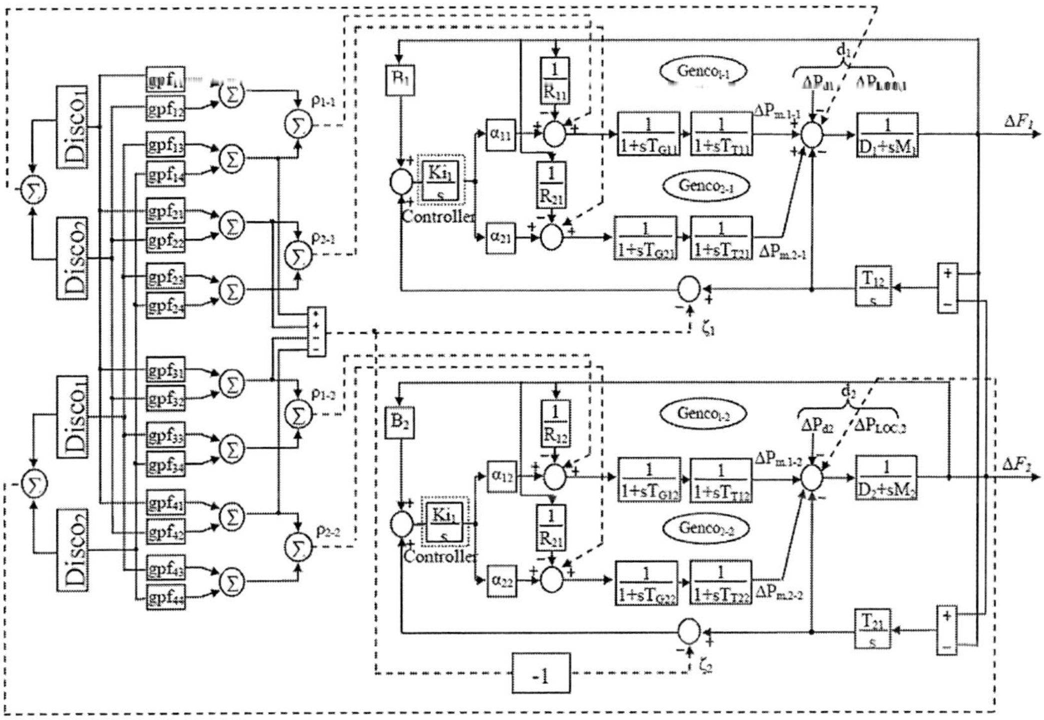

Figure 8. A two-area restructured power system.

Free LFC

As there is no contract between GENCOs and DISCOs for load following matching, the load disturbance in each area is taken up by the GENCOs in the same area. Thus, in this structure, the AGPM matrix is a zero matrix as follows:

$$AGPM = 0_{4\times4}$$

Assume that DISCO 1 in one area demands 0.1 puMw. The power system responses are shown in Figure 9. It can be seen that the LFC task is equally performed by two GENCOs in each area. Since there are no contracts between areas, the scheduled steady state power flow according to Eq. (6) over the tie-line is zero.

Charged LFC

In this structure it is assumed that by ISO coordination, GENCO 1 in one area is participated in LFC task. Thus, the AGPM structure is given by:

Frequency and generation power of GENCOs are shown in Figure 10 which is given by Eq. (9) in the steady state case as follows:

$$\begin{aligned}\Delta P_{m,1-1} &= gpf_{11}\Delta P_{L1-1} + gpf_{21}\Delta P_{L2-1} + gpf_{21}\Delta P_{L2-1} + gpf_{22}\Delta P_{L2-2} + \alpha_{11}(\Delta P_{LU1-1} + \Delta P_{LU2-1}) \\ &= 1\times 0.1 + 1\times 0 + 1\times 0 + 1\times 0.1 + 0.5\times(0+0) = 0.2\ PuMW\end{aligned}$$

and similarly:

$$\Delta P_{m,2-1} = 0\,;\ \Delta P_{m,2-1} = 0\,;\ \Delta P_{m,2-2} = 0$$

The tie-line power flow deviation in the steady state (Eq. (6)) is given by:

$$\begin{aligned}\Delta P_{tie,12} &= gpf_{13}\Delta P_{L1-2} + gpf_{14}\Delta P_{L2-2} + gpf_{23}\Delta P_{L1-2} + gpf_{24}\Delta P_{L2-2} \\ &\quad -(gpf_{31}\Delta P_{L1-1} + gpf_{32}\Delta P_{L2-1} + gpf_{41}\Delta P_{L1-1} + gpf_{42}\Delta P_{L2-1}) \\ &= 1\times 0 + 1\times 0.1 + 0\times 0 + 0\times 0.1 - (0\times 0 + 0\times 0 + 0\times 0 + 0\times 0) = 0.1\ PuMW\end{aligned}$$

Bilateral LFC

In this case, DISCOs have the freedom to have a contract with any GENCO in their and other areas. This structure is investigated for the three Scenarios.

A. Scenario1: Poolco based transactions

In this scenario, GENCOs only participate in load following control of their areas. It is assumed that a large step load 0.1 pu is demanded by each DISCOs in area 1 and DICOs 1 and 2 in second are demand step load 0.08 and 0.05 puMW, respectively.

$$AGPM = \begin{bmatrix} 1 & 1 & 1 & 1 \\ 0 & 0 & 0 & 0 \\ 0 & 0 & 0 & 0 \\ 0 & 0 & 0 & 0 \end{bmatrix}$$

Assume that DISCO 1 in one area and DISCO 2 in second area demand 0.1 puMw, respectively.

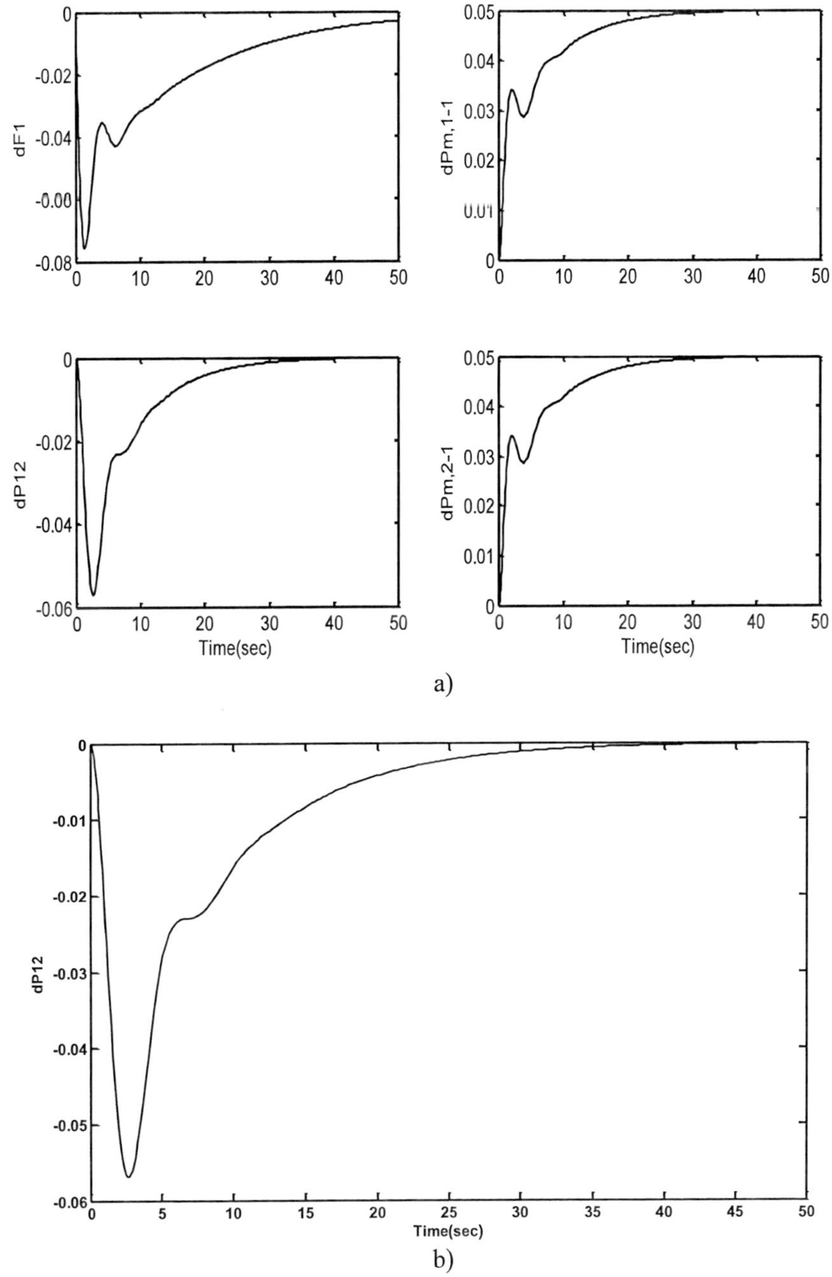

Figure 9. Power system responses in the free LFC a) Frequency deviation and GENCOs power in area 1 b) Tie line power deviation/

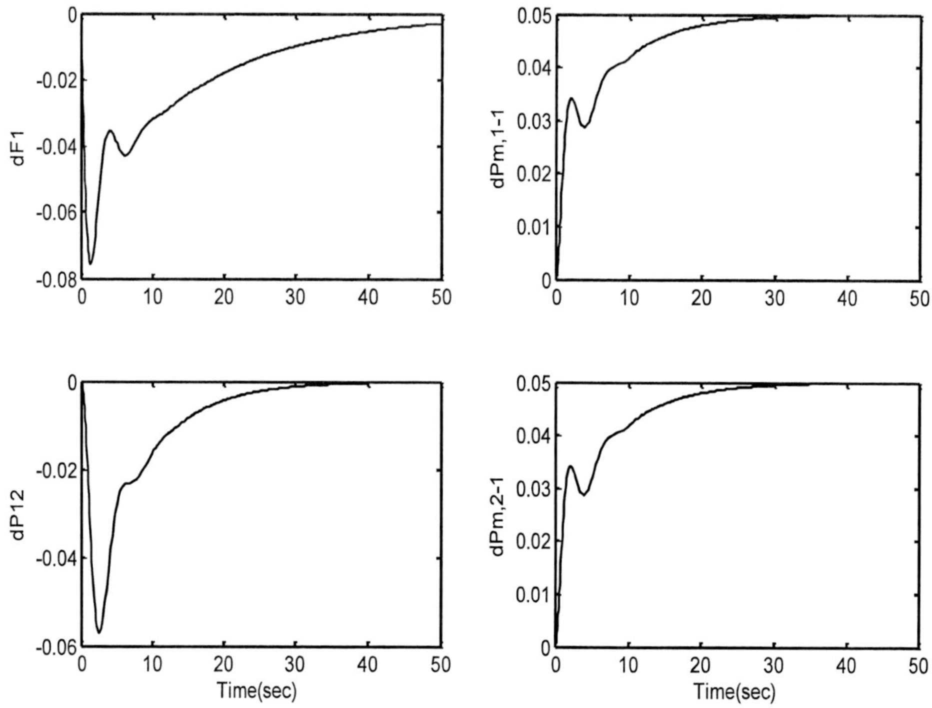

a)

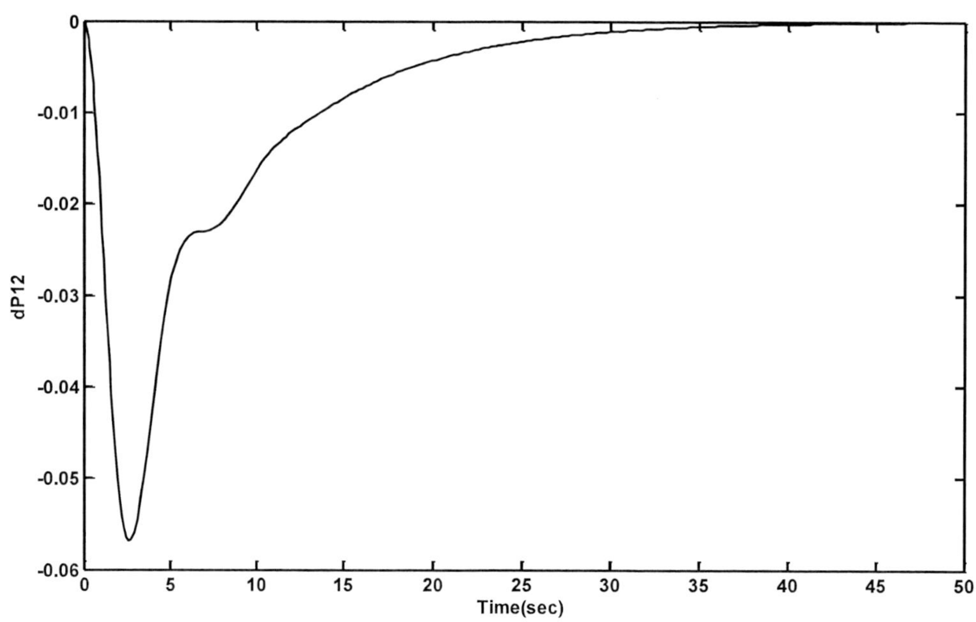

b)

Figure 10. Power system responses in the charged LFC a) Frequency deviation and GENCOs power in area 1. b) Tie line power deviation.

A case of Poolco based contracts between DISCOs and available GENCOs is simulated based on the following AGPM.

$$AGPM = \begin{bmatrix} 0.4 & 0.6 & 0 & 0 \\ 0.6 & 0.4 & 0 & 0 \\ 0 & 0 & 0.5 & 0.5 \\ 0 & 0 & 0.5 & 0.5 \end{bmatrix}$$

The power system responses are depicted in Figure 11. It can be seen that the actual generated power of GENCOs reaches to the desired values as given by Eq. (9). i.e.:

$$\Delta P_{m,1-1} = 0.1 ; \Delta P_{m,2-1} = 0.1 ; \Delta P_{m,2-1} = 0.07 ; \Delta P_{m,2-2} = 0.07 \quad PuMW$$

Since there are no contracts between areas, the scheduled steady state power flow according to Eq. (6) over the tie-line is zero.

B. Scenario 2: Combination of Poolco based and bilateral transactions

In this case, in addition to participation of GENCOs in their areas for matching load demands of DISCOs, they can be participated for matching the load demands of DISCOs in other areas.

Consider the all the DISCOs contract with the available GENCOs for power as per the following AGPM:

$$AGPM = \begin{bmatrix} 0.5 & 0.25 & 0 & 0.3 \\ 0.2 & 0.25 & 0 & 0 \\ 0 & 0.25 & 1 & 0.7 \\ 0.3 & 025 & 0 & 0 \end{bmatrix}$$

It is assumed that each DISCO demands 0.1 puMw power from GENCOs. The power system responses in this case are shown in Figure 12.

It can be seen that, the actual generated powers of GENCOs reach the desired values in the steady state as given by Eq. (9). i.e.:

$$\Delta P_{m,1-1} = 0.105 ; \Delta P_{m,2-1} = 0.045 ; \Delta P_{m,2-1} = 0.195 ; \Delta P_{m,2-2} = 0.055 \quad PuMW$$

Also, the tie-line power flow properly converges to the specified value Eq. (6) in the steady state. i.e.: $\Delta P_{tie,12,sch}$=-0.05pu.

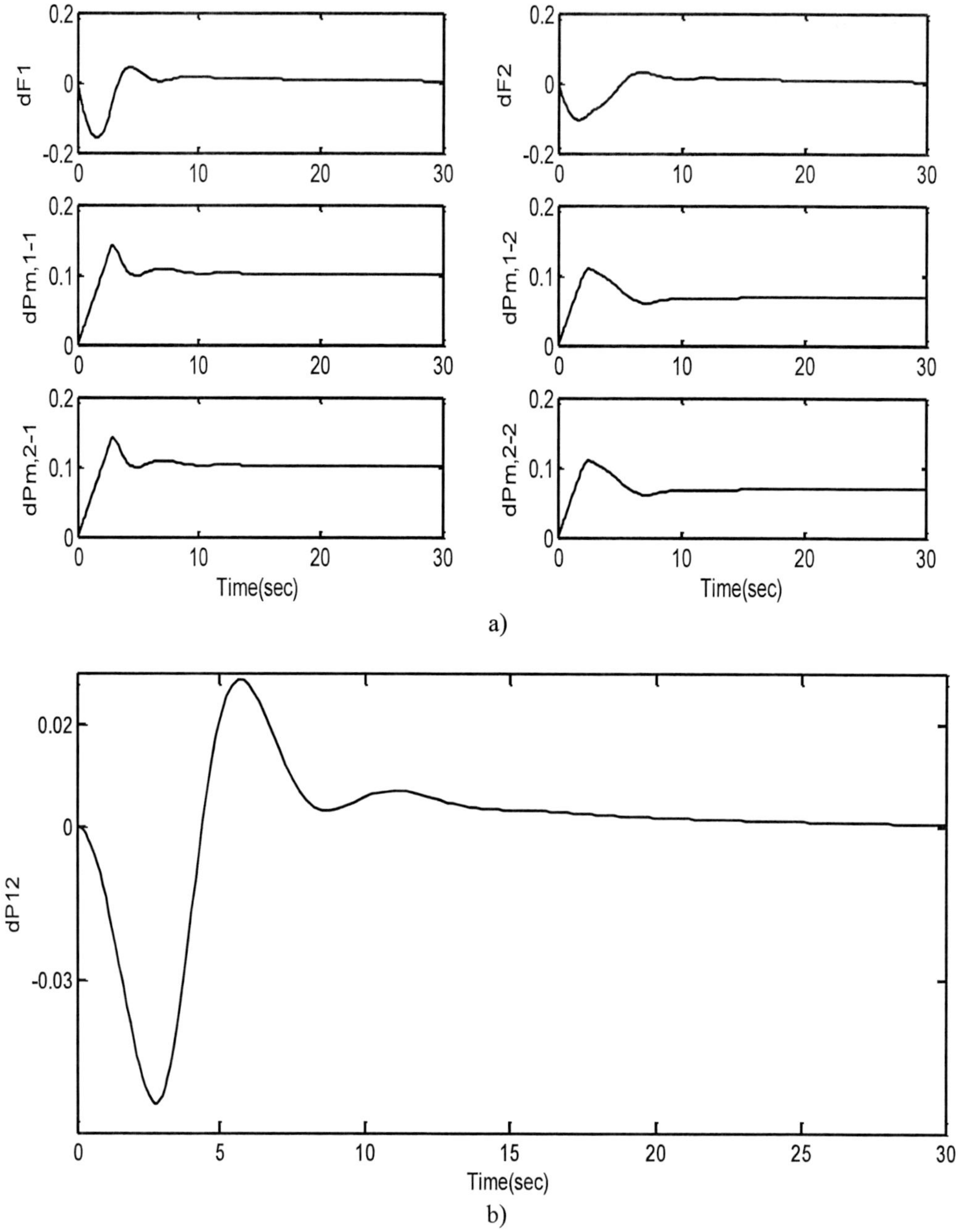

Figure 11. Power system responses in the charged LFC a) Frequency deviation and GENCOs power in areas b) Tie line power deviation.

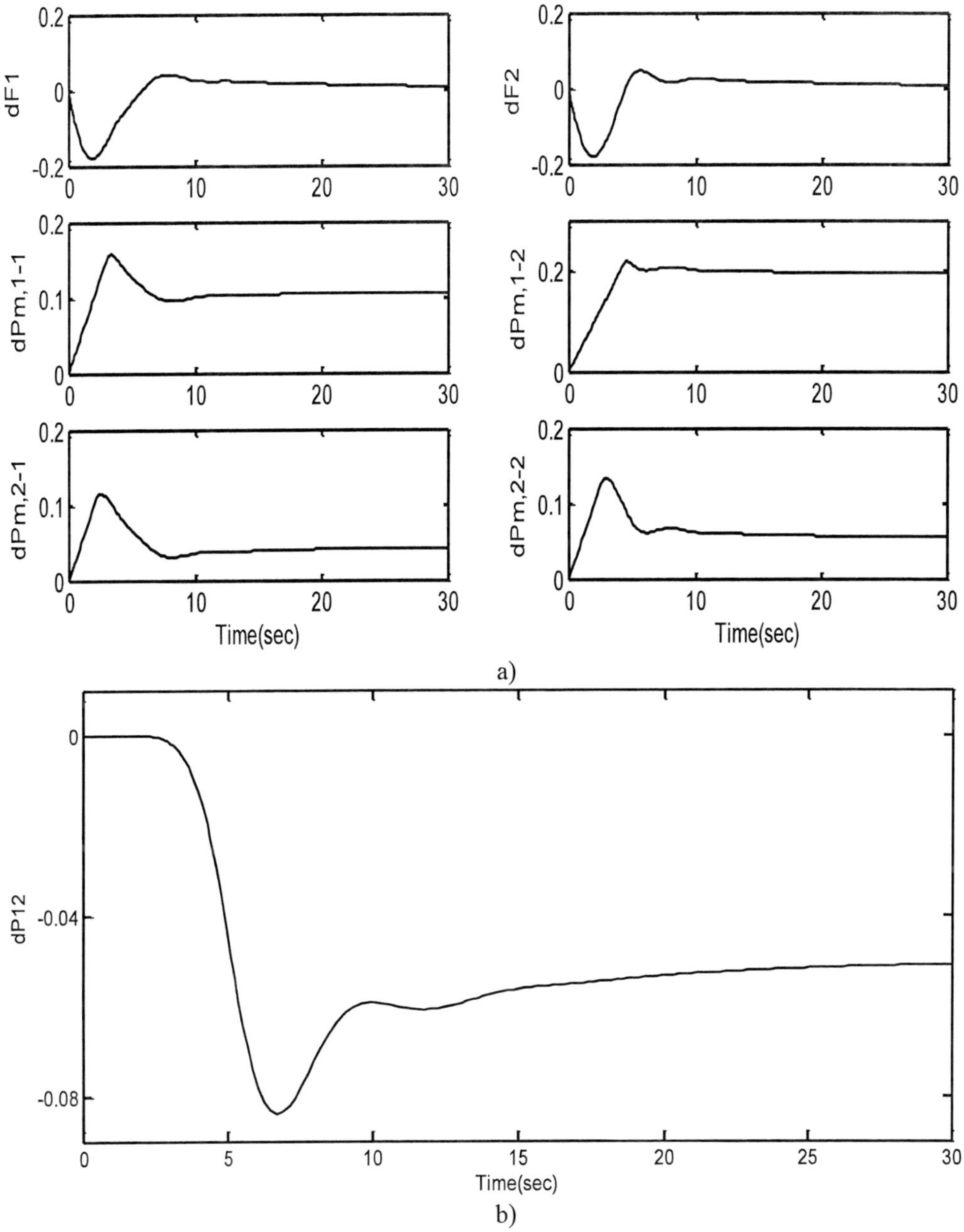

b)

Figure 12. Power system responses in the charged LFC a) Frequency deviation and GENCOs power in areas b) Tie line power deviation.

Control Scheme

For solution of the LFC problem many control schemes based on centralized control strategy have been reported. The main limitation of centralized control strategy For LFC is the need to exchange in information from control areas spread over distantly connected geographical territories along with their increased computational and storage complexities.

The decentralized LFC concept appeared in the power system control scenario to deal with such problems, very effectively and consequently, many research papers using this concept with continuous and discrete time system models have been presented in the literatures [5; 6]. It should be noted that in the dynamical operation of power systems, it is usually important to aim for decentralization of control action to individual areas. This aim should coincide with the requirements for stability and load frequency scheduling within the overall system. In a completely decentralized control scheme, the feedback controls in each area computed on the basis of measurements in that area only. This implies that, no interchange of information among areas is necessary for the LFC task. The advantage of this operating philosophy is apparent in providing cost saving in date communications and in reducing the scope of the network monitoring.

Due to these, the design of decentralized load frequency controllers based on structured singular value and H_∞ norm [5; 16; 17; 18]. They have demonstrated that when the frequency response-based diagonal dominance cannot be achieved, the structured singular values and H_∞ norm can be applied to design decentralized LFC to achieve the desired system dynamic performance, respectively in such away the stability of the overall system with the decentralized controllers was guaranteed. The effectiveness of the proposed control strategy based on ITAE performance index against parametric uncertainties on a two-area or four area power system has been confirmed.

In [19] using the Lyepunov function illustrates that the overall system was asymptotically stable for all admissible plant parametric uncertainties, when all local controllers were working together. Kazemi, Karari and Menhaj introduced a suitable transformation matrix which transformed the initial reference model to an equivalent reference model, such that the convergence of the output errors was guaranteed. An appropriate adaptive law was derived for adjusting this transformation matrix.

Decentralized biased PI dual mode controller for LFC of the interconnected power systems considering GDB and GRC nonlinearities have been reported by Velusami and Chidambaram (2007) [20]. Using an integral of square error and maximum stability margin criteria the biased PI dual mode controller was designed. The proposed control strategy has simple structure and easy to implement.

Control Design Methods

Generally, LFC control designs methodologies can be categorized as: i) classical approaches, ii) adaptive and variable structure techniques, iii) robust control theorms, iv) AI-based technique.

A. Classical Methods

The conventional control strategies for LFC problem are those which take the integral of the control error as the control signal. In classical control methods, to obtain the desired gain and phase margins, Bode and Nyquist diagrams as well as root locus are usually used. Thus, the design procedure of the classical methods for LFC problem is: straight forward, easy and amenable for practical implementation. However, the investigations carried out using these approaches reveal that it exhibits poor dynamic performance, especially in the presence of other destabilizing effects, such as parameter variations and nonlinearities [7; 21].

Modern optimal control theory based LFC schemes has been appeared in the literatures [5]. The feasibility of an optimal LFC scheme requires the availability of all state variables for generating of the feedback signals. This requirement may be met, if the system state-vector is observable from area measurements [22; 23]. However, even if the observability condition is satisfied, the resulted controllers with appropriately designed observers are normally quite complicated and therefore, these approaches are not suitable for a large scale power system where the total number of the state variables are large.

Due to practical limitations in the implementation of LFC, based on feedback of all state variables, suboptimal LFC regulator designs were considered [24; 25]. The design method employs model and singular perturbation techniques to affect decoupling of the interconnection into its subsystem components. In [26] a PID controller was tuned for AGC of hydro power system using a maximum peak resonance specification method that was graphically supported by the Nichols chart. The proposed method is efficient for controlling the stability and overshoot of the system. The controller performance was desirable against parametric uncertainties. Howevor, the GDB and GRC effects has not investigated.

B. Adaptive and Variable Structure Techniques

Adaptive control is a topic of research for more than a quarter of a century. Basically, the adaptive control systems can be classified into two categories; namely, the self tuning regulators and the model reference control systems. The task of adaptive control technique is to make the process under control less sensitive to change in plant parameters and to un-modeled plant dynamics. A review of various adaptive control techniques were reported for LFC scheme for dealing with plant parameter changes in [5]. Talag and Al-Basri [27] showed an adaptive fuzzy gain scheduling scheme for conventional PI and optimal load frequency controllers. Despite the promising results achieved by adaptive controllers, the control algorithms are complicated and require on-line system model identification. The adaptive control strategies usually require the perfect model-following condition or explicit parameters identification. These efforts seem unrealistic, since it is difficult to achieve it.

The concept of VSS to design of load frequency controller has been applied by several authors in [6]. In [28; 29], a fuzzy logic based variable structure controller was proposed for solution of the LFC problem in electric power generation systems. This method combines the salient feature of both variable structure and fuzzy systems to achieve high performance and robustness. It may be noted that the VSS controllers effectively can be improved transient response due to load disturbances in the power system by properly selecting the parameter of the controller. However, the complexity of VSS and the associated chattering problem may be the reason that these controllers were not fully appreciated in the LFC tasks. Furthermore, controllers based on the state equation of the linearized model may require estimates of inaccessible state variables. Observers can be designed for this, but it would involve the additional cost data telemetring.

C. Robust Control Theorems

Power systems subjected to different kinds of uncertainties and disturbances due to changes in system parameters and characteristics, load variation, errors in modeling and linearizing and environmental conditions. On the other hand, the operating points of a power system may changes very much randomly during a daily cycle. Because of this, an optimal LFC regulator design based on the nominal system parameter values is certainly not suitable for the LFC problem and therefore, the implementation of these regulators on the system may be inadequate to provide the desired system functioning. This could result in a degraded system dynamic performance and sometimes also in the loss of system stability. It may be noted that in robust control approaches, control objectives are to design load frequency controller to not only meet nominal stability and nominal performance requirements, but also guarantee the *robust stability* and *robust performance* in power system on LFC problem. Thus, considerable efforts have been made to design LFC controllers with better performance to cope with system uncertainties using various robust methodologies such as Riccati equation based controller, combination of the robust and adaptive control techniques, quantitative feedback theory and H_∞ control. The review of these control strategies are given in [5]. The advantages of these control techniques are considering system uncertainties and physical understudying of the system in the synthesis procedure.

Azzam [30] proposed a robust state feedback controller based on optimum finding of the close loop system eigenvalues that had minimum sensitivity to system parameter changes. The design of robust stabilizing controller based on Riccate equation method for all admissible uncertainties using the combination of 'Matching conditions' and Lyapunov stability theory has been proposed by Goshaids, Prasad and Prasad [31]. Shayeghi, Karrari and Menahj [32] have applied μ-synthesis and analysis technique for design of robust LFC considering admissible parametric uncertainties. The effectiveness of the resulting controller was compared with the proportional integral controller optimized by ISE index for wide range of plant parameter changes in the presence of GRC. The design of robust controller based on Q-parameterization theory for control of the frequency deviations was proposed by Azzam and Mohamed [33]. This approach has several advantages such as: a stable controller always exits, the closed loop poles can be located in a prescribed region for achieving of the required transient response and the resulting controller order is lower than the order of comparable robust controller such as H_∞, LQG and μ-analysis. The above robust control methods shows good dynamical responses, however, most of them are based on state feedback and require the availability of all state variables for feedback. Thus, these efforts seem unrealistic, since it is difficult to achieve it. Also, some of them have a centralized scheme and order of resulting controllers is high which is not feasible for a large scale power systems.

It may be noted that in the dynamical operation of power systems, it is usually important to aim for decentralization of control to individual areas. This aim should be coincide with requirements for stability of the overall system. The main feature of the robust decentralized scheme methods is reduction in the controller complexity and suitability for practical implementation. Thus, decentralized robust control has been suggested for multi-area power system LFC. Lim, Wang and Zhou [34] proposed a robust decentralized load frequency controller based on Riccate equation approach by embedding the local system parametric bounds in it for achieving controller robustness. Using the proposed method, the asymptotically stability of the overall power system for all admissible uncertainties was guaranteed. A decentralized LFC based on H_∞ optimal control combined with an observer has been appeared in [35]. H_∞ control was proven to show grater effectiveness of damping load disturbance over the conventional optimal control by design of control system aimed at restricting the H_∞ norm of its transfer function. The effectiveness of the proposed method was investigated on a three-area and six-area networks involving both thermal and hydraulic power plants under load fluctuations. Later, in [36] a systematic method for robust sequential decentralized load frequency controllers using μ-synthesis in a multi-area power system has been proposed. At each design step, the information about the controllers design in the previous step is taken into account in the synthesis procedure. The investigations were carried out on system dynamic performance in the presence of parametric uncertainties and step load disturbances.

It should be noted that in robust control design approaches it is possible to use the physical understanding of power systems and considering system uncertainties for synthesis procedure. But large model order, uncertain connection between subsystems, broad parameter variations and elaborate organizational structure of power systems preclude direct the application of standard robust control methodologies.

D. Ai-Based Approaches

In practice, many nonlinear plants such as power system are approximated by reduced order models, possibly linear, that clearly related by underlying plant characteristics. However, these models are only valid with in certain specific operating ranges and a different model may be required in the wake of changed operating conditions, or the control system should adopt the new system model parameters. On the other hand, due to complexity and multivariable conditions of the power system, classical and nonflexible LFC schemes do not represent good enough solutions. Thus, for the purpose of evaluating the performance of such system, a flexible method be developed. In recent years, the advent of modern intelligent methods such as ANNs, fuzzy logic, PSO and GAs has solved the above mentioned problems to great extent.

ANN Techniques

Human ability for control of complex plants has encouraged researchers to pattern on the human neural network systems. ANN technique with their massive parallelism and ability to learn any type of nonlinearities are now being used in the area of nonlinear control problems, especially when the system is operating over the nonlinear range. Reviews on the applications of ANN for solution of the LFC problem are reported in [5]. Fransione et al. [37] proposed multi-layer nonlinear networks to system control which was trained by back propagation-through-time algorithm. The proposed ANN controller in comparison with the classical PI controllers on a single area and two-area power system has achieved good performance. A new LFC scheme to incorporate the nonconforming load problem was presented by Douglas et al.[38], in which an effort had been undertaken to develop algorithms capable of discriminating between non-controllable short-term excursions and controllable long-term excursions. Out of the two techniques described, one was developed using a neural network algorithm for pattern recognition of controllable signals, and the other technique was based on the detection of the controllable signal in the presence of a noisy random load using a random signal probability model. Test results reveal that the neural network-based AGC implementation had a significant improvement over the modern AGC implementation. LFC system performance was evaluated with a nonlinear neural network controller using a generalized neural structure to yield better system dynamic performance than the individual neurons [39]. A four-area interconnected power system model with reheat nonlinearity effect of the steam turbine and upper and lower constraints for generation rate nonlinearity of hydro turbine was considered for the investigation in [40]. It has been shown in [41] that the AGC problem can be viewed as a stochastic multistage decision-making problem or a Markov Chain control problem and have presented algorithms for designing AGC based on a reinforcement learning approach.

Recently, application of ANN technique based on robust control methodologies for solution of the LFC problem in interconnected power system have appeared in the literatures [15; 42). Shayeghi and Shayanfar [42] have used the idea of H_∞ robust control technique for training of the RBF neural networks for improvement of the performance of the proposed controller under various operating conditions. In [43], the idea of μ-synthesis control techniques has been used for training ANN based LFC controller, too. These approaches combined the advantage of neural networks and robust control technique to achieve the desired level of robust performance under large parametric uncertainness and leaded to a flexible controller with relatively simple structure.

Fuzzy Based Controller

Nowadays, fuzzy logic is used in almost all sectors of the industry and science. One of them is power system control. On the other hand, their robustness and reliability make fuzzy controllers useful for solving a wide range of control problems in power systems. Unlike traditional control theory, which is essentially based on mathematical models of the controlled plants, the fuzzy control methodology tries to establish the controller directly from domain experts or operators who are controlling the plant manually and successfully. Many studies to consider various power system aspects have been reported for design of a fuzzy logic based LFC regulator in the literature [5]. Researches on LFC problem show that the fuzzy PI controller is simpler and more applicable to remove the steady error. The fuzzy PI controller is known to give poor performance in the system transient response. In view of this, Yesil, Guzelkaya and Eksin [44], have proposed fuzzy PID methods to improve the performance of LFC problem. It should be pointed out that they require a three-dimensional rule base. This problem makes the design process more difficult. The problem of decomposition of multivariable systems for the purpose of distributed fuzzy control design was proposed in [45]. This method has reduced the number of interactive fuzzy relations among subsystems. The combined intelligent technique using ANN and fuzzy logic theory to utilize the novel aspects of both designs to a single hybrid LFC system has also been presented [46].

These days, GA is the most popular and widely used algorithm of all the intelligent algorithms. GAs are search algorithms based on the mechanism of natural selection and natural genetics that operate without knowledge of the task domain and utilize only the fitness of evaluated individuals. They can be considered as a general purpose optimization method and have been widely used to solve many complex nonlinear engineering optimization problems in general and especially for the solution of LFC problems [47; 48; 49; 50; 51; 52]. In [49], optimum adjustment gain of the integral controller using GAs through performance indices ISE and ITAE are investigated. A reinforced GA has been proposed as a suitable method to automate tune of the membership function and rule sets of fuzzy gain scheduling load frequency controllers to improve the dynamic performance of multi-area power systems in the presence of system nonlinearities, such as GRC and GDB. Later, contrary to the trial-error selection of the variable structure based LFC feedback gains, a GA based method has been used for finding optimal feedback gains [48]. The test results show that not only the dynamical performance has been improved, but also the control effort is dramatically reduced. Aditya et al. [47] have presented a comprehensive study on LFC problem of an interconnected hydro power system using combined advent of ANN, fuzzy logic and GA techniques to achieve the desired level of robust performance. A higher order robust dynamic performance is achieved for LFC designs based on GA and LMIs [52]. The PI control parameters have been obtained by coordinating GA with linear matrix inequalities control toolbox for optimization of H_∞ norm based performance index. In [50], a new GA/GA-SA based fuzzy LFC scheme of a multi-area thermal generating system is developed. A function like FD has been used as the fitness function for evaluating the fitness of GA/hybrid GA-SA optimization. This function directly depends on transient performance characteristics like overshoots, under shoots, settling times and time derivation of the frequency. The hybrid GA-SA technique yields more optimal gain values than the GA method. Later, for optimization of PID gains in designing Sugeno fuzzy logic based LFC scheme, particle swarm optimization technique has been reported [51]. PSO as one of the modern heuristic algorithms is a population based evolutionary algorithm, which is motivated by the simulation of social behavior instead of survival of the fittest. The proposed PSO algorithm establish the true optimality of transient performance similar to those obtained by GA-SA based optimization technique, but it is very faster than GA-SA algorithm. The Sugeno fuzzy logic technique is very simple to implement and very fast acting than the classical fuzzy logic system. Hossieni and Etemadi [53] have presented an ANFIS based AGC, which is trained by the results of off-line studies obtained using PSO to update integral controller gains in real time according to load variations. Also, a frequency relaxation method is separately implemented using ANFIS which deactivates both the primary and secondary frequency controls whenever the magnitude of estimated frequency deviations is smaller than a threshold value. The compliance of the proposed approach with NERC recent control performance standard rules verified. Recently, various heuristic techniques such as PSO, bactria foraging and chaotic ant swarm optimizations have been applied for optimal tuning of PID gains used in Sugeno fuzzy logic based AGC scheme [54]. The objective function for off-nominal tuning of PID gains is FD. For varying off-nominal online system parameters, fast acting Sugeno fuzzy logic manipulates the nominal gains adaptively to determine transient responses. The simulation results on a ring connected three-area reheat thermal power system shows that all optimization techniques are more on less equally very effective in yielding system transient responses, But chaotic ant swarm optimization and bactria forging optimization yield much more global true optimal results. Also, PSO takes the least time.

Digital Approaches

A comprehensive review on digital based LFC scheme has been given in [6]. As the ACE representing generation mismatch in an area can be derived in discrete mode by sampling the tie-line power flow and system frequency deviations and then transferring over the telemetring links and unlinks in continuous-time system, the control vector in the discrete mode is constrained to remain constant between the sampling instants. Hari, Kothari and Nanda [55] have investigated more realistic discrete model LFC of an interconnected power system with reheat thermal plants considering a new ACE. The new ACE is derived from tie-line power deviation, frequency deviation, time error and inadvertent interchanges. Optimum gains of the integral and PI controllers using the concepts of stability margin and the ISE index based on the conventional and new ACEs, have been obtained and their dynamic performance was compared for a step load disturbance.

LFC Scheme with DC Links

The high voltage direct current transmission has emerged on a power scenario, due to its numerous technical and economic advantages, for a large chunk of power transfer over large distance. Besides, other application, the commissioning of an HDVC link in parallel with existing ac links has shown beneficial effects from the stabilization point of view of the system. Considerable research works have been appeared to investigate the damping effects of the dc link on frequency control of the interconnected power systems. Sanpei, Kakehi and Takeda [56] have developed a new dc control system based on multi-variable control for frequency control of the two ac systems optimally with maintaining their stability. A decentralized robust LFC based on Ricatte equation for a multi-area interconnected power system with AC as well as frequency controllable HVDC links, has been proposed in [57]. Kumer and Ibrahiem [58; 59] have carried out a comprehensive study for optimal load frequency controller design of a two area power systems with parallel ac/dc links. Optimal controller designs by considering the incremental dc link power flow as an additional state, as well as, control variable were analyzed. The results reveal that the system transient performance has significantly improved, considering the incremental dc link power flow as an additional state variable in comparison to that obtained when system interconnection is through the ac link only.

LFC Considering Energy Storage Devices, SSSC and DG Units

It should be noted that, most of the proposed control strategies so far for solution of the LFC problem have not been implemented due to system operational constrains associated with thermal power plants. The main reason is the non-availability of required power other than the stored energy in the generator rotors, which can improve the performance of the system, in the wake of sudden increased in load demands. Also, due to persist of the system frequency and tie-line deviations for a long duration along in the case of small load disturbances in the presence of suitable supplementary controllers, the governor system many no longer be able to absorb the frequency fluctuations due to its slow response. On the other hand, electromechanical oscillations in a power system can be effectively damped by fast acting energy storage devices, because additional energy storage capacity is provided as a supplement to the kinetic energy storage in the moving mass of the generator rotor. The energy storage devices share the sudden changes in power requirement in the load. Thus, in a power system, the instantaneous mismatch between supply and demand of real power for the sudden load changes can be reduced by the addition of active power sources with fast response such as BES, SMES and CES devices. Some researchers [60; 61] have investigated LFC performance improvement of the power systems including BES unit. A fast-acting BES can effectively dampen electro-mechanical oscillations in a power system, because they provide storage capacity in addition to the kinetic energy of the generator rotor, which can share the sudden changes in the power requirement. In [61], it has been revealed that BES is helpful in meeting sudden requirements of real power load and system transient performance is significantly improved. Also, it is effective in reducing the peak deviations of frequency and tie-line power flows. The effects of BES on LFC have been studied considering GDB and GRC. The salient feature of SMES units such as low discharge rate, fast acting, increased time required for power flow reversal and the maintenance requirement have let to the application of it as load frequency stabilizers [62]. The performance of the adaptive controlled SMES is compared with that of non-adaptive SMES, keeping the supplementary control as the conventional one with the integral controller in [63]. It has been observed that when the SMES control is adaptive, the performance is almost insensitive to control gain parameter variation. In [63], the effects of SMES on LFC problem have been studied considering GDB nonlinearity, steam reheat constraints and boiler dynamics. Parameter optimization of the controller was carried out by the second method of Lyapunov, which ensures stability of the system. The results reveal that the use of ACE for the control of SMES units considerably reduces the tie-line power deviation, and the action of SMES is localized with diminished contribution for load disturbances in the other area as compared to using frequency deviations as the control signal. Later, the feasibility of using an IGBT converter instead of the thyristor converter as a power with SMES has been achieved [64]. Demiroren [65; 66] has investigated the effects of SMES unit on LFC problem using ANN technique and fuzzy logic based supplementary controllers. The results reveal that including SMES in each control area is effective in reducing the peak deviations of frequency and tie-line power flow and improvement of system transient response. Recently, in [67] LFC of an interconnected hydrothermal power system including SMES has been investigated. The optimum value of integral controller gain in the control areas in the presence of different GRCs considering an SMES unit either thermal or hydro area, as well as, both the areas through the minimizing ISE performance index based on the frequency and tie-line power flow deviations was obtained. The results reveal that considering SMES in both the area and SMES in either of them improves the system dynamic performance significantly. However, from the point of economy, including SMES in either of the areas may be preferred. Recently, Ganapathy and Velusami [68] have proposed a multi objective algorithm based PI controller design for LFC problem including ac-dc parallel tie-lines and SMES units. The HVDC links in parallel with AC tie-lines is used to effectively damp the frequency oscillation of ac system while the SMES unit provides bulk energy storage and release, there by achieving combined benefits. Applying of the designed controller on a two-area power system reveal that it is effective in suppressing the frequency and tie-line power oscillation and achieves good robust performance against system parameter changes in the presence of GRC and GDB. In [69] a fuzzy controller considered to control SMES unit for stabilizing the frequency oscillation. To judge the ability of the control strategy, different level of load disturbance is considered in each area. The superiority of the fuzzy based SMES frequency stabilizer than the conventional PI controller using simulation results on a two-area power system was indicated.

The effects of CES units on LFC problem considering GDB nonlinearity and boiler dynamics has been studied by Tripathy [70]. It has been verified that CES provides the same kind of dynamic performance improvement of power system like SMES, but compared to SMES it is less expensive, has dielectric losses and its energy density is low. The soild-state phase shifter systems or SSSC have also been found to aid in the damping of oscillations of power system dynamic response. The new techniques of LFC regulator design, based on SSSC, have been proposed in [71; 72]. Due to high-speed performance of SSSC to the governor system, the SSSC compensates the peak value of the transient frequency deviation for a sudden load perturbation. The dynamics of governor system was eliminated in the control design process for phase shifter. In [71], a feedback signal composed of area frequency and tie-line deviations has been used to stabilize power system. The coefficient of these deviations is obtained using pole placement of reduced order system based on the aggregation of inertia center mode. In [72], parameters of lead-lag controllers considered for frequency stabilizer of SSSC are automatically optimized by a micro genetics algorithm using evaluation of the fitness function based on a linear combination of damping ratio of the inter-area mode and multiplicative stability margin for taking system uncertainties into account. The results have been confided the high robustness of this strategy against various load disturbances with changing frequency in the vicinity of inter-area mode.

Currently, small wind turbines are among the candidate systems envisioned to operate in parallel with the utilitie's generators. The combined effect of the customer's load demand and the wind turbine fluctuating power output will develop a new load-diversity curve for the utility system. As a result, the regulation or LFC requirements could differ significantly from present ones.

An AGC scheme for a wind farm in the north of Spain with 37 variable speed wind turbines was developed in [73]. The control scheme is based on two control levels: a supervisory system controls active and reactive power of the wind farm by sending out set points to all wind turbines, and a machine control system ensures that set points at the wind turbine level are reached. The advancements in wind-turbine and micro-hydro technology have made it possible to introduce decentralized hybrid electric power systems. A variable structure LFC of isolated wind-diesel micro-hydro hybrid power systems was developed and dynamic performance has been investigated by Bhatti and Kothari [74]. The static VAR compensation systems have also been found to aid in the damping of oscillations of power system dynamic responses. A new technique of AGC regulator design, based on SVC, has been suggested in [75]. A feedback signal composed of frequency deviation and reactive power variation has been used to stabilize the electrical power system. The coefficient of these deviations is a function of system and controller parameters. The influence of a PV system on LFC has also been described in [76]. Besides the other observations, it has been revealed that an electrical power system containing a 10% contribution from PV stations would require a 2.5% increase in LFC capacity over a conventional system. Moreover, BES,

SMES and CES units, a favorable effect of containing a 10% fuel cell into the power system dynamic model on power system dynamic performance has also been evident [77].

Effects of fuel cells on load frequency control in power distribution system have been investigated by Sedghisigarchi and Feliachi [78]. The design of two control loops was presented for the fuel cell to maintain an energy balance in distribution system without requiring additional power from the substation. The primary control loop is a proportional type whose fined gain was designed for fuel cell stack protection in its nominal operating condition. The secondary loop controller was designed using GA methods to regulate the distribution system error and therefore improve the frequency oscillation of the distribution system. Advanced of this control strategy considering a mix of gas turbines and fuel cells has been illustrated and was shown that fuel cells are able to follow the load for short time intervals of load demand changes.

The widespread use of various kinds of DG sources would impact the quality of the power supply with in a micro-grid power system, causing many control problems. In [79] an electrolyzer system's dynamic control method bases on fuzzy PI controller have been proposed to improve frequency fluctuation in multiple renewable energy hybrid micro-grid power systems. The motivation for using a combination of a micro-turbine, fuel cell and electrolyzer system to deal with sudden real power imbalances is that it can be offer flexible controllability functions to compensate for system's real power imbalances of micro-grid due to fast control of its power consumption by adjusting pressure in the customer piping systems. The effectiveness of the proposal fuzzy PI control scheme under the islanding and interconnected operational modes was evaluated. It was correctly shown that a power quality improvement for frequency fluctuations can be achieved by actively utilizing the load-controllable electrolyzer systems.

LFC in a deregulated environment

Nowadays, the electric power industry is in transition to a competitive energy market. In the new structure, GENCOs may not participate in the LFC task and DISCOs have the library to control with any available GENCOs in their own or other areas. Global analysis of the power system markets show that the LFC is one of the most profitable ancillary service at these systems and from the mechanism used to manage the provision of this service in ancillary markets, the bilateral contracts or competitive offers stand out [80]. Besides addressing the operational structures likely to results from deregulation, the possible approaches to LFC and associated technical issues; i.e. standards and algorithms, were described by Christen and Bose [10]. Two alternative methods to LFC of interconnected power systems of Norway and Sweden are introduced by Bakken and Grande [81]. In some of the reported strategies, attempts have been made to adapt well-tested classical LFC schemes to the changing environment of the power system operation under deregulation [83; 84]. In the recent years, several control scenarios based on linear control, robust and combined intelligent approaches have been proposed for the LFC system in the deregulated power systems.

Linear Control Technique

To obtain optimal parameters of integral controller for tuning optimal gains of the integral controller in each area, the trajectory sensitivities using a gradient Newton algorithm

[11] have been presented in deregulated environment taken of bilateral contracts into account on AGC dynamics. A two-area restructured power system was used to illustrate frequency and tie-line power flow oscillation improvement. A framework of market based optimal LFC considering the economy and security of power systems operation has been appeared in [57]. To achieve the desired value of robust system performance with the lower cost, Gausi-Newton algorithm was applied to optimize the regulation of LFC problem. A LQG based load frequency controller was designed in [84] with considering Poolco and bilateral transactions. The controller is rescheduled the GENCOS outputs according to their bids for the frequency regulation.

It should be noted that the real world power system contains different kinds of uncertainties and disturbances and coming deregulation significantly increase the severity of this problem. Under this condition, the classical control based method is certainly not suitable for LFC problem.

Robust Control Methods

As in the liberalized power markets, LFC systems are subjected to new uncertainties. Considering of these uncertainties in synthesis procedure is very important to design suitable control to provide better conditions foe electricity trading. To cope with uncertainties, a robust decentralized controller based on H_∞ control theory and mixed H_2/H_∞ control technique have been reported in [85; 86; 87] for the solution of LFC problem in a deregulated power system. To achieve decentralization, in each control area the effects of contracted signals and interfaces between control areas are treated as a set of new disturbance signal. The proposed control strategy is formulated as a general LMIs optimization problem and solved by LMI control toolbox provided by MATLAB software. The results evaluation reveals that these approaches achieve good dynamical performance, however, stability of the overall system with local controller were not discussed. Also, the order of resulting controllers is high in general. One development after deregulation of the power industry is the necessity of a communication infrastructure to support an increasing variety of ancillary services for effective implementation of LFC schemes. An article focusing on the communication network requirements for a third-party LFC services in an interconnected power system were written by Bhowmik and Tomsovic [88]. Data communication models based on queuing theory have been proposed in the study. The design of decentralized robust LFC applying H_∞ norm and structured singular value was proposed by Shayeghi and Shayanfar [87; 89]. It has shown that, when the frequency response-based diagonal dominance can not be achieved, subject to a condition based on H_∞ norm and structured singular value, each local control area can be designed independently; such that the stability of the overall closed loop system is guaranteed. The developed design strategy combines the advantage of H_2 and H_∞ control synthesis and gives a powerful multi-objectives design addressed by LMI techniques by considering the effect of plant parameter changes, system nonlinearities and disturbances [87]. A robust LFC based on mixed H_2/H_∞ control technique including SMES in the competitive electricity environment has been developed in [90]. It was shown that with use of a small capacity SMES in some area of four-area power system the dynamic performance of system such as frequency regulation and disturbances attenuation is significantly improved for a wide range of plant parameter changes.

It may be noted that the main capability of robust control methodologies based LFC schemes is in possibility of controller designed based on a more complete model of system which consider uncertainties, too. This fact is of the great importance knowing that the power system has a variable structure and subject to type of uncertainties and disturbances. However, due to complexity of actual uncertainties and large model order of the real world power system these approaches yield complex controllers whose size will be very large in general.

Computational Intelligent Based Control Scheme

According to complexity and multivariable condition of the restructured power system, nonflexible and high order robust controller do not achieve good robust performance due to geographical distribution and a large number of real system elements which makes the knowledge of all system parameters from view point of technical and economical is difficult. Recently, in order to overcome these drawbacks and keeping advantage of robust control technique, a robust decentralized neural network based LFC have appeared in the literatures [15; 89] in a deregulated power system. The motivation for using robust control strategy for training ANN technique based controller was to take large modeling uncertainties into account, cover physical constraints on control action and minimize the effects of area load disturbances. The developed approaches was combined the advantage of the neural networks and robust control technique and leaded to a flexible controller to provide robust performance. In [91] GA has been applied for optimal tuning of integral gains and bias factor for solution of the AGC problem in a three-area power system. The simulation results show that the ACE based controller parameters optimized by GA are satisfactory for three different operating cases in comparison with the case used the nominal parameter value. Fuzzy logic based AGC of TCPS based hydro thermal system under market scenario has been investigated in [92]. It was shown that with applying proposed strategy in a two-area power system, the system frequency and tie-line power oscillations is significantly improved by controlling the phase angle of TCPS which is expected to provide a new ancillary service for the future power system.

A multi-stage fuzzy PID type controller with a fuzzy switch has been proposed to solve AGC problem in a deregulator power system using the generalized LFC dynamic model [93]. This approach is a form of behavior based control where the PD controller became active only when certain conditions are met. The resulting structure is a controller using two dimensional inference engines (rule base) to perform reasonably the task of a three dimensional controller. The proposed method requires fewer sources to operator and its role in the system response is more apparent. In a fuzzy control system, optimization of membership functions plays a key role, because it is a computationally expensive combinatorial optimization problem. To reduce fuzzy system effort and cost saving, a GA-based MSF PID controller has reported by Shayeghi, Jalili and Shayanfar [21] for the solution of LFC problem. The GA works off-line and in order to improve the algorithm convergence speed and guarantee optimum value a modified GA based on the hill claiming method has been used. The salient feature of the proposed method is that it does not require an accurate model of the LFC problem and design process is less demanding compared to the other fuzzy PID controllers. The evaluation of system performance characterizes in terms of ITAE and

FD indices are revealed that it can be promising control scheme for the solution of LFC problem and ideal for the real world power systems. It should be noted that, GA seems to be good approach to solve optimization problems. When applied to problems consisting of more number of local optima, the solution from GA are just near global optimum areas. Also, it takes long simulation time to obtain the solution. Moreover, when the number of parameter is more, optimization problem is complex and coding chromosomes with more genes for increasing algorithm accuracy is caused GA convergent speed will become very slow, so that convergent accuracy may be influenced by the slow convergent speed. To overcome these drawbacks, a PSO based MSF has been presented [94]. The motivation for using the PSO technique is to reduce fuzzy system effort and take large parametric uncertainties into account. This newly developed control strategy combines the advantage of PSO and fuzzy system control techniques and leads to a flexible controller with simple stricture that is easy to implement. Simulation results on a three-area power system through ITAE and FD indices in compression with GA based MSF reveal that the proposed control strategy achieves good robust performance for wide rang of operating conditions. Recently, a PSO based fuzzy PID type controller with considering SMES was proposed for solution of the LFC problem by Shayeghi and Shayanfar [12]. Thus, to demonstrate technical and economic feasibility of them in the deregulated power system applications, a MSF controller is designed including SMES unit in some area of the power systems by PSO. It was shown that addition of the SMES units to the system significantly improves transients of frequency and tie-line power deviations against load disturbances.

CONCLUSIONS

Load frequency control is one of the important issues in power systems operation and control for supplying sufficient and reliable electric power with good quality. Especially, in the deregulated electricity market, it will serve as ancillary service and acquires a principal role to enable power exchanges and to provide better condition for electricity trading. LFC goals i.e. frequency regulation and tracking load demands, maintaining the tie-line power interchanges to specified values in the presence of modeling uncertainties, system nonlinearities, complexity and multi variable condition of power system determines the LFC synthesis as a multi-objective optimization control problem. This chapter is focused on the recent research in the area of LFC and intends to be a useful reference and search tool as well as a critical account of the up-to-date use of AI technologies in LFC problem. Emphasis has been given to categorizing various LFC strategies reported in the literature and their salient features and disadvantages.

Among discussed categories of LFC strategies, robust control and AI based methods have been shown an ability to give better performance in dealing with the system nonlinearities, modeling uncertainties and area load disturbances under different operating conditions. The main capability of robust control approaches is impossibility of controller design based on a more complete model of system which considers uncertainties and physical constraints, too. The salient feature of AI technique is that it provides a model-free description of control system and do not require an accurate model of the plant. In conclusion, we can say that, the robust and AI techniques, like all other control techniques, have relative advantages and disadvantages. There are no rules as to when a particular technique is more suitable for LFC problem. It is envisaged that this chapter will serve as a valuable resource to any further worker in this important area of research.

REFERENCES

[1] Kothari D. P. and Nagrath I. J. (2003). Modern power system analysis, 3rd ed., Singapore: McGraw-Hill.

[2] Saadat H. (1999). Power system analysis, Mc-Graw Hill, USA.

[3] Kumar J.; Hoe N. G. and Sheble G. (1997). AGC simulator for price-based operation Part I: Modeling, *IEEE Transaction on Power Systems*, 12 (2): 527 - 532.

[4] Raineri R.; Rios S. and Schiele D. (2006). Technical and economic aspects of ancillary services markets in the electric power industry: an international comparison, *Energy Policy*, 34 (13): 1540-1555.

[5] Shayeghi H.; Shayanfar H. A. and Jalili A. (2008). Load frequency control strategies: A state- of-the-art survey for the researcher, *Energy Conversion and Management*, 49: 344-353.

[6] Ibraheem; Kumar P. and Kothari D. P. (2005). Recent philosophies of automatic generation control strategies in power systems, *IEEE Transaction on Power Systems*, 20 (1): 346-357.

[7] Kundur P. (1994). Power system stability and control, New York: McGraw-Hill.

[8] Wood A. and Wollenberg B. (1996). Power generation, control and operation, 2nd ed., John Wiley and Sons.

[9] Shayeghi H. (2006). A robust control based intelligent LFC in the restructured power systems, Ph.D. Thesis, Iran University of Science and Technology.

[10] Christie R. D. and Bose A. (1996). Load frequency control issues in power system operations after deregulation, *IEEE Transaction on Power Systems*, 11 (3): 1191-1200.

[11] Donde V.; Pai A. and Hiskens I. A. (2001). Simulation and optimization in a LFC system after deregulation, *IEEE Transaction on Power Systems*, 16 No. 3, 481-489.

[12] Shayeghi H. and Shayanfar H. A. (2010). Application of PSO for Fuzzy Load Frequency Design with Considering Superconducting Magnetic Energy Storage, *International Journal of Technical and Physical Problems of Engineering*, 1 (1): 24-33.

[13] Tripathy S. C.; Balasubramanian R. and Nair P. S. C. (1992b). Effect of superconducting magnetic energy storage on automatic generation control considering governor dead-band and boiler dynamics, *IEEE Transaction on Power Systems*, 7 (3): 1266-1273.

[14] Shayeghi H. (2008). A Robust Decentralized Power System Load Frequency Control, *Journal of Electrical Engineering*, 59 (6): 281-293.

[15] Shayanfar H. A. and Shayeghi H. (2005). Decentralized load frequency control of a deregulated electric power system using ANN technique, *WSEAS Transaction on Circuits and systems*, 4 (1): 38-47.

[16] Yang T. C.; Ding Z. T. and Yu H. (2002). Decentralized power system load frequency control beyond the limit of diagonal dominance, *Electrical Power and Energy Systems*, 24 (3): 173-184.

[17] Shayeghi H. and Shayanfar H. A. (2005b). Design of decentralized robust LFC in a competitive electricity environment, *Journal of Electrical Engineering*, 56 (5): 225-236.

[18] Shayeghi H. and Shayanfar H. A. (2006b). Decentralized robust AGC based on structured singular values, *Journal of Electrical Engineering*, 57: 258-263.

[19] Lim K. Y.; Wang Y.; Guo G. and Zhou R. (1998). A new decentralized robust controller design for multi-area load frequency control via in complete state feedback, *Optimal Control and Applied Mathematics*, 19: 345-361.

[20] Velusami S. and Chidambaram I. A. (2007). Decentralized biased dual mode controllers for load frequency control of interconnected power systems considering GDB and GRC nonlinearities, *Energy Conversion and Management*, 48: 1691-1702.

[21] Shayeghi H.; Jalili A. and Shayanfar H. A. (2007). Robust modified GA-based multi-stage fuzzy LFC, *Energy Conversion and Management*, 48 (5):1656-1670.

[22] Velusami S. and Ramar K. (1997). Design of observer-based decentralized load-frequency controllers for interconnected power systems, *International Journal of Power Energy Systems*, 17 (2): 152-160.

[23] Hain Y.; Kulessky R. and Nudelman G. (2000). Identification-based power unit model for load-frequency control purposes, *IEEE Transaction on Power Systems*, 15 (4): 1313-1321.

[24] Aldeen M. and Trinh H. (1992). Combined modal and singular perturbation approach to decentralized control, *International Journal of Systems Science*, 23 (5):741-764.

[25] Aldeen M. and Trinh H. (1994). Load frequency control of interconnected power systems via constrained feedback control schemes, *International Journal of Computer and Electrical Engineering*, 20 (1): 71-88.

[26] Khodabakhshian A. and Hooshmand R. (2010). A new PID controller design for automatic generation control of hydro power systems, *Electrical Power and Energy Systems*, 32 (5): 375-382.

[27] Talaq J. and Al-Basri F. (1999). Adaptive fuzzy gain scheduling for load-frequency control, *IEEE Transaction on Power Systems*, 14 (1): 145-150.

[28] Ha Q. and Trinh H. (2001). A variable-structure based controller with fuzzy tuning for load frequency control, *International Journal of Power Energy Systems*.

[29] Kazemi A. and Andami H. (2004). A decentralized fuzzy logic enhanced variable structure controller applied to load frequency control system, *Iranian Journal of Science and Technology, Transaction B*, 28 (B3): 295-303.

[30] Azzam M. (1999). Robust automatic generation control, *Energy Conversion and Management*, 40 (13): 1413-1421.

[31] Goshaidas R.; Prasad A. N. and Prasad G. D. (1999). A new approach to the design of load frequency controller for large scale power system, *Electric Power Systems Research*, 51: 13-22.

[32] Shayeghi H.; Karrari M. and Menhaj M. B. (2000). Theory of μ synthesis for power systems load frequency control, *Journal of Electrical Engineering*, 51: 258-263.

[33] Azzam M. and Mohamed Y. S. (2002). Robust controller design for automatic generation control based on Q-parameterization, *Energy Conversion and Management*, 43 (13): 1663-1673.

[34] Lim K. Y.; Wang Y. and Zhou R. (1996). Robust decentralized load frequency control of multi-area power systems, IEE Proceedings on Generation, *Transmission and Distribution*, 143 (5): 377- 386.

[35] Ishii T.; Shirai G. and Fujita G. (2001). Decentralized load frequency control based on H∞ control, Electrical Engineering in Japan, 136 (3): 28 -38.

[36] Bevrani H.; Mitani Y. and Tsuji K. (2004a). Sequential design of decentralized load frequency controllers using μ synthesis and analysis, *Energy Conversion and Management*, 45 (6): 865-881.

[37] Franoise B.; Magid Y. and Bernard W. (1994). Application of neural networks to load-frequency control in power systems, Neural Networks, 7 (1): 183-194.

[38] Douglas L. D.; Green T. A. and Kramer R. A. (1994). New approaches to the AGC nonconforming load problem, *IEEE Transaction on Power Systems*, 9 (2): 619-628.

[39] Chaturvedi D. K.; Satsangi P. S. and Kalra P. K. (1999). Load frequency control: A generalized neural network approach, *Electrical Power and Energy Systems*, 21 (6): 405-415.

[40] Zeynelgil H. L., Demiroren A. and Sengor N. S. (2002). The application of ANN technique to automatic generation control for multi-area power system, *Electrical Power and Energy Systems*, 24 (5): 345-354.

[41] Ahamed T. P. I.; Rao P. S. N. and Sastry P. S. (2002). A reinforcement learning approach to automatic generation control, *Electric Power Systems Research*, 63: 9-26.

[42] Shayeghi H. and Shayanfar H. A. (2003); Application of ANN technique for interconnected power system load frequency control, *International Journal of Engineering, Transaction* B, 16 (3): 247-254.

[43] Shayeghi H. and Shayanfar H. A. (2006a). Application of ANN technique based on μ-synthesis to load frequency control of interconnected power system, *Electrical Power and Energy Systems*, 28: 503-511.

[44] Yesil E.; Guzelkaya M. and Eksin I. (2004). Self tuning fuzzy PID type load and frequency controller, *Energy Conversion and Management*, 45: 377-390.

[45] Gegov A. E. and Frank P. M. (1995). Decomposition of multivariable systems for distributed fuzzy control [power system load frequency control], *Fuzzy Sets Systems*, 73 (3): 329-340.

[46] Karnavas Y. L. and Papadopoulos D. P. (2002). AGC for autonomous power system using combined intelligent techniques, *Electric Power Systems Research*, 62 (3): 225-239.

[47] Aditya S. K. and Das D. (2003). Design of load frequency controllers using genetic algorithm for two area interconnected hydro power system, *Electrical Power Components and Systems*, 31 (1): 81-94.

[48] Al-Hamouz Z. M. and Al-Duwaish H. N. (2000). A new load frequency variable structure controller using genetic algorithms, *Electric Power Systems Research*, 55 (1):1-6.

[49] Chang C. S.; Fu W. and Wen F. (1998). Load frequency control using genetic-algorithm based fuzzy gain scheduling of PI controllers, *Electric Machines and Power Systems*, 26 (1): 39-52.

[50] Ghoshal S. P. (2004a). Application of GA/GA-SA based fuzzy automatic generation control of a multi-area thermal generating system, *Electric Power Systems Research*, 70 (2): 115-127.

[51] Ghoshal S. P. (2004b). Optimizations of PID gains by particle swarm optimizations in fuzzy based automatic generation control, *Electric Power Systems Research*, 70 (3): 203-212.

[52] Rerkpreedapong D.; Hasanovic A. and Feliachi A. (2003). Robust load frequency control using genetic algorithms and linear matrix inequalities, *IEEE Transaction on Power Systems*, 18 (2): 855-861.

[53] Hosseini S. H. and Etemadi A. H. (2008). Adaptive neuro-fuzzy inference system based automatic generation control, *Electric Power Systems Research*, 78: 1230-1239.

[54] Roy R.; Bhatt P. and Ghoshal S.P. (2010). Evolutionary computation based three-area automatic generation control, Expert Systems with Applications.

[55] Hari L.; Kothari M. L. and Nanda (1991). Optimum selection of speed regulation parameters for automatic generation control in discrete mode considering generation rate constraints, *IEE Proceedings on Generation, Transmission and Distribution*, 138 (5): 401-406.

[56] Sanpei M.; Kakehi A. and Takeda H. (1994). Application of multi-variable control for automatic frequency controller of HVDC transmission system, *IEEE Transaction on Power Delivery*, 9 (2): 1063-1068.

[57] Liu F.; Song Y.H.; Ma J.; Mei S. and Lu Q. (2003). Optimal load-frequency control in restructured power systems, IEE Proceedings on Generation, *Transmission and Distribution*, 150(1): 87-95.

[58] Ibraheem and Kumar P. (2003). Dynamic performance enhancement of hydropower systems with asynchronous tie-lines, *Journal of Electrical Power Components and Systems*, 31 (7): 605-626.

[59] Ibraheem and Kumar P. (2004). A novel approach to the matrix Riccati equation solution: An application to optimal control of interconnected power systems, *Journal of Electrical Power Components and Systems*, 32 (1): 33-52.

[60] Aditya S. K. and Das D. (2001). Battery energy storage for load frequency control of an interconnected power system, *Electric Power Systems Research*, 58 (3): 179-185.

[61] Lu C. F. and Liu C. C. (1995). Effect of battery energy storage system on load frequency control considering governor dead-band and generation rate constraint, *IEEE Transaction on Energy Conversion*, 10 (3): 555-561.

[62] Banerjee S.; Chatterjee J. K. and Tripathy S. C. (1990). Application of magnetic energy storage unit as load frequency stabilizer, *IEEE Transaction on Energy Conversion*, 5 (1): 46-51.

[63] Tripathy S. C.; Balasubramanian R. and Nair P. S. C. (1992a). Adaptive automatic generation control with superconducting magnetic energy storage in power systems, *IEEE Transaction on Energy Conversion*, 7 (3): 434-441.

[64] Tripathy S. C. and Juengst K. P. (1997). Sampled data automatic generation control with superconducting magnetic energy storage in power systems, *IEEE Transaction on Energy Conversion,* 12 (2): 187-192.

[65] Demiroren A. (2004). Automatic generation control using ANN technique for multi-area power system with SMES units, *Electrical Power Components Systems*, 32 (2): 193-213.

[66] Demiroren A. and Yesil E. (2004). Automatic generation control with fuzzy logic controllers in the power system including SMES units, *Electrical Power and Energy Systems*, 26 (4): 291-305.

[67] Abraham R. J.; Das D. and Patra A. (2007). Automatic generation control of an interconnected hydrothermal power system considering superconducting magnetic energy storage, Electrical Power and Energy Systems, 29: 271-579.

[68] Ganapathy S. and Velusami S. (2010). MOEA based design of decentralized controllers for LFC of interconnected power systems with nonlinearities, AC-DC parallel tie-lines and SMES units, *Energy Conversion and Management*, 51: 873-880.

[69] Hemeida A. M. (2010). A fuzzy logic controlled superconducting magnetic energy storage, SMES frequency stabilizer, *Electric Power Systems Research*, 80: 651-656.

[70] Tripathy S. C. (1997). Improved load frequency control with capacitive energy storage, *Energy Conversion and Management*, 38 (6): 551-562.

[71] Nagmroo I.; Mitani Y. and Tsui K. (2000). Application of solide-state phase shifter to stabilization of system frequency in an interconnected power system, *Electrical Engineering in Japan*, 132 (2): 19-28.

[72] Ngamroo I.; Tippayachai J. and Dechanupaprittha S. (2006). Robust decentralized frequency stabilizers design of static synchronous series compensators by taking system uncertainties into consideration, *Electrical Power and Energy Systems*, 28 (8): 513-524.

[73] Amenedo J. L.; Arnalte S. and Burgos J. C. (2002). Automatic generation control of a wind farm with variable speed wind turbines, *IEEE Transaction on Energy Conversion*, 17 (2): 279-284.

[74] Bhatti T. S. and Kothari D. P. (2002). Variable structure load-frequency control of isolated wind-diesel-microhydro hybrid power systems, *Journal of Inst. Engineering*, 83: 52-56.

[75] El-Emary A. A. and El-Shibina M. A. (1997). Application of static VAR compensation for load frequency control, *Electric Machines and Power Systems*, 25 (9): 1009-1022.

[76] Asano H.; Yajima K. and Kaya Y. (1996). Influence of photovoltaic power generation on required capacity for load frequency control, *IEEE Transaction on Energy Conversion*, 11 (1): 188-193.

[77] Paradkar A.; Davari A.; Feliachi A. and Biswas T. (2004). Integration of a fuel cell into the power system using an optimal controller based on disturbance accommodation control theory, *Journal of Power Sources*, 128 (2): 218-230.

[78] Sedghisigarchi K. and Feliachi A. (2006). Impact of Fuel Cells on Load-Frequency Control in Power Distribution Systems, *IEEE Transaction on Energy Conversion*, 21(1).

[79] Li X., Song Y. J. and Han S.-H. (2008). Frequency control in micro-grid power system combined with electrolyzer system and fuzzy PI controller, *Journal of Power Sources*, 180: 468-475.

[80] Shayeghi H., Shayanfar H. A. and Malik O.P. (2007). Robust decentralized neural networks based LFC in a deregulated power system, *Electric Power System Research*, 77: 241-251.

[81] Bakken B. H. and Grande O. S. (1998). Automatic generation control in a deregulated power system, *IEEE Transaction on Power Systems*, 13 (4): 1401-1406.

[82] Kumar J.; Ng K. H. and Sheble G. (1997). AGC simulator for price-based operation-part II: Case study results, *IEEE Transaction on Power Systems*, 12 (2): 533-538.

[83] Delfino B.; Fornari F. and Massucco S. (2002). Load-frequency control and inadvertent interchange evaluation in restructured power systems, *IEE Proceedings on Generation, Transmission and Distribution*, 149 (5): 607-614.

[84] Tyagi B. and Srivastava S.C. (2005). A LQG based load frequency controller in a competitive electricity environment, *International Journal of Emerging Electrical Power Systems*, 2.

[85] Bevrani H.; Mitani Y. and Tsuji K. (2004b). Robust decentralized LFC in a restructured power system, *Energy Conversion and Management*, 45: 2297-2312.

[86] Bevrani H.; Mitani Y.; K. Tsuji and Bevrani H. (2005). Bilateral based robust load frequency control, *Energy Conversion and Management*, 46: 1129-1146.

[87] Shayeghi H. and Shayanfar H. A. (2005a). Decentralized robust load frequency control using linear matrix inequalities in a deregulated multi-area power system, *Journal of Power Engineering Problems*, 2: 30-38.

[88] Bhowmik S.; Tomsovic K. and Bose A. (2004). Communication models for third party load frequency control, *IEEE Transaction on Power Systems*, 19 (1): 543-548.

[89] Shayeghi H. and Shayanfar H. A. (2006c). Robust decentralized LFC design in a restructured power system, *International Journal of Emerging Electrical Power Systems*, 6 (2), Article 4.

[90] Shayeghi H.; Jalili A. and Shayanfar H. A. (2008a). A Robust Mixed H2/H∞ Based LFC of a Deregulated Power System Including SMES, *Energy Conversion and Management*, 49: 2656-2668.

[91] Demiroren A. and Zeynelgil H. L. (2007). GA application to optimization of AGC in three- area power system after deregulation, *Electrical Power and Energy Systems*, 29: 230-240.

[92] Rao C. S.; Nagaraju S.S.; Raju S. N. and Sangameswara P. (2009). Automatic generation control of TCPS based hydrothermal system under open market scenario: A fuzzy logic approach, *Electrical Power and Energy Systems*, 31: 315-322.

[93] Shayeghi H., Shayanfar H. A. and Jalili A. (2006). Multi stage fuzzy PID power system automatic generation controller in the deregulated environments, *Energy Conversion and Management*, 47: 2829-2845, 2006.

[94] Shayeghi H.; Jalili A. and Shayanfar H. A. (2008b). Multi stage fuzzy load frequency control using PSO, *Energy Conversion and Management*, 49: 2570-2580.

In: Advances in Energy Research
Editor: Nicu Bizon

ISBN 978-1-61209-991-0

Chapter 9

TECHNIQUES TO MITIGATE THE FUEL CELL CURRENT RIPPLE

Nicu Bizon* and Mihai Oproescu
Faculty of Electronics, Communication and Computers
University of Pitesti, Romania

ABSTRACT

In this chapter are briefly presented the main mitigation techniques of the low frequency fuel cell current ripple: filtering by passive filters on high voltage bus and/or low voltage bus, designing of appropriate structures for inverter system powered by fuel cell, and active control of power converters from inverter system, or the active control of power interface that operate as an active filter. The designing of models used in simulation is also shown, especially for fuel cell model used to evaluate the low frequency fuel cell current ripple. The simulation results are validated by measurements performed in different experiments.

Keywords: Fuel cell, current ripple, passive filters, active control, power interfaces, energy.

INTRODUCTION

The Polymer Electrolyte Membrane Fuel Cell (PEMFC) is one of the most promising solutions to be utilized in Energy Generation System (EGS) and portable applications because of its relatively lightweight and small size, ease of construction, fast start-up and low operating temperature. Unfortunately, the relatively short PEMFC's life represents for moment a major obstacle to their commercialization [1].

Inverter current ripple represents the main factor responsible for performance degradation of PEMFC energy efficiency [2; 3; 4] and PEMFC life cycle [5; 6; 7]. The PEMFC low frequency (LF) current ripple affects in much measure the PEMFC life cycle, causes

---

* Corresponding Author: nicu.bizon@upit.ro

hysteretic losses and subsequently more fuel consumption. Recent experimental results shown that LF inverter current ripple contributes with up to 10% reduction in the available output power [8; 9]. Therefore some restrictions of the fuel cell current ripple on frequencies bands are specified. It is known that fuel cell is high intolerance for LF current ripple (for example in case of back propagation of harmonics in 50Hz load applications) or slower load transients. The fuel cell electrical equivalent model has a rather large capacitance shunting the device. This suggests that the fuel cell can in-fact tolerate high frequency (HF) current ripple. The USA National Energy Technology Laboratory (NETL) first published the following guidelines for fuel cell current ripple (NETL, 2001):

- 100/120 Hz ripple < 15% from 10% to 100% load, not to exceed 0.6 A for lighter loads;
- 50/60 Hz ripple: < 10% from 10% to 100% load, not to exceed 0.4 A for lighter loads;
- 10 kHz and above: < 60% from 10% to 100% load, not to exceed 2.4 A for lighter loads;
- >100/120 Hz to <10 kHz, limit linearly interpolated between the 120 Hz and 10 kHz limits;
- Transients below 50/60 Hz represent "load following" action of the system, and should track the maximum available current signal from the fuel cell to within 1% for purposes of both fuel cell integrity and efficiency.

The new current ripple limits are given experimentally as Ripple Factor (RF) for different frequency bands (for example, LF RF must be up to 5% from 10% to 100% load, not to exceed 0.5 A for lighter loads; HF RF must be up to 40% from 10% to 100% load, not to exceed 2 A for lighter loads). Lower values for RF are recommended to increase PEMFC performances and the use of paralleling power converters with interleaved technique may be a solution [10; 11]. Transients below grid frequency (50/60 Hz) represent "load following" action of the EGS control, and should track the Maximum Power Point (MPP) signal from the PEMFC to within 1% for purposes of both PEMFC reliability and efficiency (see figure 1) [12; 13; 14].

As it is know, on vehicle applications appear high energy demands in a short time that will cause high current slopes that will produce voltage drops (recognized as fuel starvation phenomenon). Consequently, it is necessary to add Energy Storage Devices (ESD) on Hybrid Power Source (HPS) of vehicles supplied by PEMFC stack as main energy source [15; 16; 17]. Sometime, the proper ESD's, with short time response (for example the ultracapacitors), are used as Power Dynamic Compensators (PDC). Usually batteries and ultracapacitors are used as ESD and PDC, respectively. Such devices are of bidirectional devices. They are required to absorb the energy from the regenerative braking, to compensate the fast power demand, reducing the fuel cell starvation phenomenon that is above mentioned. During these sharp power shapes and other transitory regimes (such as the fuel cell start-ups and shutdowns), the power interface control forces mixed ESD and DPC stack to supply such peaks, improving PEMFC dynamic performance and transient response [18]. For the above mentioned considerations, it is obviously that hybrid and fuel cell technologies need to be merged in vehicle HPS. Usually, a HPS combines two or more energy sources and mixed

ESD and PDS that work together to deliver power (to DC bus) or store energy in order to act as a single power delivery unit. Consequently, the challenge for designer of the power management control into the HPS is to improve the performance of all technologies that working together and to minimize fuel consumption while increasing the HPS reliability [19]. The power or current slopes are given experimentally for different PEMFC stacks, depending on their power. For example, the current slope is about 10 A/s per each kW power [9; 20; 21]. In last decade, many papers have already reported some HPS structures for vehicle applications of type: FC/ultracapacitor [22], FC/battery [23] and FC/ultracapacitor/battery [24; 25].

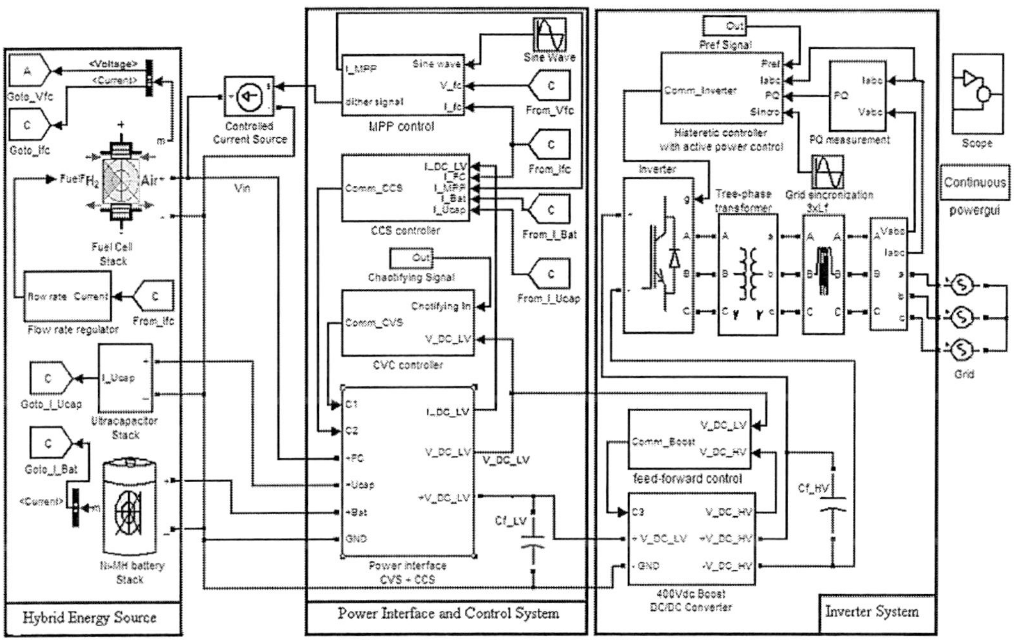

Figure 1. A typical EGS topology.

Usually, 100Hz fuel cell current ripple and other LF harmonics appear in operation of grid inverter system and HF harmonics are generated by PWM switching control of DC-DC converter. Figure 2 shows how appears the LF current ripple in the inverter system. This ripple is back propagated to the PEMFC stack. Usually, first power conversion stage in an EGS without ESD is a DC-DC converter (figure 2).

To obtain the AC output voltage from the high voltage (HV) on HV DC bus, the second power conversion stage is a DC-AC converter. Both converters form an inverter system used in figure 2 as a grid EGS. In grid EGS the HV DC bus value, V_{HVbus}, is in range of 350V to 450V.

Using, for example, a 26V/1.2kW fuel cell stack (figure 2), the DC fuel cell output voltage (which is also low voltage (LV), V_{LVbus}, on LV DC bus for EGS without power interface) is in the 25V ÷ 35V range, depending by load current. For EGS structure without power interface, input current of the DC-DC converter is the current demand from fuel cell stack. For a 230 V AC grid voltage, the boost controller maintains the HV DC voltage at the rated value of 400 V.

Fuel cell current ripple must be mitigated by passive filtering on the HV DC and LV DC buses and by an adequate active control at different energy conversion stages. Typical structure of the basic EGS is presented in figure 3. A possible structure for a low power AC load and a power demand relatively constant it is possible to use a EGS without ESD's (see figure 4, where the fuell cells stack supply directly the inverter system) or with minimum size of the ESD capacity [26]. In order to increase the inverter efficiency the number for stages of energy conversion should be minimized [27]; see figure 5). A multiport EGS topology (see figure 6) with power interface for ripple mitigation was proposed and analyzed in references [28; 29; 30; 31]. Figure 7 shows a HPS topology with Maximum Power Point (MPP) tracking control that assure a low current ripple by PEMFC stack operation near to MPP using an appropriate control of the power interface [32].

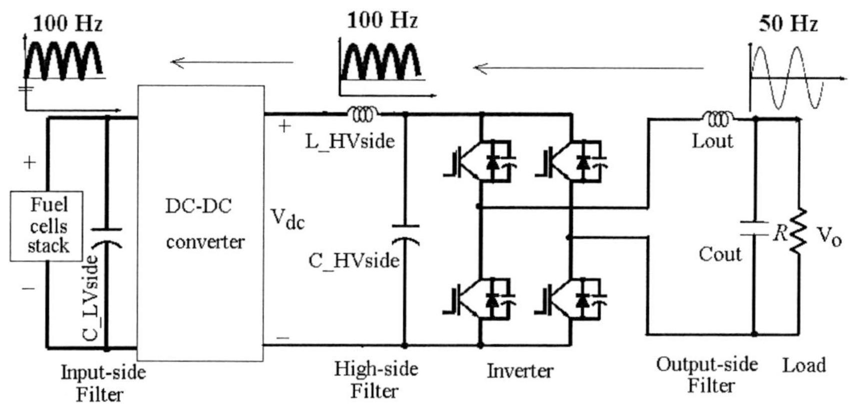

Figure 2. Back propagation of current ripple in the fuel cell inverter system.

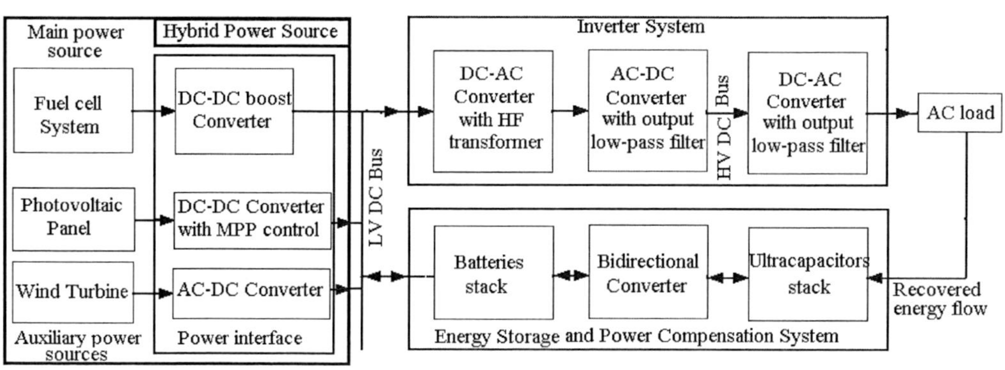

Figure 3. Basic EGS topology with the inverter system powered by a fuel cell HPS.

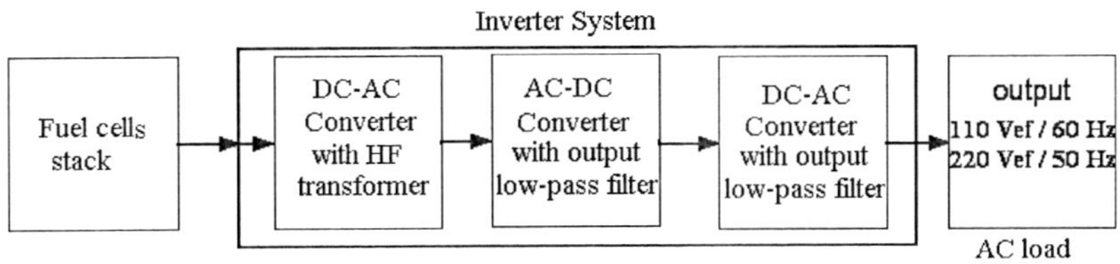

Figure 4. Inverter system powered directly from the fuel cell stack.

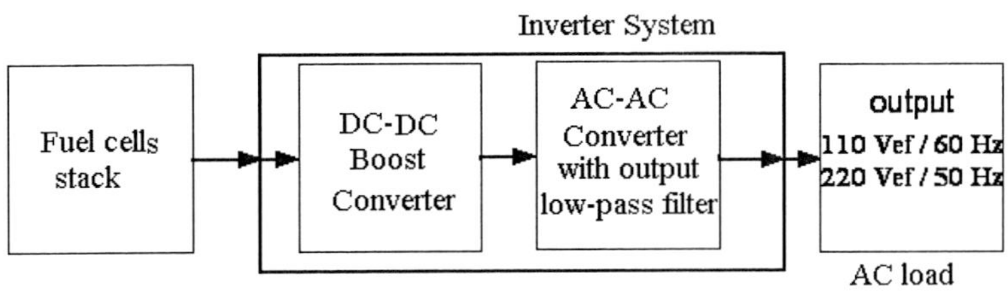

Figure 5. Two-stage inverter system powered directly from the fuel cell stack.

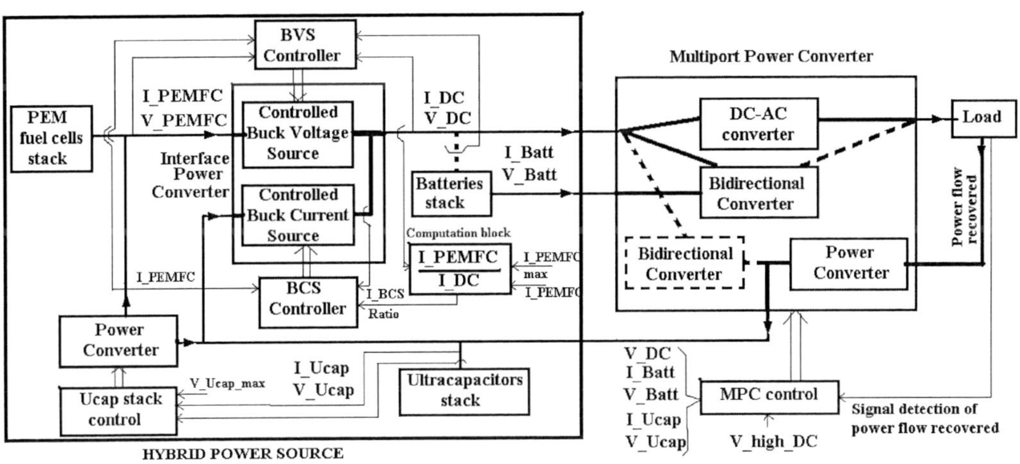

Figure 6. Multiport EGS topology with power interface for ripple mitigation (adapted from [30]).

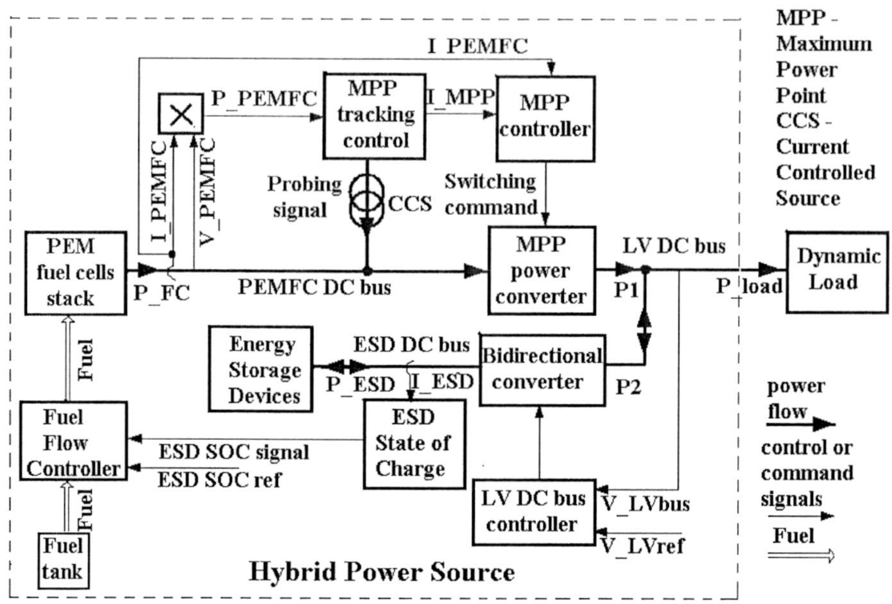

Figure 7. Hybrid power source topology with MPP tracking control.

In this chapter, the modelling analysis will be focused on designing and operating of the fuel cell HPS with passive/active ripple mitigation techniques. The remainder of the chapter is organized as follows. Section 2 presents the basic EGS topology and modelling methodology of the EGS blocks for correct evaluation of ripple in back propagation on different DC buses. Two PEMFC models, which use to model the PEMFC dynamic a transfer function and an electrical circuit, respectively, are compared in ripple mitigation simulations. The mitigation performances of passive filters connected on LV DC bus and HV DC bus are analyzed in Section 3 in order to obtain a referential for other architectural and control solutions for ripple mitigation that are presented in next sections. Experiments performed have validated the design of PEMFC models presented in Section 2. Some aspects of the appropriate control performed to a boost power interface used in fuel cell inverter system are shown in Section 4. The implemented inverter system and the obtained results when is powered by a 12V/2kW fuel cell stack are presented in Section 5. Section 6 shown some representative simulation results and analyze the mitigation performances for an active control of bi-buck interface and last section concludes the chapter.

Fuel Cell EGS System Modelling for Ripple Evaluation

The fuel cell EGS topology, which will be used to analyze the back propagation of inverter current ripple towards the PEMFC stack and to estimate the filtering effect, is show in figure 8 [11; 33]. The current ripple will be mesured by Ripple Factor (RF):

$$RF_I = \frac{I_{Max} - I_{Min}}{I_{(AV)}} \tag{1}$$

where I_{Max}, I_{Min}, and $I_{(AV)}$ are maximum, minimum and average value of the fuel cell current.

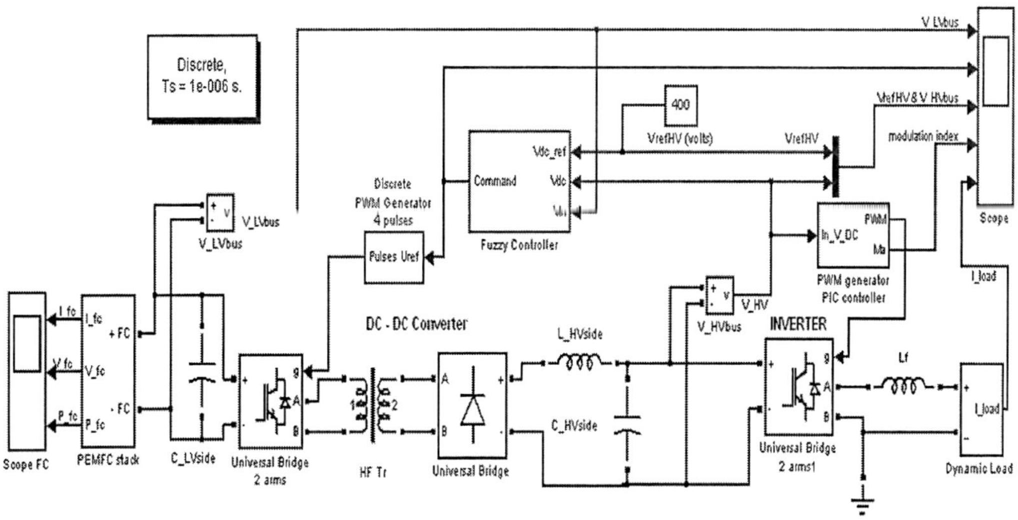

Figure 8. Matlab-Simulink diagram of the analyzed EGS.

EGS Modelling to Evaluate the Current Ripple

In this section an analysis of current ripple back propagation for a mono-phase full-bridge inverter system is made using the EGS topology show in figure 8. The current ripple harmonics on LV DC bus are shown in order to estimate the current ripple on LV DC bus by modelling the DC-DC converter. The ripple model of DC-DC converter is presented in second subsection.

Power Spectrum of the Inverter System

The command of a mono-phase full-bridge inverter system can be easy make using a full-wave switching signal, but the output voltage can't be controlled in this case. A linear control of the output voltage level by index modulation can be made using a "pure" sine PWM switching signal. Output voltage and current of the inverter are shown in figure 9 and 10, respectively, for a given index modulation.

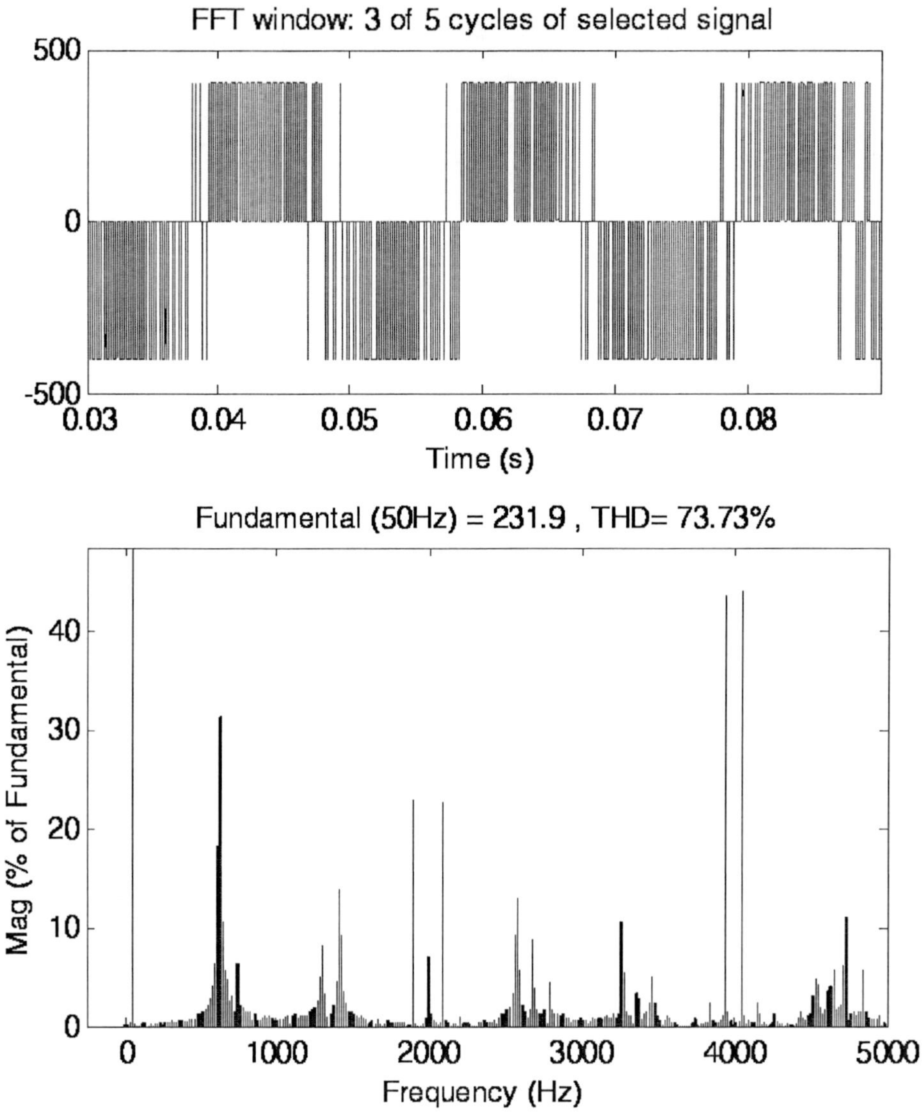

Figure 9. Output voltage of the inverter with pure sine PWM command: signal in time (top) and associate power spectrum (bottom).

Obviously, the fundamental harmonic for output signals is at 50Hz grid frequency. The output current harmonics are at odd multiples of the grid frequency. The output voltage spectrum includes high components, which depend by the used carrier frequency. Input voltage and current for the same inverter are signals on HV DC bus and it are shown in figure 11 and 12, respectively. As has been stated before, the fundamental harmonic for input signals is of twice of grid frequency.

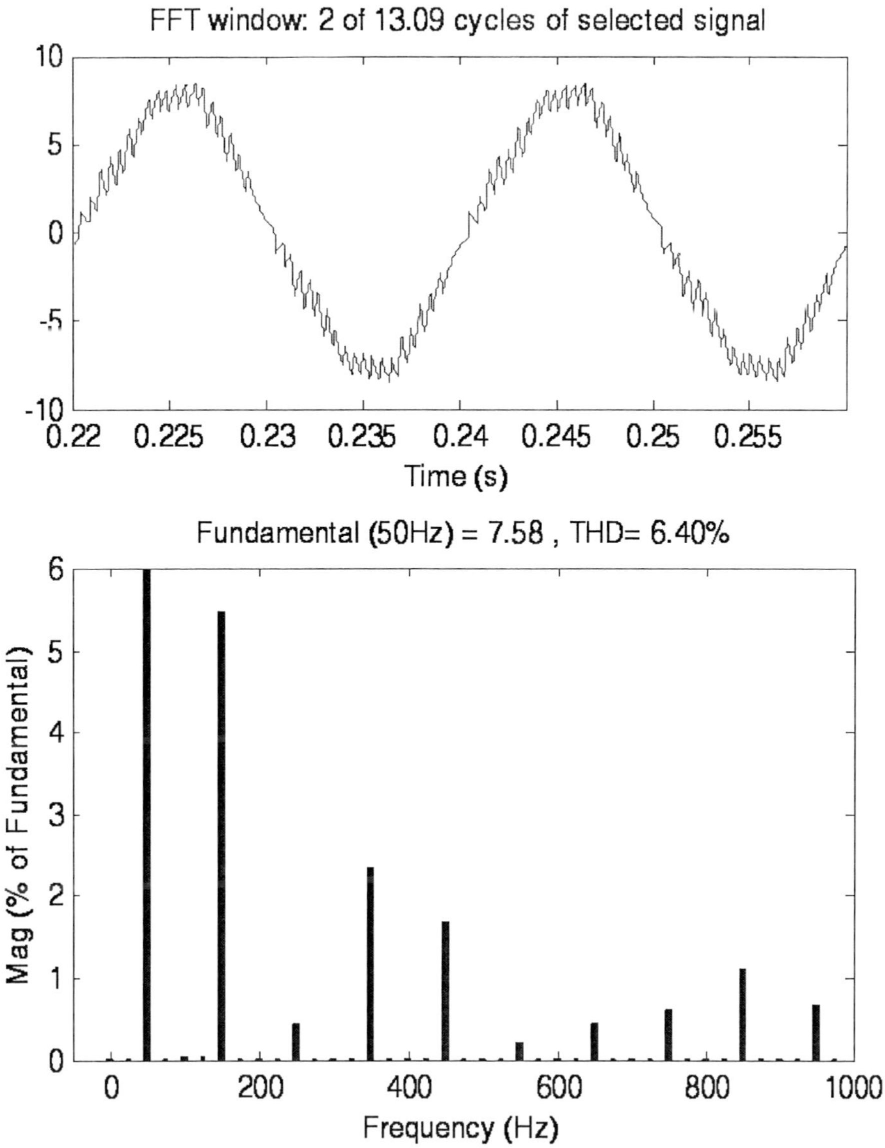

Figure 10. Output current of the inverter with pure sine PWM command: signal in time (top) and associate power spectrum (bottom).

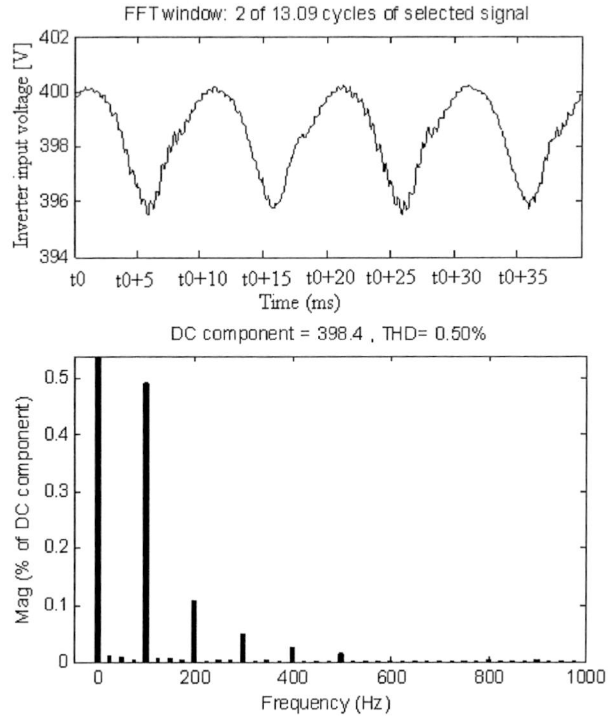

Figure 11. Input voltage of the inverter with pure sine PWM command: signal in time (top) and associate power spectrum (bottom).

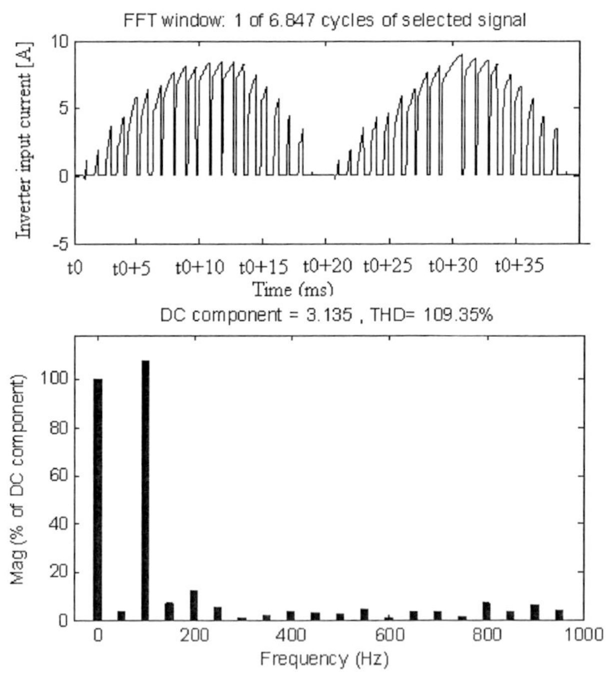

Figure 12. Input current of the inverter with pure sine PWM command: signal in time (top) and associate power spectrum (bottom).

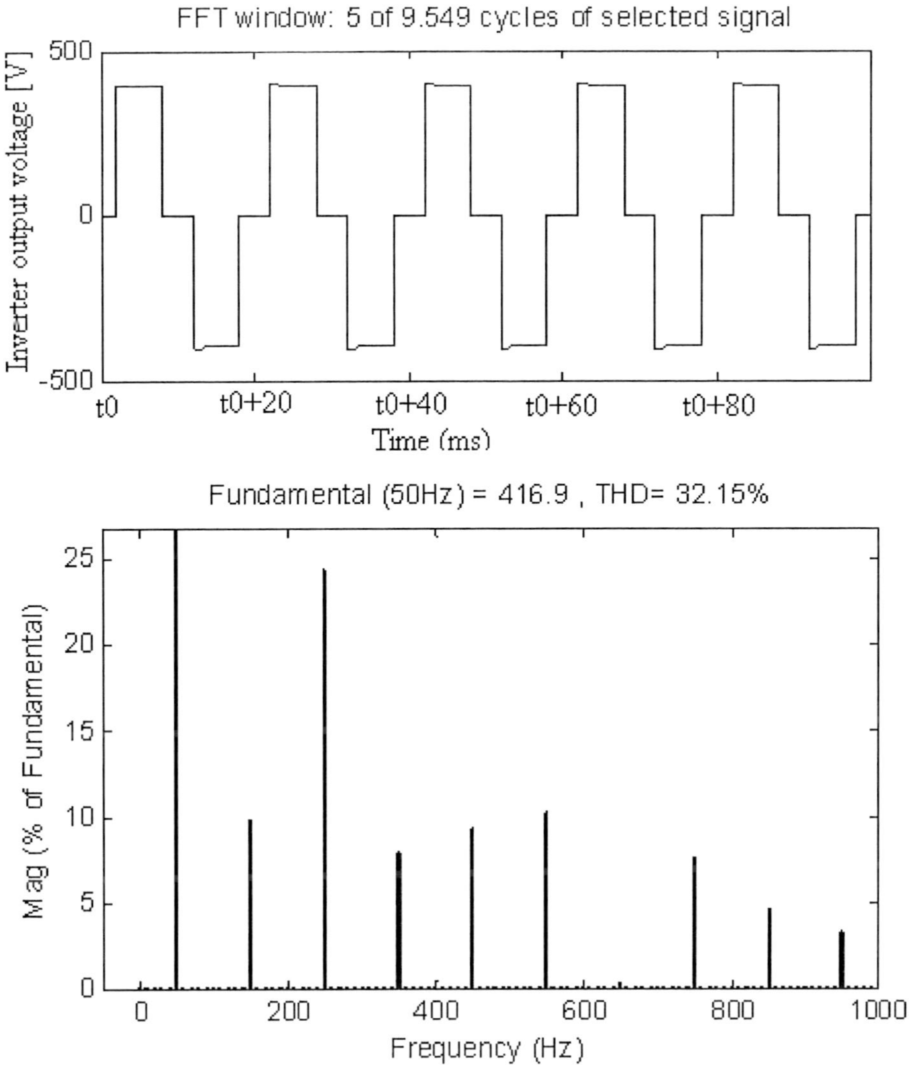

Figure 13. Output voltage of the inverter with modified sine PWM command: signal in time (top) and associate power spectrum (bottom).

A way to mitigate some harmonics is to use the phase shift command, also named as modified sine PWM command because of the shape of output voltage (figure 13). Output voltage and input current of the inverter with modified sine PWM command are shown in figure 13 and 14, respectively, for a $\pi/6$ phase shift angle.

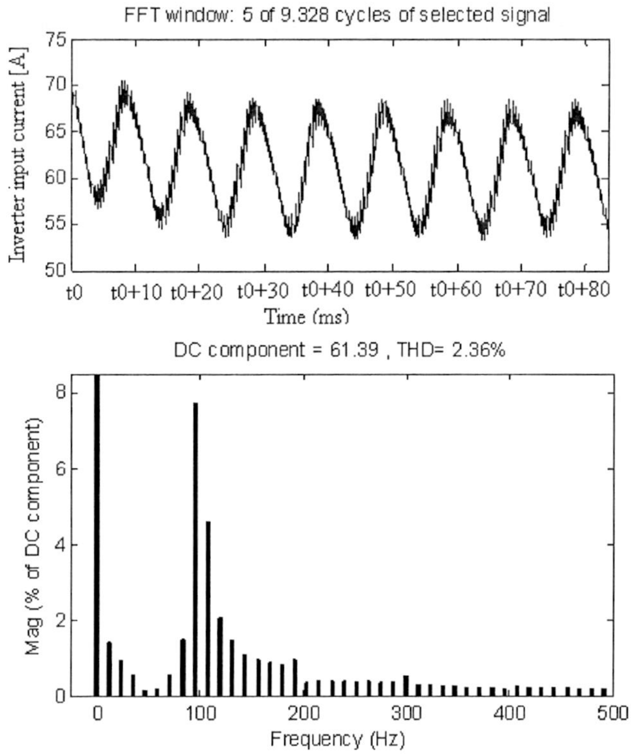

Figure 14. Input current of the inverter with modified sine PWM command: signal in time (top) and associate power spectrum (bottom).

The current harmonics level is clearly dependent by filtering capacitance value on LV-side and HV-side and the harmonics number is given by the used switching techniques: full-wave rectangular command, phase shift command or pure sine PWM command, respectively.

The current slopes for PEMFC stacks must be limited at about 10 A/s per each kW power. So, the EGS behavior under dynamic load must be analyzed. Voltage shapes on HV DC bus and LV DC bus are shown in figure 15 for a pulse load. The effect of high power demand at PEMFC stack is shown in figure 16.

Analyzing figure 16 (where PEMFC current slope exceeding 50 A/s per each kW power), we can conclude that high load pulse (which might lead to a variation from rated load to double of its - which is close of maximum of PEMFC power delivery) can cause sharp current variations that exceed acceptable limits, if are not used appropriate pasive/active filters on DC buses.

Considering the inverter current ripple level that appear on HV DC bus, an AC model of the DC-DC converter is needed to evaluate the currrent ripple that propagates back on LV DC bus by DC-DC converter. A simple model of DC-DC converter for current ripple evaluation is developed in next section.

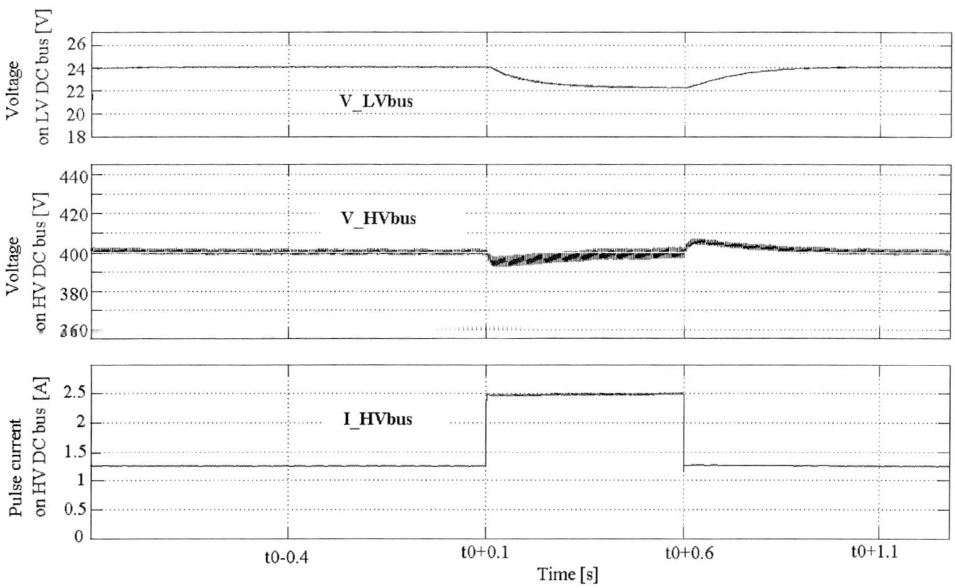

Figure 15. The effect of the current pulse, I_HVbus (bottom), on HV DC voltage, V_HVbus (middle), and LV DC voltage, V_LVbus (top).

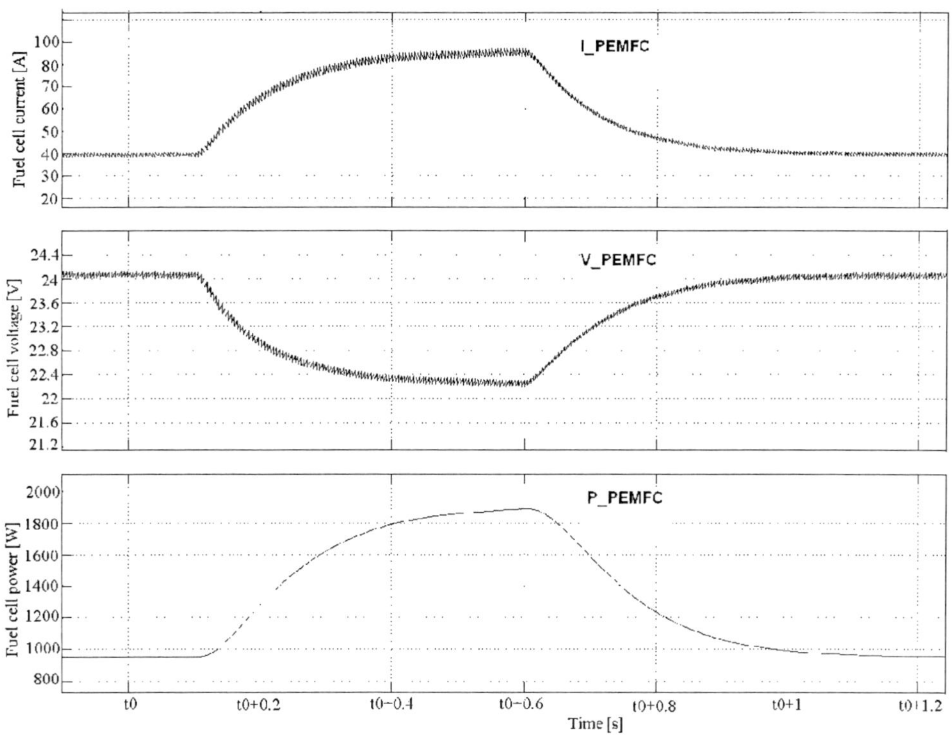

Figure 16. PEMFC stack behaviour under load pulse: power, P_PEMFC (bottom), voltage, V_ PEMFC (middle), and LV DC voltage, I_ PEMFC (top).

MODELLING OF DC-DC CONVERTER FOR BACK PROPAGATION EVALUATION OF THE INVERTER CURRENT RIPPLE

Starting from the mono-phase full bridge DC-DC converter shown in figure 17, the generic DC-DC converter topology, with passive filters at its ports of input and output, is shows in figure 18.a. The current ripple model for the DC-DC converter that will be used in the simulation are shown in figures 18.b and 18.c for case with galvanic isolation made by HF transformer and case without galvanic isolation, respectively.

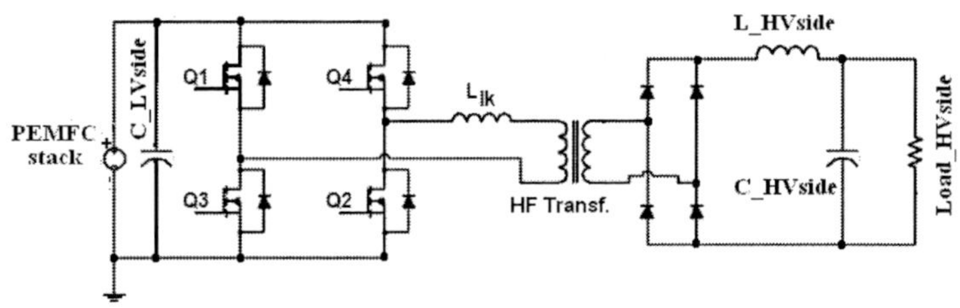

Figure 17. Mono-phase full bridge DC-DC converter topology.

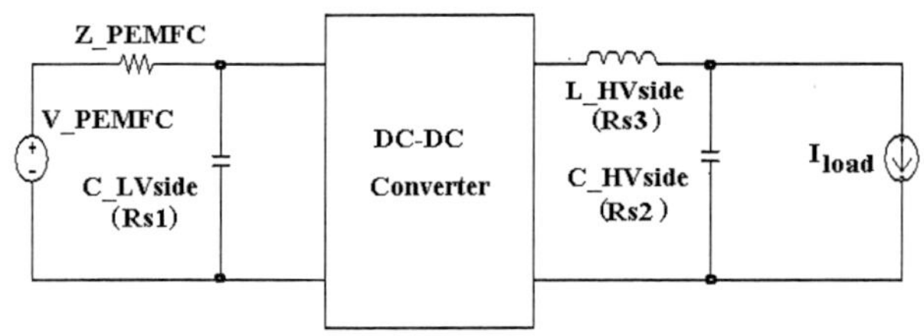

Figure 18a. DC-DC converter topology with passive filters at its input and output ports.

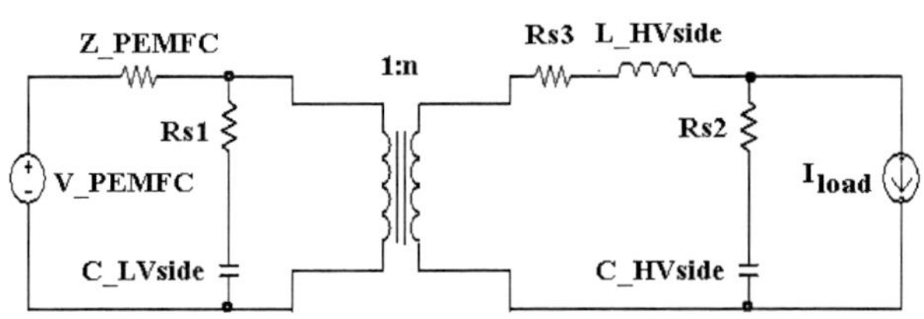

Figure 18b. Ripple model of the DC-DC converter with galvanic isolation made by HF transformer.

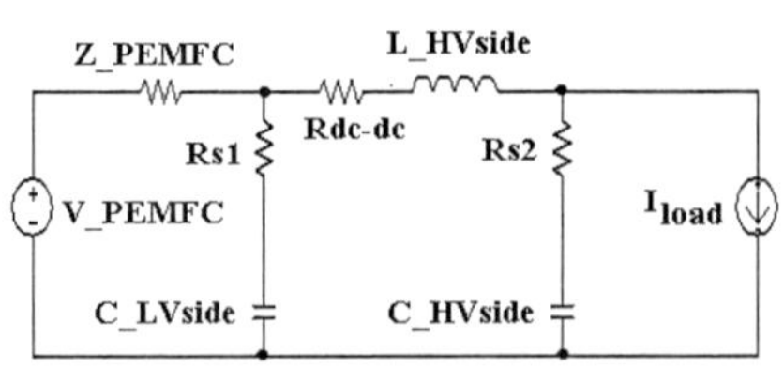

Figure 18c. Ripple model of the DC-DC converter without galvanic isolation.

The series resistances of the capacitors stack on the LV side (C_{LVside}) and on the HV side (C_{HVside}) are R_{s1} and R_{s2}, respectively. If a second order low-pass filter is used on the HV side, then the inductance, L_{HVside}, and its series resistance, R_3, may be considered in the ripple model. The harmonics mitigation is modelled by a ideal transformer or by a resistance mitigation, $R_{dc\text{-}dc}$ ($R_{dc\text{-}dc} >> R_3$) for case with galvanic isolation made by HF transformer and without galvanic isolation, respectively. Input inverter current is modeled by a controlled current source, I_{load}, which represents the ripple inverter current superimposed on the rated current. The PEMFC stack is modeled by a controlled voltage source, V_{PEMFC}, and electrical impedance, Z_{PEMFC}. Different kinds of models are proposed for modeling of PEMFC, especially for ripple current evaluation that propagates back on PEMFC stack in case of the use of an EGS topology without power interface (see figure 8) instead of one with power interface (see, for example, figure 1 or 6) that can mitigate much more the current ripple. An electrical model [34] is compared in next section with an equivalent electrical model [35] based on macromodelling.

PEMFC Modelling for Current Ripple Evaluation

In this section it demonstrated that a first order transfer function can well estimate the PEMFC dynamic [35] and the ripple current is as that reported in [34], when an electrical impedance is used in PEMFC modeling. This macro-model is simple and easy to be implemented in a PEMFC emulator. In addition, the model operates well in supplying energy to different type of power converters, instead of the electrical model of PEMFC that presents some malfunctions in connection with some type of power converters. Generally, the PEMFC voltage losses are determined in terms of three main groups: activation losses, ohmic losses and concentration losses. Figure 19 presents a typical plot of the cell voltage and the specific power density versus the current per square centimeter. By subtracting the above three loss equations from V_{cell_0} (output voltage from no-load for a single fuel cell), the cell voltage, V_{cell}, is determined in terms of the drawn current density (specified by load), $j_{fc}= j_{fc}/A$, together with a constant, j_n, determined by two phenomena's known as internal current and fuel cross over:

$$V_{cell} = V_{cell\_o} - \left(j_{fc} + j_n\right) \cdot A \cdot R_{ohmic} - A_{act} \cdot \ln\left(\frac{j_{fc} + j_n}{j_0}\right) + B_{conc} \cdot \ln\left(1 - \frac{j_{fc} + j_n}{j_{max}}\right) \quad (2)$$

The PEMFC response time, T_{fc}, usually models the PEMFC dynamic:

$$v_{fc}(t) = N_{fc} \cdot v_{cell}(u), \quad u = j(t) = \frac{i(t)}{A}, \quad \frac{I(s)}{I_{fc}(s)} = \frac{\exp(-s \cdot \tau_{fc})}{1 + T_{fc} \cdot s} \quad (3)$$

where N_{fc} is the stack cells number, A is area of membrane, i_{fc} is the PEMFC current and v_{fc} is the PEMFC output voltage. The response time, T_{FC}, and death time, τ_{fc}, is used to model the "charge double layer" phenomenon due to the build up of charges at electrode/electrolyte interface. Number of fuel cells used for a PEMFC stack is given by relation $N_{fc}=V_{PEMFC\_0}/V_{cell\_0}$, where V_{cell_0} and V_{PEMFC_0} are output voltage from no-load for a single fuel cell and PEMFC stack, respectively. Proposed PEMFC macro-model [35] that uses a transfer function for modelling the PEMFC dynamic and proposed PEMFC model [34] that uses an electrical circuit for modelling the PEMFC dynamic is shown in figure 20 and figure 21, respectively.

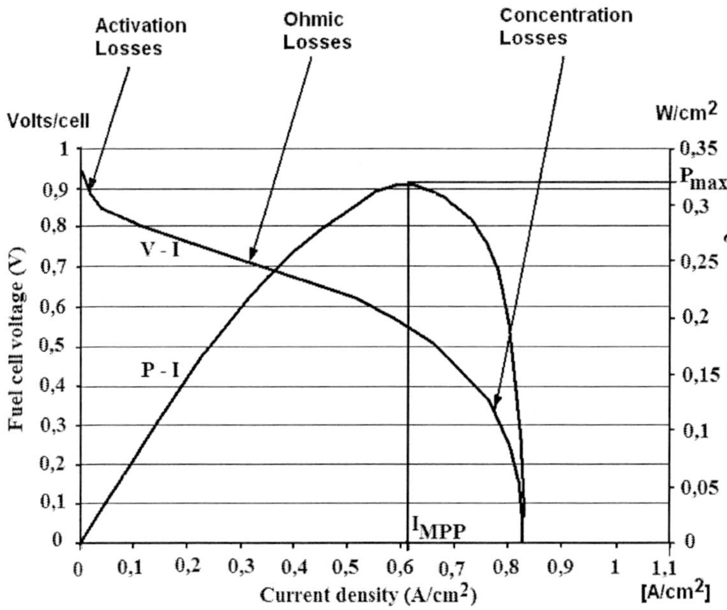

Figure 19. Polarization and power curves for a typical PEMFC [35].

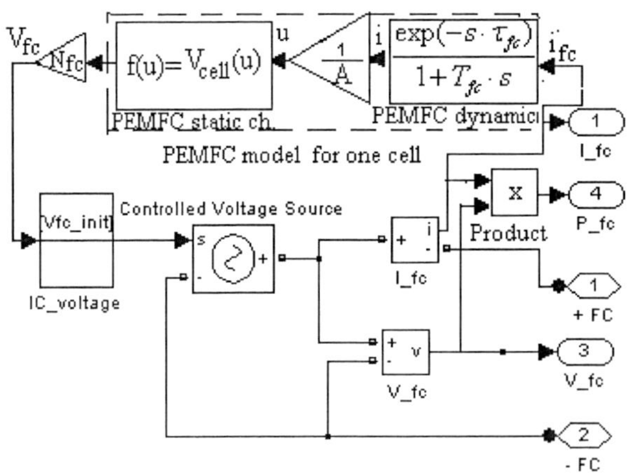

Parameter (*for λ = 1 and* P_{sys} *= 2 atm*)	Relation (2)
E° [Volts]	1.2
j_n [mA/cm²]	2
j_0 [mA/cm²]	0.067
j_{max} [A/cm²]	0.9
R_{Ohmic} [Ω]	0.03
A_{act} [Volts]	0.06
B_{conc} [Volts]	0.05

Relation (3):
f(u)=(1.2-(u+0.002)*A*0.03-0.06*log((u+0.002)/0.000067)+0.05*log(1-(u+0.002)/0.9))
where u=i/A, i[A] and A[cm²]

Figure 20. Proposed PEMFC macro-model that uses a transfer function for modelling the PEMFC dynamic [35].

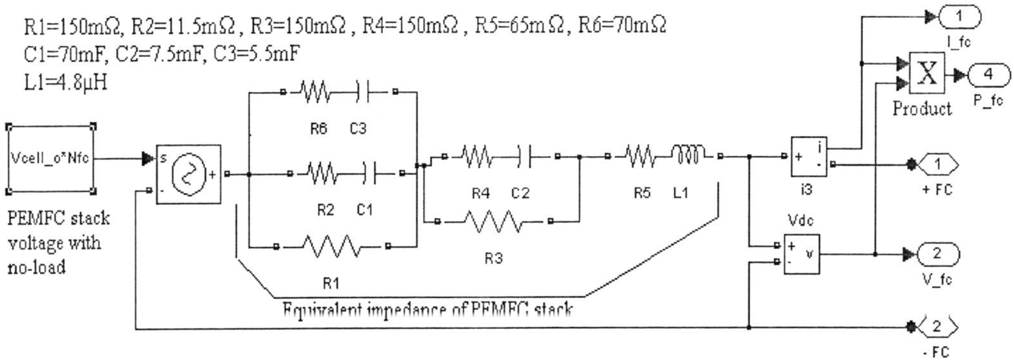

Figure 21. Proposed PEMFC model that uses an electrical circuit for dynamic modelling [35].

A detailed model of PEMFC stack will be presented in chapter dedicated to modelling of energy sources and will be compared with Matlab-Simulink® model for PEMFC stack that is included in the toolboxes. In this chapter only proposed PEMFC macro-model that uses a transfer function for modelling the PEMFC dynamic will be used.

Simulation Results

In this section, simulation results for a commercially available PEMFC stack (1.2kW/24V Ballard Nexa) are analyzed using both models of its. PEMFC parameters for electrical circuit are from [38] and are shown in figure 21. Firstly, the static performances of the PEMFC models are analyzed by drawing the U-I and P–I characteristics, and secondly the dynamic behaviour will be analyzed.

PEMFC Static Characteristics

Number of fuel cells used for PEMFC stack is given by relation $N_{fc}=V_{PEMFC\ 0}/V_{cell\ 0}$. Using the reported results for Ballard Nexa, $V_{PEMFC\ 0}$ is 42,2V and, at MPP operation for a given fuel flow rate, the power (P_{MPP}), output voltage (V_{MPP}), output current (I_{MPP}) and equivalent impedance for the stack ($R_{MPPeqv}= V_{MPP}/I_{MPP}$) are 1200W, 26.6V, 45A and 0,59Ω, respectively.

The static characteristics are clearly different (see models from figure 22 and 23 that use to model the PEMFC dynamic a transfer function and an electrical circuit, respectively). The PEMFC characteristics from figure 22 show as the real PEMFC characteristics, but not those in figure 23. As an effect of the resistive losses the static U-I characteristic from figure 23 is linear, so the P-I characteristic will be of parabolic type.

PEMFC Dynamic Characteristics

A dynamic load model is used to study the PEMFC behaviour. If a pulse load is applied to the PEMFC stack output, a time response for each model can be showed by simulation. The obtained results concerning the PEMFC time response for both PEMFC models are presented in figure 24 and 25, respectively.

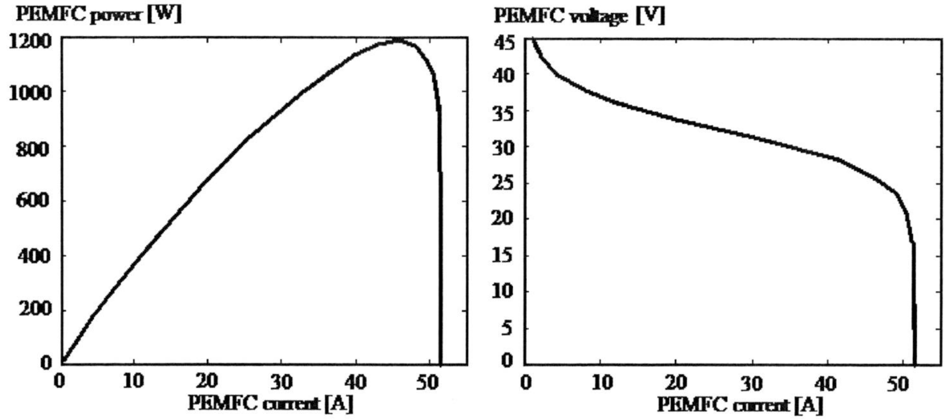

Figure 22. Nexa PEMFC characteristics for model that uses a transfer function to model the PEMFC dynamic [35].

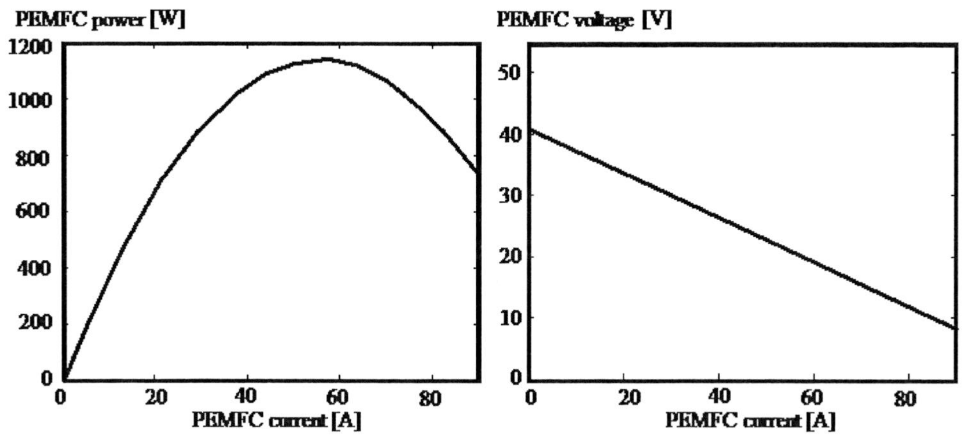

Figure 23. Nexa PEMFC characteristics for model that uses an electrical circuit to model the PEMFC dynamic [35].

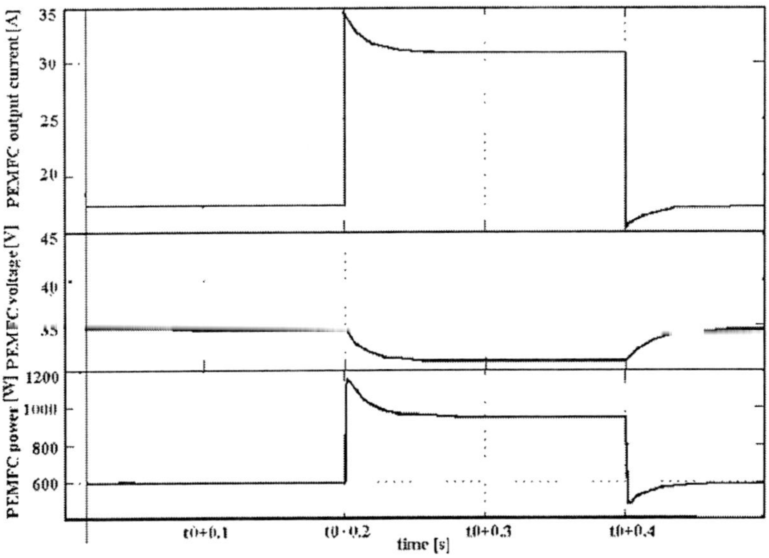

Figure 24. Behaviour under load pulse for Nexa PEMFC model that uses a transfer function to model its dynamic [35].

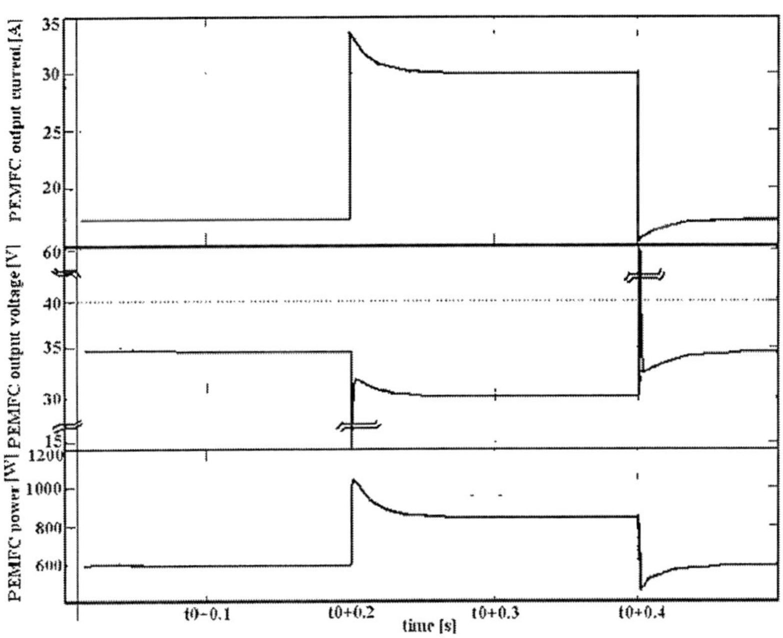

Figure 25. Behaviour under load pulse for Nexa PEMFC model that uses an electrical circuit to model the PEMFC dynamic [35].

High and low spikes appear over the PEMFC output voltage (higher and lower than V_{PEMFC_0}) for the PEMFC models that uses an electrical circuit to model the PEMFC dynamic (figure 25).

With this exception, which can be considered as normal because there are capacitors switched by load dynamic, the simulation results look almost the same for both PEMFC models. But the spikes can be very high in some power conversion circuits, distorting their proper function, as will be shown below.

PEMFC Current Ripple Evaluation

Current ripple evaluation for both PEMFC models is obtained using the Matlab diagram and a load model with LF ripple harmonics over the DC current (see figure 26).

The LF ripple harmonics are obtained by use of three sine generators with frequencies 50Hz, 150Hz and 300Hz, respectively, which are rectified by an absolute value block. The input current ripple of the mono-phase inverter system can be simulated using different magnitudes and initial phases of the sine waves.

The shape in time for the load current with LF ripple harmonics and its power spectrum are shown in figure 27 for a given set of parameters.

This simulated inverter load current includes the LF harmonics (up to 500Hz) with different level of amplitudes.

The DC-DC converter is controlled to obtain 100V on HV DC bus when on LV DC bus is connected the PEMFC stack. The controller will be presented in the chapter dedicates to power converter control.

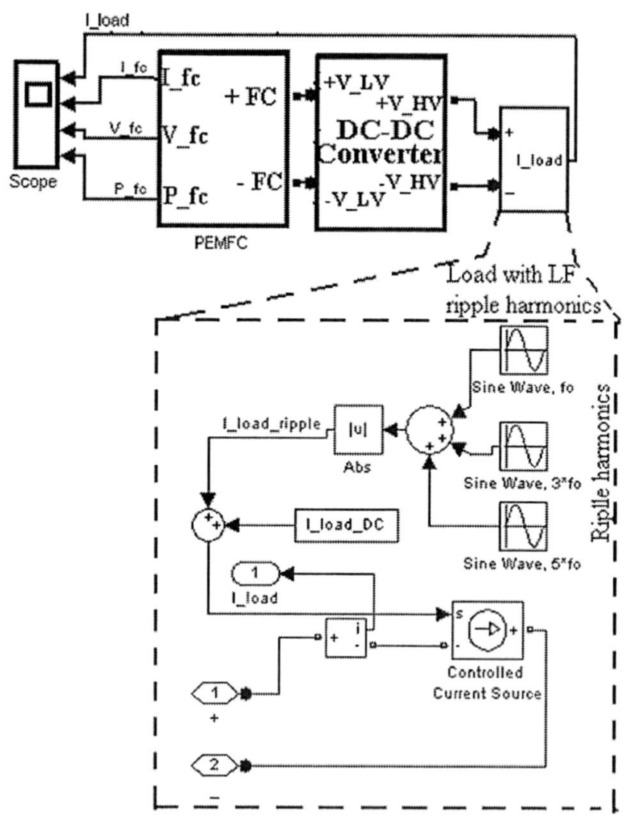

Figure 26. The Matlab diagram for PEMFC current ripple evaluation supplying a load with LF harmonics (f_0=50Hz).

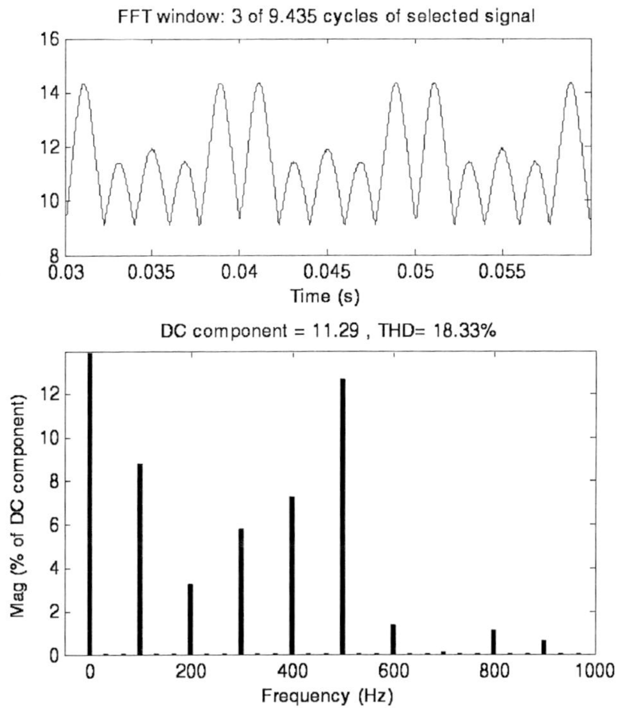

Figure 27. Shape in time for the load current with LF ripple harmonics and its power spectrum [35].

The obtained results are almost the same for both PEMFC models. The PEMFC current for both PEMFC models is presented in figure 28 and 29, respectively. The ripple factor is about of 5A/11.3A≅50% for load current that simulates the inverter system and is about of 32A/50A≅64% for PEMFC current. Consequently, the filtering techniques to mitigate this high ripple are needed to be implemented.

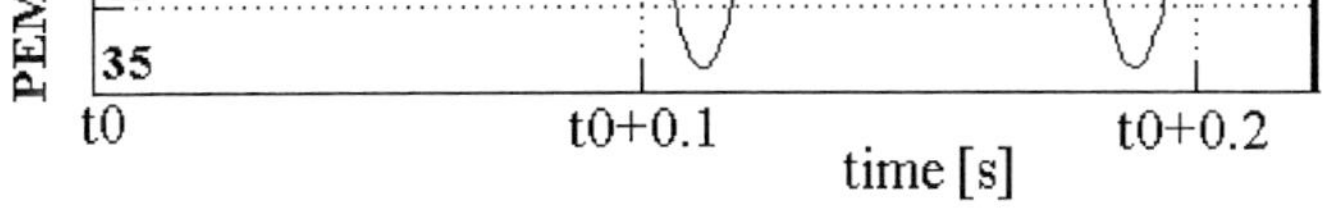

Figure 28. PEMFC current ripple for model that uses a transfer function to model its dynamic (T_{fc}=20ms).

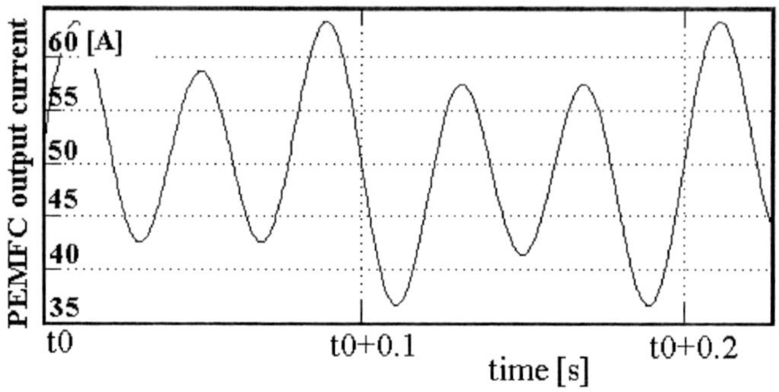

Figure 29. PEMFC current ripple for model that uses an electrical circuit to model its dynamic.

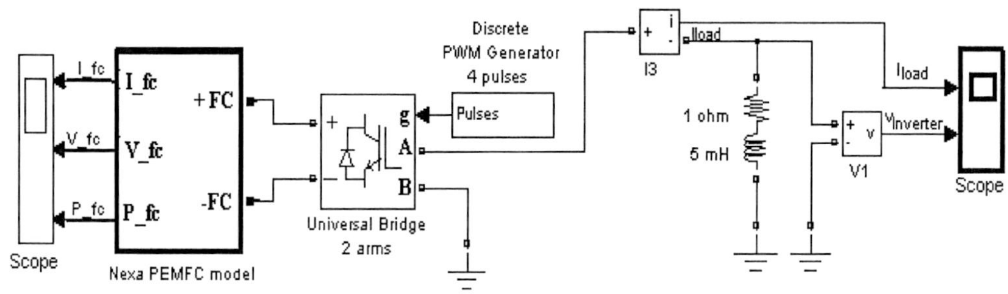

Figure 30. Fuel cell EGS (adapted from [35]).

When the inverter system is directly to the PEMFC stack (figure 30) the simulation results are clearly different for PEMFC voltage (see figure 31 and 32). Analyzing the currents of the inverter devices (IGBT's), which have high spikes for PEMFC model that uses an electrical circuit to model its dynamic, we can conclude that some resonances appear in this kind of fuel cell EGS. If changes of the resistive-inductive load may appear then much larger voltage spikes than is showed in Figure 32 can appear on PEMFC voltage. This type of behaviour was not reported in any experiment, so the model proposed in [38] can't be validated for use in simulations of some EGS's.

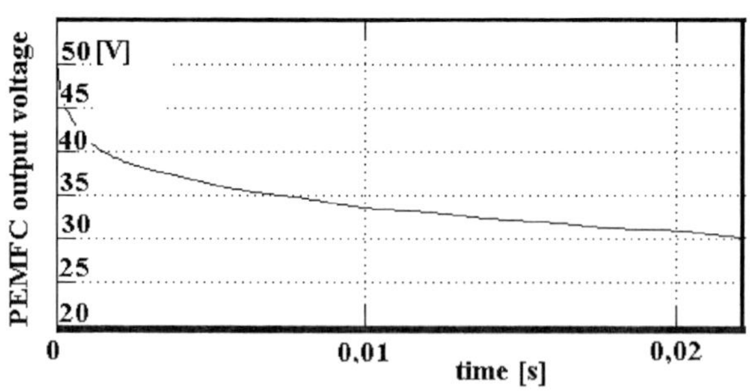

Figure 31. Behaviour at start-up for Nexa PEMFC model that uses a transfer function to model its dynamic (adapted from [35]).

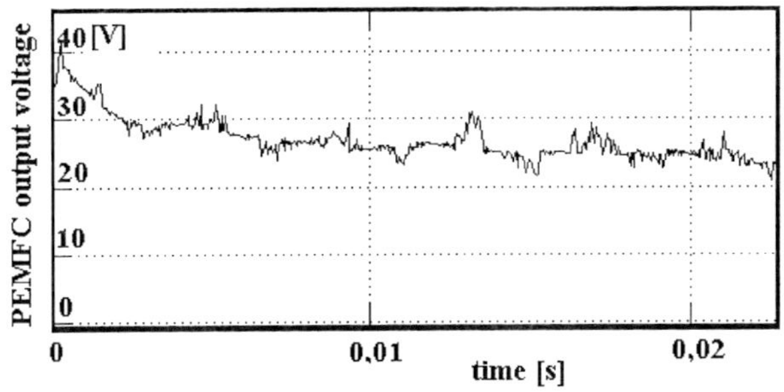

Figure 32. Behaviour at start-up for Nexa PEMFC that uses an electrical circuit to model its dynamic (adapted from [35]).

Conclusions

The proposed PEMFC model [35] that uses a transfer function to model its dynamic has a well behaviour in different Matlab-Simulink® simulations.

The proposed PEMFC model that uses an electrical circuit to model its dynamic has unexpected behaviour in many cases.

Almost the same well results are obtained using the detailed PEMFC model proposed in the Matlab-Simulink® toolboxes. Further, the static characteristic of proposed PEMFC model that uses an electrical circuit to model its dynamic are clearly different of the real PEMFC stack characteristics.

The disadvantage of the proposed PEMFC model in [35], which uses a transfer function to model its dynamic, refers to the fact that this is of macro-model type.

For example, this model can't be used for example in simulations of the output impedance of the PEMFC.

Passive Filters

Proposed state-of-the-art architectural and control solutions for inverter current ripple mitigation are the increasing of the passive filter rated capacity by adding more capacitors or adding external active filters. Because the passive filtering solution for high PEMFC output current are bulky, expensive and inherently unreliable [11; 33], active control remains an acceptable variant to reduce current ripple in the HV DC bus of the inverter system [36; 37]. Other possibility is to use a power interface between fuel cell stack and inverter system with an appropriate control to mitigate the inverter current ripple [38; 28]. If the PEMFC stack is directly connected to the inverter system, a parallel active filter on LV DC bus can be used to compensate the PEMFC current ripple [31]. In this section the performances of passive filters connected on LV DC bus and HV DC bus are analyzed to obtain a referential for other architectural and control solutions for inverter current ripple mitigation. Experiments performed have validated the modelling presented above and the simulation results obtained.

The Low Pass Passive Filter on HV DC Bus

A mono–phase inverter system contains a single-phase DC-AC converter which produces a DC current with AC ripple having the spectral harmonics at odd multiples of twice of the output frequency. Usually, a second order low pass passive filter is used on HV DC bus (see figure 2 and 8). The equivalent structure of filter is presented in figure 33 for harmonic number *n*.

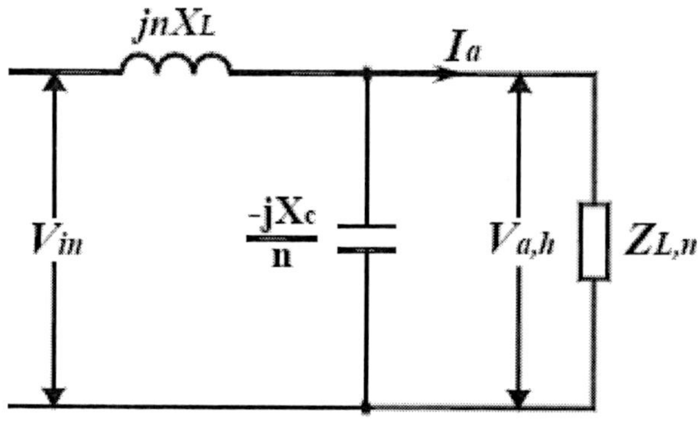

Figure 33. Low pass passive filter for harmonic number n.

Using the Matlab-Simulink® diagram shown in figure 8 and the PEMFC macro-model presented in above section, the inverter output current (before and after the filter, considering the direction of power flow propagation) is presented in figure 34. It is clearly that its power spectrum is concentrate in LF frequency band on the fundamental of 100Hz (f_{100}) and its odd harmonics. So the simulated load for the inverter system correctly models the LF harmonics of inverter current. If the effective value of the first harmonic is $I_{1(RMS)HVside}$, then the DC current on high DC bus is about of $I_{1(RMS)HVside}$ /2. To have a voltage ripple ΔV_{HVside}, the capacity for the filter capacitors stack must by higher that the value:

$$C_{HVside} = \frac{I_{1(RMS)HVside}}{4\pi f_{100} \Delta V_{HVside}} \tag{4}$$

For example, considering $I_{1(p\text{-}p)\mathrm{HVside}}$ =17A and an imposed voltage ripple value $\Delta V_{\mathrm{HVside}}$= 5V (see figure 15), these lead to a capacity, C_{HVside} , of about 2mF. For this value of capacity, the influence of the inductance value, L_{HVside} , on current ripple mitigation was tested for values in range 0.1mH-10mH. For L_{HVside} 0.2mH, C_{HVside} ==2mF and a pulse load of inverter, R=40Ω/20Ω, the simulation results are presented in figure 35 and 36.

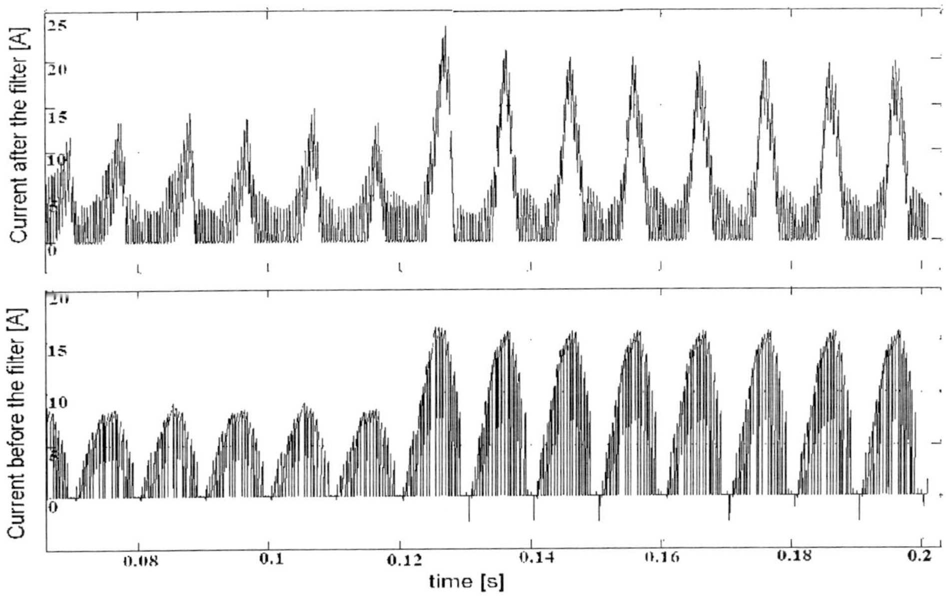

Figure 34. The current on the HV DC bus, I_{HVside}, after (top) and before (bottom) the passive filter.

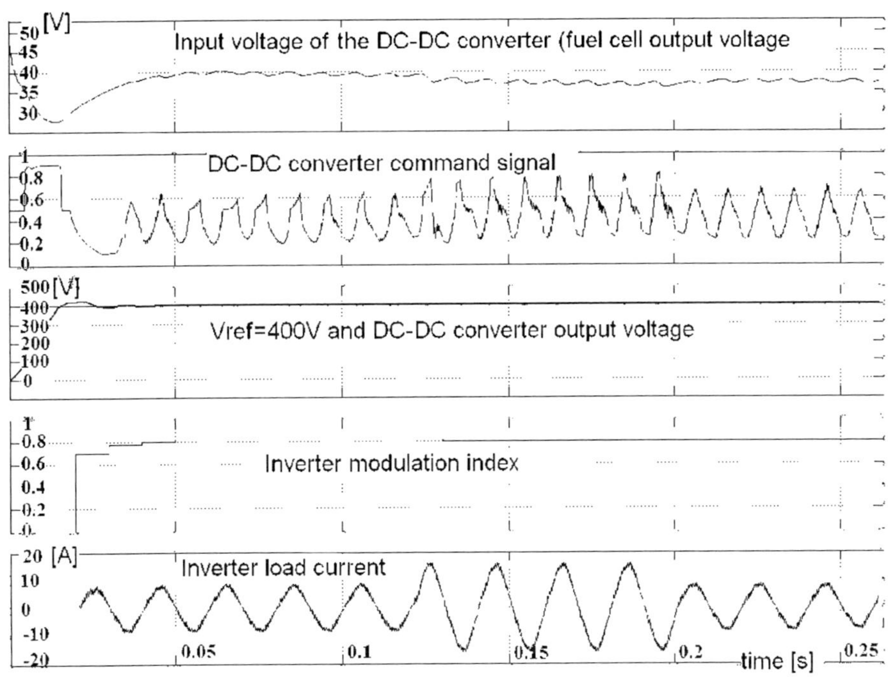

Figure 35. Significant waveforms for function of the DC-DC converter and inverter.

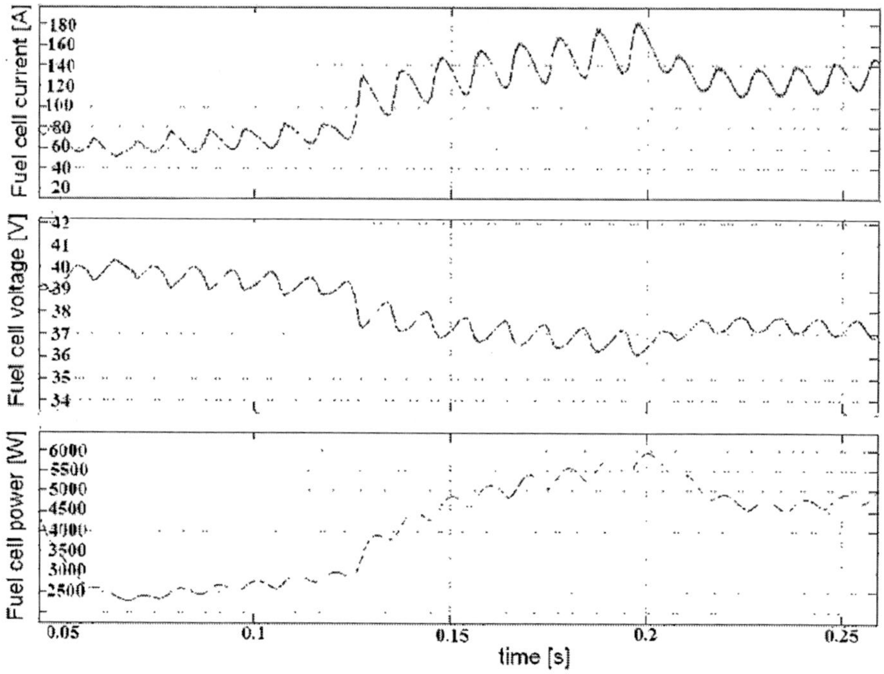

Figure 36. Fuel cell current, voltage and power.

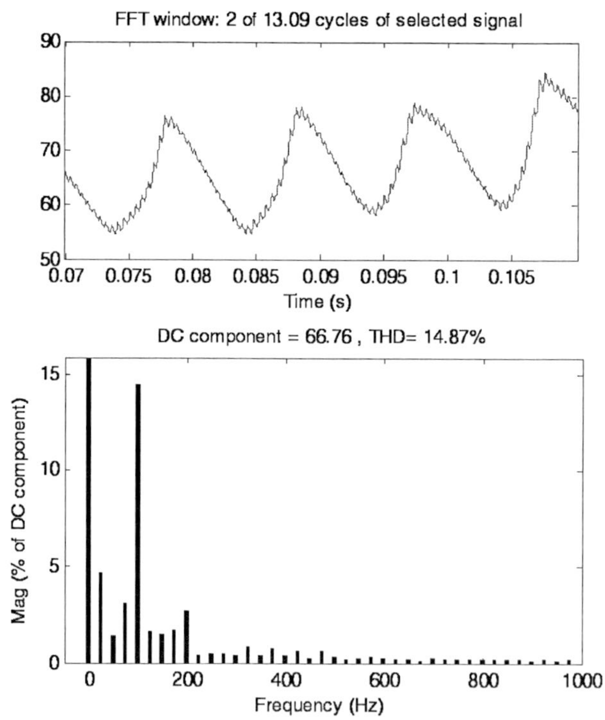

Figure 37. Fuel cell current (top) and its power spectrum (bottom) for 2.5kW load (L_{HVside} =0.2mH, C_{HVside} =2mF).

Figure 35 presents some significant wave of inverter system and figure 36 shows the behaviour of the PEMFC stack. Obviously, the LF current harmonics appears at the LV DC bus and without other power interfaces (likes boost converter or passive/active filter) represents the fuel cell current ripple, too. The PEMFC current ripple factor is about of (20A/70A)·100≅28%, much higher than the new value required (5% or even 3%).

Firstly, we will test if only a passive filter on HV DC bus (without passive filter on LV DC bus) can mitigate the current ripple to the required value of 15%. Secondly, for an optimum passive filter on HV DC bus as size and price, we find a passive filter on LV DC bus that, in conjunction with the passive filter on HV DC bus, can mitigate the current ripple to the new required value of 5%. For these simulations the power of the PEMFC stack model is increased to 6kW by tripling the area of PEMFC membrane. Power spectrum of the fuel cell current is presented in figure 37 and 38 for both values of the inverter load, 40Ω and 20Ω, respectively.

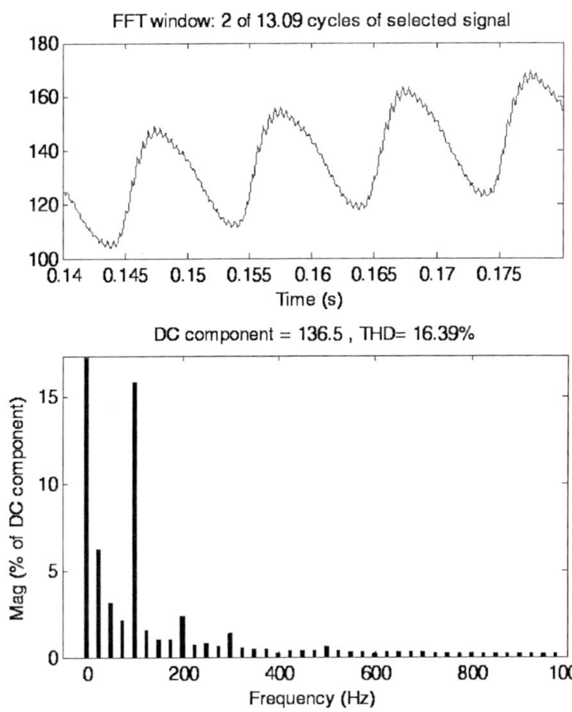

Figure 38. Fuel cell current (top) and its power spectrum (bottom) for 5kW load (L_{HVside} =0.2mH, C_{HVside} =2mF).

We can remark that 100Hz current harmonic is less dependent of the load power level. The filter inductance value is increased to 2mH in figure 39 and 40, and to 20mH in figure 41 and 42, where the power spectrum of the fuel cell current is shown for both values of the inverter load, 40Ω and 20Ω, respectively. Following these results, we note that using of a passive filter of second order type not brings major improvements compared to one of first order type (without L_{HVside}). If the inverter system power is 6kW and the voltage on HV DC bus is 400V, then the current on HV DC bus is about of 1.5A. The coil volume is high and the price of a 20mH/2A inductance is high enough, so a 0.5mH/2A inductor will be used in the experimental setup and next simulations.

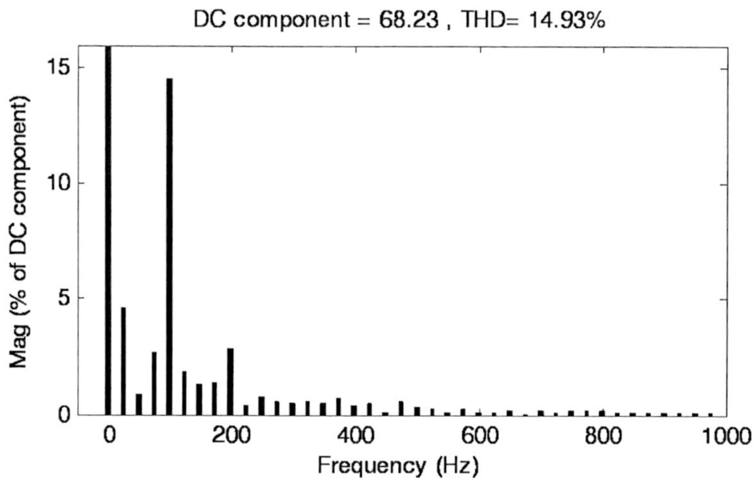

Figure 39. Power spectrum of the fuel cell current for 2.5kW load (L_{HVside} =2mH, C_{HVside} =2mF).

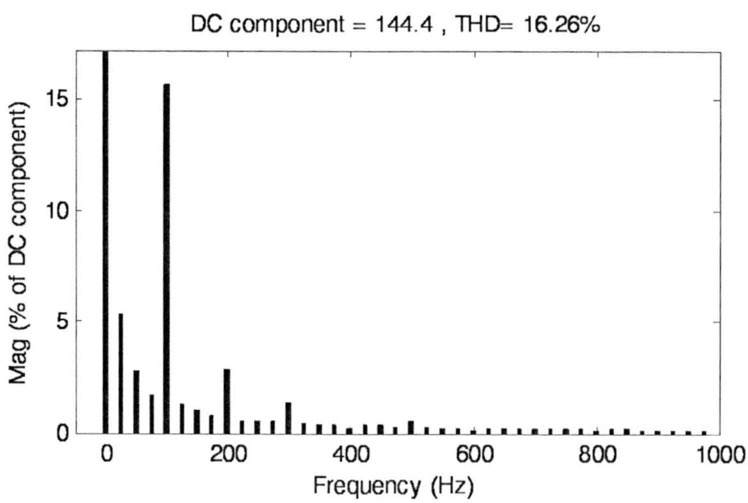

Figure 40. Power spectrum of the fuel cell current for 5kW load (L_{HVside} =2mH, C_{HVside} =2mF).

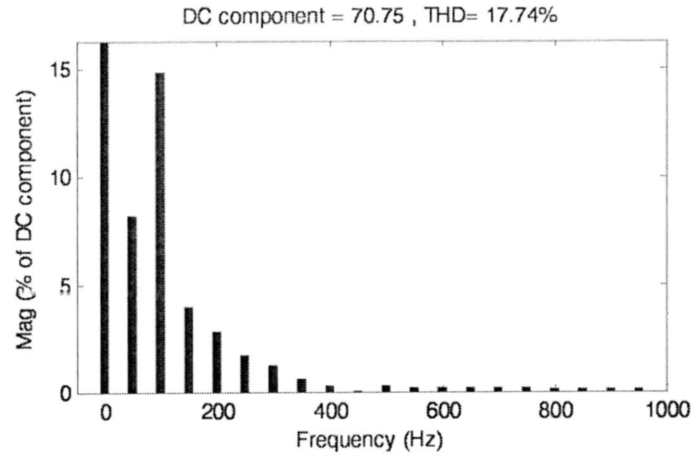

Figure 41. Power spectrum of the fuel cell current for 2.5kW load (L_{HVside} =20mH, C_{HVside} =2mF).

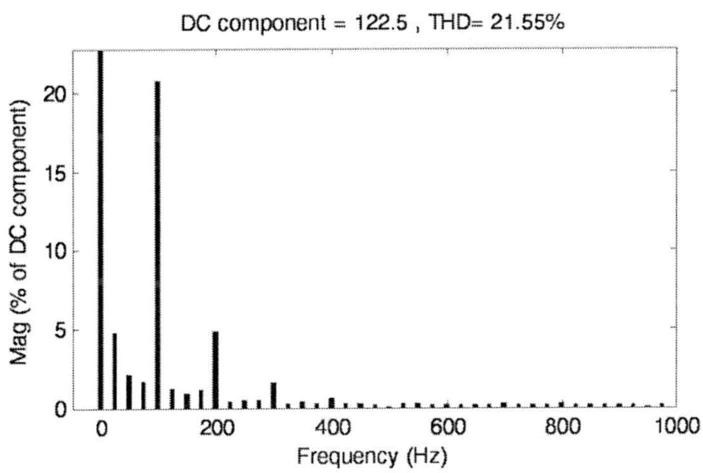

Figure 42. Power spectrum of the fuel cell current for 5kW load (L_{HVside} =20mH, C_{HVside} =2mF).

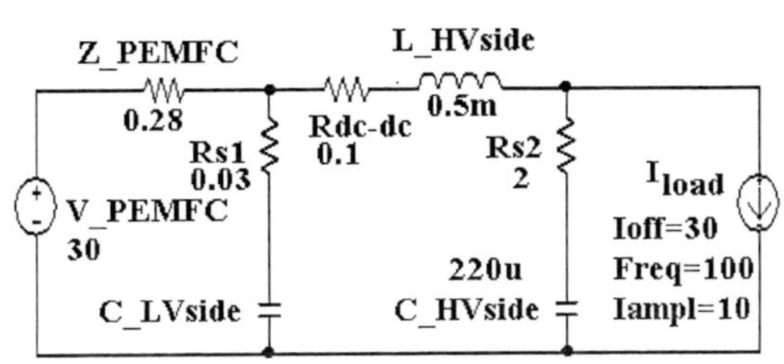

Figure 43. Current ripple model of DC-DC converter.

For L_{HVside} =0.5mH the ripple factor (RF) for PEMFC current is shown in figure 43 versus C_{HVside} . For example, a ripple factor for PEMFC current is about 15% for C_{HVside}

=0.22mF or about 5% for C_{HVside}=2.2mF. Consequently, a passive filter structure (L_{HVside} =0.5mH and C_{HVside}=0.22mF) will be considered on the HV DC bus when it is investigated the possibility to mitigate the fuel cell current ripple at the requested value by a low pass passive filter on LV DC bus.

The Low Pass Passive Filter on LV DC Bus

Simulation results for different capacity values of capacitor on LV DC bus, C_{LVside}, are made in order to compare the results when the Matlab-Simulink® diagram shown in figure 8 and the ripple model of DC-DC converter (shown in figure 18) are used, respectively. The used ripple model of DC-DC converter is shown in figure 43 and the obtained results are presented in figures 44 and 45.

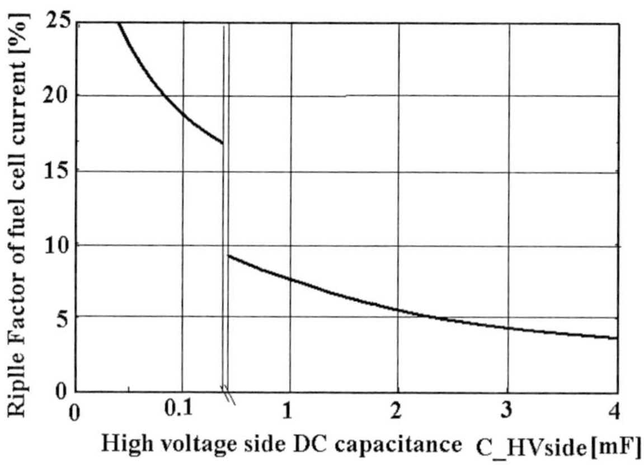

Figure 44. RF of PEMFC current vs. C_{HVside}.

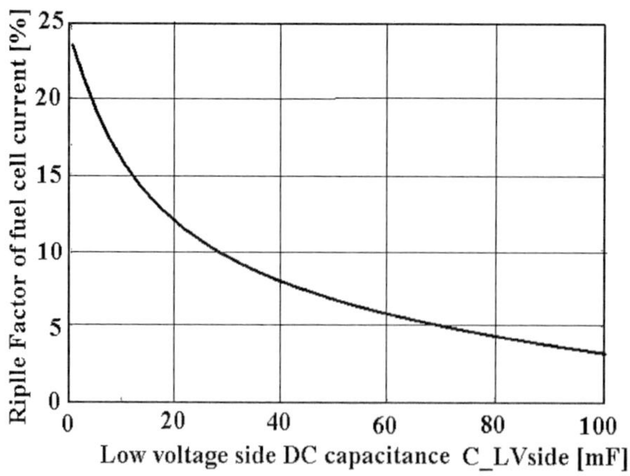

Figure 45. RF of PEMFC current vs. C_{LVside}.

For example, a ripple factor for PEMFC current is about 5% for L_{HVside}=0.5mH, C_{HVside} =0.22mF and C_{LVside}=70mF. Almost the same results are obtained using the Matlab-Simulink® diagram shown in figure 8. New technologies to achieve higher capacities (such as that of ultracapacitors) allow obtaining a reduced size for C_{LVside} stack. Also, the simulated results (which are shown in Figure 44 and 45) are similar with the reported results [7] and are validated by experiment in next section.

Experimental Results

In this section experimental results concerning the current ripple mitigation are presented for the inverter systems with different command: full wave (figure 46), modified sine PWM (figure 47) and pure sine PWM (figure 48), respectively.

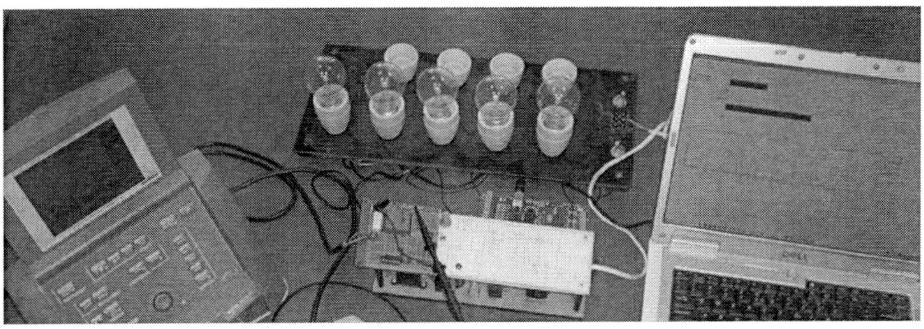

Figure 46. Experimental setup for inverter systems with full wave command.

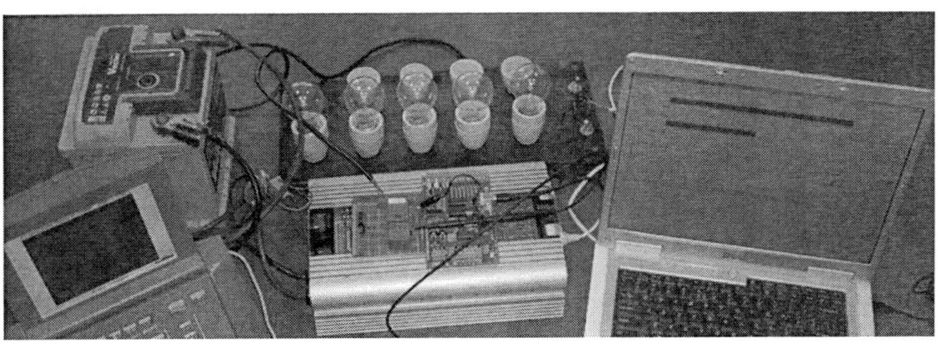

Figure 47. Experimental setup for inverter systems with modified sine PWM command.

Figure 48. Experimental setup for inverter systems with pure sine PWM command.

Laboratory tests were performed using as DC power a car battery (instead of PEMFC stack) and a variable resistive load, with maximum power of 1 kW. Car battery has characteristics of operation similar to PEMFC stack, because both DC sources are based on specific electro-chemical processes. To validate the results, experiments were repeated on a 2kW PEMFC stack, available at National Research and Development Institute for Cryogenics and Isotopic Technologies - ICSI Rm.Valcea (see appropriate section and chapter from this book).

Experimental results for above inverters system without passive filter to input are following presented in figure 49, 50 and 51 for a command type of full wave, modified sine PWM and pure sine PWM, respectively. Transfer factor of the Hall effect sensor is mention in each case (100mV/1A and 10mV/1A in figure 50.a and 51.a, respectively). The output voltage and its associated power spectrum is shown in figure 49 (a - plot and b - plot, respectively) for a full wave command of the inverter system.

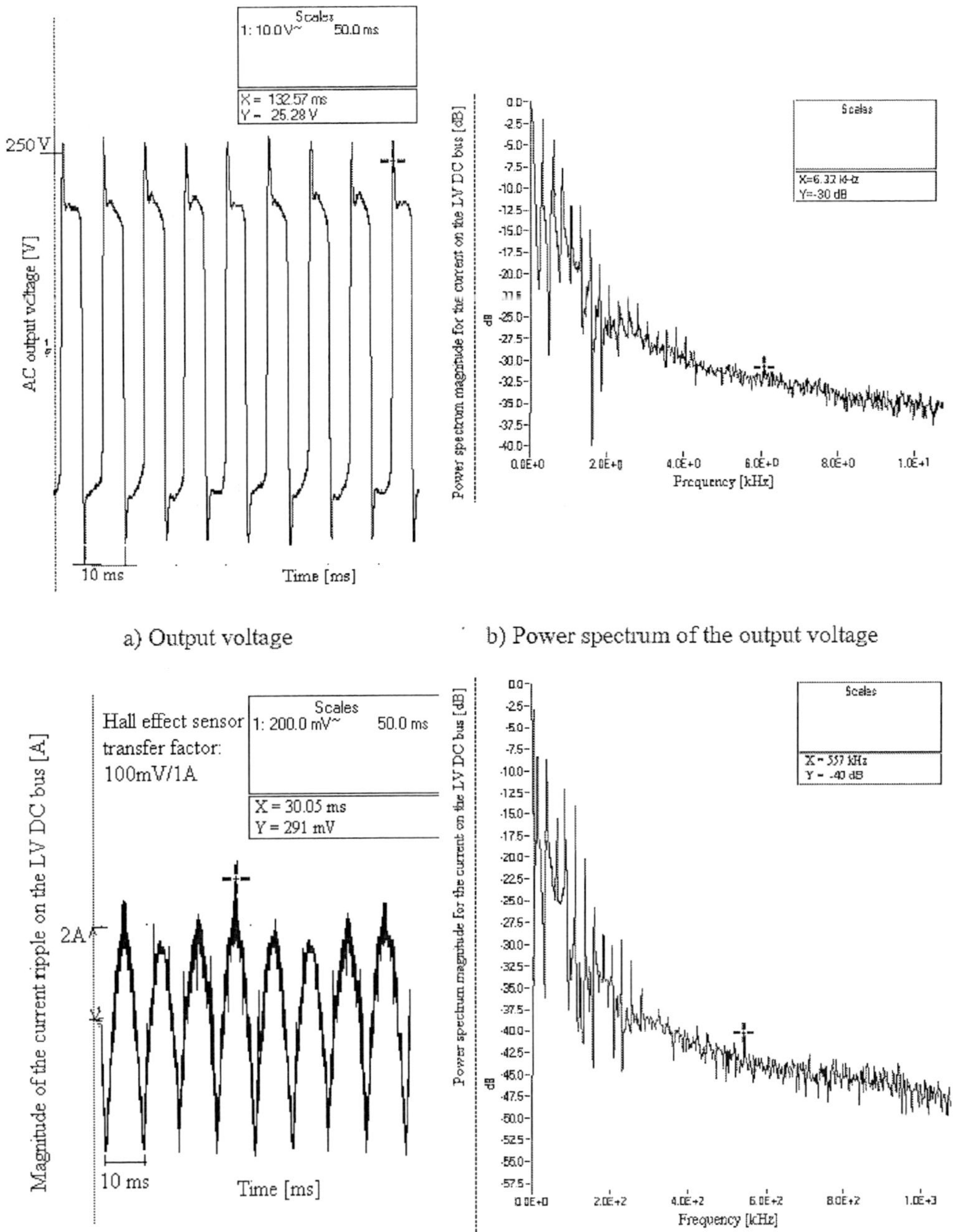

a) Output voltage b) Power spectrum of the output voltage

c) Input current ripple on LV DC bus d) Power spectrum of the input current

Figure 49. Inverter system operation with full wave command.

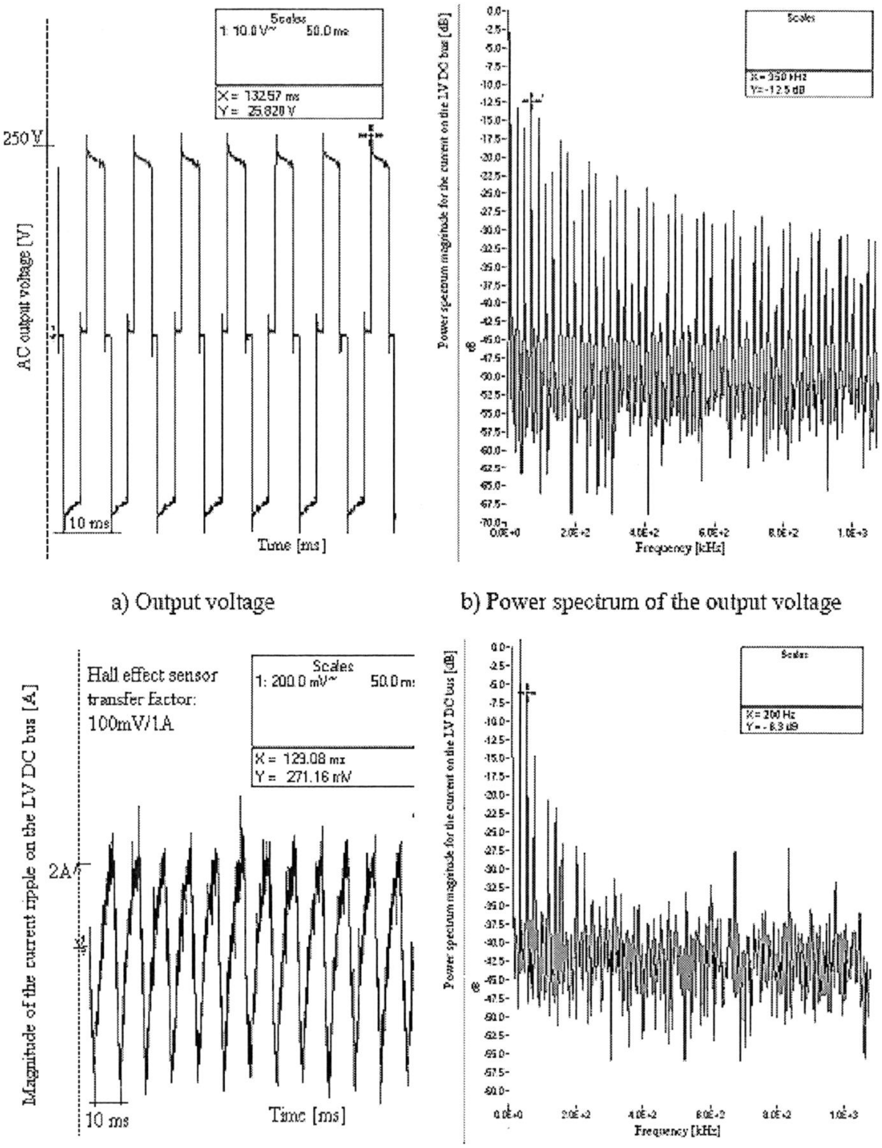

a) Output voltage

b) Power spectrum of the output voltage

c) Input current ripple on LV DC bus

d) Power spectrum of the input current

Figure 50. Inverter system operation with modified sine PWM command.

The input current ripple of the inverter system (the current ripple on LV DC bus) and its associated power spectrum (see figure 49.c and 49.d, respectively) is similar as of inverter with modified sine PWM command, which are shown in figure 50 (a - plot and b - plot, respectively).

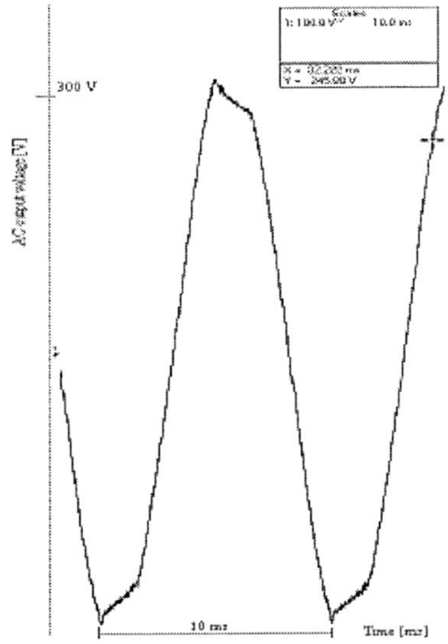

a) Output voltage

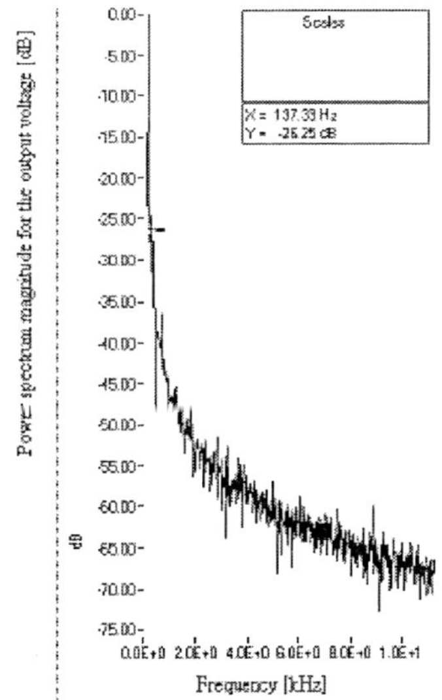

b) Power spectrum of the output voltage

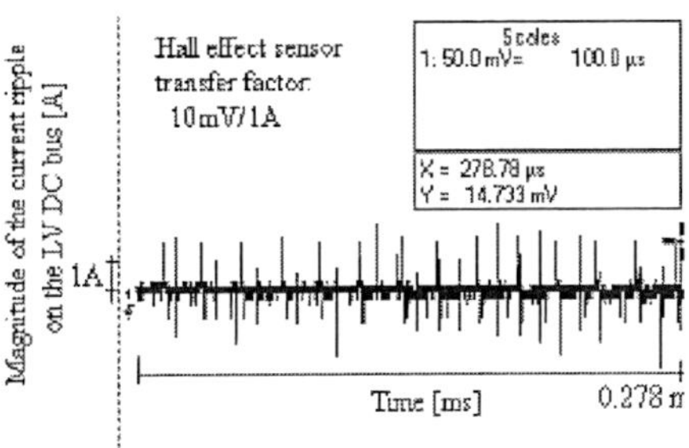

c) Input current ripple on LV DC bus (a zoom)

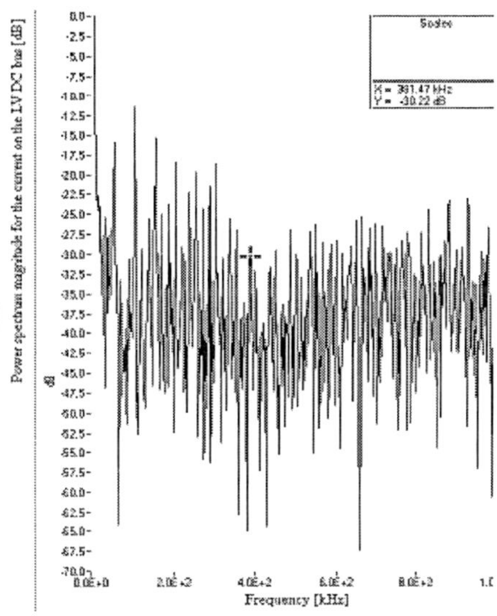

d) Power spectrum of the input current

Figure 51. Inverter system operation with pure sine PWM command.

The output voltage and its associated power spectrum are shown in figure 50 (c - plot and d - plot, respectively) for a modified sine PWM command of the inverter system, too. The input current (current on LV DC bus) and its associated power spectrum, and the output voltage and its associated power spectrum are shown in figure 51 (a - plot and b – plot, and c - plot and d - plot, respectively) for a pure sine PWM command of the inverter system. Because the LF power spectrum it was showed before, the power spectrum plots shown on the right of figures 49, 50 and 51, for the output voltage and the input current, respectively, represent the HF power spectrum (from 1kHz to 10kHz or from 1kHz to 1000kHz).

An experimental test was performed to validate the simulation results shown in figure 45. It is performed an analysis of the LF harmonics magnitude for PEMFC current in case of =0.5mH =0.22mF and different C_{LVside} values (see table 1). For all experiment the power of load is 100W and the ripple factor is around 6% (close enough to the 5% value, which was estimated by simulation). The obtained results are shown in figure 52. Plot 52.a shows that, in all cases, the fundamental harmonic has the main contribution to the current ripple. Plot 52.b shows that the harmonics mitigation on LV DC bus is significant in all cases for C_{LVside} values higher than 1mF.

In case of passive filters with structure (L_{HVside}=0.5mH and C_{HVside}=0.22mF) C_{LVside} =10mF was performed a test for different power of the load. The measured LF harmonics magnitudes are shown in table 2 and figure 53. Plot 53.a shows that for different level of load power (excluding the case with no-load) the fundamental harmonic has the main contribution to the current ripple. Plot 52.b shows that starting from the seventh harmonic (H_700) the harmonics mitigation is well in all cases. This was determined in choosing of the only the first three harmonics to modeling the inverter system by a dynamic load with current ripple as a weighted sum of their (see figure 26).

Table 1. The LF harmonics magnitude for PEMFC current in case of C_{HVside}=0.22mF and different C_{LVside} values

The harmonic amplitude of current [mA]					
C_{LVside} [mF]	Harmonics [Hz]				
	100	300	500	700	1500
0,1	364,295	110,601	64,678	47,84	21,598
0,6	363,188	104,259	55,618	36,18	10,915
1	358,665	94,721	45,693	27,855	7,2716
3	312,563	53,64	21,005	11,829	3,2614
6	232,428	30,772	12,196	7,1776	2,5022

Table 2. The LF harmonics magnitude for PEMFC current for different power values of load

	LF harmonic magnitude value [mA]				
Load [W]	H_100 Hz	H_300 Hz	H_500 Hz	H_700 Hz	H_1500 Hz
0	38	18.4	14.6	1.39	0.49
100	223	26.1	25.9	3.30	2.22
200	282	16.0	38.1	1.69	1.96
300	384	31.1	33.7	2.42	2.61
400	396	49.4	47.9	3.79	2.2
500	774	31.2	31.9	6.16	5.21

Conclusions about the Passive Filtering Technique and Further Work

Ripple factor for fuel cell is recommended to be less than 5% for an efficient and safe operation of it. An analysis of PEMFC current ripple mitigation by using passive filters was made in this section by modeling and simulation. The obtained results were validated by experiment. As a general conclusion we can say that with passive filters on HV DC bus and on LV DC bus it is possible to obtain any limit imposed in design, but this using bulky capacitors if is not used a ultacapacitor stack. On the other hand, for an EGS powered by a variable DC power source (fuel cells stack), it is recommended to be use a boost interface instead of a passive filter on LV DC voltage bus. In this case, obviously we will have a better filtering of the fuel cell current. The mitigation of first three LF harmonics increases the ripple factor in mostly, and the PEMFC life time, too. The filter on the HV DC bus must be redesigned to mitigate the current ripple under imposed limit if it is not allowed to be used the boost interface for reasons of efficiency.

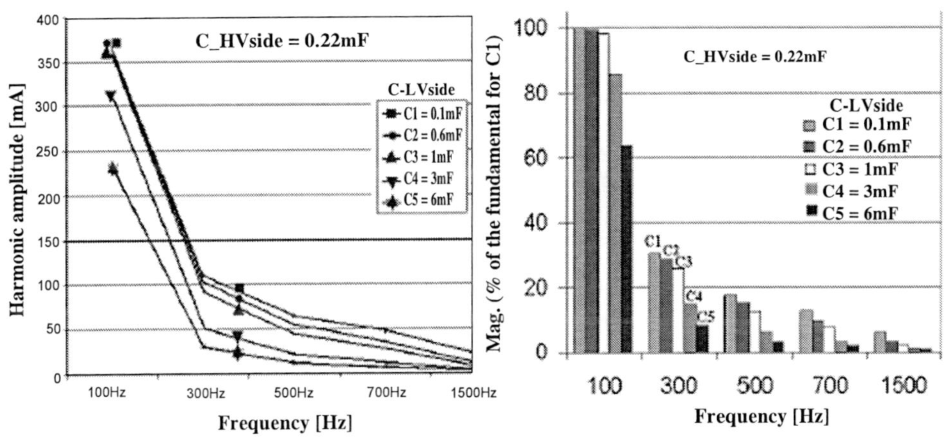

a) LF harmonic magnitude value [mA]

b) Normalized LF harmonic magnitude (% of the 100Hz harmonic for case C1=0.1mF)

Figure 52. The LF harmonic magnitude of PEMFC current for different C_{LVside} values (C_{HVside} =0.22mF).

The HV DC bus can be obtained by connecting (in parallel at input and in series to output) of several DC-DC converters (full bridge or half bridge structure) with galvanic isolation, followed by a bridge rectifier with fast diodes. This kind of power topology for inverter system will be presented in section 5. Three types of switching command (pure sine PWM, modified sine PWM and full-wave) for inverter system are analyzed as power spectrum of the output voltage and input current ripple.

Inverter systems with a rectangular full-wave switching command or a modified sine PWM control are of the grid-less type or can powered a load which accept high levels of power harmonics. For grid connected inverter system is necessary to use a pure sine PWM switching command synchronized and in phase with the grid voltage wave.

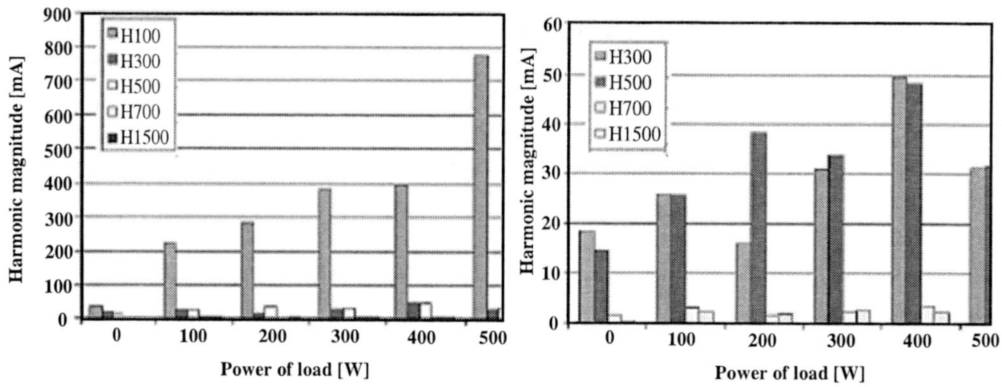

LF harmonics magnitude compared with the fundamental

LF harmonics magnitude (without fundamental)

Figure 53. The LF harmonic magnitude of PEMFC current for different power of load.

As it is mention in introductory section, for the fuel cells it is recommended a ripple factor for PEMFC current smaller than 5%. Further mitigation of the PEMFC current ripple can be achieved by other methods.

One of them is to add a boost interface between PEMFC stack and the LV DC voltage bus and will be briefly presented in next section. Other method (which will be presented in section 6) is to compensate the PEMFC current ripple by an active control of the injected current in the LV DC bus as anti-ripple to mitigate the LF harmonics of the inverter current ripple that are back propagated.

The Boost Power Interface

In this section we investigate the performances of ripple mitigation using a boost power interface between the PEMFC stack and the inverter system (see figure 3). The main goals of the boost converter are to regulate the voltage on LV DC bus and to mitigate the current ripple.

Consequently, voltage on LV DC bus and PEMFC current must be control variables. Different control techniques may be used to optimize the control performances [38] and here only two peak control techniques are presented for demonstration the ability to mitigate the current ripple by use a power interface (see figure 54).

Peak current control (PCC) laws are implemented by a fuzzy controller and a nonlinear controller shown in figure 55.a and 55.b, respectively. Peak current reference, $I_{pk(ref)}$, may be chose close to MPP current. Voltage on LV DC bus will be boosted to reference value, $V_{LV(ref)}$, by the boost converter. The ripple current is modelled by a rectified sine wave in order to compare the mitigation performances for both PCC techniques, Fuzzy Control of the PEMFC Peak Current and voltage on LV DC (FCPCLV) and Nonlinear Control of the PEMFC Peak Current and voltage on LV DC (FCPCLV). Load dynamic is modelled by a current controlled source.

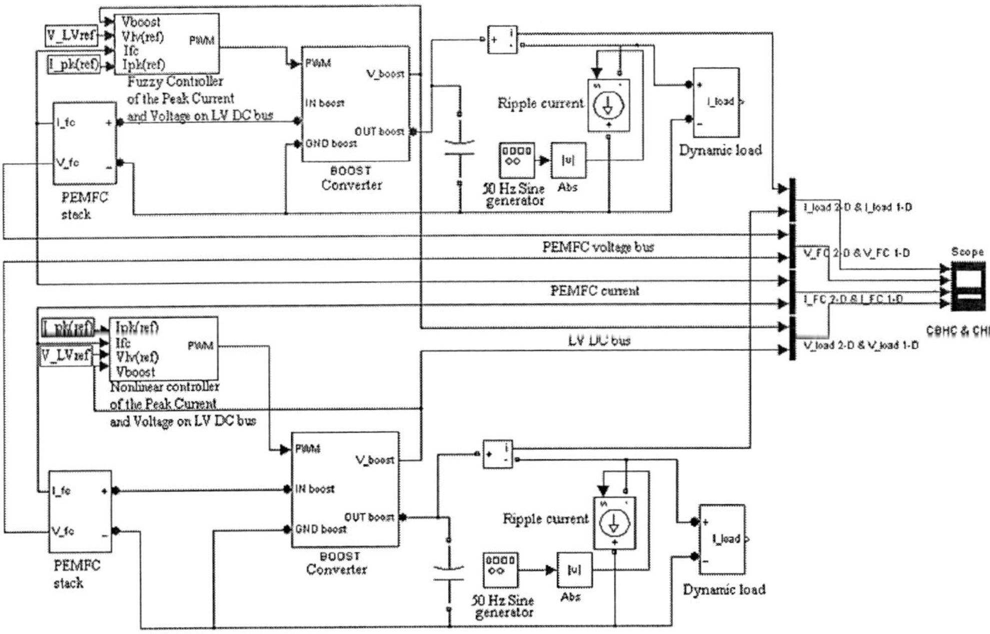

Figure 54. Boost interfaces with nonlinear and fuzzy control of the PEMFC peak current and voltage on LV DC bus.

The advantage of using a 2-D control surface for FCPCLV controller is that both control variables, fuel cell current and voltage on LV DC, directly control the time-on of the switching period. This operation mode is determined by the rules base of the Fuzzy Logic Controller (FLC).

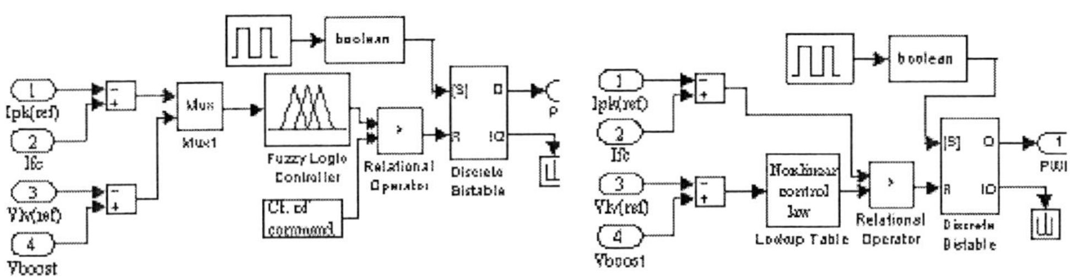

a) Fuzzy Control of the PEMFC peak current and voltage on LV DC bus (FCPCLV)

b) Nonlinear Control of the PEMFC peak current and voltage on LV DC bus (NCPCLV)

Figure 55. Control of the PEMFC peak current and the voltage on the LV DC bus.

In figure 56 is shown the fuel cell voltage ripple (bottom) and current ripple (middle) for the same load current ripple (top) using a boost power interface with peak current control of fuzzy (thin line) and nonlinear (thick line) type. The 1-D nonlinear control law was chosen for an imposed PEMFC ripple current (ΔI_{PEMFC}) from the 2-D fuzzy control surface projections obtained for a given ΔI_{PEMFC}-cut. The voltage boost factor is almost equal with ratio of DC component for PEMFC and load:

$$\eta_{Boost} = \frac{I_{(AV)load} V_{LVref}}{I_{(AV)PEMFC} V_{(AV)PEMFC}} \cong 1 \Rightarrow \frac{V_{LVref}}{V_{(AV)PEMFC}} \cong \frac{I_{(AV)PEMFC}}{I_{(AV)load}} \quad (5)$$

So, for a voltage boost factor in range 2÷5 and a PEMFC current of about 40A (see figure 54) the PEMFC voltage is about 30V (see figure 22) and the load current is in range 20A÷8A, respectively.

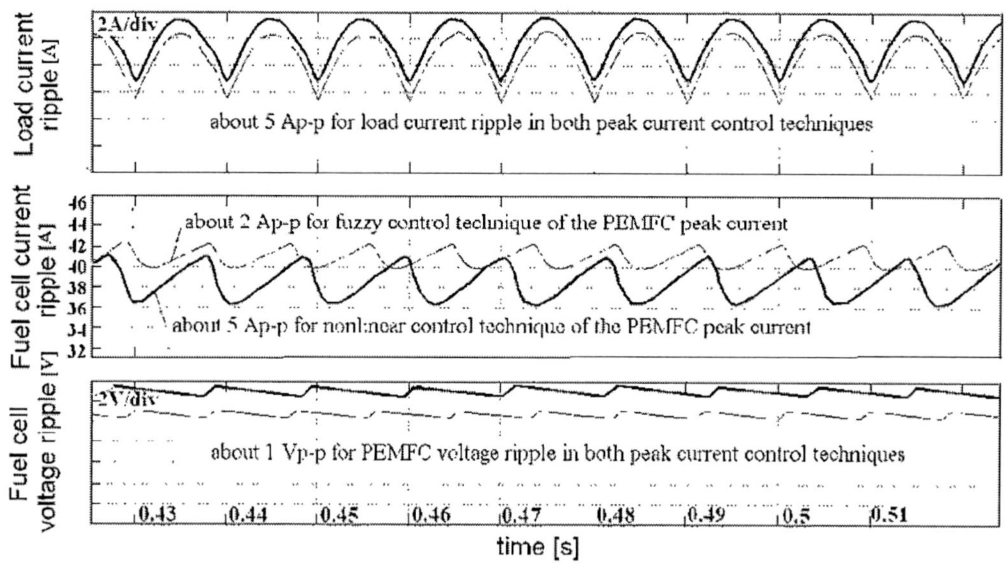

Figure 56. Fuel cell voltage ripple (bottom) and current ripple (middle) for the same load current ripple (top) using a boost power interface with peak current control of fuzzy (thin line) and nonlinear (thick line) type.

The constant of command adjusts the time-on for lowest current ripple. Instead of this 2-D control surface, the NCPCLV controller uses a 1-D control surface that gives a current ripple reference to be compared with fuel cell current ripple. Control is less flexible and thus performances of ripple mitigation are relatively low in this case. The ripple factor is about 2A/40A=5% and 5A/40A=7.5% for case of using the FCPCLV controller and NCPCLV controller, respectively. Better performances of ripple mitigation are obtained if the power interface is of multiport type (two input and one output in figure 6) that use at least an auxiliary energy source to supply the Controlled Current Source (CCS) that generates an anti-ripple. The anti-ripple that compensate the PEMFC LF current ripple is generated by an active control in tracking loop of gained PEMFC current ripple by the CCS current shape. The nonlinear gain is designed to best mitigate this ripple. Some results are presented in one before last section. Next section presents an efficient way to mitigate the HF current ripple by interleaving switching technique applied at level of the DC-DC converter.

The Inverter System with Modified Sine PWM Command

The implemented inverter system structure with modified sine PWM command has the main electrical schemes shown in figures 57 and 58 and all electrical schemes are presented in [38]. The DC-DC converter is shown in figure 57 and DC-AC converter in figure 58.

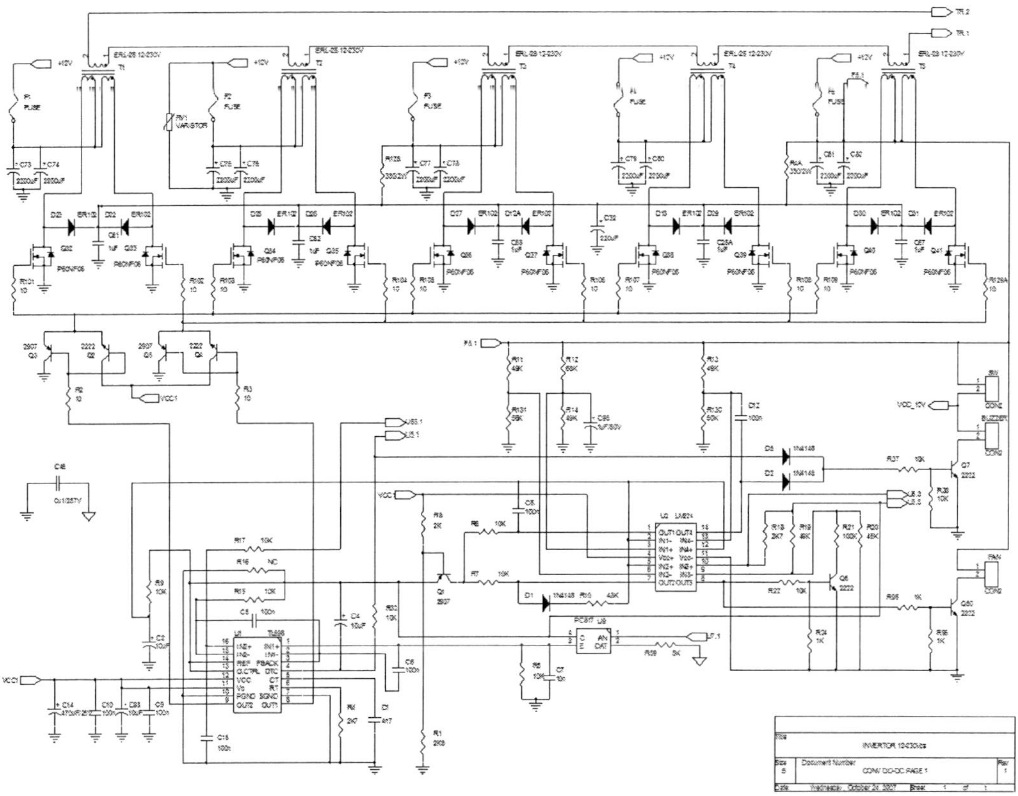

Figure 57. DC-DC converter (the power circuit part without: rectifier bridge, interleaved PWM command circuit, protection circuit and lighting indicators).

Simulation Results

The Matlab-Simulink® diagram from figure 59 represents the DC-DC converter that is obtained as in figure 55 by input paralleling of five low power DC-DC converters, which connect in series to the outputs of HF transformers in order to be rectified by one single bridge [39]. The load current ripple is modelled by a current controlled source and the used DC energy source is modelled by a 12V/2kW fuel cell stack. The interleaved switching technique is simple implemented by delaying the PWM pulses with $T_{sw}/5$, where T_{sw} is the switching period.

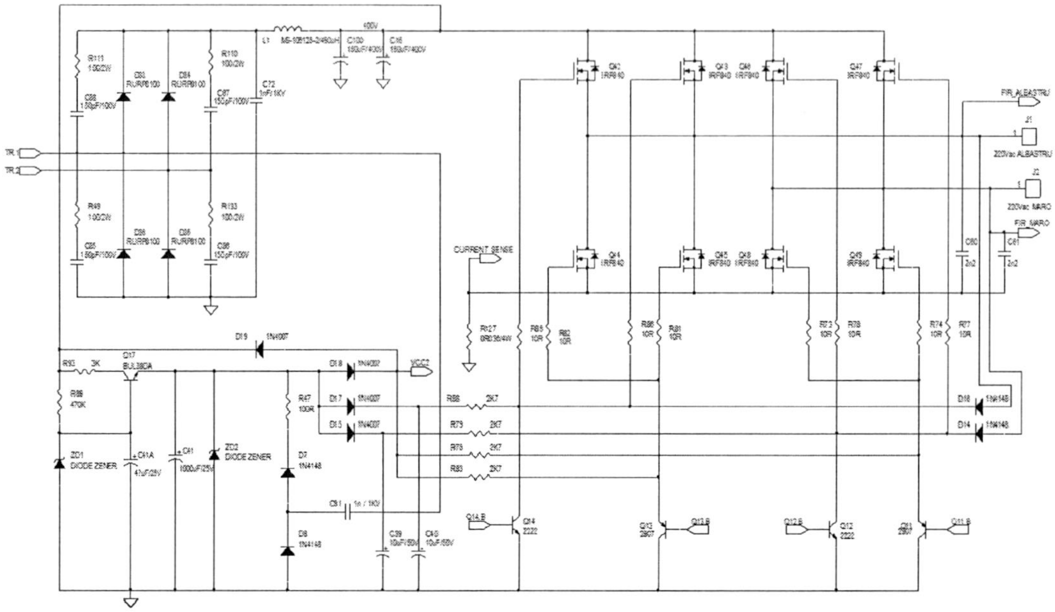

Figure 58. DC-AC converter (power circuit part including rectifier bridge and passive filter on HV DC bus, drivers for IGBT's and auxiliary power DC source for drivers; but without modified sine PWM command circuit.

The DC-DC converter behaviour is shown in figure 60 and 61 at level of HV DC bus and LV DC bus, respectively, for a load dynamic type pulse (320Ω/160Ω/320Ω) and passive filters' parameters L_{HVside}=0.5mH, C_{HVside}=0.2mF, C_{HVside}=100mF. The $1A_{p-p}$ load current ripple is modelled by a rectified 100Hz sine wave that controls a current controlled source. The DC-DC converter is powered by 2kW/12V PEMFC stack and the fuel cell behaviour is shown in figure 61 (signals of current, voltage and power from top to bottom). The fuel cell LF current ripple is about $10A_{p-p}$ (peak to peak) and $7A_{p-p}$ for PEMFC current of about 100A and 40A, respectively. It results a ripple factor of 10A/100A=10% (< 15%) and 7A/40A=17,5% (> 15%) from 75% and 25% of ratted load, respectively. The fuel cell current shape and its power spectrum is shown in figure 62 for light load (where light load means 25% of ratted load, so about 0,5kW).

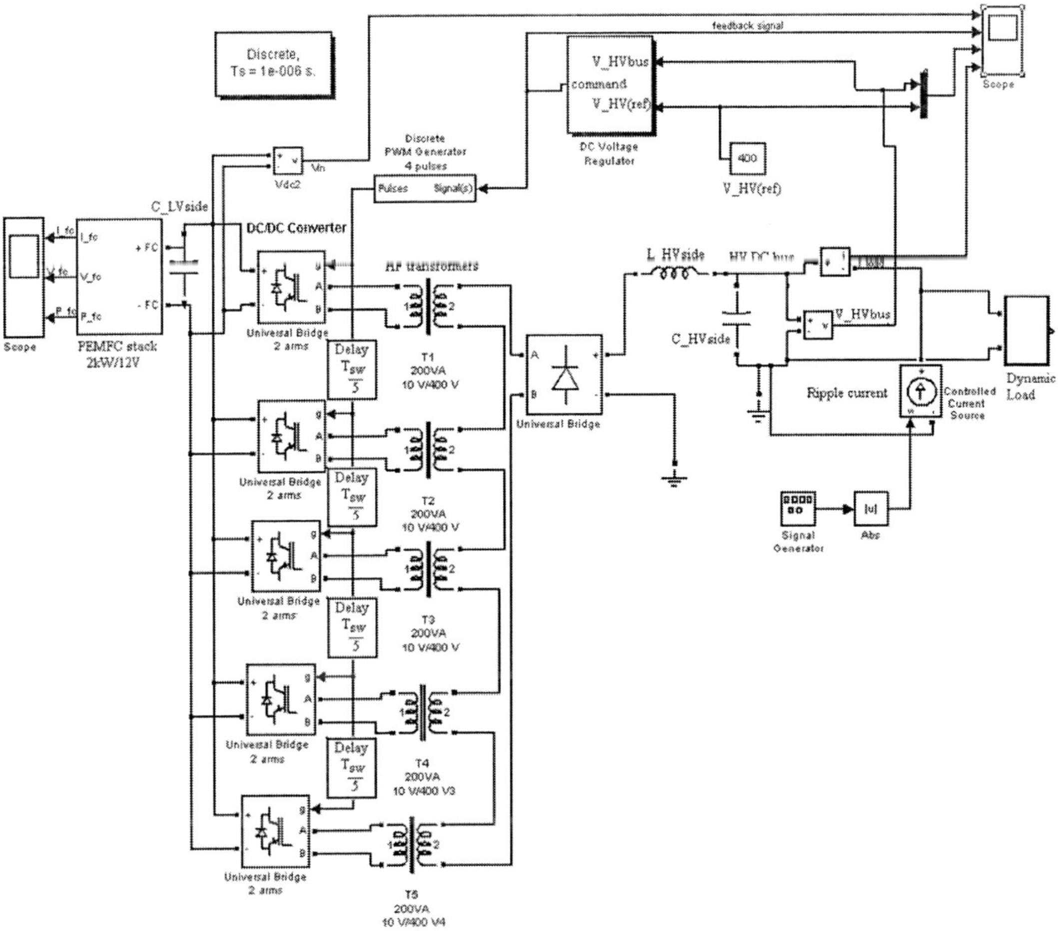

Figure 59. Matlab diagram of the DC-DC converter.

Experimental Results

The experimental fuel cell current ripple and inverter output voltage are shown in figure 63 and 64, respectively, for an inverter system without boost power interface, but with interleaving switching commands technique. Transfer factor of the Hall sensor is 500mV/1A. The PEMFC ripple factor is $RF_{I(PEMFC)} \cong 15A/100A = 15\%$ for 40A DC current delivery by PEMFC stack. These results validate the simulation results obtained with the PEMFC model proposed for ripple evaluation.

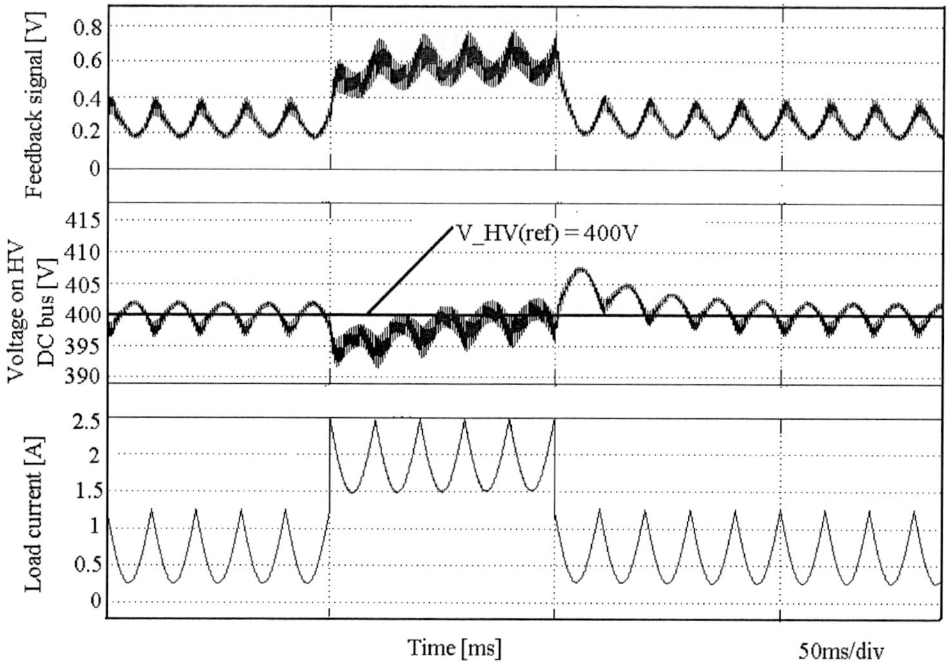

Figure 60. The DC-DC converter behaviour for pulse load.

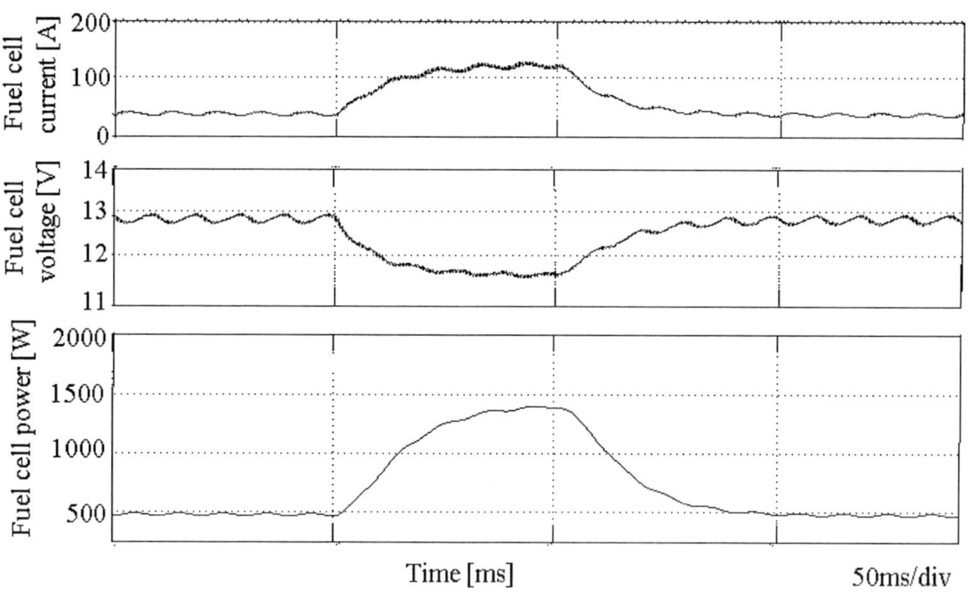

Figure 61. Fuel cell current (top), voltage (middle) and power (bottom) for pulse load.

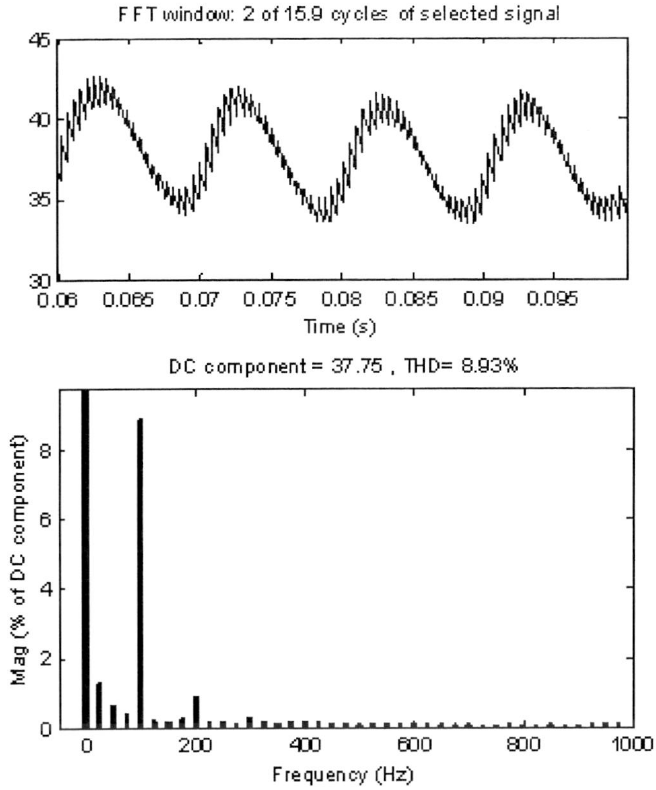

Figure 62. Fuel cell current (top) and its power spectrum (bottom) for light load of 0.5kW.

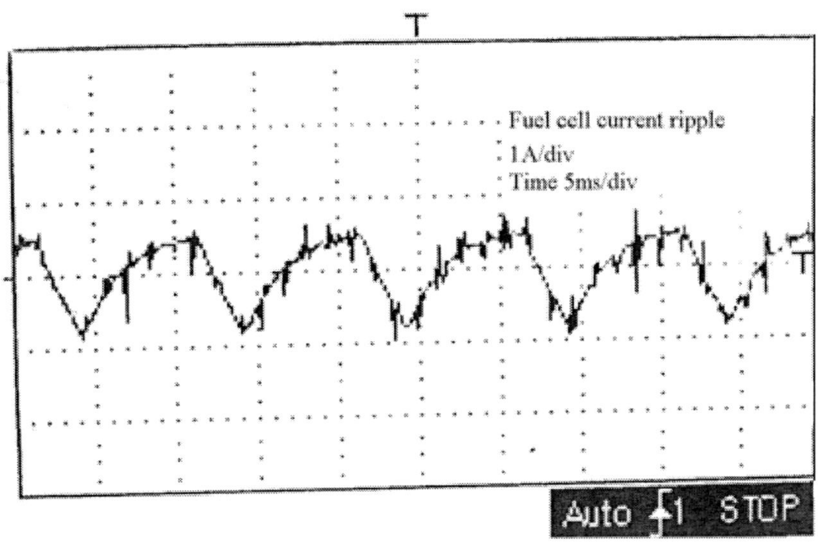

Figure 63. Fuel cell current ripple.

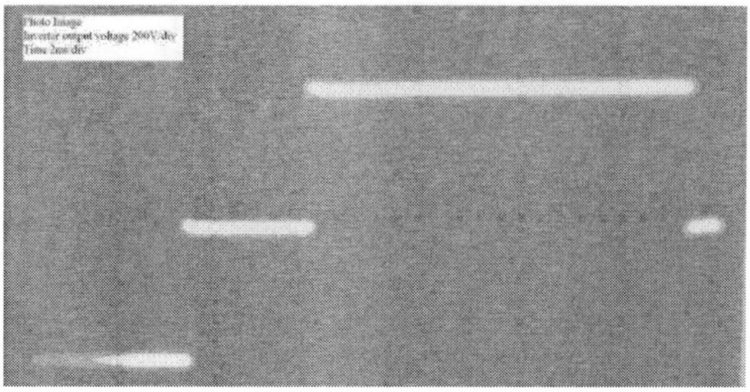

Figure 64. Inverter output voltage.

Simulation Results for Active Filtering

The new recommended limit of 3% for LF current ripple isn't easy to be satisfied using the filtering techniques that were presented in the previous sections. Active control of a tracking loop to generate an anti-ripple that can best compensate the inverter current ripple may be o solution for PEMFC current ripple mitigation [28; 29]. The proposed PEMFC power interface is a bi-buck topology shown in figure 65.

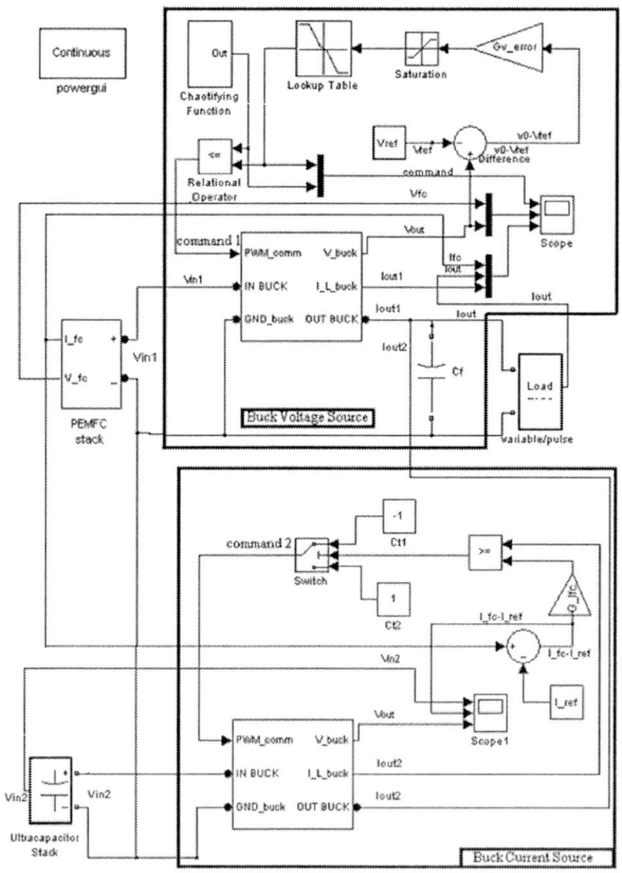

Figure 65. Bi-buck topology – Simulink diagram (adapted from [28]).

The Buck Voltage Source, BVS, is controlled by the reference voltage, V_{ref}, and the Buck Current Source, BCS, is controlled by the PEMFC gained current ripple, $G_{Ifc} \cdot (I_{FC}-I_{ref})$. The BCS try to follow the PEMFC LF current ripple, ΔI_{PEMFC}, in the control loop if $G_{Ifc} \cdot \cong \Delta I_{out}/\Delta I_{PEMFC} \cong I_{out}/I_{PEMFC} = (\eta_{BVS} \cdot V_{PEMFC})/V_{out}$, where V_{PEMFC} is the input voltage for BVS. Consequently, BCS current, I_{out2}, is a estimation of the load current ripple, ΔI_{out}, and $I_{out1(AV)} \cong I_{out(AV)}$.

If the PEMFC reference current (I_{ref}) is close to MPP then PEMFC LF current ripple is much smaller.

The load current shape that simulates inverter current is a superposition of the first three 100Hz inverter harmonics over nominal load current (see figure 26). The 2.2kW/36V PEMFC stack powered the BVS and second energy source that powered the BCS may be a low energy source (for example, can be an ultracapacitors stack). It is shown that proposed current control is independent by the V_{in2} source level. The BVS control is of nonlinear type and for spread the HF power spectrum a chaotifying techniques is used. For best performance of ripple mitigation the BCS control may be of nonlinear type, too. Further details about nonlinear control of the BCS is given in [30; 31] and other results are presented in book chapter dedicated to control field. Below, in figure 66, is shown a simulation of the bi-buck topology in order to estimate the ripple factor of PEMFC current.

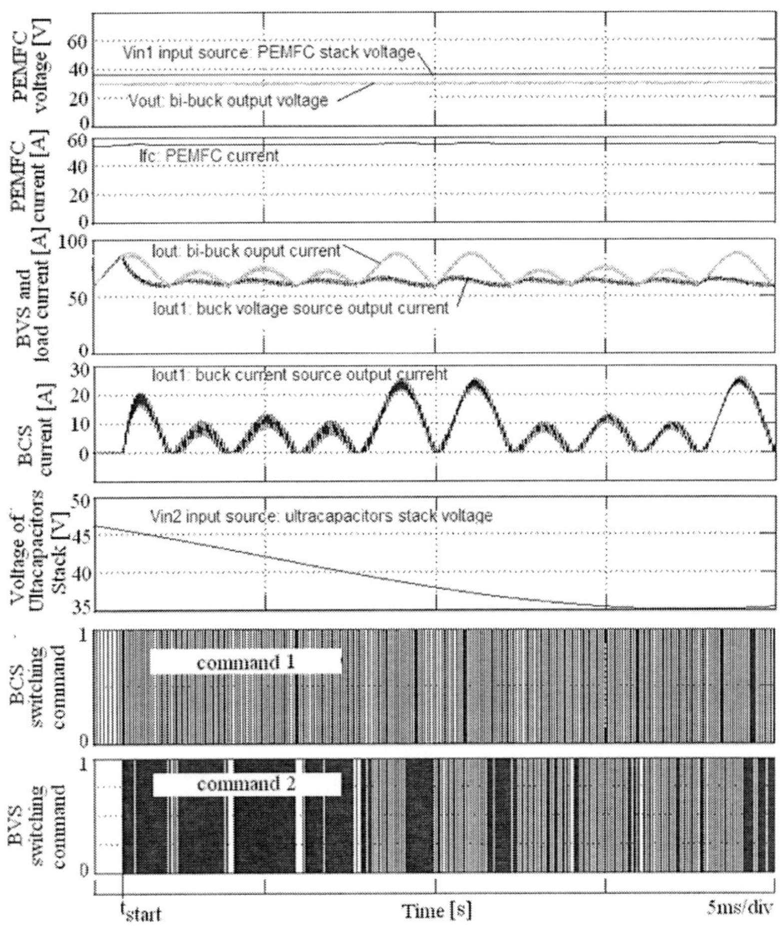

Figure 66. Bi-buck interface behaviour (adapted from [28]).

The BCS control loop start to track the PEMFC current ripple at t_{start} time. From this moment the load current ripple is compensated by the anti-ripple generated by BCS and the BVS output current ripple becomes small. Consequently, the PEMFC current ripple is mitigate by this active control of tracking loop.

The current shapes (top plots) and power spectrums (bottom plots) for load and PEMFC stack are shown in figure 67.a and 67.b, respectively, for light load. The reduction of the harmonics ratio, Rk=(load current harmonic level)/(PEMFC current harmonic level), is shown in table 3. In order to compare the bi-buck interface performance of ripple mitigation, the table 4 presents the PEMFC current harmonics with and without buck current source as % of DC component. Last row present the Mitigation Performance Ratio (MPR) of these harmonics with and without BCS. MPR is calculated with relation:

MPR = (PEMFC current harmonic level without BCS) / (PEMFC current harmonic level with BCS).

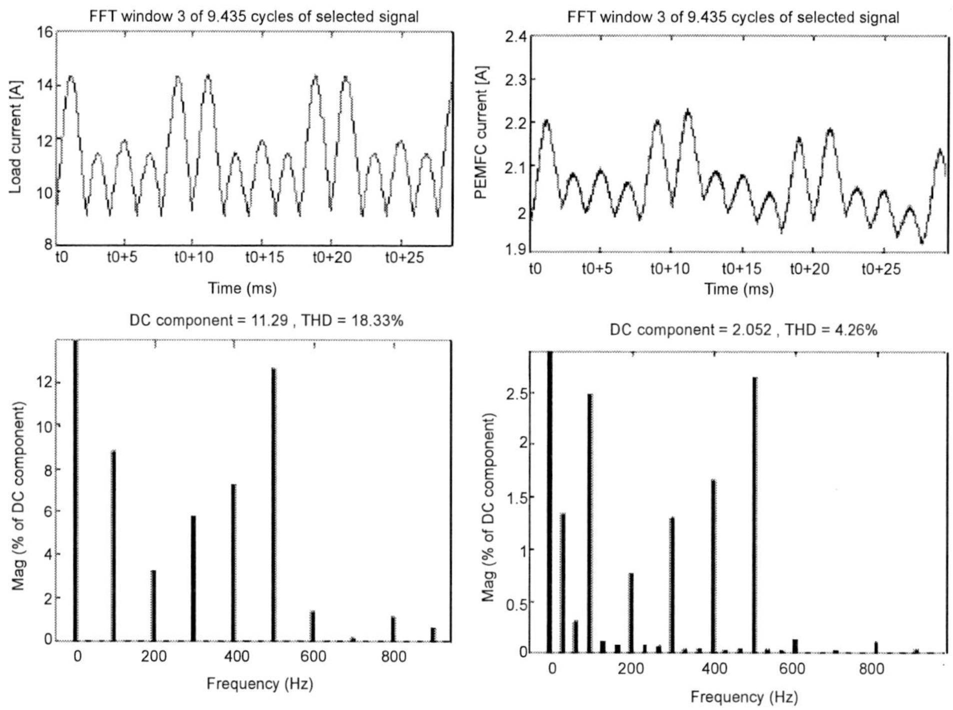

a) Load current (top) and its power spectrum (bottom)

b) PEMFC current (top) and its power spectrum (bottom)

Figure 67. Current shapes (top plots) and power spectrums (bottom plots) for: a) load and b) PEMFC stack (adapted from [28]).

The LF harmonics are mitigated by R_k value from about 18 to 27.

The results shown in table 4 reveal more clearly the mitigation performance when a BCS is used in active compensation of the inverter current. For inverter LF current harmonics the reduction ratio are in range 1.6÷2.

Also, in case of light load, the ripple factor, RF, for load current and PEMFC current are:

$RF_{load\_current} \cong (14.5\text{-}9)A/11.29A \cong 50\%$;

$RF_{PEMFC\_current} \cong (2.2\text{-}2)A/2.052A \cong 10\%$.

Better results are obtained for load power levels near to the ratted load power. The current shapes (top plots) and power spectrums (bottom plots) for load and PEMFC stack are shown in figure 68.a and 68.b, respectively, for rated load.

Table 3. The Rk value for LF harmonics (adapted from [28])

Harmonic frequency [Hz]	DC	100	200	300	400	500
R_k relation	$R_0 = \frac{11.29}{2.052}$	$\frac{8.4}{2.5}\times R_0$	$\frac{3.2}{0.73}\times R_0$	$\frac{5.7}{1.3}\times R_0$	$\frac{7.2}{1.53}\times R_0$	$\frac{12.8}{2.6}\times R_0$
R_k value	5.5019	18.49	24.12	24.12	25.89	27.09

Table 4. The MPR value for LF harmonics (adapted from [28])

Harmonic frequency [Hz]	100	200	300	400	500
Harmonic level with BCS [% of DC component]	2.5	0.73	1.3	1.53	2.6
Harmonic level without BCS [% of DC component]	4.1	1.3	2.5	1.8	3.1
MPR	1.64	1.82	1.93	1.16	1.18

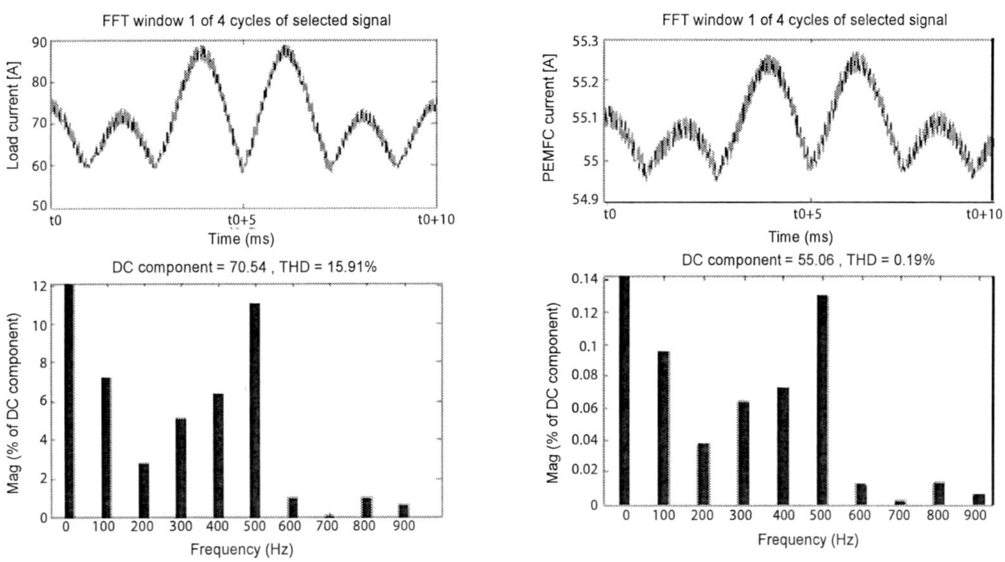

a) (top) and its power spectrum (bottom

b) PEMFC current (top) and its power spectrum (bottom)

Figure 68. Current shapes (top plots) and power spectrums (bottom plots) for load and PEMFC stack (adapted from [28]).

In case of rated load the ripple factor, RF, for load current and PEMFC current are of about:

$RF_{load\_current} \cong (88\text{-}60)A/71.25A \cong 40\%$;

$RF_{PEMFC\_current} \cong (55.25\text{-}54.95)A/55.08A \cong 0.55\%$.

To maintain the low RF across the range power of load a nonlinear controller must replace the fixed gain G_{Ifv} in tracking BCS loop.

Final Conclusion

The fuel cell current ripple can be mitigated by using a proper power interface in the low DC bus, between the PEMFC stack and inverter system. If the voltage on high DC bus is robustly controlled, then the BVS may be removed for high efficiency of the fuel cell EGS. The mitigation of the inverter ripple on high DC bus may be obtained with a cheap passive filter, but then the most part of remained ripple is back propagated on the low DC bus. This ripple can be mitigated by an appropriate active control of tracking BCS loop. For higher efficiency the PEMFC stack must operate close to MPP, so the HPS architecture is recommended, too.

Acknowledgments

The CEEX Grant #226 and CNCSIS Grant #570 of the National Research Council (MEC) have supported part of the research for this book chapter. Also, the author wishes to thank the research team for his effort in implementing the results in these contracts, especially to OPROESCU Mihai, MAZARE Alin, RADUCU Marian, and LAURENTIU Ionescu.

References

[1] Suddhasatwa, B. (2007), Recent Trends in Fuel Cell Science and Technology, New York, USA, Springer press.

[2] Wajiha, S., Rahul, A.K., and Arefeen, M. (2006), Analysis and minimization of input ripple current in PWM inverters for designing reliable fuel cell power systems, *Journal of Power Sources* 156, 448–454.

[3] Choi, B., and al. (2007), Analyze on input filter interaction in switching power converter, *IEEE Trans. on Power Electronics* 22(2), 452 – 460.

[4] Fontes, G., and al. (2007), Interactions between fuel cells and power converters: Influence of current harmonics on a fuel cell stack, *IEEE Trans. on Power Electronics* 22(2), 670-678.

[5] Gemmen, R.S. (2001), Analysis for the effect of the ripple current on fuel cell operating condition, *ASME* 369 (4), 279–289.

[6] Woojin, C., Gyubum, J., Prasad, N.E., and Jo, W.H. (2004), An Experimental Evaluation of the Effects of Ripple Current Generated by the Power Conditioning Stage on a Proton Exchange Membrane Fuel Cell Stack, JMEPEG 13, 257-264.

[7] Schmittinger, W., and Vahidi, A. (2008), A review of the main parameters influencing long-term performance and durability of PEM fuel cells, *J. Power Sources* 180, 1–4.

[8] Liu, C., and Lai, J. S. (2007), Low Frequency Current Ripple Reduction Technique with Active Control in a Fuel Cell Power System with Inverter Load, 22(4), 1453 – 1463.

[9] Thounthong, P., Davat, B., Raël, S., and Sethakul, P. (2009), An Overview of Power Converters for a Clean Energy Conversion Technology, *IEEE Industrial Electronics magazine* 3 (1), 32-46.

[10] Hwang, J.-C., Chen, L.-H., and Yeh S.-N. (2007), Comprehensive analysis and design of multi-leg fuel cell boost converter, *Applied Energy* 84, 1274–1288.

[11] Oproescu, M., Bizon, N., and Sofron, E. (2009a), Efficiency of Current Ripple Passive Filtering Techniques for Inverter System Supplied by A Fuel Cell - Modelling and Simulations, 3rd ed. of Int. Conf. on Electronics, Computers and Artificial Intelligence - ECAI'09 2, 61-66.

[12] Zhong, Z.-D., Huo, H.-B., Zhu, X.-J., Cao, G.-Y., and Ren, Y. (2008), Adaptive maximum power point tracking control of fuel cell power plants, *J. of Power Sources* 176, 259–269.

[13] Methekar, R.N., Patwardhan, S.C., Gudi, R.D., and Prasad, V. (2010), Adaptive peak seeking control of a proton exchange membrane fuel cell, *Journal of Process Control* 20 73–82.

[14] Bizon, N. (2010a), On tracking robustness in adaptive extremum seeking control of the fuel cell power plants, *Applied Energy* 87, 3115–3130.

[15] Thounthong, P., and Davat, B. (2007), A PEM fuel cell power source for electric vehicle applications with supercapacitor or battery as auxiliary, Progress in Fuel Cell Research (Eds.: P. V. Alemo, Ed. Commack), NY.

[16] Lukic, S. M., Cao, J., Bansal, R. C., Rodriguez, F., and Emadi A., (2008), Energy storage systems for automotive applications, *IEEE Trans. Ind. Electron* 55, 2258–2267.

[17] Kim M. J., and Peng, H. (2007), Power management and design optimization of fuel cell/battery hybrid vehicles, *J. Power Sources* 165, 819–832.

[18] Tang, Y., Yuan, W., Pan, M., Li, Z., Chen, G., and Li, Y. (2010), Experimental investigation of dynamic performance and transient responses of a kW-class PEM fuel cell stack under various load changes, *Applied Energy* 87, 1410–1417.

[19] Rodatz, P., Paganelli, G., Sciarretta, A., and Guzzella, L. (2005), Optimal power management of an experimental fuel cell/supercapacitor powered hybrid vehicle, *Control Eng. Practice* 13 (1), 41–53.

[20] Corbo, P., Corcione, F.E., Migliardini, F., and Veneri, O. (2009), Experimental study of a fuel cell power train for road transport application, *J. Power Sources* 145, 610–619.

[21] Thounthong, P., Chunkag, V., Sethakul, P., Sikkabut, S., Pierfederici, S., and Davat, B., (2010), Energy Management of a Fuel Cell/Solar Cell/Supercapacitor Hybrid Power Source, *Journal of Power Sources* doi:10.1016/j.jpowsour.2010.01.051.

[22] Uzunoglu M., and Alam, M.S. (2006), Dynamic modeling, design, and simulation of a combined PEM fuel cell and ultracapacitor system for standalone residential applications, *IEEE Trans. Energy Conversion* 21, 767–775.

[23] Xu, L., Li, J., Hua, J., Li, X., and Ouyang, M. (2009), Adaptive supervisory control strategy of a fuel cell/battery-powered city bus, *J. Power Sources* 194, 360–368.

[24] Thounthong, P., Raël, S., and Davat, B. (2009), Energy management of fuel cell/battery/supercapacitor hybrid power source for vehicle applications, *Journal of Power Sources* 193 (1), 376-385.

[25] Bizon, N., Lefter, E., and Oproescu, M. (2007), Modeling and Control of the Energy Sources Power Interface for Automotive Hybrid Electrical System, 21st JUMV Int. Automotive Conf. on Science and Motor Vehicles, proceedings on CD, Paper NMV0752 – Pages 12, http://riedal-dz.com/resources/52_bizon.pdf, ed. by the Society of Automotive Engineers of Serbia under FISITA patronage, Belgrade.

[26] Bizon, N. (2007), Intelligent Integrated Control of the Power Flows into an Energy Generation System, Mediterranean Journal of Measurement and Control 3(3) 113-125.

[27] Raducu, M., and Bizon, N. (2007), Efficient Energy Generation System using a Thiristors Inverter Topology, 2nd ed. of Int. Conf. on Electronics, Computers and Artificial Intellig. - ECAI'07 2, 67-70.

[28] Bizon, N., Raducu, M., and Oproescu, M. (2008), Fuel cell current ripple minimization using a bi-buck power interface, 13th International Power Electronics and Motion Control Conference, *IEEE Catalog Number* CFP0834A-CDR, 621-628.

[29] Bizon, N. (2008), Intelligent Control of Fuel cell Energy Generation System, Research report no. 3 of CNCSIS Grant # 570 (accessed in 2 may 2010 on www.cncsis.ro or directly at http://194.102.64.7/GranturiFinalizate/faces/Projects/ProjectsList.jsp).

[30] Bizon, N., and Oproescu, M. (2009), Modelling the Bi-Buck Interface for Inverter Current Ripple Compensation, *Int. J. on Technical and Physical Problems of Eng*. 1(1), 32-37.

[31] Bizon, N., and Oproescu, M. (2010), Control Performances of the Bi-Buck Interface used for Inverter Current Ripple Compensation, *Int. J. on Technical and Physical Problems of Eng*. 1(1), 5-10.

[32] Bizon, N. (2010c), Hybrid Power Source for Vehicle Applications Operating at Maximum Power Point of Fuel Cell Stack, *Int. Conf. on Applied Electronics* 2010 - APPEL'10, pp. 43-46.

[33] Oproescu, M., Bizon, N., and Sofron, E. (2009b), Efficiency of Current Ripple Passive Filtering Techniques for Inverter System Supplied by A Fuel Cell - Experimental Results, 3rd ed. of Int. Conf. on Electronics, Computers and Artificial Intelligence - ECAI'09 2, 67-74.

[34] Woojin, C., Jo, H. W., and Prasad, E. (2006), Development of an equivalent circuit model of a fuel cell to evaluate the effects of inverter ripple current, *Journal of Power Sources* 158, 1324–1332.

[35] Bizon, N. (2010b), Development of a fuel cell stack macro-model for inverter current ripple evaluation, *Revue Roumaine des Sciences Techniques series Électrotechnique et Énergetique* 55(4) (2010), 405-416.

[36] Bizon, N., Oproescu, M. (2007), Some Aspect Concerning the Fuel Cell Current Ripple as Interaction with an Inverter System, Int. Conf. on Electronics, Computers and Artificial Intelligence - ECAI'07 2, 122-132.

[37] Bizon, N. (2009), Active Filtering Techniques for Fuel Cell Stack Current Ripple, Int. Conf. on Electronics, *Computers and Artificial Intelligence* - ECAI'09 1, 33-43.

[38] Bizon, N. (2008), Control of the Bi-Buck Power Interface used for Inverter Current Ripple Minimization, *Sci. J. of the Electrical Eng. Faculty* - Valahia Univ. 1, 7-12.

[39] Wang Y., Choi, S., and Lee, E. (2009), Efficient and ripple-mitigating dc–dc converter for residential fuel cell system, *Int. J. of Electrical Power and Energy Systems* 31(1), 43-49.

In: Advances in Energy Research
Editor: Nicu Bizon

ISBN 978-1-61209-991-0

Chapter 10

APPLICATIONS OF ARTIFICIAL INTELLIGENCE IN POWER ELECTRONICS

Ires Iskender*
Department of Electrical-Electronics Engineering,
Gazi University, Ankara, Turkey

ABSTRACT

The artificial intelligence which is an important branch of computer science or computer engineering includes expert system, fuzzy logic, artificial neural network and genetic algorithm. The application of artificial intelligence in electrical engineering, power electronics and motion control has been rapidly extended recently. The objective of this chapter is to describe how neuro-fuzzy techniques can be applied in the automatic control area, particularly in feedback control. In addition, the fuzzy logic and neural network principles are briefly studied and the applications of neural networks techniques in converter modeling and the control of power electronics converters are described. The results of simulations and experiments for different applications are also given.

Keywords: Fuzzy logic, neural network, control, high dc-voltage generator, PV panel, MPP.

INTRODUCTION

There are many power conversion applications where the output signal is controlled to satisfy the load requirements. In most applications there is no a linear relationship between the input and output parameters in terms of the control parameters. Sometimes the relationship between input and output parameters is very complex. For controlling such a complicated system a classical control method can be used, but with a number of simplifying assumptions or by designing a very complex control algorithm which requires powerful

* E-mail: iresis@gazi.edu.tr

computing, fast and precise data acquisition equipment. Designing a conventional controller presents problems since modeling the system is very difficult due to its nonlinearity. However, fuzzy technique, which has gained popularity in recent years, looks very promising for these types application.

For applications like X-ray generators the output voltage applied to the x-ray tube is controlled and kept constant at the required level. The voltage across the x-ray tube produced by the generator is a low ripple voltage with fast rising time. The operating frequency is considerably high for these type applications to reduce the weight and volume of the power supply. The tube voltage is controlled by varying the phase-shift angle of the converter as the control parameter [1]. By varying the phase-shift angle of the inverter from 180° to zero, the tube voltage changes from zero to its maximum value.

For applications like solar panels the control parameter of the converter is controlled such that to obtain maximum power from the panel. There have been reported many methods in searching for the maximum-power-point (MPP) condition. The "perturb and observe" (PandO) method is a widely used approach to MPP tracking (MPPT). In this method the operating point of the solar panel is continuously changed by increasing or decreasing the panel operating voltage and observing its effect on the panel output power. When the MPP is reached the solar panel output voltage will oscillate around the actual operating voltage. The oscillation causes a power loss that depends on the width of the perturbation. In short-current pulse-based MPP tracking approach given in [2] the operating current of the MPP is estimated using short-circuit current and the converter switch is controlled to operate at this point. In this method the MPP is detected every several minutes during which the operating point is fixed. If an increase in irradiation occurs during this time interval, the efficiency of the solar panel will not be optimum until the end of the MPPT cycle. Moreover, the time required to detect MPP from the short-circuit current of the panel may not be short, which causes a reduction in the efficiency of the photovoltaic panel. In the approach given in [3] the MPP is detected based on the multiplication of the operating voltage and current. In this approach the MPP is detected every several minutes and the method has the same limitations of the short-circuit pulse-based MPPT approach. Incremental Conductance MPPT algorithm is another method developed to avoid the limitations of the PandO approach [4]. This approach is based on the fact that the derivative of the output power of panel is equal to zero with respect to the panel voltage at the maximum power point.

The I-V characteristic of a PV panel is not linear and depends on the on the atmospheric conditions such as temperature and solar irradiance. An increase in the panel current causes a reduction in its output voltage. The fuzzy logic controller can be more adequate in cases where the system is nonlinear. This method is also effective tool to deal with disturbances and uncertainties in terms of vagueness, ignorance and imprecision [5, 6, 7]. In addition, this method is effective in applications where there is wide experience in operating the plant, due to the possibility of incorporating such knowledge in ters of qualitative rules.

A Brief Study on Fuzzy Logic

The fuzzy expert system is applicable wherever the knowledge base of expert system contains fuzziness. Fuzzy control method does not need accurate mathematical model of a plant, and therefore, it suits well to a process where the model is unknown or ill-defined. Even when the plant model is known, there may be parameter variation problem.

A Fuzzy Logic Controller (FLC) may consist of three basic blocks, namely, Fuzzifications, Inference system and Defuzzification. The fuzzification interface converts the numerical values of the input variables into linguistic variables (fuzzy sets). It is a procedure to process the input variables with membership functions and determine the degree with which the input variables belong to each of the appropriate fuzzy sets via membership functions. Membership functions (MF) are used to convert each value of the input variables into a membership value between 0 and 1.

The overall performance of the system is affected by the shapes and number of the membership functions that are chosen according to the experience of expert people about the process. Just as there are an infinite number of ways to characterize fuzziness, there are an infinite number of ways to graphically depict the membership functions that describe fuzziness. Therefore, membership functions may take any arbitrary shape or form, such as triangular functions, Sigmoidal curves, Gaussian distribution curves, trapezoidal functions, exponential shapes or tables. There are a number of procedures given in [8] that can be used to build membership functions. In this study, triangular membership functions have been used as they are easier to implement and quicker to process.

In the proposed fuzzy system, there are seven fuzzy sets for each input whereas there are eleven fuzzy sets for the output. The fuzzy sets used in this study are as follows:

NVB: Negative Very Big PVB: Positive Very Big

NB : Negative Big PB : Positive Big

NM : Negative Medium PM : Positive Medium

NS : Negative Small PS : Positive Small

NVS : Negative Very Small PVS : Positive Very Small

Z : Zero

The fuzzy inference includes the process of fuzzy logic operation, fuzzy rule implication and aggregation. In the fuzzy inference system, the fuzzified input variables are processed with fuzzy operators and the IF-THEN rule implementation. The two of 49 rules of Table 1 are as:

IF error is NM AND change in error is NS THEN output is PM

IF error is NM AND change in error is Z THEN output is PS

There are a number of fuzzy reasoning methods in the literature, which can be used to obtain the inference result from a system. Those are Mamdani's, Larsen's, Sugeno's and Tsukamoto's methods [9]. The Mamdani's fuzzy reasoning strategy which is the most commonly and frequently used method has been employed in this study. The fuzzy reasoning strategy of this method is based-on the MAX-MIN or (SUP-MIN) composition. Mamdani's fuzzy reasoning is illustrated in Figure 1.

Table 1. Rule base table for converter output voltage controller

DE \ E	NL	NM	NS	Z	PS	PM	PB
NB	PVB	PVB	PB	PM	PS	PVS	NC
NM	PVB	PB	PM	PS	PVS	NC	NVS
NS	PB	PM	PS	PVS	NC	NVS	NS
Z	PM	PS	PVS	NC	NVS	NS	ZE
PS	PS	PVSS	NC	NVS	NS	PS	NB
PM	PVS	NC	NVS	NS	PS	NB	NVB
PB	NC	NVS	NS	NS	NB	NVB	NVB

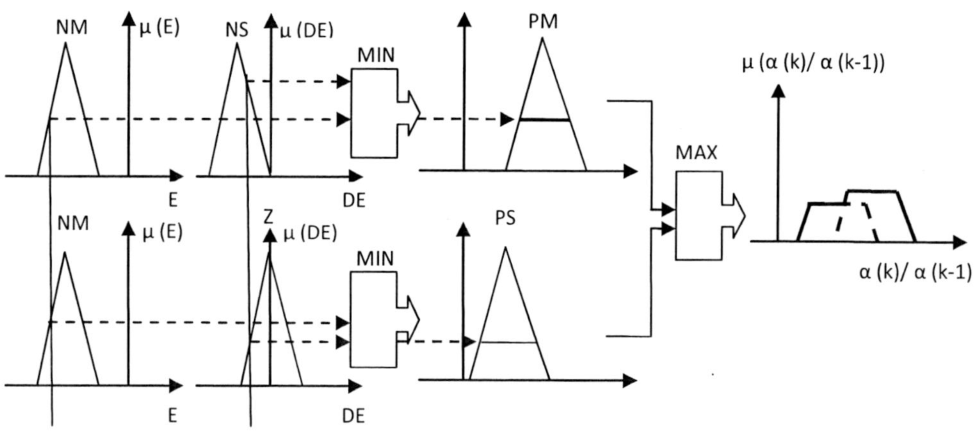

Figure 1. Fuzzy-rule-based composition method by MAX-MIN principle.

Table 1 shows the corresponding rule table for the output voltage controller. The top row and left column of the matrix indicate the fuzzy sets of the variables (E and DE), and the MFs of the output variable are shown in the body of the matrix. The proposed system has 49 rules that are built by crossing the fuzzy sets considered for each input.

Aggregation is the process by which the fuzzy sets that represent the outputs of each rule are combined into a single fuzzy set. The input of the aggregation process is the list of output fuzzy sets and the output of the aggregation process is one fuzzy set for output. In the Mamdani's fuzzy reasoning strategy used in this study, the maximum aggregate method has been used. The output of this step of FLC is a single fuzzy set obtained from the union (OR) of all the component MFs.

Defuzzification is the last step of FLC. The input for the defuzzification process is a fuzzy set obtained from the aggregation process and the output is a single value. Conversion of this fuzzy output to a crisp output is defined as defuzzification. There are a number of strategies in the literature, which can be used for performing the defuzzification. Unfortunately, there is no systematic procedure for choosing a defuzzification strategy.

Some important methods of defuzzification are:

Center-of-Area (COA) method, which determines the center of the area of the combined membership functions. The Height Method being used in this study, is the simplified state of COA. In this method the COA method is simplified to consider only the height of each contributing MF at the mid-point of the base. The Mean of Maxima (MOM) method is the further simplified model of the Height method. In this method the arithmetic mean of all values with maximum membership is considered. The fourth method, First of Maxima (FOM) uses the union of the fuzzy sets and takes the smallest value of the domain with maximal membership degree. The Last of Maxima (LOM) is another method of defuzzification which uses the union of the fuzzy sets and takes the largest value of the domain with maximal membership degree. The Middle of Maxima (MOM) method is similar to the FOM and LOM. Instead of determining the defuzzified value to be the first or the last from all values, this method takes average of these two methods. More quantitative description of the above mentioned defuzzification techniques are given in [10].

The best defuzzification method which can be used in each application is problem-dependent [8, 10]. However, there are some criteria which can be used in the selection of a suitable defuzzification method. The height method is both simple and very quick when compared with the Centroid (COA) and the steady-state error of this method is small when compared with MOM, LOM, FOM methods.

In this study, the fuzzification is realized by using membership functions of input variables, error (E) and derivative (DE) for the output voltage of the converter. The variables used for fuzzification are E(K) and DE(K).

During control (as an example in this study, the control of tube voltage of an x-ray generator) the following steps are performed:

i- First, measured tube voltage is compared with the desired value to obtain the error signal (E). The error and the derivative error (DE) are converted into per unit by dividing into the desired voltage value (V_{ref}). Using the per unit error the degree(s) of the membership function is (are) obtained (Figure 2).

ii- Derivative error (DE) is obtained by comparing the errors of the last two control steps and used in calculation of the degree(s) of the derivative error membership function.

iii- The degree(s) of output membership function is (are) calculated using the SUP-MIN method .

iv- Defuzzification is achieved using the height method.

The rules which should be activated for a specific input signal condition is determined by fuzzy controller using the fuzzy rule base and then the effective control action is obtained using the composition operation of MAX-MIN method. It is evident that for any input data E (error) and DE (derivative error), only four rules will be valid in the entire rule base given in Table 1.

The last step of fuzzy logic controller is defuzzification and can be performed by several methods, most common ones being the center of gravity (centroid) and height methods. The height method is both very simple and quick in comparison with the centroid method. The input for the defuzzification process is a fuzzy set and the output is a single value. There are a number of strategies in the literature, which can be used for performing the defuzzification [11]. Unfortunately, there is no systematic procedure for choosing a defuzzification strategy. Some important defuzzification techniques are quantitatively described in [10]. In this study, the height method is used due to its simplicity. At the end of the defuzzification step a denormalization process is performed to obtain the required increment of the duty-cycle command at the operating point.

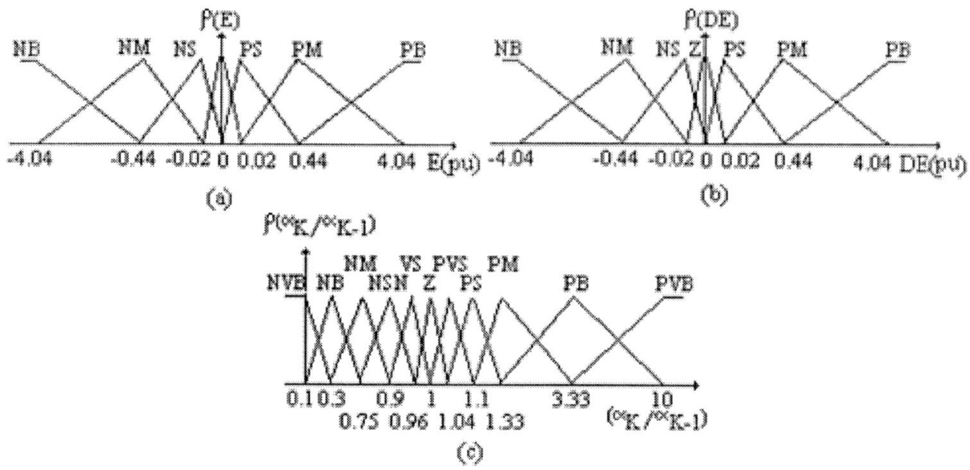

Figure 2. Membership functions for (a) Output voltage error (b) Output voltage derivative error (c) Phase-shift angle (α_K / α_{K-1}).

A Brief Study on Neural Network

Neural network performs the function of nonlinear mapping or pattern recognition. This means that for a set of data corresponding to a definite signal, the network is trained in order to give a desired pattern at the output. The input-output matching occurs by selecting appropriate weights. Initially the network is untrained or the weight values are selected at random, so the output pattern does not match the desired pattern. By comparing the actual and the desired output patterns, the weights can be adjusted by an algorithm such that the pattern matching occurs. For this case (pattern matching) the error between the actual and desired pattern is very small. The accuracy of the training depends on the number of input-output example patterns. Therefore the training should be continued with a large number of input-output example patterns.

The neural network can be trained using either on-line or batch learning algorithms. In the on-line method a pattern is presented at the input and all weights are updated before the next pattern is presented but in the second technique (batch learning algorithm) all training patterns are accumulated and the weights are updated according to the minimization of the total error function. The application of the on-line and batch learning methods depend on the problem. Some of differences among them can be summarized as:

1. The on-line method is preferred when the training patterns are not available before the training starts or when the number of training examples is very large.
2. The batch learning algorithm is used when high precision mapping is required.

Normally, the training is done off-line with the help of a computer simulator program. For the above reasons (availability of training examples and required high precision mapping) the batch learning algorithm is preferred in this study.

At the end of the training, the network is capable to recall the trained output patterns (look-up table) and also untrained output patterns by interpolating and extrapolating the trained patterns.

The training method as described before is defined as batch learning training. At the beginning, the weights of the neural model are chosen at random and they are updated by performing the following processes step by step. First the method is explained for a general neural network model (N-input, M-output model), then it is modified to be applied to the proposed single-input, single-output neural network model given as example in this study (Figure 3).

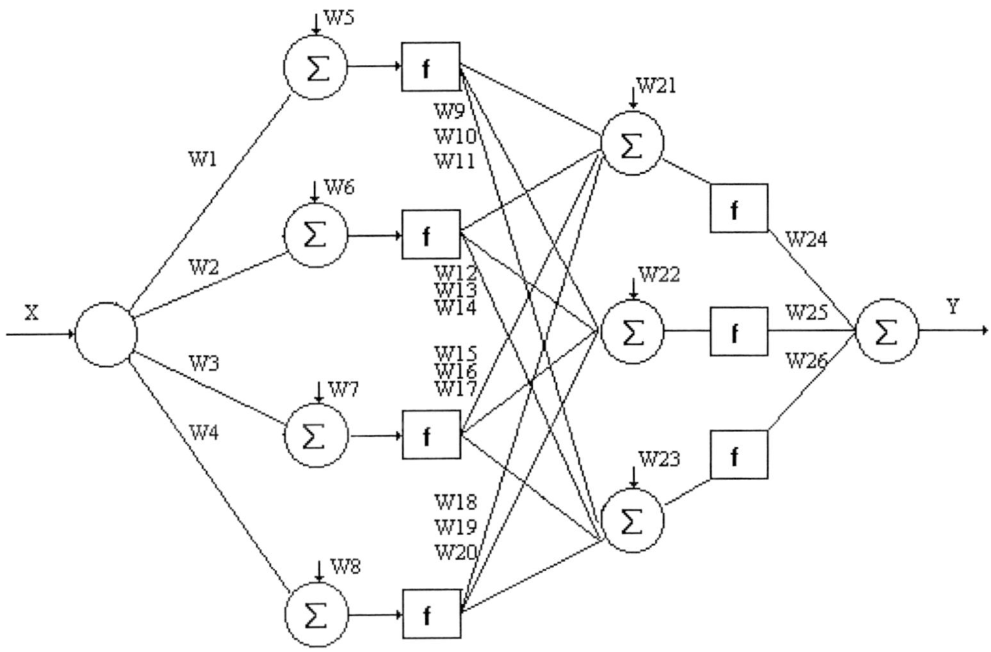

Figure 3. The neural network model.

In Figure 3, f is a sigmoid limiter function. Such a sigmoid function can mathematically be shown as:

$$Y = \tanh u\, x$$

where u is a positive constant (Figure 4).

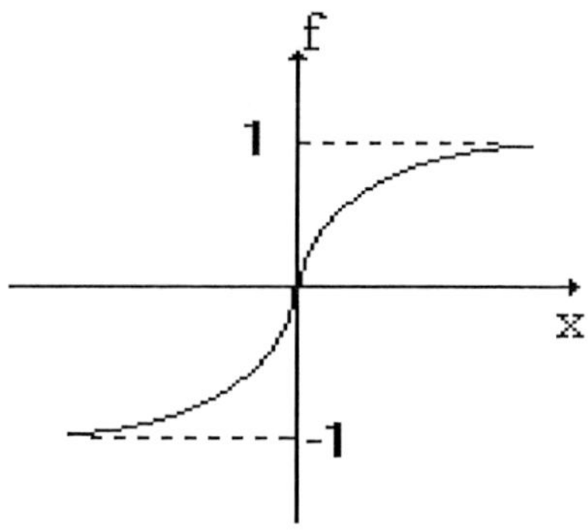

Figure 4. Sigmoid limiter.

For an N-input, M-output neural network, using the initial values of weights, the output pattern is derived for the given input pattern as:

$$V_j^p = \sum W_{ij} X_i \quad (1)$$

$$y_j^p = f_j (V_j^p) \quad (2)$$

where:

N = number of neurons feeding the j-th neuron
M = neural network output number
W_{ij} = connection weight between i-th and j-th neurons
X_i = output of the i-th neuron connected to the j-th neuron
V_j^p = output of the summing node (j-th neuron)
f_j = a sigmoidal function

The weights of the neural model are obtained following a optimization process [12].

SIMULATION

In this section, the Matlab/Simulink program is used to study the operation of two different converters. The first converter is a high dc-voltage generator supplying the x-ray tube. The phase shift angle of converter is controlled to keep the output voltage constant at the required level. Since the system is a nonlinear, the fuzzy logic control method is preferred for this application. The application of fuzzy logic to control the output voltage of a welding machine and the comparison between the machine performances corresponding to two control techniques of fuzzy logic and sliding average technique are given in [13, 14]. The second converter is related with maximum power point detecting of a solar panel. The maximum power point is achieved by operating the converter at the appropriate point. The operating point at which the output power is maximum is detected using fuzzy logic method.

HIGH DC-VOLTAGE GENERATOR

The purpose of the tube voltage control is to keep constant the output voltage of the converter at the set value by varying the phase-shift angle of the inverter. The maximum (V_{max}) and minimum values of tube voltage are obtained for α (phase-shift angle) corresponding to zero and 180°, respectively.

Figure 5 shows the Simulink model of the system. The first stage of the converter is an uncontrolled rectifier connected to the ac mains through an inductor representing the ac line inductance (L_1). The ac mains supply is considered to be 220 V (rms) at a frequency of 50 Hz. The mean value of the dc link voltage (v_{C0}) depends on both the load current and the ac line inductance. The dc link voltage has a ripple with frequency twice that of the ac mains supply.

The inverter used in simulation and experiments is a single-phase bridge operating around the resonant frequency of the system (lagging mode) so that the switching lossless of the transistors are considerably reduced [1, 15]. The high-voltage transformer was designed to have a minimum volume satisfying the load with specifications of 125 kV and 100 mA. The non-linearities of the high voltage transformer used in parallel resonant converter are incorporated in basic converter operation, rather than as parasitic elements that interfere with basic operation of the converter. The winding capacitance of the transformer is used as the resonant capacitor so that the external tank capacitor is completely eliminated and the leakage inductance of the transformer constitutes the resonant inductance. The voltage at the secondary of the high frequency transformer is once more rectified and applied to the load. The output voltage is controlled by phase-shift PWM of the power semiconductor switches. A software was developed for designing the transformer to achieve desired lumped parameters, while minimizing the volume.

An investigation is carried out to determine optimum operating frequency of the transformer. The parameters obtained in transformer design are used in both analytical simulation and laboratory experiments.

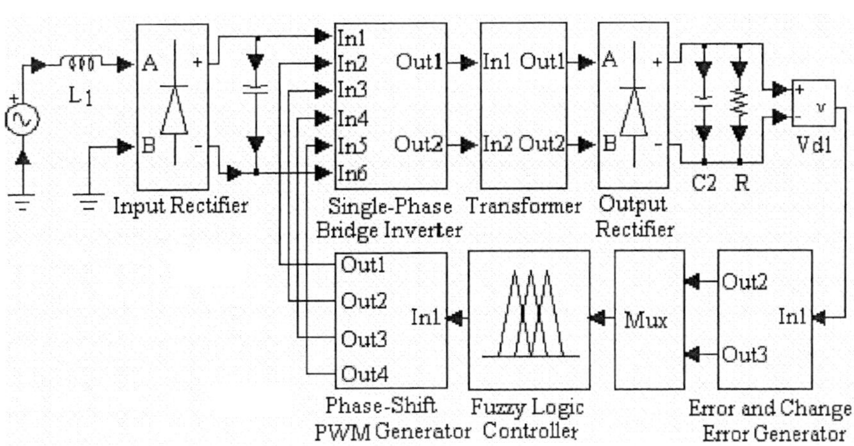

Figure 5. Simulink model of the converter.

For verifying the control strategies discussed above, the combination of mathematical model and fuzzy controller is simulated for the following operating conditions. The dc link capacitor has an initial arbitrary voltage of 140 V and the converter is connected to ac mains (220 V).

The ac line inductance is 4.5 mH and the load is considered as a resistor with resistance value equal to ratio of the tube voltage to its current (125000 / 0.1) referred to the primary side. The sampling time used in feedback control is 33.33 μs.

Figure 6 (a) shows the uncontrolled referred tube voltage for the conditions where the initial voltage of the dc link capacitor is 140 V. It is shown that the tube voltage has a ripple with peak to peak value of 20 V and frequency of 100 Hz, respectively.

Figure 6 (b) shows the tube voltage referred to the primary side, dc link voltage and the phase-shift angle of the inverter for the case as in Figure 6 (a). It is shown that the phase-shift angle changes according to fuzzy control method in such a way to keep the tube voltage constant at the set value, 160 V. This is the dc-link voltage value at the primary of the transformer corresponding to the rated output voltage value, 125000 V.

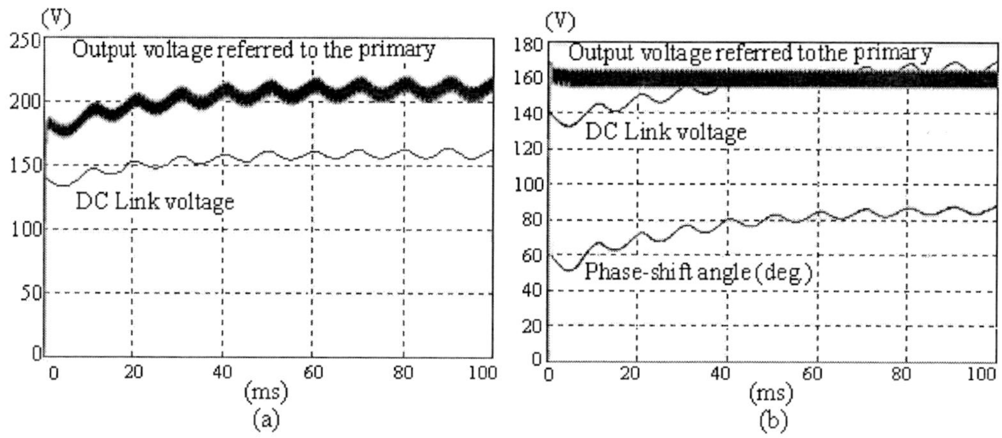

Figure 6. (a) Referred tube voltage with no control $v_{C0}(t=0) = 140$ V (b) Referred tube voltage, dc link voltage and phase-shift angle with fuzzy control. $v_{C0}(0) = 140$ V.

SOLAR PANEL MODELING AND MPP DETECTION

To analyze the overall system, firstly the neural network model of the photovoltaic panel is created in the MATLAB/SIMULINK and then combined with other blocks existing in the system. The operating point of the panel is set at MPP by adjusting the duty-cycle of the boost converter using fuzzy logic control method [6].

The objective of the control in the MPPT system is to adjust the duty-cycle of the converter in such a way to obtain maximum output power from the solar array in spite of disturbances such as changes in the solar irradiance level or temperature caused by varying the atmospheric conditions.

The operating point of the solar panel changes due to the energy radiation variations during the plant operation caused by the daily solar cycle. For a specific operating point it is possible to design a particular controller such as PID controllers that allows an acceptable behavior. However, poor transient performance is achieved when the operating point of the system changes [16]. Therefore, it is difficult to obtain a satisfactory performance over the total operating range using conventional control methods.

The relationship between the solar panel output power and the duty-cycle of the converter is nonlinear due to I-V characteristic of the solar panel. In addition, the relationship between the input and output voltages of the dc-dc converter is also nonlinear. Fuzzy logic control (FLC) technique which has gained popularity in recent years looks very promising for this application. Unlike the other conventional control schemes, FLC is a suitable method for controlling nonlinear systems and is a model free controller. Another advantage of this method is the possibility of converting linguistic control strategy based on experience and export knowledge into an automatic control strategy.

In this study, triangular membership functions were used in fuzzification of the input variables, as they are easier to implement and quicker to process. The variables used for fuzzification are the change in the panel output power ($\Delta P_k = P_k - P_{k-1}$) and the change in duty-cycle of the converter at the previous step ($\Delta D^{in}_k = \Delta D^{out}_{k-1}$).

The fuzzy sets considered in the proposed fuzzy system are as follows:

NL : Negative Large PL : Positive Large

NM : Negative Medium PM : Positive Medium

NS : Negative Small PS : Positive Small

ZE : Zero Equal

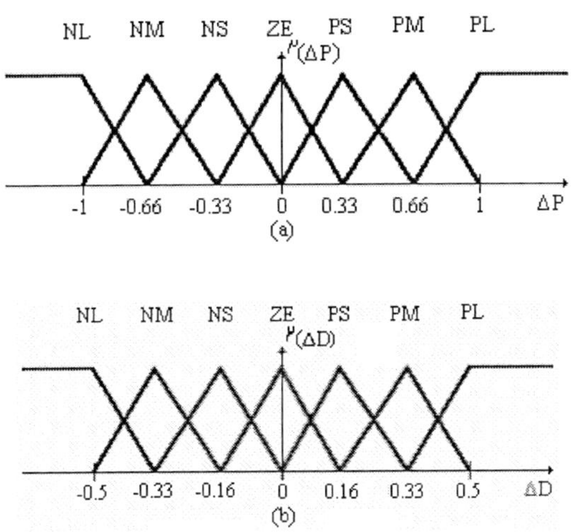

Figure 7. Membership functions for input variables (a) Change in panel output power, ΔP_k (b) Change in duty-cycle obtained at the last step, ΔD^{in}_k ($\Delta D^{in}_k = \Delta D^{out}_{k-1}$).

Figure 7 (a) shows the membership function plots of the input variables (ΔP_k, ΔD^{in}_k). The membership function of the output (change in duty-cycle, ΔD^{out}_k) is similar to that of the Figure 7 (b). Before fuzzification the input variables are normalized by dividing by the respective base values. The power base value of the panel used in this study is 42 W. The duty-cycle of the converter is normalized itself and there is no need to be converted ($D = t_{on} / T$).

In the fuzzy inference system, the fuzzified input variables are processed with fuzzy operators and the IF-THEN rule implementation. Table 2 shows the corresponding rule table for the proposed MPPT system. The top row and left column of the matrix indicate the fuzzy sets of the variables ΔP_k and $\Delta D^{in}{}_k$, respectively, and the MFs of the output variable (ΔD^{out}) are shown in the body of the matrix. The proposed system has 49 rules that are built by crossing the fuzzy sets considered for each input. The two of 49 rules of Table 2 are as:

IF ($\Delta D^{in}{}_k$ is NM) AND (ΔP_k is NM) THEN ($\Delta D^{out}{}_k$ is PM)

IF ($\Delta D^{in}{}_k$ is NS) AND (ΔP_k is NM) THEN ($\Delta D^{out}{}_k$ is PS)

Figure 8 shows the controller surface, where the horizontal axes are ΔP and ΔD^{in}, and the vertical axis is the output signal ΔD^{out}. The surface indicates that ΔD^{out} is positive large when both ΔP and ΔD^{in} are negative large or both ΔP and ΔD^{in} are positive large.

Table 2. Rule base table for the PV panel output power controller

$\Delta D^{in} \setminus \Delta P$	NL	NM	NS	ZE	PS	PM	PL
NL	PL	PL	PM	PS	NM	NL	NL
NM	PL	PM	PS	PS	NS	NM	NL
NS	PM	PS	PS	ZE	NS	NS	NM
ZE	ZE	ZE	ZE	ZE	ZE	ZE	ZE
PS	NM	NS	NS	ZE	PS	PS	PM
PM	NL	NM	NS	NS	PS	PM	PL
PL	NL	NL	NM	NS	PM	PL	PL

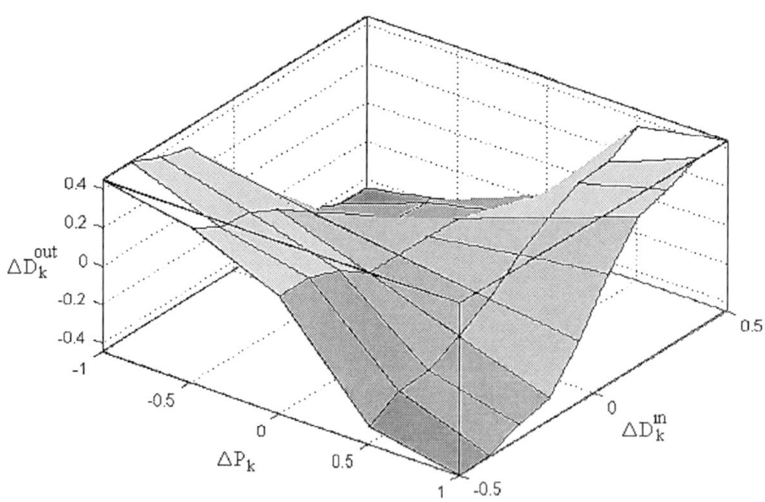

Figure 8. Control surface of fuzzy controller used in this study.

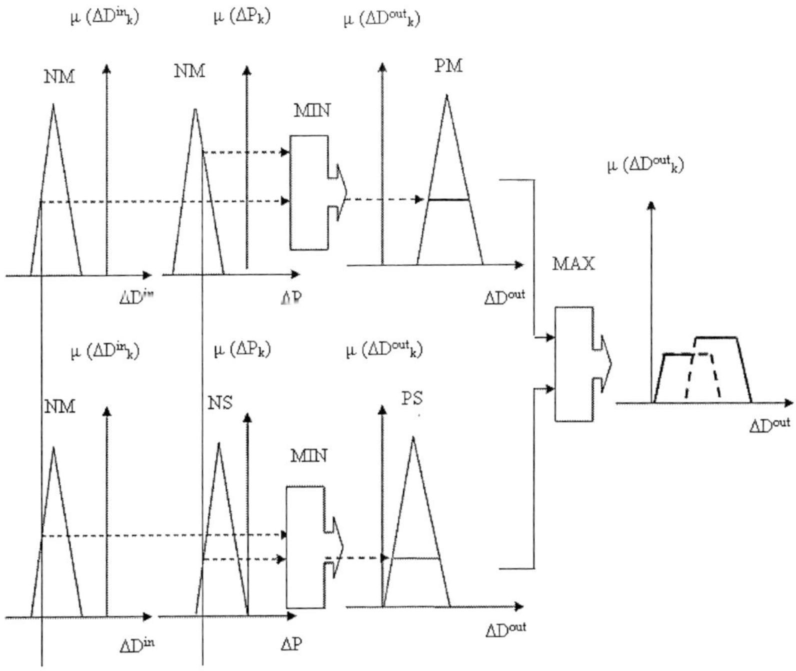

Figure 9. Two-rule fuzzy maximum power point tracking control principle.

The I-V characteristics of the solar panel are not linear and depend on the irradiance and temperature of the circumstances in which the test is carried out. There are different types of "diode model" given in literature, which can be used in PV panel performance analyzing. The model given in [17] consists of a current source supplying current to an electric circuit including a junction resistor, an internal parallel resistance and an internal series resistance. In [18] there is given another model in which the electrical characteristic of a PV cell is given by the familiar double-exponential model. The electrical characteristics of these models depend on the values of four or five parameters according to the selected model. These parameters, themselves, depend on two environmental variables being solar irradiance and temperature. A lot of experimental data and a rather complicate curve fitting process are needed to determine these parameters.

Neural network is a powerful method that is used in applications where formal analysis would be difficult or impossible. It performs the function of nonlinear mapping or pattern recognition. This means that for a set of data corresponding to a definite signal, the network is trained in order to give a desired pattern at the output. The input-output matching occurs by selecting appropriate weights. Initially the network is untrained or the weight values are selected at random, so the output pattern does not match the desired pattern. By comparing the actual and the desired output patterns, the weights can be adjusted by an algorithm such that the pattern matching occurs. For this case (pattern matching) the error between the actual and desired pattern is very small. The accuracy of the training depends on the number of input-output example patterns. Therefore the training should be continued with a large number of input-output example patterns [19, 20].

The neural network model of the PV panel is an alternate model that can be used in performance analyzing of the panel. MATLAB/SIMULINK program includes the Neural Network Toolbox which is one of the commonly used and powerful software tools for the development and design of neural networks. The software is user-friend and used in the MATLAB environment. It permits flexibility and convenience in interfacing with other toolboxes in the same environment to development a full application. Since the simulation of the panel-boost converter was carried out using the MATLAB/SIMULINK program, the neural network model was preferred to be used in the system simulation.

The proposed model in this study consists of three inputs, a single output and two hidden layers. There are three neurons in the first hidden layer and four neurons in the second one. The inputs are temperature T ($^{\circ}$C), the solar irradiance G (kW/m^2) and the panel current. The output parameter of the model is the panel voltage. The neural network of the panel is shown in Figure 10. There are 25 weights and 8 biases in this model.

Choosing a large number of hidden layer neurons may increase the accuracy, but implies larger computational time for convergence and execution time for the microprocessor.

The appropriate weights of the neural network model were obtained by training the model for more than 80 input-output patterns. The input data and their ranges which were used in the training of the panel are: i- solar irradiance with values between 0.3 kW/m^2 and 1 kW/m^2 and ii- panel output current in the range of 0-2.95 A. The temperature was constant and equal to 35° C during the experiments carried out to derive the data to be used in the training of the neural network of the panel. The back-propagation method was used in training the proposed model.

The neural network model of the PV panel is an alternate model that can be used in performance analyzing of the panel. MATLAB/SIMULINK program includes the Neural Network Toolbox which is one of the commonly used and powerful software tools for the development and design of neural networks. The software is user-friend and used in the MATLAB environment. It permits flexibility and convenience in interfacing with other toolboxes in the same environment to development a full application. Since the simulation of the panel-boost converter was carried out using the MATLAB/SIMULINK program, the neural network model was preferred to be used in the system simulation.

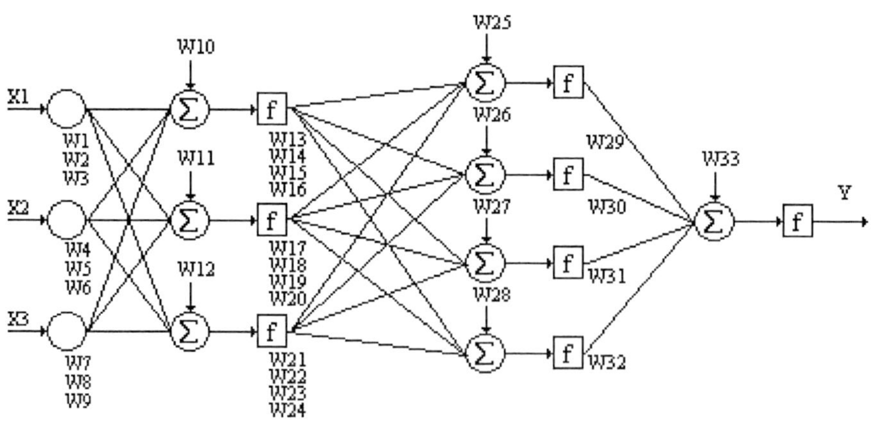

Figure 10. Structure of the proposed neural network model of the PV panel.

A neural network can be trained by using either on-line or batch-learning algorithms. In the on-line method a pattern is presented at the input and all weights are updated before the next pattern is presented, but in the second technique (batch-learning algorithm) all training patterns are accumulated and the weights are updated according to the minimization of the total error function [13].

In the simulation, the operating frequency of the converter switch is 15 kHz and the load is a 15 Ω resistor. The control unit adjusts the duty-cycle of the converter by moving the operating point of the panel to the MPP corresponding to the existing temperature and irradiance.

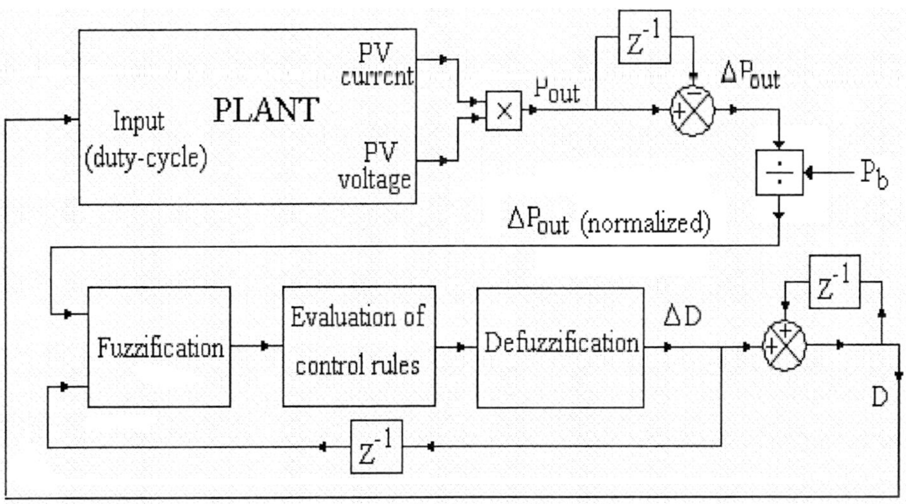

Figure 11. Block diagram of the plant and fuzzy controller.

In Figure 12, there are three periods of operation (0-30ms, 30ms-60ms, 60ms-) corresponding to irradiance values of 0.5 kW/m^2, 0.7 kW/m^2 and 1 kW/m^2, respectively. During each period, the controller adjusts the duty-cycle of the converter (In this study, duty-cycle is defined as the ratio of the turn-on duration of the switch to its switching period, $D= t_{on} / T$) to produce maximum power corresponding to the irradiance of the given period. In the first period the irradiance value is 0.5 kW/m^2 and the MPP of this period is obtained at duty-cycle of 6.9% adjusted by the controller. The maximum power obtained in the first period is 17.8 W. At instant of 30 ms the irradiance is changed to 0.7 kW/m^2 and the controller varies duty-cycle from 6.9% to 18% to find the respective MPP. The maximum power of this period is 28 W. At instant of 60 ms the irradiance is changed to 1 kW/m^2. The controller again adjusts and changes duty-cycle from 18% to 31% to produce maximum power of the respective period which is 39 W. This verification confirms that the controller works properly.

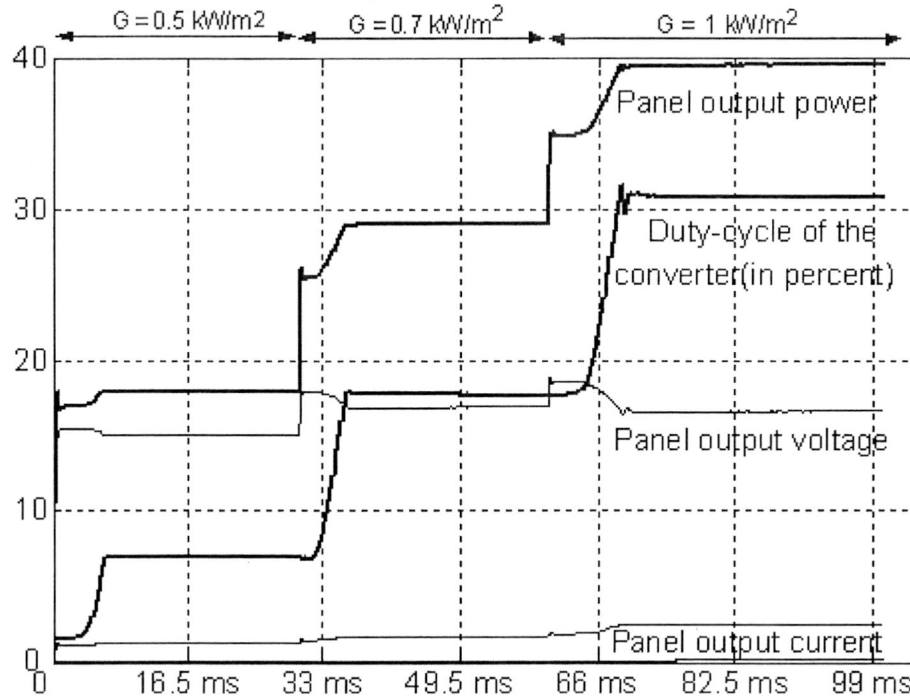

Figure 12. The performance of the controller in adjusting the duty-cycle to produce the maximum power of the respective periods.

Figure 13 shows the effectiveness of the controller in detecting the MPP. This figure is corresponding to the operation where the load, solar irradiance and temperature values are 15 Ω, 0.7 kW/m^2 and 35° C, respectively. In this operation, the panel-boost converter is operating at point (P_1) on the panel I-V characteristic under steady-state condition before the instant of 0.005 Sec..

The panel power corresponding to this point (P1) is 24 W and the duty-cycle is zero. At instant of 0.005 S the controller starts to operate and to adjust the duty-cycle under supervision of fuzzy logic controller. In Figure 13 (a) the controller varies the duty-cycle from zero (initial value) to 17.8% (final value) to increase the panel output power. It is shown that the panel power increases and reaches its maximum value, 28 W. Figure 13(b) shows the panel I-V characteristic and the variation of the panel output power. During control process the panel operating point moves from P1 to P2 on the panel I-V characteristic and the panel power increases from 24 W to 28 W which are corresponding to the operating points of P1 and P2, respectively. These results show that for a solar irradiance value of 0.7 kW/m^2 the panel maximum power is 28 W and it is achieved at duty-cycle of 17.8% when the load is a 15 Ω resistor.

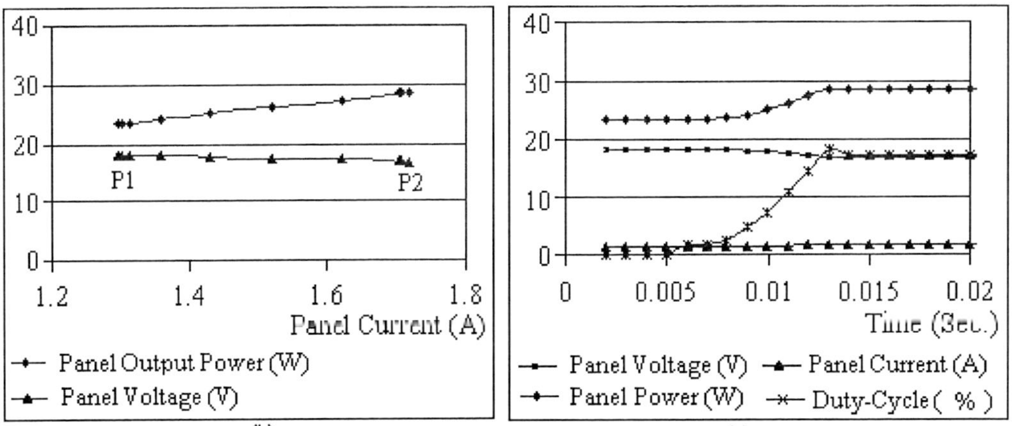

(b) (a)

Figure 13. (a) Variation of the panel current, voltage and power during control process to find MPP, (b) Motion of the panel operating point towards the MMP on the I-V characteristic.

EXPERIMENTS

In this section the results obtained from the experiment carried out on two different converters presented in Simulation Section are given.

High DC-Voltage Generator

The block diagram of the hardware used in this study is shown in Figure 14. In this figure, L_t , C_t and r_t are the leakage inductance, secondary layer capacitance and transformer winding resistance referred to the primary, respectively. The signals generated by microcontroller are sent to the inverter switches through an interface circuit. The switches are Intelligent power Module dual type (IPM) manufactured by Mitsubishi Electric (PM200DSA060). The dc link voltage is measured and converted into digital signals using DS1101 board. The measured tube voltage is compared with the reference level value to find a new phase-shift angle according to the fuzzy control method.

Using the above mentioned control algorithm, the tube voltage and output voltage of the inverter were recorded for the converter operating with the load as the referred X-ray tube equivalent resistance and the desired tube voltage of 15 V. All the signals and parameters represented as the variable in the DSP's data memory can be recorded and graphically displayed using the D1101 digital signal processor board, which contains a TMS320C14. TRACE14 software module is capable of saving the traced time histories on a readable ASCII file format. The real time traces of the tube voltage, the phase-shift angle and dc link voltage were saved on a disk and the results are shown in Figure 15. It is shown that while the dc link voltage varies the phase-shift angle changes in order to keep the output voltage constant at the desired value. For a sampling period of 320 μs, there are about 32 sampling periods in a time interval of 10 ms (ripple period of the dc link voltage) during which the phase-shift angle is constant.

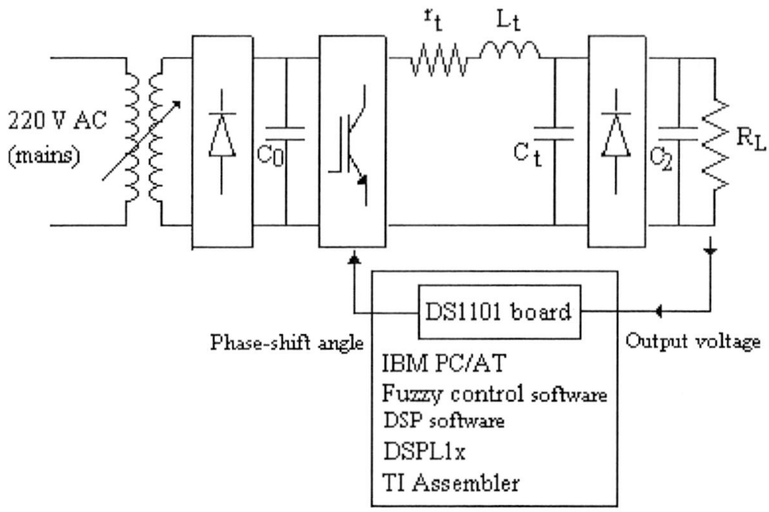

Figure 14. The blocks diagram of the converter, equivalent circuits of the transformer and the load.

The rise- and fall-time of the tube voltage depend on the tube voltage smoothing capacitance. For the smoothing capacitance of 200 μF used in the experiments, they are determined as 750 μs and 1.4 ms.

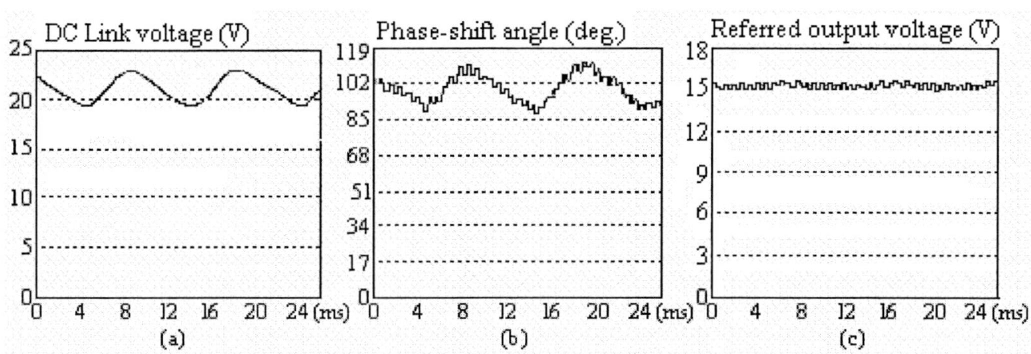

Figure 15. The real time traces of (a) dc-link voltage (b) phase-shift angle (deg.) (c) converter output voltage.

Solar Panel MPP Detection

Figure 16 shows the experimental set up used in the laboratory. The experimental study was carried out on a 42W PV panel. The semiconductor switch of the converter is a MOSFET (IRF540) operating at 15.8 kHz. The panel output current is measured on a 0.1 Ω resistor and the level of this voltage is increased using the OP-AMP, KA358 and then is applied to one of input terminals of the multiplier integrated circuit, MPY634. The panel output voltage is also measured and applied to another input of the multiplier. The panel output power is taken from the multiplier output terminal and then applied to the ADC terminal of the controller passing through the Opto-Coupler, 4N35.

The controller used in the experiments is the eZdsp^{TM} F2812 which is a stand-alone module, permitting users to evaluate certain characteristics of the TMS320F2812 digital signal processor (DSP) to determine processor applicability to design requirements. Evaluators can create software to execute onboard or expand the system in a variety of ways. The module has an operating speed of 150 MIPS. The use of a Zener diode at the input of ADC port of the controller is to protect the controller board against voltages with the level higher than 3.3 V.

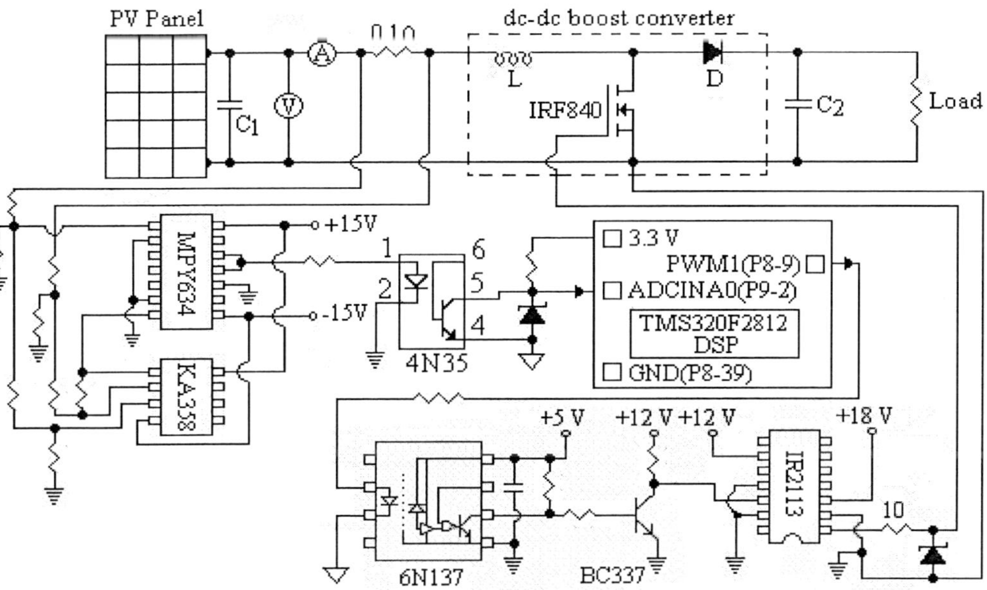

Figure 16. Experimental set up.

The duty-cycle of the converter is adjusted by the controller based on the fuzzy logic control approach. The output PWM signal of the controller is applied to the IR2113 through the Opto-Coupler, 6N137 which is used to isolate controller module from the converter and power MOSFET. The IR2113 is an integrated circuit used to drive the MOSFET. The amplitude of the controller output signals is 3.3 V. The output signals amplitudes of the Opto-Coupler and IR2113 are 5 V and 18 V, respectively.

The I-V characteristics of the panel for different irradiance values and constant temperature of 35° C are shown in Figure 17 (a). To obtain the I-V characteristic corresponding to each irradiance value a variable resistor is connected across the converter output terminals. The resistor is varied step by step from maximum (open circuit) to its minimum value (short circuit). The panel output power, voltage and current are measured and recorded for each step and are used in deriving the figure 17 (a). The P-V and I-V characteristics of the panel corresponding to the irradiance value of 0.7 kW/m^2 are shown in Figure 17 (b). The procedure used to derive this figure is similar to that of Figure 17 (a). The maximum output power for this case of operation (G = 0.7 kW/m^2) is 28 W. Figure 18 shows the panel output power, voltage and current for the case where the load is constant (R_{load} = 140 Ω) and irradiance and temperature values are 0.7 kW/m^2 and 35° C, respectively. In this test the duty-cycle of the converter is changed from 50% to 90% step by step and the panel parameters (panel power, panel voltage and current) are measured and recorded for each step.

The maximum power is 28 W and that is corresponding to a duty-cycle of 81%. Figure 19 shows the steady-state panel output voltage and current waveforms for the same atmospheric conditions of Figure 18. It can be observed from Figure 19 that the duty-cycle of the converter switch is 81% when the system is operating at steady-state condition under supervision of controller. The output power measured at the panel output for this case of operation is 28 W. This matches with the experiment results shown in Figure 18 where the maximum power is obtained at duty-cycle of 81%. It is also shown in Figure 19 that the panel current decreases with increasing the panel voltage and vice versa, and this also matches with the I-V characteristics of the panel. The oscillations shown in the panel voltage and current waveforms are generated when the switch is turned off (Figure 19). The main reason of these oscillations is due to the leakage inductance and stray capacitance of the connection wires and may be minimized by compacting the experimental set up and using snubber circuits.

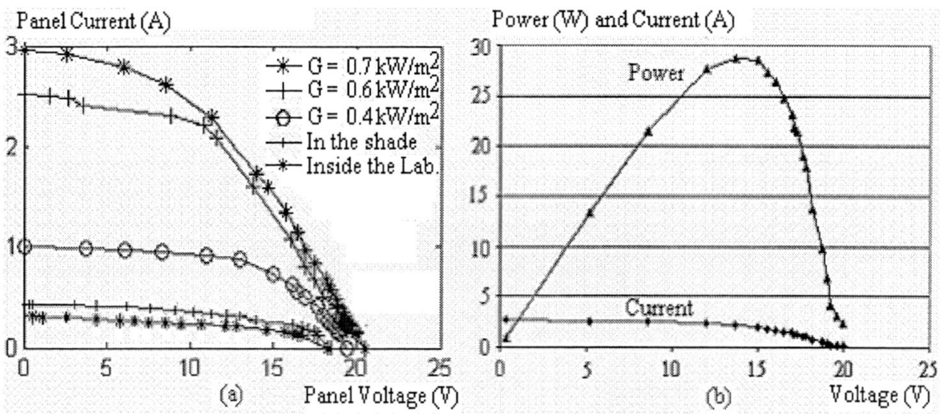

Figure 17. (a) The panel I-V characteristics for different irradiance values (b) P-V and I-V characteristics of the panel for G = 0.7 kW/m^2.

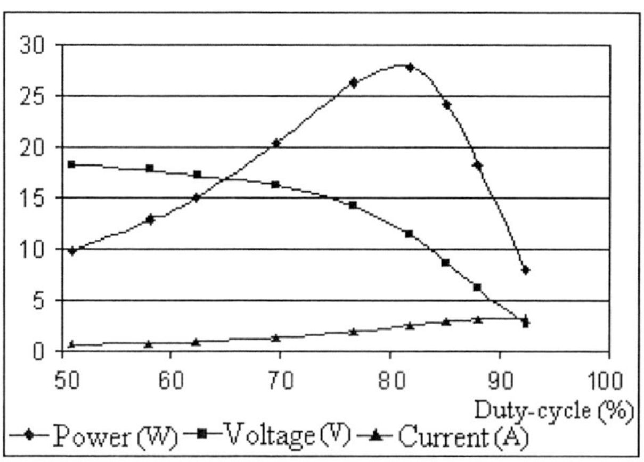

Figure 18. Variation of the panel output power, voltage and current against the converter duty-cycle corresponding to G = 0.7 kW/m^2 and R_{load} = 140 Ω.

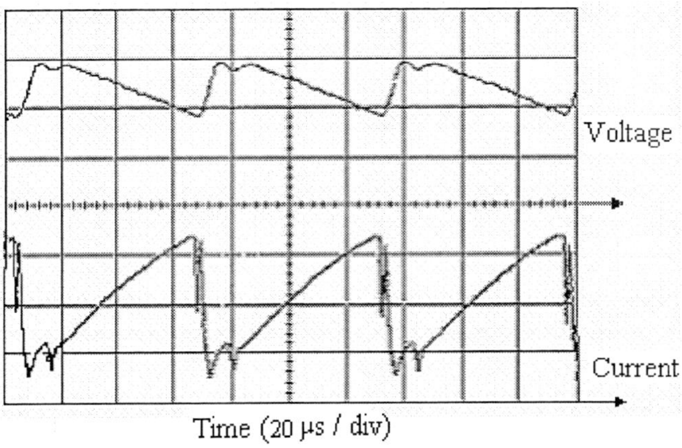

Figure 19. The panel output voltage and current waveforms when the converter is operating at MPP corresponding to $G = 0.7\ kW/m^2$ and $R_{load} = 140\ \Omega$.

CONCLUSION

In this study, the fuzzy control approach is used to adjust the output voltage of the dc-dc converter using phase-shift PWM technique. The results, which are obtained from the software and the hardware implemented in the laboratory, indicate that the fuzzy control performs well in terms of voltage regulation, rise-and fall-time of the output waveform.

Simulation and experimental results illustrated that the control system is capable of eliminating the 100 Hz ripple in the output voltage and maintaining a voltage level set by the user only with a ripple at inverter frequency. The amplitude of the ripple at inverter frequency can be decreased by using a filter at the output of the output rectifier. The rise time which is an important criterion for the quality of the output voltage can be minimized by initially charging the dc-link capacitor to a desired level before the system starts operating. The findings here suggest that fuzzy control is a good choice for controlling a dc-dc converter to obtain a high quality output voltage waveform. The chief advantage of designing and implementing the proposed controller is that controller design does not require explicit knowledge of the converter dynamics, which is a useful feature when dealing with converter parameter uncertainties.

A neural network model consisting of three inputs, a single output and two hidden layers was proposed for the PV panel. The training of the model was accomplished using back-propagation method existing in the Neural Network Toolbox in the MATLAB/SIMULINK® environment. The system model was derived by combining the models of the blocks existing in the system.

The microcontroller based solar energy conversion modular unit was designed and implemented in the laboratory. The experimental results obtained using the FLC approach are verified by Figure 18. The experimental results confirm the theoretical study and the simulation results in terms of controlling the panel output power by varying the converter duty-cycle and the effectiveness of the fuzzy logic control approach in tracking the MPP of the panel. The chief advantage of designing and implementing the proposed fuzzy controller is the ease of the design such that controller design does not require explicit knowledge of the

PV panel/converter/load dynamics, which is a useful feature when dealing with parameter and load uncertainties.

REFERENCES

[1] Iskender I., Uctug, I. and Ertan, B.,"DSP microcontroller-based fuzzy control of a DC/DC parallel resonant converter using phase-shift PWM technique" *The International Journal for Computation and Mathematics in Electrical and Electronic Engineering*, 25 (4), 827-840, 2006.

[2] Noguchi T., Togashi S. and Nakamoto R. "Short-Current Pulse-Based Maximum-Power-Point Tracking Method for Multiple Photovoltaic-and-Converter System", *IEEE Trans. on Industrial Electronics*, vol. 49, Iss. 1, pp. 217-223, Feb, 2002.

[3] Huang F., Zhimin G., Forughian T. and Tien D. "A New Microcontroller Based Solar Energy Conversion Modular Unit", *Proc. of the Power Conversion IEEE Conference*, 2, 697-700, 1997.

[4] Hussein K. H., Muta I. and Hoshino T. and Osakada M. "Maximum Photovoltaic Power Tracking: an Algorithm for Rapidly Changing Atmospheric Conditions", *IEEE Proc. of Generation, Transmission and Distribution*, vol. 142, No. 1, pp. 59-64, Jan, 1995.

[5] Iskender I. "A Fuzzy Logic Power Electronic System for Maximum Power Point Detection of a Solar Energy Panel" *The International Journal for Computation and Mathematics in Electrical and Electronic Engineering*, Vol 24, No 4, pp. 1164-1179, 2005.

[6] Iskender I. "On the Comparison of the Performance of Feed-Back and Feed–Forward Fuzzy Control Techniques of a PWM Resonant Converter Applied for an X-Ray Generator" FSSCIMIE'02, International Conference on Fuzzy Systems and Soft Computational Intelligence in Management and Industrial Engineering, pp. 185-192, Istanbul, Turkey, 29-31 May, 2002.

[7] Bose, B. K.,"Modern Power Electronics and AC Drives", Prentice-Hall, Inc. Upper Saddle River, NJ 07458, 2002.

[8] Timothy J. R."Fuzzy Logic with Engineering Applications", John Wiley and Sons, October, 2004.

[9] Patyra, M. J., "Design Consideration of Digital Fuzzy Logic Controller" in Patyra, M. J. and MLYNEK, D. M., Fuzzy Logic: Implementation and Applications, Wiley and Sons Ltd., England, 1996.

[10] Hellendoorn H. and Thomas, C. " *Defuzzification in Fuzzy Controllers, Intelligent and Fuzzy System*", vol. 1, pp. 109-123, 1993.

[11] Lin C. T. and Lee C. S. G."Neural Fuzzy Systems: A Neuro-Fuzzy Synergism to Intelligent Systems", Prentice Hall, NJ, 1996.

[12] Haykin S. "Neural Networks: A Comprehensive Foundation", Prentice Hall, Upper Saddle River, NJ 07458, 1994.

[13] Iskender I. and Karadeniz M.,"Fuzzy Logic Control of a Welding Machine Using Phase-Shift Pulse-Width-Modulation Technique" Proceedings of the 10th IFSA World Congress, Istanbul, pp. 352-356, June 29-July 2003.

[14] Iskender I. and Karaarlan A.,"The Comparison of Fuzzy Logic and State Space Averaging based Sliding Control Methods Applied on an Arc Welding Machine" World Academy of Science, Engineering and Technology, pp. 100-105, 2005.

[15] Ang S. and Oliva A., "Power Switchıng Converters" Taylorand Francis, 2005.

[16] Tang K. L. and Mulholland R. J."Comparing Fuzzy Logic With Classical Controller Designs", *IEEE Trans. On Systems, Man, Cybernetics*, vol. 17, pp. 1085-1087, Nov./Dec., 1987.

[17] Salas V. and Olias E., Barrado A. and Pleite J."Novel Algorithm Applied to Maximum Power Point Trackers (MPPT) Measuring Only One Photovoltaic Variable", 1st SWH International Conference: Solar-Wind-Hydrogen/Fuel Cells (Renewable Energies), Segovia, Spain, 2003.

[18] Gow J. A. and Manning C. D. "Development of a Model for Photovoltaic Arrays Suitable for Use in Simulation Studies on Solar Energy Conversion Systems", IEE Proc. Of 6th International Conference on Power Electronics and Variable Speed Derive, No. 429, pp.69-74, Nottingham, UK, 23-25 Sept, 1996.

[19] Madan M. G., Liang J. and Noriyasu H."Static and Dynamic Neural Networks from fundamentals to advanced theory" John Wiley and Sons, IEEE Press 2003.

[20] Iskender I., Demirkutlu E., Kılıç Ü. E. and Öztürk O. "Using Neural Network of a Solar Energy Panel in Maximum Power Point Detection." Proceedings of the İnternational Twelfth Turkish Symposium on Artificial İntelligence and Neural Networks TAINN-03, Vol. E-1, pp. 87-91, Çanakkale, Turkey, July 2-4, 2003.

In: Advances in Energy Research
Editor: Nicu Bizon ISBN 978-1-61209-991-0

Chapter 11

CONTROL OF THE FUEL CELL INVERTER SYSTEM

Nicu Bizon*
Faculty of Electronics, Communication and Computers
University of Pitesti, Romania

1

ABSTRACT

In this chapter is analyzed a fuel cell inverter system of multistage type. The mono-phase inverter is designed in the first subsection. The boost converter is supplied by a 35 V/2 kW fuel cell stack. The proposed voltage-mode control of the voltage boost stage, which has as fuzzy control variables the voltage on low voltage bus and the fuel cell current, stabilizes the low voltage and mitigates the current ripple, too. Obviously, the main control goal of the full-bridge DC-DC converter is the stabilization of the voltage on high voltage bus, if the stage of boosting voltage is removed. For the mono-phase inverter is proposed a feed-forward control of the switching command circuit that generates a pure sine PWM command as function of the voltage on high voltage bus in order to maintain the RMS level of the AC output voltage. The proposed fuzzy control is compared with classical control techniques. The simulation results confirm the robustness of the proposed control for the fuel mono-phase cell inverter system. The three-phase inverter variant is analyzed, too. Finally, the experimental setup is shown and the obtained results are compared with the simulation results.

Keywords: control, energy conversion, multistage inverter, fuel cell inverter.

MULTI-STAGE INVERTERS

Usually, in a two-stage power inverter the first stage (DC-DC converter) boosts the input voltage at about 400V and assures the electrical isolation, too. The second stage (inverter) realizes the DC-AC conversion. Each stage can be separately controlled or synchronous [1, 2,

---

[1] Corresponding Author: nicu.bizon@upit.ro

3, 4]. For implementation of the buck-boost transfer function in a multi-stage inverter, different types of energy conversion topologies were adopted: with high voltage (HV) intermediary bus of DC type or of high frequency (HF) AC type [5, 6, 7]. For buck or boost operation the DC-DC conversion stage is usually realized with a DC-DC converter or with a DC-AC-DC converter. The inverter system must have an output low-pass filter if inverter topology type has a HF HV AC bus. If it is with HV DC bus, then a low-pass filter on HV DC bus must be used to reduce the 100 Hz harmonics that are propagated back to the fuel cell stack. Using of a low voltage (LV) bus (where are connected the energy storage devices (ESD), such as mixed stack of batteries and ultracapacitors) and a boost stage help to mitigate the inverter current ripple [8] (figure 1).

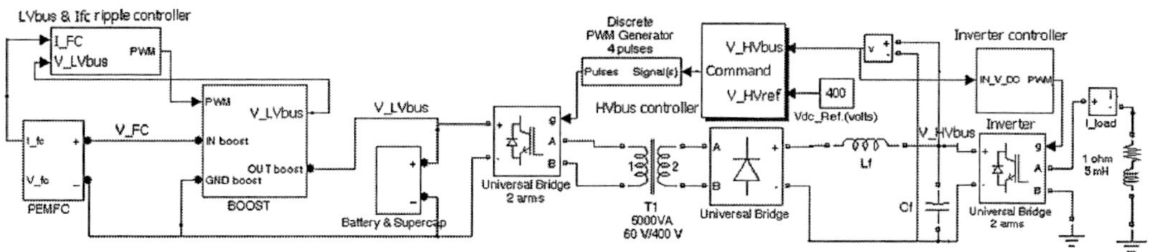

Figure 1. Matlab diagram for a mono-phase inverter system topology with HV DC bus and LV DC bus.

VOLTAGE CONTROL ON THE LV DC BUS

The main control goal of the LVbus and Ifc ripple controller is stabilization of the voltage on the LV DC bus at the set voltage, $V_{lv(ref)}$. The specific control goal is the mitigation of the inverter current ripple. Two types of controllers are used for control the boost converter in order to compare its control performances (see figure 2 and 3). The inverter current ripple is modeled by a Controlled Current Source (CCS). Only the fundamental harmonic of 100 Hz is taken in consideration to prove the mitigation effect at back propagation of this current ripple. The controllers are based of the peak current control (PCC) mode. The reset signal is obtained by processing both control variables (fuel cell current, I_{fc}, and output boost voltage, V_{boost}, that is also the voltage on the LV DC bus) in relation with theirs references ($I_{pk(ref)}$ and $V_{lv(ref)}$, respectively) by:

- a Fuzzy Logic Controller (FLC) with a 2-D control law (figure 3.a);
- a nonlinear controller with a 1-D control law (figure 3.b).

The control performances are improved comparing with the knowing performances of the classical voltage-mode control, also called duty-cycle control, which contains a single feedback loop and adjusts the duty cycle directly in response to output voltage changes. On the other hand, the proposed fuzzy control isn't of the classical current-mode control type that is a multiple-loop control method, too, but containing two loops: an inner current loop and an outer voltage loop.

The proposed control is of feed-back type regarding the voltage variable and of feed-forward type regarding the current variable.

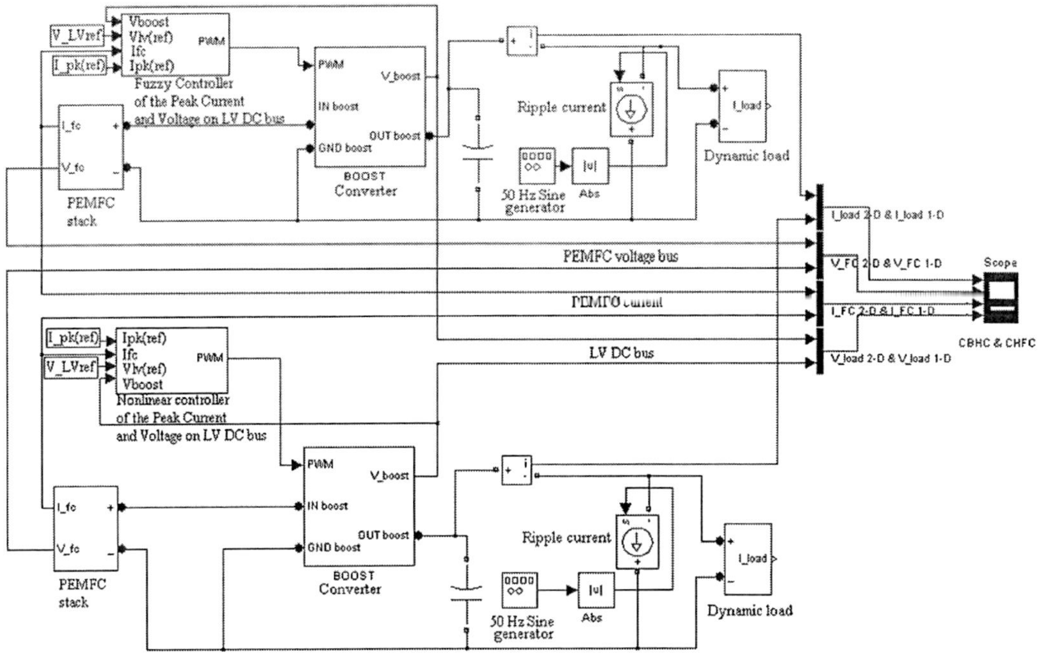

Figure 2. Boost interfaces with nonlinear and fuzzy control of the PEMFC peak current and voltage on LV DC bus.

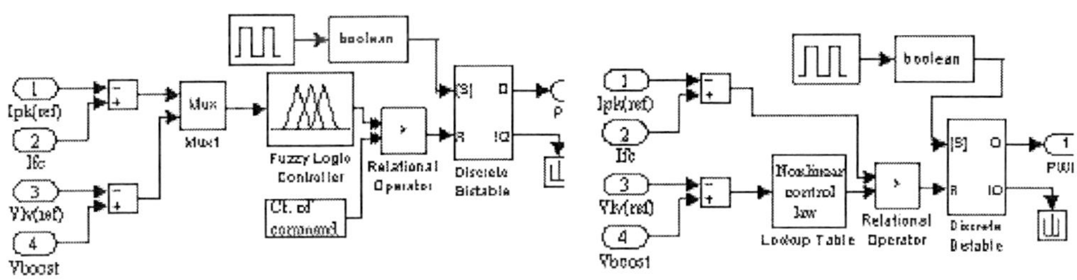

a) Fuzzy Control of the PEMFC peak current and voltage on LV DC bus (FCPCLV)

b) Nonlinear Control of the PEMFC peak current and voltage on LV DC bus (NCPCLV)

Figure 3. Control of the PEMFC peak current and the voltage on the LV DC bus.

The 1-D nonlinear control law is obtained by projection of the 2-D control law of the FLC in the $V_{boost} - I_{fc}$ plane. Further information about this kind of control, including the case when an Energy Storage Devices (ESD) stack is used, is presented in [9, 10]. There are analyzed more controller structures as level of control performances. Obviously, by adding as ESD a batteries stack, which is placed at the boost converter output port, the control law must be changed in order to avoid the “boiling” of the batteries stack. So, the required battery charging parameters must be considered in the designing of the control law. Usually a boundary control law (named *boundary control with current taper-BCCT)* is considered. This is also a function of both fuel cell current limits and battery voltage, so the control variables can be fuel cell current, I_{fc}, and output boost voltage, V_{boost}, too, but not as difference variables relative to their references.

If the ESD is connected at the boost converter output port via a bidirectional DC-DC converter then the task of correct charging/discharging of the ESD is implemented by its controller. The concept of Integrated Fuzzy Control (IFC) is proposed and analyzed in [11]. The proposed IFC structure integrates both controllers into one, in order to be cheap (only two voltage transducers and one current transducer are used) and flexible (the IFC control law can be easily redesigned for other type of stacks of PEMFCs, batteries and ultracapacitors, and different load parameters).

VOLTAGE CONTROL ON THE HV DC BUS

The main control goal of the HVbus controller is stabilization of the voltage on the HV DC bus at the set voltage, $V_{HV(ref)}$ [12]. The specific control goal can be the mitigation of the inverter current ripple, if the stage of boosting voltage is removed. Two types of controllers are used for control the boost converter in order to compare its control performances (see figure 2 and 4):

– a classical controller with a proportional – integral (PI) control law (figure 4.a);
– a fuzzy controller with a 2-D control law (figure 4.b).

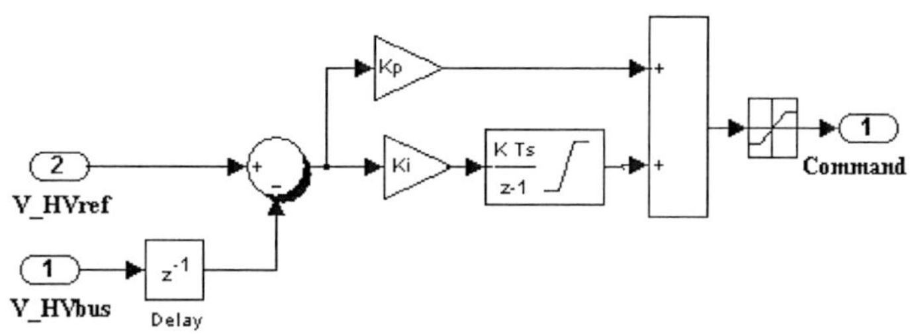

a) PI control of the voltage on HV DC bus

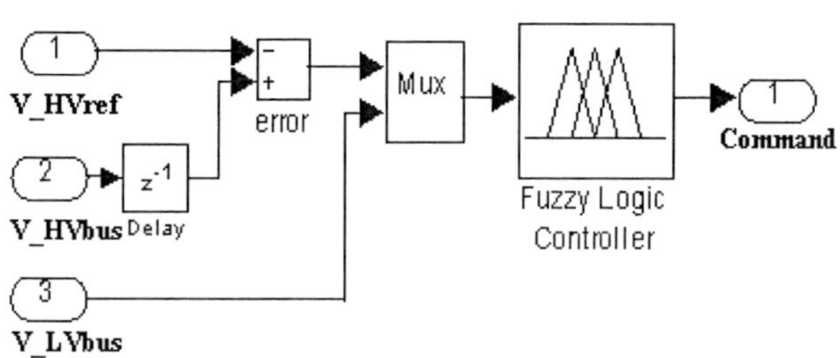

b) FLC control of the voltage on HV DC bus

Figure 4. Control of the voltage on HV DC bus.

The PI regulator constants (the proportional, k_P, and integral, k_I, constants) are designed to generate the command signal for the phase shift controller that best stabilizes the voltage on the HV DC bus at the set voltage, $V_{HV(ref)}$. (figure 5).

The dynamic behavior is presented in next figures for the DC-DC power stage using different values for the PI controller parameters. When the boost stage is removed, the DC-DC power stage is directly powered from the fuel cell stack. As it is known, some changing in the input voltage appears in this case as effect of the load dynamic. The input voltage, V_{LVbus}, can be simulated by a Controlled Voltage Source (CVS). The PI controller parameters must be designed to assure control robustness in any situation. Plotted variables and the used PI parameters are mention in every figure.

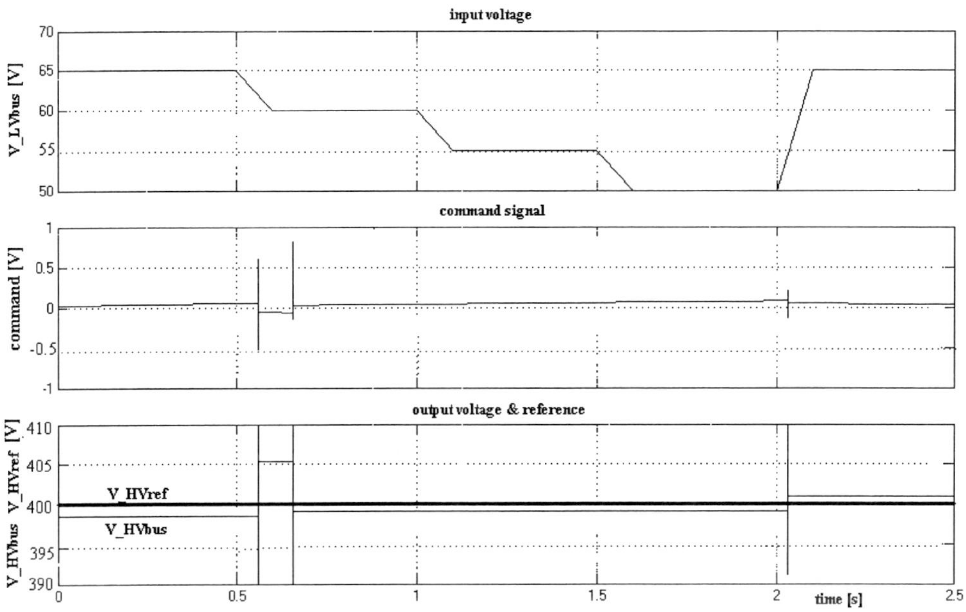

Figure 5a. DC-DC converter behavior for k_P=0.015 and k_I=0.04.

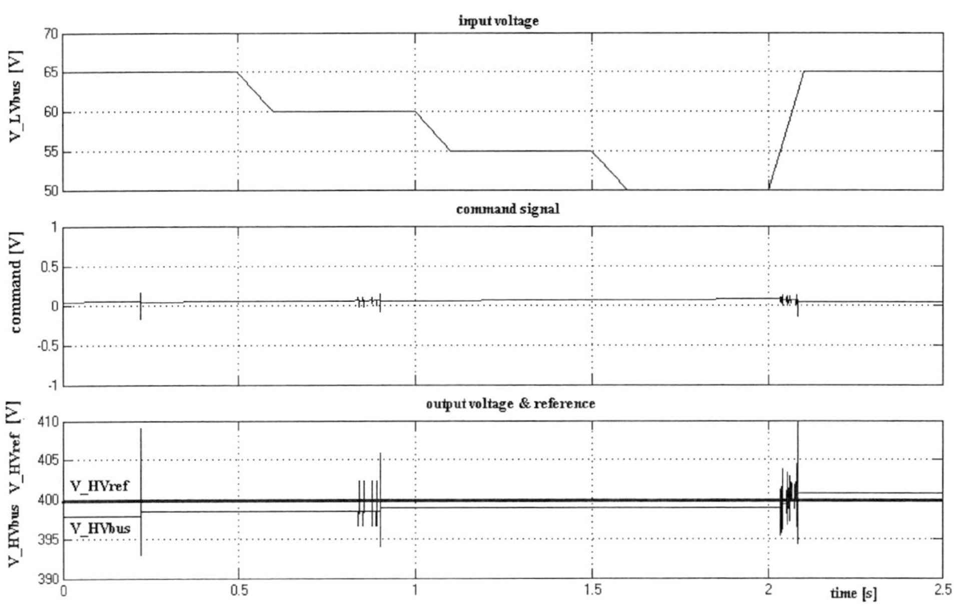

Figure 5b. DC-DC converter behavior for k_P=0.02 and k_I=0.02.

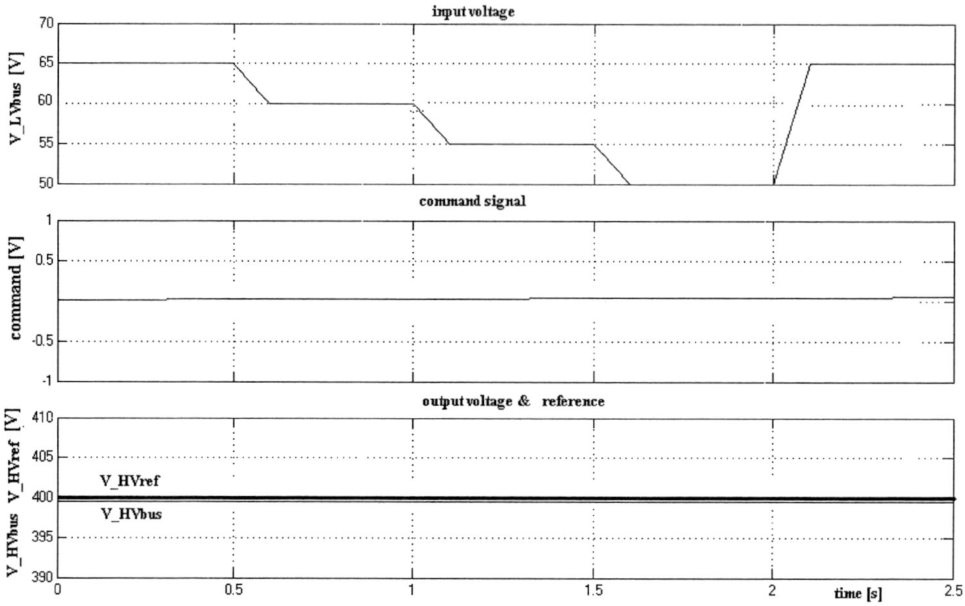

Figure 5c. DC-DC converter behavior for k_P=0.04 and k_I=0.03.

The best results are obtained for the k_P=0.04 and k_I=0.03 (figure 5.c).

The FLC for the voltage on HV DC bus has two inputs variables (the output voltage error, V_HVbus – V_HVref, and the input voltage, V_LVbus) and one output (the command signal).

The memberships functions for output voltage error (V_HVbus – V_HVref), the input voltage (V_LVbus) and command signal (command) are shown in figure 6, plot a, b and c, respectively. Three membership's functions are defined for both input variables and output variable. They are named as N=Negative, ZE=Zerro Equal and P=Positive for first input variable, L=Low, N=Normal and H=High for the second input variable, and S=Small, M=Medium and B=Big for the output variable. As it can be see in figure 6.b, the triangular membership functions for input voltage have the peaks located at values close of the rated voltage of the used PEMFC stack.

The rules base is shown in table 1.

Table 1. Rules base of the FLC for the voltage on HV DC bus

Command signal		output voltage error		
		N	ZE	P
input voltage	L	B	B	M
	N	B	M	S
	H	M	S	S

The Mamdani implication and center of gravity defuzzification method are used. The obtained 2-D control law is shown in figure 6.a.

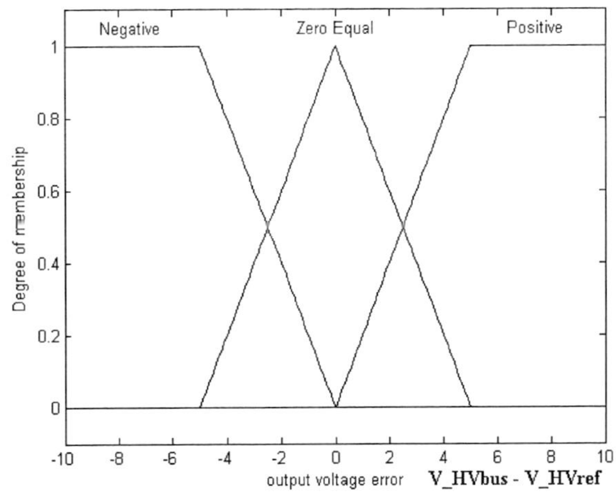

a) Output voltage error memberships

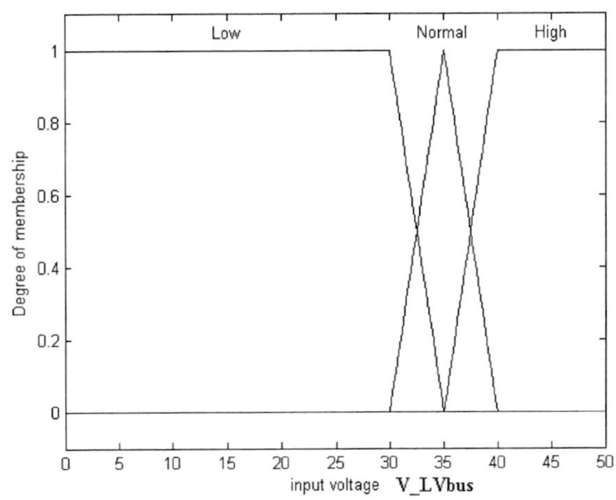

b) Input voltage memberships

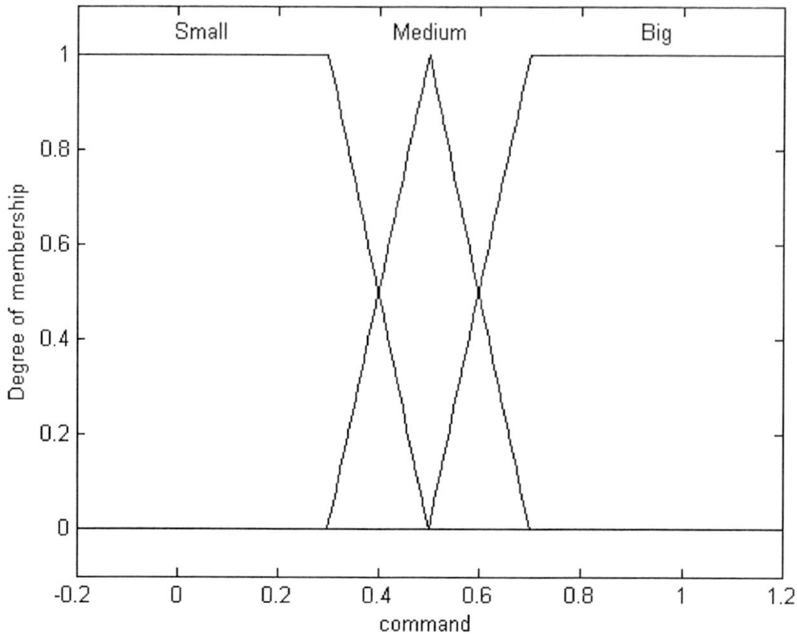

c) Command signal memberships

Figure 6. (Continued).

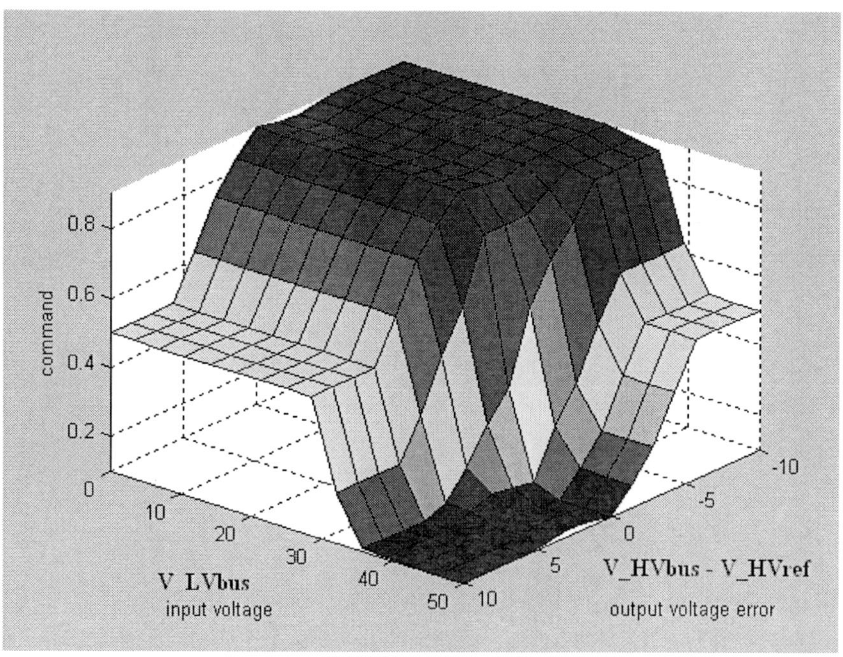

d) FLC control surface

Figure 6. Designing of the FLC and its control surface.

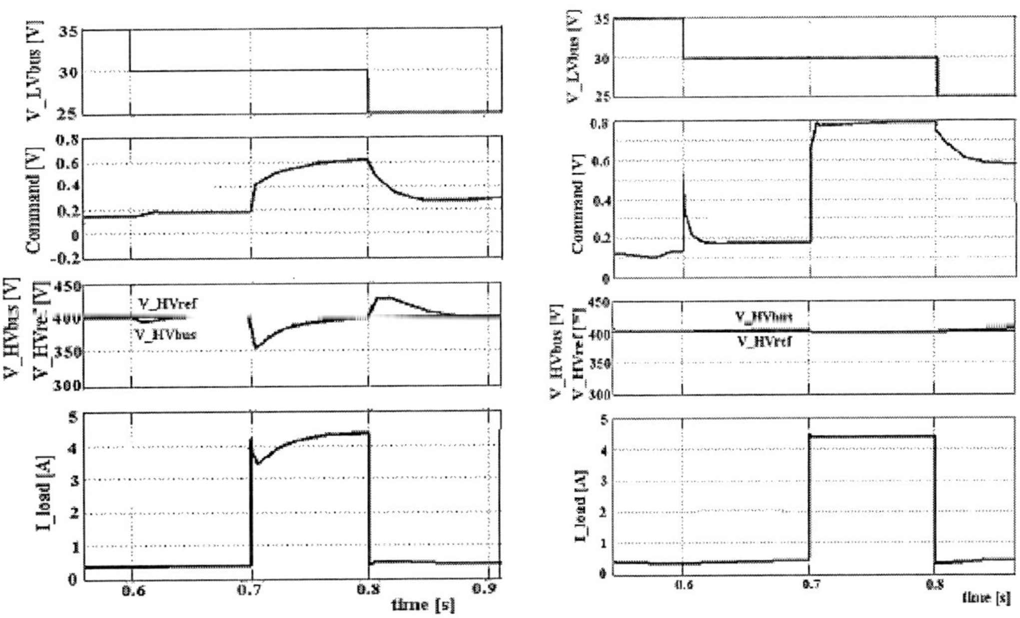

a) using a PI controller with k_P=0.04 and k_I=0.03 b) using the proposed FLC controller

Figure 7. Behavior of the DC-DC converter for a step-down input voltage and a pulse load.

The behavior of the DC-DC converter for a step-down input voltage and a 5 A pulse load using a PI controller with k_P=0.04 and k_I=0.03 and the proposed FLC controller is shown in figure 7 (plot a and b, respectively). It can be note that voltage on the HV bus has large variations under load pulses if the PI controller is used. Also, these variations appear at changing of the input voltage on the LV bus, but the level is much small than that mentioned before. All these variations of voltage on HV bus are reduced by the proposed FLC control (see the level of command signal in both cases).

The better robustness of the FLC is shown by simulation, too.

Voltage Control on the AC Bus

The control goal of the inverter controller is to stabilize the level of AC output voltage that is generated by inverter stage (DC-AC converter) powered directly from an appropriate PEMFC stack as voltage level and rated power [13]. In order to understand how is designed the forward control of pure sine PWM command circuit for a mono-phase inverter (figure 8), which will be presented in next section, the pure sine PWM control method is briefly explained below. The PEMFC stack is modeled by a CVS.

The main purpose of the Pulse-Width-Modulation (PWM) technique is to command the bridge of switches to obtain an AC output voltage from the DC input voltage. The pure sine PWM command is usually used to obtain a power spectrum of the AC output voltage with fewer harmonics. If the DC level, V_HVbus, is variable with load power, then the amplitude modulation index (see relation 3), m_a, must be controlled in order to stabilize the RMS level of the AC output voltage. In a nutshell, PWM is a way of digitally encoding for the analog signal levels. Through the use of high-resolution counters, the duty cycle of a square wave is changed to encode a specific analog signal level. Obviously, any analog signal can be encoded with pure sine PWM techniques if a high sample frequency is used.

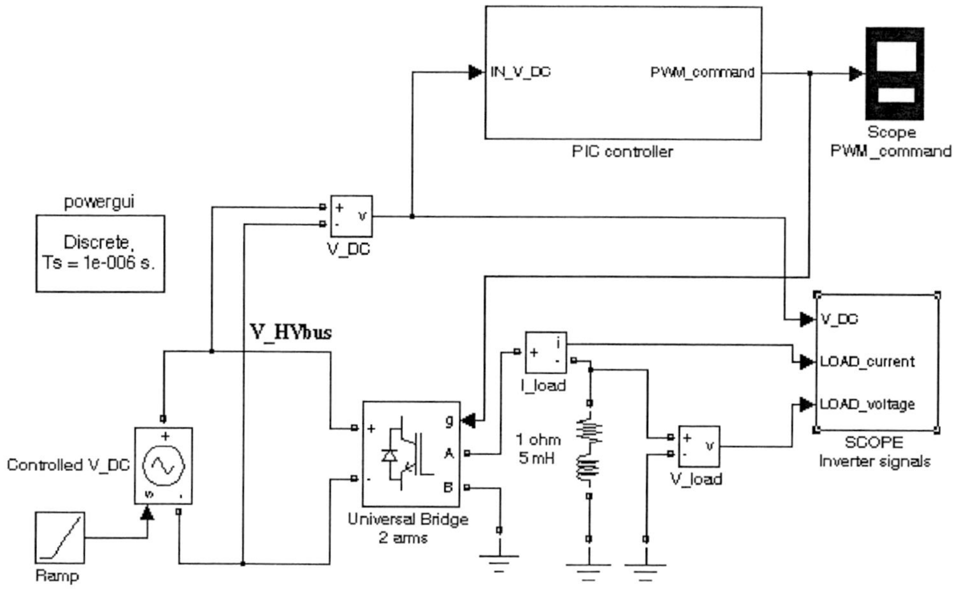

Figure 8. The structure of the mono-phase inverter.

One of the advantages of the PWM switching command is that the signal is digital from the processor to the controlled system, with different amplitudes for the digital levels. So, no digital-to-analog conversion is necessary.

By keeping the signal digital, noise effects are minimized. Noise can affect a digital signal only if it is strong enough to change one logical level into the second used level. Consequently, the increased noise immunity is yet another benefit of choosing PWM instead of an analog control.

PURE SINE PWM TECHNIQUE

Many PWM techniques have been developed in addition to pure sine PWM. Several of them are applicable to high power converters, with some modification or designed for particular configuration.

The pure sine PWM command is usually obtained with a microcontroller from PIC family (optional with dedicated PWM channels) and its programming is done to the timer channel level for a certain frequency and a controlled index of amplitude modulation. The choice of the frequency modulation index is correlated with the switching frequency, f_c. This frequency can be chosen in appropriate frequency band, where the produced EMI interferences don't affect the functionality of other equipments. If the grid frequency is f_0, then the frequency modulation index is defined as $m_f=f_c/f_0$.

The amplitude modulation index is defined as ratio between the amplitude of the AC voltage and the DC value of the input voltage, U_i. It must be less of 1 for a linear dependence of the AC amplitude as a function of this ratio. If sinusoidal command signal is (see figure 9):

$$u_C = U_{CM} \sin 2\pi f_c t \tag{1}$$

For an IGBT's half-bridge and m_a<1, the fundamental of power spectrum for the inverter output voltage, u_0, is given by relationship:

$$u_{01} = \frac{U_{CM}}{U_{TM}} \frac{U_i}{2} \sin 2\pi f_c t = m_a \frac{U_i}{2} \sin 2\pi f_c t \tag{2}$$

This fundamental harmonic of the power spectrum will be double for an IGBT's full-bridge:

$$u_{01} = \frac{U_{CM}}{U_{TM}} U_i \sin 2\pi f_c t = m_a U_i \sin 2\pi f_c t \tag{3}$$

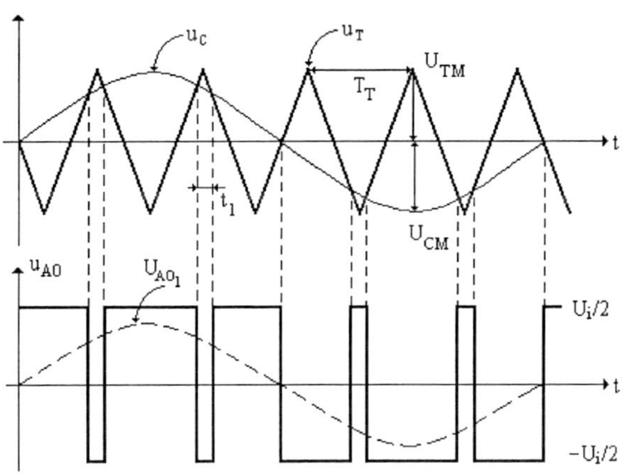

Figure 9. Pure sine PWM switching command technique.

The control of the RMS value for the inverter output voltage can be made by changing the amplitude modulation index. In a digital implementation, this supposes the change of the on/off durations that are programmed in the counters. Control loop can be of feedback type or of feed-forward type, or of mixed type. A simple and efficient forward control loop is proposed in the next section, having the advantage of a short response time.

Feed-Forward Control of the Sine Inverter

The nominal value for V_HVbus is of 400 V and the 220 V RMS value for output voltage of the mono-phase inverter is obtained for an amplitude modulation index, m_a, of 0.8.

The voltage of the PEMFC stack is variable due to load variation. So, the range for V_HVbus is around of this value. The range of 369 V to 681 V is considered. A more sensitive selection of PWM vectors is made for the input voltage close to the reference voltage, V_HVref=400V. Consequently, the voltage step is of 2 V in range [389 V, 411 V]. Outside this range the step voltage is of 10V for three ranges: 369 V<V_HVbus<379 V, 379 V<V_HVbus<389 V and 411 V<V_HVbus<421 V. The step voltage is of 20 V for 421 V<V_HVbus<681 V. Values of V_HVbus<369V can appear only to start-up. So, a single start-up PWM vector is considered for this range.

Also, at start-up can appear over-voltages (V_HVbus>421 V). The PWM vectors are named as pwm0_xx, where m_a=0.xx, and are shown in table 2 for each range of the input voltage, V_HVbus.

A simple program (e.g implemented in Matlab or Math – see figure 10) allow the calculation of the pure sine PWM command vectors for a mono-phase full-bridge of IGBTs. Some of the obtained values for m_f=25 are shown in Table 3.

Firstly, to validate the proposed feed-forward control is simulated in Matlab-Simulink®. A controlled PWM generator is implemented in figure 11, having the PWM vector selected by the V_HVbus value [14].

Table 2. Pure sine PWM command vectors for each range of the input voltage

Range [V]	PWM vector	Range [V]	PWM vector	Range [V]	PWM vector
[0, 369]	pwm1_00	(401, 403]	pwm0_79	(481, 501]	pwm0_50
(369, 379]	pwm0_95	(403, 405]	pwm0_78	(501, 521]	pwm0_45
(379, 389]	pwm0_90	(405, 407]	pwm0_77	(521, 541]	pwm0_40
(389, 391]	pwm0_85	(407, 409]	pwm0_76	(521, 561]	pwm0_35
(391, 393]	pwm0_84	(409, 411]	pwm0_75	(561, 581]	pwm0_30
(393, 395]	pwm0_83	(411, 421]	pwm0_70	(581, 601]	pwm0_25
(395, 397]	pwm0_82	(421, 441]	pwm0_65	(601, 621]	pwm0_20
(397, 399]	pwm0_81	(441, 461]	pwm0_60	(601, 641]	pwm0_15
(399, 401]	pwm0_80	(461, 481]	pwm0_55	(641, 681]	pwm0_10

Table 3. Some PWM command vectors

PWM1_00[25]={129,162,193,219,240,252,255,252,239,219,192,161,128,95,64,38,18,5,1, 5,18,38,65,96,129};
PWM0_95[25]={129,160,190,215,234,246,250,246,234,214,189,160,128,97,68,42,23,11,7 ,11,23,43,68,97,129};
PWM0_90[25]={129,159,186,210,228,240,244,240,228,210,186,158,128,98,71,47,29,17,1 3,17,29,47,71,99,129};
PWM0_85[25]={129,157,183,206,223,234,237,233,223,205,183,156,128,100,74,51,34,23,

20,23,34,52,74,101,129};

PWM0_80[25]={129,155,180,201,217,228,231,227,217,201,179,155,128,102,77,56,40,30, 26,30,40,56,77,102,129};

PWM0_75[25]={129,154,177,197,212,221,224,221,211,196,176,153,128,103,80,61,45,36, 33,36,45,61,81,104,129};

PWM0.70[25]={129,152,174,192,206,215,218,215,206,192,173,151,128,105,84,65,51,42, 39,42,51,65,84,105,129};

PWM0_65[25]={129,150,170,188,201,209,212,209,200,187,170,150,128,107,87,70,56,48, 45,48,56,70,87,107,129};

PWM0_60[25]={129,149,167,183,195,203,205,203,195,183,167,148,128,108,90,74,62,54, 52,54,62,74,90,109,129};

PWM0_55[25]={129,147,164,178,190,197,199,196,189,178,164,147,128,110,93,79,67,60, 58,61,68,79,93,110,129};

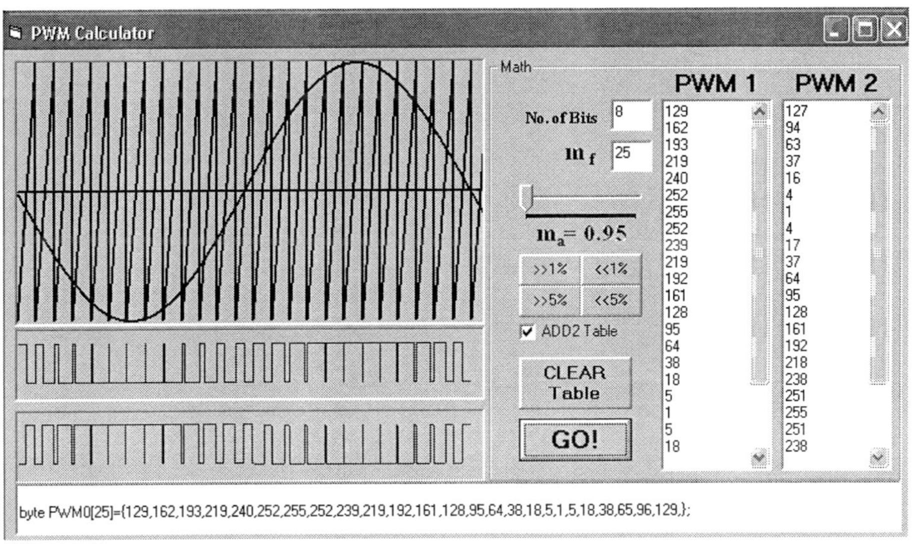

Figure 10. PWM vectors calculation.

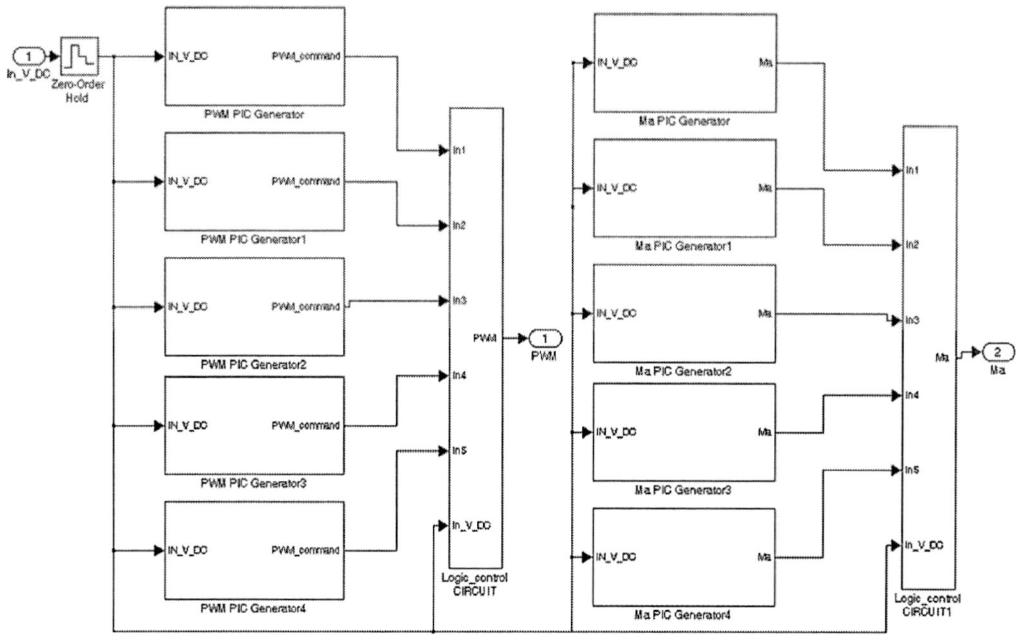

Figure 11a. The controller PWM generator structure.

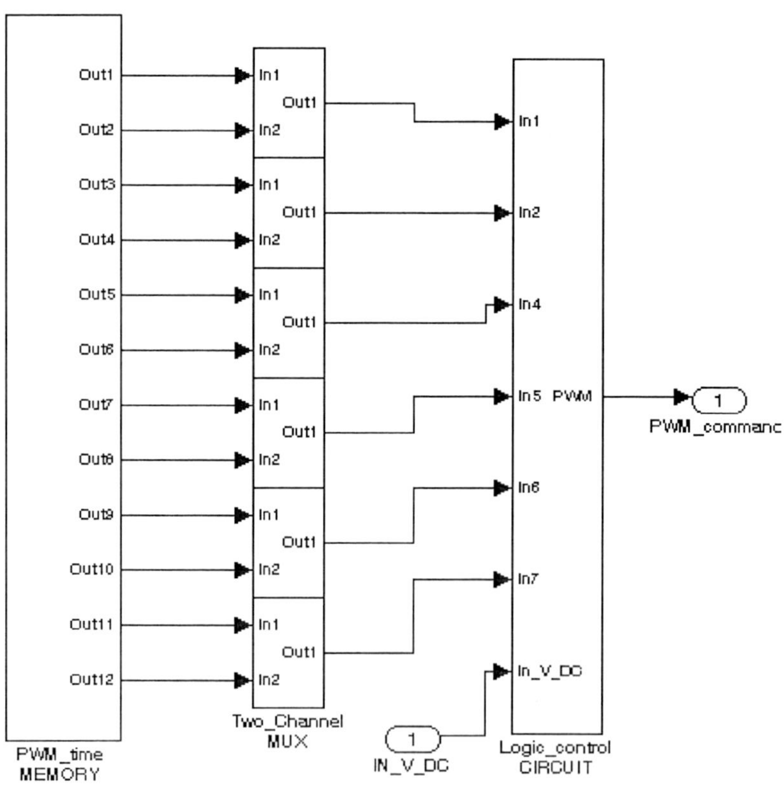

Figure 11.b. The structure of one PWM generator for a given V_HVbus range – the PWM PIC Generator block.

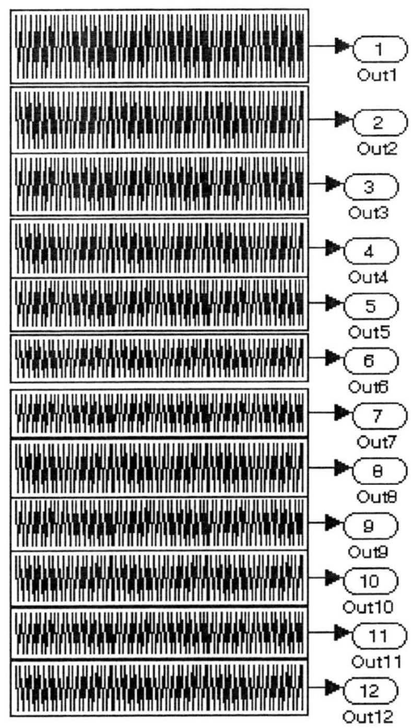

Figure 11c. PWM generators are loaded in the MEMORY block.

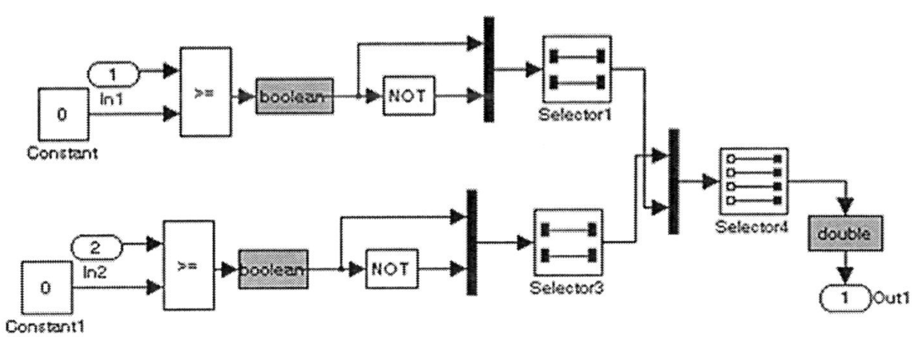

Figure 11d. PWM command circuit for the IGBTs of the full-bridge inverter – Two channel MUX block.

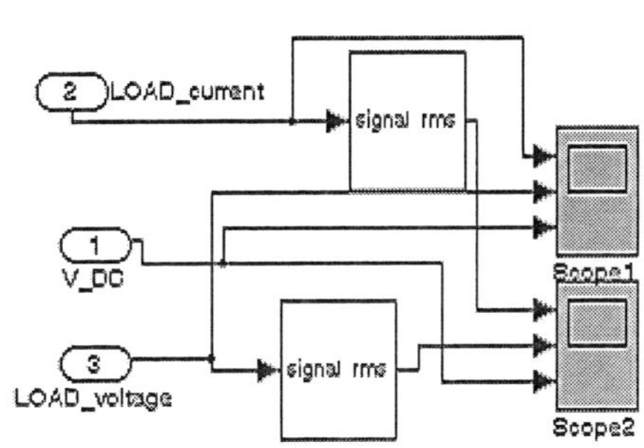

Figure 11e. The circuit that compute the RMS value of output voltage and current.

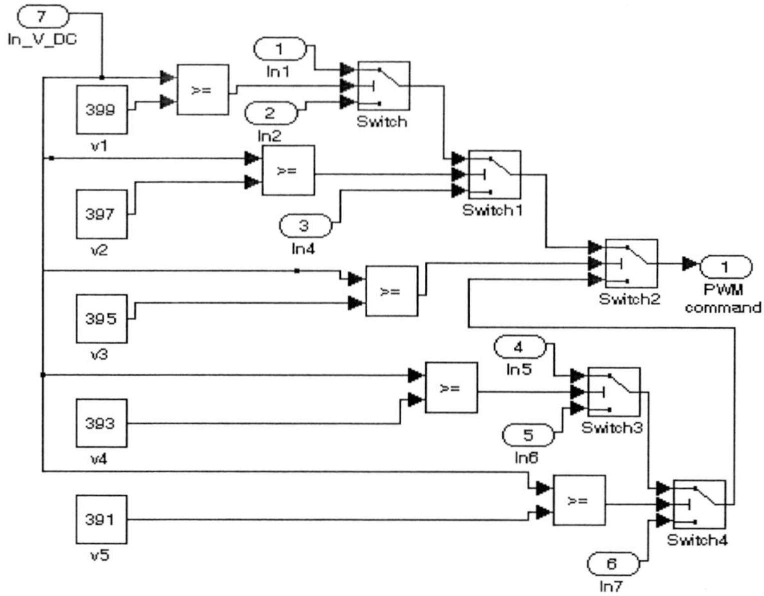

Figure 11f. Circuit for logic control and selection of the PWM switching command as function of the DC input voltage, V_HVbus – Logic control CIRCUIT.

Two RMS blocks from Signal library are used to display the RMS values of the output voltage and current (figure 11.f). The *Ma PIC Generator* block generates the m_a value that is used in each moment to be display and it is organized in the same manner as *PWM PIC Generator* block.

SIMULATION RESULTS

The obtained results by simulation are presented in this section. The robustness of the feed-forward control using the input voltage as control variable is shown in next figures using a step-up (step-down) or a ramp signal that control the CVS input control. In figure 12 is presented in time the shape of the load current, I_Load, and of the inverter output voltage, V_Inverter, Load under a linear variation of the CVS voltage, V_HVbus. In figure 13 is presented the RMS value of the output voltage from inverter, V_inverter_RMS, under a linear variation of the CVS voltage, V_HVbus. It can be observed that RMS value variation of the output voltage from inverter is up to 5% from its rated value obtained with a low-pass filter.

In figure 14 is presented the shape for load current, I_Load, and inverter output voltage, V_Inverter, under a linear variation of the CVS voltage, V_HVbus., sampled at 20 ms (synchronously with grid voltage). The variation of the amplitude modulation index to control the RMS level of the inverter output voltage is shown, too. It can be observed that RMS value variation of the output voltage from inverter is up to 1% from its rated value when the input voltage is in the range [389 V, 411 V] and close to this. The RMS value variation of the output voltage from inverter is shown in figure 16 and 17 for V_HVbus greater and lower than 400 V, respectively. By analyzing these figures it can be observed that V_inverter_RMS is well controlled for V_HVbus in range [370 V, 500 V].

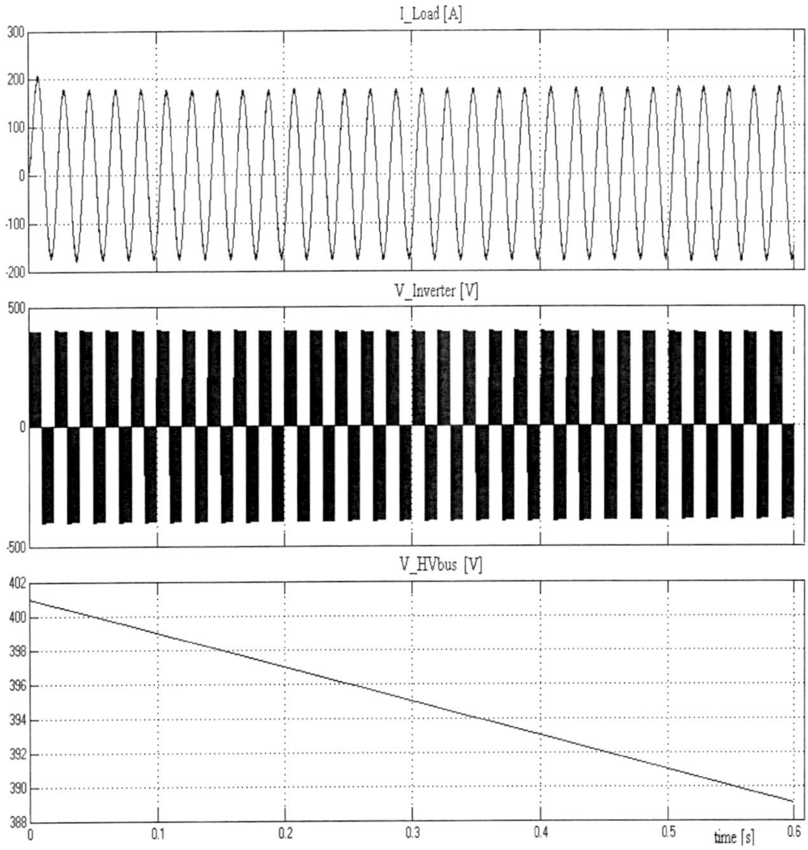

Figure 12. Load current and inverter voltage under linear variation of the CVS voltage.

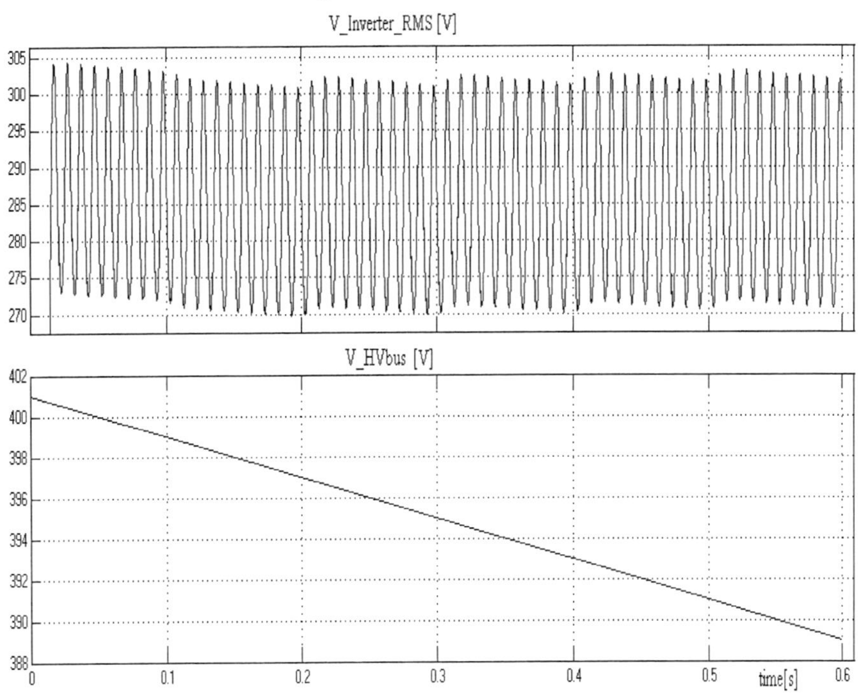

Figure 13. RMS value of the load voltage under linear variation of the CVS voltage.

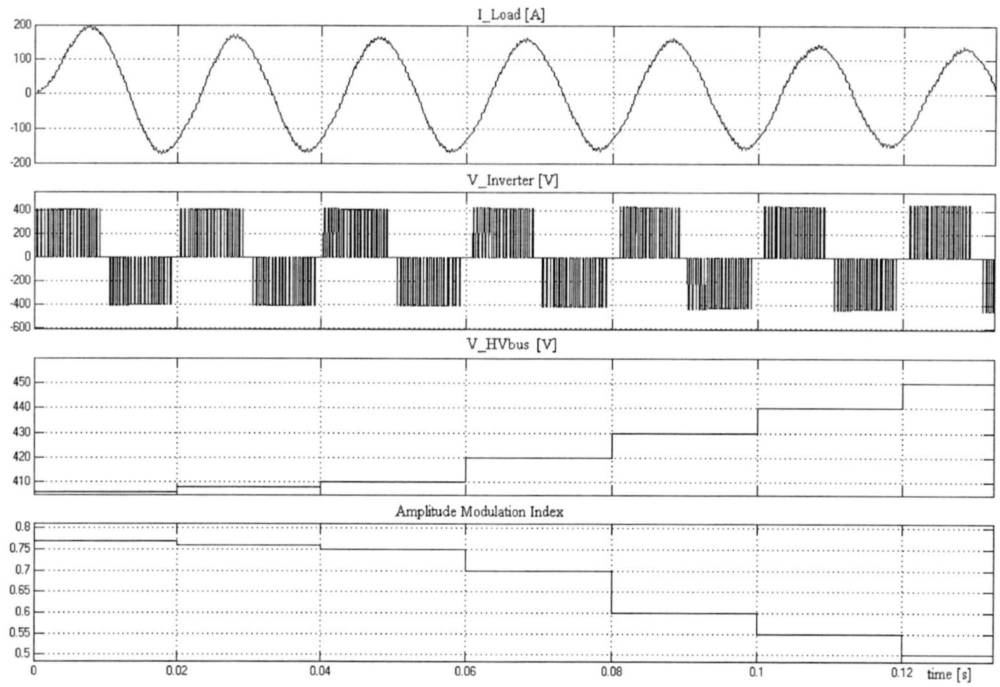

Figure 14. Load current and voltage, and amplitude modulation index variation with input voltage sampled at 20 ms.

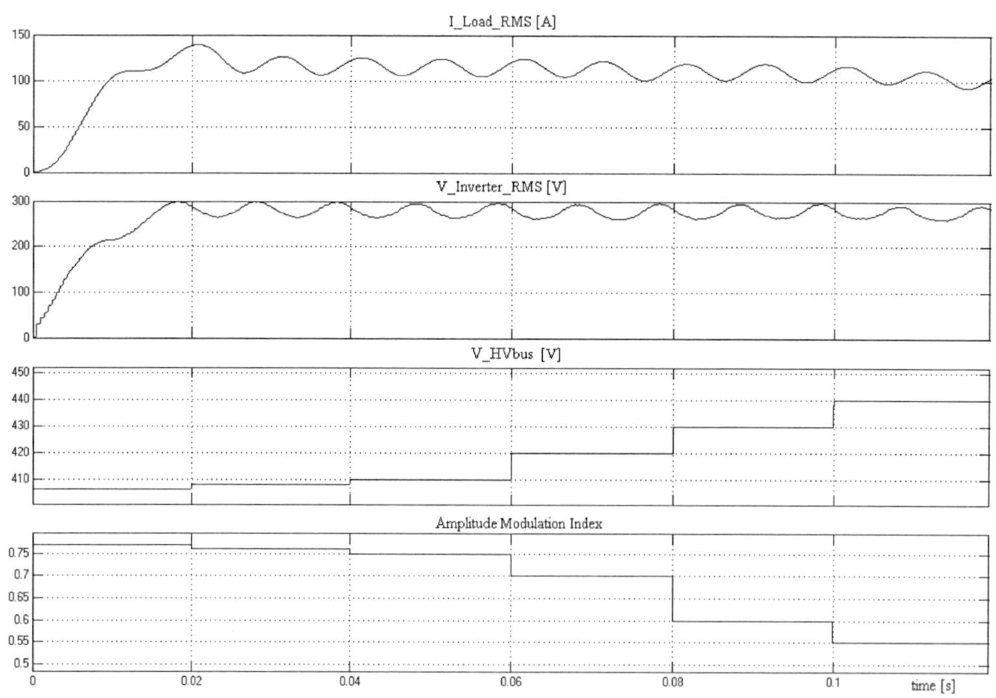

Figure 15. RMS value of the load current and voltage, and m_a variation with input voltage sampled at 20ms.

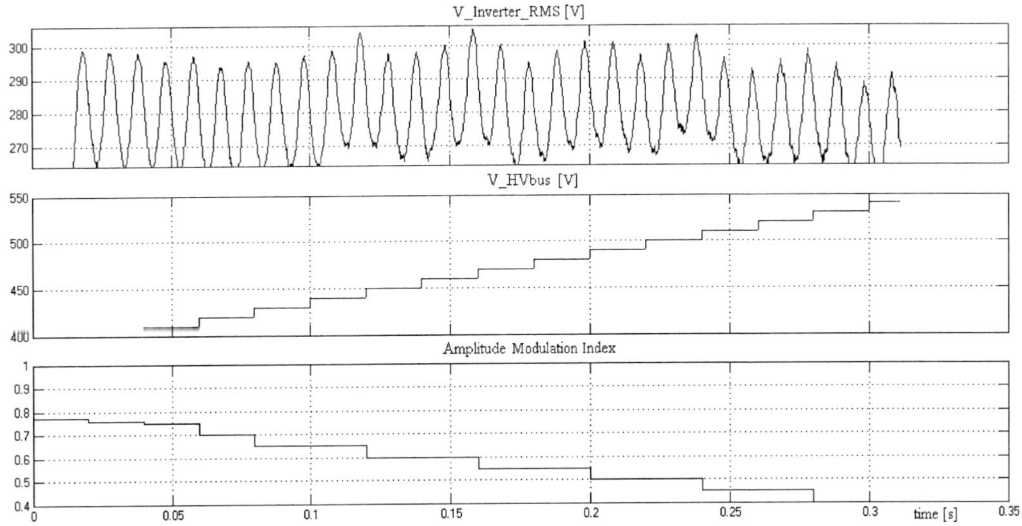

Figure 16. V_inverter_RMS for V_HVbus > 400 V.

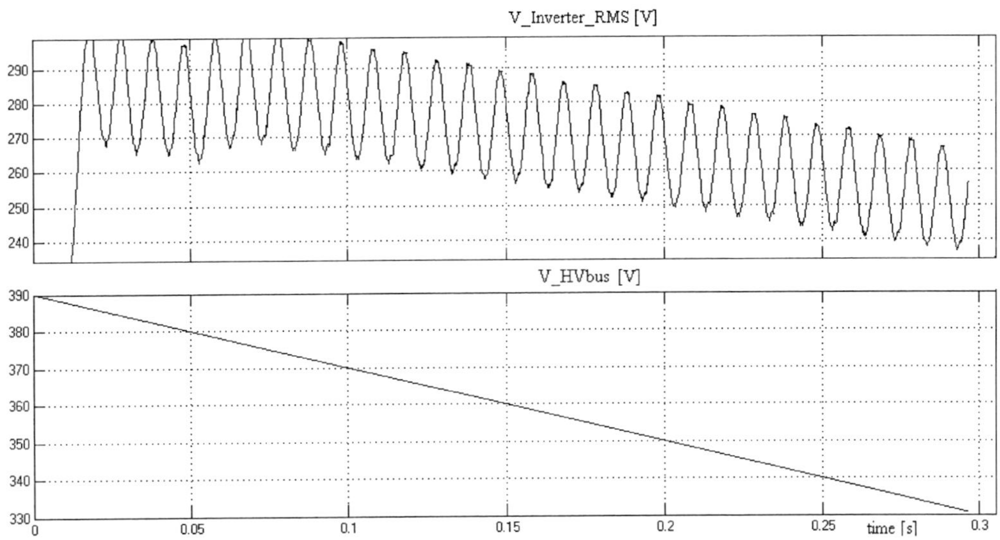

Figure 17. V_inverter_RMS (V_DC) for V_DC < 400 V.

The synchronization of the input voltage sampling with the grid voltage (at 20 ms or at 10 ms) produces a soft switching for the IGBT's of the full-bridge inverter (figure 18). A zoom is shown in figure 19.

For asynchronous sampling it is possible to appear switching over-voltages and over-currents for the IGBT's.

In figure 20 is shown the Matlab diagram for a three-phase inverter system topology with a HV DC bus and LV DC bus.

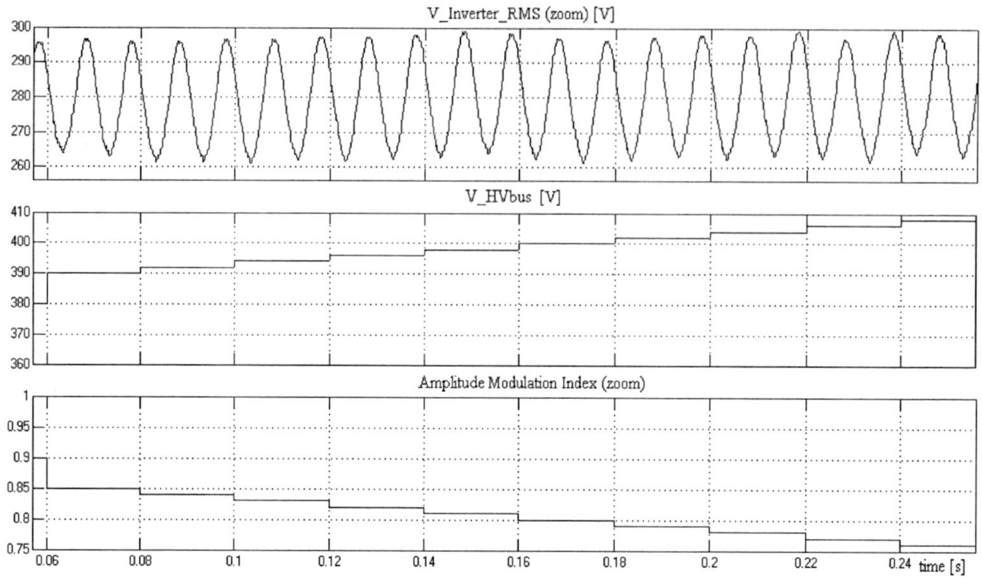

Figure 18. V_inverter_RMS variation with the V_HVbus sampled at 20 ms in range [370 V, 410 V].

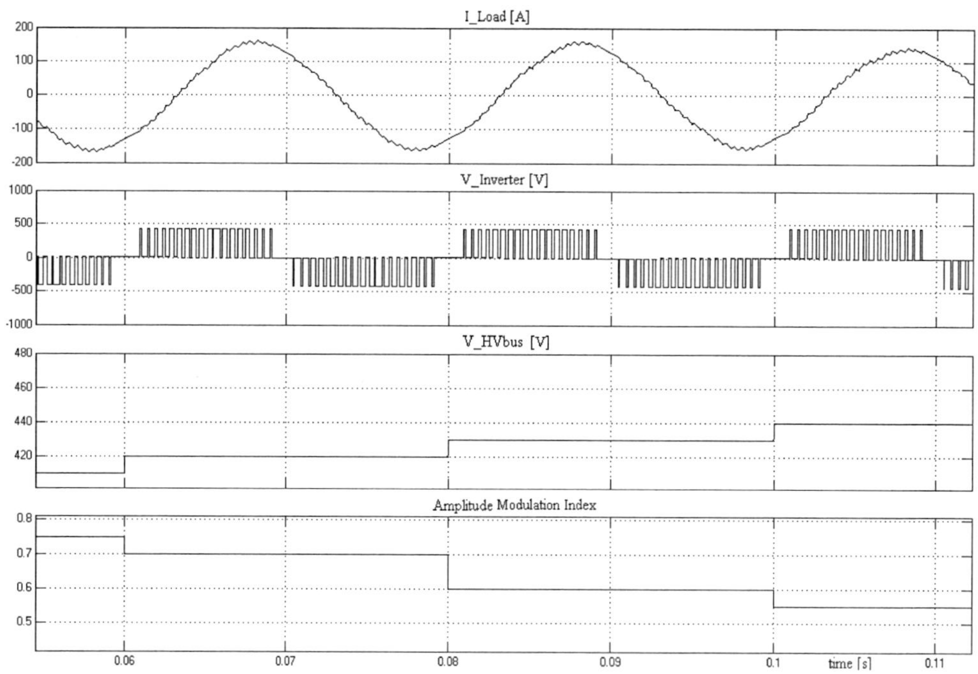

Figure 19. A zoom of the inverter behavior when the input voltage is sampled at 20 ms.

The command and control blocks are adapted form the AC-DC-AC PWM Controller demo included in the demonstrative programs of the Matlab-Simulink® library (figure 21). The amplitude modulation index is changed by the proportional-integrator (PI) controller after the voltage error. The d-q transformation of the three phase output voltages is used, before of computing of voltage error (see figure 21).

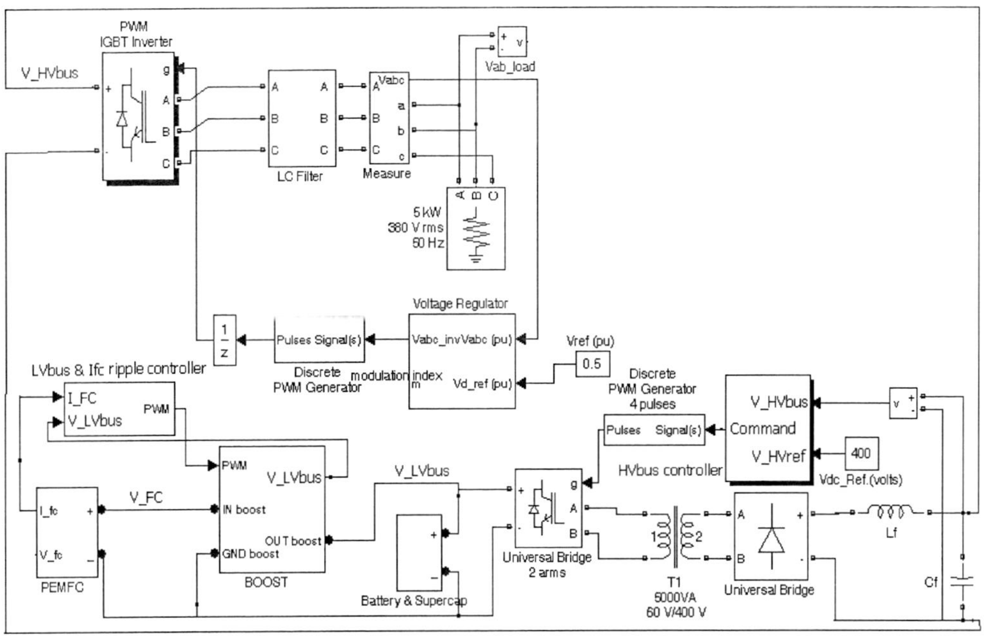

Figure 20. Matlab diagram for a three-phase inverter system topology with a HV DC bus and LV DC bus.

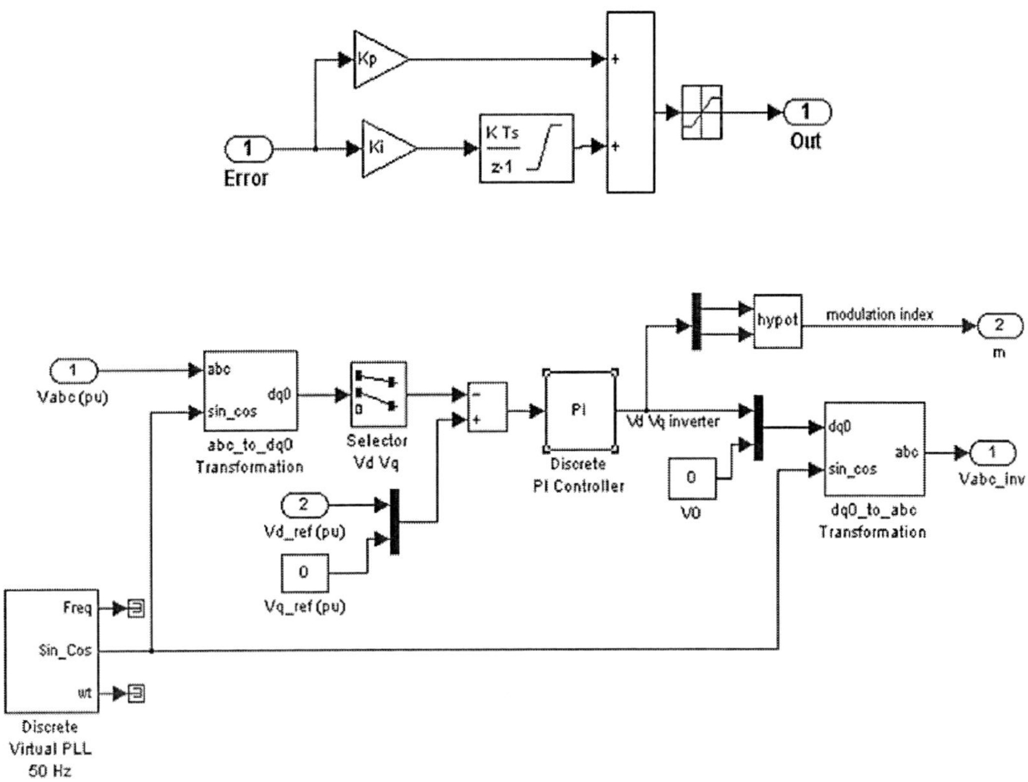

Figure 21. The diagrams of the voltage regulator (bottom) and discrete PI controller (top).

COMPARATIVE RESULTS

The control algorithm, which is purposed for the first time in [13], is compared here with the existing control presented in the Matlab- Simulink® library of demonstrative programs. The Matlab-Simulink diagram is presented in figure 22.

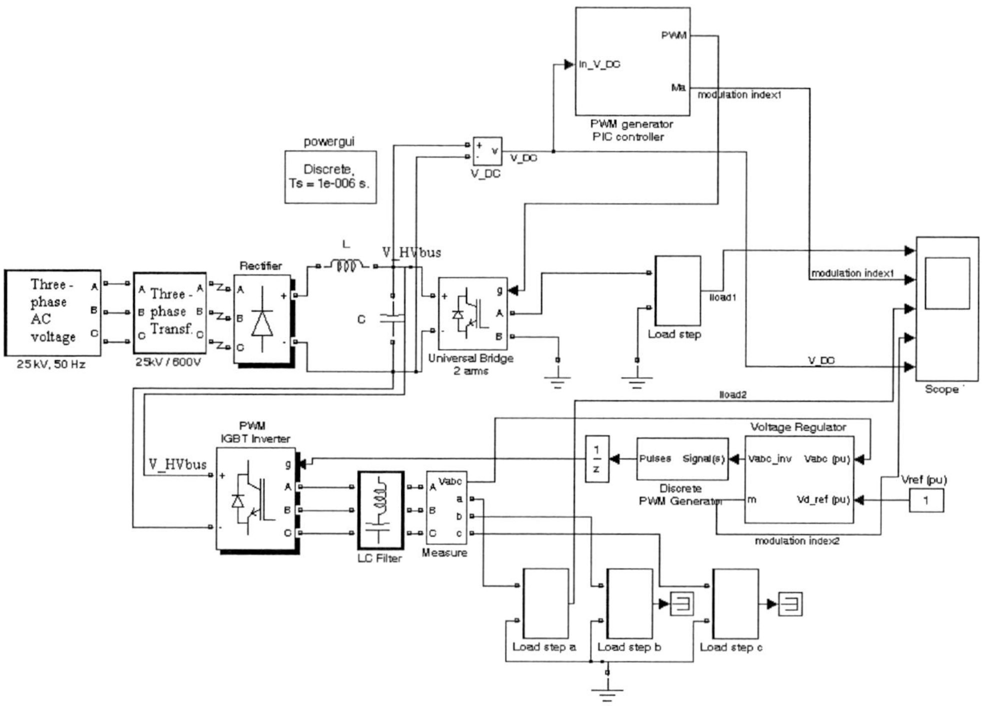

Figure 22. Matlab-Simulink diagram used to test the feed – forward control.

Both inverters are supplied from the same HV DC bus, having the rated voltage of about 400 V. The V_HVbus is obtained by rectification and low-pass filtering of the three-phase voltage. In the top side of the diagram is shown the proposed feed-forward control. The amplitude modulation index is changed by the input DC voltage. Value of amplitude modulation index, m_a, is of 0.8 for a input DC voltage of 400 V. The PIC controller generates the pure sine PWM commands for each amplitude modulation index. Variations of the load power or of the source will change the V_HVbus value.

In the bottom side of the diagram is shown the three-phase inverter. First phase is displayed to be compared with the mono-phase at level of AC voltage stabilization. The amplitude modulation indexes that are generated by the proposed controller (figure 11) and the existing controller (figure 21) are displayed, too.

In next figure are presented the variation of the amplitude modulation indexes at start-up under constant load and a zoom of it (figure 23). The inverters behaviors under a light load pulse (half of the nominal value) and under a high load pulse (double of the nominal value) applied from 0.1 ms to 0.16 ms are shown in figure 24 and figure 25, respectively.

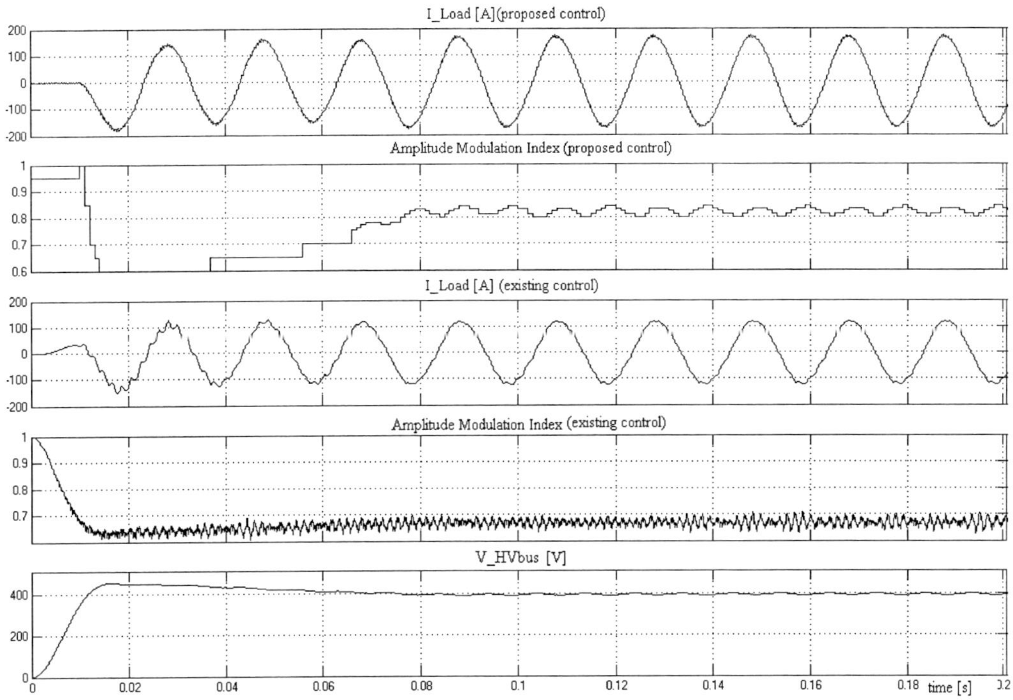

Figure 23a. Variation of the amplitude modulation indexes at start-up under constant load.

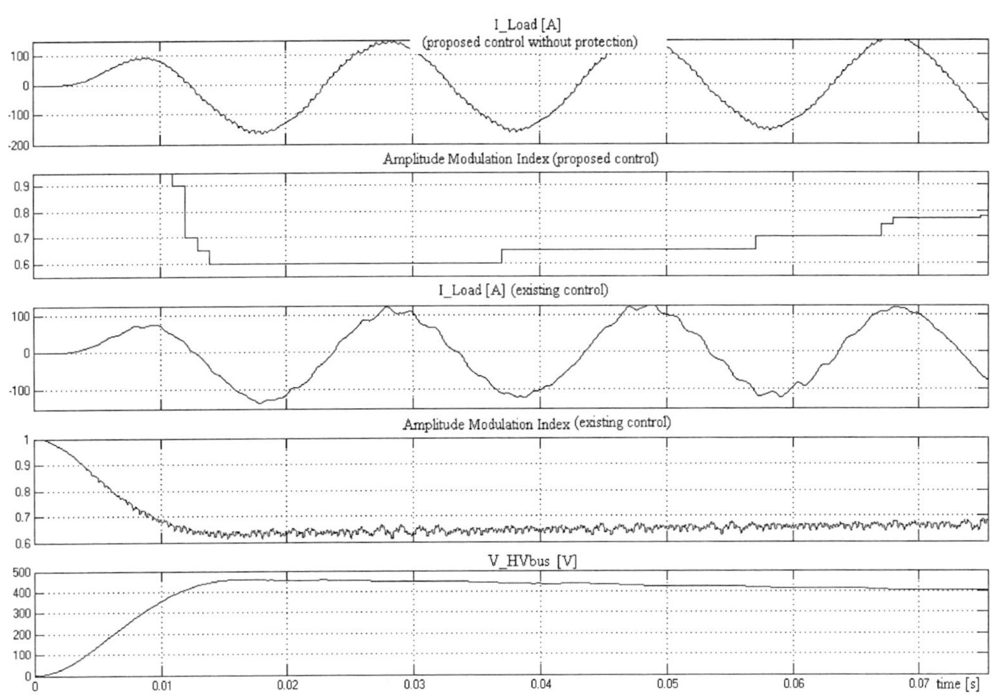

Figura 23b. Zoom of the variation of the amplitude modulation indexes at start-up under constant load.

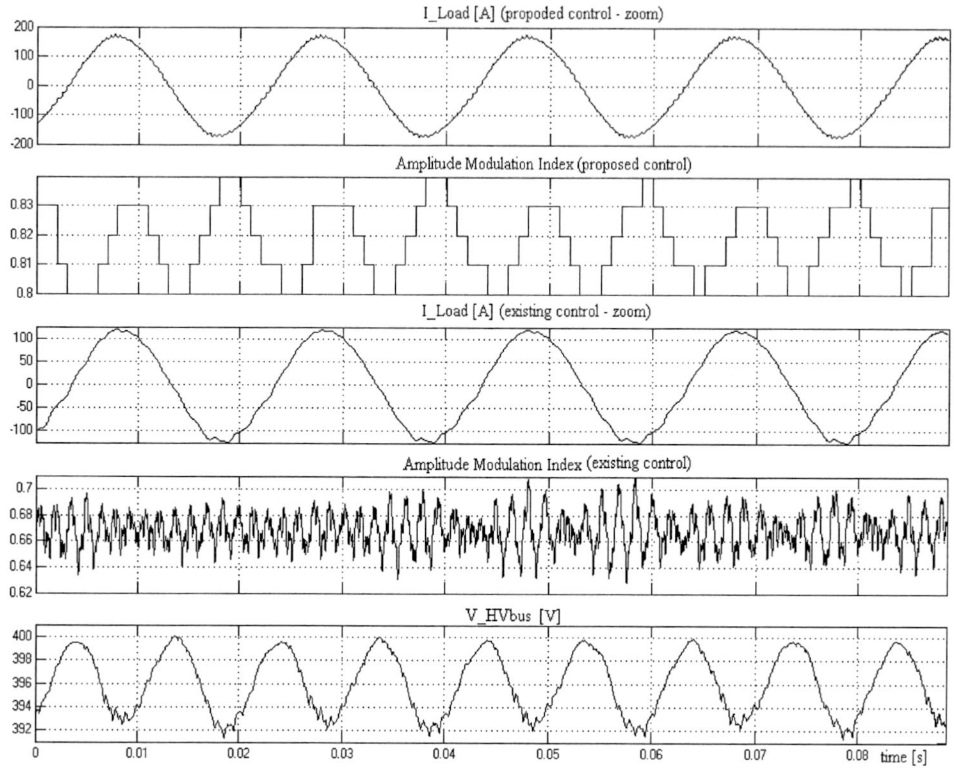

Figure 23c. Zoom of the variation of the amplitude modulation indexes under constant load.

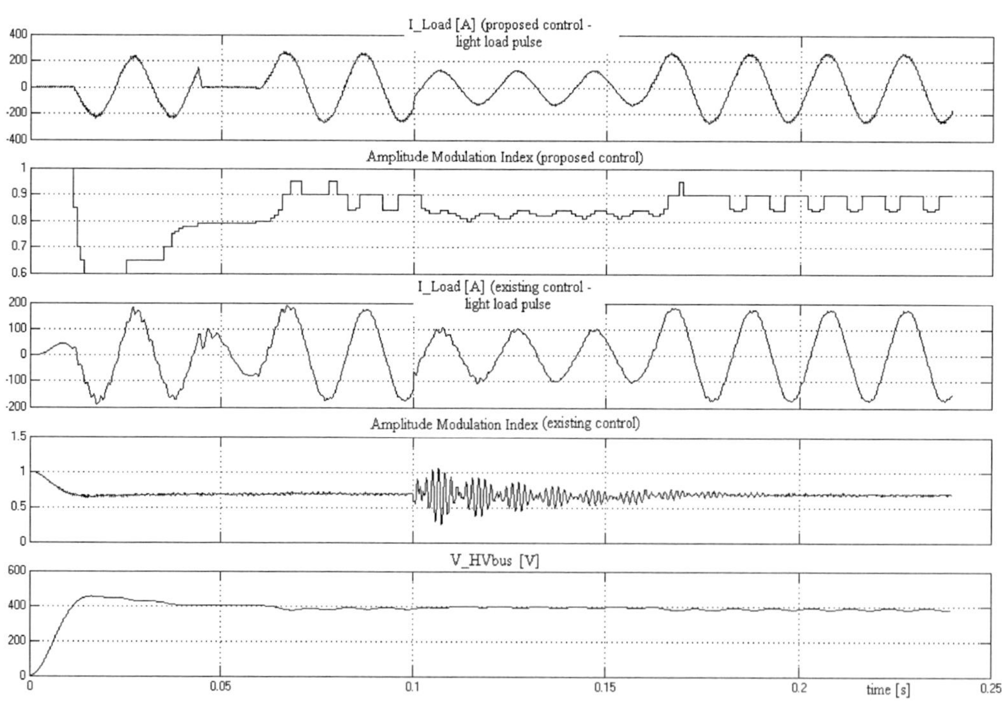

Figure 24. The inverters behavior under a light load pulse applied from 0.1 ms to 0.16 ms

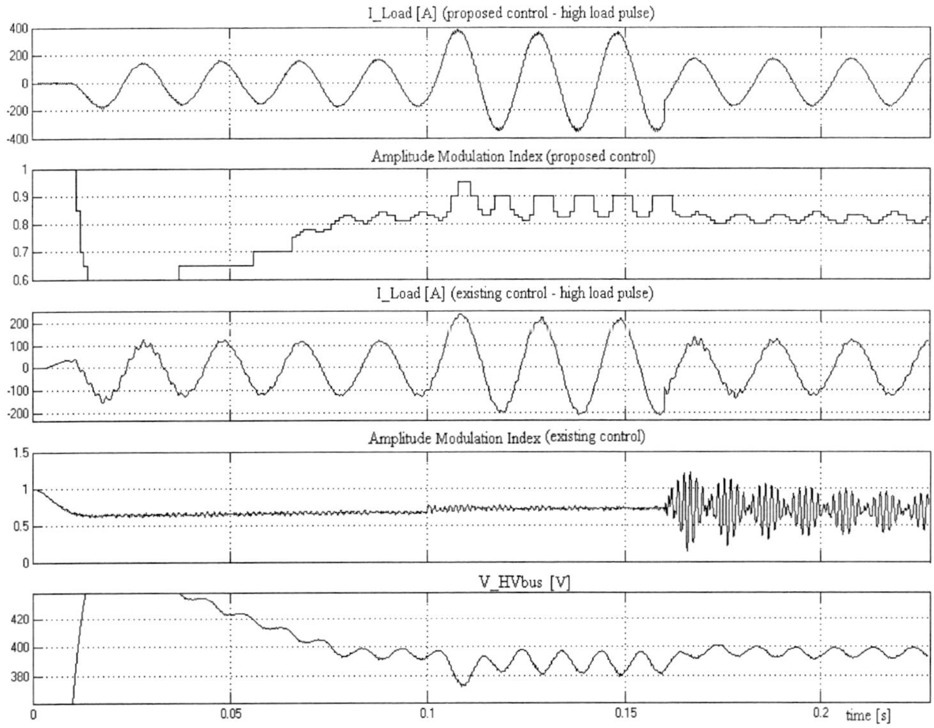

Figure 25. The inverters behavior under a high load pulse applied from 0.1 ms to 0.16 ms.

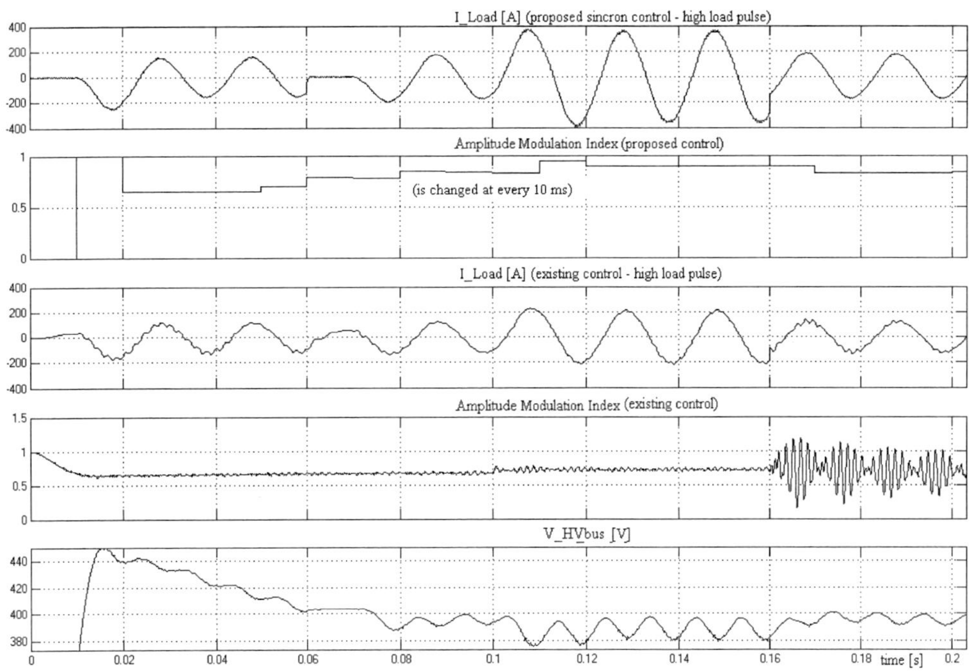

Figure 26. Mitigation of oscillations for the amplitude modulation index using the proposed feed-forward control synchronized with grid voltage at every 10 ms.

The mitigation of oscillations for the amplitude modulation index, using the proposed feed-forward control that is synchronized with grid voltage at every 10 ms, is shown in figure 26.

Taking in account the presented results in below figures it can be concluded that both control algorithms assure a RMS value of the inverter output voltage relatively stabilized to the input voltage variation or to the load power pulses. Best performances are obtained with the proposed feed-forward control. For the existing feed-back control, some oscillations of high amplitude appear for the amplitude modulation index at load power variations. The proposed feed-forward control is less sensitive to load variations. The m_a value is changed only if these load variations produce a significant variation in the input voltage (over to the voltage step considered). However, the mitigation of these oscillations can be obtained by synchronization of m_a changing with grid voltage at every 10 ms (figure 26). In this case the m_a control can not compensate the input voltage ripple of low frequency (LF) (of 100 Hz), as it is showed in figure 23.c.

Figure 27 shows that proposed feed-forward control is only dependent by input voltage, V_HVbus. If load variations didn't produce a variation of the input voltage over to the voltage step considered (2 V), then the m_a value is constant. A lower voltage step improves the control performances of the proposed feed-forward control. This step value must be higher then voltage ripple of high frequency (HF).

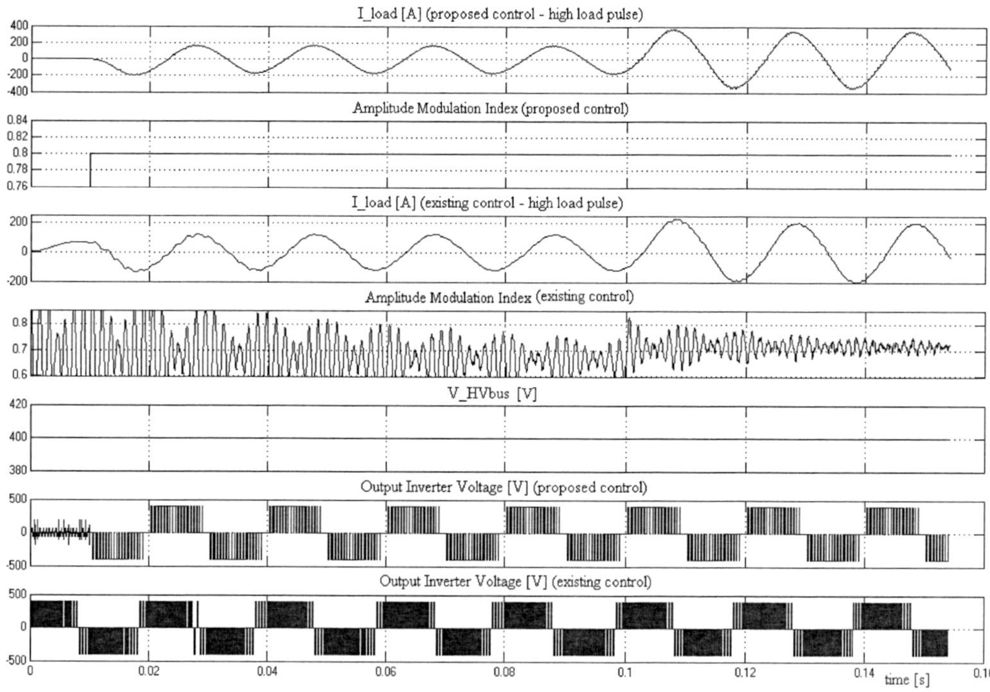

Figure 27. Output voltage of inverters supplied by a constant input voltage, V_HVbus=400 V.

A changing in the load power dynamic produces m_a changing of HF and LF (of 100 Hz) or of 50 Hz (see figure 28).

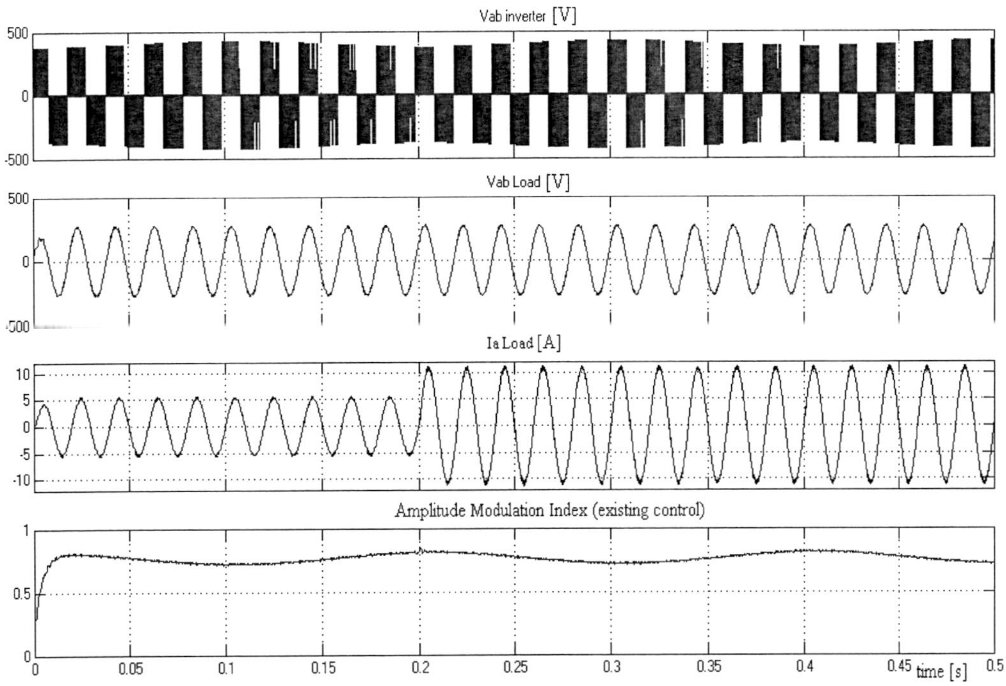

Figure 28. Inverter simulation results for PI controller parameters of k_P=0.4 and k_I=500.

In figure 29 are shown the power spectrums of output voltage for inverters with purposed feed-forward control (left plot) and with existing control from Matlab- Simulink® (right plot).

The power spectrums of load current for inverters with purposed feed-forward control (left plot) and with existing control from Matlab- Simulink® (right plot) are shown in figure 30.

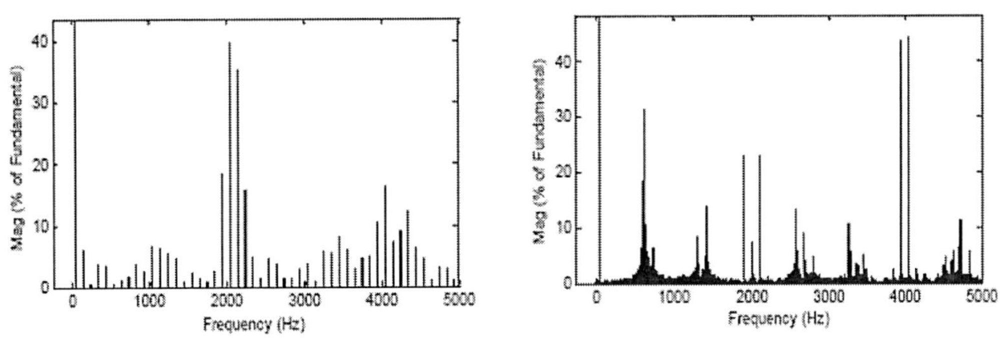

a) with purposed feed-forward control

b) with existing control from Matlab- Simulink®

Figure 29. The power spectrum of output voltage for inverters.

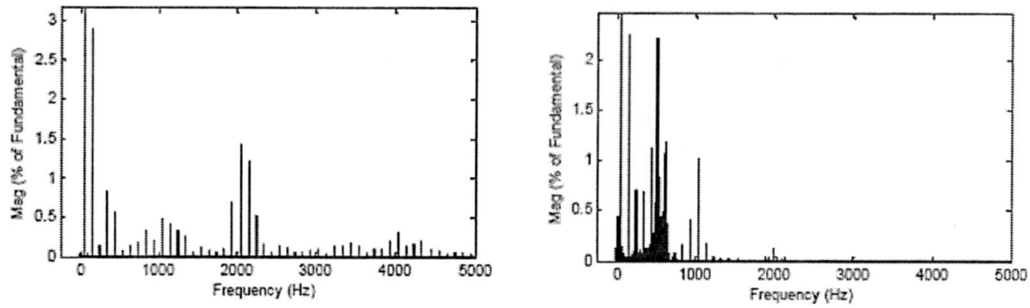

a) with purposed feed-forward control

b) with existing control from Matlab-Simulink®

Figure 30. The power spectrum of load current for inverters.

Experimental Results

In the next figure (figure 31) is presented the electrical schematic of hardware implementation for the feed-forward control. The PIC firmware flowchart is shown in figure 32.

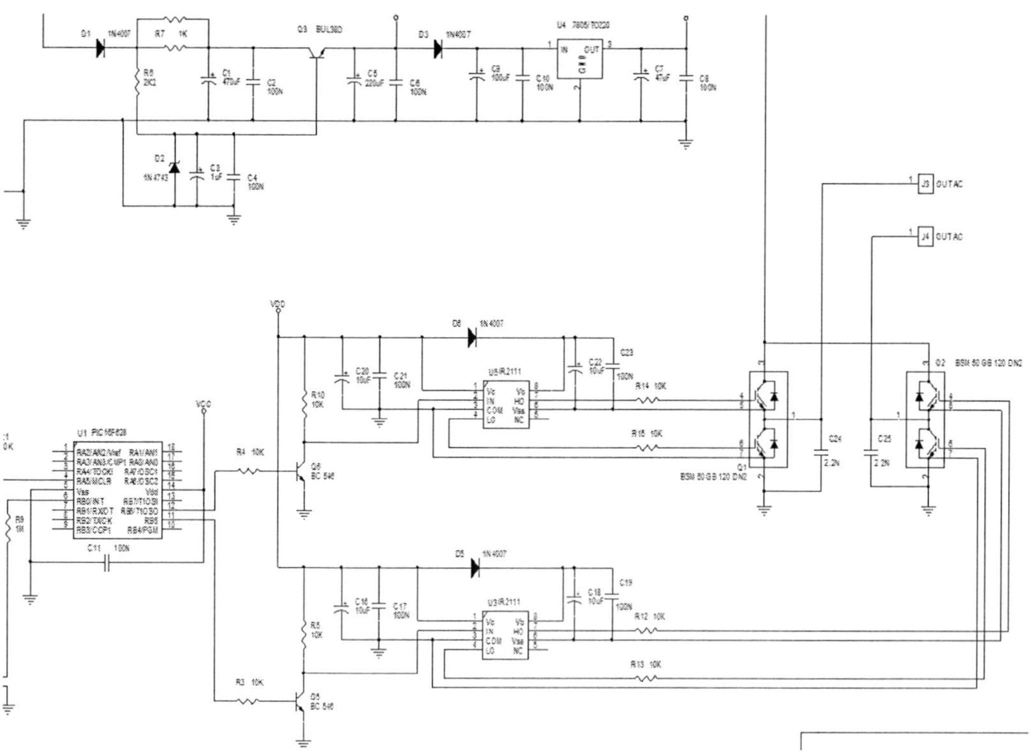

Figure 31. Schematic of hardware implementation with PIC microcontroller.

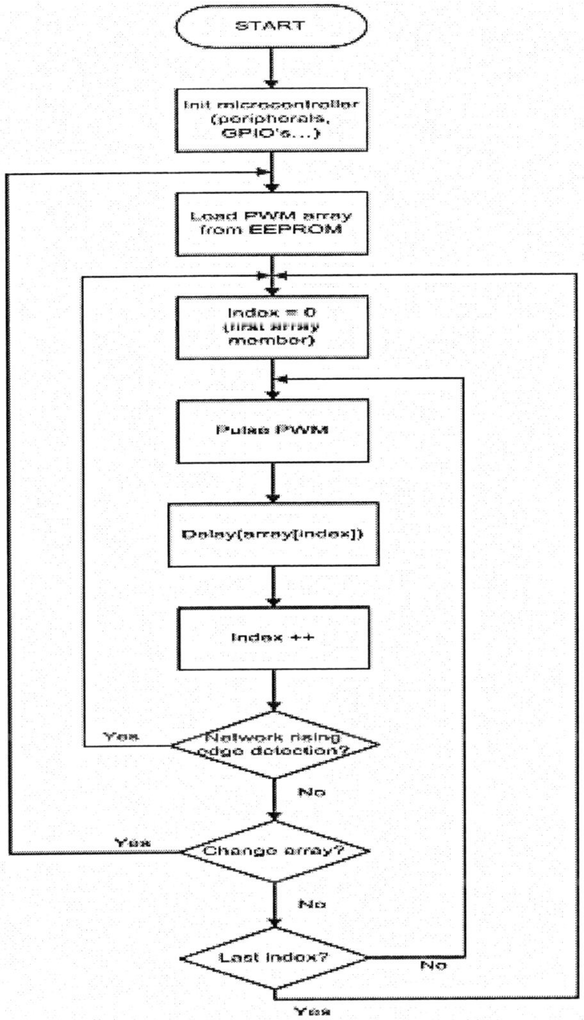

Figure 32. PIC firmware flowchart.

To generate pure sine PWM command is used a PIC16F877 from Microchip. This processor has enough memory to store the look up table shown in Table 2. The look up table has 18 vectors for the PWM command selected by the value of input voltage. The input voltage, V_HVbus, is measured using an analog input.

The microprocessor generates a synchronous PWM signals with grid frequency (50 Hz) by setting the time period and selecting the vectors from look up table at each 10 ms. The grid frequency can be easy changed by the m_f value.

Some selected of experimental results are presented in next figures. Figure 33 shows the experimental setup.

The pure sine PWM switching signal is shown in figure 34. The output voltage of the mono-phase inverter using two low-pass filters at output port is shown in figure 35 and figure 36.

The power spectrum diagram of the inverter output voltage is shown in figure 37.

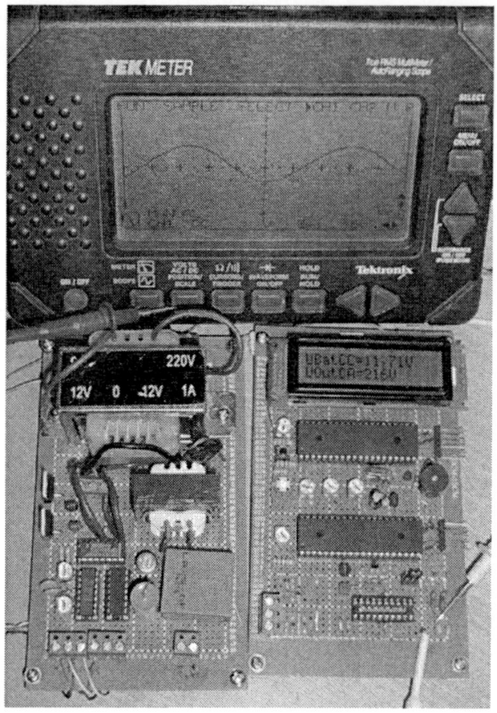

Figure 33. Experimental setup.

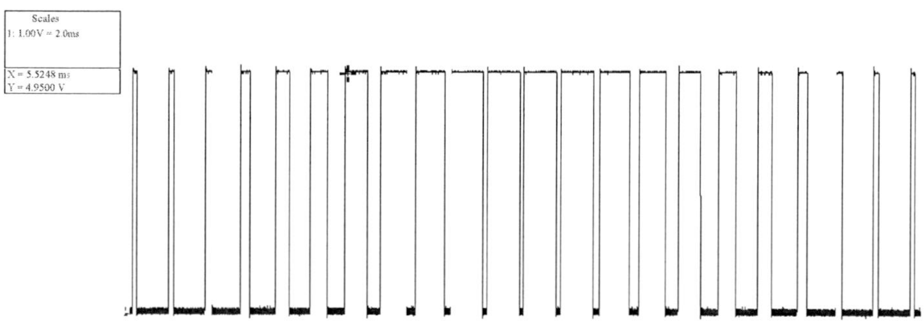

Figure 34. Pure sine PWM switching signal.

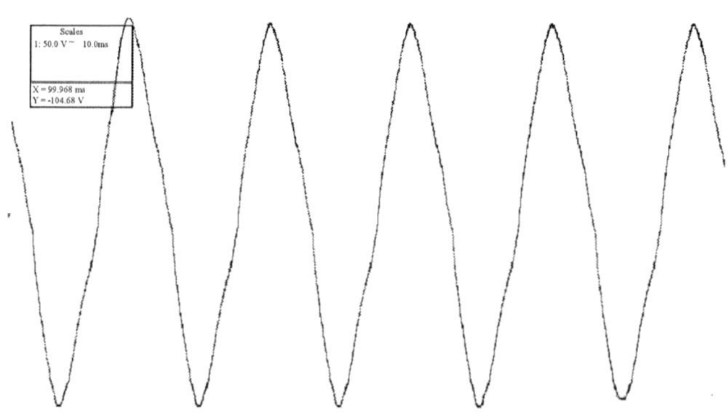

Figure 35. Output voltage of the mono-phase inverter.

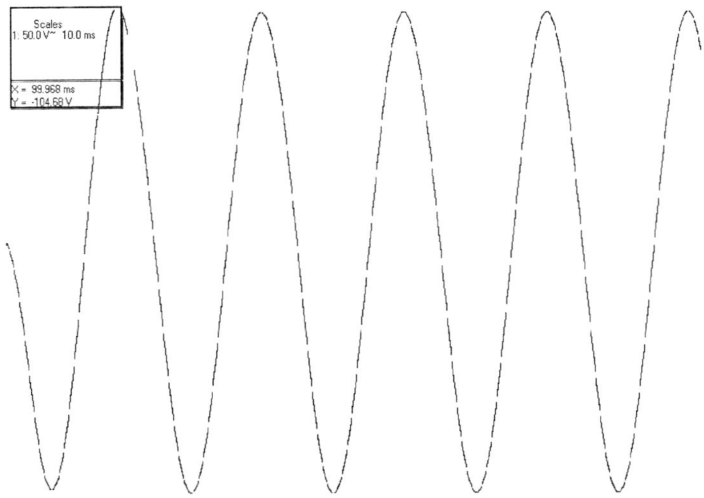

Figure 36. Filtered output voltage of the mono-phase inverter.

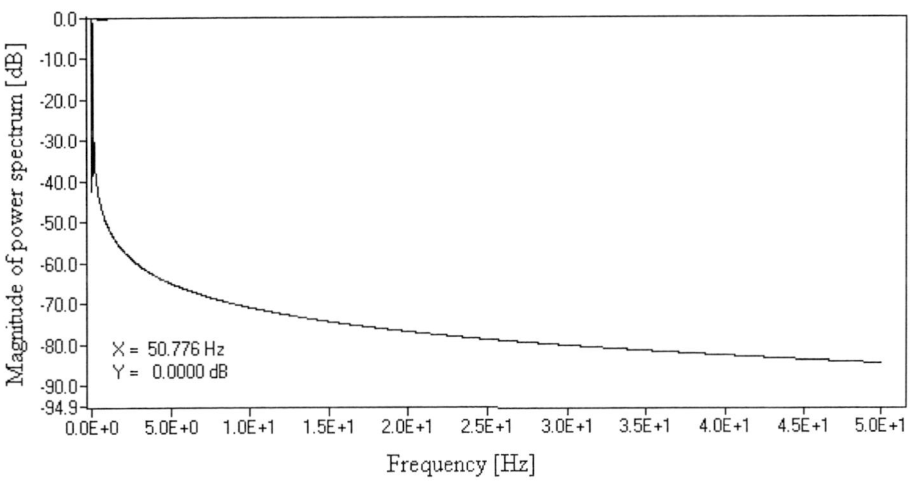

Figure 37. Power spectrum diagram of the inverter output voltage.

The dead time produces nonlinearity in the theoretical linear dependence of the RMS output voltage vs. the DC input voltage. The tests made by simulation and experiment were showed that can be obtained a RMS output voltage, V_inverter_RMS, almost constant for a DC input voltage, V_HVbus in range [370 V, 500 V]. In conclusion, the feed-forward control can be used for grid connected inverter systems.

SIMULATION OF THE INVERTER SYSTEM SUPLIED BY THE PEMFC STACK

Behavior of the fuel cell inverter system that use a PI controller for the V_HVbus and a feed-forward control for the inverter output AC voltage is shown in figure 38 [14]. The voltage ripple of V_HVbus is about of 5 V and 10 V for a load of 40 Ω and 20 Ω, respectively, using a low-pass filter having L_{filter}=3mH and C_{filter}=0.3mF.

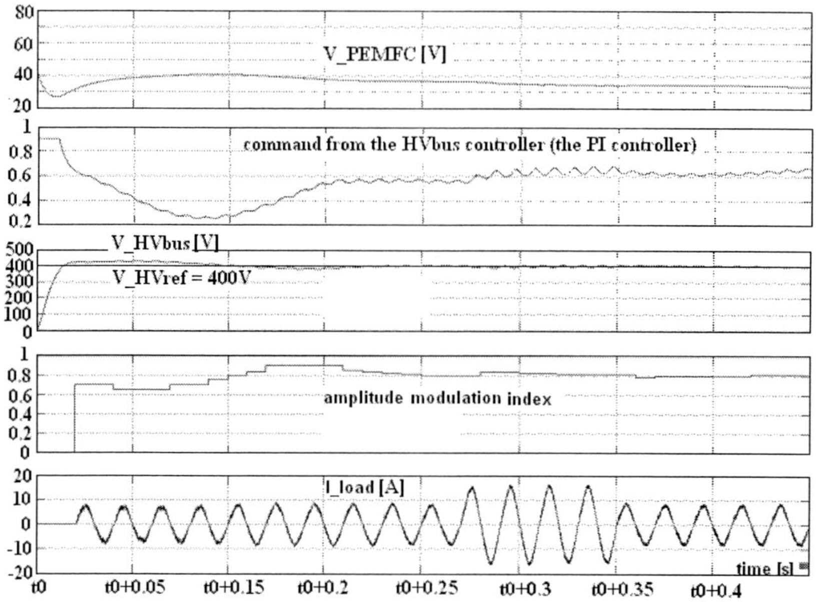

Figura 38. Behavior of the fuel cell inverter system that use a PI controller for the V_HVbus and a feed-forward control for the inverter output AC voltage.

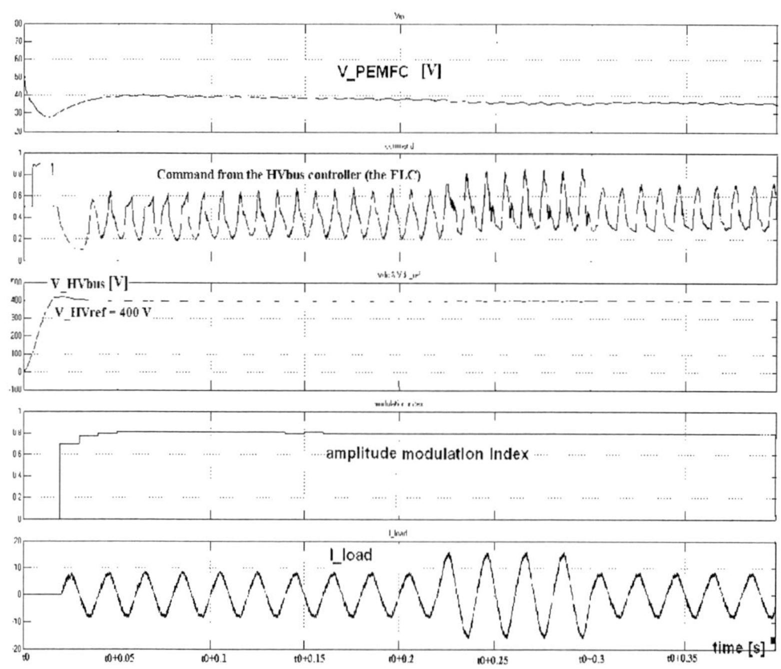

Figure 39. Behavior of the fuel cell inverter system that use a FLC for controling the V_HVbus and a feed-forward control for controling the inverter output AC voltage.

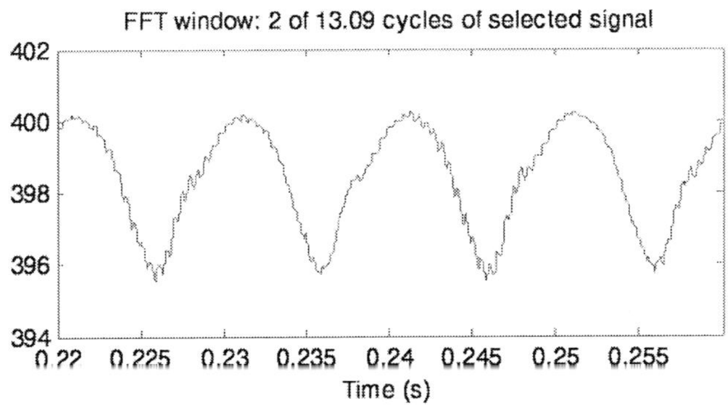

a) the ripple of DC input voltage

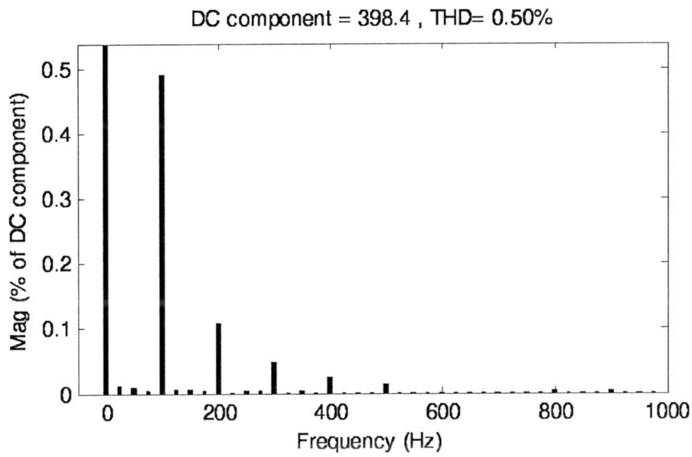

b) the power spectrum

Figure 40. The DC input voltage of inverter and its power spectrum.

Behavior of the fuel cell inverter system that use a FLC for control of the V_HVbus and a feed-forward control for control of the inverter output AC voltage is shown in figure 39. The voltage ripple of V_HVbus is about of 2 V and 4 V for a load of 40 Ω and 20 Ω, respectively, using the same low-pass filter (L_{filter}=3mH and C_{filter}=0.3mF).

An example of the ripple level of the DC input voltage of inverter and its power spectrum are shown in figure 30.

CONCLUSION

In this chapter is presented an analysis of the proper control for a fuel cell inverter system. A multi-stage inverter structure is considered. Two control techniques are considered for each energy conversion stage. The advantages of the fuzzy control are showed and the obtained results are compared with those of the classical PI control. The experimental results of a feed-forward control for the inverter stage is shown, too.

ACKNOWLEDGMENTS

The CEEX Grant #226 and CNCSIS Grant #570 of the National Research Council (MEC) have supported part of the research for this book chapter. Also, the author wishes to thank the research team for the effort in implementing the results in these contracts, especially to OPROESCU Mihai and ANGHENE Adrian.

REFERENCES

[1] Agrawal, J. P. (2001), Power electronic systems: theory and design, Upper Saddle River, N.J.: Prentice Hall.

[2] Batarseh, I. (2004), Power electronic circuits, Hoboken, NJ: John Wiley.

[3] Neacsu, D. (2006), Power Switching Converters : Medium- and High-Power, Taylor and Francis Group, LLC.

[4] Erickson, R.W., and Maksimovic, D. (2004), Fundamentals of Power Electronics, Kluwer Academic Publishers.

[5] Bizon, N., Sofron, E., Oproescu, M., and Raducu, M. (2007), Multi-stage Inverter Topologies for an Energy Generation Systems, 13th Int. Symposium on Modeling, Simulation and System's Identification - SIMSIS 2007, proceedings on CD.

[6] Bizon, N., and Oproescu, M. (2007), Power Converters for Energy Generation Systems (in romanian language: Convertoare de Putere utilizate in Sistemele de Generare a Energiei), Publishing house of the University of Pitesti, Romania.

[7] Xue, Y., Chang, L., and Bækhoj Kjær, S. (2004), Topologies of Single-Phase Inverter for Small Distributed Generators: an Overview, *IEEE Trans. on Power Electronics*, 19 (5), 1305-1313.

[8] Freris, L., and Infield, D. (2008), Renewable energy in power systems, John Wiley and Sons, Ltd.

[9] Bizon, N. (2008a), Modelling of Inverter Systems supplied by Fuel Cells (in romanian language: Modelarea Sistemelor Invertor alimentate de la Pile de Combustie), Publishing house of the University of Pitesti, Romania.

[10] Metaxiotis, K. (2009), Intelligent Information Systems and Knowledge Management for Energy: Applications for Decision Support, Usage and Environmental Protection (chapter 2, Bizon N., Intelligent control of the Energy Generation System), IGI Global, USA.

[11] Bizon, N. (2007), Intelligent Integrated Control Of The Power Flows Into An Energy Generation System, *Mediterranean Journal of Measurement and Control*, 3 (3), 59-71.

[12] Bizon, N., Mazare, A., Laurentiu, I., Oproescu, M., and Raducu, M. (2007), Some Designing Aspects of 1kw Inverter System: DC- DC Converter, *Int. Conf. on Electronics, Computers and Artificial Intelligence* - ECAI'07, 2, 147-156.

[13] Bizon, N., Laurentiu, I., Mazare, A., and Oproescu, M. (2007), Analyze of the Feed-Forward Control for a Pure Sine Inverter, *Int. Conf. on Electronics, Computers and Artificial Intelligence* - ECAI'07, 2, 71-79.

[14] Bizon, N. (2008b), Optimized Systems for Green Power Conversion (in romanian language: Sisteme Optimizate pentru Conversia Energiei Curate), MatrixROM, Bucharest, Romania.

In: Advances in Energy Research
Editor: Nicu Bizon
ISBN 978-1-61209-991-0

Chapter 12

MODELLING OF ENERGY GENERATION SYSTEMS – A BRIEF OVERVIEW OF USED MODELS

Nicu Bizon[1*], Mihai Oproescu[1], Farrokh Fattahi[2], Naser Mahdavi Tabatabaei[3] and Hamed Seddighe Abbasi[2]

[1]Faculty of Electronics, Communication and Computers
University of Pitesti, Romania
[2]Azarbaijan Higher Education and Research Complex, Iran
[3]Seraj Higher Education Institute, Iran

ABSTRACT

This chapter present a brief overview of models used in this book. Supplementary information about the relationships of models is here given, which complements the data already presented in the chapters where they were used.

First section shows the used models for Polymer Electrolyte Membrane Fuel Cell (PEMFC) stack and Energy Storage Devices (ESD). Two PEMFC models that were implemented in Matlab Simulink® are presented: a detailed model and its simplified variant when some PEMFC parameters are set to their rated or custom values. The generally accepted models for stacks of batteries or ultracapacitors are briefly shown, too. Some tests that validate the proposed models are presented at the end of this section.

Second section shows generally accepted models for wind and gas turbine. A model of horizontal axis wind turbine with convoluted flow of the wings is presented. Usually, the gas turbine drives a synchronous generator. The model of the synchronous generator is shown, too. Both models are used in simulations presented in chapter named "Case studies concerning DG system implementations".

In the third section are analyzed the multi-stage power topologies that are utilized in the Energy Generation Systems (EGS). There are presented, compared and evaluated different proposed topologies of multi-stage energy conversion for unidirectional or bidirectional power flow from energy source to the load.

* Corresponding Author: nicu.bizon@upit.ro

Keywords: energy generation system, modelling, energy conversion topology, energy source, fuel cell, wind and gas turbine, energy storage device, ultracapacitor, multi-stage converter, inverter grid connected.

MODELLING OF ENERGY SOURCES AND ENERGY STORAGE DEVICES

Fuel Cell Model

A Polymer Electrolyte Membrane Fuel Cell (PEMFC) is a high efficiency electro chemical energy conversion device, which can generate electricity and produce heat with the help of a catalyst. The PEMFC model has a rather large capacitance shunting the device. The equivalent electrical circuit for this model is shown in figure 1, where R_e is the electrolyte and contact resistance (R_{ohm}), R_{ct} is the charge transfer resistance and models the activation loss, C_d is the double charge layer capacitance, Z_D is the diffusion impedance (also called the Warburg impedance).

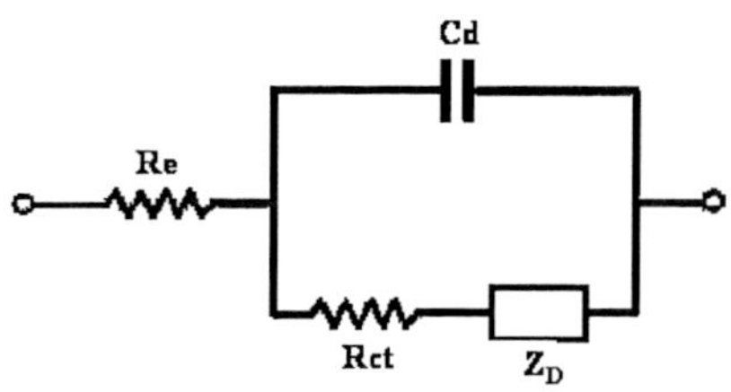

Figure 1. The PEMFC equivalent electrical circuit.

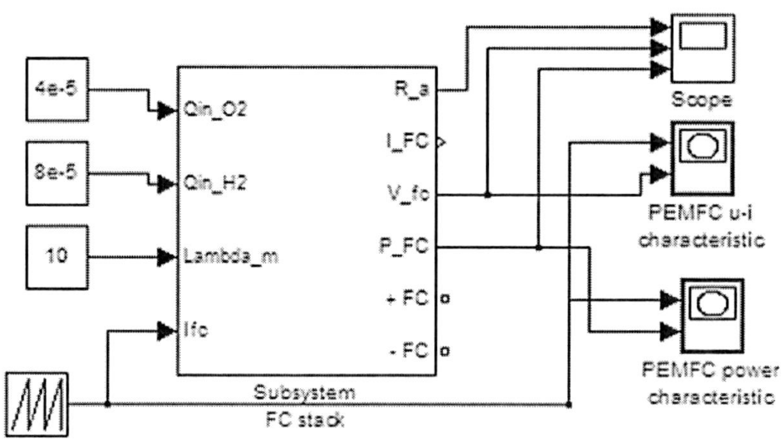

Figure 2. Simulink test diagram to plot the PEMFC characteristics of the used model [10].

The simplified model of PEMFC stack is based on this assumption. This circuit suggests that the fuel cell can in-fact tolerate inverter current ripple if its frequency if high enough.

The detailed model proposed here it is based on well know analytical description [1; 2; 3]. The simplified model is obtained for a given set of PEMFC parameters and is refined by a genetic algorithm to find the best values of model coefficients in order to fit the experimental data set. Consequently, the PEMFC macro modeling is the main goal of this section in order to understand the simulation results presented in other chapters of the book. Other detailed PEMFC models are now available in literature. See, for example, [4; 5; 6; 7; 8; 9]. A model (which combines in well manner the analytical description by function relationships) was implemented in Matlab Simulink® (Figure 2). The relationships for each block are briefly explained in next section.

Simplified PEMFC Model

Generally, the PEMFC voltage losses are determined in terms of three main groups: activation losses, ohmic losses and concentration losses (see Figure 3). Figure 3 presents a typical plot of the cell voltage and the specific power density versus the current per square centimeter.

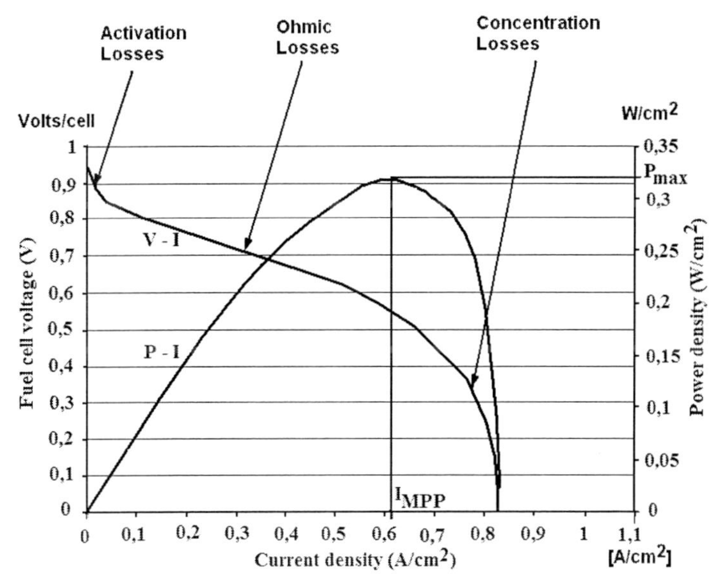

Figure 3. The polarization and power curves for a typical PEM Fuel Cell [10].

Activation Losses

The Tafel equation has been derived from experimental results and includes the losses for the reaction itself:

$$V = A_{act} \ln\left(\frac{i}{i_0}\right) \tag{1}$$

The exchange current, i_0, is a threshold value from which the voltage drop starts to fall. This means that this equation is only true when $i > i_0$. For a low temperature fuel cell, i_0, and the constant, A_{act}, is about 0.1 mA and 0.06 V, respectively.

Ohmic Losses

Ohmic losses occur due to the internal resistance of the fuel cell electrodes and the membrane resistance to the flow of ions. So, it is linearly dependent by the current and can be expressed as (ohms law):

$$V = i \cdot R_{ohmic} \tag{2}$$

In general, the biggest resistance appears in the membrane, but the connector plates and interface areas between the cells add significant resistance, too. The conductivity of the membrane is high dependent by the membrane water content:

$$R_{ohmic} = R_{electric} + R_{ionic} = R_{electric} + t_m/\sigma_m \tag{3}$$

where

- σ_m [m/Ω] is the ion conductivity related to the membrane;
- t_m [m] represents the membrane thickness;
- $R_{electric}$ [Ω] represents, mostly, the electrical resistance of the bipolar plates.

The ion conductivity can be expressed by next equation:

$$\sigma_m = (0.005193 9\lambda_m - 0.00326) \cdot \exp\left(1268 \cdot \left(\frac{1}{303} - \frac{1}{273 + T_{FC}}\right)\right) \tag{4}$$
$$\lambda_m = 0.043 + 17.18 \cdot R_H - 39.85 \cdot R_H^{\ 2} + 36 \cdot R_H^{\ 3}$$

where T_{FC} is the temperature of the fuel cell stack and R_H is the average relative humidity on the anode and cathode side.

Concentration Losses

The voltage drop is associated with the concentration loss that appears at bigger current density and can be derived from the Nernst equation. A decrease in the concentration represents a decrease in the partial pressure (after compression), P_2, which gives a voltage drop according to Nernst equation:

$$\Delta V = \frac{RT}{2F} \ln\left(\frac{P_2}{P_{REF}}\right) \tag{5}$$

Assuming that this partial pressure, P_2, decreases linearly with the current (i) that increases, and that partial pressure at the catalyst layer is zero when current is maximum (i_{max}), the equation can be rewritten as:

$$\Delta V_{conc} = -B_{conc} \ln\left(1 - \frac{i}{i_{max}}\right) \tag{6}$$

where B_{conc} is a constant (having a value which depends only by the reactants).

By adding these three loss equations and subtracting from V_{cell_0} (output voltage from no-load for a single fuel cell), the cell voltage, V_{cell}, is determined in terms of the drawn current density (specified by load), $j_{fc}= j_{fc}/A$, together with a constant, j_n, determined by two phenomena's known as internal current and fuel cross over [9]:

$$V_{cell} = V_{cell\_o} - \left(j_{fc} + j_n\right) \cdot A \cdot R_{ohmic} - A_{act} \cdot \ln\left(\frac{j_{fc} + j_n}{j_0}\right) + B_{conc} \cdot \ln\left(1 - \frac{j_{fc} + j_n}{j_{max}}\right) \tag{7}$$

where A is area of membrane

The PEMFC response time, T_{fc}, usually models the PEMFC dynamic, so:

$$v_{fc}(t) = N_{fc} \cdot v_{cell}(u),\ u = j(t) = \frac{i(t)}{A},\ \frac{I(s)}{I_{fc}(s)} = \frac{\exp(-s \cdot \tau_{fc})}{1 + T_{fc} \cdot s} \tag{8}$$

where N_{fc} is the stack cells number, A is area of membrane, i_{fc} is the PEMFC current and v_{fc} is the PEMFC output voltage. The response time, T_{FC}, and death time, τ_{fc}, is used to model the "charge double layer" phenomenon due to the build up of charges at electrode/electrolyte interface.

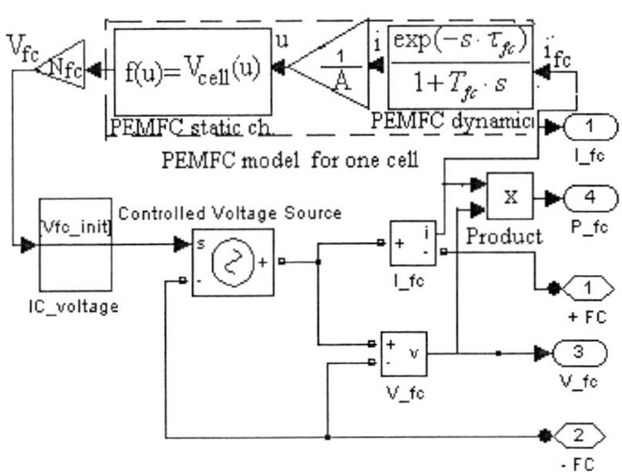

Parameter (for $\lambda = 1$ and $P_{sys} = 2$ atm)	Relation (7)
E° [Volts]	1.2
j_n [mA/cm²]	2
j_0 [mA/cm²]	0.067
j_{max} [A/cm²]	0.9
R_{Ohmic} [Ω]	0.03
A_{act} [Volts]	0.06
B_{conc} [Volts]	0.05

Relation (7):
f(u)=(1.2-(u+0.002)*A*0.03-0.06*log((u+0.002)/0.000067)+0.05*log(1-(u+0.002)/0.9))
where u=i/A, i[A] and A[cm²]

Figure 4. Proposed PEMFC macro-model that uses a transfer function for modelling the PEMFC dynamic.

Number of fuel cells used for a PEMFC stack is given by relation $N_{fc}=V_{PEMFC\_0}/V_{cell\_0}$, where V_{cell_0} and V_{PEMFC_0} are output voltage from no-load for a single fuel cell and PEMFC stack, respectively. Proposed PEMFC macro-model [11] that uses a transfer function for modelling the PEMFC dynamic is shown in figure 4.

A modified relation can be used to fit much better the experimental data set [12]:

$$V_{cell} = V_{cell\_o} - \left(j_{fc} + j_n\right) \cdot A \cdot R_{ohmic} - A_{act} \cdot \ln\left(\frac{j_{fc} + j_n}{j_0}\right) + B_{conc} \cdot \ln\left[\lambda^{k1} \cdot P^{k2} \cdot \left(1 - \frac{j_{fc} + j_n}{j_{\max} \cdot \lambda^{k3} \cdot P^{k4}}\right)\right] \quad (7')$$

where k_i parameters, i=1-4, can be find using, for example, a genetic algorithm.

Detailed PEMFC Model

The Simulink diagram of the PEMFC detailed model is presented in figure 5 [10, 13].

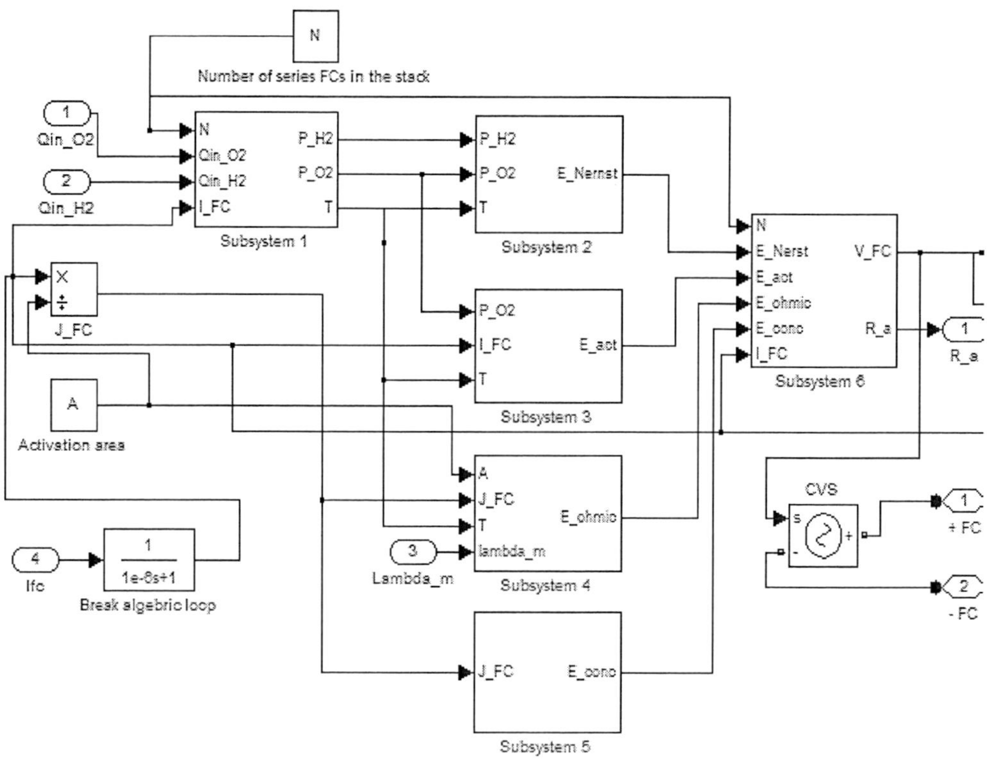

Figure 5a. Dynamical PEMFC model represented as coupled subsystems [10].

The variation of the PEMFC temperature (T) in time (t) depends by fuel cell current (I_{FC}) and can be calculated as:

$$T = 273 + T_0 + \left(T_0 - T_{rt} + T_{ic} \cdot I_{FC}\right) \cdot \left(1 - \exp\left(-\frac{t \cdot I_{FC}}{T_{it}}\right)\right) \quad (9)$$

where the parameters are mentioned in figure 5.b, next to the subsystem 1 figure.

PEMFC subsystem in figure 5.a	Dependence relationships between output - input variables and used parameters for each PEMFC subsystem
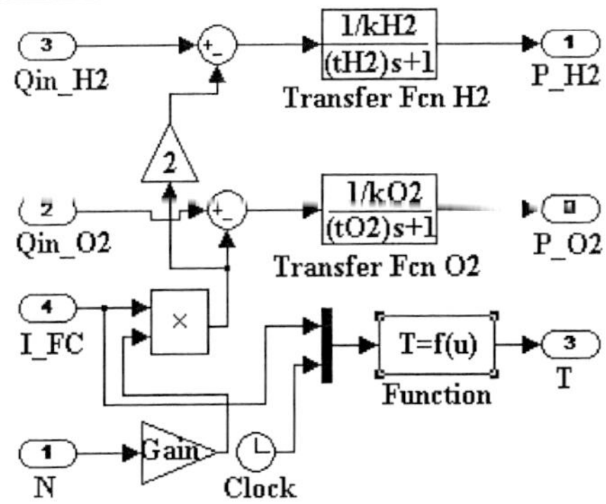 Subsystem 1: Partial pressure of hydrogen and oxygen and stack temperature depending on fuel and air flow rate, fuel cell current and number of cells in series in the stack	$Q_{in(H2)}$ - Fuel flow rate (l/min); input of the model; $Q_{in(O2)}$ - Air flow rate (l/min); input of the model; I_{FC} - Fuel cell current (A); input of the model; N - Number of cells in series in the stack; is set to 52; Gain = $1/4FU$; where: F is Faraday constant (C/kmol); U – Utilization factor for H2 (%); is set to 80%; T - Temperature of the stack (°K); $T = 273 + T_0 + (T_0 - T_{rt} + T_{ic} \cdot I_{FC}) \cdot \left(1 - \exp\left(-\frac{t \cdot I_{FC}}{T_{it}}\right.\right.$ where: T_0=28°K , T_{rt}=20 °K , T_{ic}=0.7 °K/A , T_{it}=4000s/A , P_{H2} - Partial pressure of hydrogen inside the stack (atm); t_{H2} - Hydrogen time constant (s); is set to 3.36s; k_{H2} - Hydrogen valve molar constant [kmol (atm s)-1]; set to 4.22x10^{-5}; P_{O2} - Partial pressure of oxygen inside the stack (atm); t_{O2} - Oxygen time constant (s); set to 6.47s; k_{O2} - Oxygen valve molar constant [kmol (atm s)-1]; is set to 2.11x10^{-5}.
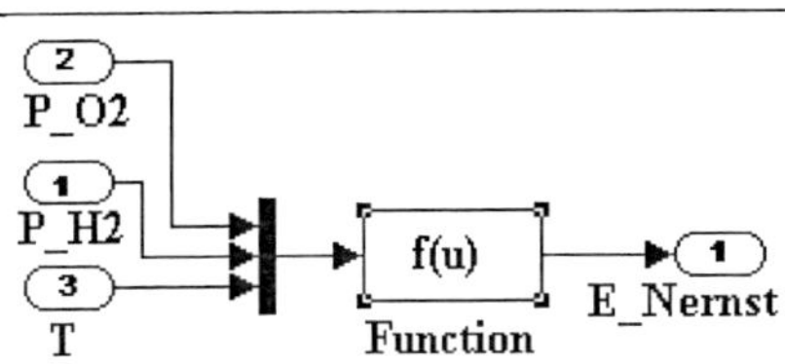 Subsystem 2: Nernst voltage depending on partial pressure of hydrogen and oxygen and stack temperature	E_{Nerst} - Nernst voltage (V): $E_{Nerst} = 1.229 - 8.5 \cdot 10^{-4} \cdot (T - 298) + 4.308 \cdot 10^{-5} \cdot T \cdot \ln\left(P_{H2} + P_O\right.$ where: P_{H2} - Partial pressure of hydrogen inside the stack (atm); P_{O2} - Partial pressure of oxygen inside the stack (atm); T - Temperature of the stack (°K);
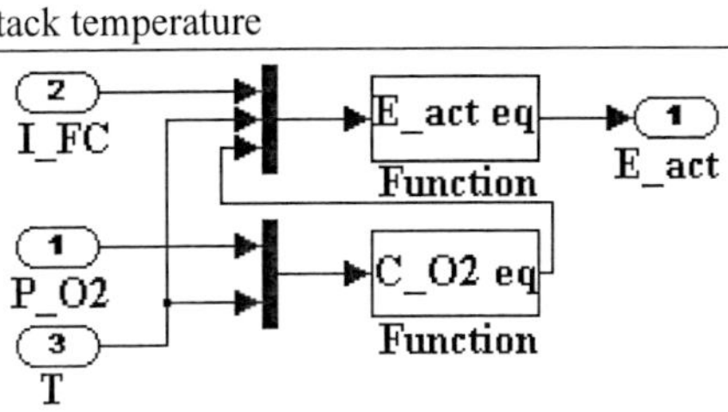 Subsystem 3: Activation voltage loss depending on fuel cell current, partial pressure of oxygen and stack temperature	E_{Act} = Activation voltage loss (V): $E_{Nerst} = 1.229 - 8.5 \cdot 10^{-4} \cdot (T - 298) + 4.308 \cdot 10^{-5} \cdot T \cdot \ln\left(P_{H2} + P_O\right.$ where: e_1=-0.9514 , e_2=0.00312 , e_3=7.5·10^{-5} , e_4= -1.87·10^{-3} and C_{O2} -Dissolved oxygen concentration in the interface of the cathode catalyst: $C_{O2} = \frac{P_{O2}}{5.08 \cdot 10^6 \cdot \exp\left(-498/T\right)}$

Figure 5b. (Continued).

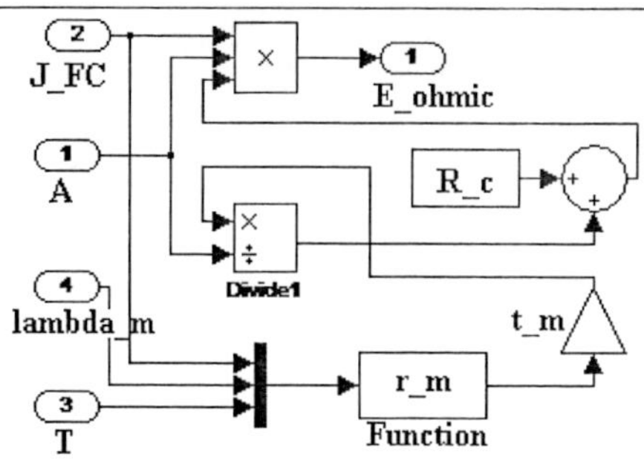

Subsystem 4: Ohmic voltage loss depending on fuel cell current density, activation aria, water content of the membrane and stack temperature

E_{ohmic} - Ohmic voltage loss (V):
$E_{ohmic}=J_{FC}\cdot R_{ohmic}$; $J_{FC}=I_{FC}/A$; $R_{ohmic}=R_m+R_c$; $R_m=r_m\cdot t_m/A$
J_{FC} – Current density (A/cm^2);
R_{ohmic} - Ohmic resistance (Ω);
R_m - Echivalent membrane resistance (Ω); R_c - Contact resistance between membrane and electrodes (Ω); is set to 0.2mΩ
t_m - membrane thickness (cm); is set to 0.018cm;
A – Activation aria (cm^2); is set to 56 cm^2;
r_m - Resistivity of Nafion membrane (Ωcm^{2}m^{-1}):

$$r_m=\frac{181.6\cdot\left[1+0.03\cdot J_{FC}+0.062\cdot(J_{FC})^{2.5}\cdot\left(T/303\right.\right.}{\left[\lambda_m-0.0634-3\cdot J_{FC}\right]\cdot\exp\left[4.18\cdot\left(T-303/T\right.\right.}$$

λ_m – water content of the membrane ; input of the model.

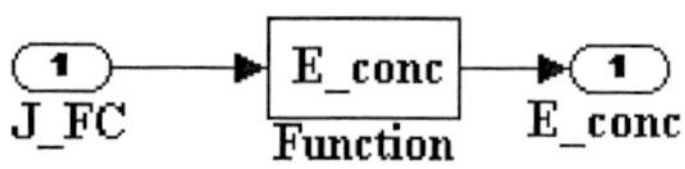

Subsystem 5: Concentration voltage loss depending on fuel cell current density

E_{conc} - Concentration voltage loss (V)
B – Modeling constant (V); set to 0.016V;
J_{max} – Maximum current density (A/cm^2); is set to 0.95A/cm^2

$$V_{conc}=B\cdot\ln\left(1-J_{FC}/J_{max}\right)$$

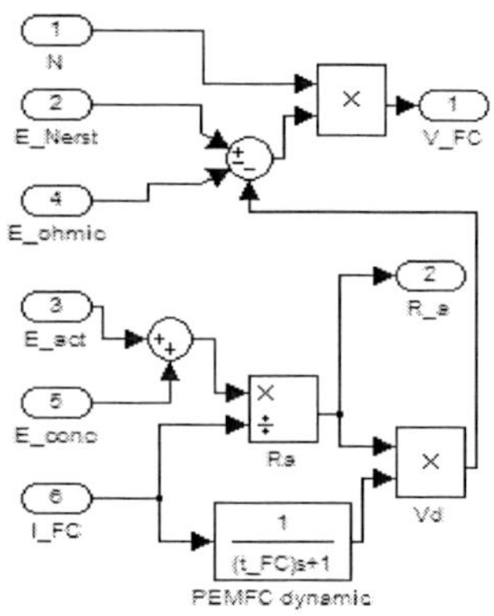

Subsystem 6: Fuel cell voltage and PEMFC dynamic

V_{FC} – Fuel cell voltage (V); output of the PEMFC model which is also used as control signal for Controlled Voltage Source:
$V_{FC}=N\,(E_{Nerst}-E_{ohmic}-V_d)$
V_d – Voltage over R_a resistance (V);

$$V_d=\frac{I_{FC}\cdot R_a}{t_{FC}\cdot s+1}$$

t_{FC} – Fuel cell time constant (s); is set to 2 s;
R_a - Equivalent resistance representing the sum of activation and concentration resistances (Ω);); auxiliary output of the PEMFC model;

$$R_a=\frac{E_{act}+E_{conc}}{I_{FC}}$$

Figure 5b. Subsystems of PEMFC model and implementing equations.

Using the basic electrochemical relationships, equation that relates molar flow of gases (hydrogen and oxygen) through the valve with its partial pressure (P_{H2} and P_{O2}) inside the channel and the ideal gas equation we obtain (after the derivation of the both sides):

$$\frac{d}{dt}p_{H2} = \frac{R \cdot T_{op}}{V_{anode}} q_{H2} = \frac{R \cdot T_{op}}{V_{anode}} (Q_{H2}^{in} - Q_{H2}^{out} - Q_{H2}^{r}) \Rightarrow$$
$$\frac{d}{dt}p_{H2} = \frac{R \cdot T_{op}}{V_{anode}} (Q_{H2}^{in} - k_{H2} \cdot p_{H2} - 2\frac{N}{4FU} I_{FC}) \quad (10)$$

where (see also figure 5.e):

- R is universal gas constant ((1 atm) (kmol K)$^{-1}$);
- F - Faraday constant (C/kmol)
- V_{anode} - Volume of the anode (m^3);
- T_{op} - PEMFC temperature operating point (about 343°K);
- Q_{H2}^{in} - Hydrogen input flow (kmol s^{-1} or l/min);
- Q_{H2}^{out} - Hydrogen output flow;
- Q_{H2}^{r} - Hydrogen flow that reacts;
- k_{H2} - Hydrogen valve molar constant [kmol (atm s)$^{-1}$];
- U – Utilization factor for H2 (%);
- N - Number of cells in series in the stack.

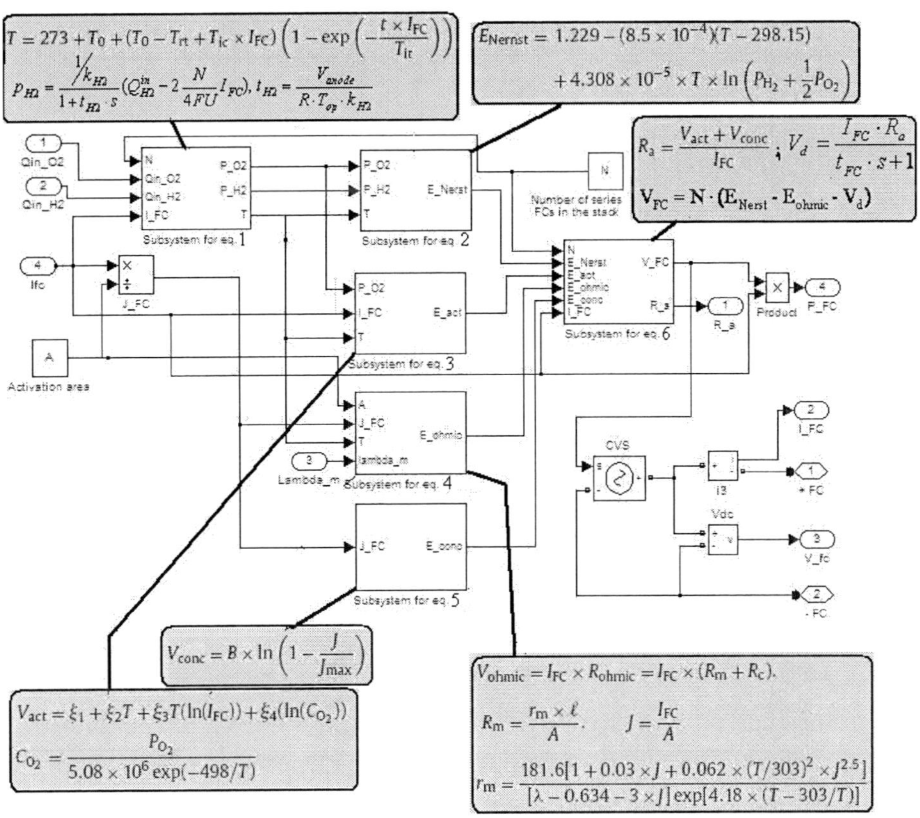

Figure 5c. The diagram of the PEMFC detailed model (adapted from [13]).

Taking the Laplace transforms of both sides of relation 10 results:

$$p_{H2} = \frac{{}^{1}\!/_{k_{H2}}}{1+t_{H2}\cdot s}(Q_{H2}^{in} - 2\frac{N}{4FU}I_{FC}),\ t_{H2} = \frac{V_{anode}}{R\cdot T_{op}\cdot k_{H2}} \tag{11}$$

where t_{H2} is hydrogen time constant (measured in seconds).

A	Activation area [cm^2 (cell^{-1})]
B	Constant utilized in modeling of concentration overvoltage (V)
C	FC double-layer equivalent capacitance (F)
C_{O_2}	Dissolved oxygen concentration in the interface of the cathode catalyst
E_{Nernst}	Nernst instantaneous voltage (V)
F	Faraday constant [C (kmol)$^{-1}$]
I_{FC}	FC current (A)
J	Current density (A (cm)$^{-2}$)
J_{max}	Maximum current density [A (cm)$^{-2}$]
N	Number of series FCs in the stack
P_{H_2}	Hydrogen partial pressure (atm)
P_{O_2}	Oxygen partial pressure (atm)
Q_{H2}^{in}	Amount of hydrogen flow required to meet the load change [kmol (s)$^{-1}$]
r_m	The resistivity of Nafion series proton exchanging membrane [Ω (cm)^2 m^{-1}]
R_a	Equivalent resistance representing the sum of the activation and concentration resistances (Ω)
R_c	The contact resistance between the membrane and electrodes (Ω)
R_m	Equivalent membrane impedance (Ω)
R_{ohmic}	FC internal ohmic resistance (Ω)
T	FC temperature (°K)
$T_0, T_{rt}, T_{ic}, T_{it}$	Empirical parameters utilized for modeling the variation of FC temperature
U	Utilization factor
V_{act}	Activation overvoltage (V)
V_d	Potential drop on R_a (V)
V_{conc}	Concentration overvoltage (V)
V_{ohmic}	Ohmic overvoltage (V)
V_{FC}	FC Stack voltage (V)
λ_m	The water content of the membrane
$\zeta_1, \zeta_2, \zeta_3, \zeta_4$	Constants utilized in modeling of activation overvoltage

Figure 5e. The parameters of PEMFC model used.

Similarly, oxygen partial pressure can be obtained:

$$p_{O2} = \frac{{}^{1}\!/_{k_{O2}}}{1+t_{O2}\cdot s}(Q_{O2}^{in} - \frac{N}{4FU}I_{FC}),\ t_{O2} = \frac{V_{catode}}{R\cdot T_{op}\cdot k_{O2}} \tag{12}$$

where:

- t_{O2} is oxygen time constant (s);
- $V_{cathode}$ - Volume of the cathode (m^3);
- - Oxygen input flow (kmol s^{-1} or l/min);
- k_{O2} - Oxygen valve molar constant [kmol (atm s)$^{-1}$];

Relations (7) – (10) are used for implementing the subsystem 1, where hydrogen and oxygen input flows are inputs for PEMFC model, too.

The Nernst's voltage can be written as:

$$E_{Nerst} = 1.229 - 8.5 \cdot 10^{-4} \cdot (T - 298) + 4.308 \cdot 10^{-5} \cdot T \cdot \ln\left(P_{H2} + {P_{O2}}/{2} \right) \tag{13}$$

and is implemented in the subsystem 2.

The activation voltage loss can be written as:

$$E_{act} = e_1 + e_2 \cdot T + e_3 \cdot T \cdot \ln(I_{FC}) + e_4 \cdot \ln(C_{O2}) \tag{14}$$

where the e_i parameters, i=1-4, are mentioned in figure 5.b, next to the subsystem 3 figure, and C_{O2} represents the dissolved oxygen concentration in the interface of the cathode catalyst and is computed using relation:

$$C_{O2} = \frac{P_{O2}}{5.08 \cdot 10^6 \cdot \exp\left(-{498}/{T}\right)} \tag{15}$$

Relations (14) and (15) are used for implementing the subsystem 3.

The ohmic voltage loss can be computed using relations:

$$E_{ohmic} = J_{FC} \cdot R_{ohmic}$$

$$J_{FC} = I_{FC} / A; \; R_{ohmic} = R_m + R_c; \; R_m = r_m \cdot t_m / A \tag{16}$$

where (see also figure 5.e):

R_{ohmic} is the ohmic resistance (Ω);

J_{FC} – Current density (A/cm^2);

R_m - Echivalent membrane resistance (Ω);

R_c - Contact resistance between membrane and electrodes (Ω);

t_m - membrane thickness [cm];

A – Activation aria (cm^2);

r_m - Resistivity of Nafion membrane (Ωcm^{2}m^{-1}).

The resistivity of Nafion series proton exchange membrane in relation (16) can be calculated as:

$$r_m = \frac{181.6 \cdot \left[1 + 0.03 \cdot J_{FC} + 0.062 \cdot (J_{FC})^{2.5} \cdot \left({T}/{303}\right)^2 \right]}{\left[\lambda_m - 0.0634 - 3 \cdot J_{FC} \right] \cdot \exp\left[4.18 \cdot \left(T - {303}/{T} \right) \right]} \tag{17}$$

where λ_m represent the water content of the membrane and is a input of PEMFC model.

Relations (16) and (17) are used for implementing the subsystem 4.

The concentration voltage loss can be written as:

$$V_{conc} = B \cdot \ln\left(1 - {J_{FC}}/{J_{\max}}\right) \tag{18}$$

where B is a modeling constant (V), J_{max} represents the maximum current density (A/cm^2), and this relation is implemented in the subsystem 5.

The PEMFC output voltage can be calculated as follows [14]:

$$V_{FC} = N\,(E_{Nerst} - E_{ohmic} - V_d) \tag{19}$$

where V_d is voltage over R_a resistance and R_a is equivalent resistance representing the sum of activation and concentration resistances:

$$R_a = \frac{E_{act} + E_{conc}}{I_{FC}} \tag{20}$$

The differential equation related with the voltage drop on R_a in parallel with double layer capacitor (C) can be written as:

$$I_{FC} = \frac{V_d}{R_a} + C \cdot \frac{d}{dt} V_d \tag{21}$$

Taking the Laplace transforms of both sides of relation 21:

$$V_d = \frac{I_{FC} \cdot R_a}{t_{FC} \cdot s + 1}, \quad t_{FC} = C \cdot R_a \tag{22}$$

where t_{FC} represents the fuel cell time constant.

The response time t_{FC} is used to model the "charge double layer" phenomenon due to the build up of charges at electrode/electrolyte interface. Relations (19) – (22) are used for implementing the subsystem 6.

All subsystems are coupled as in Figure 5.c to obtain the dynamical PEMFC model. The basic simulation conditions of the fuel cell are set for input variables: hydrogen input flow [kmol s^{-1}], oxygen input flow [kmol s^{-1}], and membrane water content (when this parameter is not used as parameter to change the MPP).

The above PEMFC model relations can be computed for a fuel cell operating point at about 70°C (which usually has been chosen to be the steady state set point) and given water content of the membrane. The result is a voltage function (f in Figure 4) depending by PEMFC current density (set by the membrane area, A) and stack cells number, N_{fc}. Number of fuel cells used for PEMFC stack is given by relation $N_{fc}=V_{PEMFC0}/V_{cell0}$, where V_{cell0} and V_{PEMFC0} are output voltage from no-load for a single fuel cell and PEMFC stack, respectively. Using the reported results for Ballard Nexa PEMFC's, V_{PEMFC0} is 42,2V, so N_{fc}=52, and A=56 cm^2. At maximum power point, obtained for full-load, the power (P_{base}), output voltage (V_{base}), output current (I_{base}) and base impedance ($Z_{base} = V_{base}/I_{base}$) are shown in Table 1.

Table 1. Base parameters for Ballard Nexa PEMFC

PEMFC type	V_{PEMFC0} [V]	V_{base} [V]	I_{base} [I]	P_{base} [W]	Z_{base} [Ω]
Ballard Nexa	42,2	26,6	45	1200	0,59

The obtained Ballard Nexa PEMFC static characteristics are presented in figure 6.

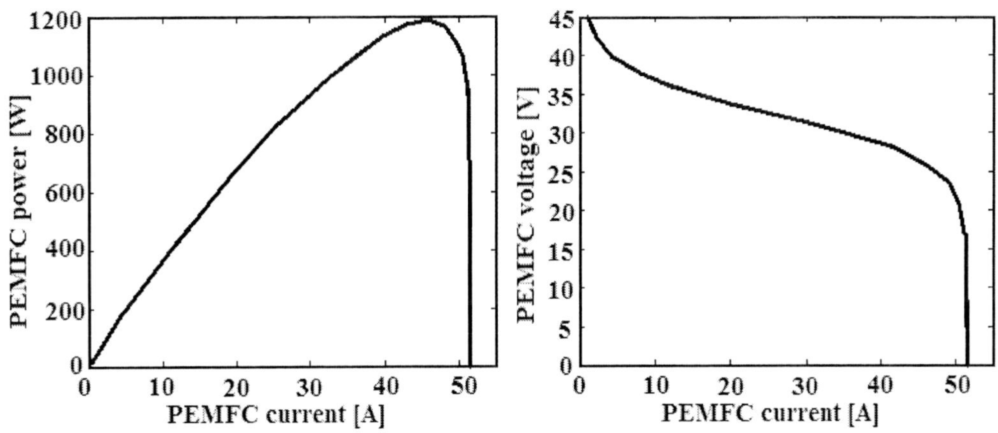

Figure 6. Ballard Nexa PEMFC static characteristics.

Modeling of the Energy Storage Devices

A generic battery model is used in this book and this model is presented in figure 7 [13; 15].

Exponential zone dynamics is modeled by relationship (see also the battery discharge characteristic shown in Figure 8):

$$E_{exp}(q) = A \cdot \exp(-B \cdot q) \tag{23}$$

where:

$q = i_{batt} \cdot t$ is current value of the battery capacity (Ah);

i_{batt} = Battery current (A);

A = Exponential voltage (V);

B = Exponential capacity $(Ah)^{-1}$.

Battery operating mode is modeled by S signal, output of the comparator

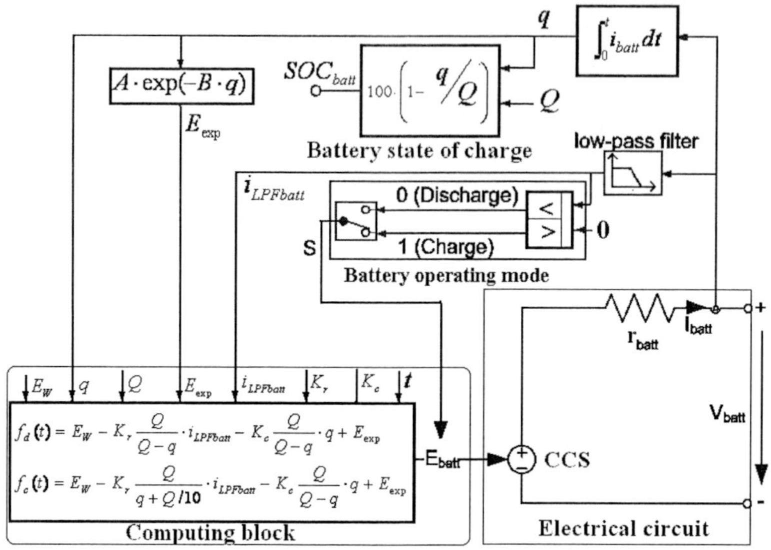

Figure 7. Battery model (adapted from [13]).

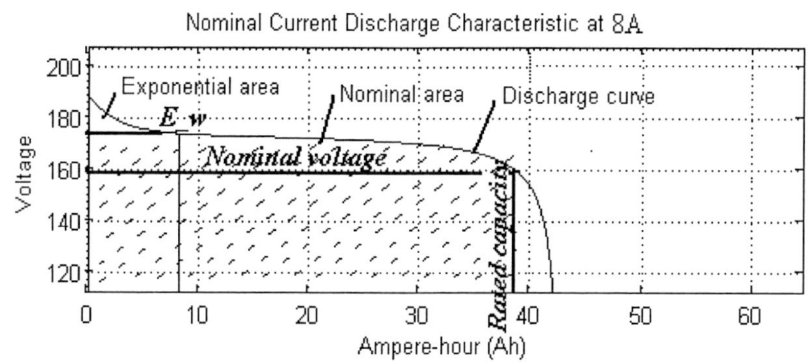

Figure 8. Battery discharge characteristic [13].

The low-pass filtered battery current, $i_{LPFbatt}$, is compared with zero reference and S = 0 during battery discharge ($i_{LPFbatt}$ >0) and S = 1 during battery charging ($i_{LPFbatt}$ <0). So, the battery voltage, E_{Batt}, is given by charge (f_c) or discharge (f_d) function as:

$$
\begin{aligned}
f_d(t) &= E_W - K_r \frac{Q}{Q-q} \cdot i_{LPFbatt} - K_c \frac{Q}{Q-q} \cdot q + E_{\exp} \\
f_c(t) &= E_W - K_r \frac{Q}{q+Q/10} \cdot i_{LPFbatt} - K_c \frac{Q}{Q-q} \cdot q + E_{\exp}
\end{aligned} \tag{24}
$$

where:

E_w is the working battery voltage (V);

K_c = Polarization constant V/(Ah);

K_r = Polarization resistance (Ω);

$i_{LPFbatt}$ = Low frequency battery current dynamics (A);

Q = Maximum battery capacity (Ah).

Battery state of charge, SOC_{batt}, is computed as:

$$SOC_{batt} = 100 \cdot \left(1 - \frac{\int_0 i_{batt} dt}{Q} \right) \tag{25}$$

The parameters of the equivalent circuit can be modified to represent a particular battery type, based on nominal voltage, rated capacity and its discharge characteristics. For example, the 160V/38Ah battery parameters are: A = 15V, $B=0.375(Ah)^{-1}$, K_c = 0.03 V/(Ah), K_r = 0.03Ω, E_w = 163V, and Q=43Ah. The nominal current discharge curve at 8A is composed of three sections, as is showed in the Figure 8. The battery internal resistance, r_{batt}, can be set to about 0.04 Ω and the battery output voltage, V_{batt}, is computed as:

$$V_{batt} = E_{batt} - r_{batt} \cdot i_{batt} \tag{26}$$

For the ultracapacitors stack a first order model is used in this book (see Figure 9. Series resistance, R_s, and parallel resistance, R_p, can be set to about of 0.01 Ω and 1 kΩ, respectively.

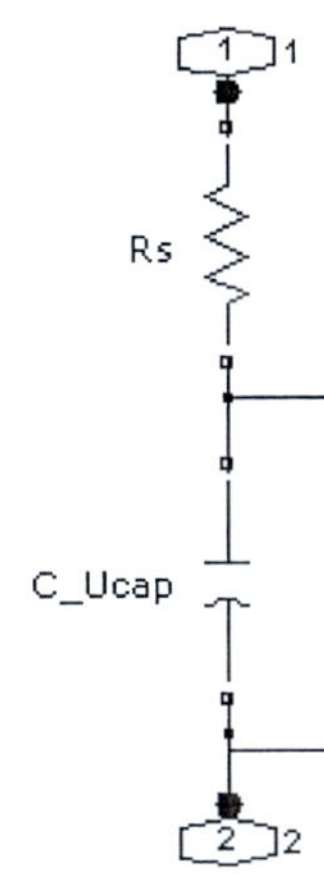

Figure 9. Ultracapacitors stack model.

Simulation Results

The PEMFC characteristics for Ballard Nexa PEMFC's stack (detailed model) are shown in figure 10, together with experimental data set for 65^0C [16]. The experimental data set are given in the range of current from 0 to 45A with step of 5A. The simulated PEMFC characteristics fit well the experimental date set. Comparing the fitted curve of the experimental data with the simulated curves, it can be seen that there is a good agreement between them, with a maximum percentage difference between the values obtained from the fitted experimental curves and simulated curves of up to 3%. Because the fitted experimental curves are too close to the simulated curves they aren't shown in figure 10.

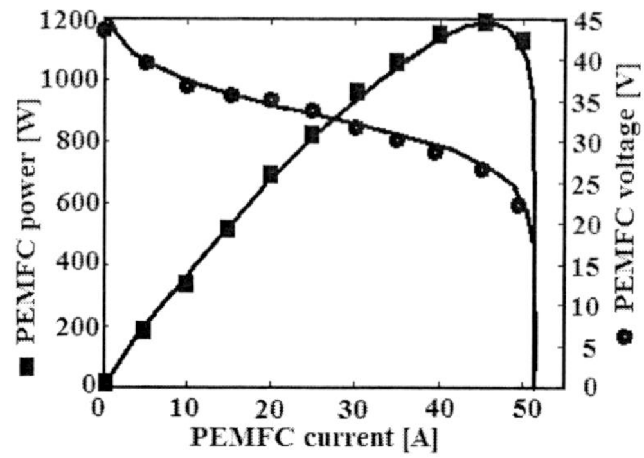

Figure 10. PEMFC characteristics for used PEMFC's stack model and experimental data set (• - voltage; ■ – power).

Consequently, the proposed PEMFC's stack model can be successfully used in simulation of the proposed HPS architecture. The same results are obtained with simplified model of Ballard Nexa PEMFC's stack. The battery discharge characteristic and its discharge in time are shown in figure 8 and figure 11, respectively.

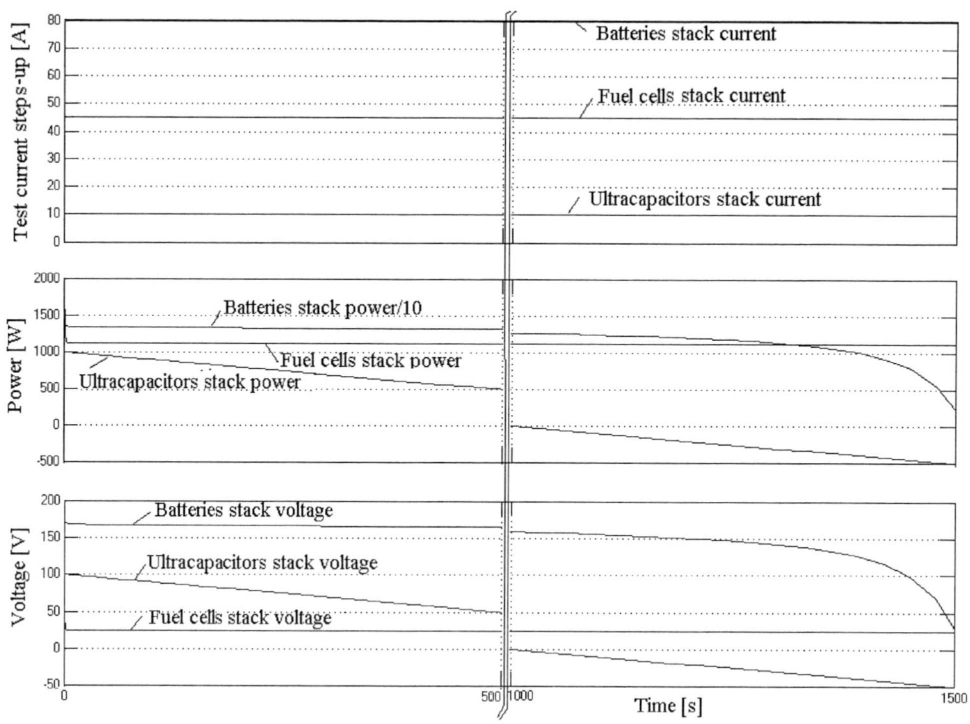

Figure 11. Behavior of the used models for fuel cells, batteries and ultacapacitors stack (adapted from [13]).

To speed-up the battery discharge, it is used a current of ten times greater than that used in drawing of discharge characteristic shown in figure 8. Anyway, the battery stack can deliver the load current for time up to 15000 s or roughly four hours (see figure 11). The behavior in separate tests under current steps-up of the used models for fuel cells, batteries and ultacapacitors stack is shown in figure 11.

These models can be used to implement an emulator of the PEMFC stack behaviour using a DC-DC converter [17].

Conclusion

This section is dedicated to explaining the used models for stacks of PEMFCs, batteries and ultracapacitors, respectively. The relationships are given for each subsystem that form the model and subsystems connections are shown, too. The proposed models are tested individually and compared with reported results or available models in Matlab – Simulink®. The results successfully validate the used models for stacks of PEMFCs, batteries and ultracapacitors, respectively.

MODELLING THE TURBINE

Introduction

The ideal model for the wind turbine was introduced by BETS in 1926. The main assumptions for this model are:

- There is no fractional tension.
- The number of the wings is unlimited.
- The driving force is equal everywhere in the rotor room.
- Convoluted flow of the wings has not been taken into account.

Taking the above assumptions into account, no work will be done on either sides of the disk, therefore the Bernoulli function will remain constant on both sides and we have [18; 19; 20]:

$$P_1 + \frac{1}{2}\rho U_1^2 = P_2 + \frac{1}{2}\rho U_2^2$$

$$P_3 + \frac{1}{2}\rho U_3^2 = P_4 + \frac{1}{2}\rho U_4^2$$

Assuming the input and output air pressure (P_1, P_2, P_3, P_4) and also air velocity (U_1, U_2, U_3, U_4) of two sides of the wing to remain constant we have:

$$T = A_2(P_2 - P_1) \Rightarrow T = \frac{1}{2}\rho A_2(U_1^2 - U_4^2)$$

$$\dot{m} = A_2 U_2 \Rightarrow U_2 = \frac{U_1 + U_4}{2}$$

$$a = \frac{U_1 - U_2}{U_1} \Rightarrow U_2 = U_1(1-a)$$

$$P = \frac{1}{2}\rho A_2(U_1^2 - U_4^2)U_2 = \frac{1}{2}\rho A_2 U_2(U_1 + U_4)(U_1 - U_4) = \frac{1}{2}\rho A_2 U_1^3[4\alpha(1-\alpha)^2]$$

$$C_F = 4a(1-a)^2$$

$$P_{out} = \frac{1}{2}\rho A_2 U_1^3(\eta_{mech}.C_F)$$

An ideal rotor is on its maximum power when its operation is such that the air velocity on rotor is $\frac{2}{3}$ of the original velocity of the wind ($a = \frac{1}{3}$). In such conditions, the power factor will be 0.5926.

Model of Horizontal Axis Wind Turbine with Convoluted Flow of the Wings

In this model, we assume that the angular velocity of the air flow ω is small compared to angular velocity of the rotor (Ω); therefore we can assume that when the axial elements of the velocity remain constant, angular velocity of air will be increased from Ω to $\Omega + \omega$ in proportion with the velocity of the wings [21; 22; 23; 24]. Figure 12 shows the parameters of blade of the wind turbine.

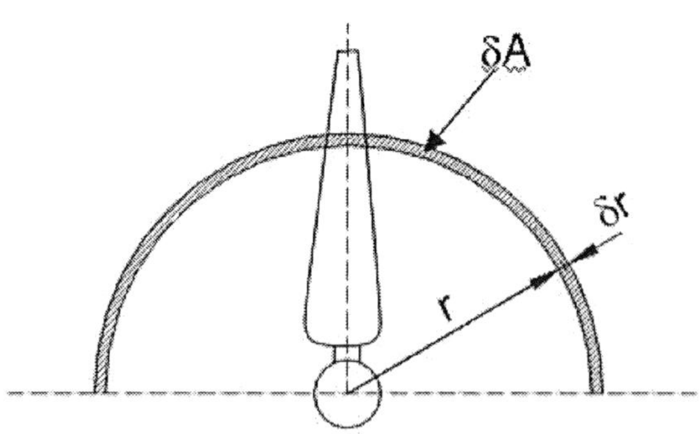

Figure 12. The parameters of wind turbine blade.

We have:

$$P_2 - P_3 = \rho(\Omega + \frac{1}{2}\omega)\omega r^2$$

$$dT = (P_2 - P_3)dA = \left[\rho(\Omega + \frac{1}{2}\omega)\omega r^2\right].2\pi r dr$$

We define the rotational inductance factor as follows:

$$a' = \frac{\omega}{2\Omega} \quad \Rightarrow \quad dT = 4a'(1 + a').\frac{1}{2}\rho\Omega^2 r^2 .2\pi r dr$$

$$\frac{a(1-a)}{a'(1+a')} = \frac{\Omega^2 .r^2}{U^2} = \lambda_r^2$$

$$\lambda_r = \frac{\Omega .r}{U} = \frac{\lambda .r}{R}$$

where dT is the partial rotational driving force and λ_r is defined as local tip velocity. The rotor's torque is equal to instantaneous variation of the convoluted flow and in partial mode it will be:

$$dQ = d\dot{m}(\omega r)r = \rho U_2 .2\pi r dr(\omega r)r$$

$$dQ = [4a'(1-a)]\frac{1}{2}\rho U\Omega r^2 .2\pi r dr$$

The power generated by the partial element of the torque is:

$$dP = \frac{1}{2}\rho A U^3\left[\frac{8}{\lambda^2}a'(1-a)\lambda_r^3 d\lambda_r\right]$$

For power factor we have:

$$dC_P = \frac{aP}{\frac{1}{2}\rho A U^3}$$

$$C_P = \frac{8}{\lambda^2}\int_0^{\lambda} a'(1-a)\lambda_r^3 d\lambda_r$$

We will discuss a situation in which we have maximum aerodynamics:

$$a' = -\frac{1}{2} + \frac{1}{2}\sqrt{1 + \frac{4}{\lambda_r^2}a(1-a)} = \frac{(1-3a)}{(4a-1)}$$

$$\lambda_r^2 = \frac{(1-a)(4a-1)^2}{(1-3a)}$$

$$C_{p,\max} = \frac{24}{\lambda^2}\int_{a_1}^{a_2}\left[\frac{(1-a)(1-2a)(1-4a)}{(1-3a)}\right]^2 da$$

$$= \frac{8}{729\lambda^2}\left\{\frac{64}{5}x^5 + 72x^4 + 124x^3 - 63x - 12\ln x - \frac{4}{x}\right\}_{x=1-3a_2}^{x=0.25a_1}$$

Gas Turbine and Model of the Synchronous Generator

In the gas turbine, there is a continuous flow of the working fluid. This working fluid is initially compressed in the compressor. It is then heated in the combustion chamber. Finally, it goes through the turbine. The turbine converts the energy of the gas into mechanical work. Part of this work is used to drive the compressor. The remaining part is known as the net work of the gas turbine. Gas turbines can be used to deliver power, heat or thrust. Therefore, different gas turbine types exist named as shaft power gas turbine and jet engine gas turbine V_t [25; 26; 27].

The shaft power gas turbine is mainly deliver shaft power and is often used in industrial applications and also electricity production. The shaft power mainly depends on the temperature drop in the power turbine (P_T).

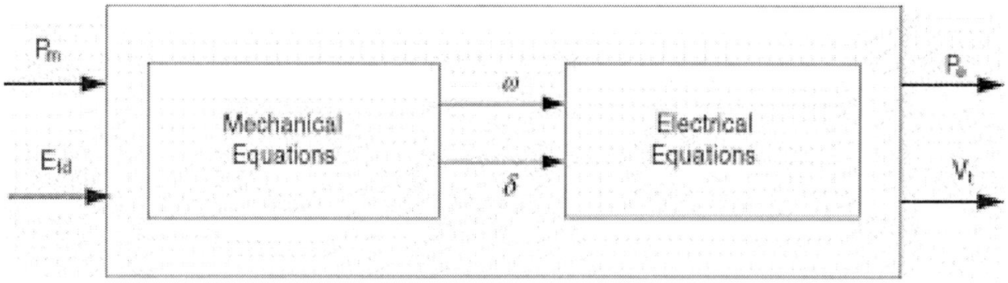

Figure 13. Block diagram of a synchronous generator.

The gas turbines are in range of sizes from 1 MW to several hundred MW. The three main components of a gas turbine are compressor, combustor, and the turbine. The compressor is used to pressurize the air before entering the combustor. Injected fuel is mixed with the compressed air in the combustor and the mixture is ignited. Mechanical power is produced when the hot combustion gases flow and expand through the turbine. The turbine drives a synchronous generator. A portion of power produced in the turbine is utilized for driving the air compressor while the rest is converted to electric power in the generator. Figure 13 shows a simplified block diagram of a synchronous generator. The inputs to the generator are mechanical power P_m and field voltage E_{fd} and outputs are the active power P_e and the terminal voltage.

Mechanical Equations

The mechanical motion of the rotating mass of a gas turbine is modeled by the dynamic equations are given below:

$$\frac{H}{\pi f_0}\frac{d\omega}{dt} + D\omega + P_e = P_m$$

$$\frac{d\delta}{dt} = \omega - \omega_0$$

where H is the inertia constant, f_0 (ω_0) the synchronous frequency in Hz (radian/s), D the damping co-efficient, ω the angular frequency of the combined rotating part, P_e the output electrical power and P_m is the mechanical power on the shaft of the generator.

Electrical Equations

A two-axis model is assumed for the gas turbine. This model is given in the form of a phasor diagram as shown in Figure 14. In this diagram, E is the voltage induced by the field current i_f and E' is the net voltage induced by i_f and the armature reaction. The terminal characteristics of the gas turbine in d-q frame of reference is obtained from the phasor diagram as follows:

$$E' = V_q + rI_q - X'_d I_d$$

$$0 = V_d + rI_d - X_q I_q$$

The above steady-state equations may also be used for dynamic cases with some approximations. The dynamic equation is related to the field winding and is given by

$$T'_{do}\frac{dE'}{dt} + E' - (X_d - X'_d)I_d = E_{fd}$$

where T'_{do} is direct axis open circuit transient time constant and E_{fd} is the field voltage.

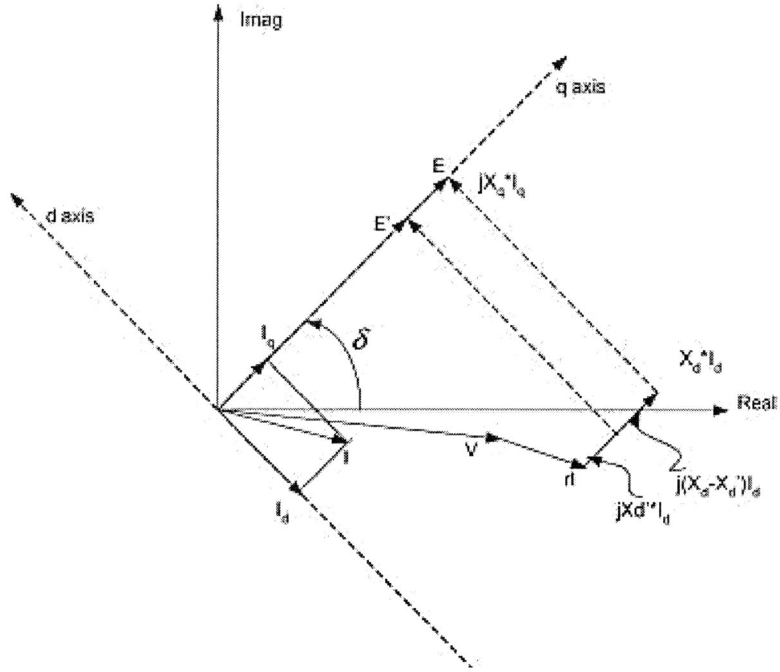

Figure 14. Phasor diagram of a synchronous generator.

Conclusion

The models used in this book for wind turbine and gas turbine are presented in this section. The turbine structure and the main operating modes are briefly showed. The mathematical background is presented, too.

Multi-Stage Inverter Topologies for Energy Generation Systems

Introduction

In multi-stage power inverter topologies, in the first stage (DC-DC converter) the DC voltage is boosted to set voltage and then is converted to AC voltage in the last stage (inverter). Usually, the electrical isolation of DC energy source to AC load is made in the first stage, too. Each stage can be controlled separately using a synchronization method for switching command of all stages [28; 29; 30]. For energy generation systems (EGS) implementation, different types of energy conversion topologies with intermediate DC or AC stage were proposed [31; 32; 33; 34; 35; 36; 37]. For voltage boost operation the first conversion stage DC-DC is usually implemented with a DC-DC converter or a DC-AC-DC converter. If the structure type with intermediate AC bus of high frequency (HF) is chosen, the EGS must contain a frequency cyclic-converter with multiple-carrier PWM sinusoidal

(MCPWMS) control [38; 39; 40; 41]. A review of the main topologies is shown in the next sections.

DC-DC-AC Power Conversion Topologies

For a DC-DC-AC boost converter with two stages (Figure 15), the use of a boost converter and buck inverter combination is usually in the case of wind energy generators or fuel cell EGSs. For this topology type results after the first stage a continuous voltage with small current ripple. The second stage (full-bridge buck inverter) use a pure sine PWM command. Usually, it isn't necessary to synchronize the conversion stages command signals.

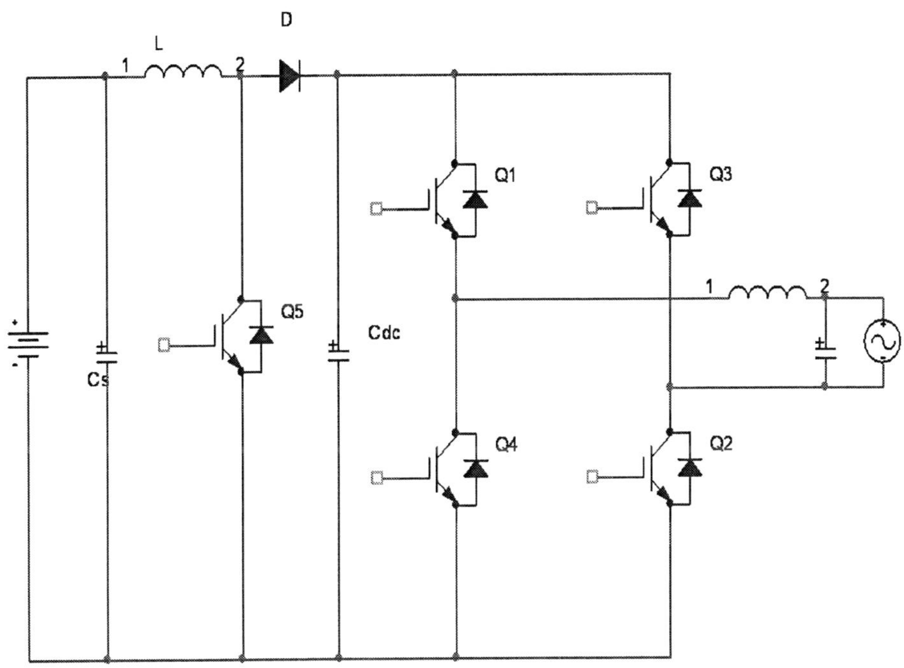

Figure 15. DC-DC-AC inverter boost.

If the MCPWMS control is used for the same topology, it is necessary to synchronize the conversion stages command signals. The HF waveform is converted by an appropriate control into alternative voltage. This lead to high energy conversion efficiency (thru reduction commutation loss) and a cheap price, too.

In figure 16 is shown the buck-boost inverter topology proposed by Saha and Sundarshing [30]. The topology is recommended for photovoltaic sources without isolation connection. For user security reasons is utilized a input DC voltage up to 100V. Actually simple, this EGS structure permits a limited range for continuous voltages variations and a limited voltage conversion rate, too.

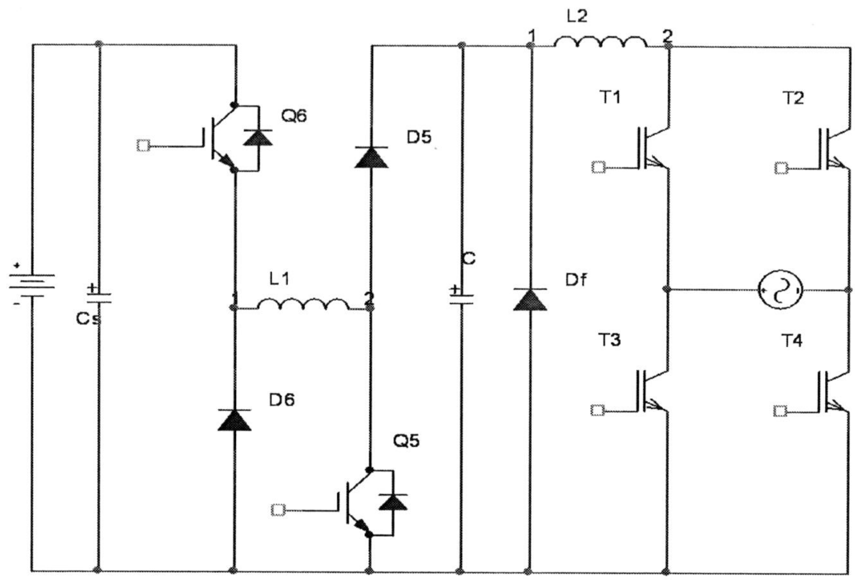

Figure 16. Buck-boost inverter without galvanic isolation (adapted from [30]).

The inverters topologies with galvanic isolation, witch utilizes HF transformers, can generate power from a renewable energy source even when the continuous input voltage is relative small. Figure 17 presents a topology that utilizes a flyback converter for electrical isolation between load and energy source.

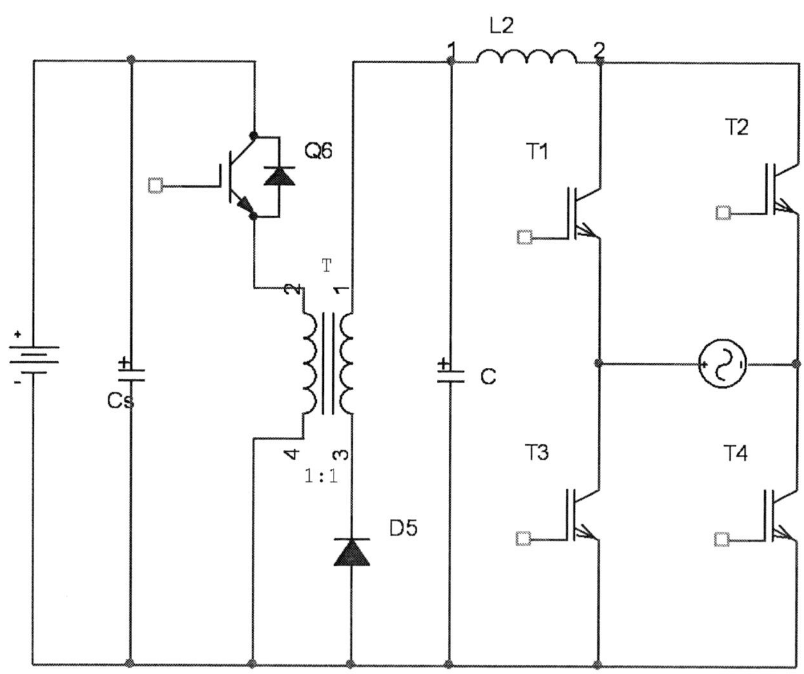

Figure 17. Flyback-boost inverter with galvanic isolation (adapted from [30]).

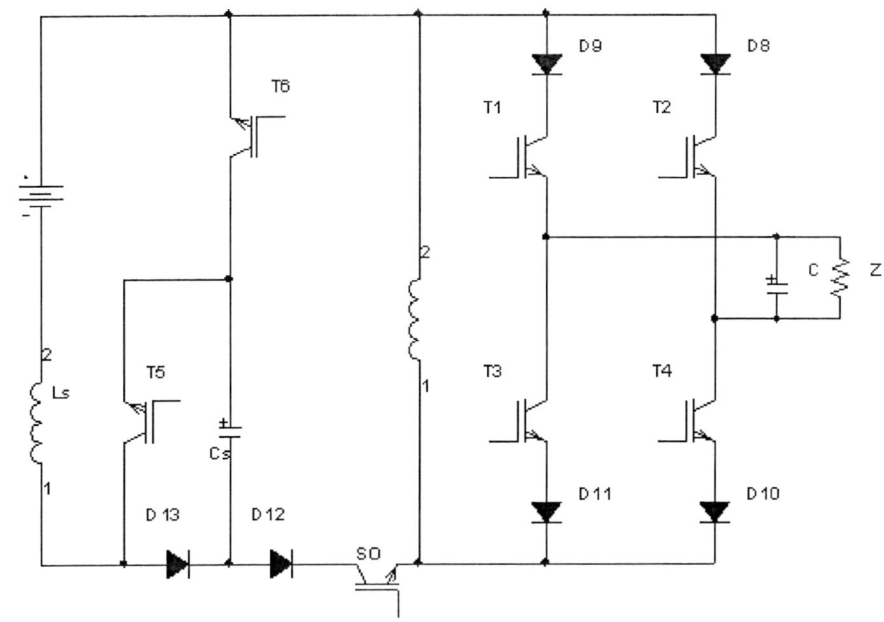

Figure 18. Buck-boost inverter with ESD (adapted from [42]).

To increase the input voltages range (without using a HF transformer) a C_s capacitor is added in the loading input circuit as energy storage device (ESD), resulting the buck-boost inverter proposed by Funabiki [42] (see Figure 18). This inverter can work normally with an energy source having a level that can exceed the amplitude of output voltage. Consequently, the proposed inverter must utilize auxiliary switches and supplementary control sequences that can be cause of some supplementary costs.

Another EGS topology with galvanic isolation between input and output (of buck-boost flyback type) is proposed by Shimizo and Kjaer [14] (see figure 19).

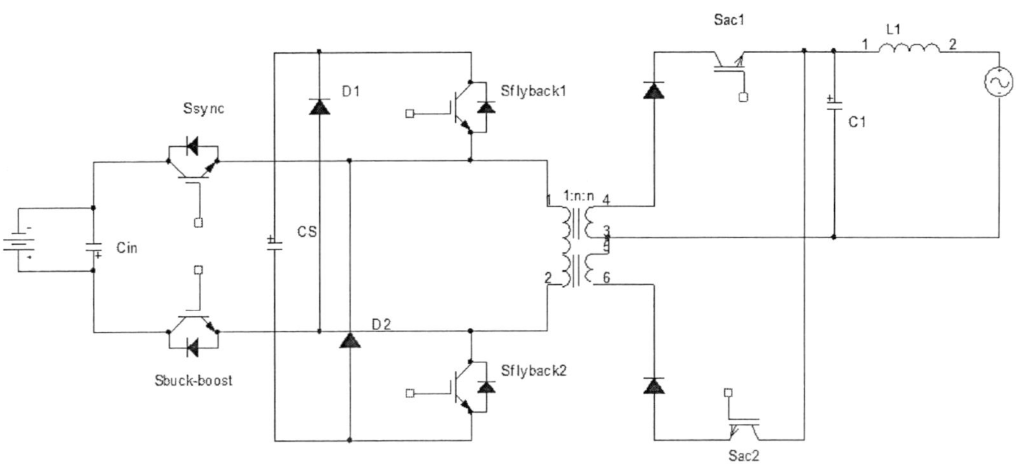

Figure 19. Flyback inverter (adapted from [14]).

The constant power mode requested by the maximum power point (MPP) control of photovoltaic modules is made in the first stage and the power balance is realized using an intermediate C_s capacitor. Finally, the storage energy is transferred via HF transformer in the secondary side. The advantage of this topology is the relative small value of the ESD capacitance. The C_s capacitor replaces the capacitors stack used for filtering the high voltage DC bus. This replacement improves the inverter's life time and gives a lower production price, too.

DC-AC-DC-AC Power Conversion Topologies

The multi-stage inverter that requires a high conversion rate (measured by the ratio of output DC voltage to the input DC voltage) is generally composed by a HF converter DC-AC-DC coupled with a HF inverter or a line current inverter system. These structure types are differenced by the DC bus type:

Authentic DC bus between these two converters

A traditionally topology with an similar bus was presented in figure 17, where the control of a DC bus constant voltage is obtained thru a HF switching command, followed by a full bridge rectifier and a passive filter. Both converters operate at a HF, so switching losses appear. Also, this topology has an expensive price.

Pseudo-bus DC between these two converters

In figure 20 is presented a multi-stage boost inverter topology. As an advantage, in comparison with topology from figure 17, the passive filter isn't used here. The last conversion stage is a line current inverter system that transforms the HF impulses train into AC voltage.

A low-pass filter is requested at the load output in order to assure, with acceptable value for THD, a sine shape for output voltage. The topology without a low-pass output filter is shown in figure 21, too.

Load AC current is in phase with system voltage without using a low-pass output filter. This topology was commercialized for a large range of power (1-100 kW) of the photovoltaic systems that are grid connected.

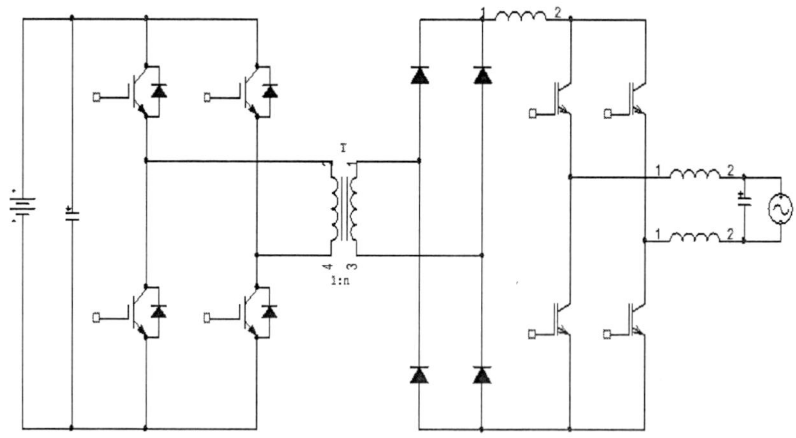

Figure 20. Boost inverter with DC pseudo-bus for line current inverter.

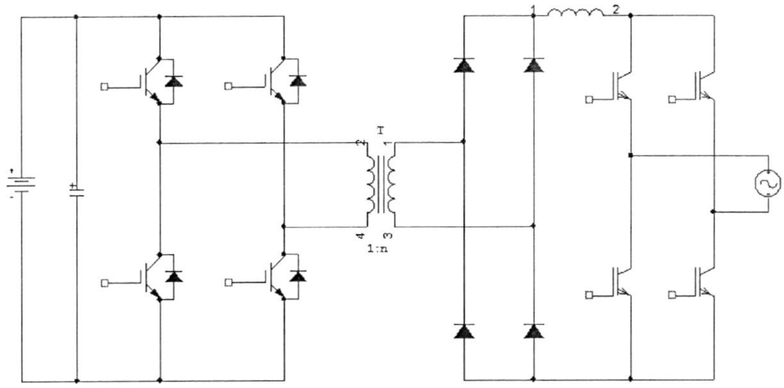

Figure 21. Boost inverter with DC pseudo-bus without a low-pass output filter.

DC-AC-AC Power Conversion Topologies

For stand-alone topologies or residential EGS are required a bidirectional control of the inverter power flows. In this case, the recycled energy will be storage in the ESD. Different proposals were made for this kind of objective. The Beristain proposal [43] is presented in the figure 22, where the second stage is a bidirectional AC-AC converter supplied from the HF AC voltage bus (the secondary side of the HF transformer).

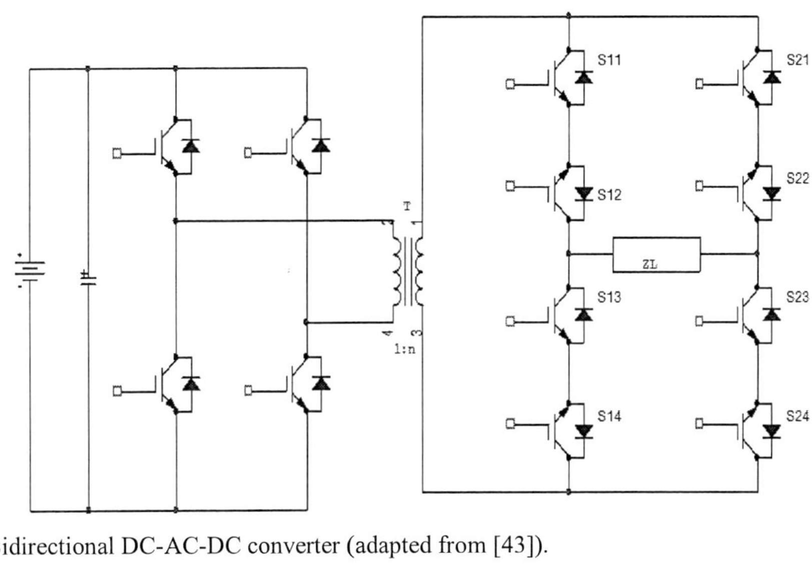

Figure 22. Bidirectional DC-AC-DC converter (adapted from [43]).

Generally, two problems appear:

- at low frequency (LF) and high load power, the LF PEMFC current ripple can be over to the imposed value [41]. Also, if the load power is over the specific limit, the control variables can't catch the control law. Consequently, bifurcation and chaos appear [44].

- at high frequency and light load power, the HF PEMFC ripple is usually up to the imposed limit, but the EGS energy efficiency is reduced dramatically because of the commutation power losses that have values increased in all power stages.

Conclusions

From the before section results some benefits of utilization of a HF transformer in the first power conversion stage and necessity of a grid inverter as a final power conversion stage. Table 2 shows a comparison of the low and medium power converters presented above using the following criterions:

- Energy storage devices used;
- Filtering technique in intermediate stage;
- Floating ground (floating source ground or floating grid load);
- Electrical isolation between energy source and AC load;
- Grid connection facility or the residential system functionality.

For on EGS topology the energy efficiency is computed for every power stage and finally the obtained value is the product of all values. For example, in case of a multi-stage DC-AC-DC-AC EGS topology, the energy efficiency is given by relationship:

$$\eta_{EGS} = \eta_{PEMFC-ESD} \cdot \eta_{DC-AC} \cdot \eta_{AC-DC} \cdot \eta_{DC-AC}$$

More cheap and energy efficient solutions is to use a multi-stage DC-AC-AC EGS topology, with an energy efficiency over to 90% at rated load power.

The same converters presented above are compared in the table 3 using the following criterions:

- minimum input voltage required to generate an AC voltage that comply the American / European energy quality standard;
- output power range;
- working frequency (fs);
- energy efficiency (η);
- recommended EGS applications;
- number of passives components (of type: transformer-T, LF-network transformer, HF-high frequency transformer, L-inductor, C_F-film capacitor, C_E-electrolytic capacitor) and power electronic devices (S-switch; D-diode).

As a final conclusion, in future years is necessary to develop efficient EGS topologies and appropriate control strategies for them [45].

Table 2. A brief comparison of the power converters presented in this chapter

Figure nr.	Type of topology	Storage energy device used / filtering technique used in intermediate stage	Floating ground /electrical isolation	Grid connection / work as residential system
15	Multi-stage DC-DC-AC	High value input capacitor and intermediary capacitor on the DC bus	yes / non-isolated	both
16	Multi-stage DC-DC-AC	High value electrolytic capacitor for input and intermediary inductor on c.c. bus	no / non-isolated	Only grid connection
17	Multi-stage DC-DC-AC	High value electrolytic capacitor for input and intermediary inductor on c.c. bus	yes / isolated	Only grid connection
18	Multi-stage DC-DC-AC	High value electrolytic capacitor for input and intermediary inductor on c.c. bus	no / non-isolated	Only able to operate as residential system
19	Multi-stage DC-DC-AC	Low value for input and intermediary capacitor	yes / isolated	Only grid connection
20	Multi-stage DC-AC-DC-AC	High value input electrolytic capacitor	yes / isolated	Only grid connection
21	Multi-stage DC-AC-DC-AC	High value input electrolytic capacitor	yes / isolated	Only grid connection
22	Multi-stage DC-AC-AC	High value input electrolytic capacitor	yes / isolated	both
23	Multi-stage DC-DC-AC-DC-AC	Low value for input and intermediate capacitor	yes / isolated	both

Table 3. Performances comparison of the power converters presented in this chapter

Figure nr.	V-Minimum input US/EU standard	Maximum output power	f_s	η	Application EGS	Number of components		
						T	L/C_F (C_E)	S/D
15	34V / 65V	1KW	-	92 %	-	No	2/1(2)	5/6
16	42V / 81V	2kW	36 kHz	-	Photovoltaic	No	2/1(1)	6/5
17	Large value range	2kW	36 kHz	-	Photovoltaic	HF	1/1(1)	5/2
18	9v / 16v	640W	5.3-17 kHz	-	Wind proof and PV	No	2/1(1)	7/6
19	Large value range	160W	-	86 %	Photovoltaic	HF	1/3(0)	6/8
20	Large value range	-	-	-	Wind proof	HF	2/1(1)	8/12
21	Large value range	4kW	10-16 kHz	90 %	Photovoltaic	HF	1/0(1)	8/8
22	Large value range	1kW	50 Hz	-	Photovoltaic	HF	1/1(1)	12/12
23	Large value range	1-10kW	-	80 %	Any EGS application	HF	3/1(1)+Ucap	5/(10)

Note. T – transformer, L – inductor, C_F – film capacitor, C_E –electrolytic capacitor , LF –grid transformer, HF – high frequency transformer , fs – switching frequency, η - efficiency.

REFERENCES

[1] Gou, B., Na, W.K., and Diong, B. (2010), Fuel cells: modeling, control, and applications, CRC Press, (chapter 3) 13-52.

[2] Larminie, J., and Dicks, A. (2003), Fuel Cell Systems Explained- Second Edition. Chichester, West Sussex, England: John Wiley and Sons Ltd.

[3] Spiegel, C. (2008), PEM fuel cell modeling and simulation using MATLAB, Elsevier Press.

[4] Erdinc, O., Vural, B., and Uzunoglu, M. (2009), A wavelet-fuzzy logic based energy management strategy for a fuel cell/battery/ultra-capacitor hybrid vehicular power system, *Journal of Power Sources* 194 (1), 369-380.

[5] Kisacikoglua, M.C., Uzunoglu, M., Alam, M.S. (2009), Load sharing using fuzzy logic control in a fuel cell/ultracapacitor hybrid vehicle, *Int. journal of hydrogen energy* 34,1497–1507.

[6] Kunusch C., Puleston, P.F., Mayosky, M.A., and Moré, J.J. (2010), Characterization and experimental results in PEM fuel cell electrical behaviour, *International Journal of Hydrogen Energy* 35 (11), 5876-5881.

[7] Marcos, M.V., and Gisele, E.-S. (2009), A practical model for evaluating the performance of proton exchange membrane fuel cells, *Renewable Energy* 34, 1734–1741.

[8] Tang, Y., Yuan, W., Pan, M., Li, Z., Chen, G., and Li, Y. (2010), Experimental investigation of dynamic performance and transient responses of a kW-class PEM fuel cell stack under various load changes, *Applied Energy* 87 (4), 1410-1417.

[9] Tirnovan, R., Giurgea, S., Miraoui, A., and Cirrincione, M. (2008), Surrogate modelling of compressor characteristics for fuel-cell applications, *Applied Energy* 85 (5), 394-403.

[10] Bizon, N. (2010a), On tracking robustness in adaptive extremum seeking control of the fuel cell power plants, *Applied Energy* 87(10) (2010) 3115–3130.

[11] Bizon, N. (2010b), Development of a fuel cell stack macro-model for inverter current ripple evaluation, *Revue Roumaine des Sciences Techniques series Électrotechnique et Énergetique* 55(4) (2010) 405-416.

[12] Bizon, N. (2008), Intelligent Control of Fuel cell Energy Generation System, Research report no. 1 of CNCSIS Grant # 570 (accessed in 2 may 2010 on www.cncsis.ro or directly at http://194.102.64.7/GranturiFinalizate/faces/Projects/ProjectsList.jsp).

[13] Bizon, N. (2010c), Hybrid Power Source for Vehicle Applications Operating at Maximum Power Point of Fuel Cell Stack, *Int. Conf. on Applied Electronics* 2010 - APPEL'10, ISSN: 1803-7232, ISBN: 978-1-4244-6818-8, INSPEC: 11579461 (2010) 3-10.

[14] Kaer, S. B., and Blaabjerg, F. (2003), Design optimization of a single phase inverter for photovoltaic applications, *IEEE PESC'03*, 1191-1196.

[15] Tremblay, O.; Dessaint, L.-A.; Dekkiche, A.-I. (2007), A Generic Battery Model for the Dynamic Simulation of Hybrid Electric Vehicles, *IEEE Vehicle Power and Propulsion* Conference - VPPC 2007, 284-289.

[16] Zhu, W.H., Payne, RU, Cahela, D.R., Tatarchuk, B.J. (2004), Uniformity analysis at MEA and stack Levels for a Nexa PEM fuel cell system. *Journal of Power Sources* 128 (2), 231-238.

[17] Marsala, G., Pucci, M., Vitale, G., Cirrincione, M., and Miraoui, A. (2009), A prototype of a fuel cell PEM emulator based on a buck converter, *Applied Energy* 86 (10), 2192-2203.

[18] Hansen, M.O.L., Sørensen, J.N., Voutsinas, S., Sørensen, N., and Madsen, H.Aa. (2006), State of the art in wind turbine aerodynamics and aeroelasticity, *Progress in Aerospace Sci.* 42 (4), 285-330.

[19] Kishinami, K., Taniguchi, H., Suzuki, J., Ibano, H., Kazunou, T., and Turuhami, M. (2005), Theoretical and experimental study on the aerodynamic characteristics of a horizontal axis wind turbine, *Energy* 30(11-12), 2089-2100.

[20] Xiong, L., Yan, C., and Zhiquan, Y. (2007), Optimization model for rotor blades of horizontal axis wind turbines, *Frontiers of Mechanical Engineering in China*, 2 (4), 483-488.

[21] Fu, P., and Farzaneh, M. (2010), A CFD approach for modeling the rime-ice accretion process on a horizontal-axis wind turbine, *Journal of Wind Engineering and Industrial Aerodynamics* 98 (4-5), 181-188.

[22] Gebhardt, C.G., Preidikman, S., and Massa, J.C. (2010), Numerical simulations of the aerodynamic behavior of large horizontal-axis wind turbines, *Int. J. of Hydrogen Energy* 35 (11), 6005-6011.

[23] Grant, I., Mo, M., Pan, X., Parkin, P., Powell, J., Reinecke, H., Shuang, K., Coton, F., and Lee, D. (2000), An experimental and numerical study of the vortex filaments in the wake of an operational, horizontal-axis, wind turbine, *J. of Wind Engineering and Industrial Aerodynamics* 85 (2), 177-189.

[24] Lanzafame, R., Messina, M. (2010), Horizontal axis wind turbine working at maximum power coefficient continuously, *Renewable Energy* 35 (1), 301-306.

[25] Jagaduri, R.T., and Radman, G. (2007), Modeling and control of distributed generation systems including PEM fuel cell and gas turbine, *Electric Power Systems Research* 77 (1), 83-92.

[26] Sanaye, S., and Tahani, M. (2010), Analysis of gas turbine operating parameters with inlet fogging and wet compression processes, *Applied Thermal Engineering*, Volume 30, Issues 2-3, February 2010, Pages 234-244.

[27] Yee, S.K., Milanović, J.V., and Hughes, F.M. (2010), Damping of system oscillatory modes by a phase compensated gas turbine governor, *Electric Power Systems Research* 80 (6), 667-674.

[28] Bizon, N., and Oproescu, M., (2006a) Control of the Inverter used into Energy Generation System, *Int. conf. Progress in cryogenics and isotopes separation*, 169-172.

[29] Bizon, N., and Oproescu, M., (2006b) Control of the DC-DC Converter used into Energy Generation System, *Int. conf. Progress in cryogenics and isotopes separation*, 173-177.

[30] Saha, S., and Sundarsingh, V.P. (1996), Novel grid-connected photovoltaic inverter, Proceedings of the IEE proceeding Generation, Transmission, Distribution 143 (2), 219-224.

[31] Agrawal, J.P., (2001) Power electronic systems: theory and design, Upper Saddle River, N.J.: Prentice Hall.

[32] Batarseh, I., (2004) Power electronic circuits, Hoboken, NJ: John Wiley Press.

[33] Chen, Y., and Smedley, K. M. (2004), A cost-Effective Single-Stage Inverter with Maximum Power Point Tracking, *IEEE Transaction on Power Electronics* 19 (5), 1289-1294.

[34] Mohan, N., Undeland, T. M., and Robbins, W. P. (2002), Power Electronics: Converters, Applications, and Design - 3rd BkandCdr ed., Wiley Press.

[35] Rashid, M. H. (2003), Power Electronics: Circuits, Devices and Applications (3rd Edition), Prentice Hall Press.

[36] Tse, C.K. (2003), Complex Behavior of Switching Power Converters, CRC Press.

[37] Xue, Y., Chang, L., and Kjær, S.B.- (2004), Topologies of Single-Phase Inverter for Small Distributed Generators :an Overview, *IEEE Trans. on Power Electronics* 19 (5), 1305-1313.

[38] Balog, R.S., and Krein, P.T., (2005) Commutation technique for high-frequency link cycloconverter based on state-machine control, *IEEE Transaction on Power Electronics* 3 (3), 101–104.

[39] Bizon, N., Sofron, E., Oproescu, M., and Raducu, M. (2007), Multi-stage Inverter Topologies for an Energy Generation Systems, 13th International Symposium on Modeling, *Simulation and System's Identification - SIMSIS* 2007, CD proceeding.

[40] Krein, P. T. (2004), Systems and Methods for Pulse Width Modulation, U.S. Patent 6 700 803.

[41] Krein, P. T., Balog, R. S., and Xin, G., (2004), High-frequency link inverter for fuel cells based on multiple-carrier PWM, *IEEE Transaction on Power Electronics* 19 (5), 1279–1288.

[42] Funabiki, S., Tanaka, T., and Nishi, T. (2002), A new buck-boost-operation-based sinusoidal inverter circuit, *IEEE PESC'02*, 1624-1629.

[43] Beristain, J., Bordonau, J., Gilabert, A., and Alepuz, S., (2004) A New AC/AC Multilevel Converter for a Single-Phase Inverter with HF Isolation, *IEEE Power Electronics Specialists Conference - PESC*, 1191-1196.

[44] Banerjee, S., Verghese, G. (Eds.), (2000) Nonlinear phenomena in power electronics, IEEE Press.

[45] Kirubakaran A, and all, A review on fuel cell technologies and power electronic interface, *Renewable and Sustainable Energy Reviews* (2009), doi:10.1016/j.rser.2009.04.004.

In: Advances in Energy Research
Editor: Nicu Bizon
ISBN 978-1-61209-991-0

Chapter 13

POWER QUALITY IMPROVEMENT USING ACTIVE AC/DC PFC CONVERTERS

Ires Iskender*[*,1] and Naci Genc[2]
[1]Department of Electrical-Electronics Engineering,
Gazi University, Ankara, Turkey
[2]Department of Electrical-Electronics Engineering,
Yuzuncu Yil University, Van, Turkey

ABSTRACT

Most modern electrical and electronic apparatus called as nonlinear loads use AC/DC power supply within their architecture. These devices draw discontinuous current from the AC network. The discontinuous currents include harmonics and cause problems for the power distribution systems. In addition, the harmonic contents of the discontinuous current cause additional losses and dielectric stresses in capacitors and cables resulting in increasing currents in windings of rotating machinery, transformers and electromagnetic interference. With the steadily increasing use of equipments including nonlinear loads, line current harmonics have become a significant problem. So, several international standards such as EN 61000-3-2 now exist limiting the harmonic content due to line currents of equipment connected to AC networks.

In this chapter, parallel to power factor correction (PFC) study, several passive and active topologies applied to single phase power factor correction are analyzed. The major advantages and disadvantages of the topologies are highlighted. In addition, the commonly used control methods for active power factor correction are discussed and compared. The switching losses and soft switching methods of the boost type active power factor correction circuits are also reviewed and discussed.

In the several active power factor correction techniques used to reduce current harmonics and to satisfy the requirements given in international standards, the boost type converters operating in continuous conduction mode have been widely adopted as power factor correction pre-regulators. For this purpose, the boost type power factor correction circuit is simulated and the related waveforms are given to show the importance of active power factor correction methods.

[*] E-mail: iresis@gazi.edu.tr ; nacigenc@yyu.edu.tr

Keywords: Power quality, power factor correction, AC/DC.

INTRODUCTION

Poor AC current phase angle is generally the result of an inductive load such as an induction motor, power transformer etc. In its simplest form, poor power factor caused by reactive linear circuit elements results as the current either leads or lags the voltage, depending on whether the load looks capacitive or inductive. This type of poor power factor is easily corrected by adding a reactive component of opposite sign in parallel with load to cancel the reactive term.

On the other hand, most electronic equipments are supplied by 50 Hz or 60 Hz utility power supplies, and almost in all of them power is processed through some kinds of power converters such as AC/DC. AC/DC conversion of electric power is widely used in adjustable-speed drives (ASDs), switch-mode power supplies (SMPSs), uninterrupted power supplies (UPSs), power supplies for telecommunication systems, measurement and test equipments etc. Most of these equipments which have nonlinear characteristics use AC/DC converter within their architecture. The input stage of these power supplies consist of a bridge rectifier followed by a large filter capacitor. These devices draw pulses of current from the ac network as shown in Figure 1.

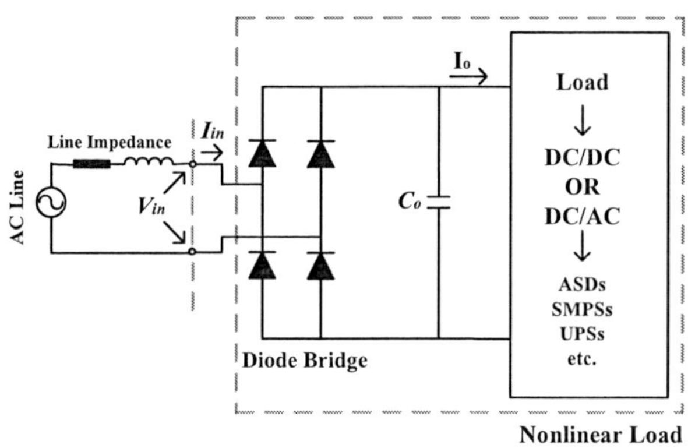

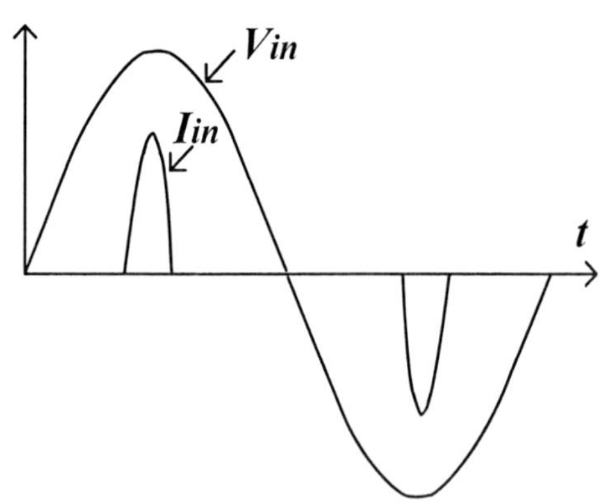

Figure 1. Nonlinear loads and the input ac line current.

Current is drawn from the line only when the line peak voltage exceeds the voltage on the filter capacitor and this causes the line current to be discontinuous. The generated periodic discontinuous current involves fundamental and odd harmonics components. These harmonics cause problems for the power distribution systems. In addition, the harmonic contents of the pulsating current cause additional losses and dielectric stresses in capacitors and cables resulting in increasing peak currents in windings of rotating machinery, transformers, and bringing about unwanted operations of relays and problems in firing of the semiconductor devices and electromagnetic interference (EMI) [1-4].

DEFINITIONS

The current of a nonlinear load supplied by an ideal sinusoidal input voltage source, $V_{in}(t)$, with period of T is given in equation (2) [1].

$$V_{in}(t) = V_{1rms} \cdot \sin(\omega t) \tag{1}$$

$$I_{in}(t) = I_0 + \sum_{n=1}^{\infty} I_n \cdot \sin(n\omega t + \theta) \tag{2}$$

where, V_{1rms} is the rms value of fundemental input voltage. For an ideal sinusoidal voltage, the rms value of the voltage is the same as fundemantal rms value ($V_{1rms} = V_{rms}$). I_o, is the average value and I_n ise n^{th} component value of the input current. θ, is the phase angle between the input voltage and input current.

Power factor (pf) is the proportion of active power (P) and appearent power (S) drawn from an AC grid.

$$pf = \frac{P}{S} \tag{3}$$

$$P = \frac{1}{T}\int_0^T V_{in}(t) \cdot I_{in}(t) \cdot dt = V_{rms} \cdot I_{1rms} \cdot \cos\theta \tag{4}$$

$$S = V_{rms} \cdot I_{rms} \tag{5}$$

For an ideal sinusoidal input voltage (Figure 2), the power factor is obtained from multiplying of displacement factor, k_θ and distortion factor, k_d. These parameters are defined by the following equations [5].

$$pf = \frac{V_{rms} \cdot I_{1rms} \cdot \cos\theta}{V_{rms} \cdot I_{rms}} = \frac{I_{1rms} \cdot \cos\theta}{I_{rms}} = k_d \cdot k_\theta \tag{6}$$

$$k_d = \frac{I_{1rms}}{I_{rms}} \tag{7}$$

$$k_\theta = \cos\theta \tag{8}$$

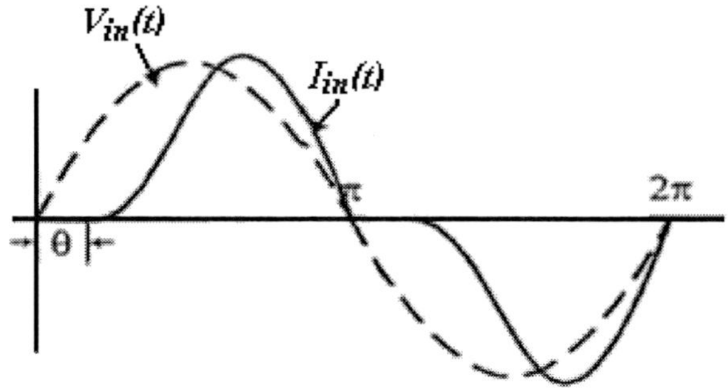

Figure 2. Input voltage and current with phase angle.

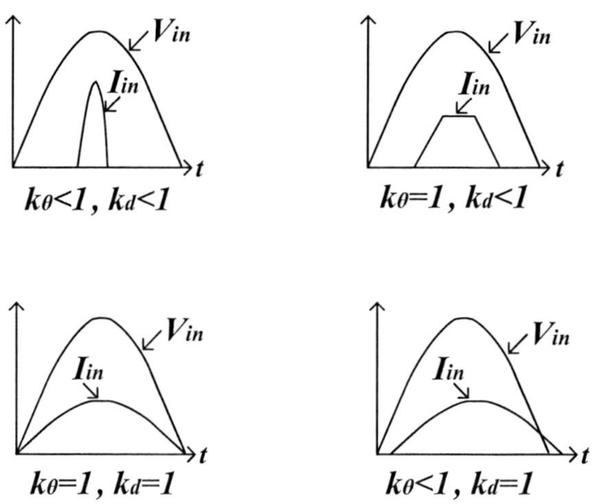

Figure 3. Voltage and current waveforms with different displacement and distortion factors.

A smaller k_θ means a larger apparent current for the same load and a small k_d means a large amount of harmonics in the current. This result in poor power quality and high harmonic contents that adversely affect other users fed from the same grid (Figure 4).

Assuming an ideal sinusoidal input voltage source, the power factor (*pf*) and Total Harmonic Distortion (*THD*) of a power electronics converter is obtained from equations (9) and (10).

$$I_{rms} = \sqrt{\sum_{n=0}^{\infty} I_{nrms}^2} \tag{9}$$

$$THD(\%) = 100x\sqrt{\frac{I_{rms}^2 - I_{1rms}^2}{I_{1rms}^2}} = 100x\sqrt{\frac{I_h^2}{I_{1rms}^2}} = 100x\sqrt{\frac{1}{k_d^2} - 1} \quad (10)$$

If the current is in phase with the voltage ($cos\theta=1$), k_θ will be 1 and therefore the current THD value can be given with equation (11).

$$THD(\%) = 100x\sqrt{\frac{1 - pf^2}{pf^2}} \quad (11)$$

From equation (11), it is seen that there is an inverse relationship between *pf* and the current THD values.

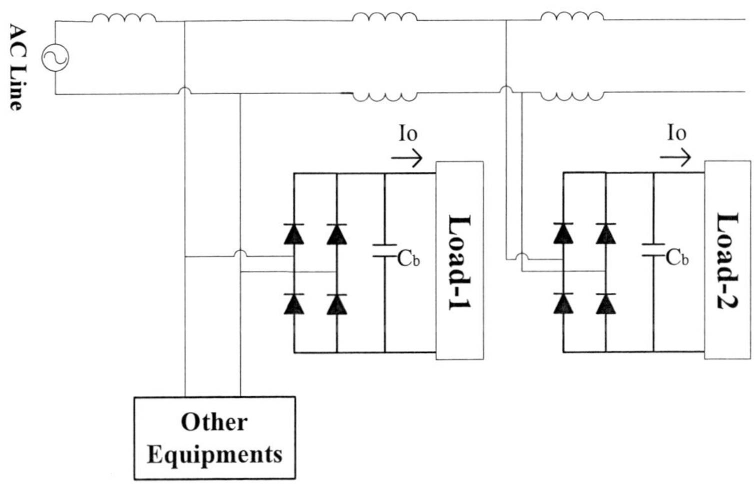

Figure 4. Different nonlinear loads on the same AC grid.

For example, for given *pf* values of 0.995 and 0.95 the corresponding THD values are 10% and 32.9% respectively. Though in some applications the *pf* of a converter may be acceptable and higher than a specified value, the THD of the corresponding converter may be far from the expected value. Thus it is clear that specifying limits for each current harmonics will help to remove the input current pollution. The process of shaping this input current is commonly called as "power factor correction," the measure of its effectiveness towards complying with international regulations is the amount of reduction in the harmonic content of the input current.

Need of Power Factor Correction

The input stage of any AC-DC converter comprises of a full-bridge rectifier followed by a large filter capacitor. The input current of such a rectifier circuit comprises of large discontinuous peak current pulses that result in high input current harmonic distortion. The high distortion of the input current is due to the fact that the diode rectifiers conduct only for a short period. This conduction period corresponds to the time when the grid instantaneous voltage is greater than the capacitor voltage. Since the instantaneous grid voltage is greater than the capacitor voltage only for very short periods of time large current pulses are drawn from the line during this short period of time. The input current harmonic distortion for this kind of rectification is very high and the typical power factor value is about 0.6. Figure 5 shows the schematic of a typical single phase diode rectifier filter circuit. Figure 6 shows the output voltage simulation results of a conventional AC/DC converter for 300 W output power with different output capacitor values. Figure 7 shows the input voltage and current waveforms.

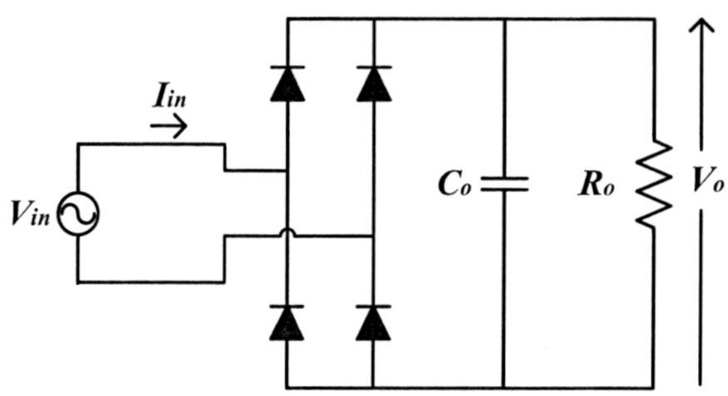

Figure 5. Typical single phase conventional diode rectifier AC/DC circuit.

It can be seen that the input peak current is very high and is about 25 A. The corresponding input voltage peak is also distorted due to the drop in the line impedance caused by this peak current.

The harmonic components and THD values of input currents for different output capacitors are given in Figs. 8 and 9. It is seen that the input current has very high THD value.

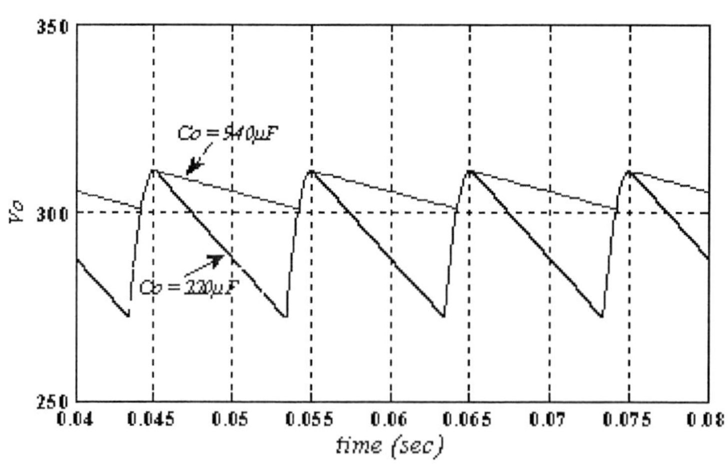

Figure 6. Output voltage waveforms for conventional diode bridge AC/DC circuit (V_{inrms} =220V, R_o = 300Ω).

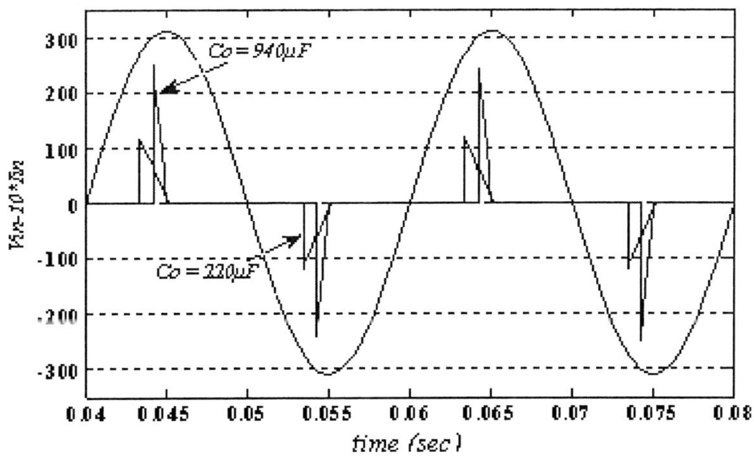

Figure 7. Input voltage and 10*input current waveforms for conventional diode bridge AC/DC circuit (V_{inrms} = 220 V, R_o = 300 Ω).

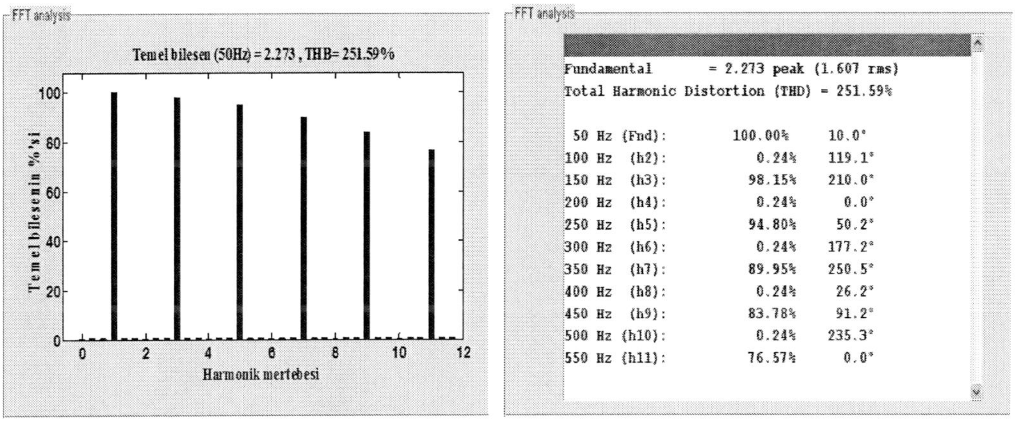

Figure 8. Input current harmonic analysis of the conventional AC/DC circuit for C_o = 940 μF, (V_{inrms} = 220 V, R_o = 300 Ω).

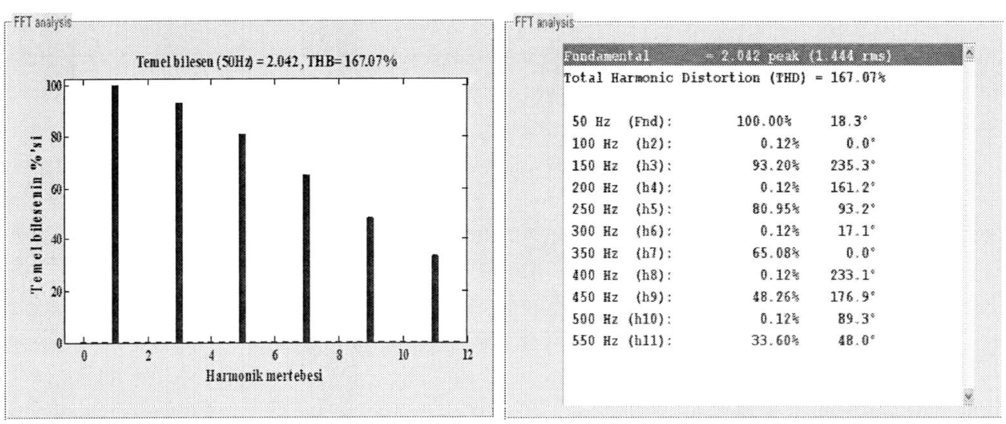

Figure 9. Input current harmonic analysis of the conventional AC/DC circuit for C_o = 220 μF (V_{inrms} = 220 V, R_o = 300 Ω).

Conventional AC rectification is a very inefficient process, resulting in waveform distortion of the current drawn from the grid. This result in poor power quality and high harmonic contents that adversely affect other users fed from the same grid. Another problem is that the power utility line cabling, the installation and the distribution transformer, must all be designed to withstand these peak current values resulting in higher electricity costs for any electricity utility company.

Thus, summarizing, conventional AC rectification has the following main disadvantages:

- It creates harmonics
- It has poor power factor.
- It produces high losses.
- It reduces maximum power capability from the line.

STANDARDS FOR LINE CURRENT HARMONICS

The negative effects of line current distortion have prompted a need for setting limits for the line current harmonics of equipment connected to the electricity distribution network. Standardization activities in this area have been carried out for many years. As early as 1982, the International Electrotechnical Committee - IEC published its standard IEC 555-2, which was also adopted in 1987 as European standard EN 60555-2, by the European Committee for Electrotechnical Standardization - CENELEC. Standard IEC 555-2 has been replaced in 1995 by standard IEC 1000-3-2, also adopted by CENELEC as European standard EN 61000-3-2. Standard IEC 1000-3-2 applies to equipment with a rated current up to 16 A_rms per phase which is connected to 50 Hz or 60 Hz, 220-240 V_rms single-phase, or 380-415 V_rms three-phase mains. Items of electrical equipment are categorized into four classes (A, B, C and D), for which specific limits are set for the harmonic content of the line current. The standard has been revised several times and a second edition was published in 2000 with an amendment in 2001[4]. Any equipment connected to the public utility grid in the European Union is covered by these four classes. Each class has its own set of limits for harmonic currents. Briefly, these different limits are as under:

Class A equipment: Balanced three-phase equipment; household appliances. All equipment that does not fit into the other three classes is categorized as Class A equipment. This class set the absolute maximum current values allowed for each harmonic and these are given in Table 1. Limits do not change with equipment rating.

Class B equipment: All portable tools are categorized as Class B equipment. Harmonic current limits are absolute maximum values. As power tools are used infrequently/for short periods, Class B limits are the least restrictive and set the absolute maximum current allowed for each harmonic to 1.5 times the Class A limits.

Class C equipment: All lighting products, including dimming devices, with an active input power higher than 25 W are categorized as Class C equipment. There are limits on the second harmonics and also on all odd harmonics. The limits are expressed in terms of the fundamental current's percentage. The maximum current percentage allowed for each harmonic, to meet Class C limits, is shown in Table 2.

Class D equipment: All equipment like computers and television sets that are frequently used for longer periods are categorized as Class D equipment. The limits depend on the power rating of the equipment. The current limit for Class D (Table 1) is expressed in terms of mA per Watt of the power consumed.

Table 1. EN 61000-3-2 standards for A and D class equipments [6]

	Class A	Class D	
	Absolute limit (No power limit)	Relative limit (600 W ≥ Power > 75 W)	Absolute limit (600 W ≥ Power > 75 W)
Harmonic order (n)	Maximum permissible harmonic current (A)	Maximum permissible harmonic current per Watt (mA/W)	Maximum permissible harmonic current (A)
Odd Harmonics			
3	2.30	3.4	2.30
5	1.14	1.9	1.14
7	0.77	1	0.77
9	0.40	0.5	0.40
11	0.33	0.35	0.33
13	0.21		
15≤n≤39 (Class A)	Use following equations		
13≤n≤39 (Class D)	2.25/n	3.85/n	2.25/n
Even Harmonics			
2	1.08		
4	0.43		
6	0.30	Not applicable	
8≤n≤40 (Class A)	1.84/n		

Table 2. EN 61000-3-2 standards for C class equipments [6]

Harmonic number (n)	Maximum permissible harmonic current expressed as a percentage of the input current at the fundamental frequency (%)
2	2
3	30 x circuit power factor
5	10
7	7
9	5
8≤n≤40	3

Besides standard IEC 1000-3-2, there are also other documents addressing the control of current harmonics. Standard IEEE 519-1992 (Table 3.) gives recommended practices and requirements for harmonic control in electrical power systems, for both individual consumers and utilities. The limits for line current harmonics are given as a percentage of the maximum demand load current I_L (fundamental frequency component) at the Point of Common Coupling – PCC at the utility (Figure 10). They decrease as the ratio I_{sc}/I_L decreases, where I_{sc} is the maximum short-circuit current at the PCC, meaning that the limits are lower in weaker grids.

Table 3. IEEE 519-1992 standards [7]

I_{sc}/I_L	n<11	11≤n<17	17≤n<23	23≤n<35	35≤n	THD (%)
<20	4.0	2.0	1.5	0.6	0.3	5.0
20<50	7.0	3.5	2.5	1.0	0.5	8.0
50<100	10.0	4.5	4.0	1.5	0.7	12.0
100<1000	12.0	5.5	5.0	2.0	1.0	15.0
1000>	15.0	7.0	6.0	2.5	1.4	20.0

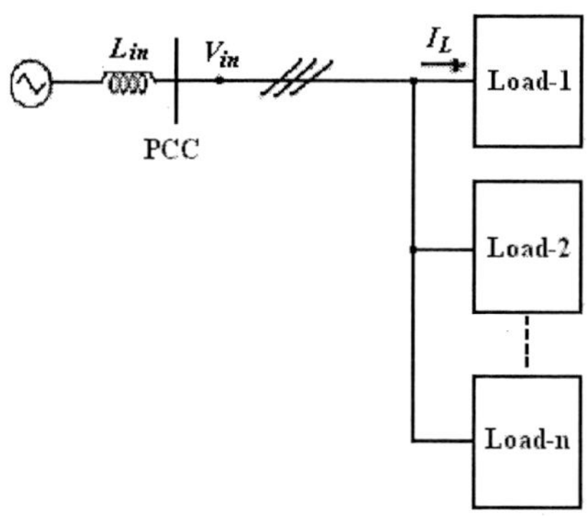

Figure 10. PCC of different loads.

Power Factor Correction

As mentioned before, the diode bridge rectifier (conventional AC/DC converter) has nonsinusoidal line current. This is because most loads require a supply voltage with low ripple, which is obtained by using a correspondingly large capacitance of the output capacitor. Consequently, the conduction intervals of the rectifier diodes are short and the line current consists of narrow pulses with an important harmonic content. Reduction of line current harmonics is needed in order to comply with the standards.

Most of the research on PFC for nonlinear loads is actually related to the reduction of the harmonic content of the line current. There are several solutions to achieve PFC depending on whether active switches (controllable by an external control input) are used or not. PFC solutions can be categorized as passive or active. In passive PFC, only passive elements are used in addition to the diode bridge rectifier, to improve the shape of the line current.

Obviously, the output voltage is not controllable. For active PFC, active switches are used in conjunction with reactive elements in order to increase the effectiveness of the line current shaping and to obtain controllable output voltage. The switching frequency of active PFC is considerably higher than the line frequency.

REVIEW OF PASSIVE PFC METHODS

Passive PFC methods use additional passive components and the diode bridge rectifier. One of the simplest methods is to add an inductor at the AC side of the diode bridge, in series with the AC line as shown in Figure 11-a. The inductor can be also placed at the DC-side, as shown in Figure 11-b. The inductor current is continuous for a large enough inductance *L*. In the theoretical case of near infinite inductance, the inductor current is constant, so the input current of the rectifier has a square shape and the power factor is *pf* = 0.9. However, operation close to this condition would require a very large and impractical inductor. Figure 12 and 13 show the simulated results for the rectifier with AC side inductor and DC side inductor respectively. The simulation studies have been done for different inductor values for comparison. The output voltage is not regulated and the output DC voltage ripple increases with load. For lower inductance *L*, the period of inductor current becomes shorter. Therefore, larger value of inductor should be used to improve the power factor [8-9].

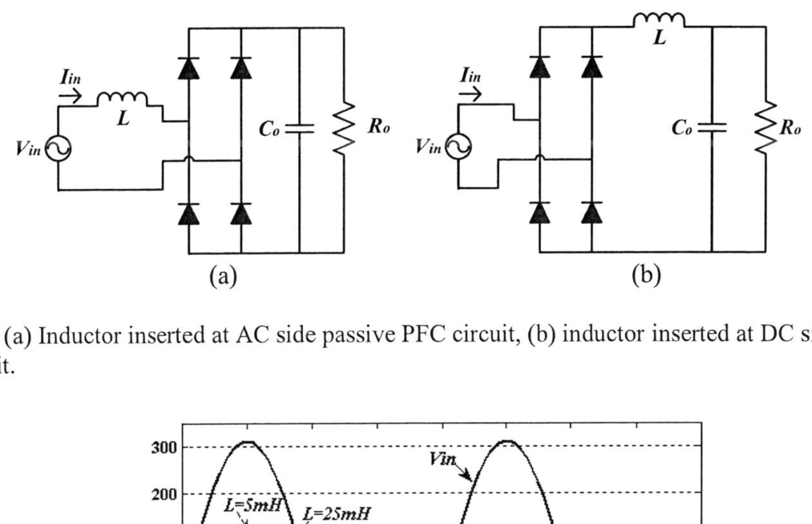

Figure 11. (a) Inductor inserted at AC side passive PFC circuit, (b) inductor inserted at DC side passive PFC circuit.

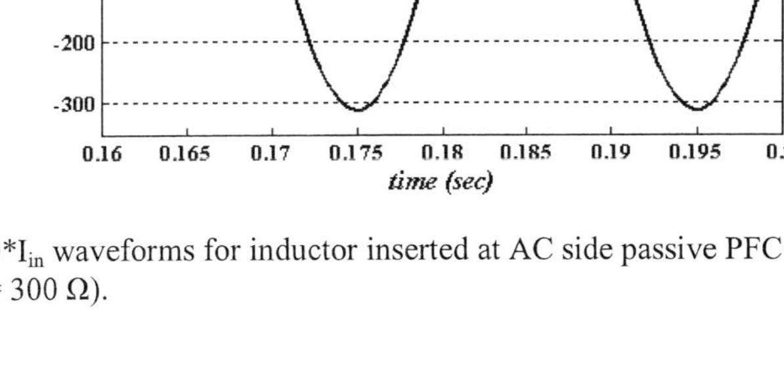

Figure 12. V_{in} and $20*I_{in}$ waveforms for inductor inserted at AC side passive PFC circuit; (V_{inrms} = 220 V, C_o = 220 μF, R_o = 300 Ω).

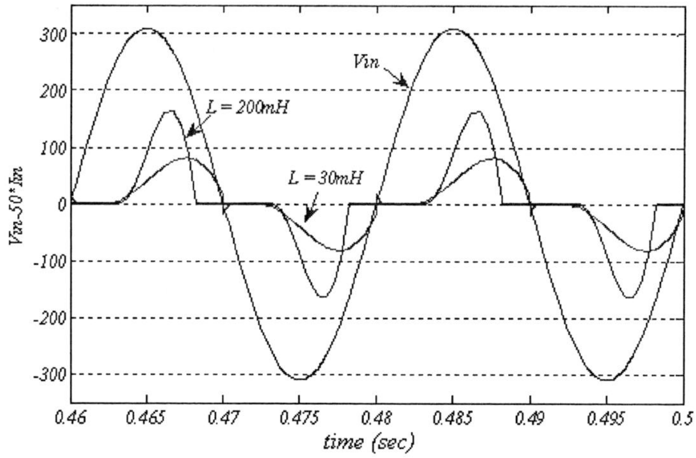

Figure 13. V_{in} and $50*I_{in}$ waveforms for inductor inserted at AC side passive PFC circuit; (V_{inrms} = 220 V, C_o = 220 μF, R_o = 300 Ω).

The solution using a band-stop filter of the parallel-resonant type [4] is presented in Figure 14. The filter is tuned at the third harmonic, hence it allows for lower values of the reactive elements. Except for third harmonic component, the other harmonic components of input current still exist. The simulated waveforms of this type are shown in Figure 15.

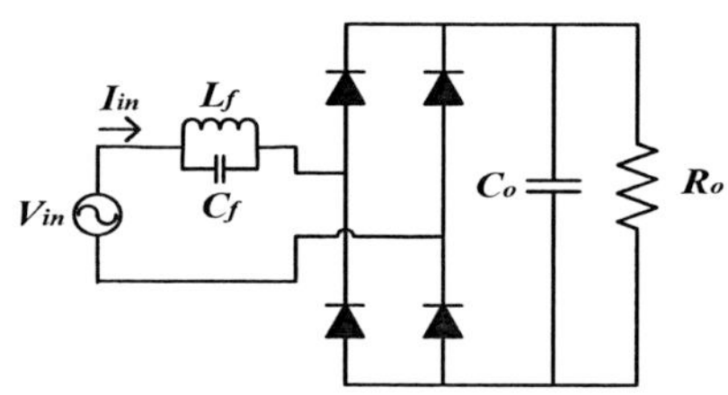

Figure 14. AC side parallel resonant circuit inserted passive PFC converter.

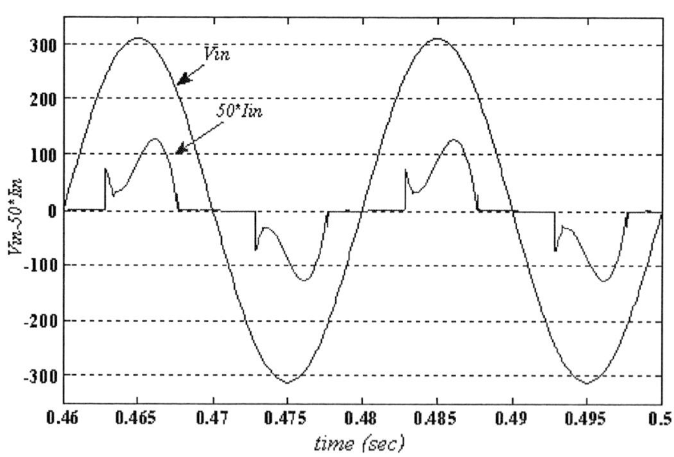

Figure 15. V_{in} and $50*I_{in}$ waveforms for AC side parallel resonant circuit inserted passive PFC converter; (V_{inrms} = 220 V, C_o = 220 μF, C_f = 47 μF, L_f = 25 mH, R_o = 300 Ω).

The AC/DC converter with an additional inductor, capacitor, and diode (LCD) is shown in Figure 16 while the simulated waveforms of this type are shown in Figure 17. The added reactive elements have relatively low values. The circuit changes the shape of the input current and only a limited reduction of the harmonic currents can be obtained [4].

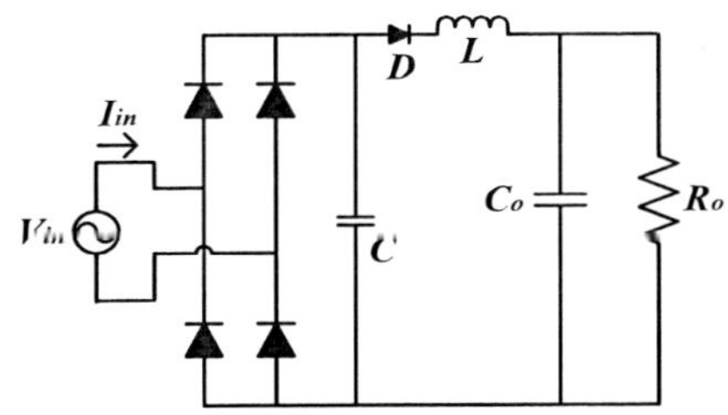

Figure 16. LCD filter inserted at DC side passive PFC circuit.

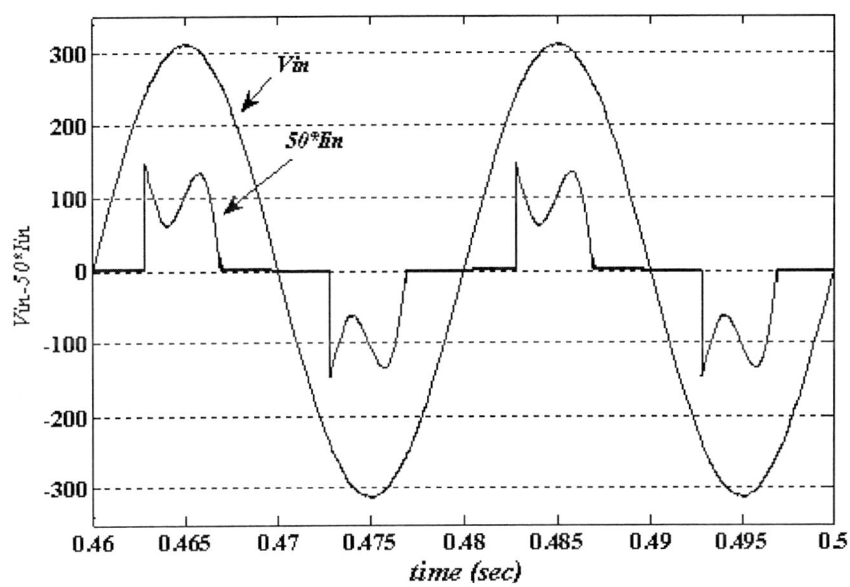

Figure 17. V_{in} and $50*I_{in}$ waveforms for LCD filter inserted at DC side passive PFC circuit; (V_{inrms} = 220 V, C_o = 220 μF, C = 4.7 μF, L =10 mH, R_o = 300 Ω).

Another solution that is often used is a harmonic trap filter. This consists of a series resonant network, connected in parallel to the AC source and tuned to the harmonic frequency that must be attenuated. Usually two harmonic traps are used and these are tuned at the dominant 3rd and 5th harmonics. Figure 18 shows the schematic of this scheme with a constant power load of 300 W. The output voltage is not regulated and the output DC voltage ripple increases with load.

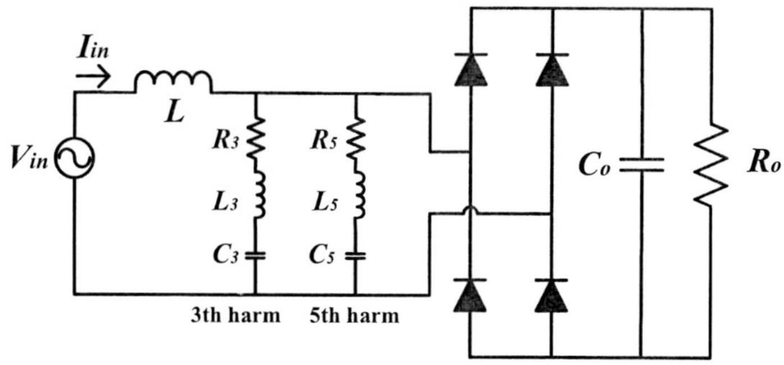

Figure 18. Harmonic trap filter inserted passive PFC circuit.

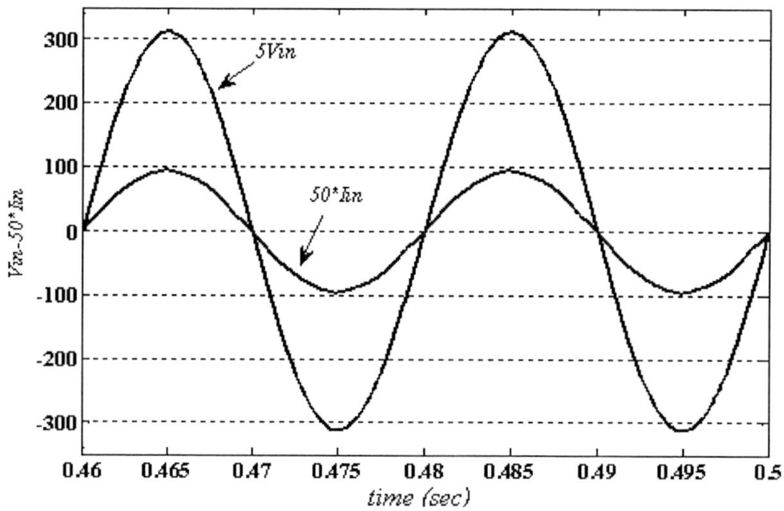

Figure 19. V_{in} and $50*I_{in}$ waveforms for harmonic trap filter inserted passive PFC circuit; (V_{inrms} = 220 V, C_o = 470 μF, L =400 mH, L_3 = 220 mH, L_5 = 100 mH, C_3 = 5.6 μF, C_5 = 4.04 μF, $R_3 = R_5$ = 0.1 Ω, R_o = 300 Ω).

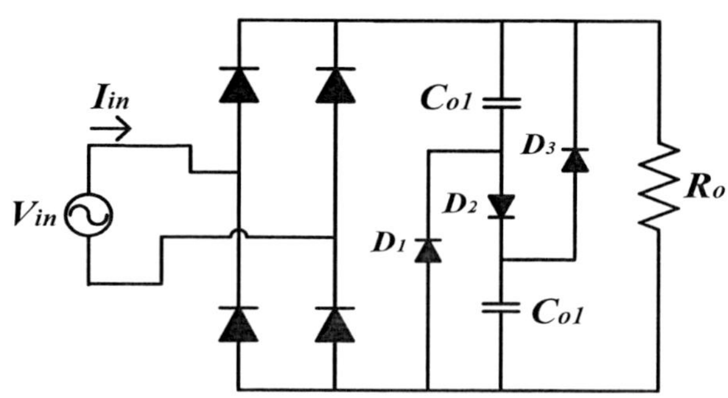

Figure 20. Valley-fill AC/DC passive PFC circuit.

This scheme is requires large inductors and is frequency sensitive. Simulated input current and input voltage waveforms for a harmonic trap filter based PFC scheme are shown in Figure 19. It can be seen that this scheme gives excellent power factor and input current wave shape at the expense of large reactive components [8].

Figure 20 represents the schematic for a 300W output power load valley-fill rectifier. The circuit reduces the harmonic content of the line current but the output voltage varies significantly because of the large output ripple voltage. This solution is often used for lighting ballast circuits [8].

Passive power factor correctors have certain advantages, such as simplicity, reliability and ruggedness, insensitivity to noise and surges, no generation of high-frequency EMI and no high frequency switching losses. However, the bulky size of these filters, their poor dynamic response, complexity and high cost, the lack of voltage regulation and their sensitivity to line frequency, limits their use to applications with lower power. The inductors used in passive correctors are large and bulky, since they work at the mains frequency. A practical passive PFC reduces the harmonic currents and improves the power factor substantially, but it does not completely eliminate the problem. Even though line current harmonics are reduced, the fundamental component may show an excessive phase shift resulting in reduction in power factor. For this purpose, active power factor correction methods have been improved. Better characteristics are obtained with active PFC circuits, which are reviewed in the following subchapter.

ACTIVE PFC AND REVIEW OF ACTIVE PFC CIRCUITS

The high frequency active PFC circuit can be realized by using a diode bridge and a DC/DC converter with a switching frequency much higher than the line frequency. The operation of the DC/DC converter by a suitable control method shapes the input current (Figure 21). For all converter topologies, the output voltage ripple is twice the line-frequency and the output DC is usually regulated.

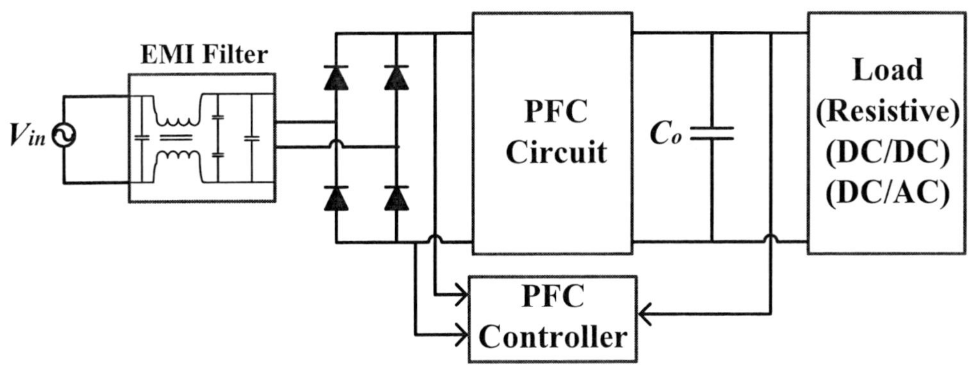

Figure 21. Block diagram of active PFC circuit.

The active switch in the DC/DC converter is controlled by an external control input. In a practical realization, this switch would be implemented by a MOSFET or an IGBT. The second semiconductor which is diode *D*, is indirectly controlled by the state of the active switch and other circuit conditions. The switching cell also contains a storage element, which is the inductor *L*. The basic Buck, Boost and Buck-Boost converters are generally used as DC/DC converter.

The PFC output voltage can be higher or lower, depending on the type of DC/DC converter being used. With a buck converter the output voltage will be lower, for a boost converter the output voltage will be higher, while for a buck-boost converter the output voltage can be higher or lower than the maximum amplitude of the input voltage. The inductor current in these converters can be either continuous or discontinuous. In the continuous conduction mode (CCM) the inductor current never reaches zero during one switching cycle while in the discontinuous conduction mode (DCM), the inductor current is zero during intervals of the switching cycle. However, though the inductor current can be continuous in all the three types of converters, the high frequency switching current components of the AC input current can be continuous only in the case of the boost converter. This is because for the buck and the buck-boost converter, the converter switch interrupts the input current in every switching cycle.

A buck converter based PFC circuit that steps down the input voltage is shown in Figure 22. Figure 23 shows its associated waveforms. Since the buck converter has step-down conversion ratio, it is possible to obtain an output voltage lower than the amplitude of the input voltage. However, since the converter can operate only when the instantaneous input voltage $V_{in}(t)$ is higher than the output voltage V_o, there is no current flow from the AC input during the period t_1 and t_2. Hence, the line current of a power factor corrector based on a buck converter has crossover distortions, as illustrated in Figure 23. Moreover, even if the inductor current is continuous, the input current of the converter is discontinuous. Thus, the input current has significant high frequency components that increase EMI and filtering requirements.

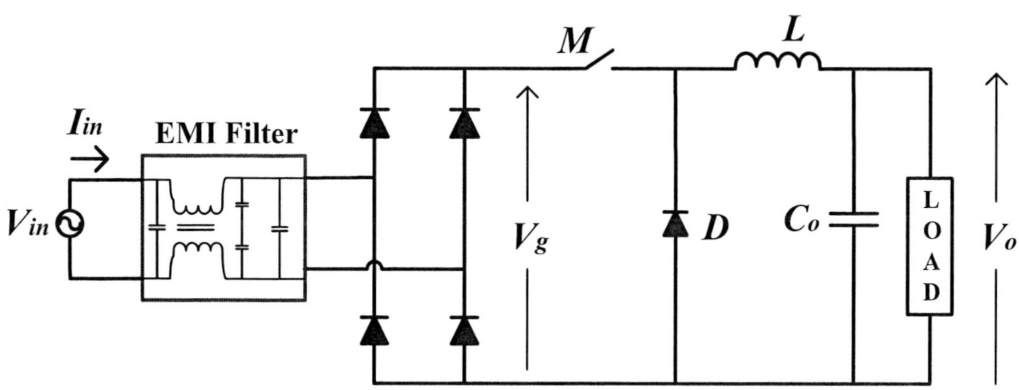

Figure 22. Buck type active PFC circuit.

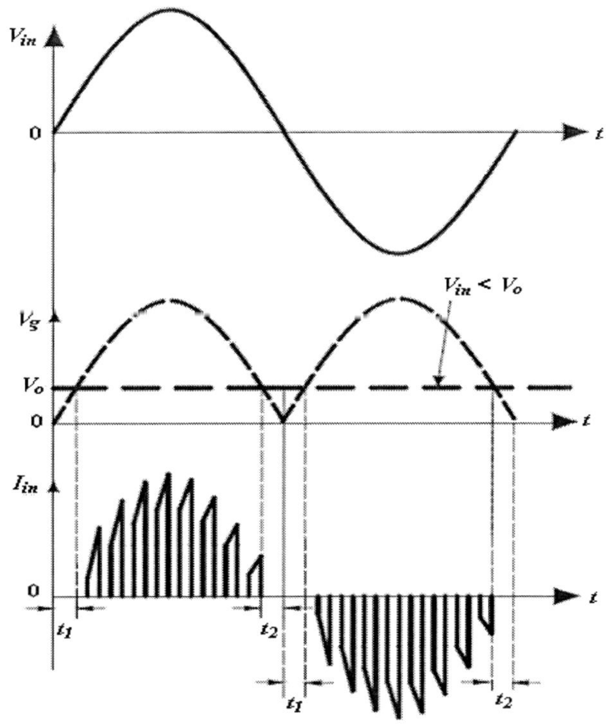

Figure 23. Voltage and current waveforms of buck type active PFC circuit.

The buck-boost converter, shown in Figure 24, can operate either as a step-down or a step-up converter. The output voltage is inverted, which translates into higher voltage stress for the switch. Since the operation of the converter is throughout the line cycle, the input current does not have crossover distortions. This gives the line current envelope no distortion near the input voltage zero crossing. Similar to the buck converter, the input current is discontinuous with significant high frequency content, as illustrated in Figure 25. Thus, the input current has significant high-frequency components that increase EMI and filtering requirements. In addition to this basic converter, the two-switch buck-boost converter [2], [10] is an interesting solution. The positive aspect of this solution is that, due to its non-inverted output voltage, the voltage stress of the switches is lower than in the classical single-switch buck-boost converter. However, this topology has an increased number of switches which leads to higher cost and conduction losses.

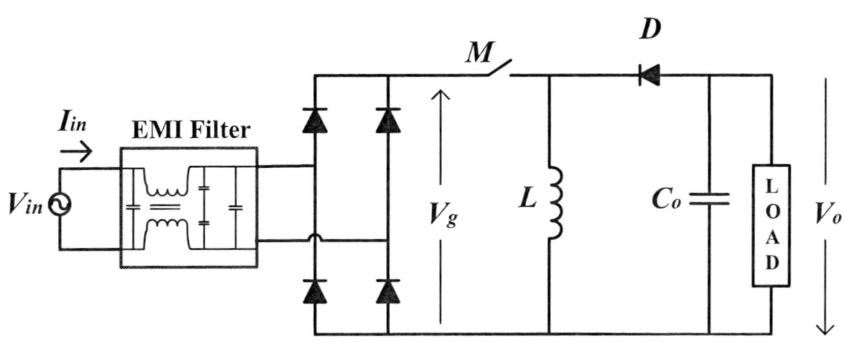

Figure 24. Buck-boost type active PFC circuit.

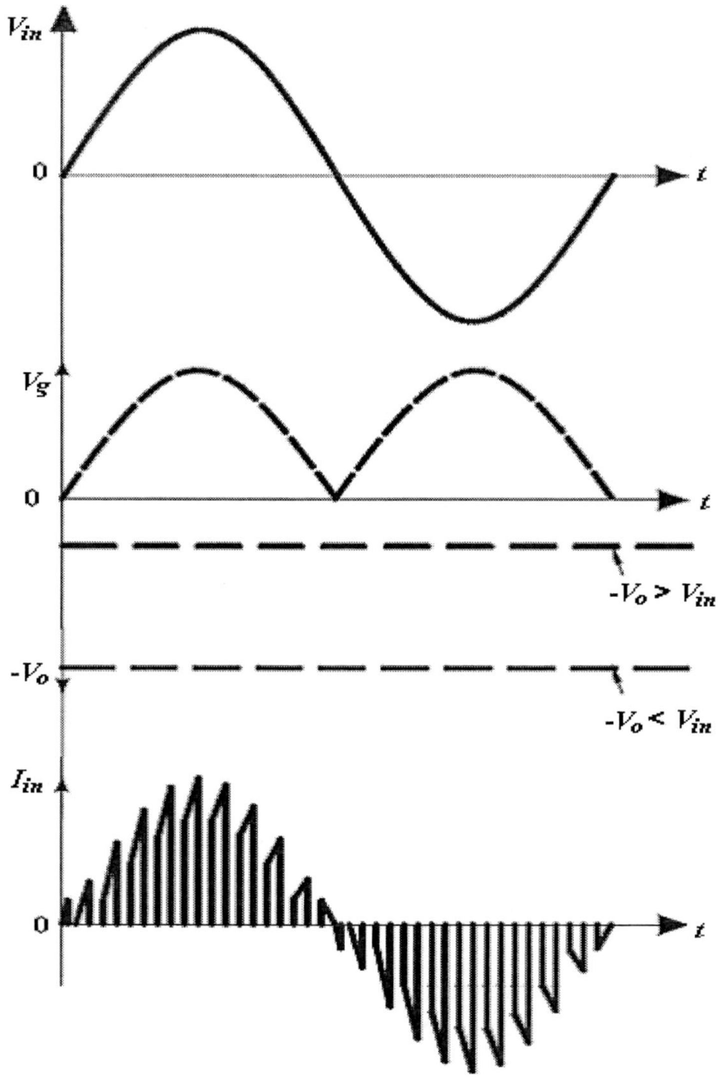

Figure 25. Voltage and current waveforms of buck-boost type active PFC circuit.

The Boost converter is shown in Figure 26. It has a step-up conversion ratio; hence the output voltage is always higher than the amplitude of the input voltage. The boost converter, the most common topology used for power factor correction, can operate in two modes, CCM and DCM. The transition mode control, also referred to as critical conduction mode (CRM) or boundary conduction mode, maintains the converter at the boundary between CCM and DCM by adjusting the switching frequency. Operation is possible throughout the line cycle so the input current does not have crossover distortions as illustrated in Figure 27. The input current can be continuous, because the inductor is placed in series at the input. Thus, the input current has less high-frequency components resulting in lower EMI and reduced filtering requirements when the converter operating in CCM. Therefore, the boost converter operating in CCM is widely used for PFC.

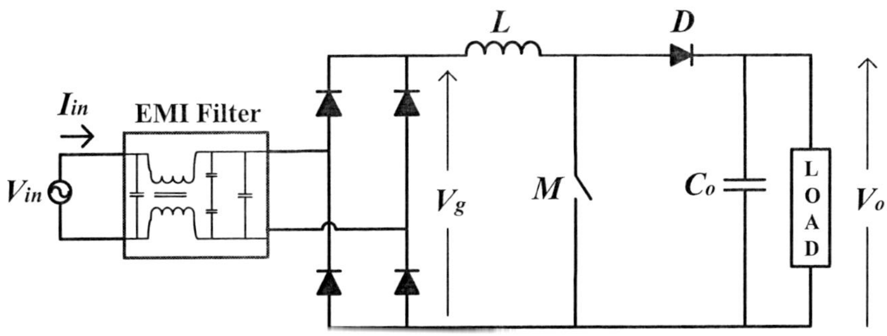

Figure 26. Boost type active PFC circuit.

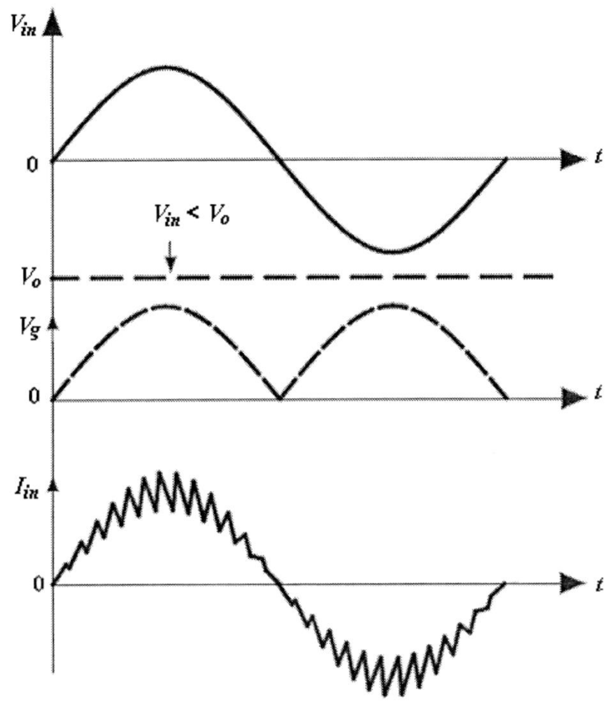

Figure 27. Voltage and current waveforms of boost type active PFC circuit.

Except for the above converters, some basic isolated converters such as flyback, forward, etc. are also used for power factor correction. The advantages of the flyback type topology are that the output voltage can be higher or lower than the input voltage and input-output isolation can be provided by the transformer. The disadvantages of this topology are higher voltage and current ratings of device and high EMI filter requirement. Therefore, it requires careful design of input filter due to discontinuous pulse waveform of the input current which makes it difficult to use current mode control.

The boost type topology is the most popular configuration because; the input current is the inductor current and is therefore easily programmed by current mode control. The boost inductor is in series with the ac power line so that the input current has smooth waveform especially at CCM resulting in much less EMI and reduced input filtering requirements.

Another advantage of this converter is the driving of the switch which has a common ground connection. Due to these advantages, the boost type topology has mostly been proposed in the literature for PFC applications.

Since the conventional boost PFC circuit includes only one switch, this switch has to be capable of carrying the load current and this makes this topology unsuitable at higher power ratings. Therefore, improved boost type PFC circuits (Figure 28 and 29) have been proposed in the literature to meet the requirements of applications and to improve power quality at input AC mains in terms of high power factor and low THD with well-regulated output dc voltage.

The topologies shown in Figure 28 are called unidirectional boost type configurations since the power flow in these configurations is in one direction. The converter shown in Figure 28-a is a well known bridgeless topology and is used in medium power applications. The main advantage of this topology compared to the conventional single-switch boost topology is that the front end diode bridge of the conventional type is eliminated. Therefore, the number of the switches carrying current at any time is reduced. However, using two controllable switches in this type topology increases the cost compared to the conventional type.

The interleaved boost type topology (Figure 28-b) is used to improve the power rating of the conventional boost converter and to reduce the higher frequency ripple of the input current which is an important advantage of this type topology. However, using more controllable switches increases the cost and causes the control circuit to be more complicated compared to the conventional type.

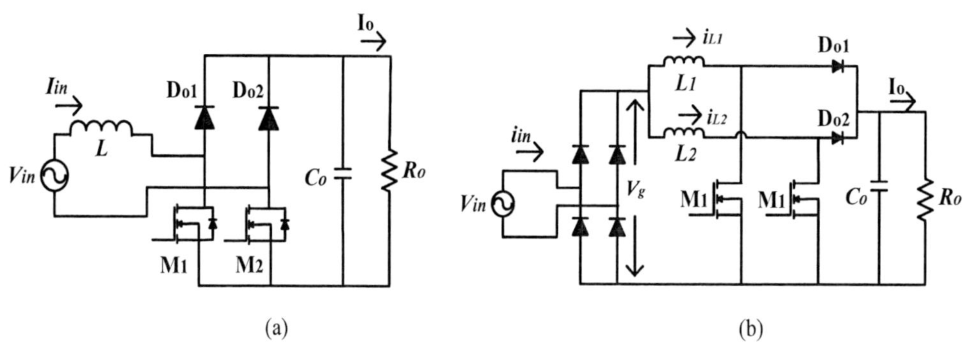

Figure 28. Unidirectional boost type circuits; (a) bridgeless, (b) interleaved boost.

In high power applications, interleaved operation (the parallel connection of switching converters) of two or more boost converters has been proposed to increase the output power and to reduce the current ripple (Figure 29). This technique consists of a phase shifting of the control signals of several cells in parallel operating at the same switching frequency. As a result, the input and output current waveforms exhibit lower ripple amplitude and smaller harmonics content than in synchronous operation modes. The resulting cancellation of low-frequency harmonics allows the reduction of size and losses of the filtering stages [11-13].

Bidirectional boost rectifiers (Figure 30) are widely used in high power factor correction applications. The advantages of these topologies compared to the conventional boost cell topology are the elimination of the diode bridge cell and reducing number of switches at any time.

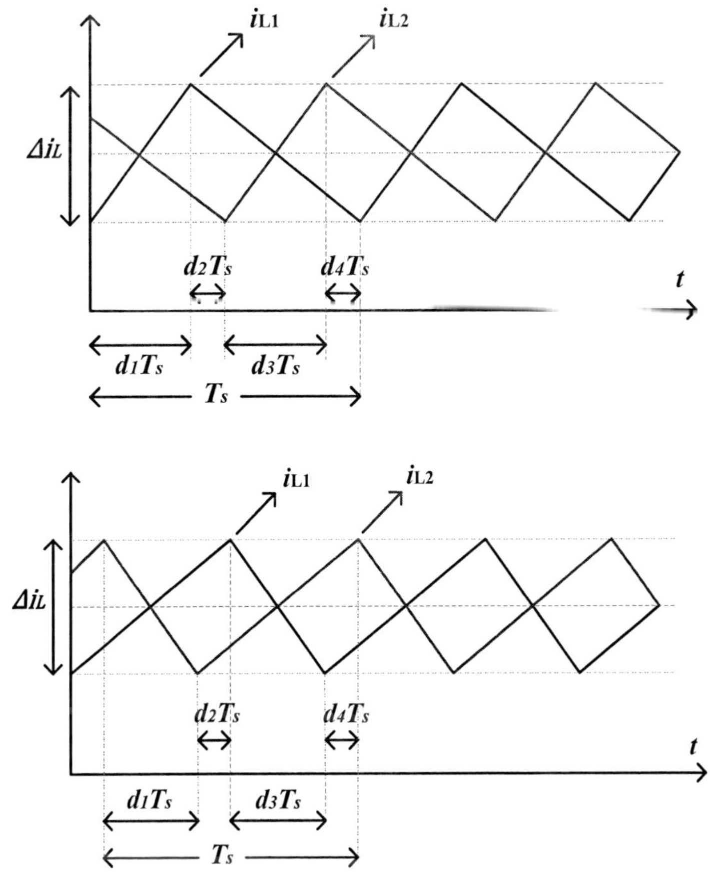

Figure 29. Inductor current waveforms of interleaved boost type circuit (for d<0.5 and d>0.5).

The main disadvantage of the half bridge and full bridge rectifier is the need for a large inductor to reduce the input current ripple and the current stress on the switches. In the half bridge topologies (half bridge and interleaved half bridge) the output voltage V_o is at least twice the peak value of the input voltage. Thus, for a 230 V (rms) input voltage range, the output voltage will be greater than 650 V. For this reason, high-voltage rated semiconductor components and passive elements are needed. Therefore, the half bridge topologies (Figure 30-a and Figure 30-c) are suitable for low ac voltage range (50-110 V rms). Another disadvantage of the half bridge topologies is that the voltage across the capacitors C_{o1} and C_{o2} are not always equal in steady-state. This inequality causes imbalance and affects the performance of the topology. The full bridge and interleaved full bridge topologies (Figure 30-b and Figure 30-d) which use only one capacitor at the output are more suitable for high power applications than the half bridge one.

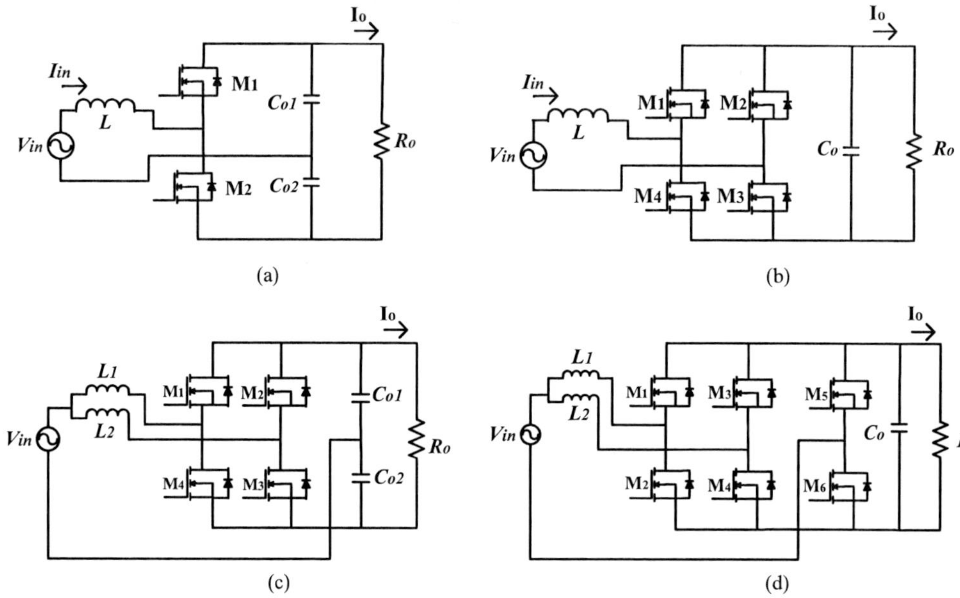

Figure 30. Bidirectional boost type circuits; (a) half-bridge, (b) full-bridge, (c) interleaved half-bridge, (d) interleaved full-bridge.

SOFT SWITCHING METHODS APPLIED TO PFC CIRCUITS

Since the PFC converters are pre regulators, it is desired to obtain a PFC converter with lossless and lower volume. In PFC circuits, higher power density and lower volume can be achieved by increasing switching frequency. Therefore, in PFC circuits, high frequency IGBT (Insulated Gate Bipolar Transistor) and MOSFET (Metal-Oxide Semiconductor Field-Effect Transistor) switches are used. As it is known, MOSFETs are faster than IGBTs. So, using MOSFETs in these active circuits are more suitable. However, higher switching frequency causes increase in switching losses and a serious EMI problem in hard switched PFC converters. The switching losses of PFC converters make a significant amount of power dissipation. Therefore, the switching losses of a converter should be minimized to increase the efficiency and power density by using soft switching techniques. The losses in an active converter caused by the switching are given below.

The typical voltage and current waveforms of an M switch are shown in Figure 31.

From the figure, t_{on} represents ON time and t_{off} represents OFF time of the switch, respectively. The voltage across the switch, the switch current, the average power loss and the maximum power loss are indicated as V_m, i_m, P_{avg} and P_{max} respectively.

The switching losses equations can be defined as follows [2]:

$$W_{on} = \frac{1}{2} \cdot V_{off} \cdot i_{on} \cdot t_{on} \tag{12}$$

$$W_{off} = \frac{1}{2} \cdot V_{off} \cdot i_{on} \cdot t_{off} \tag{13}$$

$$W_{con} = V_{on} \cdot i_{on} \cdot (T_s - t_{on} - t_{off}) \tag{14}$$

where, W_{on} and W_{off} are energy spent in ON and OFF time periods respectively. W_{con} is the energy of the conduction time period. The average power of the switch, P_{avg} is;

$$P_{avg} = \frac{1}{T_s} \int P(t) \cdot dt = (W_{on} + W_{con} + W_{off}) \cdot f_s \tag{15}$$

Since the conduction loss is very low compared to the switching losses, it can be ignored. The switching loss, P_{sw} can be given by equation (16).

$$P_{sw} = \frac{1}{T_s} \left[\int_0^{t_{on}} P(t) \cdot dt + \int_{T_s - t_{off}}^{T_s} P(t) \cdot dt \right] = (W_{on} + W_{off}) \cdot f_s \tag{16}$$

Except for active switches, all active PFC converters include fast power diodes (passive semi conductor components). If a converter is operating with hard switching, these fast power diodes have serious reverse recovery losses.

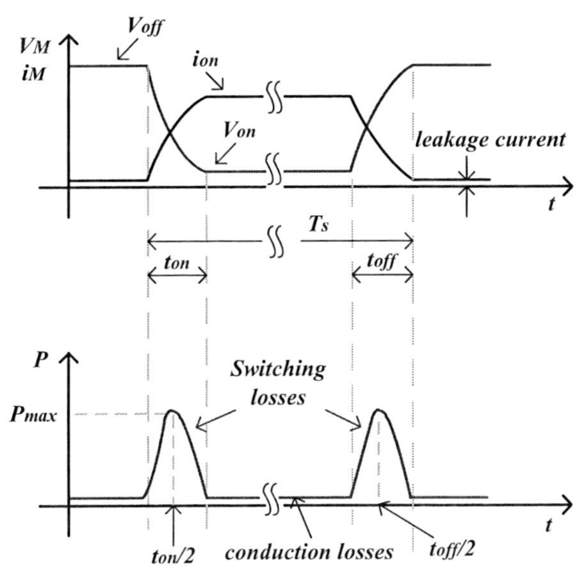

Figure 31. Voltage, current and power of a switch within a switching period.

The reverse recovery loss (Figure 32) decreases efficiency of the converter and causes voltage and current stresses on the active switches. From the figure, anode-cathode voltage is defined as V_{AK} while diode current is defined as i_D. The peak value of the diode reverse current and reverse recovery charge value (shaded area) cause the diode reverse recovery losses. Since the frequency is very low, these losses can be ignored for 50-60 Hz rectifier diodes. The parameters of the reverse recovery losses of the fast power diodes can be given as [14];

$$Q_{rr} = 0{,}5 \cdot I_{RM} \cdot t_{rr} \tag{17}$$

$$I_{RM} = \left(\frac{di_D}{dt}\right) \cdot t_1 \tag{18}$$

$$t_{rr} = \frac{2Q_{rr}}{(di_D / dt)(t_1)} \tag{18}$$

where, Q_{rr}, is the charge value of the shaded area, I_{RM} is the peak value of the reverse current, and t_{rr} is defined as the reverse recovery period and it is the sum of t_1 and t_2. Assuming, $t_1 = t_2 = t_{rr} / 2$ equations (19) and (20) are obtained:

$$t_{rr} = \left(\frac{4Q_{rr}}{(di_D / dt)}\right)^{0,5} \tag{19}$$

$$I_{RM} = \left[Q_{rr}\left(\frac{di_D}{dt}\right)\right]^{0,5} \tag{20}$$

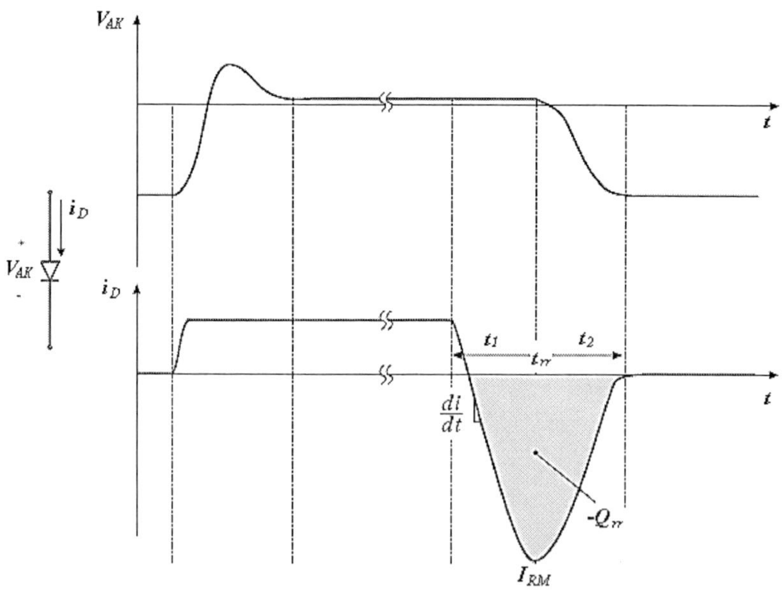

Figure 32. Voltage and current waveforms of a power diode.

The above switching losses analysis is done for an *M* switch which has a typical voltage and current waveforms. However, different types of active switches have different voltage and current waveforms. For example, Figure 33 shows the waveforms of a power MOSFET and Figure 34 shows the waveforms of a power IGBT switch respectively.

The intrinsic capacitors in the power MOSFETs (C_{gs} and C_{gd}) should be taken into account when a MOSFET is switched. As seen from Figure 33, because of these capacitors, the turn on current of a power MOSFET has a big peak value and causes higher ON loss compared to the OFF loss.

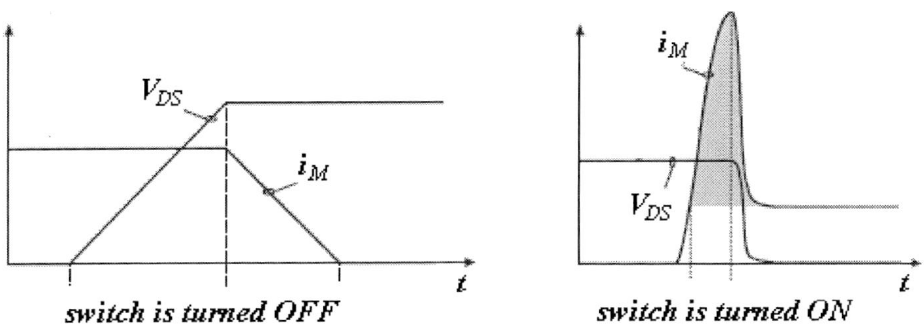

Figure 33. Switching voltage and current waveforms of a power MOSFET.

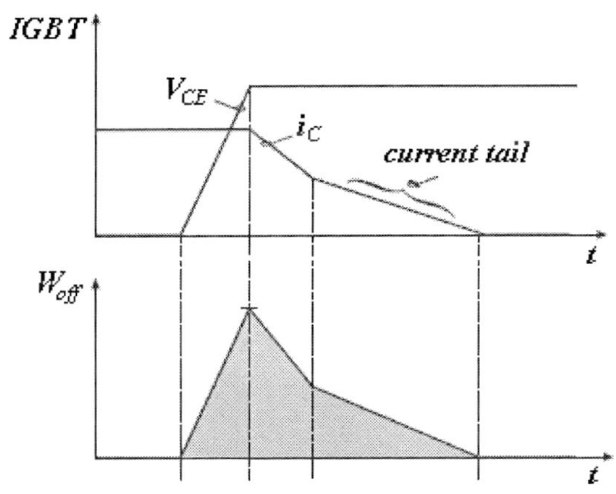

Figure 34. Switching voltage and current waveforms of a power IGBT.

It is seen from Figure 34, a power IGBT has lower ON time loss compared to power MOSFET. However, since a power IGBT has an extra current tail at the turn off, the turn off loss of a power IGBT is very high compared to that of power MOSFET. Power IGBTs do not include intrinsic capacitors. Therefore, peak currents do not occur at turn on when a power IGBT is switched.

Various kinds of soft switching techniques have been proposed in the literature to minimize switching losses of the boost converters. These techniques are implemented by passive or active snubber circuits. Converters operating at soft switching with passive snubbers are attractive, since there is no need for extra active switches and also the control scheme is simpler. The main problem with these kinds of converters is that the voltage stresses on the power switches are too high and the converter is bulky. For example, the studies presented in [15], [16] and [17], use passive methods to achieve zero-voltage-switching (ZVS) for the main switch. These studies have some disadvantages such as; the auxiliary inductor used in the circuit is very big (nearly half of the main inductors) and this results in a bulky circuit, the main switch operates in ZVS but the auxiliary diodes are under hard switching condition and the reverse recovery of the auxiliary diodes causes parasitic oscillations and increases the voltage stresses etc. Figure 35 shows the study presented in [17]. In this study, *L* and *C* prove ZVS for the main switch, but the converter draw bigger value of current compared to the hard switched operation. This cause using more rated MOSFET and therefore increases the cost.

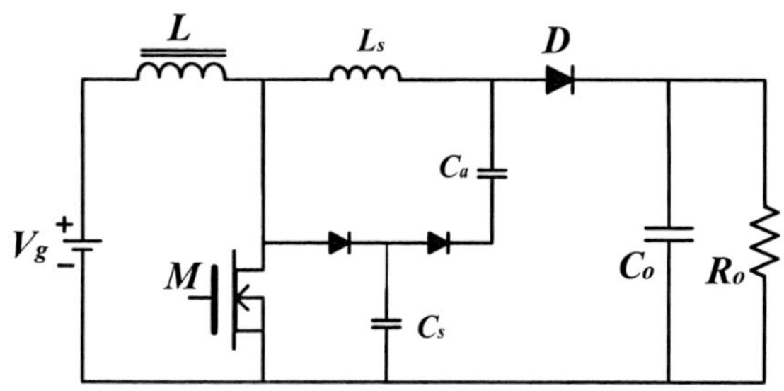

Figure 35. ZVS based boost circuit presented in [17].

Active auxiliary snubbers are also developed to reduce switching losses of boost type converters without having the disadvantages of passive auxiliary circuits. These active snubbers have additional gate circuits for the auxiliary switches to generate their gate pulses and to synchronize them with the main switch.

The most preferred active auxiliary schemes used for boost type converters are given in Figure 36-38 [18-22]. These schemes provide soft switching condition for the main switches without increasing the voltage stress of the main and auxiliary switches. However, some of these converters have some disadvantages such that the auxiliary switch operates under hard switching condition and this increases the EMI noise level and decreases the efficiency of the converter. Also, using more components cause a bulky converter and complicate the control circuit.

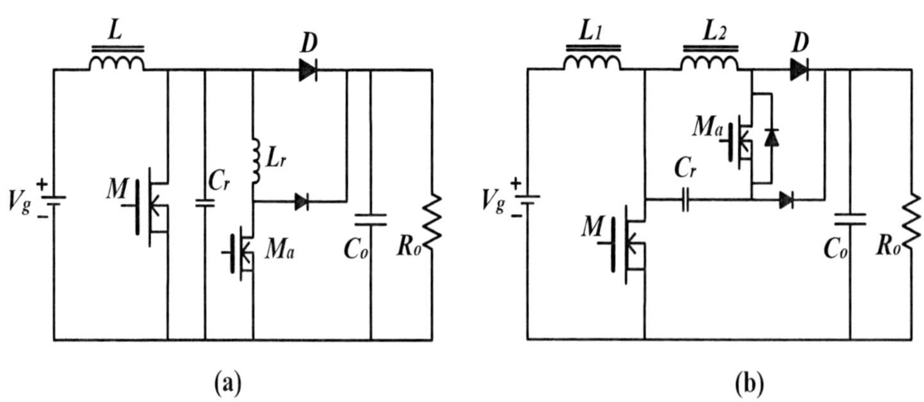

Figure 36. ZVT active auxiliary based boost circuits presented in: (a) [18], (b) [19].

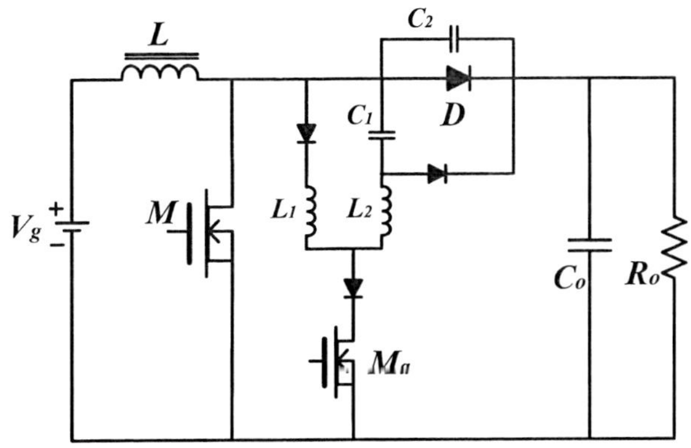

Figure 37. ZVT active auxiliary based boost circuits presented in [20].

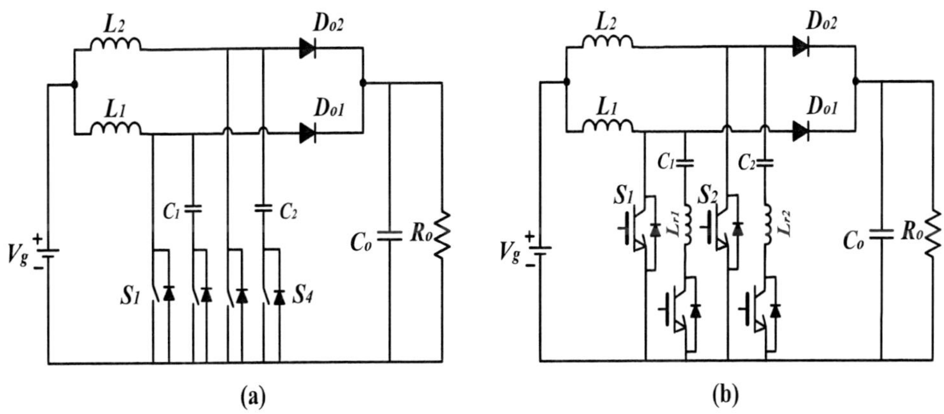

Figure 38. Active auxiliary circuit based interleaved boost converters presented in: (a) [21], (b) [22].

The review of literature shows that the zero-voltage-transition (ZVT) technique provides basically a perfect turn on process for the main switches of a converter by using a quasi resonant active circuit.

No overlap between the voltage and the current of the main switches and so no switching losses take place at turn on process. Unfortunately, the turn off process of a ZVT converter is not perfect. Since the turn off loss of MOSFETs is very low compared to the turn on loss, using MOSFETs in a converter with ZVT auxiliary circuit is more suitable. In addition, the cost and losses of an active auxiliary circuit are important subjects that must be taken into account during designing of the converter.

The auxiliary circuits which are used for soft switching should not increase the size or cost of the converter considerably.

The experimental results of a ZVT boost converter (Figure 39) are given in Figure 40.

It is seen the above results, ZVT condition is achieved for the main switch at turn on. While the main switch is switched with ZVT, the auxiliary switch is operated under soft switching.

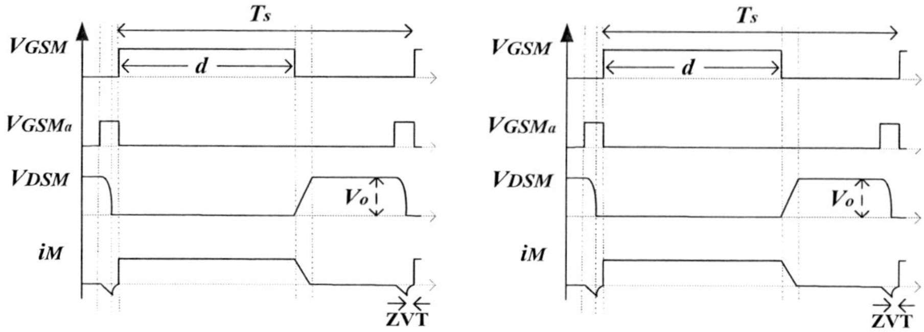

Figure 39. A ZVT active auxiliary based boost circuit and switching waveforms.

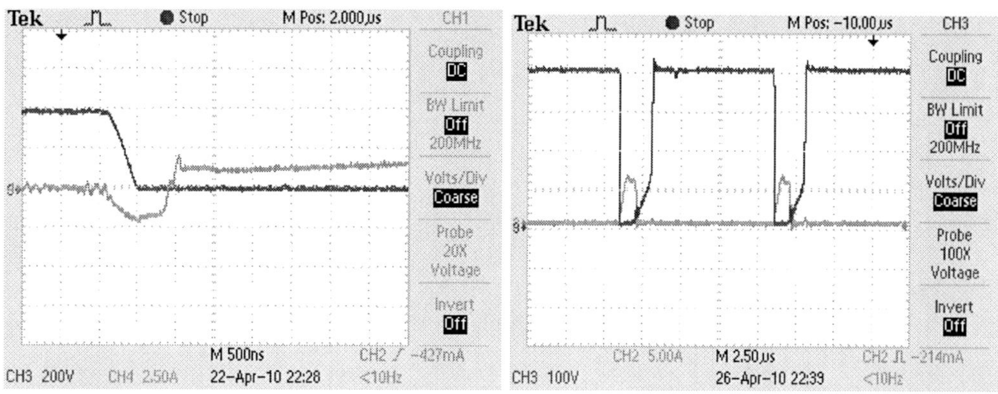

Figure 40. Experimental results of voltage and current waveforms of the ZVT boost circuit; (a) for main switch, (b) for auxiliary switch.

Commonly Used Control Methods For PFC Circuits

The classical control methods of PFC circuits are based on the inductor current shape. Therefore, before introducing the control methods, discussing the inductor current shape will be useful. The control scheme can force the inductor current to be either continuous or discontinuous related to design parameters. The DCM converter operates at fixed frequency and has switching current discontinuities in comparison to the CCM or CRM techniques. Due to the large peak currents and EMI associated with the DCM converter, it is rarely or never used. These large peak currents are due to the dead time needed at certain instantaneous input voltages to remain discontinuous over all input line variations. On the other hand the CRM converter typically uses a variation of hysteretic control with the lower boundary equal to zero current. It is a variable frequency control technique that has an inherently stable input current control while eliminating reverse recovery rectifier losses. The power stage equations and the transfer functions of the CRM converter are the same as the CCM converter. Transition mode forces the inductor current to operate just at the border of CCM and DCM. The current profile is also different and affects the component power loss and filtering requirements. As seen from Figure 41, the peak current in the CRM boost converter is twice the amplitude of the CCM boost converter leading to higher conduction losses. The peak-to-peak ripple is twice the average current, which affects MOSFET switching losses and AC losses in the boost inductor. The main trade-off in using the CRM boost converter is lower losses due to no reverse recovery in the boost diode at the expense of higher inductor ripple and peak currents. The inductor current for a 100 W converter operating in the CCM and CRM mode, are compared in Figure 41 [8]. The peak inductor current is shown as I_{max} while the average inductor current is shown as I_{avg}.

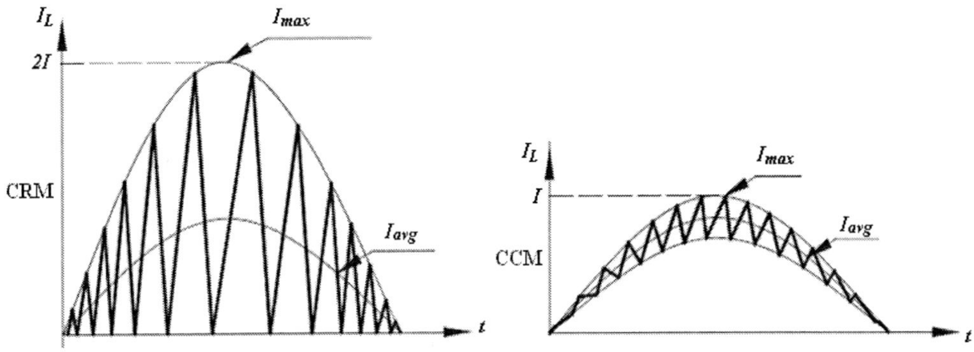

Figure 41. CCM and CRM inductor current waveforms.

Various control strategies can be implemented to PFC circuits. In the following, the most popular control techniques are reviewed and compared in order to highlight advantages and drawbacks of each solution. The multiplier and voltage-follower approaches are two traditional control strategies used in PFC circuits. While voltage follower approach is preferred for DCM converters multiplier approach is preferred for CCM converters. DCM operation is generally used for low power applications but CCM is suitable for high power applications. The multiplier approach requires a multiplier in its current loop circuit and this makes the control circuit to be more complicated. The commonly control techniques in the literature used for PFC are discussed below and shown in Figure 42 [23-31].

Discontinuous Current Mode Control Method

This method can also be called automatic input current shaper. In this approach, the internal current loop is completely eliminated (Figure 42-a) and the converter operates at DCM with fixed switching frequency. The input line current will automatically follow the input voltage. No active control of line current is required. However, the currents with high peak value increase device stresses and conduction losses. In this strategy, the input current is discontinuous with triangular waveform which requires big value of inductor to decrease the current ripple and distortion factor.

Critical Boundary Current Mode Control Method

In this control approach (Figure 42-b), the switch is turned on when the inductor current falls to zero, so that the converter operates at the boundary between CCM and DCM. The drawbacks of this method are the same as those of discontinuous current mode control method. The frequency in this method is also variable which makes it unsuitable for high-power applications.

Hysteretic Current Mode Control Method

Figure 42-c shows this type of control in which two sinusoidal current references i_{max} *and* i_{min} are generated, one for the peak and the other for the valley of the inductor current. According to this control technique, the switch is turned on when the inductor current goes below the lower reference (i_{min}) and is turned off when the inductor current goes above the upper reference (i_{max}), resulting in a variable frequency control which is an important disadvantage.

Peak Current Mode Control Method

By this control method (Figure 42-d), the switch is turned on at constant frequency by a clock signal, and is turned off when the sum of the positive ramp of the inductor current and the external ramp reaches the sinusoidal current reference.

The converter operates in CCM. This means that devices current stress as well as input filter requirements is reduced. The problem of this method is requirement of slope compensation to stabilize the control system.

Average Current Mode Control Method

Another control method, which allows a better input current waveform, is the average current control shown in Figure 42-e. Here the inductor current is sensed and filtered by a current error amplifier whose output drives a PWM modulator. In this way the inner current loop tends to minimize the error between the average input current i_L and its reference. Since the converter works in CCM, it has the same advantages that of the peak current control method.

Therefore, this method is suitable for medium and high power applications. The demerit of this technique is the current control system which is more complicated and difficult to analyze and synthesize.

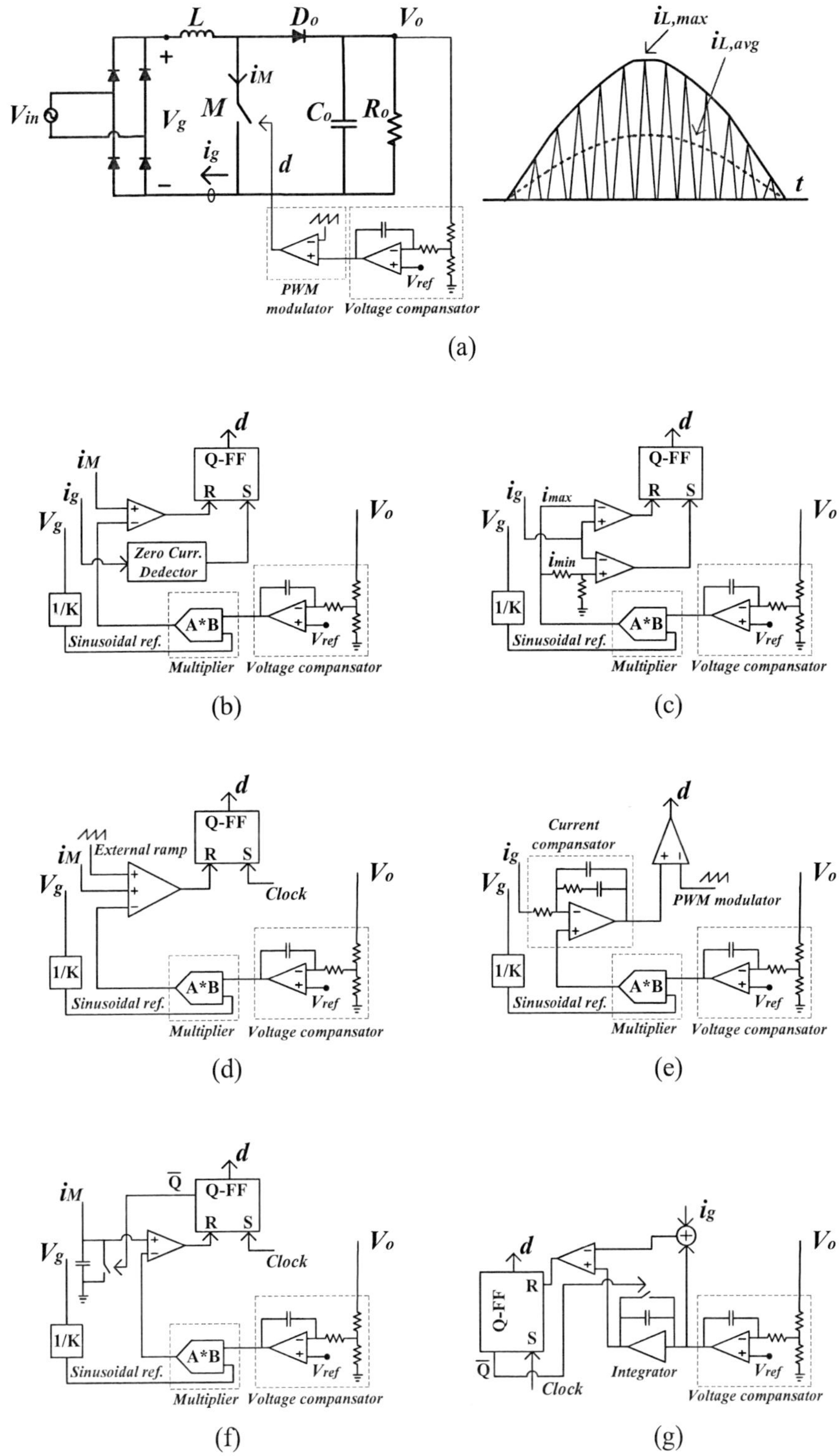

Figure 42. Commonly used control techniques; (a) discontinuous, (b) critical boundary, (c) hysteresis, (d) peak current, (e) average current, (f) charge, (g) one cycle control methods.

Charge Control Method

In this technique (Figure 42-f), the operating frequency is constant. The input current, i_s is sensed and integrated by capacitor C_T. When the voltage across C_T, which is proportional to the charge fed by the input generator in a switching period, reaches the sinusoidal reference, the switch is turned off and capacitor C_T is reset in order to be ready for the next commutation cycle.

In this way, the average input current is forced to follow the sinusoidal reference even if the converter works in CCM. This very simple approach is convenient for low-power flyback applications operating in CCM.

One-Cycle Control (OCC) Method

In this technique (Figure 42-g), the output of the voltage error amplifier is integrated over the switching cycle to produce a ramp voltage. This ramp voltage is compared with the voltage reference generated by the sum of inductor current and the error voltage and the result is then compared with the PWM comparator to determine the duty cycle of the converter power switch.

There are also other control methods introduced in the literature that have been improved and studied such as sliding mode control and adaptive control methods. These methods can be applied to the PFC circuits using analog (analog ICs) or digital (via DSP, FPGA etc.) approach.

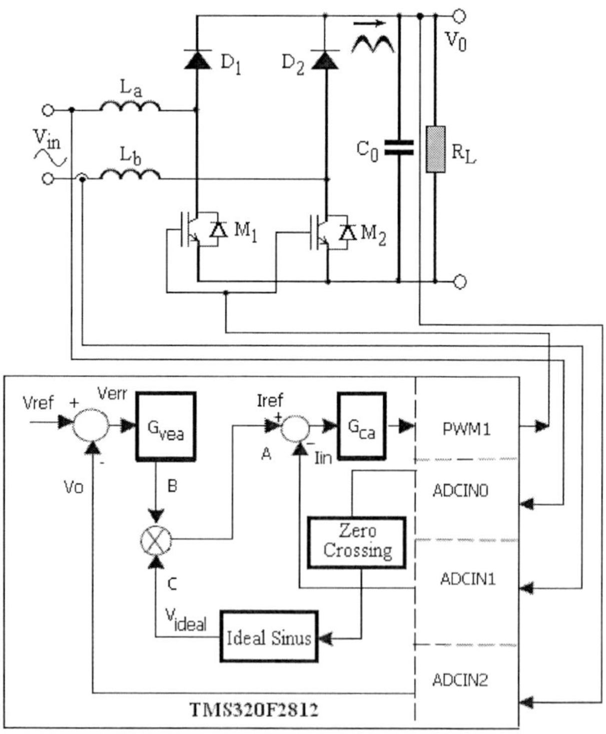

Figure 43. DSP based bridgeless PFC circuit with average current control method.

A DSP based average current mode controlled bridgeless PFC circuit is given in Figure 43 [32]. In this application a pure sinusoidal signal is generated via DSP by detecting the zero cross points of the input voltage. The AC line current is controlled to trace the generated sinusoidal signal. By this way the input current is not affected by the distorted input voltage. This method is preferred to be used in applications where the input AC voltage is not sinusoidal but distorted.

In light of the knowledge given above, it can be said that boost types PFC circuits with CCM inductor current wave shape is the most popular circuit used for medium and high power applications. Also, the average current mode controller has been introduced as the most preferred method in the literature. Therefore, the conventional boost PFC circuit with average current control method (Figure 44) has been simulated to show the related waveforms.

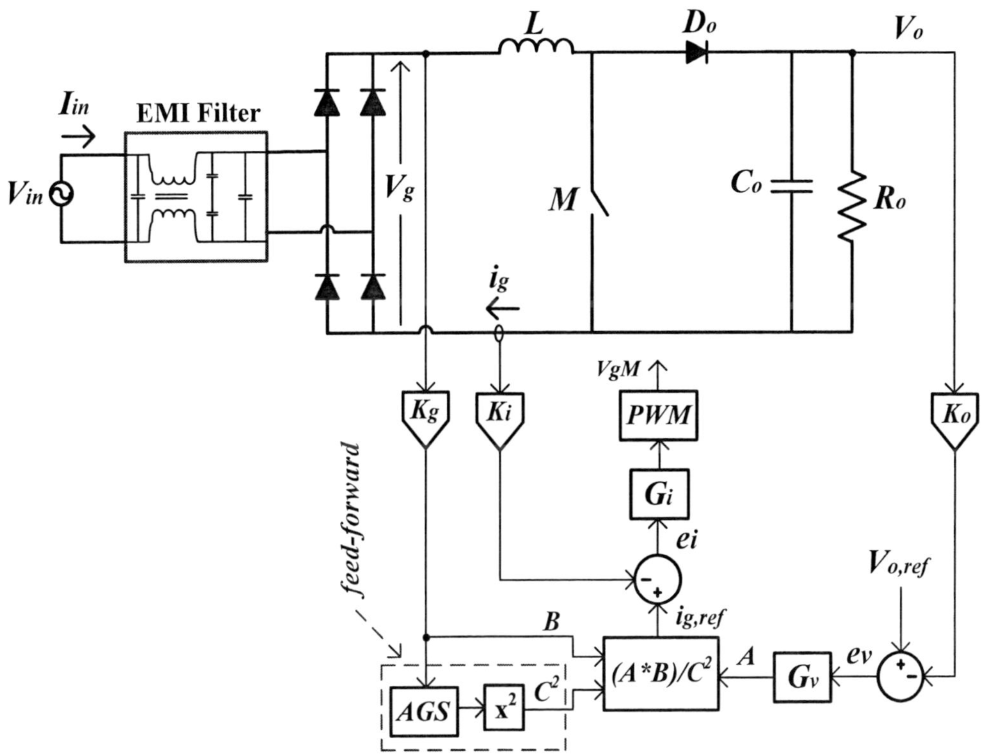

Figure 44. Boost PFC circuit with average current control method.

The simulated waveforms of the conventional boost PFC circuit are given in Figure 45 and 46.

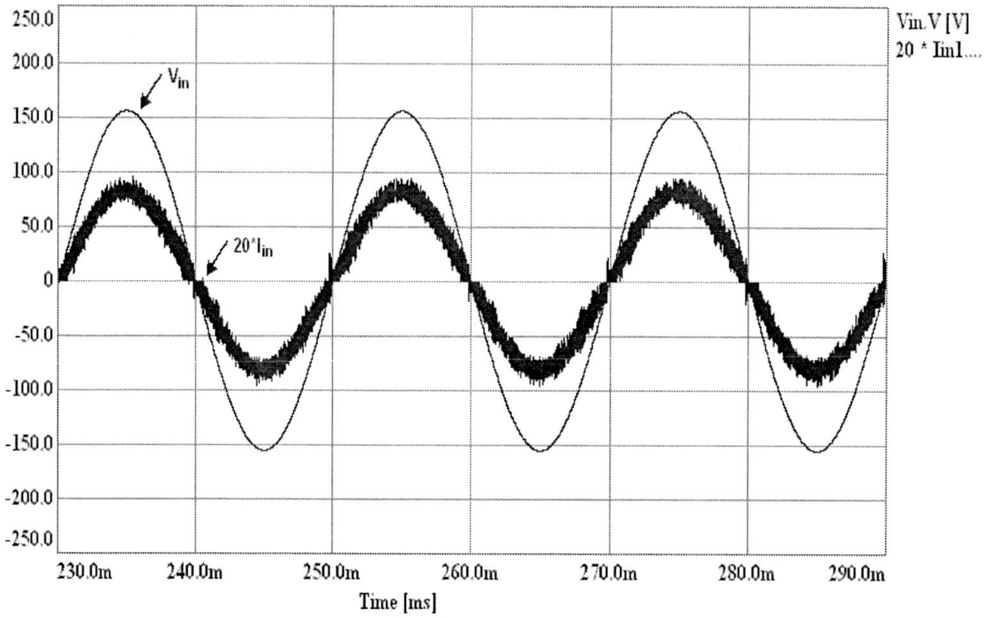

(a) input voltage and 20*input current (Vin-rms=110 V, Iin-rms=2.83 A);

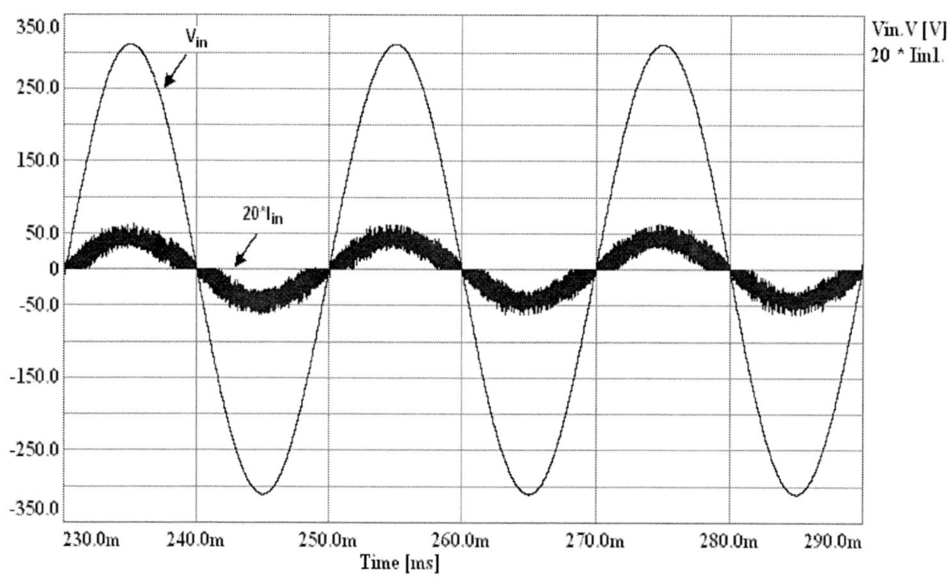

(b) input voltage and 20*input current (Vin-rms=220 V, Iin-rms=1.49 A) waveforms.

Figure 45. Simulation results of average current controlled boost PFC circuit.

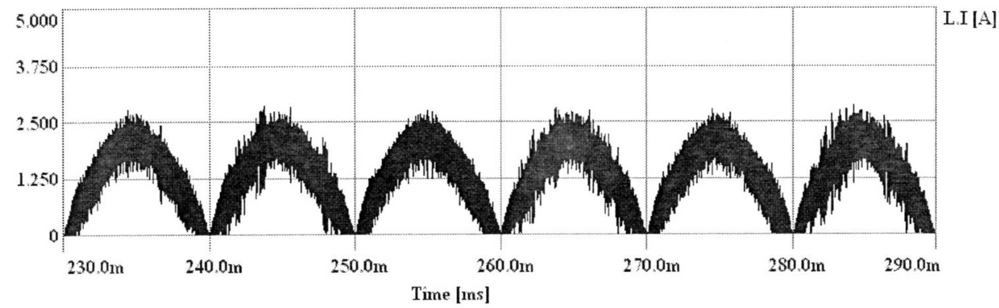

(a) inductor current,

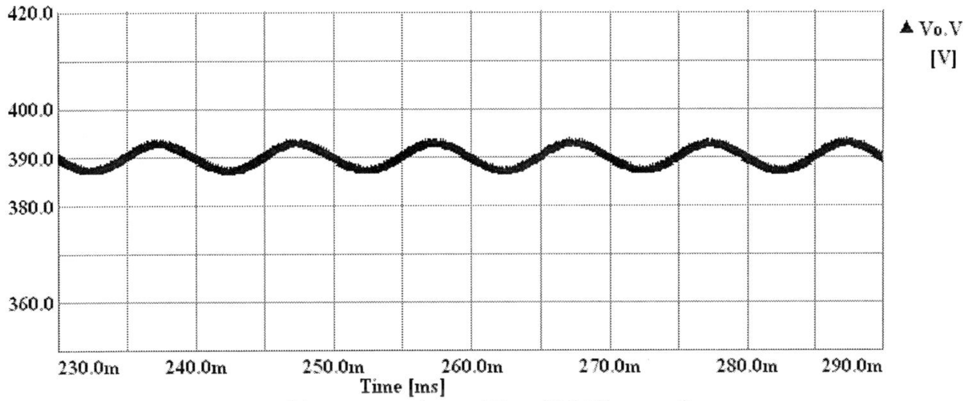

(b) output voltage (V_o = 399 V) waveforms

Figure 46. Simulation results of average current controlled boost PFC circuit.

CONCLUSION

The generally used topologies and control methods for single phase PFC circuits are discussed. The power quality phenomenon and the methods of improving power quality are analyzed.

The topologies are compared and their major advantages and disadvantages are highlighted.

In addition, the switching losses and soft switching methods of the boost type active power factor correction circuits are reviewed.

A single phase boost converter with unity power factor is simulated and the related waveforms are obtained. It is expected that, this work will be useful for students and researchers who are interested in studying and researching power quality and single phase PFC circuits.

REFERENCES

[1] Rashid, M. H. "Power Electronics : Circuits, Devices and Applications", Prentice Hall Inc., NJ, USA, 68-120 (2004).

[2] Ericson, R. W., “Fundamentals of Power Electronics”, Chapman and Hall, New York, 125-140 (1997).

[3] Qiao, C., and Smedley, K. M., ‘‘A Topology Survey of Single-Stage Power Factor Corrector with a Boost Type Input-Current-Shaper”, *IEEE Trans. on Power Electronics*, 16(3), 360-368 (2001).

[4] Grigore, V., “Topologıcal Issues In Sıngle-Phase Power Factor Correctıon”, Ph.D. thesis, Helsinki University of Technology, Espoo, Finland, 17-28 (2001).

[5] Ray, W. F. and Davis, R. M., “The definition and importance of power factor for power electronics converters”, *Proc. European conference on Power Electronics and Applications* (EPE), Europe, 799-805 (1988).

[6] EN 61000-3-2, Electromagnetic compatibility (EMC) – Part 3-2: Limits – Limits for harmonic current emissions (equipment input current ≤ 16 A per phase).

[7] IEEE Std 519-1992, IEEE recommended practices and requirements for harmonic control in electrical power systems, New York, (1992).

[8] Basu, S., “Single Phase Active Power Factor Correction Converters”, Ph.D. thesis, Chalmers University Of Technology, Göteborg, Sweden, 9-37 (2006).

[9] Kelley, A. W. and Yadusky, W. F., “Rectifier design for minimum line-current harmonics and maximum power factor”, *IEEE Trans. on Power Electronics*, 7(2): 332-341 (1992).

[10] Ghanem, M. C., Al-Haddad, K. and Roy, G., “A new single phase buck-boost converter with unity power factor”, IEEE IAS’93, Europe, 785-792 (1993).

[11] Elmore, M. S., ‘‘Input current ripple cancellation in synchronized, parallel connected critically continuous boost converters’’, *IEEE APEC’96 Cnf., San Jose*, CA, 152–158 (1996).

[12] Lee, P. W., Lee, Y. S., Cheng, D. K. W., and Liu, X. C., ‘‘Steady-state analysis of an interleaved boost converter with coupled inductors’’, *IEEE Trans. on Ind. Electronics*, 47(4): 787–795 (2000).

[13] Braga, H. A. C. and Barbi, I., ‘‘A 3-kW unity-power-factor rectifier based on a two-cell boost converter using a new parallel-connection technique’’, *IEEE Trans. on Power Electronics*, 14(1): 209–217 (1999).

[14] Gürdal, O., “Güç Elektroniği”, Nobel Yayın Dağıtım, Ankara, 50-70 (2000).

[15] Hsieh, Y. C., Hsueh, T. C. and Yen, H. C, ‘‘An interleaved boost converter with zero-voltage transition’’, *IEEE Transaction on Power Electronics*, 24(4): 973–978(2009).

[16] Streit, R. and Tollik, D., ‘‘High efficiency telecom rectifier using a novel soft-switched boost based input current shaper’’, IEEE INTELEC Cnf. Proc., Kyoto, 720-726 (1991).

[17] Liu, K. H., Lee, F. C., ‘‘Zero-voltage switching technique in dc-dc converters’’, *IEEE PESC Conf. Proc.*, 58-70 (1986).

[18] Hua, G., Leu, C., Lee, F. C., ‘‘Novel zero voltage transition PWM converters’’, *IEEE PESC Conf. Proc.*, 55-61 (1992).

[19] Bassett, J. A., ‘‘New zero voltage switching high frequency boost converter topology for power factor correction’’, *IEEE INTELEC Conf. Proc.*, Hague, 813-820 (1995).

[20] Wang, C. M., ‘‘Novel zero-voltage-transition PWM DC-DC converters’’, *IEEE Transactions on Ind. Electronics,* 53(1): 254-262 (2006).

[21] Yao, G., Chen, A. and He, X., ‘‘Soft switching circuit for interleaved boost converters’’, *IEEE Transaction on Power Electronics*, 22(1): 80–86 (2007).

[22] Oliveira Stein, C. M., Pinheiro, J. R., and Hey, H. L., ''A ZCT auxiliary commutation circuit for interleaved boost converters operating in critical conduction mode'', *IEEE Transaction on Power Electronics*, 17(6): 954–962 (2002).

[23] L. Rossetto, G. Spiazzi, P. Tenti, "Control Techniques For Power Factor Correction Converters", Proc. of Power Electronics, Motion Control (PEMC), September 1994.

[24] K. M. Smedley, S. Cuk, "One-Cycle Control of Switching Converters", IEE*E Trans. on Power Electronics*, Vol. 10, No. 6, November 1995.

[25] S. Buso, G. Spiazzi, D. Tagliavia, "Simplified Control Technique for High-Power-Factor Flyback Cuk and Sepic Rectifiers Operating in CCM", *IEEE Trans. on Ind. Applications*, Vol. 36, No. 5, September/October 2000.

[26] D. Maksimovic, Y. Jang, R. W. Erickson, "Nonlinear-Carrier Control forHigh-Power-Factor Boost Rectifiers", *IEEE Trans. on Power Electronics*, Vol. 1 I , No. 4, July 1996.

[27] M. Fu and Q. Chen, "A DSP based controller for power factor correction (PFC) in a rectifier circuit," *IEEE Applied Power Electronics Conf.*, pp. 144 –149, 2001

[28] Z. Lai, K. M. Smedley, Y. Ma, "Time Quantity One-Cycle Control for Power-Factor Correctors", *IEEE Trans. on Power Electronics*, Vol. 12, No. 2, March 1997.

[29] C. Silva, "Power factor correction with the UC 3854," Application Notes, Untroide Integrated Circuits.

[30] Genc, N., Iskender, I., ''Analysis of critical mode interleaved boost PFC converter'', ECAI 2009, 3rd International Conference on Electronics, Computers and Artificial Intelligence, Piteşti, 55-60 (2009).

[31] Genc, N., Iskender, I., ''An improved zero-voltage-transition interleaved boost converter with high power factor'', ELECO 2009, 6th International Conference on Electrical and Electronics Engineering, Bursa, 432-436 (2009).

[32] Karaarslan, A., Iskender, I., ''Pure sinusoidal input voltage based bridgeless PFC converter using TMS320F2812 digital signal processor'', ELECO 2009, 6th International Conference on Electrical and Electronics Engineering, Bursa, 234-238 (2009).

In: Advances in Energy Research
Editor: Nicu Bizon
ISBN 978-1-61209-991-0

Chapter 14

SWARM OPTIMIZATION BASED AUTOMATIC GENERATION CONTROL IN A HYDROTHERMAL POWER SYSTEM INCLUDING SMES UNITS

M. Cengiz Taplamacioglu* and Haluk Gozde
Gazi University, Electrical and Electronics Engineering Department,
Ankara, Turkey

ABSTRACT

The Automatic Generation Control (AGC) mechanism is one of the main control requirements for power generation in an interconnected power system. The most electric power consumers such as AC motors and electrically operated clocks depend on a certain system frequency. Actually, a malfunction of these equipments negatively affects our lives. Hence, AGC is carried out to the power system in order to maintain the constant and the stable system frequency. In this chapter, the main principle of AGC is investigated and its application based on swarm intelligence optimization for a hydrothermal power system is presented. After the explanation of AGC, the gain scheduling Proportional-plus-Integral (PI) control technique and the scheduling algorithm based on swarm intelligence, which is used in this application are introduced. The control technique with tunable weighted cost function is carried out to the power system which includes two control areas having a steam turbine and a hydro turbine connected together through tie lines. Both areas also include Superconducting Magnetic Energy Storage (SMES) units in order to improve the performance of the control. To optimize the gains of PI-controller and the parameters of the proposed cost function, Constriction Factor Approach Particle Swarm Optimization (CFA-PSO) algorithm is used. The usage of this algorithm to optimize the gains of the PI-controllers and the tunable weight coefficients of the cost function was detailed explained with the parts of its program in order to help the researchers which study on optimal AGC for multi area power system. The result of the control was also investigated and discussed. At the end of the simulations, it is seen that the proposed AGC technique improves the dynamic performances of the power system.

* Corresponding Author: taplam@gazi.edu.tr

Keywords: Automatic generation control, Hydrothermal power system, SMES unit, Gain scheduling PI-controller, Particle Swarm optimization.

INTRODUCTION

An electrical energy is one of the world's most commonly used types of energy. It has many desirable features such as clean, convenient, relatively easy to transfer from point of source to point of use, and highly flexible in its use [1]. An electrical energy is supplied to the consumers through huge electric power systems. The aim of such kind of systems is to generate, transport and distribute electric energy with nominal system frequency and terminal voltage. For these purpose, an electric power system includes generation, transmission and distribution units. While the transmission units are responsible for maintaining all transmission equipment to facilitate the transport of electricity, the distribution units are responsible for providing the physical wire system to connect the end users to the transmission system [1]. However, the main task of the generation units is to maintain constant and stable system frequency and terminal voltage. In the huge power systems, one or more coherent generation units may be connected to each other so as to maintain more reliable and stable system frequency via common frequency reference. These connections between neighbor generation units also provide more effective load sharing depending on agreements. These systems called "interconnected power system", and some advantages of interconnection are the more economic usage of large generators, emergency assistance to neighboring areas, reduced spinning reserves, lower overall generation costs [2]. In an interconnected power system, some control mechanisms are carried out to the generation units in order to maintain desired values of frequency and terminal voltage. While the control loop utilized to provide desired terminal voltage is called the Automatic Voltage Regulation (AVR), the control loops utilized to provide nominal system frequency are called the Automatic Generation Control (AGC) loop. The AGC loop is more complex and slower control mechanism than the AVR loop. Actually, the AGC mechanism includes two main control loops called primary and secondary control loops. The primary control loop provides the response to a change in frequency as fast as possible. The secondary control loop behaves as an integral controller and works in a slow reset mode to eliminate the small frequency errors which remain after the primary control [3].

The previous works in secondary control loop of AGC was initiated by Cohn [4]. However, a modern optimal control concept for AGC designs of interconnected systems is put forward by Elgerd and Fosha [5].

They suggested a proportional controller and a different feedback form to develop optimal controllers for secondary control loop. Until the present day, lots of different control strategies such as conventional, adaptive, variable structure, robust and based on artificial intelligence have been reported in literature [6]. But, a gain scheduling control can be distinguished from the other control techniques because it makes the process under control less sensitive to changes in process parameters and in particular, it is also simpler to implement than the other modern control techniques [8]. For these reasons, in literature it has been frequently applied to the AGC system.

Figure 1. Electric power generation, transmission and distribution.

When the related literature is investigated, it is seen that one of the first gain scheduling control methods for AGC of interconnected power system was proposed by Lee and coworkers in 1991 [9]. Their controller provided better control performance for a wide range of operating conditions than the performances obtained so far. Later on, Rubaai and Udo presented a multi-variable gain scheduling controller by defining a cost function with a term representing the constraints on the control effort and then minimizing that with respect to the control vector [10]. Since the conventional gain scheduling methods may be unsuitable for some operating conditions because of complexity of the power systems such as nonlinear load characteristics and variable operating points, a usage of artificial intelligence based methods were preferred by researchers from the beginning of these dates. In 1997, Chang reported the usage of fuzzy logic based gain scheduling method for power system [11]. As different from the usage of two fuzzy rules for integral and proportional gains in PI controller by Chang, Çam and Kocaarslan improved the performance of this approach in 2005 by which the rules for the gains are chosen to be identical [12]. Talaq and Al-Basri suggested an adaptive fuzzy gain scheduling method for conventional PI controller in 1999 [13]. Then, Pingkang optimized the gains of PI and PID controllers through real coded genetic algorithm in a two area power system in 2002 [14]. After one year from Pinkang's study, Abdel-Magid and Abido proposed a usage of PSO for the same purpose [15]. In 2004, Yeşil suggested the self tuning fuzzy PID type controller for AGC [16]. At the same year, Juang put forward the genetic algorithm based fuzzy gain scheduling method for PI controller [17]. In this study, to reduce both the fuzzy system design effort and the number of fuzzy rules, the fuzzy system that is used for gain scheduling is designed automatically by genetic algorithms. In 2006, Massiala used two layered fuzzy gain scheduling controller for improving the dynamic performance of AGC in a two area reheat thermal power system with generation rate constraints [18]. Taher composed a hybrid PSO algorithm for the gain scheduling of PI controller in a two area thermal power system in 2008 [19]. The hybrid PSO algorithm contains evolutionary operators like selection, crossover and mutation as in genetic algorithms or DE algorithm. In the year 2009, like as working by Abdel-Magic and Abido,

Gözde proposed the usage of CRAZYPSO so as to optimize AGC for a two area thermal power system with governor dead-band [20].

In the same year, Nanda suggested a maiden application of bacterial foraging optimization based gain scheduling PI controller in a multi area AGC [21]. A bacterial foraging optimization is the recent and powerful computational intelligence optimization technique that is proposed by Passino in 2002 [22].

With this algorithm, he tuned not only gains of the controller but also speed regulation parameters and frequency bias parameters, and then he compared the results with genetic algorithm and classical methods.

In 2010, Gözde suggested the cost function with tuneable weight coefficients for AGC in a two area thermal power system, and optimized the parameters of the controllers and the cost function by PSO algorithm [23].

On the other hand, an instantaneous imbalance between produced and consumed real powers may be reduced by the addition of fast acting SMES unit to the power system. Because of the frequency fluctuations caused by power imbalance is usually lie in the frequency range of 0.2 to 1 Hz., the addition of a small capacity SMES unit to the power system may decrease the frequency deviation and the tie-line power change against to these small disturbances. At the related literature, Tripathy and coworkers initiated to investigate an effect of SMES unit on the AGC for the power system including governor dead-band and boiler dynamics in 1992 [24].

In 1998, Mitani improved the dynamic performances of the power system by using SMES unit [25]. After three years, Tripathy carried out the SMES unit to the three area interconnected thermal power system [26]. Then, Demirören used a fuzzy logic controller for AGC in the power system including SMES unit in 2004 [27]. Later on, Abraham investigated the effect of generation rate constraints in a two area hydrothermal power system with SMES units in 2007 [28]. One year after, Shayeghi applied a robust mixed H_2/H_∞ based AGC to a similar power system [29].

In the same year, Pothiya used a fuzzy logic PID controller for load frequency control of the power system including SMES units, and the tuned the gains of PID controller by using Tabu Search algorithm [30]. In 2009, Gözde suggested the Constriction Factor Approach Particle Swarm Optimization (CFA-PSO) in order to optimize the parameters of the cost function and the PI-controller for AGC in a two area hydrothermal power system including SMES units [31].

In this chapter, the application of gain scheduling Automatic Generation Control based on artificial intelligence was studied for a hydrothermal power system including SMES units. Before the investigation of the simulation study, the basic theories of the AGC and the gain scheduling PI-controller based on artificial intelligence optimization were explained. The CFA-PSO was used as an artificial intelligence optimization method in order to schedule the gains of PI-controllers and the parameters of the tunable weighted cost function. The usage of this algorithm was detailed explained with the parts of its program in order to help the researchers which study on optimal AGC for multi area power system. As an output of this study, the result of the control and the effects of the SMES unit were investigated and discussed.

AUTOMATIC GENERATION CONTROL (AGC)

Basic Concept

An interconnected electric power system generates transports and distributes electric energy. The aim of such kind of systems is to supply electric energy with nominal system frequency and terminal voltage, values and tolerances of those are defined by some power quality standards. Since the frequency depended loads such as AC motors, generator turbines and electrically operated clocks are covered a high percentage of entire consumers, the maintaining constant and stable system frequency has been one of the main problems in an electric power systems.

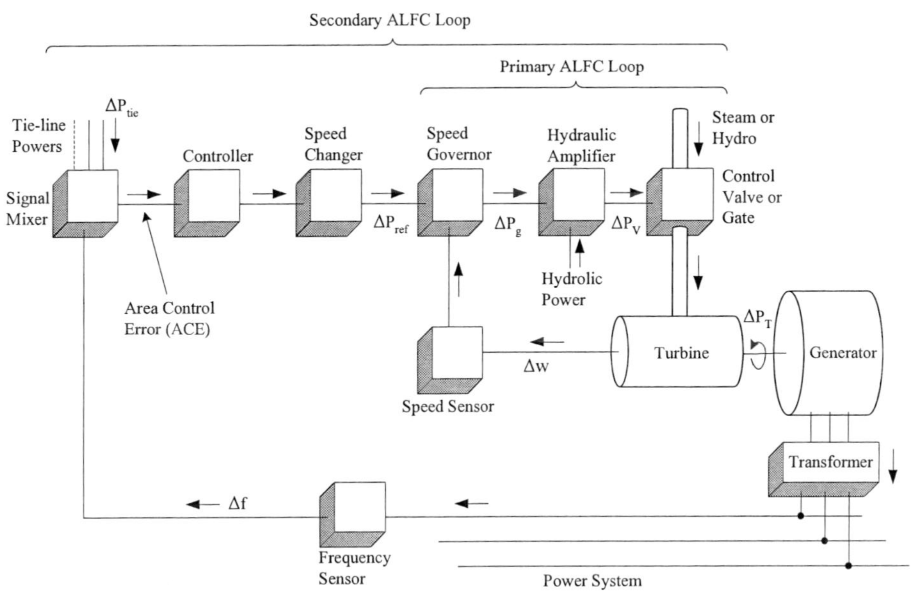

Figure 2. Automatic Generation Control Loops [3].

According to power system control theory, a nominal system frequency depends on the balance between generated and consumed real powers. Since the electric energy is transmitted at the speed of light and it cannot be stored as an electrical form, the amount of generated electric energy must equal to the amount of consumed electric energy. When there is a difference between them, this energy is added to or subtracted from the kinetic energy of the power system. Because of the kinetic energy is closely related to the speed of the generator, the difference between generated and demanded powers causes change of nominal system frequency at the normal state [3, 43]. Nevertheless, if it is taken into consideration that a properly operation of the frequency depended loads connected to the power system mostly depends on quality of electric energy; this balance must be kept in order to maintain the steady-state frequency error between acceptable values. In practice, the amount of generation of the synchronous generators is made sense for the frequency deviations occurred in the power system in order to maintain that balance. If the amount of generated power is less than the demanded amount, speed and frequency of the generator units begin to decrease, and vice versa. When the frequency deviates from the nominal value because of increasing or decreasing load, the speed of the turbine is controlled by the closed loop control system. At the result of that control, the nominal system frequency will be brought to desired level back. This control process is named Automatic Generation Control (AGC) or Automatic Load-Frequency Control (ALFC). The aims of AGC are that the steady state error of the system frequency deviations following a step load change is made zero error, and also to maintain the power exchange between the areas in an interconnected power system.

The AGC contains two different control loops which are named the primary control loop and the secondary control loop. The primary control loop includes a speed governor mechanism and serves in all generation systems locally. The purpose of this control loop is to achieve the fastest possible adjustment of the turbine power in response to a change in frequency. In this control loop, a speed governor mechanism continuously measures the speed of the turbine and produces a power command for the control valve as depicted in Figure 2. This command is fed into a hydraulic amplifier which provides a position change of the steam control valve or the control gate in the case of a hydro generator [3]. The hydraulic amplifier contains a few amounts of pilot valves and a main hydraulic piston in order to provide a required mechanical force which activates the steam valve or the water gate. The amount of steam or water which determines the rotational speed of the turbine is adjusted with these valve or gate.

In such a control system, the amount of opening of the steam control valve (or the gate) is determined with the slope of the load-frequency droop characteristic which is depicted in Figure 3(a). The load-frequency droop characteristic is the curve that indicates the amount of the change in frequency according to the load demand in MW. In practice, a few amount of the valve gap is maintained as a reserve in the generation unit. According to the droop characteristic, when the load demand is P_0 MW in the normal state, it means that the balance is maintained between generated and consumed real powers and the nominal system frequency of f_0 Hz is also kept as the desired value. If the load demand increases to P_1 MW instantly, this power difference will be tried to pay from the kinetic energy which has been accumulated in the rotating parts of the generators, at first. This mechanism provides a reaction against the excessive drop of the frequency. After that, a required power for the balance of the real powers will be met from that reserve power through control action of the speed governor. At the end of the primary control, despite the demanded power is met, the system frequency decreases to f_1 Hz as depending on the slope of the load-frequency droop characteristic. Hence, the steady-state error of f_1-f_0 Hz will occur on the system frequency. Consequently, the primary control action continues for a few ten seconds from beginning of the load change.

The secondary control loop includes wide area controls, and works in a slow reset mode to eliminate these small frequency errors which still remain after the actions of the primary control loop [3]. The measurements of the system frequency are continuously taken from the power network by a frequency sensor. When there is some deviation at the frequency, the needed power to eliminate this change in frequency is determined, and is also added to the tie-line power through the signal mixer.

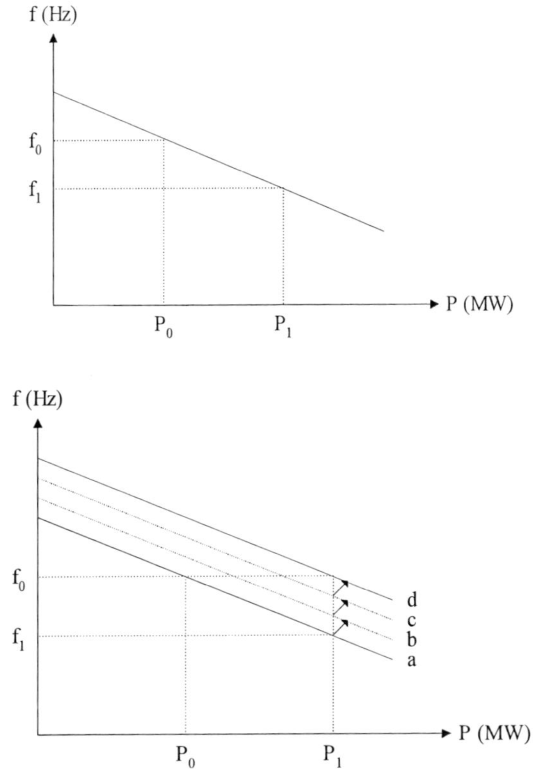

Figure 3. (a) Load-frequency droop characteristic, (b) Secondary control effect on the droop characteristic.

As a result, the Area Control Error (ACE) signal which consists of the information about frequency deviation and tie-line power change is produced in order to supply to the controller. ACE is a measurement that provides the minimum performance criteria which defines the acceptable control performance of the power system. It measures the balance of generation and demand for electricity, and the contract adherence between control areas [1]. Basic frequency responses of such a power system are represented in Figure 4.

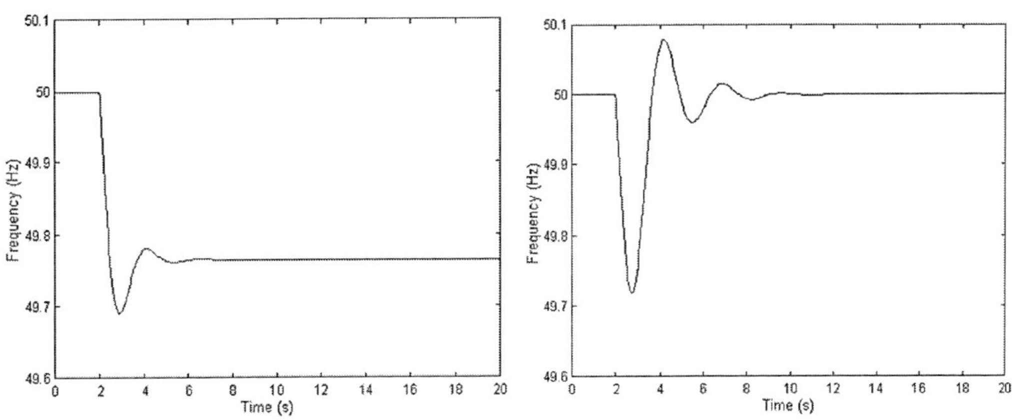

Figure 4. The frequency responses, (a) at the end of the primary control, (b) at the end of the secondary control.

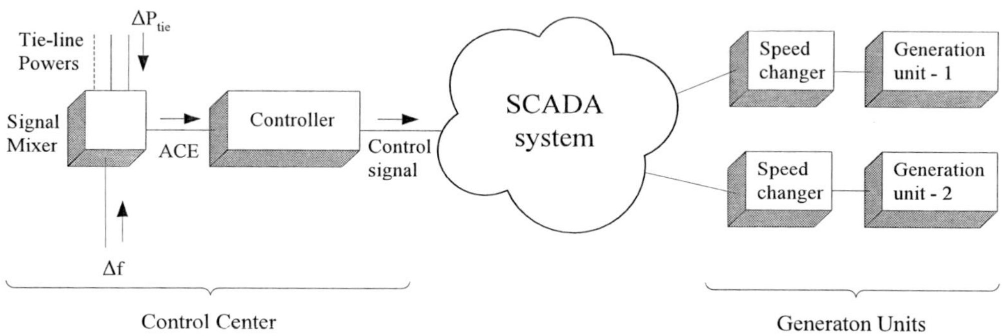

Figure 5. Transmitting the control signal to the generation units through SCADA system [2].

Another mission of the secondary control is to provide load sharing between control areas in an interconnected power system. Because the generation control cannot be carried out to the power system by the primary control in a power system where the secondary control is not used, each control area increases the amount of its power generation in proportion to its size so as to meet the demand of total load. But that is an undesirable behavior for the power system, because it is anticipated that each generation area meets its load demand without power exchange with other areas in an interconnected power system. Hence, the secondary control also arrangements the power exchange between the control areas by which the tie-line powers are fed into the control center by sampled data technique. Also, it is added to the biased frequency error so that the ACE signal may be obtained [3; 43]. The ACE is basically minimized by an integral controller in order to realize the secondary control action. Ideally, the ACE should always be zero. At the end of the secondary control, the produced control signal is send to the generation units via a Supervisory Control and Data Acquisition (SCADA) system as depicted in Figure 5.

Actually, a position of the load-frequency droop characteristic is tuned to the new position where the actual system frequency is achieved, as depicted in Figure 3(b). Thanks to the control signal taken from the SCADA system, this adjustment is realized by adjusting the load reference setting through manipulation of the speed changer motor. The usage of the speed changer motor enables the steam input and the power output at a given frequency to be changed as required [2]. The secondary control action continues for few minutes from end of the primary control action.

Modeling of Interconnected Power System

In an interconnected power system, the neighbor power networks which have one or more coherent generation units are interconnected in order to maintain more reliable and stable system frequency via common frequency reference, and to provide more effective load sharing depending on agreements. These generation units are called “control areas” and hence all synchronously connected to each other through tie-lines, they all have the same frequency [2].

In order to produce the model of such an interconnected power system for the simulation study, a small signal analysis may be used because of the relatively small variations of the system parameters and load change.

Actually, the AGC will maintain to control only during normal (small and slow) changes in load and frequency [3]. From this point of view, first a small signal model of one area power system is composed as a control area. Following, since a two area power system is being the simplest model of an interconnected power system, a model of two area power system is composed by connecting the one area systems with tie-lines.

The main control element of the primary control loop is a speed governor which senses a speed deviation and produces a power command. For the primary control loop, the inputs of the speed governor are the reference power setting, ΔP_{ref}, adjusted by the speed changer and the speed deviation of the turbine, $\Delta\omega$ (also the frequency deviation Δf of the generator) sensed by the speed governor as depicted in Figure 6.

The output of the governor ΔP_g may increase with increasing the reference power setting ΔP_{ref} or decreasing with the frequency deviation Δf. Therefore, these relationships can be written in Laplace form as below.

$$\Delta P_g(s) = \Delta P_{ref}(s) - \frac{1}{R}\Delta f(s) \tag{1}$$

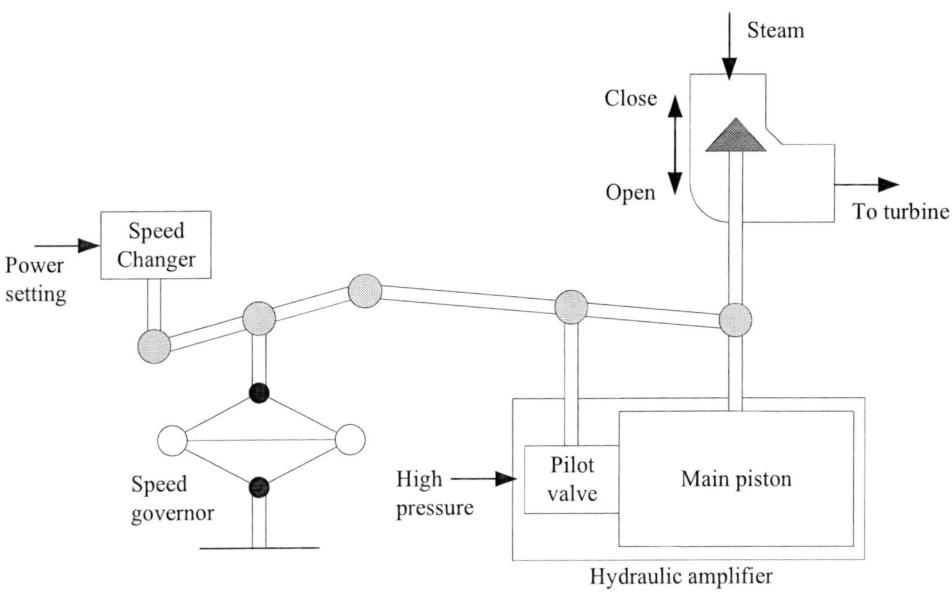

Figure 6. The small signal model of the speed governor [3].

The constant R in this formula represents the load-frequency droop characteristic and has dimension Hz/MW. Actually, this droop characteristic determines the amount of frequency change in accordance with the change in valve position or the power output as represented in Figure 3(a). The small signal model of the speed governor is depicted in Figure 7.

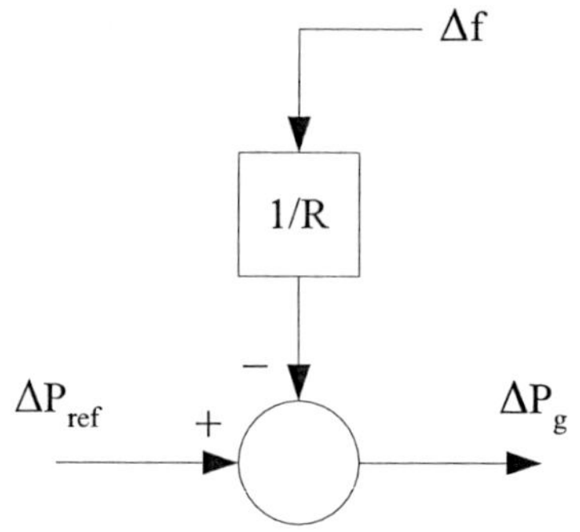

Figure 7. The small signal model of the speed governor.

The output of the governor ΔP_g supplies to the input of the hydraulic amplifier. This amplifier provides a required mechanical force in order to activate the steam valve (or water gate) of the turbine. The relationship between input ΔP_g and output ΔP_V of the hydraulic amplifier in Figure 2 can be written as a transfer function of the model. In this model which is depicted in Figure 8, T_H is named as the hydraulic time constant, and the typically value of this constant is approximately 0.1 seconds.

$$G_H(s) = \frac{\Delta P_V(s)}{\Delta P_g(s)} = \frac{1}{1 + sT_H} \tag{2}$$

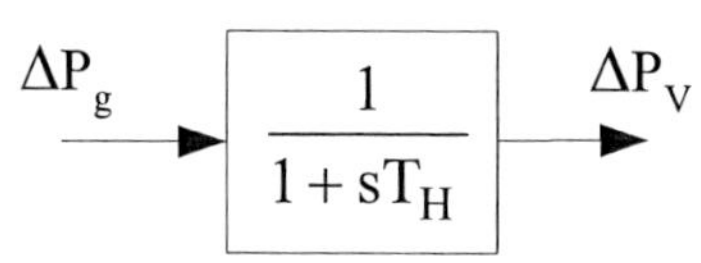

Figure 8. The small signal model of the hydraulic amplifier/

The turbine power increment ΔP_t depends on the valve power increment ΔP_V and response characteristic of the turbine [3]. Although, there are a lot of different types of turbine in literature, the transfer function model of the non-reheat steam turbine which is the simplest one is represented as Figure 9. In this model, T_T is named the turbine time constant, and explains a passing duration of steam through the valve. A typical value of this constant may be several seconds.

$$G_T(s) = \frac{\Delta P_T(s)}{\Delta P_V(s)} = \frac{1}{1 + sT_T} \tag{3}$$

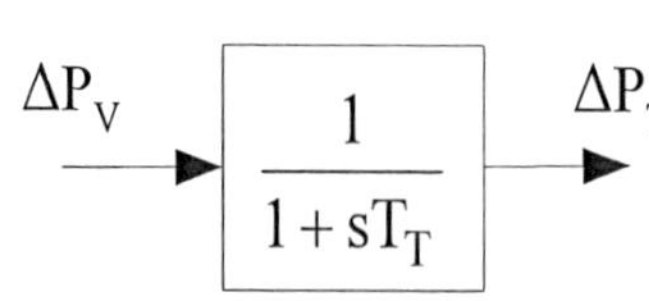

Figure 9. The small signal model of the non-reheat steam turbine.

Apart from this turbine, the reheat turbine and the water turbine will be used in this study in order to compose the model of two area hydrothermal power systems. A reheat turbine includes some re-heaters between several turbines layers as which represented in Figure 10.

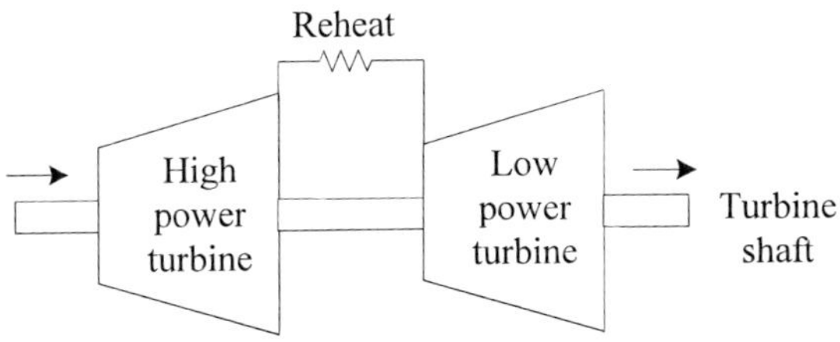

Figure 10. Block diagram of reheat steam turbine.

The steam is reheated and reenergized by these heaters. Therefore, reheat turbines increases efficiency and generally are used for large electric generation foundations. The transfer function models of water and reheat steam turbines are depicted in Figure 11. In this model of water turbine, T_w is water turbine constant, and its value is approximately 1 second. Then in the model of reheat steam turbine, T_T is turbine time constant which is determined from non-reheat steam turbine model. T_r is reheat time constant and K_r is reheat gain. Their values are generally taken 10 seconds and 0.3, respectively [32].

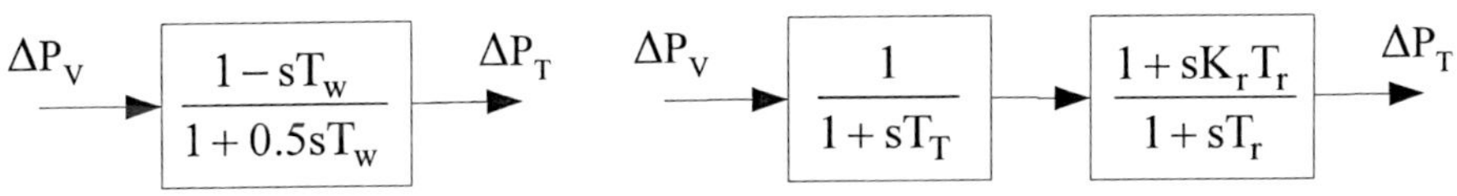

Figure 11. The small signal model of (a) the water turbine, and (b) the reheat steam turbine.

In order to create a small signal model of the control area, its inputs and its desirable outputs may be investigated. The demanded power is met from three sources in an interconnected power system consisted many control areas.

The stored area kinetic energy W_{kin} which has been accumulated in the rotating parts of the generators of the control area. ("0" represents nominal value of the parameter.)

$$\frac{d}{dt}W_{kin} = 2\frac{W_{kin}^0}{f^0}.\frac{d}{dt}(\Delta f) \tag{4}$$

The increased load consumption: Especially, for the most of frequency depended loads are being inductive; the needed power P_L on account of these loads decreases due to the decreasing frequency. The load damping constant D which simulates this behavior is expressed as follows:

$$D = \frac{\partial P_L}{\partial f} \qquad \text{MW/Hz} \tag{5}$$

The export power ΔP_{tie} via tie-line.

Here the equation of balanced powers for the control area is written as below.

$$\Delta P_T - \Delta P_L = 2\frac{W_{kin}^0}{f^0}.\frac{d}{dt}(\Delta f) + D.\Delta f + \Delta P_{tie} \tag{6}$$

In this equation, since all terms have value of MW, for simplicity they can be divided to total rated area power P_r in MW. In this case, all powers will be per unit in P_r.

$$\Delta P_T - \Delta P_L = 2\frac{H}{f^0}.\frac{d}{dt}(\Delta f) + D.\Delta f + \Delta P_{tie} \tag{7}$$

where, H is inertia constant, and the numerical range is 2-8 seconds.

$$H = \frac{W_{kin}^0}{P_r} \quad \text{Seconds} \tag{8}$$

When the Laplace transformation is carried out to (7) and the new parameters are obtained as below, the equation and the small signal model of the control area is represented as (9),

$$[\Delta P_T(s) - \Delta P_L(s) - \Delta P_{tie}(s)]\frac{K_p}{1 + sT_P} = \Delta f \tag{9}$$

where, T_p is the area time constant and K_p is the area gain as depicted below, and both define the entire control area.

$$T_P = \frac{2H}{f^0 D} \quad \text{Seconds} \tag{10}$$

$$K_P = \frac{1}{D} \quad \text{Hz/puMW} \tag{11}$$

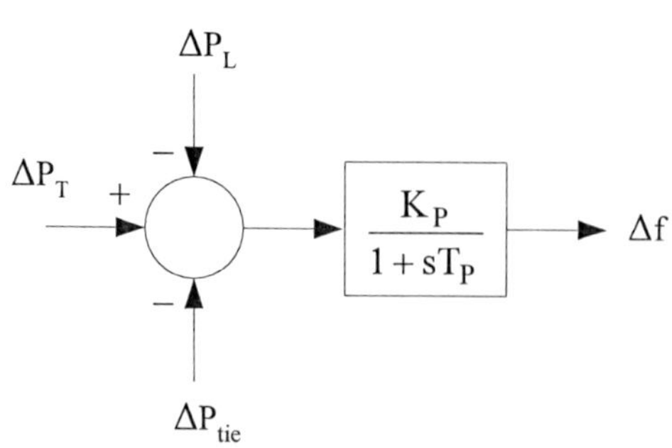

Figure 12. The small signal model of the control area according to (9).

The transfer function models obtained so far may be combined with together in order to create a model of one area power system which is depicted in Figure 13. This model represents a basic one area power system which works in a single or isolated mode. When the power system works in an isolated mode, a stability of its frequency and terminal voltage largely depends on the load changes. In this point of view, an interconnection of these one area power systems has some advantages as explained at the beginning of this chapter.

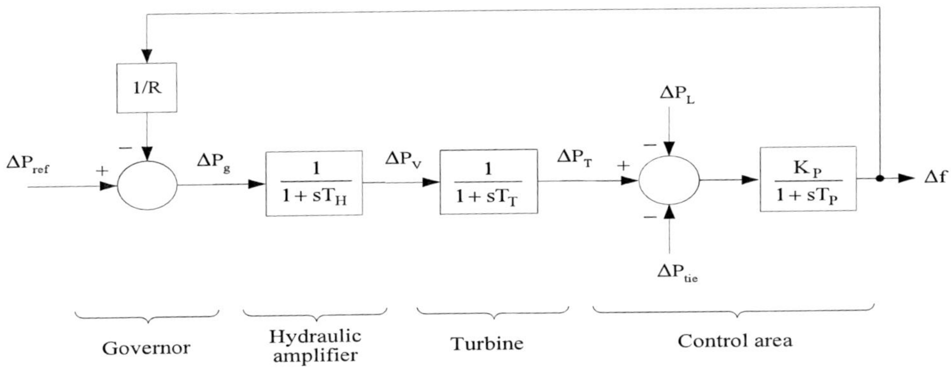

Figure 13. The small signal model of one area power system.

In order to be able to investigate the dynamic and static behaviors of an interconnected power system, the simplest model is a two area power system where the areas are connected together through relatively weak tie-lines. So such a system as depicted in Figure 14 may be investigated. In this figure, X_{12} is a tie-line reactance and P_{12} (or P_{net}) is a net power interchange between control areas.

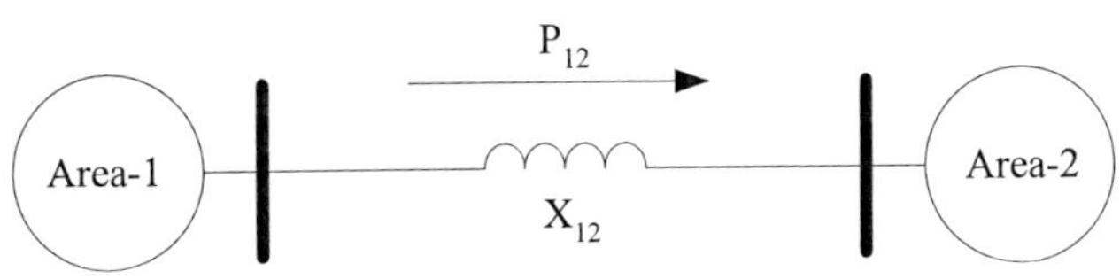

Figure 14. Two area interconnected power system.

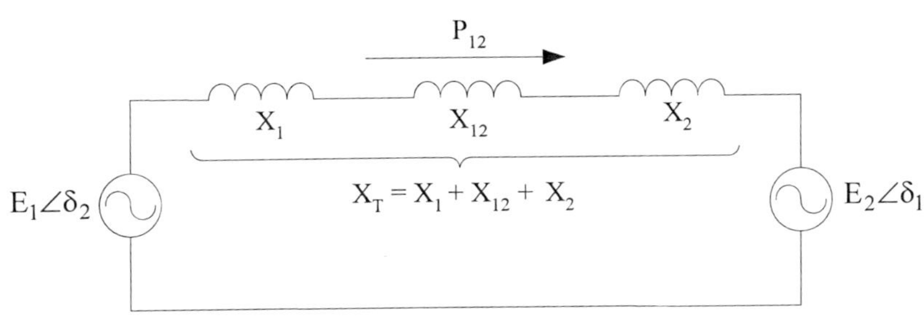

Figure 15. Equivalent circuit of two area interconnected power system.

A small signal model of tie-line power interchange can be constituted via the equivalent circuit in Figure 15. Each control area is represented as an AC power source which has the terminal voltage E, the angle δ and the equivalent reactance X. So, X_T is the total reactance which appears in this circuit. At this circuit, an equation of the net power interchange P_{12} (or P_{net}) is written as below.

$$P_{12}=\frac{E_1.E_2}{X_T}\sin(\delta_1-\delta_2) \quad \text{MW} \tag{12}$$

The power difference due to a change in the angles δ_1 and δ_2 is written as (13). In this equation, T is called stiffness or synchronizing power coefficient. It is proportional to power exchange for a given change of the phase angle.

$$\Delta P_{12}=\frac{E_1.E_2}{X_T}\cos(\delta_1-\delta_2)(\Delta\delta_1-\Delta\delta_2)=T.(\Delta\delta_1-\Delta\delta_2) \tag{13}$$

The tie-line power interchange explained in (13) can be written as (14) in time-domain and (15) in s-domain, because the frequency deviation Δf depends on the angle deviation Δδ. Finally, a small signal model of the tie-line power interchange is composed in Figure 16 in accordance with these equations.

$$\Delta P_{12}=2\pi T.\left[\int\Delta f_1 dt-\int\Delta f_2 dt\right] \tag{14}$$

$$\Delta P_{12}(s)=\frac{2\pi T}{s}.\left[\Delta f_1(s)-\Delta f_2(s)\right] \tag{15}$$

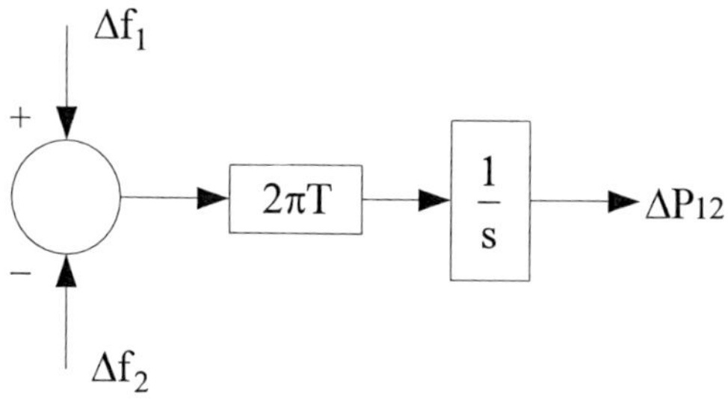

Figure 16. The small signal model of tie-line power interchange.

As a result, the small signal model of two area thermal power system may be composed by using above models. Such a model is represented in Figure 17.

The area control errors, ACE_i, for each area include the information about tie-line power error and the information about frequency deviation related to its area. The frequency deviations are weighted with the frequency bias parameters, B_1 and B_2, for all areas. The ACEs for each control area are achieved as depicted in (16) and (17).

$$ACE_1 = B_1 \Delta f_1 + \Delta P_{tie} \tag{16}$$

$$ACE_2 = B_2 \Delta f_2 - \Delta P_{tie} \tag{17}$$

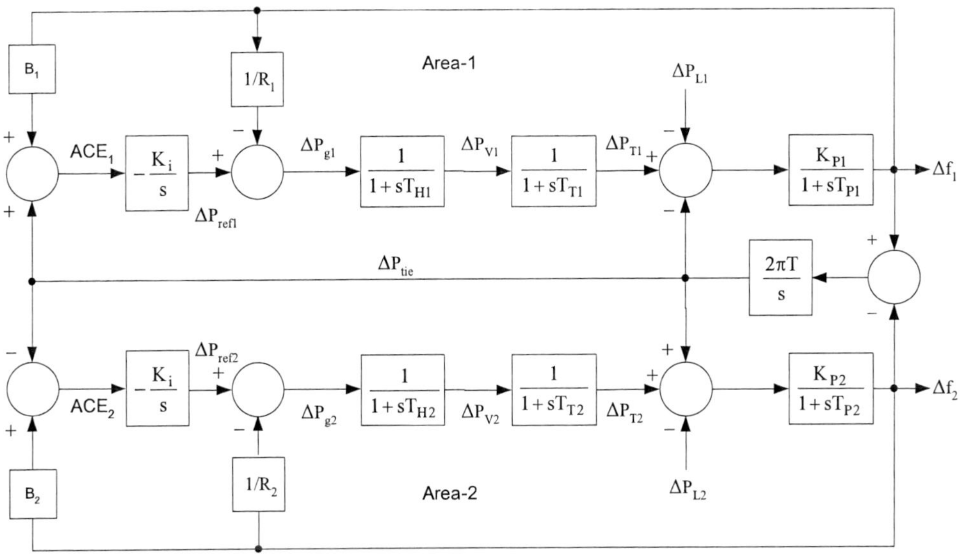

Figure 17. The small signal model of a two area power system.

When the ACE signal is fed back to the secondary control mechanism in the case of two area power system, the four control conditions which are defined in Table 1 appear. As explained above, the secondary control action needs two pieces of information about the system, the frequency deviation and the tie-line power change. The controller evaluates this information and produces the control signal. For instance, when the load demands in area-1 increases, the system takes the information about decreasing frequency and decreasing net power change leaving form area-1. Then, it increases the output of the generator, ΔP_{g1} in area-1 as a resulting control action. On the other hand, when the load demands in area-2 increases, the system takes the information about decreasing frequency and increasing net power change leaving form area-1. Then, it increases the output of the generator, ΔP_{g2} in area-2 as a resulting control action.

Table 1. The tie-line power control conditions in a two area power system

Δf	ΔP_{tie}	ΔP_{L1}	ΔP_{L2}	Resulting control action
-	-	+	0	increase ΔP_{g1} in area-1
+	+	-	0	decrease ΔP_{g1} in area-1
-	+	0	+	increase ΔP_{g2} in area-2
+	-	0	-	decrease ΔP_{g2} in area-2

The speed changer commands in accordance with the above equations are constituted by the integral controller as depicted below.

$$\Delta P_{ref1} = -K_{i,1} \int (B_1 \Delta f_1 + \Delta P_{tie}).dt \tag{18}$$

$$\Delta P_{ref2} = -K_{i,2} \int (B_2 \Delta f_2 - \Delta P_{tie}).dt \tag{19}$$

After the step load change, the new static equilibrium of the power system is achieved by the speed changer commands in the control system. In this case, it requires that the terms of integral in (18) and (19) be zero. It can be clearly seen that when this conditions can be met, Δf_0 and $\Delta P_{tie,0}$ must be zero as in (20) and (21) [3].

$$B_1 \Delta f_0 + \Delta P_{tie,0} = 0 \tag{20}$$

$$B_2 \Delta f_0 - \Delta P_{tie,0} = 0 \tag{21}$$

Finally, $K_{i,1}$ and $K_{i,2}$ are the integrator gains which have to be chosen small enough not to cause instability in the control mechanism. Here they are tuned by the gain scheduling control technique based on the artificial intelligence optimization method.

Gain Scheduling PI-Controller

The PI-controller is one of the most commonly used controller type in various industrial applications, today. Fundamentally, the PI-controller uses the difference between measurement from output of the system and the desired reference input, so as to produce the control signal. This difference is called error signal of the control system. During control process, while the proportional part changes the system output proportional to the error signal, the integral part of the controller integrates the error signal in order to achieve the sum of the previous errors. The integrator adds a pole to the transfer function and increases the order of the system. The output of system is manipulated by the controller until the error signal reaches to zero. As a result, PI-controller eliminates the steady-state error of the control system [33].

$$u = K_P \left(e + \frac{1}{T_i} \int e.dt \right) \tag{22}$$

The above equation defines the mechanism of PI-controller. In this equation, u is the control signal produced by the PI-controller. e is the error signal. K_P and T_i are the proportional gain and the integral time constant of the controller. K_p/T_i equals to the integral gain K_i.

These gains should be tuned so as to obtain the desired dynamic response. However, various classic tuning methods are available in literature; a gain scheduling control technique may be carried out to the PI-controller in order to tune K_P and K_i gains.

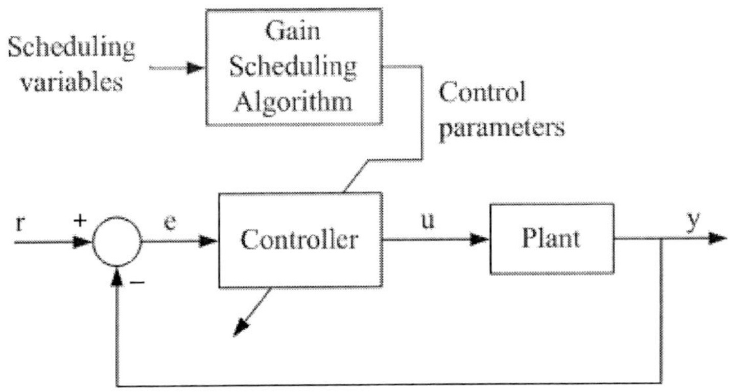

Figure 18. Block diagram of the gain scheduling control system.

A gain scheduling control is a control technique that changes some control parameters of the controller according to scheduling variables related to different operating regions which the plant works. This control technique deals with particularly nonlinear processes with time variations or situations where the requirements on the control that change with the operating conditions. The main advantage of this control is that the controller parameters can be adjusted very quickly in response to changes in the plant dynamics. It is also simpler to implement than the other adaptive control techniques [8]. A typically block diagram of this system is depicted in Figure 18.

In this figure, the scheduling variables can be the measured signal, the control signal or an external signal. The control parameters of the controller are determined with a gain scheduling algorithm. The algorithm is run for all different operating conditions by automatic tuning. In this chapter, the swarm intelligence based methods are investigated as a gain scheduling algorithm because of their some advantages such as simpler implementing, better performance of converging and less run times.

Particle Swarm Optimization (PSO) For Scheduling the Gains of PI-Controller

The Particle Swarm Optimization is swarm intelligence based optimization algorithm which was first introduced by Kennedy and Eberhart in 1995 [34]. They were called "Particle Swarm Optimization", because it is based on the food search behavior of the bird flocks or the fish schools (Figure 19). The Particle Swarm Optimization can obtain high quality solutions within shorter calculation time and stable convergence characteristics.

The algorithm uses particles which represent potential solutions of the problem. Each of M particles is a point which flies in D-dimensional search space at a certain velocity which can be adjusted in light of preceding flight experiences. That means, all individuals benefit from the experiences of itself and whole swarm. The positions of the particles are represented by D-dimensional vectors.

The number of dimension of the search space equals to the number of parameters in the solution.

$$x_i = \begin{pmatrix} x_{i1} & x_{i2} & x_{i3} & \cdots & x_{iD} \end{pmatrix}, \ i = 1, 2, 3, \ldots, M \tag{23}$$

The difference in distance between new and old positions of each particle is called velocity, and they are also represented by D-dimensional vectors for each particle.

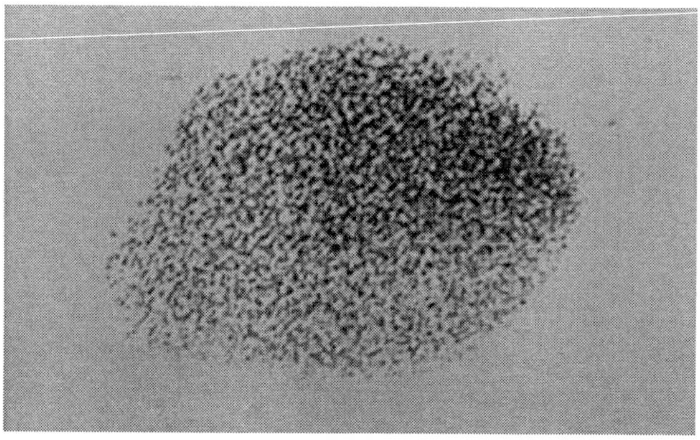

Figure 19. A bird flock.

$$v_i = \begin{pmatrix} v_{i1} & v_{i2} & v_{i3} & \cdots & v_{iD} \end{pmatrix}, \ i = 1, 2, 3, \ldots, M \tag{24}$$

Consequently, the positions and the velocities of the M particles which fly in D-dimensional search space are indicated by the following matrixes.

$$X = \begin{bmatrix} x_{11} & x_{12} & x_{13} & \cdots & x_{1D} \\ x_{21} & x_{22} & x_{23} & \cdots & x_{2D} \\ \cdots & \cdots & \cdots & \cdots & \cdots \\ x_{M1} & x_{M2} & x_{M3} & \cdots & x_{MD} \end{bmatrix} \tag{25}$$

$$V = \begin{bmatrix} v_{11} & v_{12} & v_{13} & \cdots & v_{1D} \\ v_{21} & v_{22} & v_{23} & \cdots & v_{2D} \\ \cdots & \cdots & \cdots & \cdots & \cdots \\ v_{M1} & v_{M2} & v_{M3} & \cdots & v_{MD} \end{bmatrix} \tag{26}$$

Limiting of the positions and the velocities of the particles is an important issue. There are different approaches about this subject in literature. According to Chen and co-workers, all positions of particles on each dimension are clamped in limits which are specified by the user. Then, the velocities are clamped to the range $[v_{min}, v_{max}]$ given as follows [35].

$$v^{max} = k.x^{max}, \ 0.1 \le k \le 0.5 \tag{27}$$

$$v^{min.} = -v^{max} \tag{28}$$

In PSO algorithm, the particles tend to go to their best positions obtained so far in the search space. These positions are the local best positions p_i of the particles and express the previous experiences of these particles in reality. Those positions are indicated as below.

$$p_i = \left(p_{i1} \quad p_{i2} \quad p_{i3} \quad \cdots \quad p_{iD}\right), \; i = 1, 2, 3, \ldots, M \tag{29}$$

$$P = \begin{bmatrix} p_{11} & p_{12} & p_{13} & \cdots & p_{1D} \\ p_{21} & p_{22} & p_{23} & \cdots & p_{2D} \\ \cdots & \cdots & \cdots & \cdots & \cdots \\ p_{M1} & p_{M2} & p_{M3} & \cdots & p_{MD} \end{bmatrix} \tag{30}$$

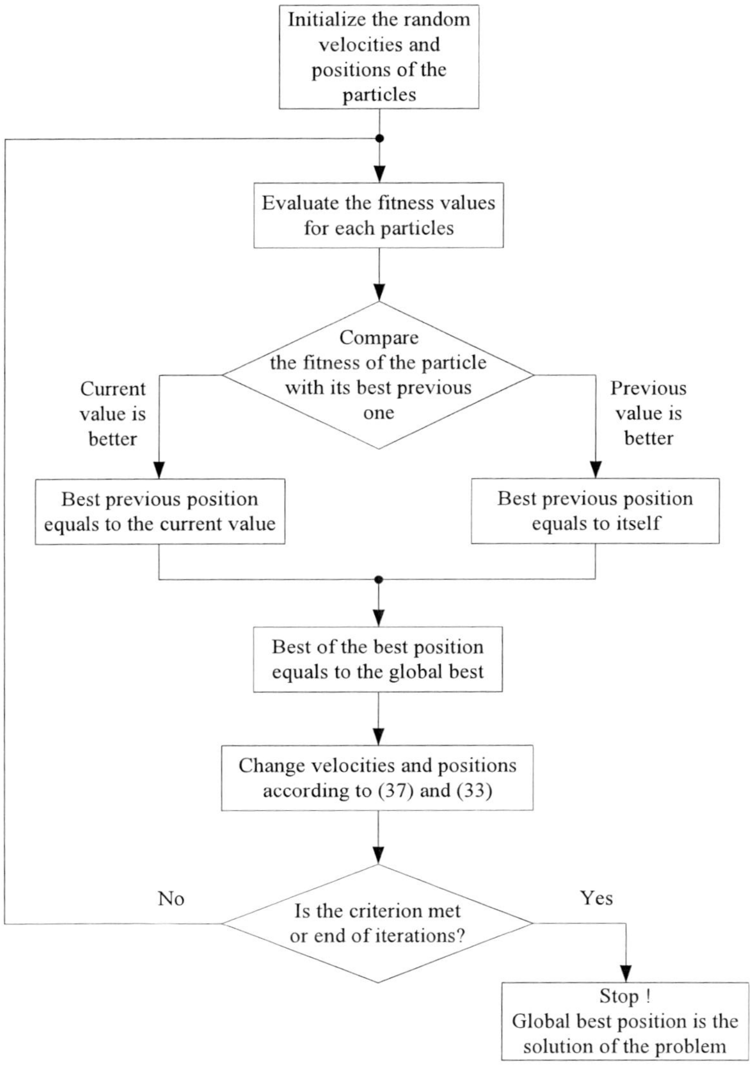

Figure 20. A basic flowchart of PSO algorithm.

The best position in the matrix P is determined as the best position g_i of the whole swarm, and is called the global best position. It expresses the previous experiences of the swarm. Especially, the final global best position vector indicates the solution of the problem.

$$g_i = (g_1 \quad g_2 \quad g_3 \quad .... \quad g_D), \; i = 1, 2, 3, ..., M \tag{31}$$

In the algorithm, the positions and the velocities of the particles are updated by the following equations developed by Kennedy and Eberhart in light of Reynolds' [36] and Hepper's [37] studies are related to the simulation of the bird flocks.

$$v_{iD}^{t+1} = v_{iD}^{t} + c_1 r_1 \left(p_{iD}^{t} - x_{iD}^{t}\right) + c_2 r_2 \left(g_i^{t} - x_{iD}^{t}\right) \tag{32}$$

$$x_{iD}^{t+1} = x_{iD}^{t} + v_{iD}^{t+1} \tag{33}$$

where, i = 1, ..., n and n is the size of the swarm, c_1 and c_2 are positive constants, r_1 and r_2 are random numbers which are uniformly distributed in [0, 1], t determines the iteration number, p_i represents the best previous position of the ith particle and g indicates the best particle among all the particles in the swarm. In 1998, Eberhart and Shi used the inertia coefficient w in (32) so as to improve the balance between global and local search speeds [38].

$$v_{iD}^{t+1} = w.v_{iD}^{t} + c_1 r_1 \left(p_{iD}^{t} - x_{iD}^{t}\right) + c_2 r_2 \left(g_i^{t} - x_{iD}^{t}\right) \tag{34}$$

In 1999, Clerc developed the constriction factor K computed as below, instead of the inertia coefficient w in order to increase the convergence performance of the optimization [39].

$$K = \frac{2}{\left|2 - \varphi - \sqrt{\varphi^2 - 4\varphi}\right|} \tag{35}$$

$$\varphi = c_1 + c_2 > 4 \tag{36}$$

Accordingly, the positions of the particles are updated by the following equation. This is called Constriction Factor Approach Particle Swarm Optimization (CFA-PSO). A basic flowchart of PSO algorithm can be represented in Figure 20.

$$v_i^{t+1} = K.\left(v_i^{t} + c_1 r_1 \left(p_i^{t} - x_i^{t}\right) + c_2 r_2 \left(g_i^{t} - x_i^{t}\right)\right) \tag{37}$$

The PSO algorithm has been used in power system researches such as AGC, automatic dispatch control, optimal power flow control, automatic reactive-power/voltage control, power system stabilizers etc. since the beginning of 2000 [6, 7].

APPLICATION OF AGC IN A HYDROTHERMAL POWER SYSTEM INCLUDING SMES UNITS

The Model of Hydrothermal Power System

The model of the power system used in this study is composed as two area hydrothermal power system. The area-1 comprises a reheat thermal power system (Figure 21.a) including governor dead-band and boiler effect. These effects are added to the model in order to increase the system nonlinearity.

The other control area comprises a hydro power system (Figure 21.b). Both areas have SMES units. The model is depicted as follows.

Figure 21. (a) Thermal power system, (b) Hydro power system.

Thermal or hydro power systems generate electrical energy by benefiting from the either heat energy of the underground resources or mechanical energy of the moving water. Basically, steam or water turns the prime mover which drives an electric generator. The prime mover is a steam turbine in the case of thermal power system and a water turbine in the case of hydro power system.

By controlling the amount of the steam or the water, the rotational speed of the generator is controlled. In the interconnected power system model depicted in Figure 22, u_1 and u_2 are the control inputs obtained from controllers that provide the secondary control action. ΔP_{L1} and ΔP_{L2} are the step load perturbations of 1% of the nominal loading in either of the areas.

Δf_1 and Δf_2 are the frequency deviations and ΔP_{tie} is the tie-line power change. The governor dead-band, the boiler effect and the SMES units used in this model are explained as follows. The values of the model parameters are defined in Table 2.

A Model of Governor Dead-Band:

A linear model of governor dead-band is represented through describing function approach. The describing function approach was introduced by Kochenburger in 1950 [40]. It is generally used in order to investigate the oscillation in nonlinear feedback systems. Since the governor dead-band nonlinearity tends to produce a continuous sinusoidal oscillation of natural period of about $T_0 = 2$ s., this approach is being used to linearize the governor dead-band in terms of change and rate of change in the speed.

Table 2. The parameters of the power system model [30]

Parameter	Definition	Quantity
T_g	Speed governor time constant	0.2 s
T_t	Turbine time constant	0.3 s
T_r	Reheat time constant	10 s
K_r	Reheat gain	0.3
$K_{p1,2}$	Power system gain	120 Hz/puMW
$T_{p1,2}$	Power system time constant	20 s
$K_{SM1,2}$	SMES gains	0.12
$T_{SM1,2}$	SMES time constant	0.03 s
T_1	Hydro governor time constant	41.6 s
T_2	Hydro governor time constant	5 s
T_3	Hydro governor time constant	0.513 s
T_W	Water time constant	1 s
T_{12}	Synchronization coefficient	0.0707 MW/rad
K_1	Steam boiler gain	0.85
K_2	Steam boiler gain	0.095
K_3	Steam boiler gain	0.92
K_{IB}	Pressure controls gain	0.03
T_{IB}	Pressure controls time constant	26 s
T_{RB}	Pressure controls time constant	69 s
T_D	Fuel system time constant	0 s
T_F	Fuel system time constant	10 s
C_B	Boiler storage coefficient	200
$B_{1,2}$	Frequency bias factor	0.425 puMW/Hz
$R_{1,2}$	Regulation constant	2.4 Hz/puMW

The governor dead-band is generally defined as the total magnitude of a sustained speed change within which there is no change in valve position [41]. The nonlinearity of this hysteresis is defined as;

$$y = F(x, \dot{x}) \tag{38}$$

In this function, x is taken as a sinusoidal oscillation with $f_0 = 0.5$ Hz.

$$x = A \sin w_0 t \tag{39}$$

If the backlash nonlinearity tends to give continuously sinusoidal oscillation with a natural period of about 2 s, such assumption will be quite realistic. Then, the F function can be evaluated as a Fourier series as follows,

$$F(x, \dot{x}) = F_0 + N_1 x + \frac{N_2}{w_0} dx / dt + \ldots \tag{40}$$

For an approximation, it is enough to consider the first three terms in (40). As the backlash nonlinearity is symmetrical about the origin, and then F_0 is equal to zero,

$$F(x, \dot{x}) = N_1 x + \frac{N_2}{w_0} dx / dt = DBx \tag{41}$$

where, DB denotes the dead-band. In this work, the backlash of approximately 0.5% is chosen and the Fourier coefficients are obtained as $N_1 = 0.8$ and $N_2 = -0.2$ [26].

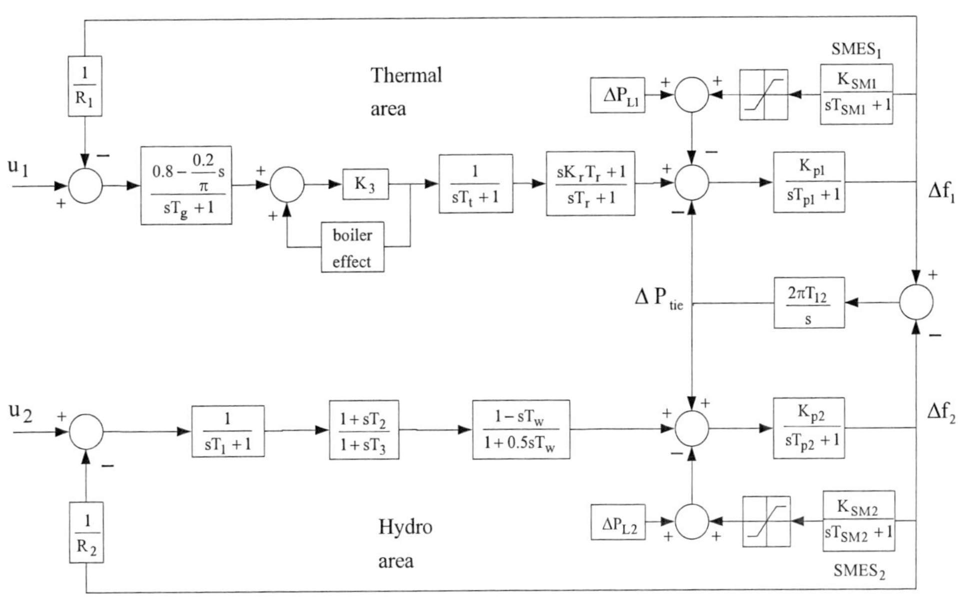

Figure 22. Transfer function model of the power system including SMES units.

As a result of the analysis, the transfer function of the governor with dead-band nonlinearity can be expressed in (42) [30];

$$G_g = \frac{0.8 - \frac{0.2}{\pi}s}{1 + T_g s} \qquad (42)$$

The Model of Boiler Effect:

The steam is produced in the boiler. The thermal energy of the steam is converted to the mechanical energy in the various sections on the turbine.

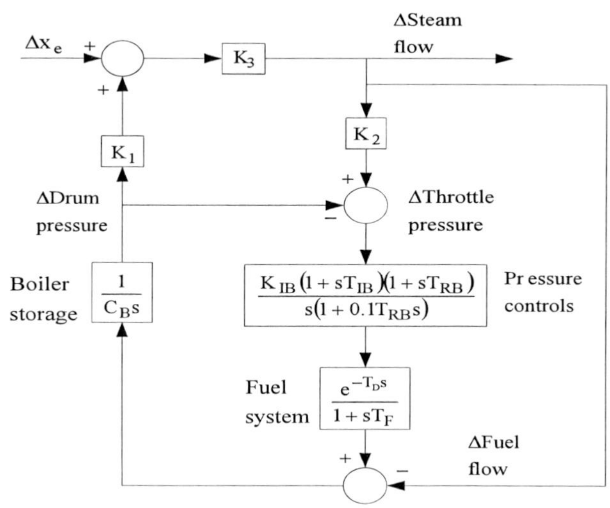

Figure 23. The model of steam boiler [42].

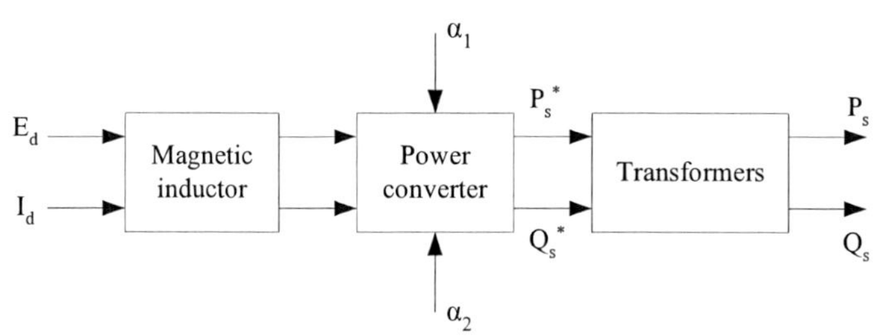

Figure 24. A SMES unit [44].

Boiler designs fall into two categories: (a) drum type boilers, and (b) once-through boilers. In this study, a boiler effect was modeled as a drum type boiler. The drum type or recirculation boilers basically rely on natural or forced circulation of drum fuel to absorb energy from the hot furnace walls for generating steam [43]. The linear model of the boiler effect in the thermal area is represented in Figure 23. In this model, the long-term dynamics of water and steam flow on boiler drum pressure are included. In this steam unit, changes in generation are initiated by turbine control valves and the boiler controls respond with necessary immediate control action upon sensing changes in steam flow and deviations in pressure [42].

The Model of Superconducting Magnetic Energy Storage (SMES) Unit:

One of the methods that improving the transient response of AGC system is to add a SMES unit to the power system. An instantaneous imbalance between produced and consumed real powers may be reduced by the addition of fast acting SMES units to the power system. Actually, the SMES unit is a device that stores energy in the magnetic field in a superconducting coil and supplies to the power system when it is needed. Nowadays, although SMES units have only operated on small scale power systems, it is preferred because of relatively short response time and less energy loss per hour than the other energy storage systems. It basically includes a magnetic inductor, power converters, transformers and a refrigerator depicted in Figure 24 [45].

In this figure, E_d and I_d are SMES voltage and current, α_1 and α_2 are the firing angles, P_s* and Q_s* are the active and reactive power which contain the harmonics, and P_s and Q_s are the SMES active and reactive power [44]. Magnetic inductor is a superconducting coil cooled until its superconducting critical temperature. The stored energy in the coil is basically computed by (43):

$$E_d = \frac{1}{2} L.I_d^2 \tag{43}$$

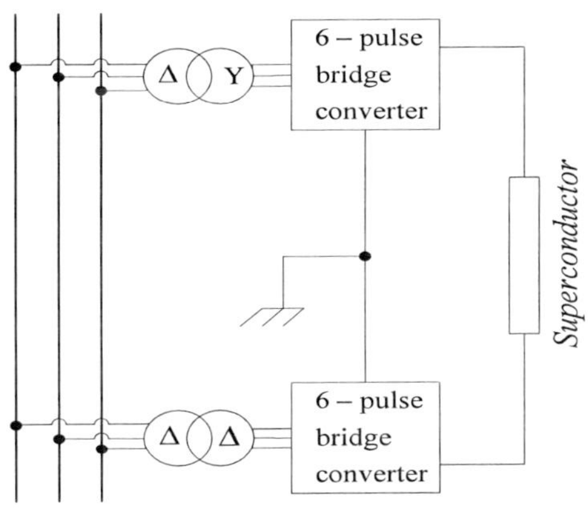

Figure 25. A basic configuration of SMES unit.

The block diagram of power converter includes two 6-pulse AC/DC converters connected to the 3-phase AC power system via transformers. These transformers are Y-Δ and Δ-Δ step down transformers and they are utilized in order to cancel the AC side 5th and 7th harmonic currents and DC side 6th harmonic voltage [46]. A basic SMES unit configuration as explained above is depicted in Figure 25. In this configuration, once the superconducting coil charged, it conducts current. When there is a sudden rise in the demand of load, the stored energy in the coil is almost immediately released through the converter to the power system. As the control mechanisms set the power system to the new equilibrium condition, the coil will be charged back to its initial value [25].

In this study, the SMES unit is modeled as a limiter and a first order transfer function.

The limiter is equipped at a power output of the transfer function and holds the power of SMES unit, ΔP_{SM}, between -0.01 and 0.01 puMW [30].

The Control Configuration

The proposed control configuretion which is composed in this study is depicted in Figure 26. In this configuration, the PI-controllers are used in order to realize the secondary control action in AGC for the power system. The obtaining optimal solution of the control inputs is taken as an optimization problem.

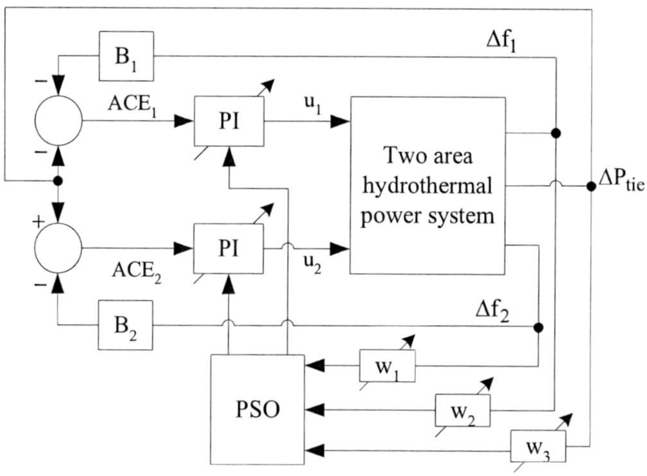

Figure 26. The proposed control configuration of the power system.

Optimum gains of the PI-controllers are tuned with CFA-PSO algorithm in accordance with the gain scheduling control method. The tunable weight cost function, J, represented in (44) is derived so as to improve the convergence of the optimization algorithm. In this function, all the observed system parameters Δf_1, Δf_2 and ΔP_{tie} are weighted with w_1, w_2 and w_3 respectively. These weights are scheduled with the gain scheduling algorithm.

$$J = \int_0^t t\left[(w_1\Delta f_1)^2 + (w_2\Delta f_2)^2 + (w_3\Delta P_{tie})^2\right]dt \quad (44)$$

The control inputs, u_1 and u_2, which are supplied to the hydrothermal power system are obtained with the PI-controllers as depicted in (45) and (46). K_p and K_i are proportional and integral gains of the PI-controller, respectively. These gains are also scheduled with the gain scheduling algorithm.

$$u_1 = K_{p1}ACE_1 + K_{i1}\int ACE_1 dt \quad (45)$$

$$u_2 = K_{p2}ACE_2 + K_{i2}\int ACE_2 dt \qquad (46)$$

The controlling of the power system is realized in case of 0.01puMW of step load change, ΔP_{L1}, in area-1. Optimization procedure of the power system by PSO algorithm is explained in the next section.

Power System Control Optimization with PSO Algorithm

In order to optimizing the system parameters, the Constriction Factor Approach PSO algorithm was utilized in this chapter. The PSO algorithm had been explained in the previous section, and the usage of the algorithm is explained with some parts of the MATLAB codes in this section. For the simulation study, the MATLAB/Simulink model of the hydrothermal power system was created as depicted in Figure 27.

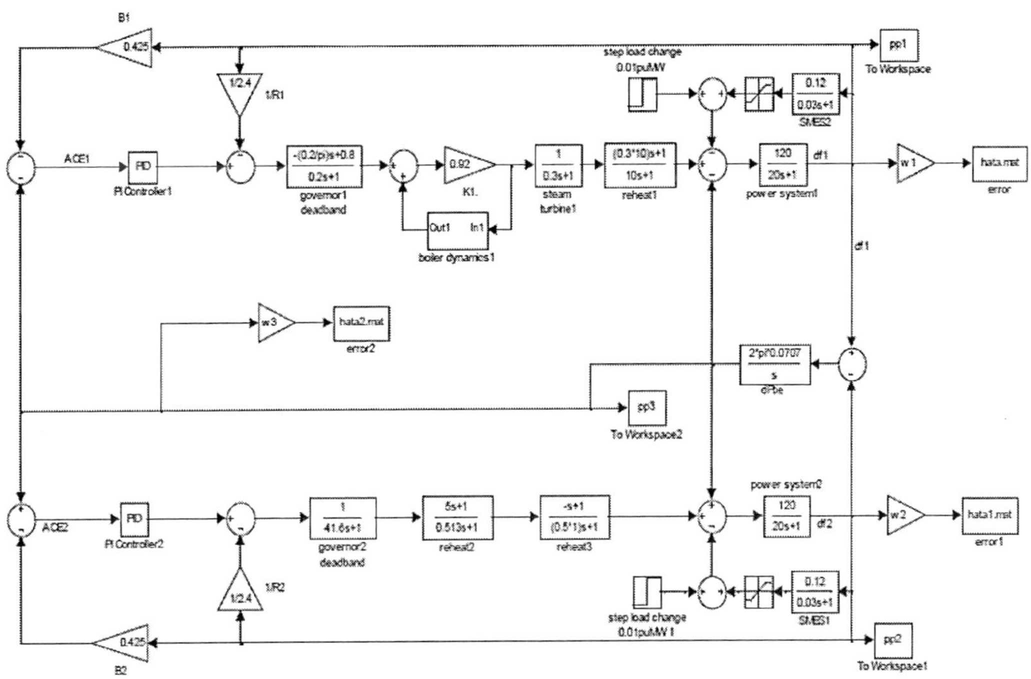

Figure 27. The MATLAB/Simulink model of the hydrothermal power system.

At the beginning of the program, the initial values and the limits of some parameters related to the PSO algorithm, PI-controller and cost function were chosen. The number of parameters, N, has been chosen 5 for each particle. These parameters are K_p, K_i, w_1, w_2 and w_3, and this number also determines a dimension of the search space. The number of particles in the swarm which flies in the 5-dimensions search space has been chosen 10, and the iteration numbers have been chosen 20. Because of the less number of parameters to be optimized, the number of particles and iterations were assessed as adequate.

The parameters are optimized and also limited as follows for the control stability.

$$-1 \leq K_p \leq 1$$

$$0 \leq K_i \leq 1$$

$$0.5 \leq w_1, w_2, w_3 \leq 1.5$$

The limits of the velocities of the particles were computed by the equations as represented in the following part of the algorithm. Then, the constriction factor, K, was also computed below as depicted in (35). Both acceleration constants, c_1 and c_2, which determine the step size of each particle to the optimal position have been chosen 2.05 in accordance with the literature.

```
%**********************************************************************
N = 5;                                          % the number of parameters
M = 10;                                         % the number of particles
T = 20;                                         % the number of iterations

x_min = [-1 0 0.5 0.5 0.5];                     % min. limits of the parameters
x_max = [1 1 1.5 1.5 1.5];                      % max.limits of the parameters

for dd=1:N,
v_max(1,dd) = (x_max(1,dd)-x_min(1,dd))*0.15;   % the computation of the limits
v_min(1,dd) = -v_max(1,dd);                     % of the velocities of the particles
end
c1 = 2.05;                                      % acceleration constants
c2 = 2.05;
F = c1+c2;
K = 2/abs(2-F-sqrt(F^2-4*F));                   % computation of the constriction
factor
%**********************************************************************
```

The initial positions of the particles have been computed in the algorithm. The initial values of the velocities, fitness values of the positions and fitness values of the best positions for each particle have been chosen zero, since these will be assigned with the first iteration. In addition to this, the first values of the best position matrix, p, have been made equal to the values of the position matrix, x. Thus, the initial settings have been completed.

```
%**********************************************************************
for m = 1:M,
for n = 1:N,
x(m,n) = x_min(1,n)+(x_max(1,n)-x_min(1,n))*rand(1,1);% initial values of the positions
v(m,n) = 0;                                     % initial values of the velocities
end
fx = zeros(M,1);                                % initial values of the fitness's of
the positions
```

```
fp = zeros(M,1);                              % initial values of the fitness's of
the best positions
p = x;
fg = 0;                                       % initial value of the fitness of the
                                              best position
end                                           % of the entire swarm

%*********************************************************************
```

Then, the updating of the velocities and the positions of all particles has been realized by the following part of the algorithm. These are the update equations that have been modified according to constriction factor approach. In this part of the program, the limits of the search space have been also determined in accordance with the minimum and the maximum values of the parameters that will be optimized, which had been chosen previously. In this equation, r_1 and r_2 are the normally distributed random numbers in the range of [0, 1]. They increase likelihood of the convergence to the desired solution by reassigning randomly in all iteration.

```
%*********************************************************************
for y = 1:M,
for z = 1:N,
v_yn(y,z) = K*(v(y,z)+c1*r1*(p(y,z)-x(y,z))+
c2*r2*(g(1,N)-x(y,z)));                       % updating of the velocities
if v_yn(y,z) > v_max(1,z),                    % limiting of the velocities
v_yn(y,z) = v_max(1,z);
end
if v_yn(y,z) < v_min(1,z),
v_yn(y,z) = v_min(1,z);
end
x_yn(y,z) = x(y,z)+v_yn(y,z);
if x_yn(y,z) > x_max(1,z),                    % limiting of the positions
x_yn(y,z) = x_max(1,z);
end
if x_yn(y,z) < x_min(1,z),
x_yn(y,z) = x_min(1,z);
end
x(y,z) = x_yn(y,z);
v(y,z) = v_yn(y,z);
end
end
%*********************************************************************
```

The following part of the algorithm has been used in order to determine the fitness values of the positions of the particles for consecutive iterations. The meaning of SSE is sum of squared error obtained by which the optimized parameters were applied to the power system control, and $f_x(i)$ is the fitness values of the positions of the particles. Due to obtain SSE, the gains of the controller and the weight parameters of the cost function were applied to model of the power system by running the “sim” command. Later on, the values vectors of the outputs of the power system are obtained as the vectors "fark", "fark1" and "fark2".

```
%************************************************************************
for i = 1:M,
Kp = x(i,1);
Ki = x(i,2);
w1 = x(i,3);
w2 = x(i,4);
w3 = x(i,5);

sim('PSO_2Bv1');
load('hata.mat');
f1 = fark(2,:);                          % frequency deviation of the area-1
time        = fark(1,:);
load('hata1.mat');
f2   = fark1(2,:);                       % frequency deviation of the area-2
load('hata2.mat');
ptie = fark2(2,:);                       % tie-line power change
SSE = 0;
for k = 1:length(f1),
SSE = SSE+(time(k)*((f1(k)^2)+(f2(k)^2)+(ptie(k)^2)));
end
fx(i) = 1/SSE;                           % fitness values of the positions
end
%************************************************************************
```

The fitness values of the best positions of each particles and the fitness value of the best position of the entire swarm have been updated with the following parts of the CFA-PSO program. In these parts of the program, if the fitness values of the recent positions, $f_x(i)$, are bigger than the fitness values of the previous best positions, $f_p(i)$, the previous best positions and their fitness values replace the recent positions and their fitness's.

```
%************************************************************************
if fp(i) < fx(i),                  % updating of the fitness values of the positions
fp(i) = fx(i);
for q = 1:N,
p(i,q) = x(i,q);
end
end
%************************************************************************
if fg < fp(i),                     % updating of the fitness value of the best position
fg = fp(i);                        % of the entire swarm
for r = 1:N,
 g(1,r) = p(i,r);
 end
end
%************************************************************************
```

Finally, the algorithm was completed at the end of the last iteration. The particle which has the best position of the entire swarm has been taken as optimum result of the problem. The parameters of this particle have been carried out to the PI-controllers and tunable weighted cost function in the AGC system, and the optimum results for this application have been achieved. These results are investigated in the next section.

RESULTS AND DISCUSSION

The optimum parameters which are obtained at the end of the CFA-PSO algorithm are shown in Table 3. The parameters achieved from the model of power system; without SMES unit, with SMES unit only in the thermal area, with SMES unit only in the hydro area and with SMES units in both areas, separately. The optimization algorithm is run on the power system in case of the step load change, $\Delta P_{L1} = 0.01$ puMW, only in the thermal area.

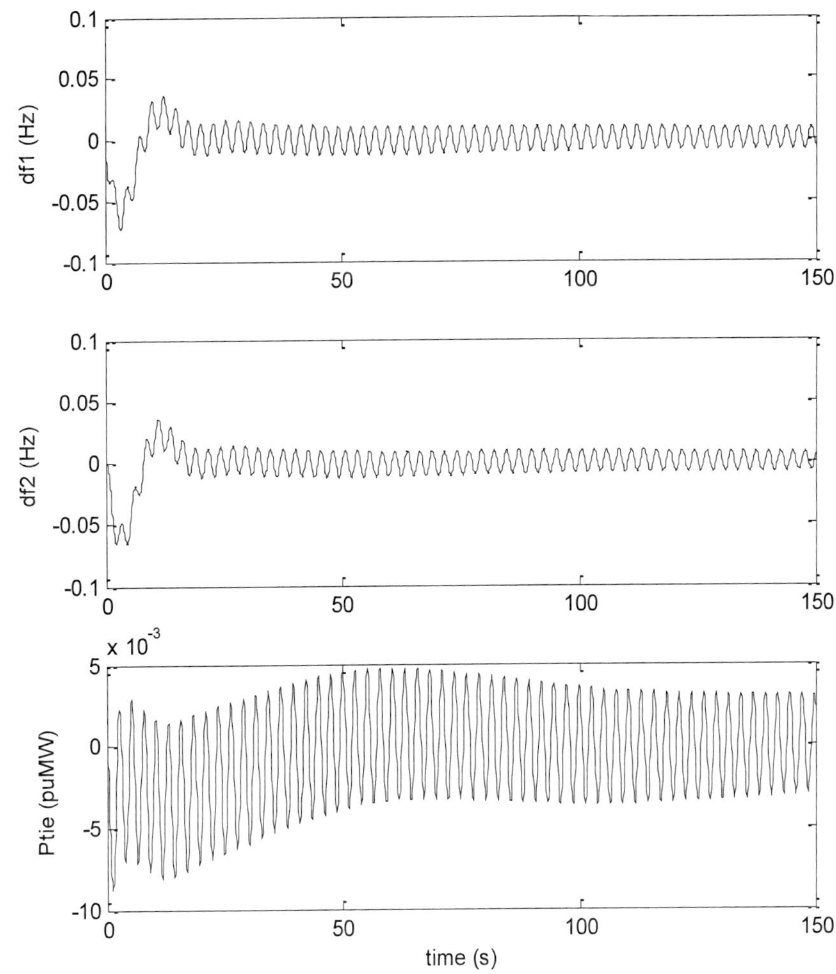

Figure 28. The curves of Δf1, Δf2, and ΔPtie in case of without SMES unit.

Table 3. The optimized parameters of the power control system

	K_p	K_i	w_1	w_2	w_3
Without SMES	-0.16	0.10	0.87	0.82	0.80
With SMES in thermal area	0.67	0.62	0.63	0.70	1.10
With SMES in hydro area	0.13	0.63	0.73	1.04	1.43
With SMES in both area	0.71	0.87	1.01	0.93	1.39

When these parameters are substituted in the PI-controllers and the cost function with tunable weights in the model of the power system, the optimal curves of the frequency deviations Δf_1 and Δf_2, and the tie-line power change ΔP_{tie} are obtained in the following figures for above situations of SMES units and load change. Figure 28 shows these curves for the model of power system without SMES unit.

Figures 29 and 30 show the curves for the models of power system within the thermal area and within the hydro area. Finally, Figure 31 shows the curves for the model of power system with in both areas. In Figure 28, the frequency deviations and tie-line power change curves have some fluctuations in steady state because of the nonlinear characteristics of the power system.

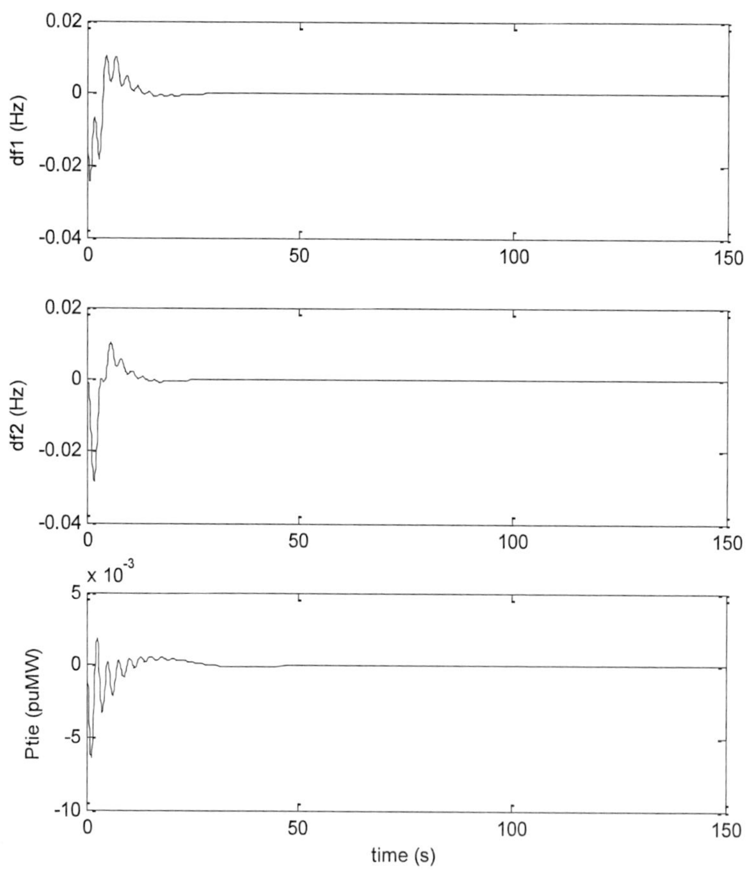

Figure 29. The curves of Δf_1, Δf_2, and ΔP_{tie} in case of the SMES unit in thermal area.

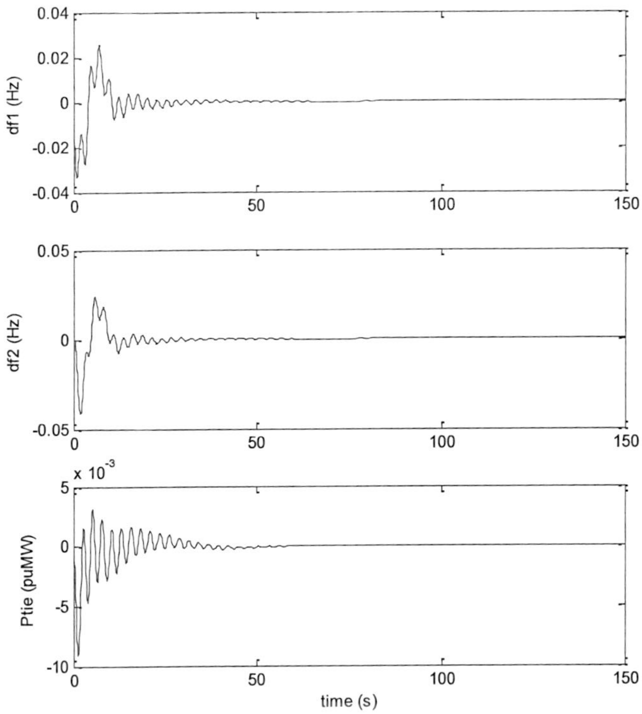

Figure 30. The curves of Δf_1, Δf_2, and ΔP_{tie} in case of the SMES unit in hydro area.

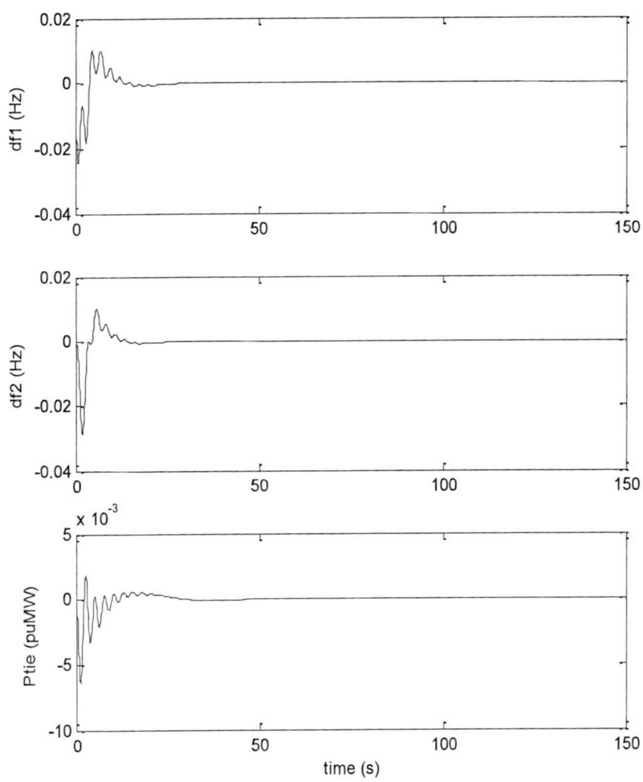

Figure 31. The curves of Δf_1, Δf_2, and ΔP_{tie} in case of the SMES unit in both area.

One of the reasons many fluctuations, the tuning range of the PI-controller is small, so they are in the range of [-1, 1] for K_p and [0, 1] for K_i. As a result of this, the PI-controller may be therefore insufficient. However, those tuning ranges of the gains of the controller may be increased. But, in this case, the controller will be able to go to instability.

As a result of the addition of SMES unit to any area, the fluctuations are eliminated as depicted in Figures 29 and 30. Especially, the curve of tie-line power change is largely smoothed. In addition to this, the effect of SMES unit is more prominent when it is connected to the area where the load change is applied. Otherwise, if the SMES unit is added to the area without load, the effect of the SMES unit decreases. When the SMES units are connected to the both control areas, the fluctuations are also eliminated such as the previous applications. But in this case, regardless of where the load is applied, all SMES units which are added to each area support the power system, and reduce the fluctuations. Thus, the effect of the usage of optimal controller gains and the usage of SMES units are observed for AGC in an interconnected power system. Especially, adding the SMES units to the power system is positively contributed to the dynamic responses of the power generation units.

Conclusion

This chapter has been prepared for helping to the researchers which study on optimal Automatic Generation Control (AGC) for multi area power system, and has been also presented the effect of SMES unit which has been added to the power system. The proposed method has been provided high quality and reliable electrical energy. This control system is also very simple and easy to implement since it does not require many information about system parameters.

At the beginning of this chapter, the introduction to the AGC system is made, and the related literature is given to the readers in order to reveal the studies on this issue so far. After that, the basic concept of the AGC mechanism and its control loops are explained. The transfer function model of an interconnected power system is created by examining and combining the parts of the power system. Then, the gain scheduling PI-control technique and the scheduling method based on Constriction Factor Approach Particle Swarm Optimization (CFA-PSO) algorithm are investigated in order to provide the optimal control of such a power system. The control configuration and the cost function with tunable weight coefficients, which are developed in this study in order to improve the dynamic performance of the power system, are also explained. The usage of the optimization algorithm to optimize the gains of PI-controllers and the tunable weight coefficients of that cost function is explained in detail with the parts of its MATLAB model and codes. Finally, the results of the control are investigated and discussed in detail.

At the end of this study, it is seen that the CFA-PSO algorithm is an effective algorithm to obtain the optimal AGC in a two area hydrothermal power system with nonlinearity, and it is seen that choosing suitable cost function is also quite important for performance of the convergence to the best solution. However, it is clear that the researchers are used different optimization algorithms based on artificial intelligence and different cost functions which derived from the control system. The results obtained with the proposed control system showed that the dynamic performance of the power system is improved.

ACKNOWLEDGMENTS

The authors would like to thank Prof.Dr. İlhan KOCARSLAN for his suggestions and support.

REFERENCES

[1] Bergen, R., and Vittal, V. (2000). *Power system analysis*, USA: Prantice Hall, Second Edition.

[2] Weedy, B.M. (1987). *Electric power systems*, USA: John WileyandSons, Third Edition Revised.

[3] Elgerd, O.I. (1982). *Electric energy systems theory*, New York, USA: Mc Graw-Hill Press.

[4] Cohn, N. (1957). Some aspects of tie-line bias control on interconnected power systems. American Institute of Electrical Engineering Transactions, 75, 1415-1436.

[5] Elgerd, O.I., and Fosha, C.E. (1970). Optimum megawatt-frequency control of multi-area electric energy systems. *IEEE Transactions on Power Apparatus and Systems*, 89(4).

[6] Kumar, I.P., and Kothari, D.P. (2005). Recent philosophies of automatic generation control strategies in power systems. *IEEE Transactions on Power Systems*, 20(1), 346-357.

[7] Yang, B., Chen, Y., and Zhao, Z. (2007). Survey on applications of particle swarm optimization in electric power systems. IEEE, International Conference on Control and Automation - ICCA 2007: INSPEC Accession Number 9807935 (pp. 481-486). China: IEEE Press.

[8] Aström, K.J., and Hagglund, T. (1995). PID Controllers, New York, USA: Instrument Society of America.

[9] Lee, K.A., Yee, H., and Teo, C.Y. (1991). Self-tuning algorithm for automatic generation control in an interconnected power system. *Electric Power Systems Research*, 20(2), 157-165.

[10] Rubaai, A., and Udo, V. (1994). Self tuning LFC: Multilevel adaptive approach. Proceeding of Institution of Electrical Engineering Generation, Transmission, Distribution, 141(4), 285-290.

[11] Chang, C.S., and Fu, W. (1997). Area load frequency control using fuzzy gain scheduling of PI controllers. *Electric Power Systems Research*, 42, 145-152.

[12] Cam, E., and Kocaarslan, İ. (2005). A fuzzy gain scheduling PI controller application for an interconnected electrical power system. *Electric Power Systems Research*, 73(3), 267-274.

[13] Talaq, J., and Al-Basri, F. (1999). Adaptive fuzzy gain scheduling for load frequency control. *IEEE Transactions On Power Systems*, 14(1), 145-150.

[14] Pinkang, L., Hengjun, Z., and Yuyun, L. (2002). Genetic Algorithm Optimization for AGC of Multi-Area Power Systems. IEEE, IEEE Region 10 Technical Conference On Computers, Communications, Control And Power Engineering - TENCON 2002: IEEE Catalog Number 02CH37368 (pp. 1818-1821). China: IEEE Press.

[15] Abdel-Magid, Y.L., and Abido, M.A. (2003). AGC Tuning of Interconnected Reheat Thermal Systems with Particle Swarm Optimization. IEEE, 10th IEEE International Conference on Electronics, Circuits and Systems – ICECS 2003: IEEE Catalog Number 03EX749 (pp. 376-379). United Arab Emirates: IEEE Press

[16] Yeşil, E., Güzelkaya, M., and Ekşin, I. (2004). *Self-tuning fuzzy PID type load and frequency conversion and Management*, 45(3), 377-390.

[17] Juang, C-F., and Lu, C-F. (2004). Power System Load Frequency Control By Evolutionary Fuzzy PI Controller. IEEE, IEEE International Conference on Fuzzy Systems: INSPEC Accession Number 8262584 (pp. 715-719). Hungary: IEEE Press.

[18] Massiala, M., Ghribi, M., and Kaddouri, A. (2006). A Two-Layered Self-Tuning Fuzzy Controller For Interconnected Power Systems. IEEE, Canadian Conference on Electrical and Computer Engineering CCECE/CCGEI - 2006: INSPEC Accession Number 9364960 (pp. 1086-1089), Ottawa: IEEE Press.

[19] Taher, S.A., Hemati, R., Abdolalipour, A., and Tabie, S.H. (2008). Optimal decentralized load frequency control using HPSO algorithms in deregulated power systems. *American Journal of Applied Sciences*, 5(9), 1167-1174.

[20] Gözde, H., Taplamacioğlu, M.C., and Kocaarslan, İ. (2009). A Swarm Optimization Based Load Frequency Control Application in A Two Area Thermal Power System. IEEE, International Conference on Electrical and Electronics Engineering ELECO - 2009: (pp. 124-128). Turkey: Uludağ University Press.

[21] Nanda, J., Mishra, S., and Saikia, L.C. (2009). Maiden application of bacterial foraging based optimization technique in multiarea automatic generation control. *IEEE Transactions on Power Systems*, 24(2), 602-609.

[22] Passino, K.M. (2002). Biomimicry of bacterial foraging for distributed optimization and control. *IEEE Control Systems Magazine*, 22(3), 52-67.

[23] Gözde, H., Taplamacıoğlu, M.C., Kocaarslan, İ., Şenol, M.A. (2010). Particle swarm optimization based PI-controller design to load-frequency control of a two area reheat thermal power system. Isı Bilimi ve Tekniği Dergisi (*Journal of Thermal Science and Technology*, 30(1), 13-21.

[24] Tripathy, S.C., Balasubramania, R., and Chanramohanan, P.S.N. (1992). Adaptive automatic generation control with superconducting magnetic energy storage in power system. *IEEE Transactions On Energy Conversion*, 7(3), 434-441.

[25] Mitani, Y., Tisuji, K., and Murakami, Y. (1998). Application of superconducting magnet energy storage to improve power system dynamic performance. *IEEE Transactions On Power System*, 3(4), 1418-1425.

[26] Tripathy, S.C., and Bak-Jansen, B. (2001). Automatic Generation Control Of Multi Area Power System With Superconducting Magnetic Storage Unit. IEEE, Porto Power Technology Conference: INSPEC Accession Number 7219796 (pp. 6). Portugal: IEEE Press.

[27] Demirören, A., and Yeşil, E. (2004). Automatic generation control with fuzzy logic controllers in the power system including SMES units. *Electrical Power and Energy Systems*, 26, 291-305.

[28] Abraham, R.J., Das, D., and Patra, A. (2007). Automatic generation control of an interconnected hydrothermal power system considering superconducting magnetic energy storage. *Electrical Power and Energy Systems*, 29, 571-579.

[29] Shayeghi, H., Jalili, A., and Shayanfar, H.A. (2008). A robust mixed H2/H∞ based LFC of a deregulated power system including SMES. *Energy Conversion and Management*, 49, 2656–2668.

[30] Pothiya, S., and Ngamroo, I. (2008). Optimal fuzzy logic-based PID controller for load frequency control including superconducting magnetic energy storage units. *Energy Conversion and Management*, 49, 2833-2838.

[31] Gözde, H., and Taplamacioğlu, M.C. (2009). A Swarm Optimization Based Automatic Generation Control In A Hydrothermal Power System Including SMES Units. IEEE, International Conference on Electronic, Computers and Artificial Intelligence ECAI - 2009: (pp. 67-72). Romania: University of Pitesti Press.

[32] IEEE Committee Report (1973). Dynamic Models for Steam and Hydro Turbines in Power System Studies. *IEEE Transactions on Power Apparatus and Systems*, 92(6), 1904-1915.

[33] Ogata, K. (2002). Modern Control Engineering, USA:, Prentice-Hall Inc., Fourth Edition.

[34] Kennedy, J., and Eberhart, R.C. (1995). Particles Swarm Optimization. Proceeding *IEEE International Conference on Neural Networks*: (pp. 1942-1948). Piscataway, NJ: IEEE Service Center.

[35] Chen, J., Ren, Z., and Fan, X. (2006). Particle swarm optimization with adaptive mutation and its application research in tuning of PID parameters. 1st International Symposium on Systems and Control in Aerospace and Astronautics - ISSCAA 2006: INSPEC Accession Number: 9017226 (pp. 994). China.

[36] Reynolds, C.W. (1987). Flocks, herds and schools: A distributed behavioral model. Proceedings Of The 14th Annual Conference On Computer Graphics And Interactive Techniques, (pp. 25-34). USA, New York.

[37] Heppner, F., and Grenander, U. (1990). A stochastic nonlinear model for coordinated bird flocks. In S. Krasner (Ed.), The Ubiquity of Chaos. Washington, D.C.: AAAS Publications.

[38] Eberhart, R.C., and Shi, Y. (1998). A modified particle swarm optimizer. IEEE, Proceedings Of The International Conference On Evolutionary Computation: (pp. 69-73). Piscataway, NJ: IEEE Press.

[39] Clerc, M. (1999). The swarm and the queen: Towards a deterministic and adaptive particle swarm optimization. IEEE, Proceeding Congress on Evolutionary Computation: (pp. 1951-1957). Piscataway, NJ: IEEE Service Center.

[40] Kochenburger, R.J. (1950). A frequency response method for analyzing and synthesizing contractor servomechanisms. *AIEE Transactions*, 69(1), 270-284.

[41] Lu, C-F., Liu, C-C., and Wu, C. (1995). Effect of battery energy storage system on load frequency control considering governor dead-band and generation rate constraints. *IEEE Transactions On Power System*, 10(3), 555-561.

[42] Tripathy, S.C., and Chanramohanan, P.S.N. (1994). Automatic generation control with superconducting magnetic energy storage in power system. *Electric Power Components and Systems*, 22(3), 317-338.

[43] Kundur, P. (1993). Power system stability and control, New York, USA: Mc Graw-Hill Press.

[44] Ham, W.K. (2003). Active and reactive power control model of superconducting magnetic energy storage (SMES) for the improvement of power system stability. Ph.D.Thesis, USA: The University of Texas at Austin Press.

[45] Tripathy, S.C., Balasubramania, R., and Chanramohanan, P.S.N. (1992). Effect of superconducting magnetic energy storage on automatic generation control considering governor dead-band and boiler dynamics. IEEE Transactions On Power Systems, 3(7), 1266-1273.

[46] Mohan, N., Undeland, T., and Robbins, W. (1995). Power Electronics, Converters, Applications and Design (2nd edition), New York, USA: John Willy and Sons press.

In: Advances in Energy Research
Editor: Nicu Bizon

ISBN 978-1-61209-991-0

Chapter 15

A POINT OF VIEW REGARDING WEB SERVICES-BASED INFRASTRUCTURE FOR GEOSPATIAL INFORMATION AND IMPLEMENTING DECISION SUPPORT SYSTEM FOR COMPLEX PR*OCESS MONITORING

Angela Ionita[1*], Adrian Zafiu[2] and Radu Constantin Gogu[1,3]

[1]Research Institute for Artificial Intelligence,
Romanian Academy, Romania
[2]University of Pitesti, Romania
[3] Technical University for Civil Engineering of Bucharest, Romania

ABSTRACT

This chapter aims to provide a non-exhaustive point of view, based on the literature, regarding the premises for building Spatial Decision Support Systems for complex processes monitoring using Web Services-based Infrastructure for Geographic Information, and to identify some challenges and to highlight on the possible solutions in order to overtake the critical point in this areas: syntactic and semantic interoperability from geospatial information used in different fields of geosciences. Starting from the terminological aspects and evolution, has been presented a frame of work covering the European and worldwide initiatives in connection with GI infrastructures building and some application domains associated with the challenges coming from the intention to offer the best solution in order to solve the problems of semantic interoperability and semantic heterogeneity followed by some relevant examples.

Keywords: web services, geoweb services, portal, geoportal, geospatial information, (spatial) decision support system, interoperability, ontology.

* Corresponding Author: aionita@racai.ro

INTRODUCTION

Given the advances in information and communications technologies in general and management of Geospatial Information (GI) in particular, Spatial Decision Support Systems (SDSS) has been and will be an important subset of Decision Support Systems (DSS) applications in next future. This trend will be driven by the relevance of spatial information as a component of the information needed for a wide range of decisions. This class of DSS will make an important contribution, not because of its use of the latest technology, but because it will allow decision makers to incorporate a spatial dimension in their decision making. This spatial dimension, which is not fully accounted for in traditional DSS designs, is an important feature of many areas of DSS application. These potential areas of application in general refer complex processes monitoring. For this larger class of decision makers who might use SDSS, spatial data will be used with other types of information required originating in specialized models, often of a non-spatial nature. For this broad range of applications, Geographic Information Systems (GIS) technology alone can only make a partial contribution to decision support. Comprehensive decision support will require the effective integration of GIS and non GIS techniques. The challenge for SDSS builders is to harmonize the search and the access of the big amount of heterogeneous data from different sources and interface with heterogeneous database, drawn from the GIS and specialized domains, in order to achieve an appropriate synthesis of modeling techniques and to provide effective decision support for interest areas. The content of the big amount of heterogeneous data from different sources and of these heterogeneous databases generate a lot of problems starting from availability, accessibility, interoperability and understanding. In additional, after 12 years from Al Gore visionary's "Digital Earth" [1] describing a virtual representation of the Earth on the Internet that is spatially referenced and interconnected with the world's digital knowledge archives, supported by recently available geo-browsers (like Google Earth, Microsoft Virtual Earth, NASA Worldwind) and web applications (like Google Maps, Microsoft Live Search Maps and Yahoo Local Maps) these systems have introduced geo-services to ordinary Internet users, offering them high-resolution aerial imagery with responsive performance [2; 3]. Moreover, in order to enable a coherent and contextual use of GI and to improve the decision making in different areas, an increasing necessity to share this information between different stakeholders (e.g., departments in central and local public administration, professionals, citizens, etc.) and diverse information systems has been signaled and a new generation of Decision Support System (DSS) for complex process monitoring having as fundamental Web Services Infrastructure for Geospatial Information occurs. The integration of these aspects harnesses the decision analytic power of traditional DSS and the spatial capabilities of GIS, on the one side, and the Infrastructure for Geospatial Information and Web Services, on the other side.

The necessity of GI organization forms the basis for a number of different international and national initiatives, to set up global, international, national, local and regional infrastructures for the collection and dissemination of geographical data, including among others:

- INSPIRE: the INfrastructure for SPatial InfoRmation in Europe (INSPIRE) - an European Directive approved in 2007, based on the goal to improve the accessibility, interoperability and affordability of spatial data and information systems in Europe, specifically: "*Experience in the Member States has shown that it is important, for the successful implementation of an infrastructure for spatial information, that a minimum number of services be made available to the public free of charge. Member States should therefore make available, as a minimum and free of charge, the services for discovering and, subject to certain specific conditions, viewing spatial datasets.*" [4].

- SEIS: the Shared Environmental Information System (SEIS) - a collaborative initiative of the European Commission, the European Environmental Agency (EEA) and the member countries of the Agency. It aims to improve the availability and quality of information needed to design and implement the European Union's environment policy [5].
- GEOSS: the Global Earth Observation System of Systems (GEOSS) - being built by the Group on Earth Observations (GEO). GEOSS seeks to connect the producers of environmental data and decision-support tools with the end users of these products, with the aim of enhancing the relevance of Earth observations to global issues. The ultimate result is to provide a global public infrastructure that generates comprehensive, near-real-time environmental data and information for a wide range of users (GEOSS, 2008).
- GMES (Global Monitoring for Environment and Security) - a joint initiative of the European Commission and European Space Agency, adopted by EU Heads of State at the Gothenburg Summit in 2001, and aimed at achieving an autonomous and operational capability in the exploitation of geo-spatial information services by 2008. GMES is the European Union contribution to GEOSS [6].
- OGC (Open GeoSpatial Consortium) initiative - Open GeoSpatial Consortium's (OGC) (http://opengeospatial.com/) core mission is to deliver spatial processing interface specifications that are openly available for global use. OGC envisions world in which everyone benefits from geographic information and services made available across any network, application or platform. Currently Open GIS is involved in the development and provision of Open GIS specification, development of strategic business opportunities and promotion of demand for interoperable products.
- Global Mapping Initiative - an international cooperation initiative through voluntary participation of National Mapping Organizations (NMOs) of the world with the aim of developing digital geo-information framework datasets with a spatial resolution of 1 km. The primary objective of the Project is to contribute to sustainable development, environmental conservation and natural disaster mitigation through provision of basic geo-information framework datasets. The Project was initially proposed by Ministry of Construction of Japan (present Ministry of Land, Infrastructure, Transport and Tourism of Japan) in 1992, in conjunction with the adoption of Agenda 21 at United Nations Conference on Environment and Development (UNCED) in Rio de Janeiro. Preparatory works to implement the proposal eventually resulted in the establishment of the International Steering Committee for Global Mapping (ISCGM) in 1996.
- GSDI (Global Spatial Data Infrastructure) initiative – was formed to promote the SDI concept and the first GSDI conference with the theme "The Emerging GSDI" was

held in Bonn, Germany, September 1996. Since then the GSDI conferences which are held annually have promoted the SDI concept globally. Among its goals, the association GSDI supports the establishment and expansion of local, national, and regional spatial data infrastructures that are globally compatible and supports interdisciplinary training, education and research activities that advance spatial data infrastructure development. Both internal and external funds are used to support projects and programs that further these goals [7].

Key Development of Web Services-Based Infrastructure for Geospatial Information and Implementing Decision Support System For Complex Process Monitoring

For didactical and practical reasons, it may be useful to group key developments of relevance to this section into seven related themes: organizing geographic information, using geography as a way to organize information, contribution of supporting technology, portals and geoportals, web services, geoweb services, spatial decision support systems.

Organizing Geographic Information

This theme includes the many initiatives since the early 1990s aimed at increasing the availability and accessibility of geographic information through the development of spatial data infrastructures (SDIs). By the mid 1990s, Masser [8] identified 11 SDIs at varying stages of development spanning large countries like the USA, Canada, and Australia, small ones like the Netherlands and Portugal, and developing nations like Malaysia, Indonesia, and Qatar. That first generation of SDIs was largely led by national mapping agencies and oriented towards the completion of national spatial databases addressing topography and other key layers of general use. The documentation of existing resources via metadata, and access mechanisms via catalogues and clearinghouses were other key features of these early developments.

In Europe a major recent development has been the adoption of a legal framework in 2007 to establish a distributed Infrastructure for Spatial Information in Europe [4; 9] built on the SDIs of the 27 Member States of the European Union. In 2009, in the framework of eSDI+ project [10], the six person jury around of Ian Masser carried out a rigorous evaluation and selected 12 SDIs from 9 European countries from total number of submissions for the best practice awards and were highly commended at an International Conference in Turin on November 26th and 27th 2009 in Turin, Italy.

Since then happen a rapid diffusion of Spatial Data Infrastructures (SDIs) worldwide facilitated by the establishment in 1996 of the Global Spatial Data Infrastructures Association (www.gsdi.org) that has helped the promotion of best practice and sharing of experiences, and capacity building in the Americas, Africa, Asia and the Pacific and Europe. At the international level, it is also worth noting that in 2006 the United Nations Geographic Information Working Group (UNIGWG) [11] developed the vision, strategy and the institutional governance framework for a United Nations SDI (UNSDI) initiative. In its initial phase (2008-2010) the UNSDI initiative is project-based, defined around projects outputs that can involve non-UN partners but in the medium-term the UNIGWG membership recognizes that the UNSDI will require legislative legitimacy.

The nature of these more recent SDIs has also shifted with an increased number of stakeholder organizations engaged in the process. There is also a stronger emphasis on distributed data and processes, and the interoperability of services to discover, view, access, and integrate spatial information.

The interoperability of systems through services has been the major target of the Open Geospatial Consortium (OGC) (www.opengeospatial.org) established in 1994 (as Open GIS Consortium) as an international partnership between government agencies, industry, and academia. The OGC client-server interface specifications and the standards adopted by the International Standards Organization (ISO) have become the cornerstone of most SDIs in their current form.

Using Geography as a Way to Organize Information

This theme includes the development of geo-browsers (e.g. Google Earth, Microsoft Virtual Earth, NASA Worldwind) which use the globe as mechanism to pan, zoom, and fly over the Earth's surface to areas of interest much as in the original vision of Digital Earth. Associated to these 3D representations of the Earth are also 2D applications (Google Maps, Microsoft Live Search Maps) that also allow users to add and share information via simple Application Programming Interfaces (APIs). Beyond visualization however, geobrowsers have become client applications for accessing a more complex infrastructure behind the scenes and for fulfilling a wider goal: organizing the world's information, in part spatially [12]. As mentioned in [13], the "Digital Earth Initiative" (DEI) chaired by National Aeronautics and Space Administration (NASA) [14] has been identified as a key milestone in trying to put the Al Gore vision into practice. Between 1998 and 2001, the initiative involved a number of US federal agencies and focusing on interoperability, infrastructure and organizational issues. A major step forward [2] came from the commercial sector in 2001 with the launch of Keyhole's Earth Viewer Google acquired Keyhole in 2004 and launched its Google Earth in 2005, the same year also making available its API for Google Maps thus making it possible for anybody to add information to the Google platform. The Keyhole Markup Language (KML) has also become a powerful way to document and index information with a geographic reference, and displays it on maps or globe interfaces. The declared aim of Google is therefore not to organize geographic information (as in SDIs) but to use geography as a way to search and view information with a geographic footprint. The success of Google Earth [15] and Google Maps [16] with hundreds of millions of users, spurred Microsoft to accelerate its own developments and release at the end of 2006 its alternative products, with a stronger emphasis on 3D visualizations, particularly in urban areas. It is not to be unexpected that users focus their attention on the geobrowser client, for after all it is this which must be installed on their desktop machine and allows them to interact with geographic information.

In this chapter refer to this as the GeoWeb infrastructure to provide contrast to the traditional SDI approach.

Contribution of Supporting Technology

The fact that today geobrowsers function at very high speeds while accessing, on-line, vast qualities of global-scale geodata is due in large part to advancements in the underlying technology [2]. Given the vast quantities of information that are needed to represent Earth digitally it is impractical for each user to maintain their own local copy of a Digital Earth vision. A more practical solution is to store the information of a collection of shared servers that can be accessed over a high-speed network connection. Although multiple protocols are available in theory, the major technology providers are rapidly converging on using Web protocols. On the server side, Google (and later Microsoft) boldly illustrated the speed accruing from distributing its index and its stored data across thousands of servers around the world. These servers cache copies of the most frequently used or viewed data, including map tiles. They are also capable of processing vast quantities of information in order to respond to client queries.

Portals and Geoportals

The origins of geoportals can be traced to the early growth of the Internet. Web sites like MapQuest [17] and MapBlast [18] are examples of portals that capitalized on the advent of the Internet and the interest of the public to locate places and map them. Even after adjusting for the hype factor, this is still a significant indicator of the value of geographic technology in popular media. The US Federal Geographic Data Committee's (FGDC) Clearinghouse web sites represent one of the earliest spatial data infrastructure (SDI) web portal initiatives. The FGDC portals were driven by the first US presidential requirement to support the sharing of geographic information, the National Spatial Data Infrastructure or NSDI [19; 20].

A technical definition of the word "portal" is "a web site considered being an entry point to other web locations" [21]. Append the term "geo", and the result might be as follows. Geoportal: *a web site that presents an entry point to geographic content on the web or, more simply, a web site where geographic content can be discovered.* Many web sites, however, have some association with geographic content or functions. For example, most web sites representing businesses have, at a minimum, addresses for the business locations on their sites, and many have links to map images depicting these locations. What then is the defining characteristic that differentiates a geoportal from other web sites with some geographic content? In most sites, geographic content supports the site but is not the site's primary focus. So to refine the definition further, *a geographic portal is a web site where the discovery of geographic content is a primary focus.*

As examples of geoportals we have to mention, in order to enumerate a few:

- INSPIRE geoportal [22] will be fully operational on 2011;
- GEOportal [23] provides an entry point to access Earth Observation information and services. It connects to a system of existing portals, addressing the GEO Societal Benefit Areas globally while also providing national and regional information to enhance understanding;
- SDI-T GeoPortal launched at World Food Programme [24];

- GEO Data Portal of United Nations Environment Programme [25];
- GeoSUR La Red Geospacial de America del Sur [26];
- Mountain GeoPortal [27];
- National GeoPortal in Finland [28];
- French Geoportal launched in June 23, 2006, rival to Google Earth [29];
- an organizational Romanian Geoportal (data delivery or download portal) to provide information and access to the geo-products of the ANCPI [30];
- a hydrogeological webmapping geoportal [31].

Web Services

Web services are internet-based interoperable ICT (Information Communication and Technology) services proscribed by a formal model which support business functions within the SOA (service-oriented architectures) paradigm [32]. Because they are standardized, self describing and XML based, they can be discovered and invoked automatically, are dynamically bound at the time of use and are interoperable in multiple environments. According to a report by the Stencil Group [33] the modularization of applications fundamental to web services, allows a flexibility and agility whereby organizations can add new offerings or reorient existing functions.

Web services allow to access data from a variety of databases and internet locations. After has been accessed the data, a Web service consumer can search, mine, and then transform the data to use it with a data warehouse for further analysis.

Through modularizing applications and making the modules interoperable through standard protocols on a fee for service basis, web services reduce the need for customization and up-front investment, lower the transaction cost of commonly needed ICT services, and offer organizations flexibility and agility in it integrating and enhancing their ICT-based functions.

According to [34], the availability (and most importantly, the growing acceptance) of common, platform-independent, tools and technologies was required to support true Networked Enterprises and Web services. Web Services is a distributed computing technology that offers interaction and collaboration among vendors and customers, with the vision of providing ubiquitous computing.

According to [35] some of the early products in Web Services started appearing in 1997 when Sun announced its Jini platform and Hewlett-Packard its e-speak After that, many big players such as IBM and Microsoft joined this race. The Web Services arena picked up steam after the big players roped in and several small players also joined hands for what was perceived as the next Internet wave. Server-standard body consortiums were formed, which developed numerous standards on different aspects of Web Services. Some of the key standard bodies' consortiums are: World Wide Web Consortium (W3C) (http://www.w3.org/), Organization for the Advancement of Structured Information Standards (OASIS), (http://www.oasis-open.org/), Java Community Process (JCP) (http://jcp.org/), Object Management Group (OMG) (http://www.omg.org/), and several individual efforts by a group of companies.

"Fundamental to web services, then, is the notion that everything is a service, publishing an API for use by other services on the network and encapsulating implementation details" (OGC® White Paper [36]).

Web Services and SOA have great promise to enhance integration efforts and streamline many business processes. But despite emerging standards that should hasten their adoption, many problems still exist in special in connection with interoperability and in particular with semantic interoperability.

Two of the key problems solved by Web Services over earlier distributed systems were discussed [35]:

- Interoperability: Earlier distributed systems suffered from interoperability issues because each vendor implemented its own on-wire format for distributed object messaging. By using XML as an on-wire standard, the two camps of Java/J2EE and .NET/C# now could speak to each other.
- Firewall traversal: Collaboration across corporations was an issue because distributed systems used non-standard ports. As a result, collaboration meant punching a hole in your firewall, which was often unacceptable to ICT. Hence, this did not allow any dynamic collaboration, as it required going through a manual process for collaborating with partners. The dynamic nature of Web Services interaction offers several exciting services for the users.

On the short and according to [37], a Web Service is the externally visible part of a specific business function that is exposed to the web via a well-defined interface, and described and invoked via standard web protocols.

The problem of integration of web services on the basis of the capabilities that they provide has recently received a considerable attention.

Geoweb Services

With the arrival of major players such as Google, Microsoft, and Yahoo, the landscape of the geospatial world has changed. Google, armed with a friendly exchange format, a powerful desktop client, and an easy to use Application programming interface (API) for building web applications has successfully provided a way for users who know nothing about GIS to create maps. They have successfully marketed web mapping to the average consumer. The upshot of which has been an increase in the amount of geospatial data on the web. Central to this revolution is the notion of the 'geoweb'. The Geospatial Web or Geoweb is a relatively new term (http://en.wikipedia.org/wiki/ Geoweb) that implies the merging of geographical (location-based) information with the abstract information that currently dominates the Internet. This would create an environment where one could search for things based on location instead of by keyword only – i.e. "What is Here?"

The concept of a Geospatial Web may have first been introduced by Charles Herring [38]. Charles Herring proposed that the problem of defining the physical domain in a computer or cyber-infrastructure, providing real time and appropriate fidelity, required a cyber-spatial reference or index combining both Internet Addressing and Hierarchical Spatial Addressing. As such, the Geoweb would be characterized by the self synchronization of network addressing, time and location. The Geoweb would allow location to be used to self organize all geospatially referenced data available through the Internet.

Another early reference to the concept of the Geoweb is the Genasys Spatial Web Broker [39]. Development of this capability began in 1994 and was officially announced in 1996. This product was designed to allow any web developer to integrate GIS functions, such as routing, into portal applications, e-government applications, and e-business applications.

The interest in a Geoweb has been advanced by new technologies, concepts and products. Virtual globes such as Google Earth and NASA World Wind as well as mapping websites such as Google Maps, Live Search Maps (http://www.maps.live.com/) and Yahoo Maps (http://maps.yahoo.com/) have been major factors in raising awareness towards the importance of geography and location as a means to index information.

For the context of this chapter the connection between SDI and geoweb services is relevant. According to [40]: "*Building a Spatial Data Infrastructure (SDI) is a hard problem. Realizing the model of sharing and collaboration among the organizations which participate in a SDI is easier said than done. There are numerous obstacles and challenges that stand in the way. Perhaps at the forefront is the issue of cost-benefit. Obtaining and maintaining good data is an expensive and time consuming process. Sharing this data often defeats the purpose of gathering it in the first place as high-quality data often gives an organization an edge in the market place. In cases where sharing of data is not an issue, more often then not it such a low priority that the effort simply diminishes. Data is used to drive the decisions which solve problems. Once a problem is solved any remaining effort is considered overhead.*"

The capacity of Geospatial Web would be similar to Google Search and likely provide similar value. It is conceived that the Geospatial Web will present as a visual medium and geospatial platform for data self-organization, discovery and use. Capabilities that allow every Internet user to post to this flow of information and anyone to poll or pull the information will lead to a new commons, media or marketplace for publication, trade and commoditization of information.

Technologies and innovations revolving around the emerging geoweb make the world of geospatial an exciting place. But it is important not to lose sight of what is important. No matter how good technology gets there is no substitute for good data. It is high quality data that makes high quality maps, not the tools. It is for this reason that SDI initiatives are of crucial importance to creating a true web of geospatial information [40].

Spatial Decision Support Systems (SDSS)

A new category of Decision Support Systems (DSS), considered a subset of DSS, called Spatial Decision Support System (SDSS) has emerged which uses a spatial representation for displaying spatial data and supporting spatial modeling. Research on SDSS originated from three different sources: DSS, GIS and ICT. DSS has been an active area of research in Information Systems for many years; however, DSS researchers have always acknowledged one of its major limitations - its inability to support spatial data. On the other hand, GIS is efficient in storing and managing spatial data, but has lagged behind in providing adequate tools to facilitate managerial decision making and cooperative problem solving. The synergy between these three research groups has lead to the adoption of state of the art technical solutions and the development of sophisticated SDSS that satisfy the needs of geographers and top-level decision makers. Recently, the Web has added a new dimension to SDSS and Web-based SDSS (WebSDSS) are being developed in a number of application domains.

In a review of GIS, in [41] identified SDSSs as a growth area in the application of GIS technology. However the value of SDSS is not determined by its innovative use of technology. Instead the contribution of these applications will be determined by how well they support the need for a spatial component in decision making.

In [42] has been suggested that three categories of decision maker may find that SDSS can make a contribution to their decisions. The first group is in the traditional areas of application of GIS, in disciplines such as geology, forestry, and land planning. In these subject areas there will be growth of decision making applications in much the same way as data processing applications evolved into DSS in traditional business applications.

The second group of decision makers, for whom SDSS can make an important contribution, is in fields such as routing or location analysis. Although the spatial component of such decisions is clear, DSS design has in the past been driven predominantly by the management science models used. In the future these models will be incorporated into GIS based SDSS, providing superior interface and database components to work with the models. This synthesis of management science and GIS techniques will provide more effective decision making. Keenan [43] has argued that the use of GIS techniques can extend the range of decision support for vehicle routing problems, allowing consideration of path constraints that have not been comprehensively modelled in the past. Routesmart [44] is a good example of the use of GIS for a routing application. Another is presented in [45].

The third group of decision makers who will find SDSS important include those where the importance of both spatial data and modelling is somewhat neglected at present. In disciplines, additional possibilities for analysis are provided by the availability of increasing amounts of spatially correlated information.

CHALLENGES AND REQUESTED SOLUTIONS

Interoperability Issues

The complex processes monitoring systems based on SDI(s) must be implemented in a manner that allows easy interoperability between heterogeneous organizations, systems, databases and archives. The common frame of work is based on a generic software platform, which supports a variety of geospatial dataset types as well as comprehensive tools for data management, editing, analysis, and visualization in order to create a good support for decision making.

The Institute of Electrical and Electronics Engineers (IEEE) defines interoperability as: "the ability of two or more systems or components to exchange information and to use the information that has been exchanged".

Interoperability issues pervade both geospatial data and geographic services. Moreover, to share geo datasets, a number of geographic services have to be provided by the system. In this section, in order to better present and discuss these issues, a separate analysis [46] for geographic data (geo-data) and geographic services (geo-services) is presented. In [47] an extended definition and analysis of interoperability in the field of GI is also given.

Geo-Data Interoperability Issues

One of the key services supplied for a complex processes monitoring system based on SDI(s) is the possibility to retrieve geographical datasets provided by heterogeneous resources.

Mainly because the logical architecture of an SDI can be based on a set of different data resources and heterogeneous geographical information has to be integrated. Since each geo-data producer adopts internal rules in order to manage its geographical datasets, heterogeneity at the data level arises from a number of different reasons [48]:

- *Different syntax*: geo-datasets are retrieved from different sources that can use different data formats (e.g., ESRI shape files, Mapinfo Files, Oracle geo-DB, GRASS files).
- *Different structure*: geographical features can be represented using different geometrical and data schemas. Often the same geographic feature is represented using different geometric features (for instance, roads can be represented using polygons or lines) or by other techniques [49].
- *Different semantics*: interoperability problems due to different semantics are caused by different reasons. *Naming conflicts* occur when classes or attribute types with different semantics are given the same names (homonyms) or when classes or attribute types that are semantically the same are named differently (synonyms). Moreover, geographical datasets have specific properties, different from other types of data, including [50]:
- *Implicit linking*: in general, explicit relationships must be present to combine information in a comprehensive manner. However, geographic information enables linking without explicit references, for instance, via coordinate reference systems.
- *Massive datasets*: compared to financial, organizational information, geo information can be massive.
- *Multiple versions*: multiple versions of the same entities on the Earth's surface can differ radically in terms of data model, scale, data generalization, conceptual model, and semantics the data collectors use. The main reason is that data collectors are often represented by different institutions/organizations at different levels (e.g., regional, national and international) in different places and countries. Specifically, for the case of geo-data integration, we have also scale conflicts and different precisions/ resolutions issues.

Geo-Service Interoperability Issues

Distributed service discovery, composition and coordination have been and continue to be the main research topics in the field of web services. General issues dealing with service integration include:

- *Geo-service discovery*: GIS desktop applications provide to the user a lot of complex functions in order to perform GIS data acquisition, creation, analysis, visualization and mapping. Distributed Service Oriented Architecture (SOA) is a common framework for modern distributed information systems [51]. SOA and Open Geospatial Consortium (OGC) [52] specifications are the base technology used by an SDI in order to provide catalogue services for discovering appropriate data and services for a specific task.

- *Geo-service integration* (composition or coordination): after discovering, services can be composed or coordinated to provide complex functionalities. Although at present, the main available web service in GIS is the map request service, the trend is to supply a technological environment that provides a number of as known stand-alone GIS services. Te majority of these geo-services exist as single services. In the case of a request for a complex service a manual and static composition of a number of predefined geoservices has to be performed. The future challenge is the (semi)-automatic composition of arbitrary services in order to obtain flexible complex services based on the available web services. In practice, however, chaining geographic services is a nontrivial task, mostly because geographic data have varied differences from other types of data and also because individual web service providers use different syntactic structure just as they use different vocabulary to define web service signatures and descriptions. As it mentioned in [53], when integrating geo-services from heterogeneous sources, some specific issues have to be taken into account:
- *Maps as implicit interfaces:* everyone is familiar with reading maps, so they are a natural human–machine interface for the services interacting with the user and presenting (intermediate) results of geo-information.
- *Geometry based information*: since geo-information is geometry based, geo-service interface has always to take into account the geometric component of the data they provide and process (e.g., the bounding box of a map and the coordinate reference system of geographical layers).
- *Specific topological operations*: it is also possible to apply a whole set of common mathematical tools in geo-services to compute their topological relationships, e.g., to compute the distance between two objects, the buffer around an object, the intersection between different features, and the neighbors of a polygon. From a technological point of view, SOA for business services and OGC specifications for geographic information represent the reference framework when discovering and integrating available geo-services. However, as in the case of geo-data, also geo-services are defined using implicit or, in the best case, local semantics. At present, no standard notions are used for defining the semantics of a geographic web service. This problem is typically referred to as the need for *semantic interoperability among autonomous and heterogeneous systems*. Semantic heterogeneity (the differences in meaning) problem is an actual challenge for geographic services integration.

In order to summarize, the interoperability can be classified as:

SYNTACTIC INTEROPERABILITY -	If two or more systems are capable of communicating and exchanging data, they are exhibiting syntactic interoperability. Specified data formats, communication protocols and the like are fundamental. In general, (Extensible Markup Language (XML) or Structured Query Language (SQL) standards provide syntactic interoperability. Syntactical interoperability is required for any attempts of further interoperability.
SEMANTIC INTEROPERABILITY -	Since the early 1990s, interoperability of spatial data has been a major concern in the spatial information community (standardization bodies, research community, developers and users). However, the recent increase in the number of data sources and their widespread availability on Internet has emphasized the interoperability difficulties. One major concern is to deal with the interoperability of spatial data in their global context that is to take into consideration their intrinsic semantics in combination with the semantics of their spatial and temporal properties. Doing so requires a more global approach that has named geosemantics.

Interoperability between GI Systems is an area of increasing interest within academic institutions, not only in Geography and other geo-related areas but also in Computer Science departments. From the software-engineering point of view, a revolution may be said to have occurred in the way of design such systems, a revolution inspired by the appearance of OGC standards, guidelines and working documents.

Interoperability issues pervade SDIs, since they can connect heterogeneous organizations and systems. Research in information system interoperability is motivated by the increasing heterogeneity of the computer systems. Heterogeneity in GIS field is not an exception and may be classified as syntactic (differences in data format), structural (differences in schemas) and semantic (differences in intended meaning of terms in specific contexts) heterogeneity [54]. Beyond the ability of two or more computer systems to exchange information, semantic interoperability is the ability to automatically interpret the information exchanged meaningfully and accurately in order to produce useful results as defined by the end users of both/multiple systems. To achieve semantic interoperability, both/multiple sides must defer to a common information exchange reference model giving the possibility for common understanding.

In the emerging open and distributed environments such as complex processes monitoring systems based on SDI(s), interoperability is essential. Interoperability involves the development of methodologies, models and tools to allow sharing, exchange and control of data [54]. The objective is to allow a collection of information systems to interoperate by means of well-known interfaces. The efforts of the OpenGeospatial Consortium (http://www.opengis.org) have been realized in several specifications that describe a technical infrastructure for the interoperability of GIS systems. The OpenGeospatial Consortium, in its Web Services Architecture specification [55] includes three main types of georeferenced information access services: Web Map Server (WMS), Web Coverage Server (WCS) and

Web Feature Server (WFS). The Web Map Server (WMS) [56; 57] provides specialized capabilities that support the visualization of geospatial information. The Web Feature Server (WFS) [58] and the Web Coverage Server (WCS) [59] provide access to collections of data, also through well-known interfaces. Different services can be chained to support information production workflows.

FUTURE DIRECTION OF RESEARCH

Using Ontological Approach

The research area of semantic integration – semantic interoperability of geospatial information is relatively young. In fact, even if the concept of geo-service publication is not new, OGC specifications and ISO standards became stable only during the last years [46]. Thereafter, different geo-information providers have started to publish their geo-data and services on the web in a standardized manner. Only recently, the integration of GI became relevant and feasible because of the availability of geo web services.

The work done on the area of assuring semantic interoperability and overcoming the semantic heterogeneity could be classified in two categories: (a) geo web services-based and (b) ontology-based methods.

From the first category it is necessary to mention the examples presented in (Shengu Tu and Mahdi Abdelguerfi, 2006) where have been mentioned five examples, as follows:

- technical insight into TerraService technologies through a bottom up tour of the GIS Web service site's design and construction, as well as end-to-end descriptions of two large-scale GIS projects that use TerraService [60]:
- applying W3C Web services technologies, but to a GIS Web service site with additional complexity [61];
- model-based analysis for a specific spatial data infrastructure, a groundwater vulnerability assessment service [62];
- a prototype framework for a semiautomatic Web service chaining tool [50];
- two interesting efforts that support Web services for GIS: increasing WMS resources' availability and mediating clients' requests and available services based on domain-specific semantics [63].

Some of the most exciting examples come from the new concept named ontological geosciences. By ontological geosciences we mean geosciences that are recognizable by computers and expressed in hierarchical concrete concepts and geoscientific concept models. A geoscientific concept model is a system model that relates different geoscientific concepts and/or concept models. There are many possible relationships among geoscientific concepts and concept models. One important relationship that expresses the complex Earth system is a parent-child relationship that can refine the general concept or concept model into a more specific one. When a concept has been refined it is represented by its instances and/or data in the databases. Therefore, geoscientific data should be incorporated into the earth system model.

In [64] has been introduced a layered Earth concept data model as a basic model of the Earth system.

In [45] there is a survey of different works done in last ten years on the area of using ontologies in order to solve the problem of semantic interoperability as follows: GEOscience Network (GEON) [65] project; Semantic Web-Service Interoperability for Geospatial Decision Making (SWING) [66] project; ORCHESTRA (http://www.eu-orchestra.org /overview.shtml) project.

Evolution of Spatial Decision Support Systems (SDSS)

As mentioned in section dedicated to Spatial Decision Support Systems (SDSS), research on SDSS originated from three different sources – decision support system (DSS), geographic information system (GIS) and information communication technology (ICT). Thus, the three streams of research that lead to the development of spatial decision support systems can be characterized as geographical information based systems and decision support based systems using a Web Services-based Infrastructure for Geospatial Information. Although GIS have been in existence for the past three decades, only recently GIS technologies are being incorporated into mainstream ICT decision support solutions. While GIS has traditionally been used in areas such as utilities, environmental and urban planning, real estate, government, and natural resources management, there is a growing interest in the use of GIS technology for decision support within the business community because of analytical and visualization capabilities. Increasingly, organizations are adopting GIS-based solutions in a number of domains including customer relationship management, vehicle routing, healthcare management, complex processes monitoring: civilian protection, urban planning, forestry, roads, water, agriculture, cadastral, environment and geologic survey.

According to [67; 68] “the evolutionary path of the decision support technology from the information systems community contains four distinct stages: a) Traditional Model based DSS, b) Expert/Knowledge-based DSS, c) Web-based DSS, and d) Service based DSS. Similarly, the GIS-driven evolutionary path of decision support systems from the geosciences community contains: a) Traditional GIS, b) Spatial Decision Support System, c) WebSDSS/Distributed SDSS, d) Mobile SDSS, and e) Service-based SDSS.”

This is a good reason in order to try to improve the performance of SDSS for complex processes monitoring having as infrastructure the Web Services-based Infrastructure for Geospatial Information developed starting from the new initiatives for SDIs implementation mentioned in introductive section of this chapter and more another from local levels of different institution of public administration and private sector and trying to overtake the syntactic and semantic integration, semantic heterogeneity and semantic interoperability with new techniques and methods.

Final Remarks

The intention of this chapter has been to provide a non-exhaustive point of view, based on the literature, regarding the possibility to build Spatial Decision Support Systems for complex processes monitoring using Web Services-based Infrastructure for Geographic Information; to identify some challenges and to highlight on the possible solution in order to overtake the critical point in this areas: syntactic and semantic interoperability from geospatial information used in different fields of geosciences. Starting from the terminological aspects, has been presented a frame of work covering the European and worldwide initiatives in connection with GI infrastructures building and some application domains associated with the challenges coming from the intention to offer the best solution in order to solve the problems of semantic interoperability and semantic heterogeneity followed by some relevant examples.

There are a number of challenges that are critical to the development and use of applications for the Geosciences that utilise the semantics of information. These challenges include the need to identify appropriate standards and languages for formalising semantics, the limitations of ontologies for representing the meaning, Grid challenges and top down versus bottom up development of geosemantic standards and applications.

As mentioned in [69] one of the main drivers of research in this area is the increased expectation of users of Geosciences data to be able to access the data they require digitally and in a standard form from different suppliers, including legacy data. This requires not only the adoption of Geosciences data exchange standards but also semantic interoperability, reasoning across heterogeneous definitions of geoscience concepts in order to achieve semantic integration across disciplines.

Geospatial data and geoscientific research standards are needed to support geosemantic interoperability. There are levels of interoperability from a basic interchange between software applications, to the exchange of data and concepts, to a high level of interoperability that involves automated discovery and use of geospatial resources. Geosemantic standards need to be able to handle things such as dynamic geometric representations, discrete and continuous data, spatio-temporal things and processes. Ideas for how this might be accomplished may involve semantic languages such as Web Ontology Language (OWL) and extensions such as Semantic Web Rule Language (SWRL) and Semantic Web Services Language (SWSL), building on the work done on existing relevant ontologies such as the GEON project [70].

In Europe the INSPIRE initiative is acting as a driver for standardisation of data structures for data exchange and interoperability, such as the use of ISO/OGC standards to enable the interoperability of both marine and earthquake data. Similar standards-based interoperability initiatives are taking place in the USA, with the coordination of Geoinformatics efforts between United States Geological Survey (USGS) [71] and state geological surveys, and in Australia with the AuScope program (http://www.auscope.org.au/) aiming to integrate a wide range of earth science information. Globally, the GeoSciML initiative [72] is developing an internationally agreed geoscience data exchange format, underpinned by an agreed conceptual data model. While this conceptual data model involves developing agreed geoscience concept definitions, the semantics of these definitions are not yet being formalised, as is the case with the majority of other efforts. However the objective is that the work will lead to the development of formal ontologies, which both handle the relationships between concepts and also the relationships between the 'same' concepts defined by different agencies in slightly differing ways.

Geoscience ontologies as a basis for 'smart' data discovery and use are seen as a solution to semantic interoperability but ontologies are more concerned with 'what' than 'why' or 'how', limiting the questions they can address. Furthermore, ontologies are typically static. Ontologies could develop through usage and context with evolving concepts.

From the educational point of view it is necessary to identify the technical challenges and managerial challenges and to offer some high performing solutions.

The technical challenges refer to:

- the problem of improving performance, one of the most important limitations in the development of SDSS developed on Web Services-based Infrastructure for Geospatial Information because SDSS models may use large non-spatial datasets (attribute data) and spatial data in different formats and is time consuming activity;
- the problem of technology integration because SDSS was and will be implemented using open architectures so that new components can be easily added resulting in "plug and play" geoprocessing environments. In addition, these systems must be robust with efficient communication mechanisms via the Internet/Intranet. Another challenge is integrating the distributed SDSS with legacy systems that may still have valuable GI datasets and programs [42].
- the problem of interoperability because full integration of geospatial data and resources requires developing a common set of syntactic and semantic interoperability standards and geoweb services will need to tolerate heterogeneous frameworks because no distributed component technology will be optimal for all kinds of tasks.
- the problem of security and privacy remain a major consideration in implementing SDSS developed on Web Services-based Infrastructure for Geospatial Information because many geospatial data sets and services are proprietary and private or are under the rules regarding strategically aspects. Such SDSS will face additional security problems because of the sharing of spatial objects over the Internet, and is prone to viruses, hackers, and network jams. Moreover, the introduction of mobile spatial technologies will add an additional layer of security concerns due to the inherent risks associated with wireless technology [42].
- the problem of quality of service because in a distributed SDSS environment with commercial services, the level of service provided by the individual nodes may vary because of node failures, unreliable communication or disconnected network links. Hence, the architecture must include technical solutions to combat such disruptive events [42].

Between managerial problems it is necessary to enumerate a few in connection with: policy issues and organizational commitment because organizations from public and private sector, being at different levels: national, regional, European, local have to develop concise policies to ensure security and privacy of sensitive spatial and non-spatial data and must to develop contingency plans for disaster recovery and security breaches; and it will be necessary to redesign some of the business processes.

Regarding a research agenda for SDSS developed on Web Services-based Infrastructure for Geospatial Information, it is necessary to take into account that [73] have updated Keen's agenda [74] and articulated a new DSS research agenda for the next decade. In our opinion, this new DSS research agenda is also pertinent to SDSS developed on Web Services-based

Infrastructure for Geospatial Information and can be adapted to guide future developments. As mentioned in [68] the following aspects in order to shape future research in web-based spatial decision support (WebSDSS) technologies have been proposed:

a) identifying potential uses for such SDSS in solving business problems with spatial aspects as well as incomplete and uncertain data,
b) building intelligent such SDSS using enabling technologies,
c) providing a distributed WebSDSS environment for cooperative spatial problem solving and linking geo services,
d) facilitating interoperability of spatial data and systems, and
e) demonstrating and improving the effectiveness of WebSDSS in decision making through training and knowledge transfer.

In the opinion of authors of this chapter, the DSS research agenda - also pertinent to SDSS developed on Web Services-based Infrastructure for Geospatial Information - could be corroborated with Spatial Data Infrastructure (SDI) implementation agenda, in order to obtain a new dimension of collaborative information necessary for decision improvement in complex processes monitoring at different level: national, regional, local, global.

REFERENCES

[1] Gore, A. (1998). The Digital Earth: Understanding our planet in the 21st Century. Photogrammetric Engineering and Remote Sensing vol. 65(5), p. 528, http://www.isde5.org/al_gore_speech.htm [accessed 30 April 2010].

[2] Craglia, M., Goochild, M. F., Annoni, A., Câmara, G., Gould, M., Kuhn, W., Mark, D., Masser, I., Maguire, D., Liang, S. and E. Parsons (2008). Next-Generation Digital Earth. *International Journal of Spatial Data Infrastructures Research*, Vol. 3:146-167, available at http://ijsdir.jrc.ec.europa.eu/in dex.php/ijsdir/article/view/119/99 [accessed 30 April 2010]

[3] Tu, S. and M. Abdelguerfi (2006). "Web Services for Geographic Information Systems", *IEEE Internet Computing*, 10(5):pp. 13-15.

[4] INSPIRE (2007). Directive 2007/2/EC of the European Parliament and of the Council of 14 March 2007 establishing an Infrastructure for Spatial Information in the European Community (INSPIRE). *Official Journal of the European Union*. L108:pp. 1-14.

[5] SEIS (2008). Towards a Shared Environmental Information System (SEIS). Communication from the Commission to the Council, the European Parliament, the European Economic and Social Committee and the Committee of the Regions, at http://eurlex.europa.eu/LexUriServ/LexUriServ.do?uri =CELEX:52008DC0046:EN: NOT [accessed 30 April 2010].

[6] GMES (2008). Kopernicus – Global Monitoring for Environmental and Security (GMES) at http://www.gmes.info/, [accessed 30 April 2010].

[7] http://www.gsdi.org.

[8] Ian Masser (1999). All shapes and sizes: the first generation of national spatial data infrastructures. *Int. J. of Geographical Information Science*, 13(1):67-84, 1999.Institute

of Electrical and Electronics Engineers. (1990), IEEE Standard Computer Dictionary: A Compilation of IEEE Standard Computer Glossaries. New York, NY: 1990.

[9] http:// www.ec-gis.org/inspire/.

[10] http://www.esdinetplus.eu.

[11] http://www.ungiwg.org/.

[12] Jones, M.T. (2007). Google's geospatial organizing principle. *IEEE Computer Graphics and Applications*, July/August 2007, 8-13.

[13] Grossner KE, Goodchild MF, Clarke KC (2008). Defining a digital Earth system. Transactions in GIS 12: 145-160.

[14] http://www.nasa.gov/.

[15] http://earth.google.com/.

[16] http://maps.google.com/.

[17] http://www.mapquest.com.

[18] http://www.mapblast.com.

[19] Federal Geographic Data Committee (2003). http://www.fgdc.gov/ [accessed 30 April 2010].

[20] Longley, P. A., Goodchild, M. F., Maguire, D. J., and Rhind, D. W. (2001). Geographic information systems and science. Chichester: John Wiley and Sons, Ltd.

[21] http://www.dictionary.com.

[22] http://www.inspire-geoportal.eu/.

[23] http://www.geoportal.org/web/guest/geo_about.

[24] http://www.wfp.org/logistics/blog/first-test-version-sdi-t-geoportal-launched-wfp.

[25] http://geodata.grid.unep.ch/.

[26] http://www.geosur.info/geoportal/.

[27] http://geoportal.icimod.org/.

[28] http://www.eurogeographics.org/news/national-geoportal-finland.

[29] http://www.geoportail.fr/.

[30] Zevenbergen, Jaap, Jandirk Bulens, Marjolijn Kuyper, Henk Koerten, Michiel Jellema and Frederika Welle Donker. (2009). Geoportal network - more process catalyst then project, in Proc of. "Urban and Regional Data Management" – UDMS Annual 2009, 105 – 114.

[31] Gogu, R.C., Gaitanaru, D., Leonard, A., Enescu Iosifescu, I., Nedelcu, I. (2009). "Applied webmapping in groundwater monitoring – A hydrogeological web mapping geoportal", Proc.of the Int. Conf. on ELECTRONICS, COMPUTERS and ARTIFICIAL INTELLIGENCE – ECAI'09, pp.124-129.

[32] Joseph Bih (2006). Service Oriented Architecture (SOA) A New Paradigm to Implement Dynamic E-business Solutions, Ubiquity -- Volume 7, Issue 30 (August 8, 2006 - August 14), 2006.

[33] The Stencil Group, I. (2002). UDDI.org White Paper: The evolution of UDDI. 2002.

[34] Phil Karecki (2001). "Web Services Fundamentals: From Whence It Came. Web Services Fundamentals: From Whence It Came", Nov. 2001, http://soa.sys-con.com/node/39300 [accessed April 30, 2010].

[35] Ravi Trivedi (2003). "Web Services Tutorial: Understanding XML and XML Schema—Part 1", April 23, 2003, http://www.developer.com/xml/article.php/2195981/Web-Services-Tutorial-Understanding-XML-and-XML-SchemamdashPart-1.htm, [accessed 30 April, 2010].

[36] An OGC® White Paper May 30, (2001). Introduction to OGC Web Services, http://www.ibm.com/developerworks/webservices/library/w-ovr/ or www.opengeo spatial.org

[37] 799–818.GigaTS D2.2.9 * Web Services About to integrate distributed applications?

[38] Herring, Charles (1994). "An Architecture of Cyberspace: Spatialization of the Internet". U.S. Army Construction Engineering Research Laboratory.

[39] Carl Reed (2003). Intellectual Property, Patents and Web Mapping. *A Historical Perspective on Claiming Innovation*, in ISPRS, Vol 8, No 3, September 2003, pp 33 - 35.

[40] Justin Deoliveira (2008). GeoServer: Uniting the "GeoWeb" and Spatial Data Infrastructures, http://www.gsdi.org/gsdiconf/gsdi10/papers/TS26.4paper.pdf, [accessed 30 April, 2010]

[41] Muller, J.C. (1993). Latest developments in GIS/LIS. *International Journal of Geographical Information Systems*, 7, 4, 293-303.

[42] Keenan, P. (2004). "Using a GIS as a DSS Generator", DSSResources.COM, 12/17/2004, available at http://dssresources.com/papers/features/keenan/keenan1217 2004.html, [accessed 30 April, 2010].

[43] Keenan, P. (1995). Spatial Decision Support Systems for Vehicle Routing, Working Paper MIS 95/10, Graduate School of Business, University College Dublin.

[44] Bodin, L., Fagan, G., Levy, L., and Rappoport, H. (1992). The RouteSmart System. Proceedings of the Conference on GIS in Business and Commerce, Denver, Colorado, USA, 1992.

[45] Ionita, A., Zafiu, A., Dascalu, M., Franti, E., Visan, M. (2009). "An approach for intelligent decision support system for urban traffic monitoring", in Proc of. "Urban and Regional Data Management" – UDMS Annual 2009, Ljubljana, Slovenia, June, 24 – 26, 2009, Eds. Alenka Krek, Massimo Rumor, Sisi Zlatanova, Elfriede Fendel, CRC Press Taylor and Francis Group – A BALKEMA BOOK, ISBN: 978-0-415-55642-2, ISBN: 978-0-203-86935-2, pp. 455 – 464.

[46] Lorenzino Vaccari, Pavel Shvaiko, Maurizio Marchese (2009). "A geo-service semantic integration in Spatial Data Infrastructures", *Int. Journal of Spatial Data Infrastructures Research*, 2009, Vol. 4, 24-51, available at: http://creative commons.org/licenses/by-nc-nd/3.0/, [accessed 30 April 2010].

[47] Kuhn, W., (2005), Geospatial semantics: why, of what, and how? *Journal on Data Semantics, Special Issue on Semantic-based Geographical Information Systems*, 3534, pp. 1-24.

[48] Friis-Christensen, A., Ostländer, N., Lutz, M. and Bernard, L. (2007). Designing Service Architectures for Distributed Geoprocessing: Challenges and Future Directions. Transactions in GIS, 11(6).

[49] Parent, C., Spaccapietra, S. and E. Zimányi (2006). Conceptual Modeling for Traditional and Spatio-Temporal Applications, The MADS Approach. Springer-Verlag, Berlin.

[50] Lemmens, R., Granell, C., Gould, M., and Oosterom, P. v. (2006). Integrating semantic and syntactic descriptions to chain geographic services. *IEEE Computer Society*, 10(5): 42-52.

[51] Papazoglou, M.P. (2003). "Service-Oriented Computing: Concepts, Characteristics and Directions", Proceedings 4th International Conference on Web Information Systems Engineering (WISE), December 10-12 2003, Rome, Italy, pp. 3-12.

[52] http://www.opengeospatial.org.

[53] Stuckenschmidt, H. (2003). Ontology-Based Information Sharing in Weakly Structured Environments. PhD thesis, Vrije Universiteit Amsterdam (NL).

[54] Yétongnon K., Benslimane D., Laurini R., Leclercq E., Jouanot F., Cullot N. (2000). “GIS interoperability: An Overview”. Proceedings of Telegeo’2000, Second Internatinal Symposium on Telegeoprocessing, Sophia-Antipolis, France, May 10-12, 2000, pp. 185-204.

[55] Atkinson R., Martell R. (ed.) (2002). “OWS1 Web services Architecture”. April 2002, available at http://www.opengis.org/techno/abstract.htm [accessed 30 April 2010].

[56] de La Beaujardière J. (ed.) (2002). “Web Map Server Implementation Specification (WMS)”, January 2002.

[57] Shekhar S., Raju R., Sahay N., Burk T. E., Lime S. (2001). “WMS and GML based Interoperable Web Mapping System”. Proceedings of the Ninth ACM International Symposium on Advances in Geographic Information Systems, Atlanta, Georgia, USA, 2001, pp 106-111, ACM Press 2001.

[58] Vretanos P. A. (ed.) (2001). “OpenGIS Web Features Server Specification”, January 2001.

[59] Evans J. D. (ed.) (2002). “OWS1 Web Coverage Service (WCS)”. April 2002.

[60] Barclay T., Jim Gray, Steve Ekblad, Eric Strand, Jeffrey Richter (2006). "Designing and Building TerraService," *IEEE Internet Computing*, vol. 10, no. 5, pp. 16-25, Sep./Oct. 2006, doi:10.1109/MIC.2006.95.

[61] Roger Zimmermann, Wei-Shinn Ku, Haojun Wang, Amir Zand, Jean-Pierre Bardet (2006). "A Distributed Geotechnical Information Management and Exchange Architecture," *IEEE Internet Computing*, vol. 10, no. 5, pp. 26-33, Sep./Oct. 2006, doi:10.1109/MIC.2006.92.

[62] Marius Scholten, Ralf Klamma, Christian Kiehle (2006). "Evaluating Performance in Spatial Data Infrastructures for Geoprocessing," *IEEE Internet Computing*, vol. 10, no. 5, pp. 34-41.

[63] John T. Sample, Roy Ladner, Lev Shulman, Elias Ioup, Fred Petry, Elizabeth Warner, Kevin Shaw, Frank P. McCreedy (2006. "Enhancing the US Navy's GIDB Portal with Web Services," *IEEE Internet Computing*, vol. 10, no. 5, pp. 53-60, Sep./Oct. 2006, doi:10.1109/MIC.2006.96.

[64] Xiaoning Guo, Kangping Sun, Qiuju He, Longjian Xie (2008). Ontological geosciences: An example of a layered Earth concept data model, Proc. of 33rd Int. Geological Congress in Oslo.

[65] http://www.geongrid.org/.

[66] http://www.swing-project.org/.

[67] Sugumaran, V., Sugumaran, R. (2005). “Web-based Spatial Decision Support Systems (WebSDSS): Evolution, Architecture, and Challenges”, Third Annual SIGDSS Pre-ICIS Workshop Designing Complex Decision Support: Discovery and Presentation of Information and Knowledge December 11, 2005, Las Vegas, Nevada, available at http://mis.temple.edu/sigdss/icis05/proceedings/DSS Workshop05-Sugumaran.pdf, [accessed 30 April 2010].

[68] Sugumaran, V., Sugumaran, R. (2007). "WEB-BASED SPATIAL DECISION SUPPORT SYSTEMS (WebSDSS): EVOLUTION, ARCHITECTURE, EXAMPLES AND CHALLENGES", Communications of AIS (2007) Volume: 2007, Issue: 19, Pages: 844-875, ISSN: 15293181.

[69] Reitsma, Femke; Laxton, John; Ballard, Stuart; Kuhn, Werner; Abdelmoty, Alia. (2009). Semantics, ontologies and eScience for the geosciences. Computers and Geosciences, 35 (4). 706-709. 10.1016/j.cageo.2008.03.014.

[70] http://www.geongrid.org/.

[71] http://www.usgs.gov.

[72] http://www.geosciml.org/.

[73] Shim, J.P., M. Warkentin, J.F. Courtney, D.J. Power, R. Sharda, and C. Carlsson. (2002). "Past, present, and future of decision support technology," Decision Support Systems 33(2), 111 – 126.

[74] Keen, P. (1987). "Decision support systems: the next decade," Decision Support Systems 3 (3) pp. 253– 265.

In: Advances in Energy Research
Editor: Nicu Bizon

ISBN 978-1-61209-991-0

Chapter 16

CASE STUDIES - IMPLEMENTATION STAGE OF THE DISTRIBUTED GENERATION SYSTEM

***Zdenek Vostracký**[*1], **M.Cengiz Taplamacioglu**[3], **Farrokh Fattahi**[4], **Radu Porumb**[2], **Naser Mahdavi Tabatabaei**[5], **Jan Veleba**[1], **Fazil Kaytez**[3] and **Hamed Seddighe Abbasi**[4]*

[1]Department of Electric Power Engineering and Ecology
University of West Bohemia, Pilsen, Czech Republic
[2]University "Politehnica" of Bucharest,
Energetics Department, Romania
[3]Gazi University, ElectricalandElectronics Engineering Department,
Ankara, Turkey
[4]Azarbaijan Higher Education and Research Complex, Tabriz, Iran
[5]Electrical Engineering Department,
Seraj Higher Education Institute, Tabriz, Iran

ABSTRACT

This chapter is focused on implementation stage of the Distributed Generation (DG) system in some countries: two from the European Union (EU), one candidate to EU and one from outside of EU. If last section of first chapter (named „Stage of Integrating Renewable Sources into DG Systems of the European Union) shows some approach for accessing the available renewable energy of each EU state (that put in evidence the diversity of energy efficiency measures and implementation of renewable energy strategies), then these aspects will be detailed by case studies in this chapter.

First of all, the energy resources of each state (that is subject of a case study) are shown as level of non-renewable energy resources and renewable resources. Then the electricity generation, transportation and distribution is discussed in perspective of integration of the available renewable energy sources.

So, proposed objectives for the DG system, the rehabilitation measures and levels achieved for the implementation stages are presented. In this frame, the energy policy of

* Email: zdenekv@kee.zcu.cz

the each state is discussed, showing the priorities of this energy policy and future energy policy measures. These strategies are based on new technologies and equipments that are applied using new infrastructure (as is the smart grid, for example).

Some case studies are sustained by simulation models. Finally, some conclusions are given.

Keywords: energy policy, electricity generation, distribution network, distributed generation, smart metering, WAMS, smart grids, renewable resources, hydro, biomass, biogas, natural gas, solar-thermal systems, photovoltaic, wind, geothermal, hydropower plants, thermal power plants, cogeneration, nuclear power, micro-turbines, wind turbines, gas turbines, control, ripple control, fault, fuzzy logic controller, simulation. Distribution network, ripple control, distributed generation, Smart Metering, WAMS, Smart Grids.

Distribution Power System of the Czech Republic

For more than one hundred years, energy policy has been the topic of interest for the Czech government. Strategic decision was made to build the main transportation system (dominantly with nominal voltages 110 and 400 kV). Nowadays, new renewable energy sources are being connected to the system bringing new potentials but also new challenges. It is necessary to solve new tasks to sustain stability and reliability requirements in electric power generation, transportation and even distribution. Therefore, advanced systems of Smart Grids and Smart Cities are being currently studied showing promising results for future development of the Czech electric power system.

Introduction

Modern construction of the distribution power system in former ČSR was started rapidly after the Second World War. During that time, significant effort was already placed on the control of electric power consumption and related optimized operation of the distribution network.

To accomplish this, ripple control was also integrated into the distribution power system of West Bohemia – for the first time in Cheb in 1960 for the control of public lighting. Subsequently, signal receivers were used in hundreds of consumption points – especially in those with accumulation appliances and two-tariff metering. In 1963, transmitter of ripple control was put into operation in 22 kV substation Křimice for coverage of Pilsen and surrounding area. In following years, remaining 22 kV substations have been successively equipped with ripple control transmitters (on frequency of 1050 Hz). However, only one third of the supplied area was covered by the signal in 1980s. Important turning point for the ripple control in the region of West Bohemia was the year 1983, when the signal transmission was initiated into the 110 kV power system in substations Vítkov and Přeštice. After its implementation in substation Chrást, new transmitters with frequency of 217 Hz ensured reliable propagation of the signal to lower voltage levels in the entire area of West Bohemia.

Remarkably, the ripple control was also used in the past for calling mining rescue teams and military/safety bodies.

Nowadays, the ripple control has been provided by central automatics containing the processing up to three different telegram rasters transmitted on the same frequency, autonomous switching programs with week-long regimes, real time back-up for no more than 64 hours and many other advanced functions. However, in the days of entirely liberalized electric power market, the ripple control cannot be fully used. The reason for this is the fact, that all distributors are obligated to respect unbundling – one of the legislative steps from the EU for better economical transparency and equal entry of all traders and producers of electrical energy on the power market. Nevertheless, from the historical point of view the ripple control was undoubtedly the first level to the concept of Smart Grids.

Until 2005, eight regional joint-stock companies (REAS) existed in the Czech Republic providing the distribution of electrical energy to the customers (JČE, JME, PRE, SČE, SME, STE, VČE and ZČE). Each of these companies ensured complete distribution services for the region of their provinces. Today, only three distribution companies are operating in the Czech Republic – ČEZ Distribuce, E.ON Distribuce [1] and PRE distribuce (PRE) [2]. All of these companies are licensees for electric power distribution and distribution system operators (DSOs) according to the Energy Act (Regulation No. 458/2000 Coll.). Province areas of individual DSOs are schematically presented in Figure 1 highlighting the province of our largest DSO, joint-stock company ČEZ Distribuce – member of ČEZ Group [3], with its main dispatching centre in Děčín (North Bohemia). Major role of each DSO is to ensure secured delivery and power quality of electrical energy to all the customers by maintaining voltage values on the customer side within their permitted limits along with minimum active power losses. Electricity prices are set by the Energy Regulatory Office [4] in terms of expenses for effective performance of licensed activities, taxes depreciation and reasonable profits ensuring recoverability of all investments executed.

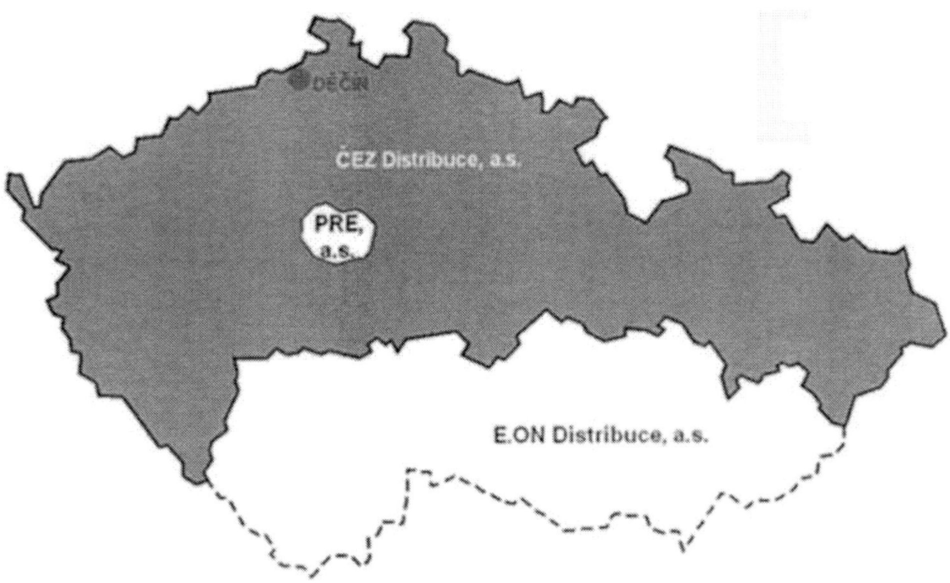

Figure 1. Provinces of individual DSOs in the Czech Republic (taken from [3], slightly modified).

Basic Data about the Czech Distribution Power System

The distribution power system is supplied from the 400 (220)/110 kV central substations of the transmission system and partially from distributed generation units, such as small industrial power plants, cogeneration units and renewable sources using water, wind, biomass and solar energy. Its fundamental HV is 110 kV and it continues through the MV levels with their values depending on the area (35 kV in Eastern Bohemia, 22 kV in the largest part of the Czech power system) and on the extent of consumption. Smaller municipalities in the 35 kV network regions are supplied on the level of 10 kV, while those in the 22 kV network regions are powered on the level of 6 kV. Low voltage is predominantly 400 V/230 V, other voltage levels – such as 500 and 690 V are used exceptionally, mostly for heavy industrial purposes (factories, mines).

The overview of the standardized set of 3-phase ac line-to-line voltages in kV operated in the Czech power system is shown in Table 1 along with their upper operational limits.

Table. 1. Standardized set of 3-phase ac line-to-line voltages in the Czech DS (taken from [5])

Low voltage	Rated	0.4 / 0.23			0.50	0.69
(LV) [kV]	Max. operating	0.42 / 0.241			---	0.73
Medium voltage	Rated	3	6	10	22	35
(MV) [kV]	Max. operating	3.6	7.2	12	25	37
High voltage	Rated	110		220	---	---
(HV) [kV]	Max. operating	123		245	---	---
Extra high voltage	Rated	400		---		
(EHV) [kV]	Max. operating	420		---		

Note: Industrial plants and electrical traction systems – 25 kV ac (south), 3 kV dc (north) – are supplied form the public distribution power system.

Figure 2. Scheme of the Czech distribution and transmission power systems on 110 to 400 kV levels – January 2009 (employed from [5], amended).

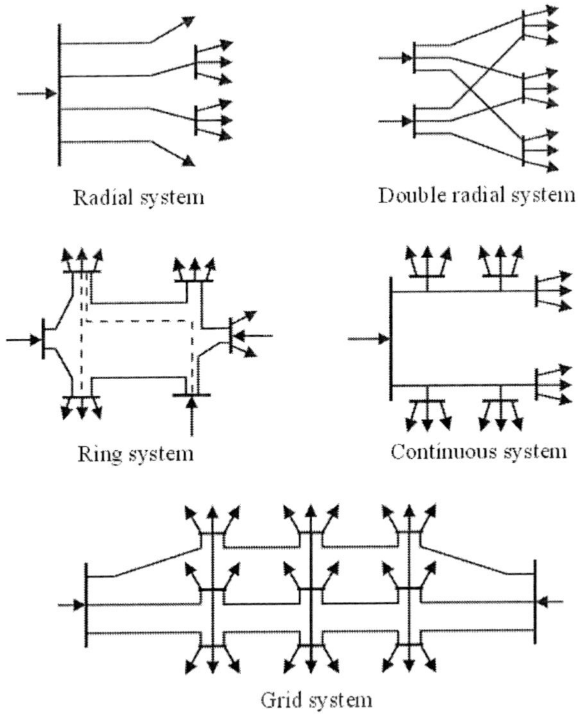

Figure 3. Types of electrical energy distribution (taken from [5]).

Delivery of electrical energy to the customers is realized either using radial, continuous, ring or grid distribution systems (Figure 3.). Selection of suitable network type depends on the operation method of the system in terms of power distribution, safety and economy. The 110 kV and MV distribution networks are operated as radial or continuous systems, although their design allows number of connections in double radial or ring configurations for possible energy back-up in case of failure. The LV distribution networks are predominantly controlled using radial or continuous designs. Although the radial distribution method is operationally the simplest and least expensive, it provides the lowest reliability. To minimize supply interruptions, several standards are defined in the Energy Act (Regulation No. 540/2005 Coll.) determining penalties for system operators if not respected. Dense LV networks are projected as grid systems.

When concerning the operation method of the transformer system point, the 110 kV and LV distribution networks (along with 400 kV and 220 kV transmission power systems) are operated with directly earthed system points enabling clear evaluation of the fault occurrence using single-pole reclosing. Due to significant fault current magnitude it is necessary to switch off the fault immediately.

In small MV networks, insulated system points are allowed for capacity currents below 20 Amps. The advantage of these networks is transparent evaluation of outlets with earth fault and possible network operation with the fault during the searching time of its location. However, for capacity current values higher than 10 Amps it is recommended to use system point earthed via arc-suppression coil (for overhead lines) or via resistance (for cable lines). In Table 2, an overview of operation methods for transformer system points is presented.

At present, the Czech distribution power system undergoes rather revolutionary change, where especially dispersed electric power sources and advanced technologies (WAMS, Smart Metering, Smart Grids, etc.) are being integrated into the grid operation. However, all of these challenges create increased burden for the DSOs to maintain secured performance and control of the distribution power system. In the following chapters, each of these problems is objectively described along with present situation in the Czech Republic.

Table 2. Operation methods for transformer system points (taken from [5])

<table>
<tr><td>400 kV</td><td rowspan="3">directly earthed system points</td><td>EHVV</td></tr>
<tr><td>220 kV</td><td rowspan="2">HV</td></tr>
<tr><td>110 kV</td></tr>
<tr><td>35 kV</td><td rowspan="4">insulated system points or earthed via arc-suppression coil/resistance</td><td rowspan="4">MV</td></tr>
<tr><td>22 kV</td></tr>
<tr><td>10 kV</td></tr>
<tr><td>6 kV</td></tr>
<tr><td>0.69 kV</td><td rowspan="2">insulated system points</td><td rowspan="3">LV</td></tr>
<tr><td>0.5 kV</td></tr>
<tr><td>0.4 kV</td><td>directly earthed (exceptionally insulated) system points</td></tr>
</table>

Dispersed Generation in the Czech Republic

In the past, the generation energy mix in the Czech Republic was strongly based on coal-fired power plants burning brown coal mined in surface mines in north-west Bohemia. This coal, containing high levels of sulfur (0.8 to 2.8 percent) and fly ash (15 to 40 percent), was the main reason for massive air pollution in 1960s to 1990s. Although all coal-fired power stations are desulphurized today, vast volumes of waste gases (especially carbon dioxide emissions) with high concentrations of heavy metals are still emitted to the surrounding area. Nowadays, when coal reserves are rapidly decreasing along with new territorial restrictions for opening new coal mines, the Czech Republic can no longer rely on this electric power source.

The solution, approved by the Czech government, lies in power production increase from nuclear and gas power stations with significant support of renewable power sources (wind, biomass, photovoltaics). Concerning reservoir power plants, stagnancy of their numbers is expected due to depletion of possible standing water localities. Building of a new pumped storage power plant (at least four planned projects) is postponed because of higher construction costs and ecological limits.

To meet the EU restrictions for emissions and updated requirements for generation from renewable sources, increased number of new biomass, wind and solar power plants have to be installed and connected to the grid.

Note: Currently, the generation from renewables is 6.5 percent. The target (in 2020) for the Czech Republic is 13 percent.

Biomass power plants have the biggest potential from all renewable sources in the Czech Republic. As the fuel, energetic plants, wooden sawdust and straw are used either in biomass power plants themselves or as an ingredient to brown coal for combined heat and power generation in coal-fired power plants and thermal power stations in many larger Czech cities. Planned increase of power production is expected from today's 4.67 percent to 13.1 percent in 2030. In spite of not entirely optimal solar and wind conditions in the Czech Republic (see Figure 4, 5), energy production by photovoltaic and wind power plants has been generously supported by the Czech government offering obligatory connection to the Czech distribution network. In case of photovoltaics, purchasing price of approximately 13 CZK/kWh (one of the highest in the EU) has been also allowed.

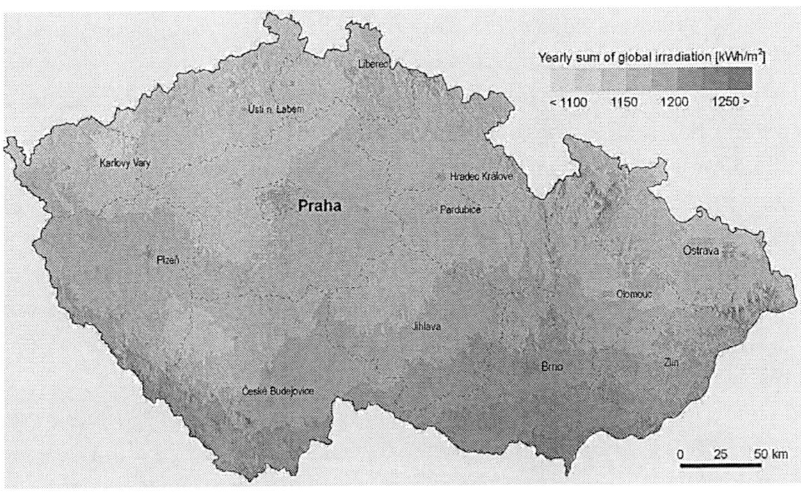

Figure 4. Map of global solar irradiation in the Czech Republic (taken from [6], modified).

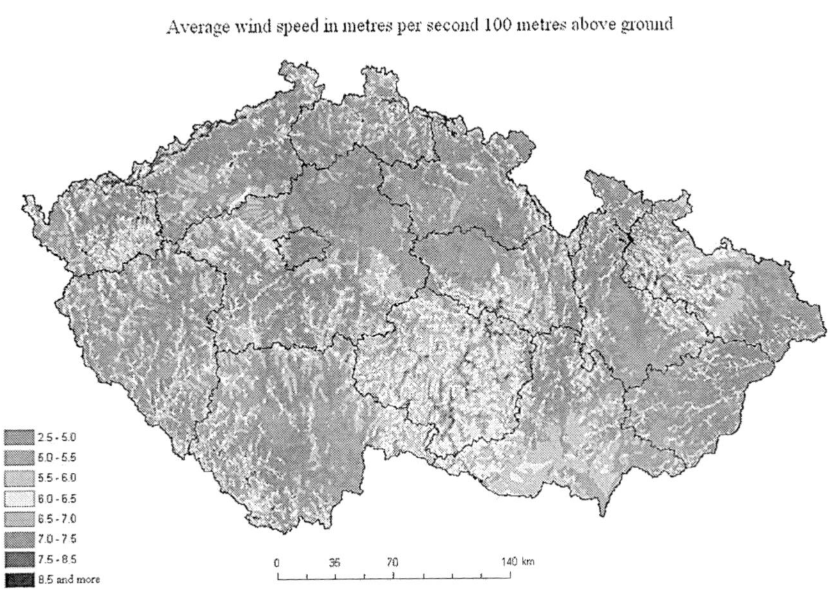

Figure 5. Wind map of the Czech Republic (taken from [7], slightly amended).

Unfortunately, in last several months an extreme situation has arisen in the Czech distribution network, where over 8000 MW of new solar power plants have been approved and built, currently waiting for being connected. Understandably, such large amount of power from an uncontrollable power source would have destructive effect on the Czech distribution and transmission power system.

Therefore, connection process of new photovoltaic sources to the grid has been provisionally suspended along with planned reduction of purchasing price to 6 CZK/kWh coming into force in 2011. This entire situation is being intensively analyzed and new technical and juridical precautions will be taken soon to successfully solve this problem.

Promising wind conditions can be found only in a few areas of the Czech Republic. For all that, wind power plants are being installed in many localities - mostly in areas, where the average wind speed does not exceed 7 meters per second. Moreover, new wind farms are in the planning or installation process. Many application studies have to be performed to assess if a new wind farm can be safely connected to the grid. Due to uncontrollable power flow from these sources, possible voltage and frequency instability of the system can occur causing the overloading of power lines which would result in a failure state of the network or even the blackout in the worst-case scenario.

Therefore, thorough investigation of possible negative impacts of each new wind farm on distribution and even transmission power system (through 110kV/MV transformers) must be carried out. To minimize these effects, automatic secondary voltage control (ASVC), static shunt compensators and tap-changing transformers are used on a mass scale for the control of voltage and reactive power.

Detailed connection study of wind farm Horní Loděnice (9x2MW) via two parallel cable lines and the 110/30 kV tap-changing transformer to the distribution system of ČEZ Distribuce in substation Šternberk has been performed by joint-stock company EGÚ Praha Engineering ([8], [9]) – see Figure 6.

Individual asynchronous generators Vestas are wired to the local substation 30/0.7 kV. In the tertiary winding of the 110 kV/MV transformer, shunt inductor of 1.5 MVAr is located as a backup for possible voltage increase in the substation above the value of 119.6 Volts. The distance between the wind farm and 110 kV substation Šternberk is approximately 13 kilometers.

From the ASVC point of view, the wind farm is a self-reliant regulating element for being connected to the relevant pilot bus of 110 kV. This is an ideal situation because no negative interactions between multiple regulating elements have to be taken into consideration. Although this wind farm contains only the local ASVC device, i.e. without any links to the ASVC device of the DS, the dispatcher can affect its operation for their mutual cooperation. However, this cooperation can be negatively influenced by voltage step change caused by connection of shunt inductor to the tertiary winding of the 110 kV/MV transformer.

Because the pilot bus consists of the substation with one single-piece busbar, it is not necessary to evaluate the topology of the substation and only measured voltages can be sufficiently taken for the ASVC requirements. Due to longer distance from the 110 kV substation, larger voltage drop can occur on interconnecting cable lines along with significant differences between reactive power generated by the wind farm and transmitted to the 110 kV substation.

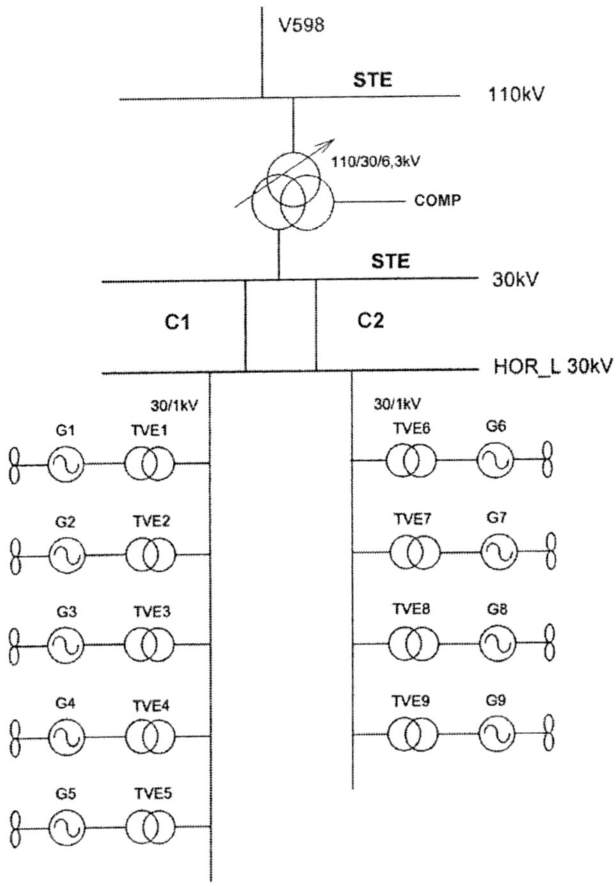

Figure 6. Connection of wind farm Horní Loděnice to 110 kV substation Šternberk (used from [8]).

Concerning the voltage stiffness, low values up to 6 MVAr/kV can lead to rather broad range of voltage control by the wind farm in 110 kV pilot bus. In case of zero generation, reactive power is not produced by the wind farm and is excluded from the ASVC. Only the 110 kV/MV transformer and connected shunt inductor can be regulated by ASVC during that time.

When massively used, renewable sources can pose significant threat to the secured operation of the distribution network. Therefore, power engineers must take reasonable actions to determine acceptable volume of electrical energy produced by these sources in combination with new strategies (Smart Grids) to avoid all possible abnormal situations in the network.

WAMS Technology

Another new trend in the distribution network operation is the synchronized metering of working conditions. The reason for this is, that the quality of measured data can be significantly influenced by asynchronism of measurement times resulting in lower accuracy of data analysis. Therefore, more precise data collection must be performed to maintain high reliability, control and running process of the distribution power system.

WAMS systems (Wide Area Monitoring, Control and Protection) provide measurements of synchronous phasors capable to handle this problem. Measurements are realized by phasor terminals PMU (Phasor Measurement Unit), where output signals of measuring voltage and current transformers are transmitted for the calculation of synchronized phasor values. After that, collected data are sent to the main server of the dispatching centre (Phasor Data Concentrator - PDC) using the GPS system, where further analysis of mutual phase angles between individual synchronous phasors is accomplished.

Note: Measuring points are mostly busbars, power lines and transformer outlets at the substation.

Based on obtained measured data, monitoring, analysis, control and protection of large power networks can be provided in real time. The WAMS technology can be applied for determination of safe manipulation processes, fault location, network stability (monitoring of network frequency and possible network oscillations) and calculation of network passive parameters (R, X, B) under load, which is vital for more accurate settings of electrical protections and network simulations (state estimation, calculation of power losses, network models, etc.). Its further use is especially to implement three-phase state estimation for minimizing errors cased by unsymmetrical network loadings and to prevent possible stability loss and voltage collapse situations.

According to the Bonneville Power Administration [10], the accuracy of WAMS technology is 300 meters (time delay of 1μs) for fault locations and ±0.1 degrees (time delay of 5.5 μs) for phasor measurements and controls of network stability.

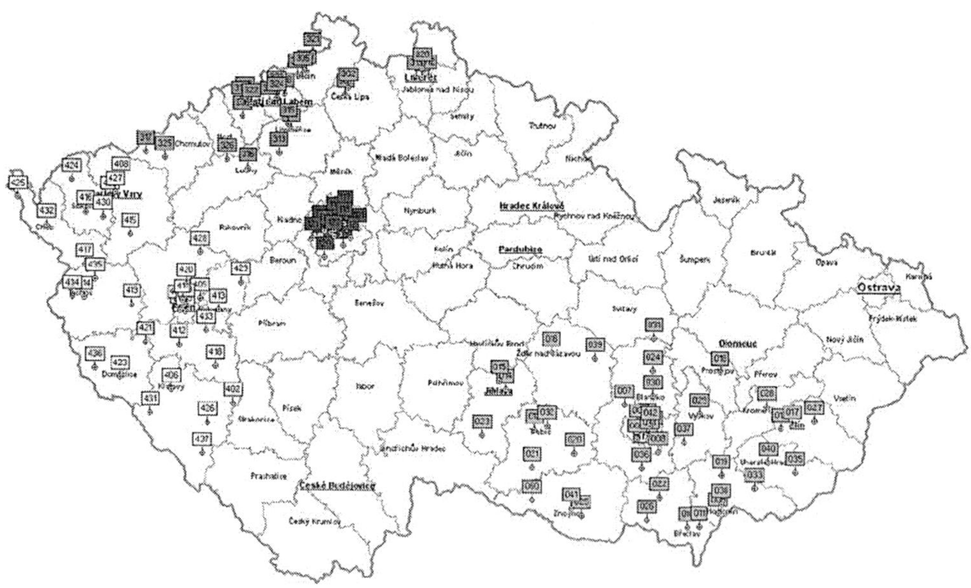

Figure 7. Implementation process of WAMS technology into the Czech DS – situation in 2009 (taken from [12]).

The most widely-used measuring system with WAMS technology in the Czech Republic is FOTEL. This product has been manufactured by non-stock corporation AIS [11], pioneer

Czech company focusing on measuring and information technology for power engineering. In three regions of PRE, almost 100 percent of 110/22 kV substations are fully equipped with FOTEL system. In 2004, the FOTEL system with WAMS technology was installed in several areas of individual DSOs in the Czech Republic – at 89 substations with synchronous phasor measurements in 297 buses. At the end of 2006, increase to 120 substations of the Czech DS was reached. From 429 metered buses, 146 buses were measured in all three phases (721 phasors in total). In October 2007, number of substation was increased to 122. Today's situation covers 128 substations with 475 buses and 767 phasors – see Figure 7 above.

It is expected, that following inclusion of WAMS technology into the Czech distribution network will continue in the future to provide essential information about the system behavior.

Smart Grids – New Concept for the Czech DS

Several factors resulted in world-wide discussion about the new concept for electric power system operation – the Smart Grids. For the 21st century, increasing prices of fossil fuels caused massive installation of new types of electric power generation units – especially renewable power sources (wind, photovoltaics). However, without any feasible control of their production, secured operation of the network can be significantly threatened. The situation is even more complex since also other dispersed sources, such as geothermal/biogas/biomass power plants, cogeneration units and small water power stations are operated in the distribution power system.

With decrease of fossil fuels and CO_2 emissions, broad use of electromobiles is expected for future traffic conditions. Using their accumulators, electromobiles would be also employed as very effective power storage devices. However, impact of high power converters on electric power quality seems to be very serious problem for today's distribution system control.

Another challenge is the integration of final consumers to the distribution network so, that they would be able to control their own electric power consumption (using Smart Meters) based on current tariff rate selection from the distributor. This would result in more advanced control of the distribution system, minimization of power losses and optimization of generation and distribution costs.

Smart Grids are intelligent and self-regulated power systems enabling the electric power delivery from both centralized and decentralized power sources to final customers. They ensure maximum use of electric power from individual power plants to meet all the requirements of the distribution power system (power quality, reliability of electric power supply) by avoiding possible abnormal network conditions, such as over- and undervoltages, branch overloadings and sudden power outages.

Smart Grids enable the integration of many partial areas, such as metering of electrical energy, customer services, consumption control, distributed generation, microregions and microgrids, electric power storage devices, islanding, stabilization of operational network parameters (voltage control, power factor compensation, etc.), control and automation, regulation of network power flows, virtual power plants, stability of distribution and transmission power systems and others.

Important feature of Smart Grids is so-called 'self-healing', which ensures the evaluation of possible power outages in particular network section and reliable return to normal operation of the network with renewed delivery of electrical energy to the customers. This can be provided using intelligent switching devices, remote sensors and other intelligent technological elements without any additional operator's involvement. This new concept would lead to minimized number of possible abnormal conditions with more effective and stable control and operation of distribution and even transmission power systems.

Smart Metering is one of crucial pillars of the Smart Grid concept. It includes the installation of new measurement technology called AMM-AMI (Automated Meter Management Infrastructure) for mutual communication between the power distributor and the customer. Distant measurement data collection with remote meter manipulations are important for further studies of the network operation, for sharing vital information with consumers and for control of electric power demands (Demand-Side-Management). However, conversion of distribution power systems to Smart Grids is rather a revolutionary way. To support this, hierarchy of control systems for the concept of Smart Grids along with all relevant standards is shown in Figure 9. In the future, only single standard should be valid for this new strategy for operation of distribution/transmission power systems.

During years 2006-2008, first pilot project was performed confirming possible large-scale implementation of the AMM technology to the distribution power system of joint-stock company ČEZ Distribuce.

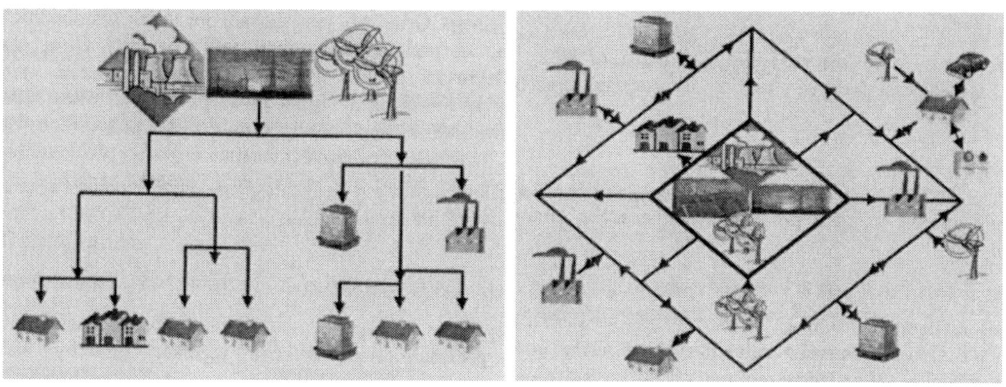

Figure 8. Today's structure of the DS (left) and future concept using Smart Grids (right) – taken from [13].

In June 2009, new initiative has been introduced by ČEZ Group called 'Futur/E/Motion' highlighting decreasing reserves of strategic resources, higher ecological requirements and legitimate needs for reduction of electricity costs and higher reliability of electric power supply. As a response, significant support of renewable power sources, electromobiles, power storage devices, technologies for power savings and higher efficiency of present generation units has been announced. In the area of Smart Grids, new pilot project called 'Smart Region' will be carried out by 2015 using the high-tech distribution technologies with strong links to the decentralized production of electrical energy. Total expenses for this particular project are estimated up to 1 billion CZK [16-17].

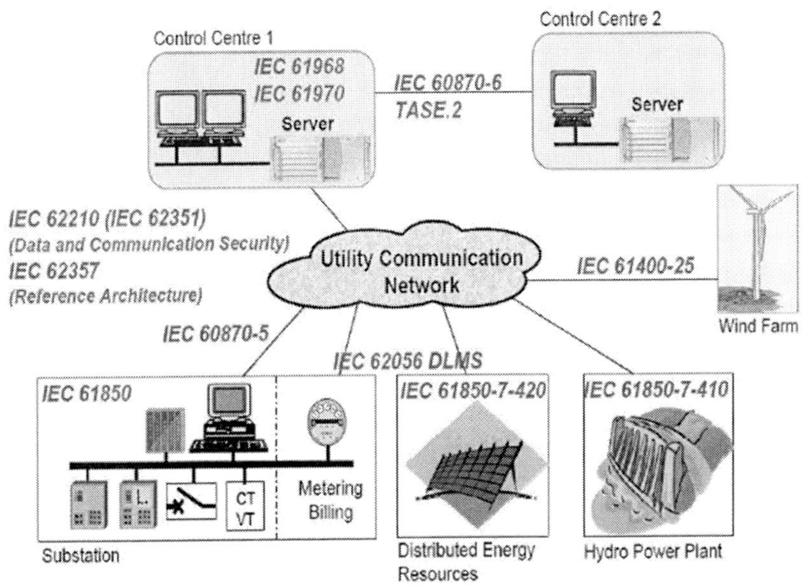

Figure 9. Hierarchy of control systems (including relevant standards) for Smart Grids (used from [14]).

Another pilot project, as a part of Technological Platform for Sustainable Energy (Smart Cities), is currently running in city Vrchlabí (North Bohemia).

Conclusion

In this chapter, brief history of electric power system development in the Czech Republic was introduced along with current situation and future challenges in the network operation in terms of distributed electric power generation and sustainable control. Classical and new technologies are briefly discussed highlighting especially new energy sources and control systems based on "smart" technology (Smart Cities, Smart Grids). In the future, these systems would be able to optimize power consumption in a great extend including final customers. However, more work must be done to solve related problems to this topic for successful integration of this technology into the distribution grid operation in the Czech Republic.

Romanian Distributed Generation System

Energy Resources of Romania

Non-Renewable Energy Resources

Known raw materials deposits are limited and there is a decrease of internal production. Nowadays, oil resources are estimated to be 73.7 mil. Tons [18]. Oil production decreased from 14.7 mil. tons in 1976 to 4.7 mil. tons in 2009 [19]. Present natural gas reserves are estimated to be 184.9 mld.m^3 and production decreased to 12.9 mld.m^3 in 2009, which represented 71.4% from annual national consume.

In 2009, coal production was roughly 35 mil. tons, of which 30 mil. tons lignite and 5 mil. tons mineral coal. Uranium reserves are estimated to be 7.5 mil. tons. Yearly production is 61.000 tons, processed at Feldioara Plant for nuclear combustible for Units I and II of CNE Cernavodă.

Upper estimations are made by taking into account following conditions:

- consume level does not change;
- new deposits will not be discovered;
- technology remains at present level and the economically exploitable reserve does not increase.

Table 3. Energy resources of Romania [20]

	Reserve			Yearly production 2009	Estimated covered period
	U.M.		Mil. t_{ep}	Mil. tons *)	Years
1	2	3	4	5	6 = 3 / 5
Coal -mineral coal -lignite	 Mil. tons Mil. tons	 721 400	 274 629	 3 28	 240 121
Oil	Mil. tons	4	72	4.7	15
Natural gas	Bil. m^3	85	159	12,9	14
Uranium	Mil. tons	7.5	107	0.061	122

*) exclusive natural gas, expressed in bil. m^3.

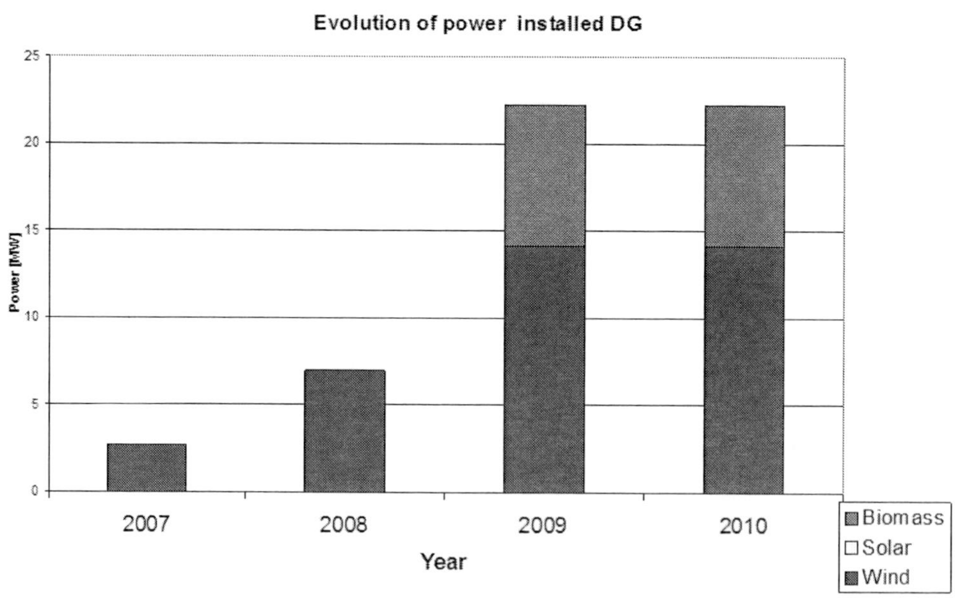

Figure 10. Evolution of DG-installed power [26].

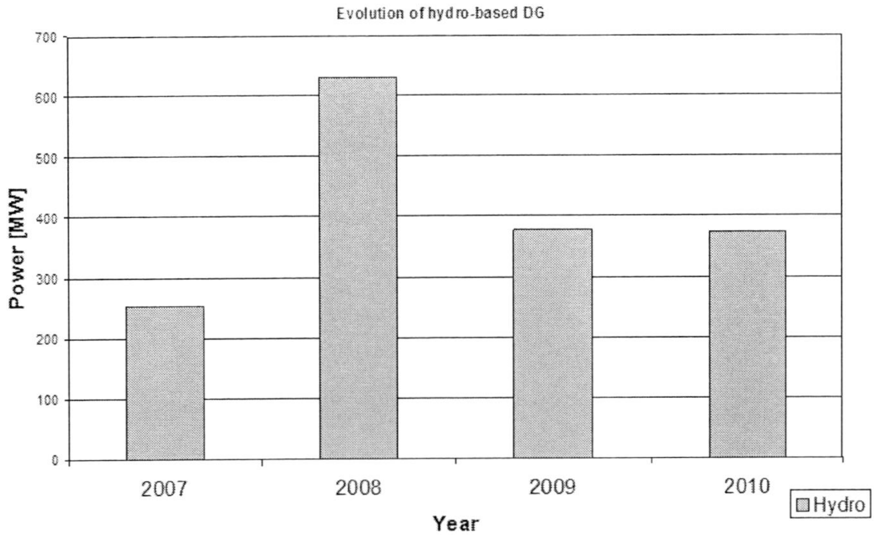

Figure 11. Evolution of hydro-based DG[26].

Renewable Resources

Technical hydro-electric potential of Romania is 36 TWh/year. Economical hydro-electric potential of Romania is estimated to be 23-25 TWh/year, corresponding of an installed electrical power of 8000 MW [20]. In 2009 the capitalization level reached approximate 80% of economical potential. There are under development hydro plants summing 600 MW, with a production potential of 1870 GWh/year. Energy potential of biomass is about 7594 thousand t_{ep}/year, of which 15.5% represents waste from forest exploitations and fire wood, 6.4% sawdust and other wooden residue, 63.2% farm waste, 7.2% house waste and 7.7% biogas [21-26]. Energy potential of solar-thermal systems is evaluated at 1434 mii t_{ep}/year, and of photovoltaic systems at 1200 GWh/year. Technical wind potential is estimated at 8 TWh/year. Geothermal low enthalpy resources is about 167 thousand t_{ep}/year; of which only 30 thousand t_{ep}/year were put to use [26-29].

Electricity Generation, Transportation and Distribution in Last Ten Years

Electricity Generation

Internal production of primary energy maintained a constant level (approx. 28 mil.t_{ep}), the increase of coal production compensate the decrease of oil and natural gas. Due to the increase in consume, the dependence of import grew from 22,5% in 2000 to about 34% in 2004.

At the end of 2008, the installed power in national electrical system was 14.714 MW. In the same year, electrical energy production was 59,7 TWh, with respect to 51,9 TWh, in 2000. At the end of 2004 the number of electrical energy consumers was 8.594.067, of which 8.036.932 home consumers.

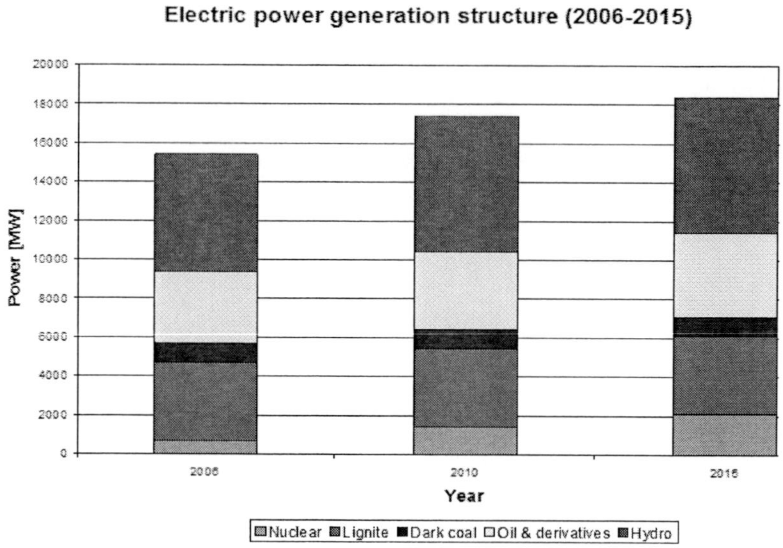

Figure 12. Forecast of electrical power installed (up to 2015)[26].

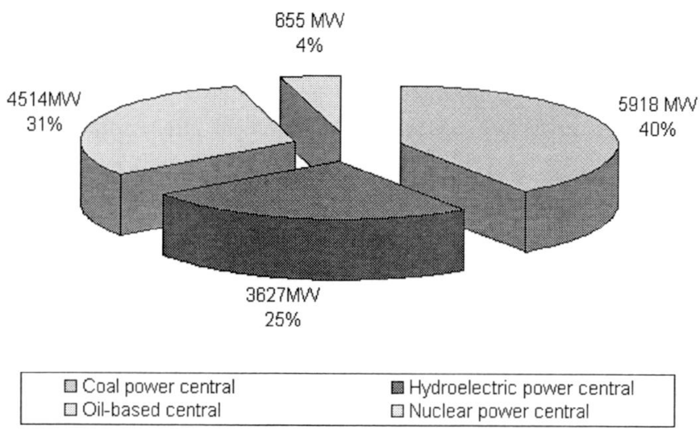

Figure 13. Structure of installed power on generating types [26].

Electric Energy Transportation

Transelectrica SA core activity is electricity transmission through the Power Transmission Grid (PTG) consisting of power substations and lines. PTG is the power grid of national and strategic interest with a rated voltage higher than 110 kV (Figure 14).

The equipment park managed by Transelectrica SA consists of 78 power substations: 1 substation of 750 kV, 35 substations of 400 kV and 42 substations of 220 kV. It also consists contains 9028.8 km of overhead lines, of which: 154.6 km 750 kV, 4740.3 km 400 kV, 4095.9 km 220 kV and 38 km 110 kV (lines of interconnection with the neighboring power systems).

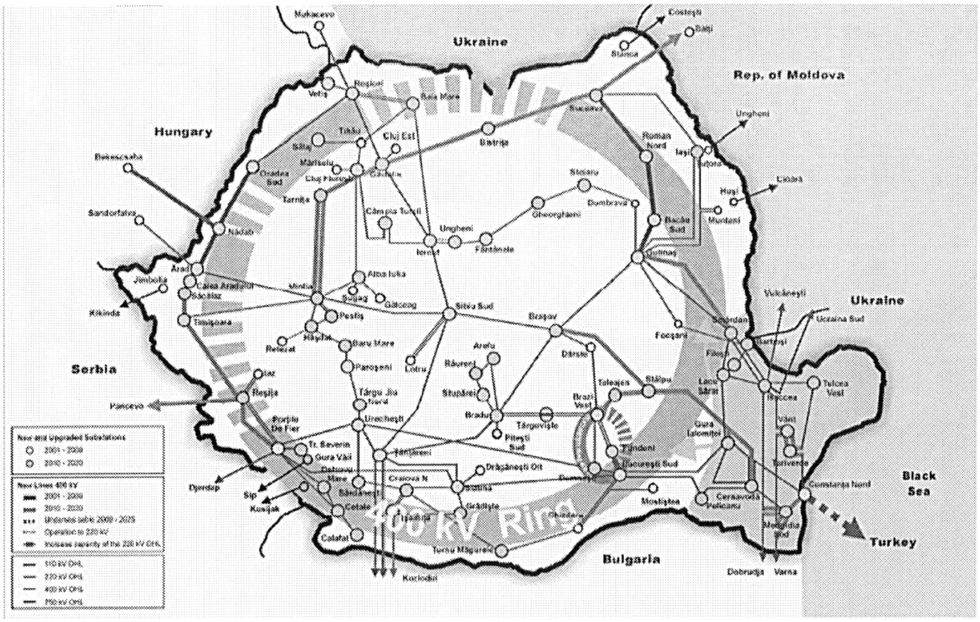

Figure 14. Romanian electricity transmission network [18].

Swot Analysis of the Romanian Power Sector and Market

Strong issues	Weak issues
Diversified transport infrastructure for electrical energy and natural gas; Institutional and legislative framework, adapted to E.U. acquis; Relatively strong interconnection of the transport systems of electrical energy and natural gas with neighbor countries.	Large part of electrical energy production installations are old and do not comply with environmental regulations; Growing dependency on raw materials import (a single natural gas supplier); Weak performances of energy companies own by Romanian State; Price distortions at consumers.
Opportunities	Treats
Favorable geographical position for pan-European oil, natural gas and electrical energy thoroughfares; towards S-E Greece, Bulgaria, Turkey ; towards West: Central Europe; Existence of physical and financial energy markets, as well as access to regional markets, inclusively with system services; Attractive investments climate for foreign investors and Romanian investors, including in privatization process; Increase of trust in Romanian capital market functioning, beginning of Stock Exchange listing for energy companies; Full liberalization of energy and natural gas markets from 2007.	Volatility of raw materials prices on international markets; Competition limitation in energy sector at European level, due to concentration tendencies in the sector; Existence of not-performed payments at stat companies, as well as old system of debt compensation; High vulnerability level of population regarding growing price of energy;

Energy Generation in the Period 2006 – 2009

Table 4. Electrical energy forecast until 2015 [26]

	UM	2005	2006	2007	2008	2009	2009/2005 [%]	2010	2015
Coal total of which:	Mil tep Mil tons	6,19 31,64	6,49 32,50	7,31 36,60	7,69 38,60	7,68 38,9	124	7,75 39,40	7,3 37,30
-Lignite	Mil tep Mil tons	5,10 28,69	5,27 29,20	5,90 32,80	6,25 34,70	6,35 35,3	122,5	6,5 36,00	6,2 34,40
-Mineral coal	Mil tep Mil tons	1,09 2,95	1,22 3,30	1,41 3,80	1,44 3,90	1,33 3,60	122,1	1,25 3,40	1,10 2,90
-Gas	Mil tep Bil m^3	10,05 12,48	9,82 12,2	9,40 11,68	9,08 11,28	8,9 11,0	88,6	8,8 10,90	8,5 10,60
Oil	Mil tep	5,21	5,12	5,12	5,21	5,2	100	5,2	5,1
Hydro energy	Mil tep TWh	1,69 19,90	1,41 16,40	1,50 17,40	1,51 17,60	1,51 17,60	*)	1,53 17,8	1,57 18,2
Nuclear Energy nuclear uranium minerals	Mil tep TWh Th t/year	1,33 5,10 61	1,32 5,10 73	1,95 7,50 122	2,65 10,20 122	2,65 10,2 122	200 200 200	2,65 10,2 122	3,96 15,3 183
Other combustible+ RER	Mil tep	2,90	2,92	2,97	3,00	3,10	106,9	3,3	4,0
TOTAL	Mil tep	27,37	27,08	28,20	29,14	29,04	106,1	29,23	30,43

*)Due to exceptional hydro.

Dependency on imports (specially natural gas) is estimated to be 39% in 2010:

Table 5. Forecast of import variation (until 2015) [26]

	UM	2004	2005	2008	2009	2010	2015
Total primary energy necessary, covered by:	Mil tep	39,018	41,36	45,00	46,30	47,65	51,50
- internal production	Mil tep	28,094	27,37	29,14	29,04	29,00	31,00
- imports	Mil tep	11,852	13,99	15,86	17,26	18,65	20,50
Imports dependency	%	30,4	33,8	35,2	37,3	39,00	40,00

Energy Policy of Romania

Priorities of the New Energy Policy

Energy sector must deal with internal and global challenges: security of energy supply, increase of economic competence and complying with environmental rules. For 2005-2008 period,

National Policy of Energy settled following strategic objectives:

- Diversifying supplying sources with energy resources and transit paths through country;
- Achievement of an increased capacity on energy markets, through full liberalization of markets, transparent and non-discriminatory treatment;
- Ensuring for energy accessibility for the most disadvantaged social categories, without affecting the capacity of energy markets;
- Institutional framework improvement regarding energy markets supervision and monitoring, for eliminating the distortions that affects the competition and harmonizing in the internal E.U. market.

Energy Policy Measures

Through the document of Energy Policy, the government initiated the following directions of action, establishing priorities of Romania policy in the energy domain:

- choose an energy mix with an accent on coal, nuclear energy and renewable energy resources;
- improving internal energy market abilities, active participation to regional markets and cross-border exchanges;
- maintaining the dependency on primary energy resources from import, at an acceptable level;
- consolidating, within European framework, collaboration with countries which are producing raw materials, as well as transit ones;
- diversifying supply sources and transport paths;
- liberalization of energy transit and ensuring permanent and non-discriminatory access to transport networks and international interconnections;
- continue with restructuring process in energy sector (privatization will be achieved with foreign investors as well with companies listing at stock exchange).

General Measures

- Continuous improvement of institutional and legislative framework, expanding concurential mechanisms;
- Drawing attention of foreign and national capital, giving priority to international cooperation projects;
- Integration of electrical energy and natural gas concurential markets into regional and European markets.

Measures for the Power Sector

- Use with high efficiency of coal, in rehabilitated and new groups (Paroşeni, Borzeşti, Brăila etc.);
- Completion of Units 3 and 4 (after 2012) at Cernavodă Nuclear Plant;
- Setting-up of new hydro facilities and re-powering of the existing ones;

- Construction of hydro plant with accumulation through pumping Tarniţa-Lăpuşteşti, in order to satisfy the demand of system services for internal and regional markets (1000 MW, 747 mil. Euro); This objective represents an advantage in day-night relationship with Ukraine;
- Continuing with transport and distribution networks renewal and increase of interconnection degree with E.U. states and states from Black Sea region;
- Separation of distribution and supply of electrical energy activities until 1.07.2007;
- Finalizing distribution companies' privatization process and starting the privatization process for production companies of electrical and thermal energy (is still under discussion a regrouping in production sector).

Investments necessary in electrical energy sector between 2006-2012 is estimated at approximate 3,6 bil. Euro, of which approximate 35% is destined for new facilities construction, as well as for plants re-powering in hydro sector, approximate 15% in energetic facilities, around 20% in thermo sector, approximate 10% in nuclear domain, around 10% in renewal and development of transport networks and approximate 10% in renewal and development of distribution networks.

During 2005-2010 period, will be put out of service, units with reduced performances and having available power of approximate 2100 MW. A power of 1100 MW will be put in use, in new units with high performances and re-powering will be performed for units with power of approximate 950 MW. From 2009 on, the weight of electrical energy obtained by raw materials was reduced at about 14% due specially to putting into service of Unit 2 from Cernavodă.

TURKEY ELECTRICITY SYSTEM

Energy Resources of Turkey

Turkey is relatively well endowed with energy and mineral resources. The extensive mountainous terrain provides numerous hydroelectric sites, although most are far from the main population and consumption centers. The country also has substantial exploitable lignite resources and small reserves of hard coal, petroleum, and natural gas. Commercially exploitable deposits of many minerals have been located, but the territory has been surveyed only partially. Exploitation of these natural resources has occurred relatively slowly.

The combined demands of industrialization and urbanization nearly tripled energy consumption in the 1960s and 1970s. An inappropriate pricing policy, especially the subsidy of petroleum that led to unduly cheap products, was one cause of shifts in the sources of energy that exacerbated shortages. In 1960 more than half of the primary energy consumed came from noncommercial sources, mainly firewood but also manure and other agricultural wastes. These noncommercial sources, plus domestic coal and lignite, accounted for more than 80 percent of all primary energy consumed; oil supplied only 18 percent. By 1980, in contrast, oil supplied about 47 percent of the primary energy consumed, coal and lignite about 21 percent, hydroelectric power 8 percent, and noncommercial sources such as firewood and animal wastes only 23 percent. By 1992, 43.5 percent of final energy came from petroleum,

31.1 percent from lignite and hard coal, 4.1 percent from hydroelectric power, 6.9 percent from natural gas, and 14.4 percent from other energy sources, including solid fuels, geothermal, solar power, and wind power.

Although Turkey's energy resources remained underdeveloped in early 1995, the country had relatively good energy production potential. One estimate places the economically feasible hydroelectric potential at around 29,500 megawatts, which would allow annual production to reach roughly 100,000 gigawatt-hours in years with normal rainfall. Lignite is the second most important potential source of energy, with proven and probable deposits put around 6.4 billion tons. However, Turkish lignite, containing high amounts of water and sulfur, is hard to burn and pollutes the air. Turkey's proven and estimated petroleum stocks are equivalent to about three years' consumption. Proven reserves are estimated at about 16 million tons, and enhanced oil-recovery techniques may allow extraction of another 30 million tons. Proven reserves of natural gas total about 12.4 billion cubic meters, and reserves of hard coal about 1 billion tons. Turkey's geothermal resources are considerable, but they have not yet been systematically explored (Metz, 1995 [30]).

Non-Renewable Energy Resources

Imports of petroleum averaged more than 15 million tons per year in the early 1980s and increased to about 23 million tons in the early 1990s. Most of Turkey's oil fields are located in southeastern Anatolia near the borders with Iran, Iraq, and Syria. Because of the country's fractured substrata, deposits are often contained in small pockets, which make exploration and extraction difficult. In 1985 exploration proved that Turkey has oil deposits at very deep levels, but it was not known how large the deposits might be. Shell Oil determined that oil at Paleozoic levels would be recoverable, and other investigations proved significant deposits in central Anatolia under the salt flats in the plain north of Konya. In 1991 British Petroleum began exploring for oil in offshore areas of the Black Sea. It is also suspected that the Aegean shelf contains considerable petroleum deposits, but as long as relations with Greece remain strained, conflicting claims to the Aegean seabed limit prospects for exploration. To speed up the exploration process, the Turkish government in 1983 eased regulations on such activities by foreign oil companies, allowing them to export 35 percent of production from fields they discovered in Anatolia and 45 percent from offshore fields. Although several foreign concerns started exploration after the liberalization package went into effect, up to the mid-1990s no major finds had been reported.

The state-owned oil company, Turkish Petroleum Corporation (TPAO), Shell Oil, and Mobil control most petroleum output, which had climbed gradually to a peak of 3.6 million tons in 1969 but declined to about 2.1 million tons in 1985 as deposits were depleted. By the early 1990s, output had increased once again to nearly 4.4 million tons. The main petroleum project during the 1980s was an attempt at secondary recovery at the Bati Raman fields in southeastern Anatolia, which were expected to produce roughly 1.5 million tons a year over a twenty-year period. Five refineries with a total capacity of about 713,000 bpd meet most of the country's need for petroleum products. Until early 1995, about 85 percent of refinery capacity was in public hands in four refineries located at Aliaga near Izmir, Kocaeli, Kirikkale, and Batman. A fifth refinery, jointly owned by Mobil, Shell Oil, British Petroleum, and a Turkish company, is located at Mersin.

Natural gas became important in the 1980s. Gas tapped in Thrace (Trakya, European Turkey) was piped to the Istanbul region and used to produce electricity, thereby reducing the need for energy imports from Bulgaria. In 1986 Turkey began construction of a pipeline to carry Soviet natural gas from the Bulgarian border to Ankara; the line was completed in the late 1980s. In 1990 government officials announced that they also desired to purchase natural gas from Algeria, a move that would help balance Turkey's large purchases from the Soviet Union.

Policy makers in the early 1970s had targeted lignite as the most abundant domestic source of hydrocarbons, and production grew rapidly from an average of about 7.9 million tons for the 1970-75 periods to more than 31 million tons in 1985. Mines operated by the state-owned Turkish Lignite Company are responsible for about two-thirds of output; private firms produce the remainder. Production of hard coal is entirely controlled by the government-owned Turkish Coal Company, which suffers from poor management and outmoded technology. Coal production is also hampered by the great depth of the country's deposits. Hard coal output fell from around 6.5 million tons in 1976 to about 3.8 million tons in 1983, and unit costs exceeded those of coal imports. As a result of these trends, Turkey is beginning to import coal for use in power plants. In 1992 Turkey produced about 12 million tons of coal and imported a net of about 4.2 million tons (Metz, 1995 [30]).

Renewable Resources

Turkey has abundant reserves of renewable energy, such as solar, wind and geothermal (Table 6). The benefits of exploiting these sources would be enormous:

Table 6. Turkey's renewable energy potential

Energy type	Usage purpose	Natural capacity	Technical	Economical
Solar energy	Electric (bilion kWh)	977.000	6.105	305
	Thermal (mtoe)	80.000	500	25
Hydro power	Electric (bilion kWh)	430	215	124.5
Wind (land)	Electric (bilion kWh)	400	110	50
Wind (off shore)	Electric (bilion kWh)	-	180	-
Wave energy	(bilion kWh)	150	18	-
Geothermal energy	Electric (10^9 kWh)	-	-	1.4
	Thermal (mtoe)	31.500	7.500	2.843
Biomass energy	Total (mtoe)	120	50	32

- By relying completely on indigenous resources, renewable would reduce reliance on imported fuels and enhance Turkey's energy security.
- The environmental impact of renewable technologies is far less than that of nuclear and fossil fuel power plants. There are no emissions of greenhouse gases or toxic wastes;
- The cost of electricity from some renewable energy sources is already competitive with many conventional technologies, and is dropping rapidly;

– Renewable technologies have no fuel costs and are virtually inexhaustible;

The real beginning for renewable energy policy was the definition of renewable energy sources in the decree of the Modification of the License Regulation in the Electricity Market in 2003. Before then, there was no national renewable energy policy and few government incentives existed to promote market deployment of renewable energy. However, the Electricity Market Licensing Regulation, in itself, is not expected to be sufficient to overcome the high investment cost, risk and lack of security associated with the entrance of renewable power plants into the electricity market.

Solar:

It has been calculated that Turkey receives sunlight equivalent to roughly 11 thousand times the amount of electricity generated in Turkey in 1996. Clearly, both photovoltaic and solar-thermal systems could be used to great effect. Use of solar thermal is already widespread. In 1995 it was calculated that solar thermal generated 52,000 tones-of-oil-equivalent of heat. Indications suggest that this could increase six-fold in the next 15 years. The potential for PV is virtually unlimited, and the price for these systems is declining rapidly.

Wind:

Wind energy is already in use in many parts of the world at a price competitive with conventional technologies. The latest technologies can produce electricity at 4.4c/kWh, comparable to many conventional sources. In Turkey, wind energy has great potential. The figures from the Organization for Economic Cooperation and Development (OECD) show that Turkey theoretically has 160 Terawatt hours a year of wind potential which is equivalent to twice the country's electricity production in 1996.

Geothermal:

Turkey has one eighth of the world's geothermal potential and is ranked 7th in the world. The cost of electricity generated from geothermal reserves is between 3-10c/kWh. The bottom end of this range is competitive with conventional systems. The Turkish Government has already recognized the potential of geothermal and in 1972 began to implement a program of construction which would have seen nearly 710 Megawatts installed by now. Unfortunately, the program was not fully implemented and this rich resource remains unexploited. Recently some local authorities have announced plans to utilize geothermal, but there is still a lack of coordinated support from the central Government (Elsam Power Pool Report, 1997 [31]).

Electricity Generation, Transportation and Distribution

Electricity Generation

As of February 2008, Turkey's total generation capacity is 40161 MW. Of this, 23356 MW is provided by Electricity Generation Company (EUAS) while 9201 MW comes from BO, BOT and TOOR type plants, with a further 6882 MW produced by auto-producers and private generation companies. Thus, state owned generation plants dominate with a 58.2% share of all generation.

As far as primary sources used for generation are concerned, 13. 063 MW is based on hydro power whereas 12.735 MW and 7851 MW are based on natural gas fired power plants and lignite fuel plants, respectively. The remaining 6.513 MW of the total capacity consists of electricity generated from petroleum products, hard coal, geothermal, wind and other renewable sources.

In 2006, annual total electricity production reached 176 billion kWh, an 8.6% increase compared to the previous year. Meanwhile, total electricity consumption was 174 billion kWh, corresponding to an 8.3% annual increase. 131 billion kWh of total production was provided by thermal plants, whereas 44 billion kWh and 344 billion kWh were provided by hydro plants and other renewable, mainly wind turbines, respectively.

Turkey's electricity generation is mainly based on thermal plants. Their share in total electricity production was 75% in 2006. The remaining quarter of total electricity is produced by hydro power plants. Wind power is negligible, amounting to less than 1% of total generation despite the priority given to renewable, especially wind power, in the governments' new energy policy. Natural gas fired power plants are the largest single source of generation, amounting to 44% of total electricity generated.

Security of supply is a priority in Turkey's energy strategy. As a result of the growth in demand, installed capacity has increased nearly 2000 times since 1923, but this has been outpaced by energy demand, which has increased more than 3000 times since 1923.

In terms of demand side management, Energy Efficiency Law no.5584, approved by the Turkish Grand National Assembly in 2007, approved a number of instruments to support energy efficiency both at the micro-and macro-level. These instruments include undertakings such as energy management, monitoring, utilization of heat and temperature control, the use of individual metering systems fort he central heating of buildings, energy performance certificates for buildings, co-generation and micro-co-generation, research and development activities as well as a number of incentives namely, projects for improving energy efficiency on existing industrial plants, co-generation and promoting central heating systems in buildings.

Electricity Transportation

Turkish Electricity Transmission Company (TEIAS) is responsible fort he construction, operation, and maintenance of the electricity transmission grid. It is also responsible, as systems operator, for dispatch, frequency control, monitoring real time system reliability and fort he purchases and provision of ancillary services. TEIAS is also responsible for the running of the electricity market in the form of Day-Ahead balancing and settlement market. Transmission grid and market activities are completely unbundled from other market and regulated activities.

The transmission system is the backbone of the whole electricity system in Turkey, since most generation is undertaken in the eastern regions where the big hydro plants are located, while most of the consumption takes place in industrialized and urbanized western cities. TEIAS performs its duties through one national and seven regional load dispatch centers. At the end of 2009, its transmission assets consist of transformers with total capacity of 92.199 MWA and 47.284 km of lines. Most lines are 380kV and 154kV transmission lines. Transmission system losses are 2.3% in 2009.

Electricity Distribution

The distribution network in Turkey in 2005 consisted of around 842.393 km of lines (approximately 60% low-voltage and 40% mid-voltage of which around 94% is airborne, with the remaining 6% underground). There are 285.175 distribution transformers with a total capacity of 90.000 MVA and around 13.5 millions poles in Turkey. Transmission losses and distribution losses amount to nearly 20.2 TWh in total which means that 17.8% of total electricity generated in Turkey has disappeared or has been stolen.

Kayseri is the only region in which a private company distributes electricity, following the transfer of operating rights to KCETAS (Kayseri Electricity Distribution Company) in 1990. It transferred to PA (Privatization Administration). Distribution Regions' geographical coverage is showed in Table 7 early. Baskent EDAS (Electricity Distribution Company) and Sakarya EDAS were transferred to private sector by PA in 2009.

Table 7. Electricity Distribution Regions in Turkey (TEDAS 2008)

Region No.	Region Name	Number of customers	Energy consumption (GWh)
1	Dicle EDAS	1.044.300	5.214,0
2	Van Golu EDAS	401.400	1.137,2
3	Aras EDAS	725.200	1.655,8
4	Coruh EDAS	989.600	2.267,7
5	Firat EDAS	663.700	2.145,2
6	Camlibel EDAS	734.700	2.087,9
7	Toroslar EDAS	2.597.400	13.904,9
8	Meram EDAS*	1.530.500	5.858,9
9	Baskent EDAS	3.075.800	11.161,5
10	Akdeniz EDAS	1.469.800	6.048,6
11	Gediz EDAS	2.331.500	13.861,9
12	Uludag EDAS	2.278.500	10.940,5
13	Trakya EDAS	767.800	5.473,2
14	Anadolu Y. EDAS	2.037.900	8.672,1
15	Sakarya EDAS	1.307.300	8.760,5
16	Osmangazi EDAS	1.277.300	5.041,7
17	Bogazici EDAS	3.832.800	18.947,6
18	Kayseri(KCETAS)	521.453	2.224,4
19	Menderes EDAS*	1.475.700	3.815,4
20	Goksu EDAS	479.800	3.302,6
21	Yesilirmak EDAS	1.466.700	4.062,7
Total	TEDAS (Turkey)	31.009.153	136.584,30

Proposed Objectives and Implementation Stage

In Turkey, electricity is produced by thermal power plants (TPPs), consuming coal, lignite, natural gas, fuel oil and geothermal energy, and hydropower plants (HPPs) (Table 8). The development of nuclear power in Turkey is so new. Present and future total final energy production and consumption in Turkey are shown in Table 9 and Table 10, respectively. Long term planning studies (from 1996 to 2010) indicate a heavy burden of investments amounting to some 68 billion US$. Turkey's funding needs for the energy sector is the highest of the southern and eastern Mediterranean countries.

Table 8. Electricity production from thermal and hydropower sources according to years in Turkey

Year	Thermal (GWh)	Hydropower (GWh)	Total (GWh)	% of hydropower
1950	759	30	789	3.80
1960	1814	1001	28`5	35.55
1970	5590	3033	8623	35.17
1780	11,927	11,348	23,275	48.75
1990	34,395	23,148	57,543	40.22
1995	50,621	35,541	86,153	41.25
1999	81,661	34,678	116,339	29.81
2000	93,934	30,879	124,813	24.74
2001	98,563	24,010	122,573	19.60
2002	71,966	44,034	116,000	38.00
2003	104,898	35,325	140,283	25.18

Table 9. Present and future of total final energy production in Turkey

Energy sources	1990	2000	2005	2010	2020	2030
Coal and lignite	12.41	13.29	20.69	26.15	32.36	35.13
Oil	3.16	2.73	1.66	1.13	0.49	0.17
Gas	0.18	0.53	0.16	0.17	0.49	0.17
Com. renewable and wastes	7.21	6.56	5.33	4.42	3.39	3.75
Nuclear	-	-	-	-	7.30	14.60
Hydropower	1.99	2.66	4.16	5.34	10.00	10.00
Geothermal	0.43	0.68	0.70	0.98	1.71	3.64
Solar/wind/other	0.03	0.27	0.22	1.05	2.27	4.28
Total production	25.86	26.71	34.12	39.22	58.20	71.68

Table 10. Present and future of total final energy consumption in Turkey

Energy sources	1990	2000	2005	2010	2020	2030
Coal and lignite	16.94	23.32	35.46	39.70	107.57	198.34
Oil	23.61	31.61	40.01	51.17	71.89	102.38
Gas	2.86	12.63	5.33	4.42	3.93	3.75
Com. renewable and wastes	7.21	6.56	5.33	4.42	3.39	3.75
Nuclear	-	-	-	-	7.30	14.60
Hydropower	1.99	2.66	4.16	5.34	10.00	10.00
Geothermal	0.43	0.68	0.70	0.98	1.71	3.64
Solar/wind/other	0.03	0.27	0.22	1.05	2.27	4.28
Total production	53.01	77.49	129.63	152.22	279.18	463.24

The needs of each energy sector are:

Electricity : 56 billion US$ (82%),
Gas : 6 billion US$ (9%),
Oil : 4 billion US$ (6%),
Solid fuels : 1 billion US$ (1%).

In 1996, 5.9 billion US$ were invested in the Turkish economy, 24% of which were in the energy sector.

Energy Policy of Turkey

Turkey is among the first five leader countries in its geothermal direct use applications [30-32]. In Turkey, the district heating system applications were started with large scale, city based geothermal district heating systems. The investigations on geothermal energy in the country gained speed in the 1970s. However, the utilization of geothermal energy could not become widespread sufficiently due to scaling problems up to the early 1980s. Since then, important developments have been recorded in geothermal energy utilization. Recently, geothermal direct use applications have reached up to 52,000 residences equivalence of geothermal heating, and engineering design of nearly 300,000 residences equivalence geothermal district heating has been completed.

Parallel to the development of geothermal energy utilization in Turkey, it is projected that by the years 2010 and 2020, the total installed capacity will increase to 3500MWt (500,000 residences equivalent, which is about 30% of the total residences in the country) and 8300 MWt (1,250,000 residences equivalent) for space heating and to 500 MWe and 1000 MWe for power production, respectively.

The investment cost for geothermal district heating systems per residence with a floor area of 100m2 is about 1500–2500 US$ (excluding heater costs in the residence), while the payback period varies between 5 and 8 years. About 30–50% of the investment costs have been paid by consumers as a connection subscription fee, like a capital investment. The heating fees (2001 heating season) were in the range of 14–29 US$.

Turkey can not perform a clear strategy concerning the renewable energy sources because of energy costs and investment costs. The State encouraged the private sector for natural gas combined circuit plants and guaranteed to buy the generated electricity with a low cost and with special conditions. State performed the strategy of build, operates and transfer (BOT) system and succeeded it. The share of the natural gas combined circuit plants increased to 20% in total primary energy supply. The state achieved the sustainability in this wise. Turkey is interested in renewable energy resources and gives effort to provide the sustainability of using these energy resources. The state encouraged the municipalities in respect of the geothermal energy and gave them the permission to behave self-governing. In Turkey, the efficiency of energy utilization is not as high as Europe yet. The state leads the private sector to the World Bank's credit in all sources of renewable energy.

The State says that it will be the guarantor for the 30-40% of the cost of the private sector's investments which are for their own needs. If the private sector can find buyer, it can sell the electricity produced in these plants.

The cost of 1 kilowatt (kW) power from a renewable energy resource is 13-15 cent. If the State buys for example 10% of the generated energy by private sector for this cost, the State will gain from that too in future. It is so recently that less energy consuming building projections have taken place. Ground sourced heating and passive heating systems are not common either. For a sustainable development of renewable energy resources and settling to the Mediterranean strategy, ground sourced and water sourced heat pump systems, wind and solar energy power plants have to be kept unobstructed always. The renewable energy Technologies and the energy quantities which are necessary for production per unit have to be kept always in the journal and policy of the country. Legal regulations have to be performed and the State has to give effort to make the public assimilate these regulations highly.

Priorities of the New Energy Policy

The most important handicap for foreign investors is Turkish bureaucracy. The permission for a foreign investor can be taken through one-year period with applying numerous different associations. New government had promised to make the permission producer easier. Hydroelectric generation, biomass combustion, solar energy for agricultural grain drying and hot water heating, and geothermal energy have been in use in the country for many years. Domestic water heating is the primary active solar technology. In Turkey, approximately 30,000 solar water heating systems have been installed since the 1980s. This is a minute fraction of the total potential. About 50% of existing dwellings could be fitted effectively with a solar water heater. If this potential were extended to 2025, the deployment of approximately 5 million systems (allowing for a rise in the Turkish housing stock) would be required. This could save an estimated 30 PJ (9.0 TWh) per year of oil, coal and gas and 2.0 TWh per year of electricity, giving a saving of 5.0 million tones of CO2 per year, or just under 1% of current Turkey CO2 production [32].

Agricultural residues have a high potential to take the place of the lignite (40 million tons) and hard coal (1.3 million tons) used in electricity production.

Biogas systems are considered to be strong alternatives to the traditional space heating systems (stoves) in rural Turkey. Geothermal heat pumps are a relatively new application of geothermal energy that has grown rapidly in recent years. On the other hand, the biggest benefit of geothermal heat pumps is that they use 25-50% less electricity than conventional heating or cooling systems. Geothermal heat pumps can also reduce energy consumption, and corresponding air pollution emissions, up to 44% compared to air source heat pumps and up to 72% compared to electric resistance heating with standard air conditioning equipment.

Conclusions

Turkey recently has the capacity to generate 80 Terawatt-hours of electricity per year. This is to be expanded as spending in this sector by state-controlled and private companies will be about $80bn.

For nearly two decades, Turkey has been one of the fastest growing power markets in the world. In terms of the energy efficiency, Turkey has made good progress by adopting the framework law on energy efficiency in order to reduce the high energy intensity of the

Turkish economy. The law does not, however, contain any objectives and the provisions on high-efficiency cogeneration do not comply with acquits.

Coal which is the most reliable domestic energy source in Turkey should be consumed more in the industry and also electricity production in order to reduce the energy production costs of Turkey and reduce the dependency on other countries. Moreover, Turkish government should improve the coal burning technologies in the thermal power plants, so the energy production rate from coal will increase and will contribute to the developing economy of Turkey.

Natural gas is an expensive energy source for Turkey. However, because of the misguided energy policies, the consumption of natural gas is high in Turkey. Moreover, the share of natural gas in electricity generation is 45% in Turkey. Because of dramatically high dependency on natural gas, Turkey will be one of the most affected countries in a possible natural gas crisis in the world. In other words, high consumption of natural gas is a disadvantage for Turkey in terms of development. As mentioned in the section on natural gas, Turkish government should use coal-fired power plants efficiently and increase the load factor of these power plants. Moreover, it is not possible to shut down all natural gas fired power plants, because of the high energy demand.

In the area of renewable energy sources, Turkey had already adopted a Law for the use of renewable energy sources for energy production and implemented a regulation on the guarantee of origin. The framework law also increases the potential for promoting the production of electricity from renewable energy sources. This legal framework still needs to be supplemented by ambitious objectives in order to exploit the vast unused potential of renewable energy sources in Turkey. Energy production from renewable energy sources should be improved in Turkey to reduce the dependency and increase the development level of the country by increasing the economic level of the country. Turkey should use its renewable energy sources efficiently and should promote new technologies and also need to use all its renewable energy potential in next future.

In 2002, the Energy Market Regulatory Authority (EMRA) began to issue licenses for various activities in the electricity sector and to approve companies' tariffs. The independence and administrative capacity of this authority need to be strengthened, including in terms of the number of qualified staff it employs. Further legislative and administrative steps need to be taken to ensure the proper functioning of a competitive electricity market in line with acquits. The dominant position of the State trading company in the wholesale market should be adjusted. Current restrictions for cross-border trading removed and the matter of existing long-term power purchase agreements should be addressed. In this connection, distribution rights have been granted to several new regional distribution companies.

The Turkish electricity sector has made remarkable progress in terms of deregulation of market-based activities and regulation of naturally monopolistic activities. At the same time, there are still some issues that need to be solved in order to provide for a sound, credible and smoothly running market system.

IRAN-AZARBAIJAN DISTRIBUTED GENERATION SYSTEM

Recently Study about Application of Distributed Generation (DG) System shows that by using of DG elements for generating of electrical energy to feed of electric consumers. We study on the Iranian Azarbaijan (Iran-Azarbaijan) Electric Network involves of more purposes to get more technical results. In any subchapter we try to get application results for using of DG System and how to use that with the best electrical and technical qualifications results.

Rehabilitation of Distributed Generation System

Application of selected DG system for electric network was possible for getting the acceptable results in damping of any electrical oscillation specially oscillation of voltage and current. These parameters were very important and the selected DG system would have been electrical capability for feeding of any electrical loads for this aim. The selected micro turbine (gas) and wind turbine are used to attention of technical situation of the network. We consider the effect of distance on DG system against Iran-Azarbaijan electrical network and explain about the methods of energy generation and renewable energy that can be study in the following subjects [33, 34].

Energy Management

Distributed energy management technologies include energy storage devices and various methods for reducing overall electrical load. Energy storage technologies are essential for meeting the levels of power quality and reliability requirements by high-tech industries. Storage can provide emergency power and peak-shaving benefits. Energy storage is important for distributed energy devices by giving them more load-following capability, and also supports renewable technologies such as wind and solar electricity by making them dispatchable. Load reduction technologies are used to reduce energy consumption and consumer energy expenses. By maximizing end-use efficiency, they can also help avoid or postpone the construction of new generating plants [35].

Industrial Gas Turbines

A gas turbine is a heat engine that uses high-temperature, high-pressure gas as the working fluid. Part of the heat supplied by the gas is converted directly into mechanical work. High-temperature, high-pressure gas rushes out of the combustor and pushes against the turbine blades, causing them to rotate. In most cases, hot gas is produced by burning a fuel in air, which is why gas turbines are often referred to as "combustion" turbines. Because gas turbines are compact, lightweight, quick-starting, and simple to operate, they are used widely in industry, universities and colleges, hospitals, and commercial buildings to produce electricity, heat, and steam [36].

In such cases, "simple cycle" gas turbines convert a portion of input energy from the fuel to electricity and use the remaining energy, which is normally rejected to the atmosphere, to produce heat. This waste heat may be used to create steam to power a separate turbine. The attached steam turbine may generate electricity or power a mechanical load. This is referred to as a combined-cycle combustion turbine because two processes or cycles are derived from one fuel input to the primary turbine.

Simple-cycle efficiencies range (without external use of exhaust heat) is from 21% to 40%. Turbines produce high-quality heat that can generate steam for combined-cycle or combined heat and power applications and significantly enhance efficiency. These systems capture and use the heat produced during the combustion process for steam, hot water, or thermally activated equipment such as absorption chillers. Taking advantage of the normally wasted heat means a tremendous gain in efficiency, nearly 90% in some cases. Domestic and global potential markets for advanced turbines are large [37].

At least half of all new power generating capacity added between now and future is likely to use gas turbines. In addition, mid-sized turbines have tremendous potential for use as base load, CHP, peaking, and standby/backup power in commercial and industrial settings.

Micro Turbines

Micro Turbines are small combustion turbines approximately the size of a refrigerator with outputs of 25-500 KW. They evolved from automotive and truck turbochargers, auxiliary power units for airplanes, and small jet engines and are composed of a compressor, combustor, turbine, alternator, recuperator, and generator.

Micro turbines offer a number of potential advantages compared with other technologies for small-scale power generation. These advantages include their small number of moving parts, compact size, light weight, greater efficiency, lower emissions, lower electricity costs, and ability to use waste fuels. They also have the potential to be located on sites with space limitations for the production of power. In addition, waste heat recovery can be used with these systems to achieve efficiencies greater than 80%.

Turbines are classified by the physical arrangement of their component parts: single-shaft or two-shaft, simple-cycle or recuperated, inter-cooled, and reheat. The machines generally rotate at more than 40,000 rpm. Bearing selection, whether the manufacturer uses oil or air, is dependent on use. Single-shaft is the more common design because it is simpler and less expensive to build. Conversely, the split shaft is necessary for machine drive applications, which do not require an inverter to change the frequency of the AC power.

Micro turbines can also be classified as simple-cycle or recuperated. In a simple-cycle, or unrecuperated, turbine, compressed air is mixed with fuel and burned under constant pressure conditions. The resulting hot gas is allowed to expand through a turbine to perform work. Simple-cycle micro turbines have a lower cost, higher reliability, and more heat available for cogeneration applications than recuperated units. Recuperated units use a sheet-metal heat exchanger that recovers some of the heat from an exhaust stream and transfers it to the incoming air stream. The preheated air is then used in the combustion process. If the air is preheated, less fuel is necessary to raise its temperature to the required level at the turbine inlet. Recuperated units have a higher thermal-to-electric ratio than unrecuperated units and can produce 30% to 40% fuel savings from preheating [37].

Compressed air is mixed with fuel and burned in a combustor under constant pressure conditions. The resulting hot gas is allowed to expand through a turbine to perform work. A sheet-metal heat exchanger is used to recover some of the heat from the exhaust stream and transfer it to the incoming air stream. The preheated air is then used in the combustion process. By preheating the air, less fuel is needed to raise its temperature to the required level at the turbine inlet. The rest of the exhaust stream is passed through a waste-heat recovery device that captures more of the heat for use in industrial processes, space heating, and water heating.

Advanced materials such as ceramics and thermal barrier coatings are some of the key enabling technologies to further improve micro turbines. Efficiency gains can be achieved with materials such as ceramics, which allow a significant increase in engine operating temperature.

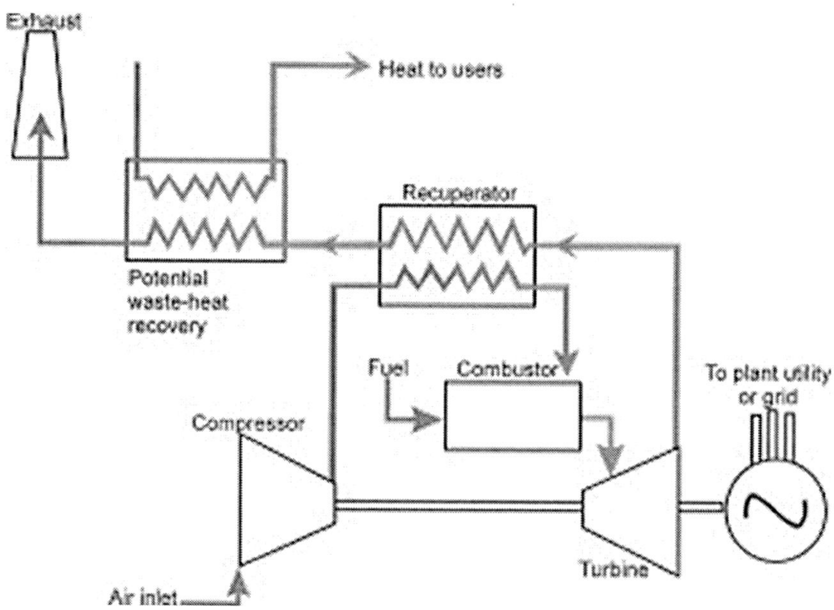

Figure 15. Micro turbine cycle [38].

Because of their compact size, relatively low capital costs and expected low operations and maintenance costs, and automatic electronic control, micro turbines are expected to capture a significant share of the distributed generation market [37].

Wind Energy

Winds are caused by the uneven heating of the atmosphere by the sun, the irregularities of the earth's surface, and rotation of the earth. Wind flow patterns are modified by the earth's terrain, bodies of water, and vegetative cover. This wind flow, when "harvested" by modern wind turbines, can be used to generate electricity.

Historical of Wind Energy

The wind energy is used in last time and different methods by human. In 200 years ago wind energy used by Iranian for mixed of corn foods. Of course Egyptian peoples were used from wind energy for driving of ships. Holland people were used wind energy for windmill. They used for any kind of applications in other countries for example in Italy, England or India.

How Wind Energy Is Generated

The term wind energy describes the process by which the wind is used to generate mechanical power or electricity. Wind turbines convert the kinetic energy in the wind into mechanical power.

This mechanical power can be used for specific tasks (such as grinding grain or pumping water) or a generator can convert this mechanical power into electricity to power homes, businesses, schools, and other facilities.

Wind Turbines

Most wind turbines turn in the moving air like aircraft propeller blades, and power an electric generator that produces an electric current. A wind turbine is essentially the opposite of a fan: instead of using electricity to make wind, like a fan, wind turbines use wind to make electricity.

The wind turns the blades, which spin a shaft, which connects to a generator and makes electricity.

Wind Turbine Types

Modern wind turbines fall into two basic groups; the horizontal-axis variety, like the traditional farm windmills used for pumping water, and the vertical-axis design, like the eggbeater-style Darrieus model, named after its French inventor. Most large modern wind turbines are horizontal-axis turbines.

Figure 16. Horizontal-axis wind turbine [39].

Figure 17. Vertical-axis wind turbine [39].

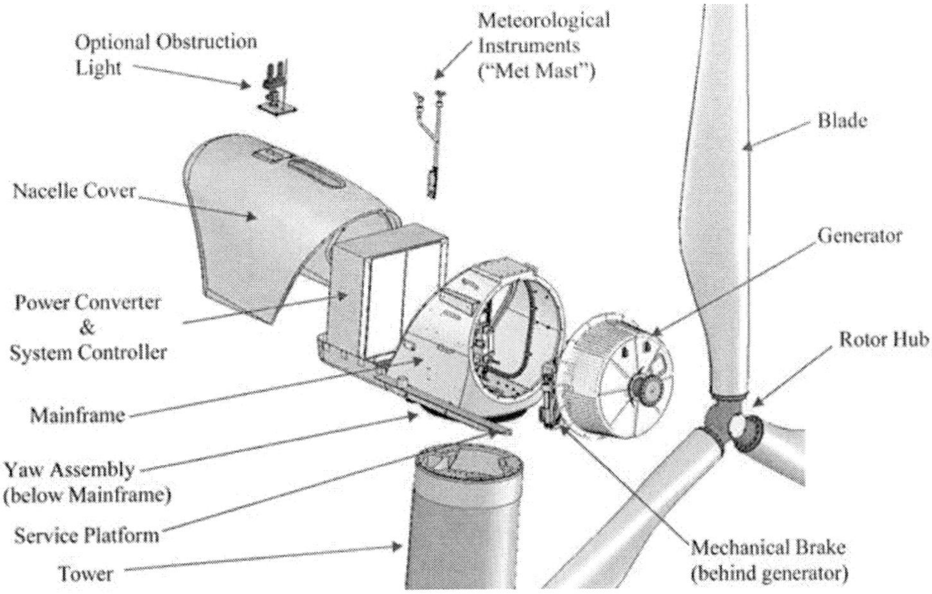

Figure 18. Wind turbine Components [40].

Horizontal turbine components include [40] (Figure 18):

- blade or rotor, which converts the energy in the wind to rotational shaft energy;
- a drive train, usually including a gearbox and a generator;
- a tower that supports the rotor and drive train; and
- Other equipment, including controls, electrical cables, ground support equipment, and interconnection equipment.

Turbine Configurations

Wind turbines are often grouped together into a single wind power plant, also known as a wind farm, and generate bulk electrical power. Electricity from these turbines is fed into a utility grid and distributed to customers, just as with conventional power plants.

Wind Turbine Size and Power Ratings

Wind turbines are available in a variety of sizes, and therefore power ratings. Commercially available land-based turbines range in capacity from 0.25 KW to 2.5 MW, while offshore wind turbines may exceed 4.5 MW. When operating at full power, the smallest and the largest wind generators can supply enough electricity to power a few light bulbs and thousands of homes, respectively.

The capacity of a turbine is determined largely by its rotor diameter. Present-day technology can be divided into three broad size ranges, briefly characterized below:

Residential: rated capacity below 30 kW, rotor diameter of 4 to 43 ft, hub height of 60 to 120 ft Intermediate: rated capacity between 30 and 500 KW, rotor diameter of 43 to 100 ft, hub height of 115 to 164 ft Commercial: rated capacity between 500 KW and 4.5 MW, rotor diameter of 100 ft to more than 325 ft, hub height of 164 to more than 260 ft.

Although wind turbine blades may appear to turn slowly, because most wind turbines have constant rotational speed, the speed with which the tip of the rotor blade moves through the air (the tip speed) may exceed 140 miles per hour, while at the center of the hub it is zero [40].

Figure 19. 450 KW wind turbine and 1.5 MW wind turbines [39].

Rated capacity is a measure of electricity generation under ideal conditions. In the real world, the wind blows intermittently, at varying speeds. Wind speed at hub height is the main determinant of actual capacity.

A Renewable Non-Polluting Resource

Wind energy is a free, renewable resource, so no matter how much is used today, there will still be the same supply in the future. Wind energy is also a source of clean, non-polluting, electricity. Unlike conventional, fossil-fueled power plants, wind plants emit no air pollutants or greenhouse gases. According to the U.S. Department of Energy, in 1990, California's wind power plants offset the emission of more than 2.5 billion pounds of carbon dioxide, and 15 million pounds of other pollutants that would have otherwise been produced. It would take a forest of 90 million to 175 million trees to offset the same amount of carbon dioxide production.

Cost Issues

Even though the cost of wind power has decreased dramatically in the past 10 years, the technology requires a higher initial investment per MW of installed capacity than fossil-fueled generators. If wind generating systems are compared with fossil-fueled systems on a "life-cycle" cost basis (counting fuel and operating expenses for the life of the generator), however, wind costs are much more competitive relative to other generating technologies because there is no fuel to purchase and minimal operating expenses.

Environmental Concerns

Although wind power plants have less impact on the environment than fossil fuel power plants, there is concern over birds and bats being killed by flying into the towers or rotors, visual impacts, and the noise produced by the rotor blades. Other impacts are associated with construction and operation of transmission lines associated with wind energy development, trenching and burying of collector lines, and construction of access roads.

Supply and Transport Issues

The major challenge to using wind as a source of power is that it is intermittent and does not always blow when electricity is needed. Wind energy is not easily stored, and not all winds can be harnessed to meet the timing of electricity demands. Further, good wind sites are often located in remote locations far from areas of electric power demand (such as cities). Finally, wind resource development may compete with other uses for the land, and those alternative uses may be more highly valued by some than electricity generation. However, wind turbines can be located on land that is also used for grazing or even farming [39].

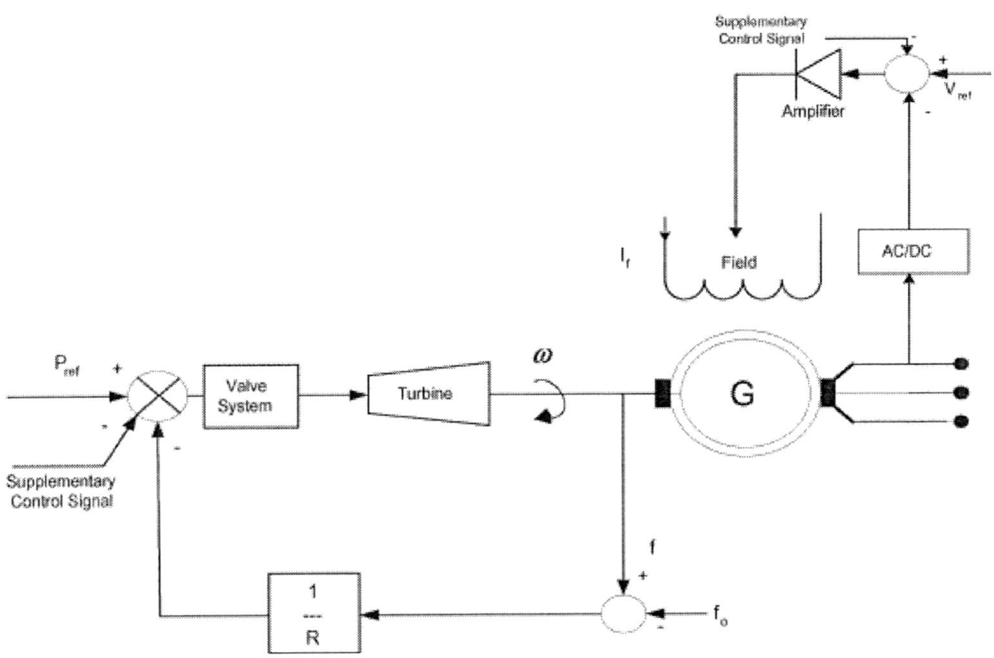

Figure 20. Synchronous generator with traditional controllers.

Control Loops

This section deals with the control of active power and terminal voltage of gas turbine [33]. A schematic of the generator together with its control loops is shown in Fig 20. The generator frequency is mostly affected by changes in power balance ($P_m - P_e$). Thus a scaled version of frequency deviation may be used for achieving power control as shown in Fig 20. The field voltage is provided by means of ac/dc converter that uses the terminal ac voltage. The two supplementary control signals shown in this figure and as well as other two supplementary control signals are generated by the proposed multivariable fuzzy logic controller.

Using above equations and ignoring resistance r, the algebraic equations of the gas turbine becomes:

$$\begin{bmatrix} I_q \\ I_d \end{bmatrix} = \begin{bmatrix} 0 & -\dfrac{1}{X_q} \\ \dfrac{1}{X'_d} & 0 \end{bmatrix} \cdot \begin{bmatrix} V_q \\ V_d \end{bmatrix} - \begin{bmatrix} 0 \\ \dfrac{E'}{X'_d} \end{bmatrix}$$

or in compact form it can be expressed as follows:

$$\boldsymbol{I}_{dq} = \boldsymbol{Y}_{dq} . \boldsymbol{V}_{dq} - \hat{\boldsymbol{E}}$$

The d-q components of gas turbine voltage and current are transformed to the system side frame of reference (real = r and imaginary = m) using the transformation matrix T given below:

$$T = \begin{bmatrix} \cos\delta & \sin\delta \\ -\sin\delta & \cos\delta \end{bmatrix}$$

Using this transformation matrix, the d-q and r-m components of current and voltage are related to each other as follows:

$$\begin{bmatrix} I_q \\ I_d \end{bmatrix} = \begin{bmatrix} \cos\delta & \sin\delta \\ -\sin\delta & \cos\delta \end{bmatrix} \cdot \begin{bmatrix} I_{re} \\ I_{im} \end{bmatrix}$$

$$\begin{bmatrix} V_q \\ V_d \end{bmatrix} = \begin{bmatrix} \cos\delta & \sin\delta \\ -\sin\delta & \cos\delta \end{bmatrix} \cdot \begin{bmatrix} V_{re} \\ V_{im} \end{bmatrix}$$

or equivalently in compact form, is expressed as

$$\boldsymbol{I}_{rm} = \boldsymbol{T}^{-1} . \boldsymbol{I}_{dq}$$

$$\boldsymbol{V}_{rm} = \boldsymbol{T}^{-1} . \boldsymbol{V}_{dq}$$

Using above equations and rearranging, the following equation is obtained for gas turbine:

$$\boldsymbol{I}_{rm} = (\boldsymbol{T}^{-1} \boldsymbol{V}_{dq} \boldsymbol{T} . \boldsymbol{V}_{rm}) - \boldsymbol{T}^{-1} \hat{\boldsymbol{E}}$$

Fuzzy Logic Controller (FLC)

The development of the fuzzy logic approach here is limited to the controller structure and design. More detailed discussions on fuzzy logic controllers are widely available [34, 41]. The proposed multivariable supplementary fuzzy logic controller (FLC) is designed to improve the stability of the system. For the design of this controller, the stability of the system is measured by the deviation of gas turbine frequency from synchronous speed (called error *e*) and gas turbine acceleration power, which is an indication of the derivative of error (called $\dot{e}$). The proposed fuzzy logic controller (FLC) is based on the fuzzy stabilizer which is used only to provide a supplementary control signal at the voltage regulator junction. The controller proposed here, on the other hand, uses the same principle and generates four supplementary control signals. These signals are injected at the summing junctions of both power- and voltage-control loops of the wind turbine and gas turbine.

The control signal of the fuzzy logic controller (FLC) is determined through a set of control rules, $\{R_i\}$. Each control rule R_i is of the form:

IF e is A_i,

AND is B_i,

THEN u is C_i,

where A_i, B_i and C_i are fuzzy sets with triangular membership functions as shown in Fig 21. The membership functions are normalized between −1 and 1. Also the following definitions apply: LP, large positive; MP, medium positive; SP, small positive; ZE, zero; SN, small negative; MN, medium negative; SN, small negative. The error and error change are classified according to these fuzzy membership functions scaled by an appropriate constant. A specific signal may have non-zero membership in more than one set. Similarly, a specific control signal may represent the contribution of more than one rule.

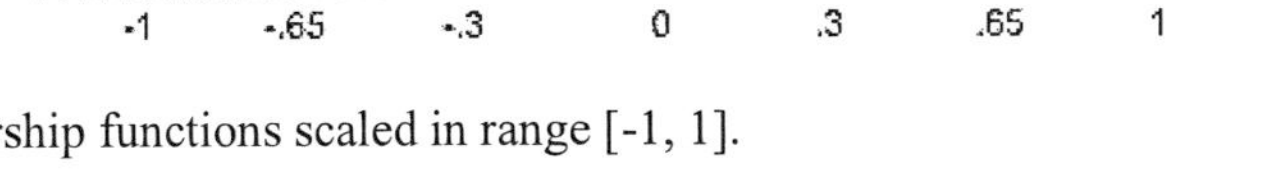

Figure 21. Membership functions scaled in range [-1, 1].

Rule conditions are joined by using the minimum intersection operator so that the resulting membership function for a rule is

$$\mu_R(e,\dot{e}) = \min(\mu_A(e), \mu_B(\dot{e}))$$

The suggested control output from rule i is the center of the membership function C_i. Rules are then combined using the center of gravity method to determine a normalized control output U, defined as follows:

$$U = \frac{\sum_{i=1}^{n} \mu_{Ri}(e:\dot{e}).U_i}{\sum_{i=1}^{n} \mu_{Ri}(e:\dot{e})}$$

A particular control design requires specification of all control rules and membership functions. The control rules are designed from an understanding of the desired effect of the controller. The complete set of control rules is shown in Table 11.

Table 11. Rules table

e / ė	LN	MN	SN	ZE	SP	MP	LP
LN	LP	LP	LP	MP	MP	SP	ZE
MN	LP	MP	MP	MP	SP	ZE	SN
SN	LP	MP	SP	SP	ZE	SN	MN
ZE	MP	MP	ZE	SN	SN	MN	MN
SP	MP	SP	ZE	SN	SN	MN	LN
MP	SP	ZE	SN	MN	MN	MN	LN
LP	ZE	SN	MN	MN	LN	LN	LN

Each of control rules represents a desired controller response to a particular situation. The control rules are meant to be symmetric under the assumption that if necessary any asymmetries could be best handled through scaling and membership functions defining. For low distortions FLC uses only some of these rules (involving the ZE membership functions and its neighborhoods), but for higher distortions may be active others. In presented simulations, the proposed controller here actually uses four of these rules.

Technical Study of Iran-Azarbaijan Electric Network

In this chapter we select Iran-Azarbaijan Electric Network in voltage range of 132 KV to 20 KV of Tabriz city, East Azarbaijan province, Iran. This part of network has six main distribution substations for distributing of energy after transformer. Of course any kind of substation has special technical specification for feeing of loads that is shown in Fig 22 [42, 43]. Figure 22 is designed for feeding of any electrical loads in distribution energy system of Tabriz city in Iran. Of course each load directly connected to 20 KV feeder in substation and the DG system can be connected to the network in substation or any side of network in distribution system to attention of specification of network and loads for getting in the best result of operation of DG system.

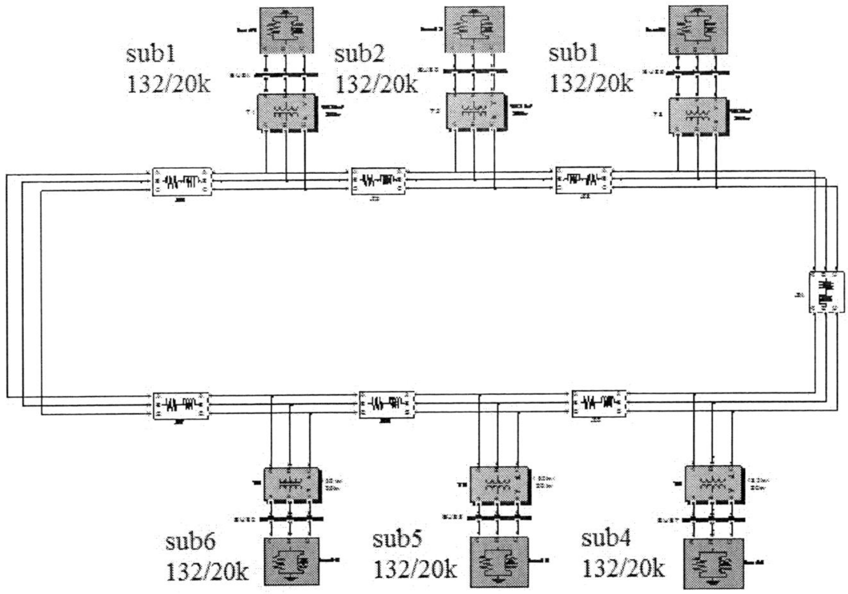

Figure 22. Sample Network Diagram of 132\20kv Iran-Azarbaijan.

Technical Data of Applied Dg in Iran-Azarbaijan Electric Network

The DG system specifications for using and impact on Iran-Azarbaijan electric network have been used by micro-turbines and wind turbines of course the specification of micro-turbines is in Table 12.

Table 12. Technical specifications for selected DG system

	Micro Turbines	Wind Turbines
General Information	-	On shore and in lend
Application Range	30 KW - 20 MW 35 KW - 1 MW Small scale application up to 1 KW	20 W - 3 MW
Electric Conversion Efficiency	20% - 30%	Not Applicable
Application	Power generation possible whit CHP added	-
Fuel	Generally uses natural gas but fare land in are biogas can also be used	Wind
Comments	-	Non predicate output capacity locator on store 20-25%

Simulation Results of DG System Operation in Iran-Azarbaijan Electric Network

The operation of DG systems is shown in Figure 23 that is connected to real network (Iran-Azarbaijan electric network). It has six stations for study of operation of DG system with main source in the network considering different distances of DG from main network. The results are shown after computation and simulation processes.

The main electric network in Figure 23 of Iran-Azarbaijan and DG system that forms by wind and gas turbines for controlling of turbines used fuzzy logic controller (FLC) [41]. Specification of blocks that used for control of main network and DG system is represented by standard symbol that named: substations (132/20 KV), model of network, gas turbine, wind turbine, model of transmission line, model of fuzzy logic control and many scopes system for simulation [44].

The technical data of gas and wind turbines and also transformers have been shown in Table 13 and Table 14, respectively.

Table 13. Technical data of gas and wind turbines

Turbine type	*P+Q*	*V*	X_G
Gas Turbine	50 MVA	20 KV	9%
Wind turbine	9 MVA	11 KV	18%

Table 14. Technical data of transformers

Transformer	*P+Q* [MVA]	V_1/V_2 [KV]	X_G
T_{GT}	50	20/63	10%
T_{WT}	50	11/63	16%
T_N	50	63/11	20%

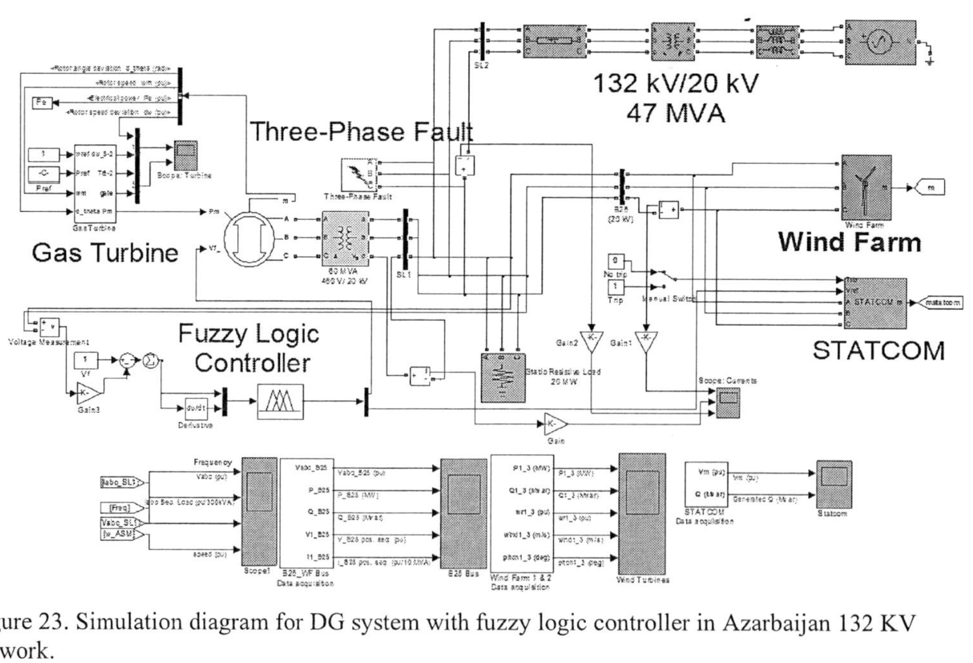

Figure 23. Simulation diagram for DG system with fuzzy logic controller in Azarbaijan 132 KV network.

There are three power sources in the simulation diagram. The first source is a gas turbine with the characteristics given in Table 14. The second source is the main network with 132 KV nominal voltage and 50 Hz nominal frequency. The third source is a wind turbine farm with the characteristics given in Tables 15.

Table 15a. Gas turbine characteristics

V_{rms} [KV]	F_n [Hz]	P_n [KVA]
13.8	50	50

Table 15b. Wind turbine farm characteristics

Base Wind [m/s]	F_n [Hz]	P_n [KW]
9	50	45

Figure 24 shows that only modeling of gas turbine and variation of voltage and current in fault duration at T=15. Figure 25 include the results of modeling of wind turbine that results are only for operation of wind turbine special at fault time in five steps.

Figure 26 shows modeling and simulation of operation of STATCOM for wind turbine when have been fault at T=15 operation of STATCOM for near distance of wind turbine to fault point diagram of oscillation would be satisfactory. Fig 27 shows the variation of current for wind and gas turbine and main network of system.

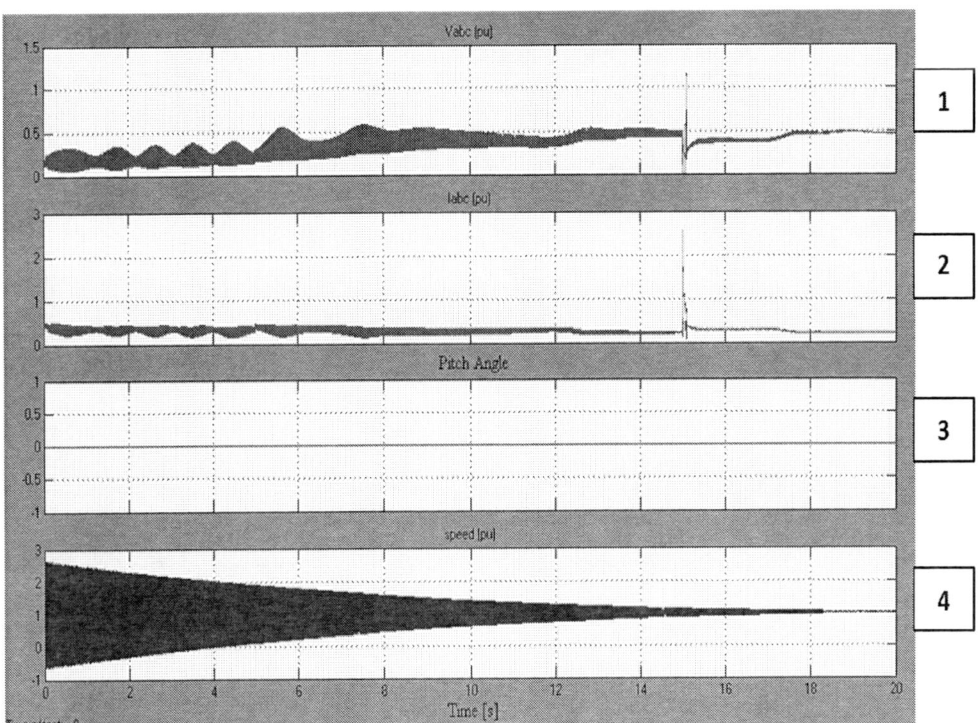

Figure 24. Gas turbine simulation results 1. Voltage 2. Current 3. Pitch Angle 4. Speed.

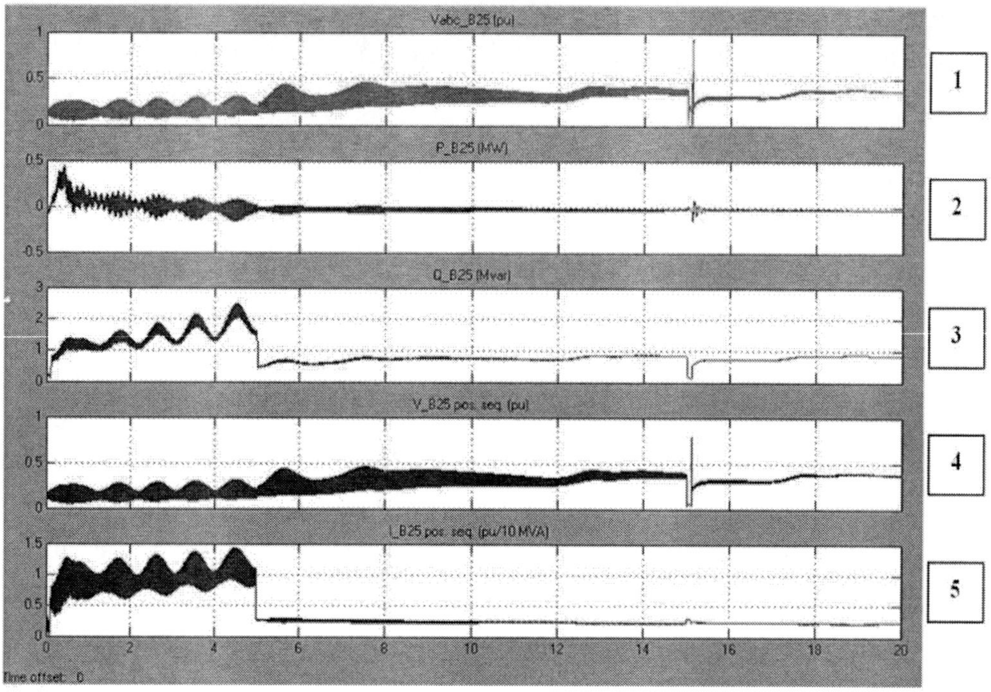

Figure 25. Wind Turbine- simulation Results 1. Voltage; 2. Active power;3. Reactive power; 4. Positive sequence of voltage; 5. Positive sequence of current.

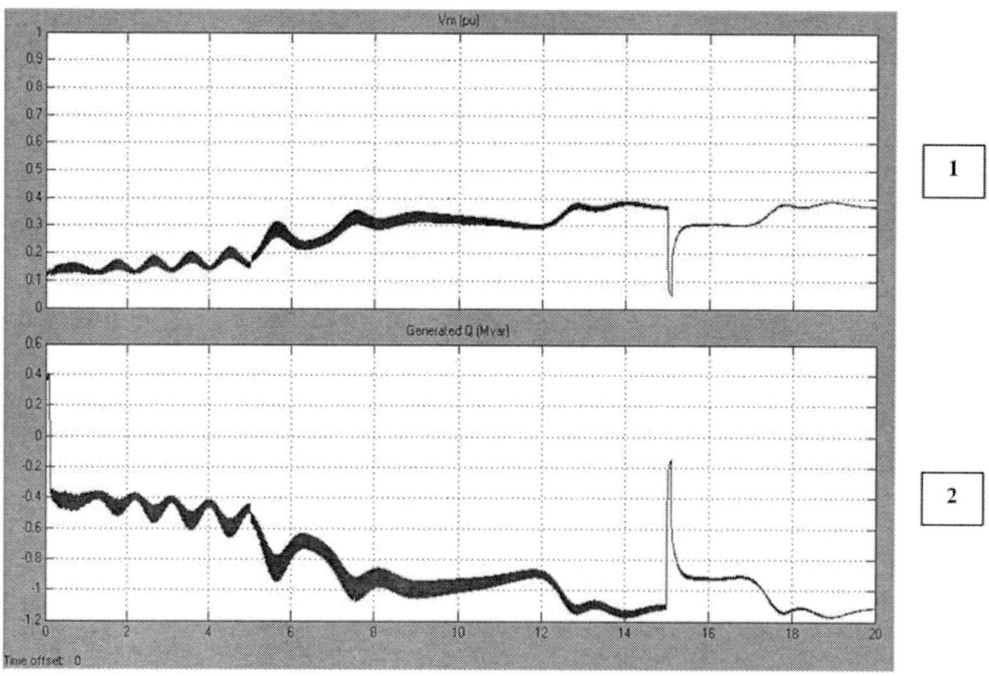

Figure 26. STATCOM specifications in simulation 1. Voltage 2. Reactive power.

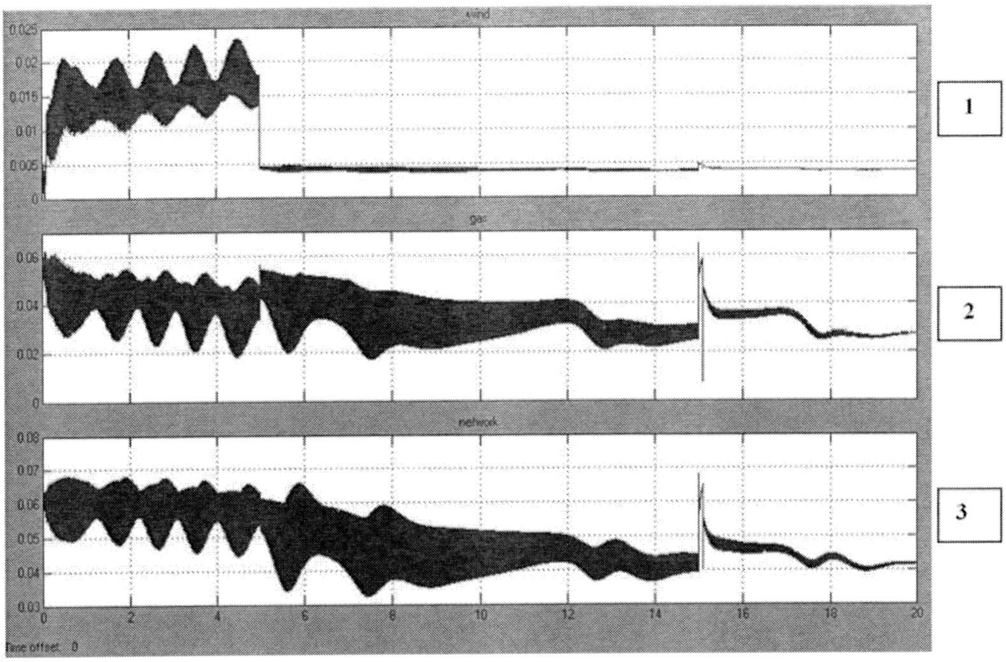

Figure 27. Simulation results of currents (gas-wind-network) 1. Wind turbine current 2. Gas turbine current 3. Network current.

The block diagram in Figure 28 shows the modeling and simulation of wind and gas turbines against the main network considering fuzzy logic controller (FLC) in Iran-Azarbaijan electric network and enables to study the effect of any fault in network with dynamic system of DG especially in fault time. The fuzzy logic controller (FLC) rules for gas turbine and wind turbine are given in Figure 18 and Figure 19, respectively.

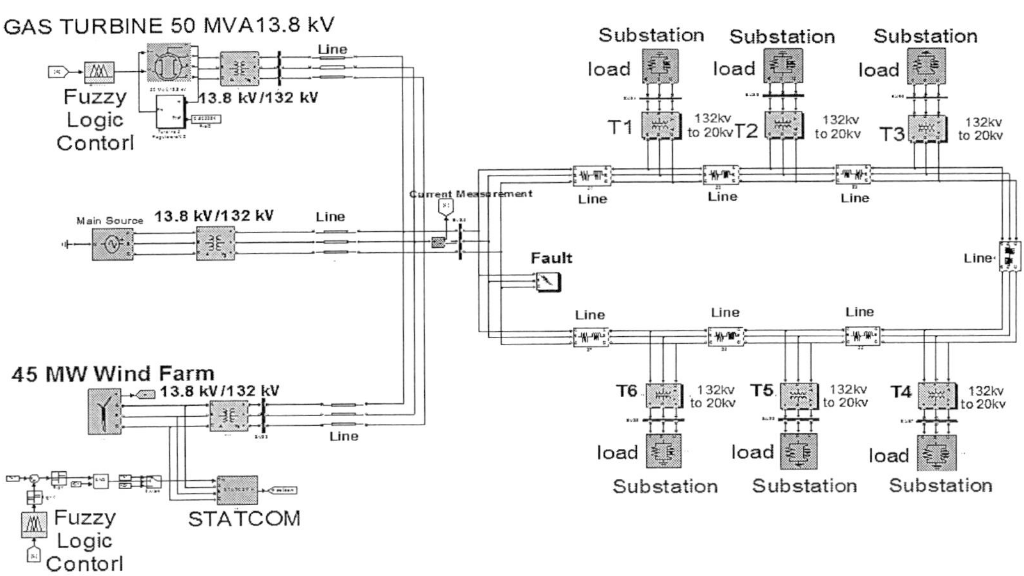

Figure 28. Simulation of operation of gas and wind turbines as DG system with fuzzy logic controller (FLC) in Iran-Azarbaijan electrical network.

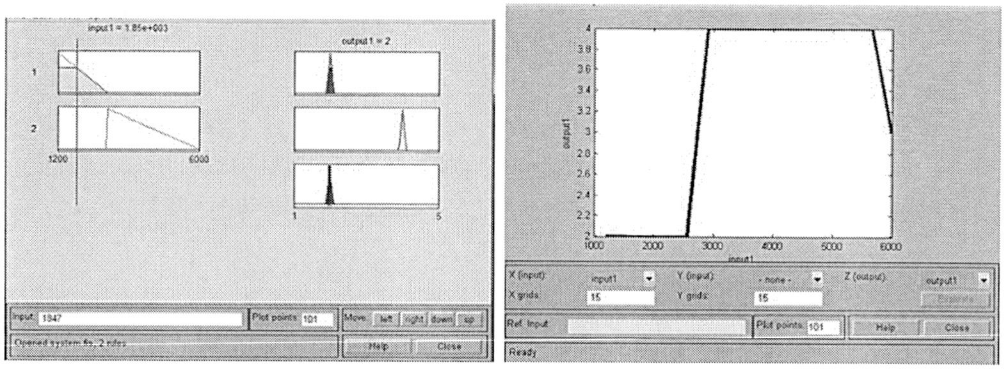

Figure 29. FLC rules for gas turbine.

The diagram in Figure 29 shows short circuit of second phase in the network and results of the system against fault considering fuzzy logic controller (FLC) for gas turbine and 5km distance between gas turbine of DG system and the main network.

Results for gas turbine of simulation for 5 km distance between of gas turbine of DG system and main network and also effect of DG system for damping of oscillation of current after fault is shown in Figure 30. The diagram in Figure 31 shows short current of second phase of network and simulation results for voltage oscillation of wind turbine.

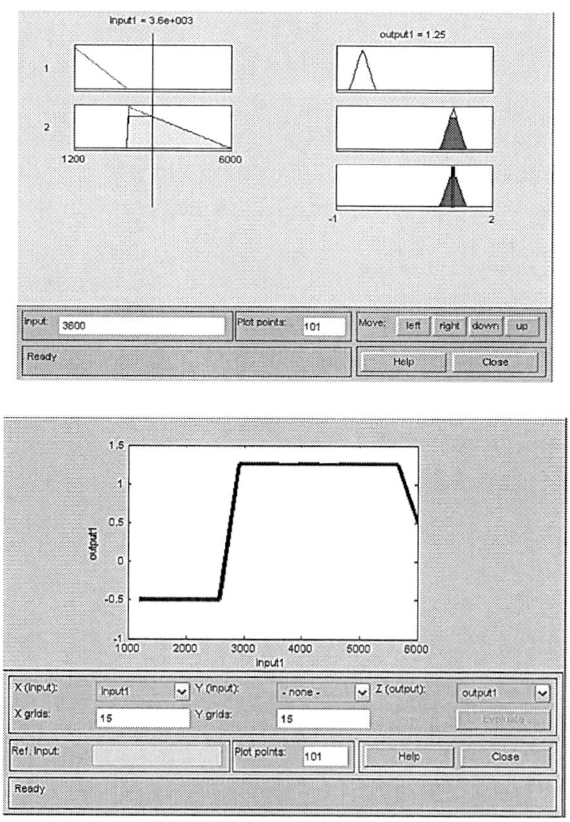

Figure 30. FLC rules for wind turbine.

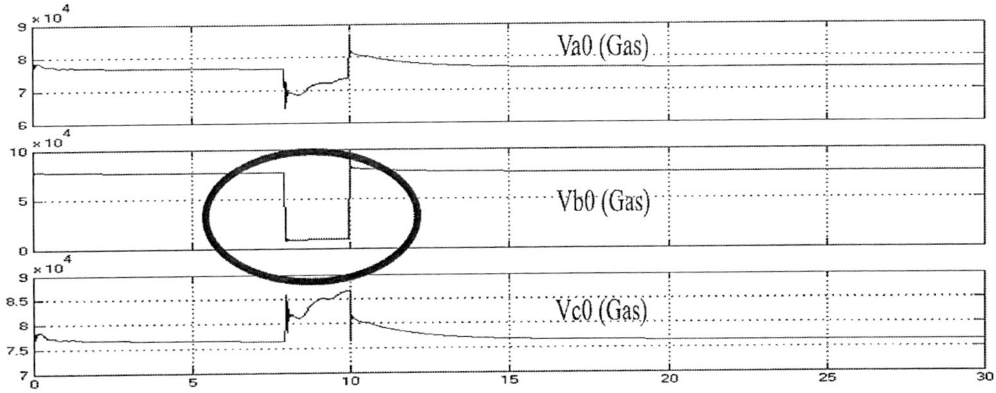

Figure 31. Variation of voltages of network with fuzzy logic controller (FLC) during fault time on one phase to ground and distance of 5km (D=5km) between gas turbine and the main network.

Figure 32 shows the oscillation currents of phases after fault (on second phase of network) for wind turbine. Figure 33 and Figure 34 show the damping of oscillation of voltage and current after fault for DG systems and network with fuzzy logic controller (FLC) after fault and *D*=5km of DG from the main network.

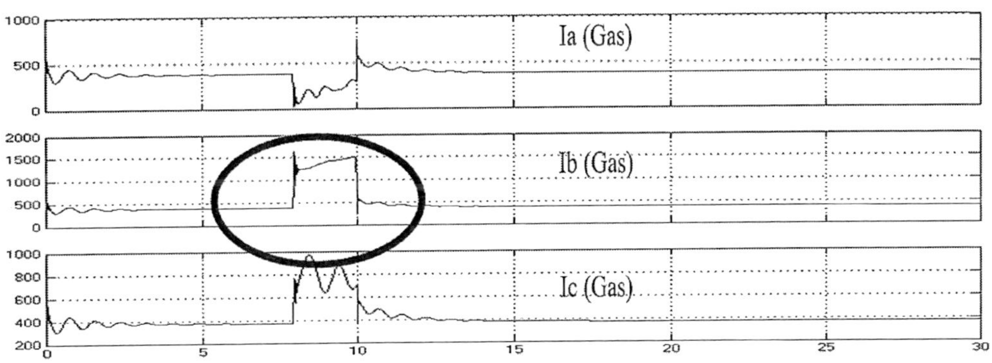

Figure 32. Variation of current of network with FLC during fault time on one phase to ground and D=5km between gas turbine and the main network.

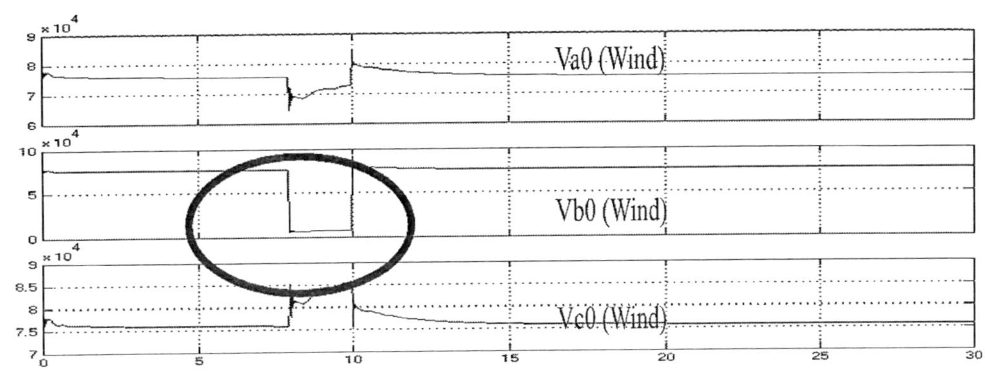

Figure 33. Variation of voltage of network with FLC during fault time on one phase to ground and D=5km between wind turbine and the main network.

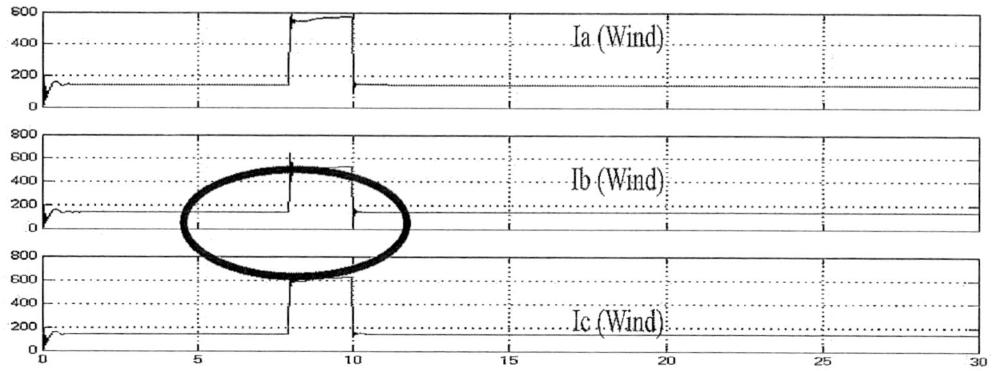

Figure 34. Variation of current of network with FLC after fault on one phase to ground and D=5km between wind turbine and the main network.

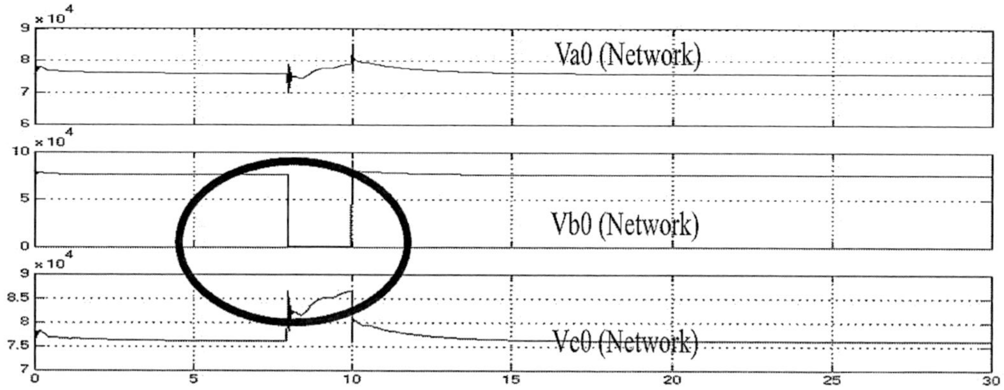

Figure 35. Variation of voltage of network with FLC after fault time on one phase to ground and D=5km between DG system and the main network.

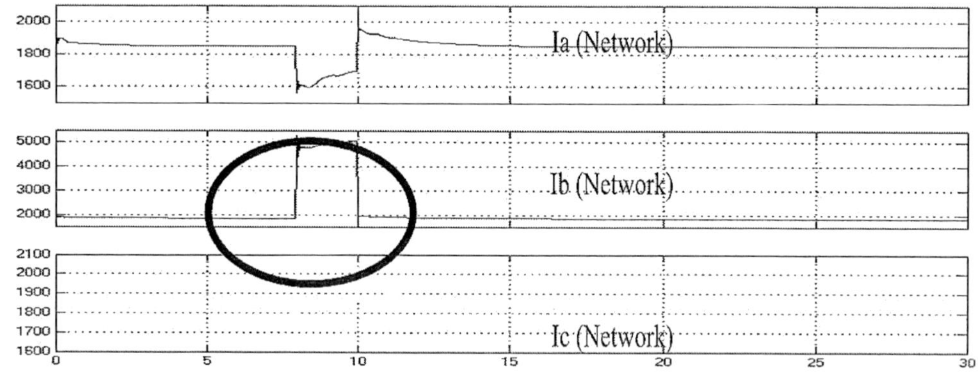

Figure 36. Variation of current of network with FLC after fault time on one phase to ground and D=5km between DG system and the main network.

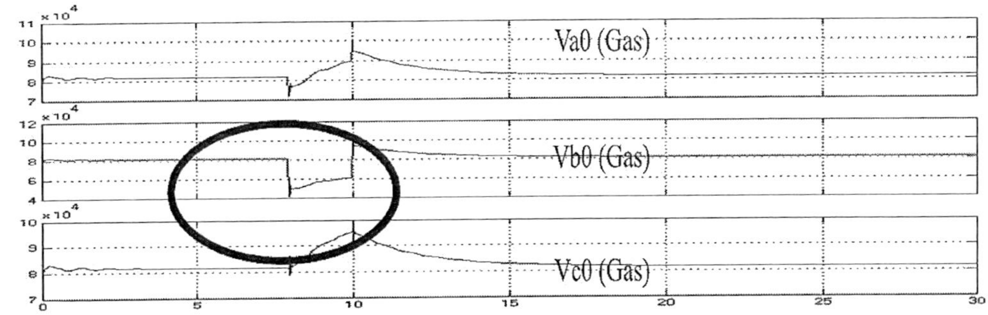

Figure 37. Variation of voltage of network with FLC during fault time on one phase to ground and D=70km between gas turbine and the main network.

Figure 37 and Figure 38 show the results of simulation for voltage and current oscillation of the gas turbine at D=70km in fault duration. It is important that oscillation rang after fault can be controlled by FLC. In the Figure 39 and Figure 40 the fault is happening on the second phase of network and the damped oscillation is detected in network by the fuzzy logic controller (FLC) at 70km distance of DG.

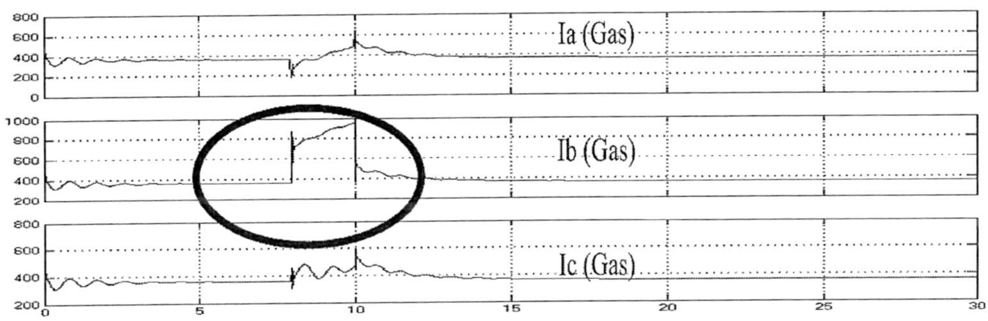

Figure 38. Variation of current of network with FLC during fault time on one phase to ground and D=70km between gas turbine and the main network.

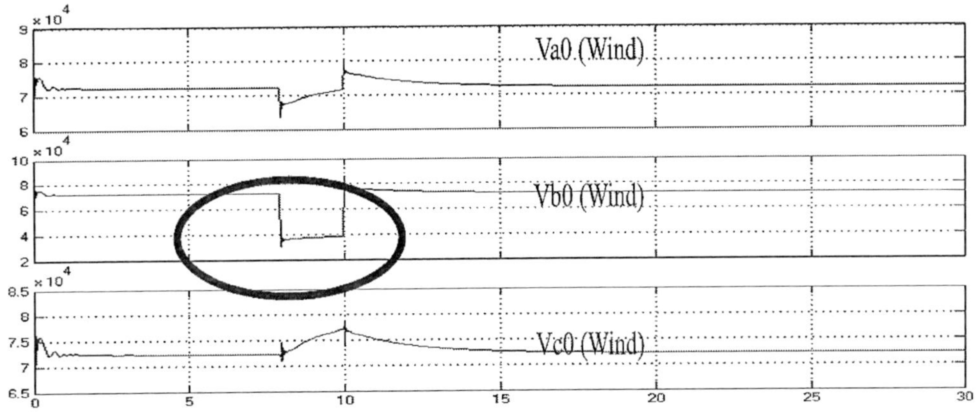

Figure 39. Variation of voltage of network with FLC during fault time on one phase to ground and D=70km between wind turbine and the main network.

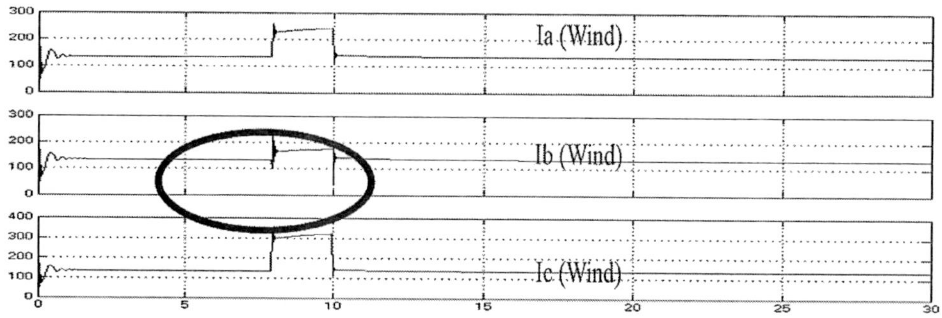

Figure 40. Variation of current of network with FLC during fault time on one phase to ground and D=70km between wind turbine and the main network.

Figure 41 and Figure 42 show the results of simulation and how to damping of oscillation after fault with DG and FLC at *D*=70km of DG system in the main network. Diagrams also show how to damping of oscillation is possible when use from DG system in nearly distance of DG from the main network. So in this study the average of distance is about 40-50km of DG from the main network.

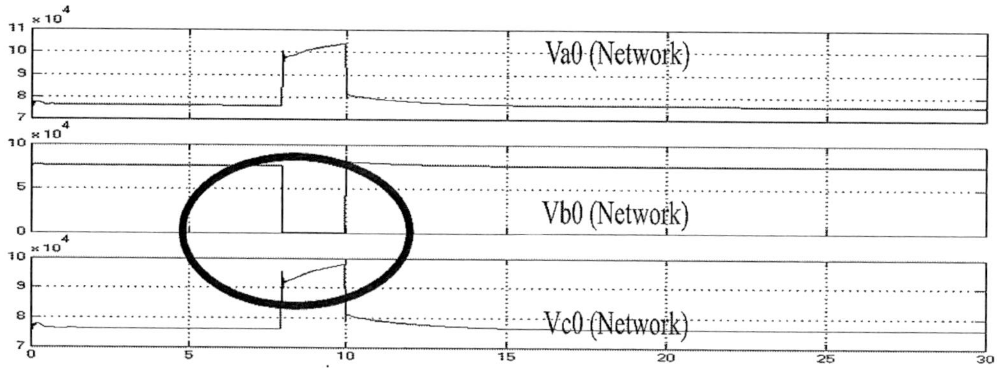

Figure 41. Variation of voltage of network with FLC after fault time on one phase to ground and D=70km between DG system and the main network.

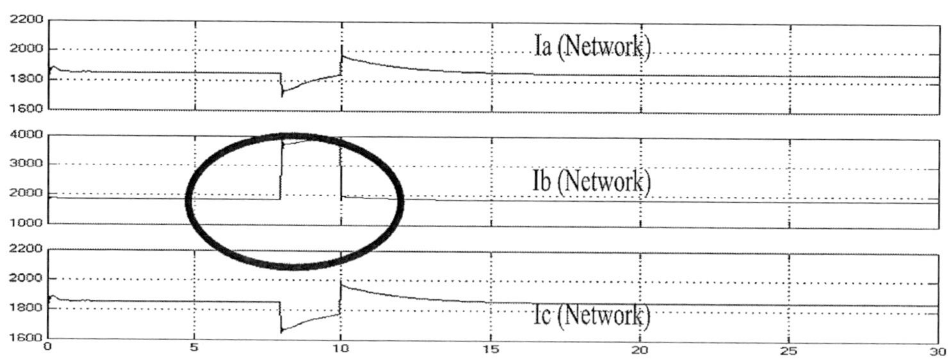

Figure 42. Variation of current of network with FLC after fault time on one phase to ground and D=5km between DG system and the main network.

CONCLUSION

This chapter is dealing with attention for development of electrical network and increase of electrical loads regarding to feed of loads with standard voltage, active power and reactive power specially feeding with minimum losses and have standard power supply and reliable power system. So, using variety of power supplying systems e.g. more power station and different DG systems are important. In this chapter we try to show the effect of DG system using wind turbine and gas turbine and the results show that gas turbine in a DG system with suitable controller can be feed to long distance of DG system in the main network [45]. The wind turbine can be feed for loads to improve the voltage profile as well as loads power quality which can be corrected specially in peak time of power system and fault time to fix up amplitude of voltage and current. This chapter is also considering the matter of using DG systems with position factor for power system and environmental conditions to reach the improvement of electrical condition of main network. The subjects are studied in Azarbaijan of Iran (Iran-Azarbaijan) and give very useful results that how to applying DG system with together in a large main network. Of course gas turbine can be held to feed for short distance network from loads and can be see the best effect with fuzzy logic controller (FLC) to supervise the output of DG system.

ACKNOWLEDGMENTS

The work done in first section was supported from Department of Electrical Power Engineering and Ecology, University of West Bohemia and funded by PTSE project No. 1M 06059 (Program Research Centre MŠMT Prague), project No. SGS-2010-018, the European Regional Development Fund and Ministry of Education, Youth and Sports of the Czech Republic - project No.Z.1.05/2.1.00/03.0094: Regional Innovation Centre for Electrical Engineering (RICE).

REFERENCES

[1] E.ON Česká Republika, s.r.o. (www.eon.cz).
[2] Group PRE (www.pre.cz, www.predistribuce.cz).
[3] ČEZ Group (www.cez.cz).
[4] ERÚ (www.eru.cz).
[5] ČEPS, a.s. (www.ceps.cz).
[6] Photovoltaic Geographical Information System (http://re.jrc.ec.europa.eu/pvgis/).
[7] Czech Wind Energy Association (www.csve.cz).
[8] Habrych, R. EGU seminar Proc. 2009, 10, 1-25.
[9] EGÚ Praha Engineering, a.s. (www.egu-prg.cz).
[10] The Bonneville Power Administration (http://www.bpa.gov/corporate/).
[11] AIS spol. s r.o. (www.ais-brno.cz).
[12] Popelka, A. CIRED conf. pres. 2009, PS03.
[13] Popelka, A. CPS conf. pres. 2008.

[14] Popelka, A.; Juřík, D. ČK CIRED conf. pres. 2007.

[15] Mertlová, J.; Noháčová, L. Elektrické stanice a vedení; ISBN: 978-80-7043-724-7; ZČU: Pilsen, CZ, 2008; Vol. 1, pp 5-15.

[16] Petrák, K.; Lisý, P. EGU seminar pres. 2010, 4, 1-20.

[17] Roubal, J. CIRED pres. 2009, PS03/1.

[18] National Energy Policy (in Romanian language: Politica Națională în domeniul Energiei).

[19] Transelectrica S.A., http://www.transelectrica.ro/EN/gestionare.php.

[20] Romania Statistic Report, 2009.

[21] Regulatory Whitepaper, ANRE code 67.2.431.0.01.04/05/00.

[22] Roadmap energy in Romania (in Romanian language: Foaie de parcurs în Domeniul Energetic din Romania), the Romanian Government, Ministry of Industry and Commerce, July 2003.

[23] Energy, Environment and Sustainable Development, Part B Energy, DECENT Final Report, October 2005.

[24] GREEN PAPER, A European Strategy for Sustainable, *Competitive and Secure Energy*, COM(2006) 105 final, Brussels, 8.3.2006.

[25] COMMUNICATION FROM THE COMMISSION TO THE COUNCIL AND THE EUROPEAN PARLIAMENT Report on progress in creating the internal gas and electricity market, COM(2005) 568 final, Brussels, 15.11.2005.

[26] Energy Regulators Regional Association, www.erranet.org/Committeee-WG/Tariff/Members.

[27] COMMISSION STAFF WORKING DOCUMENT Report on Progress in Creating the Internal Gas and Electricity Market Technical Annex to the Report from the Commission to the Council and the European Parliament, SEC(2005) Brussels 2005.

[28] Transelectrica S.A., www.transelectrica.ro.

[29] Romanian Power Market Operator – OPCOM Statistic Report 2009.

[30] Metz, H.C. (1995). Turkey: A Country Study. Washington: GPO for the Library of Congress.

[31] Elsam Power Pool Report (1997), Denmark.

[32] www.tedas.gov.tr.

[33] Jagaduri, R.T. and Radman, G.; Modeling and Control of Distributed Generation Systems including PEM Fuel Cell and Gas Turbine, *J. El. Power Systems Research*, 2007, 77, pp. 83-92.

[34] Hasan, A.R., Martis, T.S. and Ula, A.H.M.S.; Design and Implementation of a Fuzzy Controller Based Automatic Voltage Regulator for a Synchronous Generator, *IEEE Transactions on Energy Conversion*, 1994, Vol. 9 (3), pp. 550-557.

[35] Ramirez, D., Beites, L.F., Blazquez, F. and Ballesteros, J.C.; Distributed generation system with PEM fuel cell for electrical power quality improvement, *International Journal of Hydrogen Energy*, 2008, Vol. 33 (16), pp. 4433-4443.

[36] Munteanu, I., Bratcu, A.I. and Changa, E., Wind turbulence used as searching signal for MPPT in variable-speed wind energy conversion systems, *Renewable Energy*, 2009, 34, pp. 322–327.

[37] www.nrel.gov/learning/eds_energy_management.htm.

[38] http://plainswindeis.anl.gov/index.cfm.

[39] http://www.northernpower.com/technology/turbine-configuration.php.

[40] Lee, J.; On Method of Improving Performance of PI-Type Fuzzy Logic Controller, *IEEE Transactions on Fuzzy Systems*, 1993, Vol. 1 (4), pp. 298-301.

[41] R. Brehmer and P. Nemetz, Optimal Control Methods for Power System Operation, Springer Berlin, Heidelberg, 2006.

[42] Lee, K. and Shin, M.C.; Power Plants and Power System Control, Elsevier, NY, 2003.

[43] The Math Work Fuzzy Logic Toolbox Guide, MIT, NW, 2004.

[44] Khan, U.N.; Impact of Distributed Generation on Electrical Power Network, Proceedings of the 7th *Int. Conference EEEIC,* Cottbus, 2008, pp. 82-85.

[45] Hajizadeh, A., Golkar, M.A. and Feliachi, A.; Power Management Strategy of a Hybrid Distributed Generation System, *Int. J. Distr. Energy Resources*, Vol. 3 (2), 2007, pp. 139-156.

In: Advances in Energy Research
Editor: Nicu Bizon

ISBN 978-1-61209-991-0

Chapter 17

EUROPEAN RESEARCH ENVIRONMENT IN THE FIELD OF HYDROGEN AND FUEL CELL

Roxana Elena Ionete, Elena Carcadea and Mihai Culcer***
National Research and Development Institute for
Cryogenics and Isotopic Technologies,
Rm. Valcea, Romania

ABSTRACT

Public debate about hydrogen and fuel-cell technology takes place between two extremes: it is perceived as the immediate answer to all questions related to air pollution and noise, climate change and expensive fuels, but also as too dangerous even to think about. There are also obligatory references to hydrogen as the energy carrier of the (far) future, but the option hardly features in the governmental policy debates about the above-mentioned issues. The problem for hydrogen and fuel cells is that they are good for many policy objectives – air quality, climate change, security of energy supply, competitiveness /innovation – but not the best solution for any of them in the short term. It is the role of science to solve all the issues related to the real hydrogen-based energy applications.

Keywords: hydrogen, fuel-cell technology, European research.

INTRODUCTION

Within the context of sustainable development there is strongly underlined the role hydrogen could play to meet two of Europe's key policy objectives: providing energy supply security and reduction of greenhouse gas emissions and consequently mitigating the climate changes.

---

* Corresponding Author: roxana@icsi.ro

Hydrogen and electricity together represent one of the most promising ways to realize sustainable energy, whilst fuel cells provide the most efficient conversion device for converting hydrogen, and possibly other fuels, into electricity. Hydrogen and fuel cells open the way to integrate "open energy systems" that simultaneously address all of the major energy and environmental challenges, being flexible and adaptable to diverse renewable energy sources available [1,2].

On the other hand, it is recognized that the transition to a hydrogen energy system will require further technological advances in hydrogen production, conversion, storage, distribution and end use applications to achieve the desired levels of performance and cost. Effective research and development (R&D) activities are essential for these advances to take place. To have a picture of the funds allocated to the European level, Figure 1 illustrates some statistics regarding the total public H&FC research budgets including not only H&FC programme budgets but all other H&FC spending for the year 2005 [3,4,5].

There is a wide spread in magnitude of the budgets - the top three countries (Germany, Italy and France) account for more than 64% of the total (excluding the European Commission contribution) and there is a gap of €10M in research spending between France and the next biggest budget, the Netherlands. It is apparent that the majority of countries have relatively small budgets of the order of €10M or less.

In terms of evaluating research effort within Europe, it is more useful to derive indicators based on research spending as a proportion of their GDP as the countries vary greatly in size and in economic power. This changes the picture quite dramatically, with Iceland and Denmark now at the top of the list and a much smaller spread between the major players, Figure 2. In relation to GDP, research spending on H&FC in France and the UK is comparatively lower, indicating a proportionally smaller research effort in these countries. This is not due to a lack of research funding in general, however. Both countries have among the highest GERD (the most commonly used metric for comparing R&D efforts of countries) per unit GDP of the countries studied, but make hydrogen less of a research priority than in other countries. In relation to overall research spending, the countries with the greatest focus on H&FC research are Lithuania and Romania, followed by Italy, Iceland and Norway.

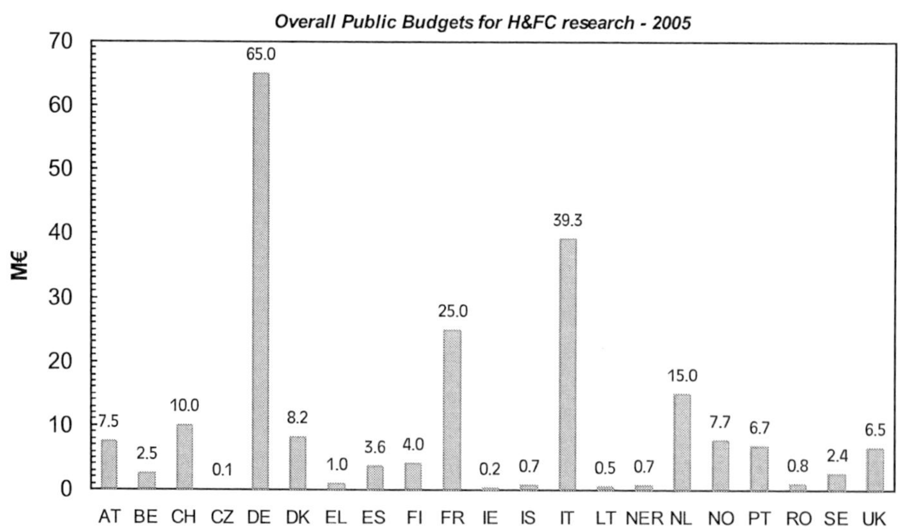

Figure 1. Research input: public budgets for H&FC research.

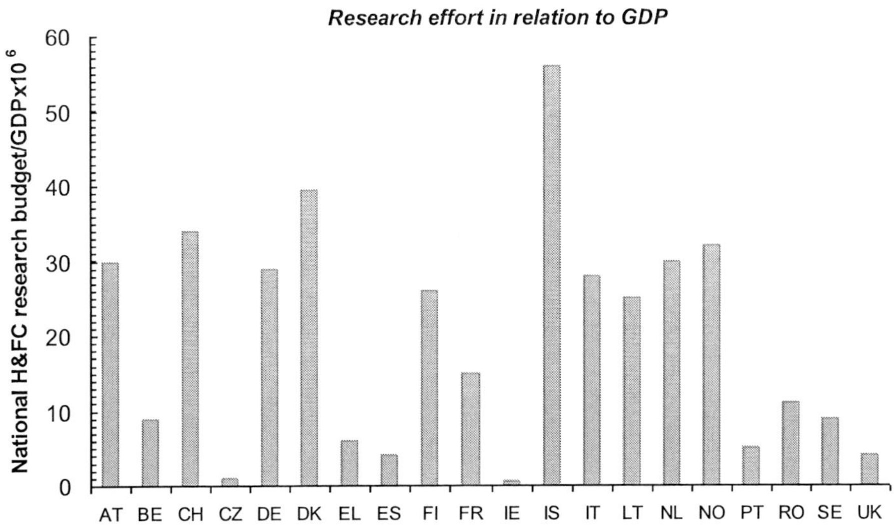

Figure 2. Research input: H&FC research budgets in relation to GDP.

While there are numerous R&D activities being undertaken in the European countries, there is increasing acceptance that international coordination of research is more efficient and leads to faster technology development through pooling of resources, providing common answers to common issues and developing common approaches to exploit synergies.

One of the key initiatives in hydrogen and fuel cells research was to establish the European Platform for Hydrogen and Fuel Cells, launched in 2004 to facilitate and accelerate the development of fuel cells and hydrogen production energy systems. The purpose of this platform was a more effective research in Europe by developing a common and consistent vision at European level, and the promotion of public and private funding in this field. The European Platform was created based on the recommendations specified in the "Vision Report" prepared by the "High Level Group for Hydrogen and Fuel Cells" with intent to accelerate and implement these key technologies in Europe.

In 2006, the European Union has set a higher level of integration of research and development in this area by establishing a Joint Technology Initiative - JTI concept as a legal entity proposed to achieve a new type of public-private partnership in the relevant field of R&D at European level. Officially launched in October 2008, the Joint Technology Initiative Undertaking on Hydrogen and Fuel Cells has a default budget sufficient to reach a critical mass for a period of six years and was created with the intent to significantly reduce the time required to achieve commercial stage.

This new institutional framework should implement basic and applied research, technology validation and demonstration projects and also should support actions based on the results obtained and appreciated by the European Platform for Hydrogen and Fuel Cells. The main participants in JTI are the European Commission and the Industrial Group consisting in major European companies with particular interest in hydrogen. The JTI members have equal rights and financial contributes with equal amounts to R&D activities. However, the European Commission included as an additional participant a research group, as consultant regarding the strategies and the ultimate goal of short and medium term research.

JTI funded research in fuel cell aims to reduce costs and improve the performance, durability and safety of fuel cell systems for stationary and transport applications, in their competition with conventional technologies. This will include processes and materials developing, the optimization and simplification of fuel cell components and sub-systems and also their modelling, testing and characterization.

We must emphasize that the transition to a hydrogen economy is on the agenda of research and development around the world, most governments investing heavily in this area, with the aim of domain developing and also to implement the solutions already existing on the market. Hydrogen and fuel cells based technologies may contribute to a sustainable energy economy. Several factors that may influence the success of these technologies are:

- the scientific and technical issues that still need resolution;
- the transition to a hydrogen economy by strategies developing.

Based on several strategies - Snapshot 2020, which correspond to those of European countries and beyond, is very likely that the small applications, less relevant for energy savings, to enter the market first. Stationary fuel cells will follow, since they do not depend on the infrastructure transition to a hydrogen economy.

Should be specified that Romania, as an intrinsic part of the European Union, can not stay away out of these activities, even if it's a minor player compared with major financial potential countries. An important feature of hydrogen technologies is its strongly interdisciplinary character, involving various fields of research: heterogeneous catalysis, absorption/desorption processes, gas physics, electrochemistry, IT, mathematical modeling, etc. In Romania, in many of these areas with links to hydrogen energy systems, were developed technologies and researches, even there was no a national specific research programme for hydrogen and fuel cell.

As a relevant example may be mentioned the experience of researchers from National R&D Institute for Cryogenics and Isotopic Technologies – ICIT Rm. Valcea, which, since 2002, established as a priority the research into hydrogen energy production and its related areas, starting with exploratory research on PEM hydrogen fuel cells. These surveys had originally a modest goal, the investigation of possibilities to implement and use these systems for various applications.

The high global interest and the achievieng of proposed initial goals for this research led to the development of a first integrated system for energy production using hydrogen fuel cells – a small-scale experimental model. This allowed the studies of hydrogen production from methane through catalytic reforming, developing methodologies for hydrogen purification, investigating new materials for catalysts and PEM fuel cell modelling, development and testing. At that time there was a premiere for Romania and allowed the entry of ICIT at the international scientific community active in hydrogen and fuel cell field.

Meanwhile, interesting results were obtained as: a pilot trial for hydrogen production, an application of fuel cells in transport - a small car moving on the basis of fuel cells, etc.

The experience and positive results achieved to ICIT convinced the policy makers to develop a major investment to Ramnicu Valcea, a National Centre for Hydrogen and Fuel Cells, as a separate entity within ICIT (Figure 3).

Newly created Centre received a substantial financial investment to construct new facilities and instrumentation acquisition, an essential prerequisite relevant research unit at the European level.

Despite the poor integration capacity, Romanian research in this area is known in Europe by several international projects and also by the quality of its membership to the European research organizations. The best example is the presence of the newly created Romanian Centre for Hydrogen and Fuel Cells as a full member to the „Research Group of the Joint Technology Initiative - Joint Undertaking on Hydrogen and Fuel Cells" [6,7,8].

The membership grants full rights to participate in decision making activities related with the defining of European strategy and the participation to the projects calls initiated under the JTI.

Figure 3. Romanian Centre for Hydrogen and Fuel Cell.

A SUCCES STORY – THE ROMANIAN HYDROGEN AND FUEL CELL CENTRE

The Romanian Hydrogen and Fuel Cell Centre (RHFCC) under the coordination of National Authority for Scientific Research has the main aim to develop and implement new technologies related to Hydrogen energy. The Centre has been created officially in 2008 through a large direct financial investment, but based on a prior experience of the National R&D Institute for Cryogenics and Isotopic Technologies – ICIT Ramnicu Valcea staff in working with hydrogen.

The declared mission of the RHFCC is to grow deeper scientific knowledge of the Hydrogen and Fuel Cell chemical and physical processes and to bring together technology and expertise in order to provide a platform for new business development into this new energy field.

Actually, ICIT Ramnicu Valcea has been involved in the Hydrogen “business” since 1970, when a new research facility has been setup in order to develop the Heavy Water production technology in Romania. Since then, the Institute was an active player on the Hydrogen isotope “market”. Several “success stories” has been developed in Ramnicu Valcea, almost all being based on Hydrogen Isotopes researches: Heavy Water production technology, Tritium Separation Pilot facility, Depleted Deuterium Water for medical applications, etc.

From 2002, ICIT has been involved in the Hydrogen Fuel Cell research projects and started to push the research and development of these technology at national level. Step by step, the knowledge and expertise of the research staff increased and the achievements of the last years showed that can pass to the next level of performance, transforming distinct and gained research projects into a more focused and coordinated work into this new organizational structure.

In 2008, a large financial investment has been granted to ICIT Ramnicu Valcea to be assigned for a new Hydrogen Research Centre. This investment includes a new building and new instrumental and testing facilities, at the same location as the Institute.

The new Centre was planned to become the main experimental facility for research and also for testing and validate new technologies related to Hydrogen for medium and large scale.

The first issue which could be raised is “Why Hydrogen in Romania?” There are several rationales for developing and implementing these new technologies in Romania:

- Romania has the scientific potential to give an original contribution into this complex area - HFC
- Is part of European Union and should integrate its research into the Framework of European Research
- Economic reasons – soon or later the oil and natural gases reserves will be exhausted and Romania should gain its energetic independence
- The cost of doing nothing to combat the issues related to the energy may be far exceed the cost of investing in new technologies to mitigate the effects of GHG emissions

Considering that Romania has to fulfill the 20-20-20 Directive, and an assumed increase of energy demand by another 20% given by economic growth, it will results that about 40% from the total energy of the future will have to come from non-CO_2 generating sources in 2020. This is not the end of the story! Hydrogen is only a secondary source of energy. The First Law of Thermodynamics being what it is, producing this quantity of Hydrogen will require even bigger quantities of primary energies. Depending of the nature of the conversion devices envisaged the efficiency of hydrogen production could be anything between 50 and 90%. So, assuming an average efficiency of 80% it would be needed in Romania even more – 50% from non-carbon sources.

RHFCC R&D ACTIVITIES

Hydrogen Production Systems

It is one of the main targets to develop and implement a portfolio of Hydrogen production technologies at small and medium scale. The Hydrogen Production activity is thought to be focused on developing hydrogen production technologies that enable the introduction and long-term viability of hydrogen as an energy carrier for transportation and stationary power.

Goal: *to investigate, develop and/or implement low-cost and efficiently high-purity Hydrogen production technologies from diverse source, putting in the forefront renewable source.*

The Hydrogen production group is planned to develop and implement technologies for producing hydrogen in a distributed manner from natural gas and by water electrolysis, but also and is developing centralized renewable production options that include water electrolysis integrated with renewable power (e.g., wind, solar, hydroelectric, geothermal), solar-driven high-temperature thermochemical water splitting cycles, direct photoelectron-chemical water splitting, and biological processes.

It must be emphasis that a methane steam reformer-based Hydrogen production pilot facility is functional at ICIT Ramnicu Valcea which could be used in the future for investigations of process optimization. Main topics which are planned to be developed:

a) Methane reforming:

- develop a low-cost methane reformer membrane reactor to deliver high-purity Hydrogen for small PEMFC stacks;
- modeling a distributed SMR-based hydrogen production network and perform experimental simulations;
- demonstrate the reliability and durability of the SMR system built on –site;
- develop an integrated Ceramic Membrane System for Hydrogen Production.

b) Water electrolysis

- explore and compare system-level integration issues surrounding multiple electrolyzers of proton exchange membrane (PEM) and alkaline technologies that also produce hydrogen gas at different pressures to gauge their efficiencies, responsiveness, and performance;
- demonstrate PEM electrolyzer operation at moderate pressure and determine the optimum operating pressure for low-cost hydrogen production;

- develop a conceptual design of a PEM electrolyzer with decreased capital costs and increased efficiency.

c) Thermochemical water splitting

- identify and characterize new semiconductor materials that have the possibility of meeting the criteria for a viable photoelectrochemical (PEC) hydrogen-producing device;
- evaluate the potential of the Copper-Chlorine(Cu-Cl) thermochemical cycle for large-scale hydrogen production using nuclear energy and perform an integrated lab-scale (ILS) experiment to demonstrate closed-loop operation of the Cu-CI cycle;
- demonstrate the feasibility of hybrid cycles base on termochemical reactions and electrolysis for hydrogen production – proof-on experiments.

d) Integrating renewable sources with water electrolysis

- model, evaluate, test and optimize the renewable electrolysis system performance for dedicated hydrogen production and electricity/hydrogen cogeneration;
- gain operational experience with a hydrogen production facility, including the compression of gas and the use of a hydrogen-fueled internal combustion engine to generate electricity during peak demand hours;
- create synergies from the co-production of electricity and hydrogen by providing consistent support of the electric grid via off-peak storage of hydrogen.

e) Cross-cutting issues

- evaluate safety systems and system controls for the safe operation of hydrogen production technologies;
- develop computer application for modeling and controlling hydrogen production systems;
- scoping standards and codes for hydrogen production.

Fuel Cell Design, Testing and Technology Validation

The experience achieved in the last years in designing single cell PEM fuel cell or small PEMFC stacks later, was a good basis to approach new and more complex project in this area. It is known that fuel cell systems are complex. They may include as many as 10 subsystems incorporating electrochemistry, heterogeneous catalysis, endo and exothermic chemical reactions, and multiple gas and fluid handling operations, all of which must be designed and engineered to interact with one another in an efficient manner. It is the foreseen task of the cooperation of electrochemistry, designing and modeling groups to try to combine new materials and designs in order to get the required performances.

The impact of design and material choices on fuel cell manufacturability and performance must be investigated through extensive modeling of the detailed systems dynamics prior to equipment or material specification to shorten design cycles and to minimize life cycle cost.

The emphasis will be on testing and technology validation phase and a wide range of characterization instrumentation and testing facility will be soon operational on-site.

The topics which are planned to be analyzed in the next future are:

a) Analysis and characterization of PEMFC

- develop a wide range of microstructural and electrochemical characterization of PEM Fuel Cell MEAs;
- simulate a close-to-reality fuel cell model in order to get a better insight view of the material requirements in various working parameter;
- elucidate membrane electrode assembly (MEA) degradation and/or failure mechanisms by conducting extensive microstructural characterization;
- develop correlations between MEA structure/composition and durability/ performance.

b) Investigate the effect of impurities on PEMFC performance

- develop analytical chemistry protocols and tools to detect the nature and fate of contaminating species within fuel cells and determine through controlled laboratory experimentation study the main drivers for voltage decay.
- develop contaminant analytical models and computer simulations that explain and predict these effects and validate contaminant models through single cell experimentation using standardized test protocols.
- develop and validate novel technologies for mitigating the effects of contamination on fuel cell performance.

c) Investigate dynamic behavior of PEMFC stacks small and medium power in various working conditions

- develop Hardware-in-Loop experiments and analysis for optimizing the dynamic control of PEMFC and PEMFC small stacks – start-up and shut-down procedures.
- investigate PEMFC behavior at extreme environmental temperature and pressure conditions
- model, design and develop a regenerative PEM fuel cell; analyze of performance in start-up/shot-down conditions

d) Demonstrative PEMFC project – Analysis of durability and reliability

- design and build a small scale integrated hydrogen processor-PEMFC stack of 1 kW for demonstrative purposes

- perform a long-term investigation of durability and reliability of the tested model based on parameter acquisition and a statistical multivariate analysis
- develop a set of codes and standards for operating a small scale PEMFC system type "hand-held" for various applications

Fuel Cell – Based Applications

Considering our mission of implementing Hydrogen-based technologies in Romania, the main part of the activity is planned to be in the area of application design, fuel cell system testing and technology validation. Taking into account the specific conditions in Romania and also the limits of our capacity to deliver breakthrough science it was clear for us that we have to find our place in a more technological area were engineering capability and knowledge of using new techniques and instrumentations could be a plus for our expertise.

Therefore, it's one of our declared aims of establishing fuel cells as a commercially viable low carbon energy technology. To support this aim, it has established a Fuel Cell-based application - Development Group, designed to support all parts of the fuel cell supply chain. The Development Group, together with its support team and extensive range of partner organizations, is supposed to offer technical support to product and project developers, future Romanian PEMFC Systems integrators and manufacturers, reducing the risks associated with the commercialization of the technology. The Group will be equipped with advanced fuel cell test stations, allowing fuel cell testing and experimentation on a wide range of fuel cells, from single cell assemblies to 50 kW systems. Dynamic load testing can be carried out on PEM, and DMFC systems, with full data logging. The Centre offers LABVIEW capability for fuel cell testing and control and CHEMCAD capability for the design and simulation of complete fuel cell systems. Filling facilities for hydride and composite hydrogen tanks will be also available in the testing room.

Few of the topics which are planned to be investigated are:

a) Model and design of renewably powered hydrogen system

- PV modeling and design for Hydrogen – based system
- Wind-machine modeling
- PV-electrolyzer-PEMFC system modeling

b) Develop on-site a totally independent demonstrative system "zero-emission-energy"

c) Perform energy, economic and environmental analysis on RET-hydrogen systems in residential buildings

d) Improve modeling, designing and building capability of using PEMFC systems for mobile applications – develop demonstrative applications

e) Develop a model for a complete reversible fuel cell system

f) Elaborate an extensive experimental study – laboratory-scale, of a decentralized electricity generation system based on the concepts on distributed Hydrogen energy resources.

This R&D activity clearly links our work to National and European specified priorities and initiatives. It will provides a framework for working effectively across the research groups within RHFCC, and working in partnership with national and European collaborators, which include universities, research and development institutes and companies, industry organizations.

EXAMPLE OF ICIT R&D ACHIEVEMENT - EXPERIMENTAL PILOT FOR ENERGY CONVERSION USING PEM FUEL CELLS

This experimental pilot plant is composed from a natural gas steam reforming unit, a metallic hydride storage system and a proton exchange membrane fuel cells low current assemble.

The first step in hydrogen production is the natural gas catalytic steam reforming process based on an endothermic reaction. In the steam reforming process, the fuel is heated and vaporized, than injected with superheated steam into the reaction vessel. An excess of steam is required to suppress carbon formation and promote the reforming reaction, so that the steam to carbon molar ratio is in our reactor 4:1.

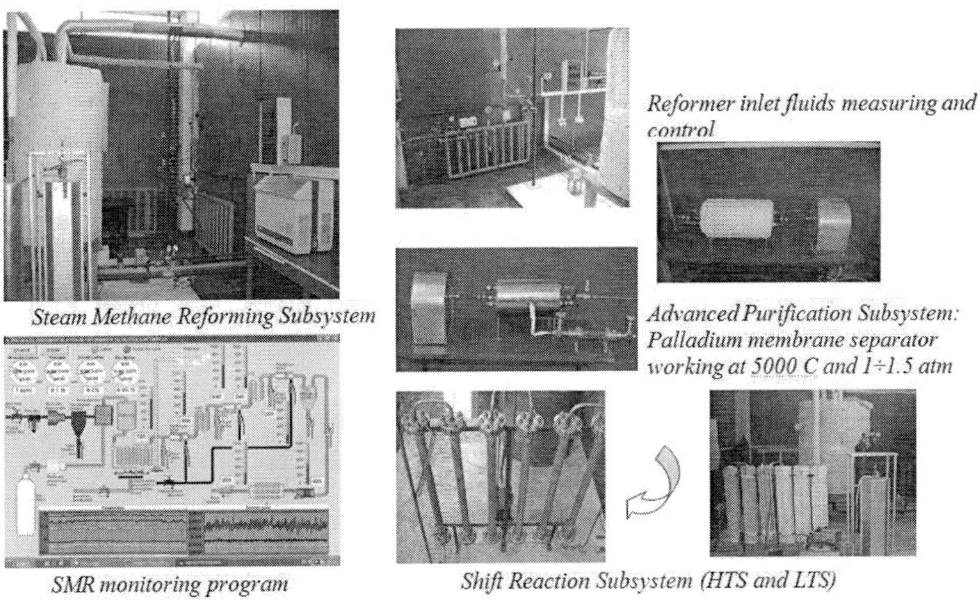

Reformer inlet fluids measuring and control

Steam Methane Reforming Subsystem

Advanced Purification Subsystem: Palladium membrane separator working at 5000 C and 1÷1.5 atm

SMR monitoring program

Shift Reaction Subsystem (HTS and LTS)

The reactor woks at 700° C and 3 atm, the steam reforming process being produced on a Ni based catalyst disposed in ten columns, annular distributed. The synthesis gas contains a mixture of $H_2/H_2O/CO_2/CO/CH_4$. After reformation the output gas is cooled and drawn into the HTS (high-temperature shift) and LTS (low-temperature shift) reactors. The water shift reaction is an exothermic reaction and takes place at 500°C and 200°C respectively. In the high-temperature shift reaction (500°C) in HTS reactor takes place in presence of Fe2O3/Cr2O3 catalyst disposed in three columns. The low-temperature shift reaction (250°C) in LTS reactor takes place in presence of CuO/ZnO catalyst disposed in three columns. The output gas contains a mixture of H_2 (aprox. 60%), CO_2, H_2O, and small amounts of CH_4 si CO.

The exchange proton membranes used as solid electrolyte for hydrogen fuel cells (PEM) can be easily poisoned with carbon monoxide. Therefore a final purification unit that consists by a Pd membrane reactor is used, to purify the carbon monoxide level lower than 10 ppm. This reactor is made by three cannular membranes on γ-alumina support and annular disposed within a stainless steel reactor.

The plant operating parameters are measured for each unit the plant being monitorized against two possible hazards, the atmospheric pollutants and hydrogen explosion.

Conclusion

In this chapter are presented some research in the field of hydrogen applications performed at National Research and Development Institute for Cryogenics and Isotopic Technologies, Rm. Valcea, Romania, by the authors and their colleagues of the research team.

Acknowledgments

The authors wish to thank to all members of the research team for their contribution in the research grants performed at National R&D Institute for Cryogenics and Isotopic Technologies – ICIT Rm. Valcea (some of them mentioned in the references list).

References

[1] Culcer, M., Iliescu, M., Enache, A., Ionete, R.E., Stefanescu, I., Duta, A., Enesca, A., *Energy Conversion Integrated System Using Renewable Sources*, Hydrogenia 2008.

[2] Carcadea, E., Stefanescu, I., Ionete R.E., Ene, H., PEM fuel cell systems performance optimization using the mathematical modeling", vol. „Stiinta moderna si energia – Producerea, Transportul si utilizarea energiei", 2006, *Risoprint Publishing House*, ISBN 978-973-751-215-4.

[3] Hugo Seymour, Francisco C. Borges, Rei Fernandes – "Indicators of European public research in hydrogen and fuel cells - An input-output analysis", *International Journal of Hydrogen Energy*, 32 (2007) 3212 – 3222.

[4] ***, "HY-CO" project, coordination action to establish a H&FC ERA-NET, EC research grant ERAC-CT-2004-0011744, 2004.

[5] ***, „HyWAYS" - the European Hydrogen Roadmap project, EC research grant SES6-502596, European Commission.

[6] ***„The Strategy of National Centre for Hydrogen and Fuel Cells", ICIT Report, 2009.

[7] ***, „Electricity Integrated Production System Based on Proton Exchange Membrane Fuel Cells", research grant.

[8] ***, „Optimizing the geometry and thermodynamics characteristics of the PEM fuel cell type assembly", research grant.

INDEX

A

B

D

E

F

G

H

I

J

K

L

O

P

Q

R

S

T

U

V

W

X

Y

Z